Lecture Notes in Physics

T0238664

Volume 871

For further volumes:
www.springer.com/series/5304

The Lecture Notes in Physics

The series Lecture Notes in Physics (LNP), founded in 1969, reports new developments in physics research and teaching—quickly and informally, but with a high quality and the explicit aim to summarize and communicate current knowledge in an accessible way. Books published in this series are conceived as bridging material between advanced graduate textbooks and the forefront of research and to serve three purposes:

- to be a compact and modern up-to-date source of reference on a well-defined topic
- to serve as an accessible introduction to the field to postgraduate students and nonspecialist researchers from related areas
- to be a source of advanced teaching material for specialized seminars, courses and schools

Both monographs and multi-author volumes will be considered for publication. Edited volumes should, however, consist of a very limited number of contributions only. Proceedings will not be considered for LNP.

Volumes published in LNP are disseminated both in print and in electronic formats, the electronic archive being available at springerlink.com. The series content is indexed, abstracted and referenced by many abstracting and information services, bibliographic networks, subscription agencies, library networks, and consortia.

Proposals should be sent to a member of the Editorial Board, or directly to the managing editor at Springer:

Christian Caron
Springer Heidelberg
Physics Editorial Department I
Tiergartenstrasse 17
69121 Heidelberg/Germany
christian.caron@springer.com

Dmitri Kharzeev · Karl Landsteiner ·
Andreas Schmitt · Ho-Ung Yee

Editors

Strongly Interacting Matter in Magnetic Fields

 Springer

Editors
Dmitri Kharzeev
Department of Physics
Stony Brook University
Stony Brook, NY, USA

Andreas Schmitt
Institut für Theoretische Physik
Technische Universität Wien
Vienna, Austria

Karl Landsteiner
Instituto de Física Teorica UAM/CSIC
Universidad Autonoma de Madrid
Madrid, Spain

Ho-Ung Yee
Physics Department
University of Illinois at Chicago
Chicago, IL, USA

ISSN 0075-8450   ISSN 1616-6361 (electronic)
Lecture Notes in Physics
ISBN 978-3-642-37304-6   ISBN 978-3-642-37305-3 (eBook)
DOI 10.1007/978-3-642-37305-3
Springer Heidelberg New York Dordrecht London

Library of Congress Control Number: 2013938882

Printed on acid-free paper

Springer is part of Springer Science+Business Media (www.springer.com)

Contents

Chapter 1
Strongly Interacting Matter in Magnetic Fields: A Guide to This Volume

Dmitri E. Kharzeev, Karl Landsteiner, Andreas Schmitt, and Ho-Ung Yee

1.1 Introduction

Electromagnetic probes have proved to be extremely important for understanding strongly interacting matter—for example, the discovery of Bjorken scaling in deep-inelastic scattering (DIS) has allowed to establish quarks as the constituents of the proton, and has opened the path towards Quantum Chromodynamics (QCD) and the discovery of asymptotic freedom. The development of QCD has quickly led to the realization that the dynamics of extended field configurations is crucial for understanding non-perturbative phenomena, including spontaneous breaking of chiral symmetry and confinement that define the properties of our world. The challenge

D.E. Kharzeev (✉)
Department of Physics and Astronomy, Stony Brook University, Stony Brook, NY 11794-3800, USA
e-mail: Dmitri.Kharzeev@stonybrook.edu

D.E. Kharzeev
Department of Physics, Brookhaven National Laboratory, Upton, NY 11973-5000, USA

K. Landsteiner
Instituto de Física Teórica UAM-CISC, C/Nicolás Cabrera 13-15, Universidad Autónoma de Madrid, Madrid 28049, Spain
e-mail: karl.landsteiner@uam.es

A. Schmitt
Institut für Theoretische Physik, Technische Universität Wien, Vienna 1040, Austria
e-mail: aschmitt@hep.itp.tuwien.ac.at

H.-U. Yee
Department of Physics, University of Illinois, Chicago, IL 60607, USA
e-mail: hyee@uic.edu

H.-U. Yee
RIKEN-BNL Research Center, Brookhaven National Laboratory, Upton, NY 11973-5000, USA

D. Kharzeev et al. (eds.), *Strongly Interacting Matter in Magnetic Fields*,
Lecture Notes in Physics 871, DOI 10.1007/978-3-642-37305-3_1,
© Springer-Verlag Berlin Heidelberg 2013

of understanding the collective dynamics in QCD calls for the study of response of strongly interacting matter to intense coherent electromagnetic fields. Such fields induce a host of interesting phenomena in QCD matter, and understanding them brings us closer to the ultimate goal of understanding QCD. Some of these phenomena (e.g. Magnetic Catalysis of chiral symmetry breaking [1],[1] and Inverse Magnetic Catalysis [2]) exist in a static equilibrium ground state, and affect the phase diagram of QCD matter in a magnetic field. About one half of this volume addresses such equilibrium phenomena, mostly in the context of QCD [3–6], and we give an overview over this part in Sect. 1.3. The other half addresses mainly anomaly-induced transport phenomena and is summarized in Sect. 1.2. These phenomena include the Chiral Magnetic and Chiral Vortical effects (reviewed in [7–14]) which require the existence of a chirality imbalance induced by the topological transitions in matter or the presence of vorticity.

Experimental access to the study of QCD plasma in very intense magnetic fields with magnitude $eB \sim m_\pi^2$ (or $\sim 10^{18}$ G) is provided by collisions of relativistic heavy ions at nonzero impact parameter. They create a magnetic field which is (on average) aligned perpendicular to the reaction plane. Somewhat weaker magnetic fields $\sim 10^{15}$ G exist on the surface of magnetars. They are possibly much larger in the interior of the star, where they may affect the properties of cold dense quark matter as reviewed in [15].

1.2 Chiral Magnetic Effect and Anomaly-Induced Transport

The Chiral Magnetic Effect (CME) is the phenomenon of electric charge separation along an external magnetic field that is induced by a chirality imbalance. In QCD matter, the source of chirality imbalance are the transitions between the topologically distinct states—the index theorem and the axial anomaly relate the resulting change in topological number to the chirality of fermion zero modes. The CME is thus a topological effect: it results from an interplay of topology of the zero mode of a charged fermion in an external Abelian magnetic field, and of the non-Abelian

---

[1]In this introduction we refer only to the contributions in the volume, and provide the corresponding arXiv references when available; many more references can be found in these individual contributions. If you would like to cite one of the contributions on a specific topic, please refer to them directly (instead of citing the entire volume), e.g.

E. D'Hoker, P. Kraus, in *Strongly Interacting Matter in Magnetic Fields*, ed. by D. Kharzeev, K. Landsteiner, A. Schmitt, H.-U. Yee. Lect. Notes Phys. **871**, 467 (2013). arXiv: 1208.1925

If you would like to refer to a broad review representing the collective work of the authors, you can refer to the entire volume as

Strongly Interacting Matter in Magnetic Fields, ed. by D. Kharzeev, K. Landsteiner, A. Schmitt, H.-U. Yee. Lect. Notes Phys. **871**, 1 (2013). arXiv:1211.6245

topology of the gluon field configurations. Because of this, the CME current is topologically protected (not sensitive to local perturbations) and non-dissipative.

At weak coupling, the quasi-particle picture is appropriate, and allows to understand the phenomenon in a very simple and intuitive way, as we now explain. An external magnetic field aligns the spins of the positive and negative fermions at the lowest Landau level in opposite directions (only the lowest Landau level matters for the CME, as the contributions of all excited levels cancel out—see [8] for details). Therefore, the electric charge, chirality and momentum of the fermion are correlated—for example, a positively charged right-handed fermion propagates along the direction of magnetic field, and a negative right-handed fermion propagates in the opposite direction. This creates an electric current, which however is usually compensated by the left-handed fermions that propagate in the opposite direction.

Let us however imagine that the fermions, apart from the Abelian $U(1)$ charge, are also charged under a non-Abelian group—for us, the most important example is provided by quarks, which carry both electric and color charges. The non-Abelian gauge theories possess a rich spectrum of topological solutions, and the axial anomaly links the topology of gauge fields to the chirality of fermions. Therefore, in a topologically non-trivial non-Abelian background, the numbers of left- and right-handed fermions will in general differ—because of this, their contributions to the electric current will no longer cancel. As a result, an external magnetic field will induce an electric current along its direction—an effect that is absent in Maxwell electrodynamics.

The absence of CME in conventional electrodynamics follows already from symmetry considerations—the magnetic field is a (parity-even) pseudo-vector, and the electric current is a (parity-odd) vector. Therefore, CME signals the violation of parity—indeed, as we discussed above, its presence requires the asymmetry between the left and right fermions.

It is well known that there are no perturbative corrections to the axial anomaly, so the CME expressions for the electric current and electric dipole moment are exact (at the operator level). Moreover, because the origin of CME is topological, it appears that the CME current at zero frequency remains the same even in the limit of strong coupling, that is accessible theoretically through the holographic correspondence—see [12, 16]. A phenomenon similar to CME arises when instead of a magnetic field there is an angular momentum (vorticity) present—this is the so-called Chiral Vortical Effect (CVE). The CVE is also caused by the quantum anomaly, but by the gravitational one [16]. In a holographic setup, this is described through a mixed gauge-gravitational Chern-Simons term.

In the absence of external charge, magnetic field in the framework of AdS/CFT correspondence induces an RG flow to an infrared $AdS_2 \times \mathbb{R}^2$ geometry [17]. This theory is the holographic description of the dimensionally reduced 2d conformal field theory describing the strongly coupled fermions on lowest Landau levels. In general, the dimensional reduction proves very beneficial to treating the CME and related phenomena—see [9] for review. In particular, in the limit of strong magnetic field one can construct an explicit solution describing the QCD instanton in magnetic background [9]—since the instanton induces an asymmetry between left- and right-handed fermions, this solution has been found to possess electric dipole moment, in accord with CME expectations. It is of great interest to investigate the dynamics of CME by considering the decay of topological objects in magnetic backgrounds—see [11] for a review.

The persistence of CME at strong coupling and small frequencies makes the hydrodynamical description of the effect possible, as reviewed in [10]. The quantum anomalies in general have been found to modify hydrodynamics in a significant way. This has a profound importance for transport, as the anomalies make it possible to transport currents without dissipation—this follows from the P-odd and T-even nature of the corresponding transport coefficients. The existence of CME and CVE in hydrodynamics is interesting also for the following reason—usually, in the framework of quantum field theory one thinks about quantum anomalies as of UV phenomena arising from the regularization of loop diagrams. However, we now see that the anomalies also modify the large distance, low frequency, response of relativistic fluids. This is because the anomalies link the chirality of fermion zero modes to the global topology of gauge fields.

The CME can be studied numerically from first principles on Euclidean space-time lattices, see [13, 14] for reviews. On the lattice, one can measure the fluctuations of electric charge asymmetry induced by the dynamical topological fluctuations in QCD vacuum and plasma in a magnetic background [13]. Alternatively, one can introduce the chiral chemical potential, and measure the CME current explicitly, testing the relation between the current, magnetic field, and the chiral chemical potential—this approach is reviewed in [14]. Note that the chiral chemical potential (unlike the baryon one) does not lead to the determinant "sign problem" and thus does not prevent one from performing lattice QCD simulations.

From the experimental viewpoint, CME makes it possible (at least in principle) to observe directly the fluctuations of topological charge in heavy ion collisions—indeed, these fluctuations in magnetic field induce the asymmetry of electric charge distributions with respect to the reaction plane. Such a study has to carefully separate the CME effects from all possible backgrounds, as reviewed in [18]. One of the CME tests discussed in [18] is the collision of two Uranium ions where the deformation of the Uranium nucleus allows to separate the CME from the backgrounds, as we now explain.

The main idea behind the UU measurement is the following: all possible backgrounds to CME should, on symmetry grounds, be proportional to the elliptic flow of hadrons. The elliptic flow stems from the ellipticity of the initial fireball produced in heavy ion collisions, and so it exists only in non-central collisions, just like the magnetic field that drives the CME. This complicates the separation of CME from possible background effects [18]. However, since the Uranium nucleus is strongly deformed, even most central UU collisions (where there is no magnetic field) produce a deformed fireball and hence a sizable elliptical flow of hadrons. Thus, if the charge separation persists in most central UU collisions, the observed effect is most likely due to background, and vice versa. Very recently, the RHIC result on UU collisions has been presented at the Quark Matter 2012 Conference by the STAR Collaboration. The presented result indicates that the signal vanishes in the absence of a magnetic field, providing a strong support for the CME interpretation.

1.3 Phase Structure in a Magnetic Field

Equilibrium properties of matter can be changed significantly by the presence of a background magnetic field. This is true even for non-interacting matter. Consider for example a Fermi sphere of charged, non-interacting fermions, say at zero temperature, that is subject to a static and homogeneous magnetic field. Instead of a continuous spectrum with respect to all three momentum directions, the system will develop discrete energy levels with respect to the momentum directions transverse to the magnetic field. Only the longitudinal momentum remains continuous. If the magnetic field is large enough, i.e. of the order of the Fermi momentum squared, the discretization of the energy levels, called Landau levels, will have an important effect on the properties of the system. For instance, upon increasing the magnetic field, the distance between the energy levels increases and as these levels "pass" the given Fermi energy, the observable quantities such as the number density change. Eventually, for sufficiently large magnetic fields, all fermions reside in the lowest Landau level, where only one spin polarization is allowed. The system has become fully polarized.

A more challenging question is how the equilibrium properties of *strongly inter-acting* matter are affected by a magnetic field. One might for instance ask whether there are still Landau levels or whether this description becomes incorrect. One might also wonder about the phase transitions induced by magnetic field—do we enter genuinely different phases upon increasing a background magnetic field in a given, strongly-interacting, system? If yes, what are the order parameters, what is the order of these phase transitions, and what are the critical magnetic fields? Or, ultimately, what is the phase diagram of a given system with one of the axes corresponding to the magnetic field? A well-studied example from condensed-matter physics is the phase diagram of Helium-3. Due to the nontrivial structure of the order parameter with respect to spin and orbital angular momentum, Helium-3 has more than one superfluid phase, and an externally applied magnetic field can induce phase transitions between different superfluid phases.

1.3.1 Phases of QCD in a Magnetic Field

In these Lecture Notes, the term "strongly interacting matter" mostly refers to the matter governed by QCD, although some chapters address questions from "ordinary" condensed-matter systems [17, 19, 20], see also the discussion about graphene in Sect. 3.3 of Ref. [1]. Usually, "the phase diagram of QCD" is drawn in the plane spanned by the temperature T and the baryon chemical potential μ. But various additional directions, i.e., higher-dimensional versions of the phase diagram, are of interest as well. First, one can imagine to change the parameters of QCD by hand, for example by adding an axis for the quark mass(es) or for the number of colors N_c to the phase diagram. The purpose of such a theoretical manipulation can be to enter a more tractable regime and/or to put the QCD phase structure into a wider context, with the ultimate goal to understand better the phase structure of real-world QCD. Second, there are external parameters of direct phenomenological interest that reflect the characteristics of matter, such as the baryon and isospin chemical potentials that become important in the context of neutron stars. A (uniform) magnetic field is both: it is of phenomenological interest but also a theoretically useful "knob" by which interesting and rich physics can be introduced which may help us to deepen our understanding of strongly interacting systems. (This volume mostly addresses the effects of a magnetic field from ordinary $U(1)$ electromagnetism; the effects of chromomagnetic fields are briefly discussed in [6].)

Phase transitions in QCD can be expected to occur at energy scales comparable to the QCD scale $\Lambda_{\text{QCD}} \sim 200$ MeV. As a consequence, we are interested in magnetic field strengths of the order of $B \sim (200 \text{ MeV})^2 \simeq 2 \times 10^{18}$ G. As mentioned above, such strong magnetic fields indeed exist in non-central heavy-ion collisions [18] (at least temporarily), and possibly in the interior of compact stars, then called magnetars [15]. These two instances are (together with the early universe) the only systems with which we can reach out "experimentally" also into the μ-T plane of the QCD phase diagram and thus probe the phase structure of QCD. Most notably, we are interested in the nonperturbative phenomena of confinement and chiral symmetry breaking. Both heavy-ion collisions and compact stars are expected to "live" in a region of the phase diagram where the transitions to deconfined and/or chirally restored matter occur and thus it is relevant to discuss the location and the nature of these transitions in magnetized QCD matter.

The various chapters addressing the phase structure of QCD in magnetic fields [1–6, 15, 21] make use of different theoretical tools. At asymptotically large energies, methods from perturbative QCD can be applied, see parts of [1, 15]. For moderate energies, however, first-principle QCD calculations can only be done on the lattice and are restricted to vanishing (vector) chemical potentials [6]. These results from QCD are complemented by model calculations, which may differ (and, judging from the results presented here, *do* differ) in important aspects from actual QCD. Nevertheless, they may be useful to get an idea about some of the physical mechanisms behind the phase structure. The models discussed here are the Nambu-Jona-Lasinio (NJL) model, including different variants with respect to the interaction terms and extensions to incorporate confinement (PNJL) [1–3, 5], the MIT bag model [4], and a quark-meson model [3, 4]. Additionally, we employ the gauge/gravity duality [2, 21] which provides us with a reliable tool for the physics in the strongly coupled limit, albeit for theories that differ more or less—depending on the model at hand—from real QCD.

The holographic calculations presented here make use of two different setups. Firstly, based on the original AdS/CFT correspondence, a background geometry given by D3-branes is discussed, where D7-branes, corresponding to fundamental (as opposed to adjoint) degrees of freedom, are introduced as probe branes [21]. And secondly, we use the Sakai-Sugimoto model, where "chiral" D8-branes are embedded into a background of D4-branes, which includes a compact extra dimension that serves to break supersymmetry completely [2, 21].

The picture that emerges from these studies is, concerning QCD, a preliminary one at best, and many questions are still open—and, in fact, are raised by these studies. Nevertheless, let us try to summarize the current picture. Before one addresses the question of how a magnetic field affects a hot and dense medium, one might ask whether and how the vacuum changes in a magnetic field. A very important part of the answer to this question—since it is of very general nature and thus not only relevant for QCD—is "magnetic catalysis", which is reviewed in great detail in Ref. [1]. This effect has been confirmed in numerous model calculations, as well as in QED-like theories and—at least at zero temperature—in QCD lattice calculations. In simple words, it says that a magnetic field favors chiral symmetry breaking. A more precise version, for instance employing the mean-field approximation of the NJL model, is as follows. For $\mu = T = 0$ and vanishing magnetic field, there is a critical coupling strength above which a chiral condensate forms, i.e., there is a phase transition from the chirally restored to the chirally broken phase as one increases the (attractive) coupling of the four-fermion interaction. In the presence of a background magnetic field, however, there is a chiral condensate for *arbitrarily small* coupling. Thus, the magnetic field has a profound qualitative effect on chiral symmetry breaking. One way to understand the physics behind magnetic catalysis is the analogy to BCS Cooper pairing. In both cases, the dynamics of the system becomes effectively $1 + 1$ dimensional at weak coupling, in the case of Cooper pairing because of the presence of the Fermi surface, in the case of chiral condensation because of the magnetic field. As a consequence, an infrared divergence occurs which is cured by a nonvanishing mass gap that shows precisely the same exponential, nonperturbative behavior in both cases.

Another possible effect of an ultra-strong magnetic field on the QCD vacuum is the condensation of ρ mesons [5]. In a weak-coupling picture, this condensation is suggested from the Landau-level structure of spin-1 bosons. Since ρ mesons are electrically charged, their condensation implies electric superconductivity. The details of this interesting idea are reviewed in Ref. [5].

How does magnetic catalysis manifest itself in the QCD phase diagram? A straightforward, but, as we now know, too naive, expectation is that the phase space region of the chirally broken phase should get larger with the magnetic field. The current picture is more complicated and can be summarized as follows.

- *Hot medium, $\mu = 0$.* Lattice studies with physical quark masses suggest that the (pseudo-)critical temperature T_c for the chiral crossover *decreases* [6], whereas the results of all above mentioned model calculations (as well as lattice results for unphysically large quark masses [6]) show a monotonically *increasing* T_c. While it is clear that none of the models can capture all features of QCD, the physical mechanism behind the lattice result is still under discussion, see for instance Sec. 3.2 in Ref. [1]. There are also slight differences in the behavior of T_c in the different models: for instance, while T_c saturates at a finite value for asymptotically large magnetic fields in the holographic Sakai-Sugimoto model [2, 21], no such saturation can be seen in the NJL-like models [2, 3].
- *Dense medium, $T = 0$.* In this case, there are currently no lattice results due to the sign problem. In the model calculations we see an interesting nontrivial effect,

termed "inverse magnetic catalysis" [2]. At strong coupling, the critical chemical potential can *decrease* with the magnetic field, which is in apparent conflict with magnetic catalysis. In contrast to the $\mu = 0$ result on the lattice, a physical explanation for this behavior is known. It can be traced back to the cost in free energy that has to be paid for chiral condensation at nonzero μ. Crucially, this cost is not independent of B and competes with the gain from condensation which, due to magnetic catalysis, increases with B.

We know that the QCD phase structure at large densities can be very rich due to various color-superconducting phases. Model calculations at nonzero B including two-flavor color superconductivity seem to confirm the effect of inverse magnetic catalysis. From first principles we know that at asymptotically large densities, the ground state of three-flavor quark matter is the color-flavor locked (CFL) phase. The CFL order parameter is invariant under a certain combination of generators of the color and electromagnetic gauge groups, resulting in a massless gauge boson (which is predominantly the original photon, with a small admixture from one of the gluons). As a consequence, the CFL phase is a color superconductor (all gluons become massive), but not an electromagnetic superconductor. An ordinary magnetic field can thus penetrate this phase. Depending on the strength of the magnetic field, certain variants of the CFL phase become favored. These phases as well as their possible astrophysical relevance are discussed in [15].

For the deconfinement crossover in a magnetic field, the results on the lattice are similar to the chiral crossover [6]. Again, with physical quark masses the temperature for the crossover seems to decrease. This behavior is not reproduced by the PNJL model, where confinement is mimicked by including the expectation value of the Polyakov loop by hand. Depending on the details of the interactions within the model, the deconfinement and chiral phase transitions do or do not split in a magnetic field, but in any case the critical temperature of both transitions seems to increase monotonically with the magnetic field [3]. This is also the case in a quark-meson model [4]. Within the MIT bag model, however, with the confined phase simply being modelled by a noninteracting pion gas, the deconfinement transition decreases with the magnetic field, in qualitative agreement with the lattice results [4]. These studies show that the mechanism behind the behavior of the deconfinement transition—just like for the chiral transition—in a magnetic field is not yet understood, and further analyses that clarify this picture are necessary.

1.3.2 Condensed Matter Systems in a Magnetic Field via AdS/CFT

The gauge-gravity or AdS/CFT correspondence plays an important role in the above mentioned studies of the phase structure of QCD. This is quite natural since in general the gauge-gravity correspondence relates a non-Abelian gauge theory at strong coupling and large N to a weakly coupled (super)gravity background with asymptotic AdS boundary conditions. The idea that strongly coupled phases of quantum field theories can be described by a gravity dual is however more general than that. In particular it has been applied to "tabletop" laboratory condensed matter systems that inherently involve strongly correlated electrons and are conjectured to be governed by an underlying quantum critical phase such as high Tc superconductors or the so called strange metals. Many condensed-matter systems are supposed to undergo a quantum phase transition at zero temperature upon varying a parameter such as the doping in a high-T_c superconductor, the pressure or the applied magnetic field. At the phase transition the system is at a quantum critical point which often turns out to be in a strongly coupled regime with an emergent Lorentz (or more generally Lifschytz) invariance. In the vicinity of the quantum critical point, e.g. at finite temperature the dynamics of the system is still governed by the degrees of freedom relevant at the critical point. One might hope that such strongly coupled quantum critical points can be described by a gravity dual just as QCD in its strongly coupled regime might be described by e.g. the Sakai-Sugimoto model. At the very least one can expect that the gauge-gravity duality helps to develop interesting toy models of quantum critical points that can lead to a better understanding of qualitative or even quantitative behavior of condensed matter systems whose theoretical description is otherwise rather elusive due to their strongly coupled nature.

A central role is played by charged asymptotically AdS black holes. Due to the underlying scaling symmetry two regimes can naturally be distinguished, one in which the temperature T is much larger than the chemical potential μ. This is the hydrodynamic regime and the gauge gravity duality allows to compute transport coefficients such as viscosities or conductivities. The other regime of interest is the opposite with $\mu \gg T$. Here the black hole is near extremality. At precisely $T = 0$ the black hole horizon becomes degenerate with profound implications for the dual field theory physics. It turns out that the entropy of an extremal charged AdS black hole is macroscopically large, scaling with some positive power of the number of the underlying microscopic degrees of freedom N. This feature is generally interpreted as pointing towards an inherent instability of such black holes. Indeed many such instabilities are known to arise upon adding additional fields, e.g. scalars (even uncharged ones) tend to condense near extremality leading to a superconducting phase transition. The charge that was hidden behind the horizon is pulled outside and the scalarfield forms the (symmetry breaking) condensate. If one replaces the scalar with a fermion a similar transition occurs where the black hole is replaced by a geometry without horizon and the charge is carried by the fermions forming an "electron star" in asymptotically AdS. The field theory interpretation of this transition is one between a "fractionalized" phase and a "mesonic" phase. This is in one-to-one correspondence to the deconfined/confined phases of non-Abelian gauge theory, only that in condensed matter it is often the mesonic or confined phase that appears as the more fundamental one. Real world fractionalization occurs for example in the separation of spin and charge degrees of freedom of electrons or the appearance of quasiparticles carrying fractions of the electron's charge. Such gravity duals of fractionalized/mesonic phases in the presence of a magnetic field are the subject of [19]. A crucial role is played by the presence of a dilaton which allows to find a rich structure of solutions including partially fractionalized phases.

The quantum critical behavior of four dimensional field theories described by a gravity dual including a $U(1)$ gauge field with an additional Chern-Simons term in a magnetic field is the subject of [17]. The authors find a quantum phase transition for large enough magnetic field beyond which the charge is completely expelled outside the horizon and the background geometry has zero entropy.

Another important theme in the condensed matter applications of the gauge gravity correspondence is the spectrum of fermions in asymptotically AdS black holes. Fermions in the gravity dual obeying a Dirac equation dual to gauge invariant fermionic operators in the field theory. One can study the holographic fermionic two point function and in particular look out for poles that identify the presence of Fermi surfaces. It is indeed well known by now that such probe fermions can show behavior consistent with Landau's theory of Fermi liquids but also more exotic possibilities such as marginal Fermi liquids in which the residue of the corresponding pole vanishes or completely non-Landau Fermi liquid behavior are realized. The Fermi level structure of such probe fermions in a four dimensional dyonic, i.e. including a magnetic field, asymptotically AdS black hole is the subject of [20]. For strong magnetic field the Fermi surface vanishes and the authors associate this with a metal to strange metal phase transition.

To summarize, this volume of Lecture Notes presents a review of the current research of strongly interacting matter in magnetic fields. Most of the applications considered here concern QCD matter, but a number of important cases from condensed matter physics is considered as well. The focus of the volume is on the theoretical results; however these results have a direct significance for experiment, in particular for heavy ion collisions that are currently under intense study at RHIC and LHC. While most of the contributions in this volume reflect the work done very recently, the field is evolving so rapidly that we expect to see a significant progress already in the near future. Because of this, we do not expect this volume to express a "final word" in any sense; instead, we view it as a snapshot of the exciting work that is being done right now. A large number of open questions has emerged as a result of this work; some of them have been mentioned in this brief overview, but we encourage the reader to read the individual contributions for an in-depth exposure. We hope that this volume will convince the reader that strongly interacting matter in a magnetic field is a rich and vibrant research area, and many more discoveries and surprises can be fully expected.

Acknowledgements   We would like to express our gratitude to all authors of these Lecture Notes for their contributions. This overview was finalized during the workshop "QCD in strong magnetic fields" at the European Centre for Theoretical Studies in Nuclear Physics and Related Areas (ECT*). We would like to thank ECT* for its kind hospitality and the organizers for a very interesting and stimulating workshop. The work of D.K. is supported in part by the US Department of Energy under Contracts DE-AC02-98CH10886 and DE-FG-88ER41723. The work of K.L. is supported by Plan Nacional de Altas Energías FPA2009-07890, Consolider Ingenio 2010 CPAN CSD2007-00042 and HEP-HACOS S2009/ESP-2473.

References

1. I.A. Shovkovy, Magnetic catalysis: a review, in *Strongly Interacting Matter in Magnetic Fields* (Springer, Berlin, 2013), p. 13. arXiv:1207.5081 [hep-ph]
2. F. Preis, A. Rebhan, A. Schmitt, Inverse magnetic catalysis in field theory and gauge-gravity duality, in *Strongly Interacting Matter in Magnetic Fields* (Springer, Berlin, 2013), p. 49. arXiv:1208.0536 [hep-ph]
3. R. Gatto, M. Ruggieri, Quark matter in a strong magnetic background, in *Strongly Interacting Matter in Magnetic Fields* (Springer, Berlin, 2013), p. 85. arXiv:1207.3190 [hep-ph]
4. E.S. Fraga, Thermal chiral and deconfining transitions in the presence of a magnetic background, in *Strongly Interacting Matter in Magnetic Fields* (Springer, Berlin, 2013), p. 119. arXiv:1208.0917 [hep-ph]
5. M.N. Chernodub, Electromagnetic superconductivity of vacuum induced by strong magnetic field, in *Strongly Interacting Matter in Magnetic Fields* (Springer, Berlin, 2013), p. 141. arXiv:1208.5025 [hep-ph]
6. M. D'Elia, Lattice QCD simulations in external background fields, in *Strongly Interacting Matter in Magnetic Fields* (Springer, Berlin, 2013), p. 179. arXiv:1209.0374 [hep-lat]
7. A.R. Zhitnitsky, P odd fluctuations and long range order in heavy ion collisions. Deformed QCD as a toy model, in *Strongly Interacting Matter in Magnetic Fields* (Springer, Berlin, 2013), p. 207. arXiv:1208.2697 [hep-ph]
8. K. Fukushima, Views of the chiral magnetic effect, in *Strongly Interacting Matter in Magnetic Fields* (Springer, Berlin, 2013), p. 239. arXiv:1209.5064 [hep-ph]
9. G. Basar, G.V. Dunne, The chiral magnetic effect and axial anomalies, in *Strongly Interacting Matter in Magnetic Fields* (Springer, Berlin, 2013), p. 259. arXiv:1207.4199 [hep-th]
10. V.I. Zakharov, Chiral magnetic effect in hydrodynamic approximation, in *Strongly Interacting Matter in Magnetic Fields* (Springer, Berlin, 2013), p. 293. arXiv:1210.2186 [hep-ph]
11. A. Gorsky, Remarks on decay of defects with internal degrees of freedom, in *Strongly Interacting Matter in Magnetic Fields* (Springer, Berlin, 2013), p. 329
12. T. Nishioka, C. Hoyos, A. O'Bannon, A chiral magnetic effect from AdS/CFT with flavor, in *Strongly Interacting Matter in Magnetic Fields* (Springer, Berlin, 2013), p. 339
13. P.V. Buividovich, M.I. Polikarpov, O.V. Teryaev, Lattice studies of magnetic phenomena in heavy-ion collisions, in *Strongly Interacting Matter in Magnetic Fields* (Springer, Berlin, 2013), p. 375. arXiv:1211.3014 [hep-ph]
14. A. Yamamoto, Chiral magnetic effect on the lattice, in *Strongly Interacting Matter in Magnetic Fields* (Springer, Berlin, 2013), p. 385. arXiv:1207.0375 [hep-lat]
15. E.J. Ferrer, V. de la Incera, Magnetism in dense quark matter, in *Strongly Interacting Matter in Magnetic Fields* (Springer, Berlin, 2013), p. 397. arXiv:1208.5179 [nucl-th]
16. K. Landsteiner, E. Megias, F. Pena-Benitez, Anomalous transport from Kubo formulae, in *Strongly Interacting Matter in Magnetic Fields* (Springer, Berlin, 2013), p. 431. arXiv:1207.5808 [hep-th]
17. E. D'Hoker, P. Kraus, Quantum criticality via magnetic branes, in *Strongly Interacting Matter in Magnetic Fields* (Springer, Berlin, 2013), p. 467. arXiv:1208.1925 [hep-th]
18. A. Bzdak, V. Koch, J. Liao, Charge-dependent correlations in relativistic heavy ion collisions and the chiral magnetic effect, in *Strongly Interacting Matter in Magnetic Fields* (Springer, Berlin, 2013), p. 501. arXiv:1207.7327 [nucl-th]
19. T. Albash, C.V. Johnson, S. MacDonald, Holography, fractionalization and magnetic fields, in *Strongly Interacting Matter in Magnetic Fields* (Springer, Berlin, 2013), p. 535. arXiv:1207.1677 [hep-th]
20. E. Gubankova, J. Brill, M. Čubrović, K. Schalm, P. Schijven, J. Zaanen, Holographic description of strongly correlated electrons in external magnetic fields, in *Strongly Interacting Matter in Magnetic Fields* (Springer, Berlin, 2013), p. 553. arXiv:1304.3835 [hep-th]
21. O. Bergman, J. Erdmenger, G. Lifschytz, A review of magnetic phenomena in probe-brane holographic matter, in *Strongly Interacting Matter in Magnetic Fields* (Springer, Berlin, 2013), p. 589. arXiv:1207.5953 [hep-th]

Chapter 2
Magnetic Catalysis: A Review

Igor A. Shovkovy

2.1 Introduction

The magnetic catalysis is broadly defined as an enhancement of dynamical symmetry breaking by an external magnetic field. In this review, we discuss the underlying physics behind magnetic catalysis and some of its most prominent applications. Considering that the ideas of symmetry breaking take the center stage position in many branches of modern physics, we hope that this review will be of interest to a rather wide audience.

In particle and nuclear physics, spontaneous symmetry breaking is commonly used in order to explain the dynamical origin of the mass of elementary particles. In this context, the idea was realized for the first time over 50 years ago by Nambu and Jona-Lasinio [151, 152], who suggested that "the nucleon mass arises largely as a self-energy of some primary fermion field through the same mechanism as the appearance of energy gap in the theory of superconductivity." As we now know, the analogy with superconductivity is not very close and the description of chiral symmetry breaking in terms of quarks may be more natural than in terms of nucleons. However, the essence of the dynamical mass generation was captured correctly in Refs. [151, 152]. In fact, with the current state of knowledge, we attribute most of the mass of visible matter in the Universe to precisely this mechanism of mass generation, which is associated with breaking of the (approximate) chiral symmetry.

The conceptual knowledge that the mass can have a dynamical origin opens myriads of theoretical possibilities that would appear meaningless in classical physics. For example, keeping in view the above mentioned mechanism of mass generation through chiral symmetry breaking, it is reasonable to suggest that the masses of certain particles can be modified or even tuned by proper adjustments of physical parameters and/or external conditions.

I.A. Shovkovy (✉)
School of Letters and Sciences, Arizona State University, Mesa, AZ 85212, USA
e-mail: Igor.Shovkovy@asu.edu

D. Kharzeev et al. (eds.), *Strongly Interacting Matter in Magnetic Fields*,
Lecture Notes in Physics 871, DOI 10.1007/978-3-642-37305-3_2,
© Springer-Verlag Berlin Heidelberg 2013

One of the obvious knobs to control the value of the dynamical mass is an external magnetic field. In addition to be a good theoretical tool, magnetic fields are also relevant to many applications. For example, they are commonly present and play an important role in such physical systems as the Early Universe [18, 30, 44, 95, 183], heavy ion collisions [123, 179], neutron stars [38, 182], and quasi-relativistic condensed matter systems like graphene [155, 189].

As we discuss in detail in this review, the magnetic field has a strong tendency to enhance (or "catalyze") spin-zero fermion-antifermion condensates. Such condensates are commonly associated with breaking of global symmetries (e.g., such as the chiral symmetry in particle physics and the spin-valley symmetry in graphene) and lead to a dynamical generation of masses (energy gaps) in the (quasi-)particle spectra. The corresponding mechanism is called *magnetic catalysis* [99].

It should be emphasized that, in a striking contrast to its role in superconductivity, the magnetic field helps to *strengthen* the chiral condensate. There are many underlying reasons for its different role. Unlike the superconductors, the ground state with a nonzero chiral condensate shows no Meissner effect. This is because the chiral condensate can be thought of as a condensate of neutral fermion-antifermion pairs, not charged Cooper pairs that can give rise to supercurrents and perfect diamagnetism. Also, in a usual Cooper pair, the two electrons have opposite spins and, therefore, opposite magnetic moments. When placed in a magnetic field, only one of the magnetic moments can minimize its energy by orienting along the direction of the field. The other magnetic moment will be stuck in a frustrated position pointing in the opposite direction. This produces an energy stress and tends to break the Cooper pair. (Note, however, that the orbital motion plays a much more important role in breaking nonrelativistic Cooper pairs.) In a neutral spin-zero pair, in contrast, the magnetic moments of the fermion (with a fixed charge and spin) and the antifermion (with the opposite charge and spin) point in the same direction. Therefore, both magnetic moments can comfortably align along the direction of the magnetic field without producing any frustration in the pair. (Also, in relativistic systems the fermion-antifermion condensate is not destroyed by the orbital motion.)

The above explanation of the role that the magnetic field plays in strengthening the chiral condensate is semi-rigorous at best and does not capture all the subtleties of the dynamics behind magnetic catalysis (e.g., completely leaving out the details of the orbital motion). It does demonstrates, however, how the magnetic field can have, at least in principle, so drastically different effects on the dynamical generation of mass on the one hand and on superconductivity on the other. (It may be curious to mention here that, in cold dense quark matter, it is possible to obtain color superconducting states, in which diquark Cooper pairs are neutral with respect to the in-medium (but not vacuum) electro-magnetism [7, 85]. In such quark matter, the in-medium magnetic field is not subject to the Meissner effect and, in fact, can enhance color superconductivity [54, 55, 62, 63, 75, 141, 154, 186].)

The early investigations of the effects of strong magnetic fields on chiral symmetry breaking in $(2 + 1)$- and $(3 + 1)$-dimensional models with local four-fermion interactions have appeared in late 1980s and early 1990s [122, 127–131, 135, 167, 181]. In these studies, it was already found that a constant magnetic field stabilizes the chirally broken vacuum state.

The explanation of the underlying physics was given in Ref. [99], where the essential role of the dimensional reduction, $D \to D - 2$, in the low-energy dynamics of pairing fermions in a magnetic field was revealed. As a corollary, it was also established that the presence of a magnetic field leads to the generation of a dynamical mass even at the weakest attractive interaction between fermions [99–102]. The general nature of the underlying physics was so compelling that it was suggested that the corresponding dynamical generation of the chiral condensate and the associated spontaneous symmetry breaking in a magnetic field are universal and model-independent phenomena. To emphasize this fact, the new term "magnetic catalysis" was coined [99].

The model-independent nature of magnetic catalysis was tested in numerous $(2 + 1)$- and $(3 + 1)$-dimensional models with local four-fermion interactions [10, 13, 29, 40, 55, 56, 80, 114, 116, 117, 132, 134, 142, 143, 157, 184, 193], including models with additional gauge interactions [119], higher dimensional models [84], $\mathcal{N} = 1$ supersymmetric models [42], quark-meson models [8, 9], models in curved space [79, 81, 118] and QED-like gauge theories [5, 6, 11, 12, 51, 53, 101, 104–107, 110, 111, 137, 138, 158, 159, 165]. The realization of magnetic catalysis was investigated in chiral perturbation theory [32, 33, 176] and in QCD [121, 146], as well as in a models with the Yukawa interaction [43, 59–61]. There are studies of magnetic catalysis using the methods of the renormalization group [74, 166, 173], lattice calculations [14–16, 24, 28, 34, 35, 52] and holographic dual models of large-N gauge theories [4, 20, 21, 45–50, 64–66, 162, 163, 187]. Similar ideas were extended to solid state systems describing high-temperature superconductivity [57, 58, 139, 172, 192, 194], highly oriented pyrolitic graphite [90, 124, 125], as well as monolayer [87, 91, 93, 98, 113, 171, 174] and bilayer [86, 88, 89, 92] graphene in the regime of the quantum Hall effect. Finally, the generalization of magnetic catalysis was also made to non-Abelian chromomagnetic fields [39, 41, 97, 133, 175, 185, 191], where the dynamics is dimensionally reduced by one unit of space, $D \to D - 1$. For earlier reviews on magnetic catalysis, see Refs. [96, 145].

2.2 The Essence of Magnetic Catalysis

As already mentioned in the Introduction, the essence of magnetic catalysis is intimately connected with the dimensional reduction, $D \to D - 2$, of charged Dirac fermions in the presence of a constant magnetic field. In this section, we discuss in detail how such a dimensional reduction appears and what implications it has for the spontaneous symmetry breaking.

2.2.1 Dimensional Reduction in a Magnetic Field

Before considering a fully interacting theory and all details of the dynamics responsible for the generation of the chiral condensate and the symmetry breaking, associated with it, let us start from a free Dirac theory in a constant external magnetic field. This appears to be a perfect setup to understand the kinematic origin of the dimensional reduction, $D \to D - 2$.

2.2.1.1 Dirac Fermions in a Magnetic Field in $3 + 1$ Dimensions

Let us start by reviewing the spectral problem for charged $(3+1)$-dimensional Dirac fermions in a constant magnetic field. We assume that the field is pointing in the positive x^3-direction. The corresponding Lagrangian density reads

$$\mathscr{L} = \bar{\Psi}\left(i\gamma^\mu D_\mu - m\right)\Psi, \tag{2.1}$$

where the covariant derivative $D_\mu = \partial_\mu - ieA_\mu^{\text{ext}}$ depends on the external gauge field. Without loss of generality, the external field A_μ^{ext} is taken in the Landau gauge, $A_\mu^{\text{ext}} \equiv (0, -\mathbf{A}^{\text{ext}})$, where

$$\mathbf{A}^{\text{ext}} = \left(0, Bx^1, 0\right), \tag{2.2}$$

and B is the magnetic field strength. By solving the Dirac equation of motion, one finds the following energy spectrum of fermions [3]:

$$E_n^{(3+1)}(p_3) = \pm\sqrt{m^2 + 2|eB|n + (p_3)^2}, \tag{2.3}$$

where $n = 0, 1, 2, \ldots$ is the Landau level index. It should be noted that the Landau level index n includes orbital and spin contributions: $n \equiv k + s + \frac{1}{2}$, where $k = 0, 1, 2, \ldots$ is an integer quantum number associated with the orbital motion, while $s = \pm\frac{1}{2}$ corresponds to the spin projection on the direction of the field. [For the orbital part of the wave functions, see (2.82) in the Appendix.] Considering that the energy depends only on n, we see that the energy of a quasiparticle in orbital state k and spin $s = +\frac{1}{2}$ is degenerate with the energy of a quasiparticle in orbital state $k+1$ and spin $s = -\frac{1}{2}$. The lowest Landau level with $n = 0$ is special: it corresponds to the lowest orbital state $k = 0$ and has only one spin projection $s = -\frac{1}{2}$. The letter, in particular, implies that the lowest Landau level is a spin polarized state.

On top of the spin degeneracy of higher Landau levels ($n > 0$), there is an additional (infinite) degeneracy of each level with a fixed n and a fixed value of the longitudinal momentum p_3. It is connected with the momentum $p_2 \in \mathbb{R}$, which is a good quantum number in the Landau gauge utilized here. As follows from the form of the orbital wave functions in (2.82), the value of $-p_2/|eB|$ also determines the location of the center of a fermion orbit in the x^1-direction. A simple analysis [3] shows that the area density of such states in the perpendicular x^1x^2-plane is $\frac{|eB|}{2\pi}$ for $n = 0$ and $\frac{|eB|}{\pi}$ for $n > 0$ (here the double spin degeneracy of the higher Landau levels is accounted for).

When the Dirac mass is much smaller than the corresponding magnetic energy scale (i.e., $m \ll \sqrt{|eB|}$), we find that the low-energy sector of the Dirac theory is determined exclusively by the lowest Landau level ($n = 0$). As we see from (2.3), the corresponding spectrum of the low-energy excitations is given by $E(p_3) = \pm\sqrt{m^2 + (p_3)^2}$, which is identical to the spectrum of a $(1 + 1)$-dimensional quantum field theory with a single spatial coordinate, identified with the longitudinal direction. This spectrum of the low-energy theory confirms the obvious kinematic aspect of the dimensional reduction, $3 + 1 \to 1 + 1$, in a constant magnetic field.

From the physics viewpoint, the dimensional reduction is the result of a partially restricted motion of Dirac particles in the $x^1 x^2$-plane perpendicular to the magnetic field. The effect can be seen already at the classical level in the so-called cyclotron motion, when the Lorentz force causes charged particles to move in circular orbits in $x^1 x^2$-plane, but does not constrain their motion along the x^3-direction. A very important new feature at the quantum level is the quantization of perpendicular orbits. Without such a quantization, the clean separation of the low-energy sector, dominated exclusively by the lowest Landau level, would not be possible.

It should be noted that the spin also plays an important role in the dimensional reduction of Dirac particles. If the spin contribution were absent ($s = 0$), the energy of the lowest Landau level would scale like $\sqrt{|eB|}$, which is not vanishingly small compared to the energy of the next Landau level $\sqrt{3|eB|}$. Then, a clean separation of the lowest Landau level into a dimensionally reduced, low-energy sector of the theory would become unjustified and meaningless.

2.2.1.2 Dirac Fermions in a Magnetic Field in 2 + 1 Dimensions

It is straightforward to obtain the spectrum of charged Dirac fermions also in $2 + 1$ dimensions. The vector potential in the Landau gauge takes the form: $\mathbf{A}^{\text{ext}} = (0, Bx^1)$. In the absence of the longitudinal direction x^3, the magnetic field B is not an axial vector, but a pseudo-scalar. Concerning the Dirac algebra in $2 + 1$ dimensions, there exist two inequivalent irreducible representations, given by

$$\gamma^0 = \sigma_3, \qquad \gamma^1 = i\sigma_1, \qquad \gamma^2 = i\sigma_2, \tag{2.4}$$

and

$$\gamma^0 = -\sigma_3, \qquad \gamma^1 = -i\sigma_1, \qquad \gamma^2 = -i\sigma_2, \tag{2.5}$$

where σ_i are the Pauli matrices. In each of these representations, the nature of the lowest Landau level is somewhat unusual: it has either only a particle state (with a positive energy $E_0 = m$) or only an antiparticle state (with a negative energy $E_0 = -m$). Such an asymmetry in the spectrum is known to induce a Chern-Simons term in the gauge sector of the theory [153, 164]. In order to avoid the unnecessary complication, it is convenient to use the following reducible representation instead:

$$\gamma^0 = \begin{pmatrix} \sigma_3 & 0 \\ 0 & -\sigma_3 \end{pmatrix}, \qquad \gamma^1 = \begin{pmatrix} i\sigma_1 & 0 \\ 0 & -i\sigma_1 \end{pmatrix}, \qquad \gamma^2 = \begin{pmatrix} i\sigma_2 & 0 \\ 0 & -i\sigma_2 \end{pmatrix}. \tag{2.6}$$

(Incidentally, in the low-energy theory of graphene, which is a real quasi-relativistic system in $2 + 1$ dimensions, such a reducible representation appears automatically [170].) The corresponding Dirac spectrum reads

$$E_n^{(2+1)} = \pm\sqrt{m^2 + 2|eB|n}. \tag{2.7}$$

As we see, this is very similar to the $(3 + 1)$-dimensional result in (2.3), except for the missing dependence on the longitudinal momentum p_3.

Repeating the same arguments as in the $(3 + 1)$-dimensional case, we find that the low-energy sector of the Dirac theory in $2 + 1$ dimensions is also determined by the lowest Landau level. Here we assume again that the Dirac mass is much smaller than the corresponding Landau energy scale ($m \ll \sqrt{|eB|}$) in order to insure a clear separation of the low- and high-energy scales.

Just like in the higher dimensional case, all Landau levels are (infinitely) degenerate. In particular, the number of degenerate states per unit area is $\frac{|eB|}{2\pi}$ in the lowest Landau level. A special feature of the $(2+1)$-dimensional theory is a discrete, rather than continuous spectrum of excitations. In the absence of the x^3-direction and the associated quantum number p_3, all positive energy states in the lowest Landau level have the same energy $E_0 = m$. Moreover, when $m \to 0$, this energy goes to zero and becomes degenerate with the negative energy states $E_0 = -m$. In this limit, there is an infinite vacuum degeneracy even if the condition of charge neutrality may favor a state with exactly half-filling of the lowest Landau level. It should be expected, however, that taking into account any type of fermion interaction will lead to a well defined ground state, in which the interaction energy is minimized. One can even make an educated guess that the corresponding ground state should be a Mott-type insulator with a dynamically generated mass/gap.

Another special and rather unusual feature of the $(2 + 1)$-dimensional Dirac fermions in a magnetic field is a spontaneous symmetry breaking, which is manifested by a nonzero "chiral" condensate $\langle \bar{\Psi}\Psi \rangle$ already in the *free* theory. To see this, let us make use of the proper-time representation of the fermion propagator in the magnetic field, see (2.101) in the Appendix. In the limit of a small bare mass ($m_0 \to 0$), we easily derive the following (regularized) expansion for the condensate:

$$\langle \bar{\Psi}\Psi \rangle \equiv -\text{tr}\big[S_{2+1}(x,x)\big] = -\frac{m_0|eB|}{2\pi^{3/2}} \int_{1/\Lambda^2}^{\infty} \frac{ds}{\sqrt{s}} e^{-sm_0^2} \coth\big(s|eB|\big)$$

$$\simeq -\frac{|eB|}{2\pi}\text{sign}(m_0) - \frac{m_0}{\pi^{3/2}}\left[\Lambda + \sqrt{\frac{\pi|eB|}{2}}\zeta\left(\frac{1}{2}, 1 + \frac{m_0^2}{2|eB|}\right)\right]. \tag{2.8}$$

It can be shown that the first term, which remains nonzero even in the massless limit, comes from the lowest Landau level. At first sight, this may appear to be a very surprising result. Upon a closer examination, one finds that this condensate is directly connected with a nonzero density of states and a nonzero spin polarization in the lowest Landau level of the free Dirac theory. The result is unambiguous only after specifying the sign of the bare mass parameter, which is also typical for spontaneous symmetry breaking.

In connection with the result in (2.8), it may be useful to recall that the chirality is not well defined in the $(2+1)$-dimensional space. However, as we will discuss in Sect. 2.2.2, the condensate $\langle \bar{\Psi} \Psi \rangle$ is still of interest because it breaks another global symmetry that has a status similar to that of the conventional chiral symmetry.

2.2.2 Magnetic Catalysis in 2 + 1 Dimensions

Now, let us consider a Nambu-Jona-Lasinio (NJL) model in $2+1$ dimensions, in which the magnetic catalysis of symmetry breaking is realized in its simplest possible form [99, 102]. When using the reducible representation of Dirac algebra, given by (2.6), one finds that the kinetic part of the massless Dirac theory is invariant under a global $U(2)$ flavor symmetry. The generators of the symmetry transformations are given by $T_0 = I$, $T_1 = \gamma^5$, $T_2 = \frac{1}{i}\gamma^3$, and $T_3 = \gamma^3 \gamma^5$, where $\gamma^5 \equiv i\gamma^0\gamma^1\gamma^2\gamma^3$. A dynamical Dirac mass will break this $U(2)$ symmetry down to the $U(1) \times U(1)$ subgroup with generators T_0 and T_3.

The NJL-type Lagrangian density, with the interaction term invariant under the $U(2)$ flavor symmetry, can be written down as follows:

$$\mathcal{L} = \bar{\Psi} i\gamma^\mu D_\mu \Psi + \frac{G}{2}\left[(\bar{\Psi}\Psi)^2 + (\bar{\Psi}i\gamma^5\Psi)^2 + (\bar{\Psi}\gamma^3\Psi)^2\right], \qquad (2.9)$$

where G is a dimensionfull coupling constant. This theory is nonrenormalizable, but can be viewed as a low-energy effective theory with a range of validity extending up to a certain ultraviolet energy scale set by a physically motivated choice of the cutoff parameter Λ.

2.2.2.1 Weak Coupling Approximation

As usual in problems with spontaneous symmetry breaking, we use the method of Schwinger-Dyson (gap) equation in order to solve for the dynamical mass parameter. We assume that the structure of the (inverse) full fermion propagator is the same as in the free theory, but has a nonzero dynamical Dirac mass m,

$$S^{-1}(x, x') = -i\left[i\gamma^0\partial_t - (\boldsymbol{\pi} \cdot \boldsymbol{\gamma}) - m\right]\delta^3(x - x'). \qquad (2.10)$$

In the mean-field approximation, the dynamical mass parameter satisfies the following gap equation:

$$m = G\mathrm{tr}\left[S(x, x)\right]. \qquad (2.11)$$

To leading order in weak coupling, $G \to 0$, this equation can be solved perturbatively. Indeed, by substituting the condensate calculated at $m_0 \to 0$ in the *free* theory, see (2.8), into the right-hand side of the gap equation (2.11), we obtain

$$|m| \simeq G \frac{|eB|}{2\pi}. \tag{2.12}$$

As we see, the dynamical mass is induced at any nonzero attractive coupling.

The massless NJL model in (2.9) is invariant under the flavor $U(2)$ symmetry. The generation of a Dirac mass m is only one of many equivalent ways of breaking this symmetry down to a $U(1) \times U(1)$ subgroup. Indeed, by applying a general $U(2)$ transformation, we find that the Dirac mass term can be turned into a linear combination of the following three mass terms: m, $\gamma^3 m_3$, and $i\gamma^5 m_5$. (In principle, there is also a possibility of the so-called Haldane mass term $\gamma^3 \gamma^5 \Delta$, which is a singlet under $U(2)$. We do not discuss it here. However, as we will see in Sect. 2.3.3, the Haldane mass plays an important role in graphene.)

In our perturbative analysis, we did not get any nonzero m_3 or m_5 because the vacuum alignment was predetermined by a "seed" Dirac mass m_0 in the free theory, see (2.8).

2.2.2.2 Large N Approximation

It is instructive to generalize the above analysis in the NJL model to the case of strong coupling. While magnetic catalysis occurs even at arbitrarily weak coupling, such a generalization will be useful to understand how magnetic catalysis is lost in the limit of the vanishing magnetic field.

At strong coupling, a reliable solution to the NJL model can be obtained by using the so-called large N approximation, which is rigorously justified when the fermion fields in (2.9) carry an additional, "color" index $\alpha = 1, 2, \ldots, N$, and N is large. Using the Hubbard-Stratonovich transformation [115, 180], one can show that the NJL theory in (2.9) is equivalent to the following one:

$$\mathscr{L} = \bar{\Psi} i \gamma^\mu D_\mu \Psi - \bar{\Psi} (\sigma + \gamma^3 \tau + i\gamma^5 \pi) \Psi - \frac{1}{2G} (\sigma^2 + \pi^2 + \tau^2). \tag{2.13}$$

Note that the equations of motion for the new composite fields read

$$\sigma = -G(\bar{\Psi}\Psi), \qquad \tau = -G(\bar{\Psi}\gamma^3\Psi), \qquad \pi = -G(\bar{\Psi}i\gamma^5\Psi). \tag{2.14}$$

Under $U(2)$ flavor symmetry transformations, these composite fields transform into linear combinations of one another, but the quantity $\sigma^2 + \pi^2 + \tau^2$ remains invariant.

The effective action for the composite fields,

$$\Gamma = -\frac{1}{2G} \int d^3x (\sigma^2 + \tau^2 + \pi^2) - i\mathrm{tr}\mathrm{Ln}[i\gamma^\mu D_\mu - (\sigma + \gamma^3\tau + i\gamma^5\pi)], \tag{2.15}$$

is obtained by integrating out the fermionic degrees of freedom from the action. It is convenient to expand this effective action in powers of derivatives of the composite fields. The leading order in such an expansion is the effective potential V (up to the minus sign). Because of the flavor symmetry, the effective potential depends on σ, π, and τ fields only through their $U(2)$ invariant combination $\rho^2 = \sigma^2 + \pi^2 + \tau^2$.

Using the proper-time regularization, one obtains the following explicit expression for the effective potential [99, 102]:

$$V(\rho) \simeq \frac{N}{\pi} \left[\frac{\Lambda}{2} \left(\frac{1}{g} - \frac{1}{\sqrt{\pi}} \right) \rho^2 - \sqrt{2} |eB|^{3/2} \zeta \left(-\frac{1}{2}, 1 + \frac{\rho^2}{2|eB|} \right) - \frac{\rho |eB|}{2} \right], \quad (2.16)$$

where we dropped the terms suppressed by the ultraviolet cutoff parameter Λ and introduced a dimensionless coupling constant $g \equiv N\Lambda G/\pi$.

The field configuration ρ that minimizes the effective potential is determined by solving the equation $dV/d\rho = 0$, i.e.,

$$\Lambda \left(\frac{1}{g} - \frac{1}{\sqrt{\pi}} \right) \rho = \frac{|eB|}{2} + \rho \sqrt{\frac{|eB|}{2}} \zeta \left(\frac{1}{2}, 1 + \frac{\rho^2}{2|eB|} \right). \quad (2.17)$$

In essence, this is the gap equation. At weak coupling, $g \to 0$, in particular, we obtain the following approximate solution:

$$m = \rho_{\min} \simeq GN \frac{|eB|}{2\pi}, \quad (2.18)$$

which is the large N generalization of the result for the dynamical mass in (2.12).

2.2.2.3 Zero Magnetic Field Limit in 2 + 1 Dimensions

Before concluding the discussion of the $(2 + 1)$-dimensional NJL model, it is instructive to consider how the above analysis of flavor symmetry breaking modifies in the zero magnetic field case. By taking the limit $B \to 0$ in (2.16), we arrive at the following effective potential:

$$V_{B=0}(\rho) \simeq \frac{N}{\pi} \left[\frac{\Lambda}{2} \left(\frac{1}{g} - \frac{1}{\sqrt{\pi}} \right) \rho^2 + \frac{1}{3} \rho^3 \right]. \quad (2.19)$$

Now, the zero-field limit of the gap equation, $dV_{B=0}/d\rho = 0$, is given by

$$\Lambda \left(\frac{1}{\sqrt{\pi}} - \frac{1}{g} \right) \rho = \rho^2. \quad (2.20)$$

This is very different from (2.17). In particular, the only solution to this equation at $g < \sqrt{\pi}$ is a trivial one, $\rho = 0$. The nontrivial solution appears only in the case of a sufficiently strong coupling constant, $g > \sqrt{\pi}$. This is in a stark contrast with the dynamical mass generation in the presence of a magnetic field, where a nontrivial solution exists at arbitrarily small values of the coupling constant g.

2.2.3 Magnetic Catalysis in 3 + 1 Dimensions

Let us now extend the analysis of the previous subsection to the case of a $(3 + 1)$-dimensional model, where the dynamics is truly nonperturbative. The Lagrangian density of the corresponding NJL model reads

$$\mathcal{L} = \bar{\Psi} i \gamma^\mu D_\mu \Psi + \frac{G}{2} [(\bar{\Psi}\Psi)^2 + (\bar{\Psi} i \gamma^5 \Psi)^2]. \tag{2.21}$$

This model possesses the $U(1)_L \times U(1)_R$ chiral symmetry. The symmetry will be spontaneously broken down to the $U(1)_{L+R}$ subgroup when a dynamical Dirac mass is generated.

2.2.3.1 Weak Coupling Approximation

In the weakly coupled limit, the gap equation in the mean-field approximation reads

$$m = G \text{tr}[S(x, x)]. \tag{2.22}$$

Formally, it is same as the gap equation in the $(2 + 1)$-dimensional model in (2.11). However, as we will see now, its symmetry breaking solution will be qualitatively different.

Let us start by showing that the chiral condensate vanishes in the free theory in $3 + 1$ dimensions when the bare mass goes to zero, $m_0 \to 0$. By making use of the proper-time representation, see (2.97) in the Appendix, we obtain

$$\langle \bar{\Psi}\Psi \rangle \equiv -\text{tr}[S(x, x)] = -\frac{m_0 |eB|}{(2\pi)^2} \int_{1/\Lambda^2}^\infty \frac{ds}{s} e^{-sm_0^2} \coth(s|eB|)$$

$$\simeq -\frac{m_0}{(2\pi)^2} \left[\Lambda^2 - m_0^2 \left(\ln \frac{\Lambda^2}{2|eB|} - \gamma_E \right) \right.$$

$$\left. + |eB| \ln \frac{m_0^2}{4\pi |eB|} + 2|eB| \ln \Gamma \left(\frac{m_0^2}{2|eB|} \right) \right]$$

$$\simeq -\frac{m_0}{(2\pi)^2} \left[\Lambda^2 + |eB| \ln \frac{|eB|}{\pi m_0^2} - m_0^2 \ln \frac{\Lambda^2}{2|eB|} + O \left(\frac{m_0^4}{|eB|} \right) \right]. \tag{2.23}$$

In the limit $m_0 \to 0$, we see that the condensate indeed vanishes. This means that we cannot apply a perturbative approach to find any nontrivial (symmetry breaking) solutions to the gap equation in (2.22).

The explicit form of the gap equation (2.22) reads

$$m \simeq G \frac{m}{(2\pi)^2} \left[\Lambda^2 + |eB| \ln \frac{|eB|}{\pi m^2} \right]. \tag{2.24}$$

Its nontrivial solution is given by

$$m \simeq \sqrt{\frac{|eB|}{\pi}} \exp\left(\frac{\Lambda^2}{2|eB|}\right) \exp\left(-\frac{2\pi^2}{G|eB|}\right). \tag{2.25}$$

When $G \to 0$ this result reveals an essential singularity. Obviously, such a dependence cannot possibly be obtained by resuming any finite number of perturbative corrections in powers of a small coupling constant. Therefore, despite the weak coupling, the result for the dynamical mass (2.25) is truly nonperturbative.

2.2.3.2 Zero Magnetic Field Limit in 3 + 1 Dimensions

It is instructive to compare the above dynamics of spontaneous symmetry breaking with case of the zero magnetic field. At $B = 0$, the chiral condensate in the free theory is easily obtained from taking the appropriate limit in (2.23), i.e.,

$$\langle \bar{\Psi}\Psi \rangle_{B=0} = -\frac{m_0}{(2\pi)^2}\left[\Lambda^2 - m_0^2\left(\ln\frac{\Lambda^2}{m_0^2} + 1 - \gamma_E\right)\right]. \tag{2.26}$$

The corresponding gap equation is

$$m \simeq G\frac{m}{(2\pi)^2}\left[\Lambda^2 - m^2\ln\frac{\Lambda^2}{m^2}\right]. \tag{2.27}$$

Because of the negative sign in front of the logarithmic term, this equation does not have any nontrivial solutions for the dynamical mass at vanishingly small coupling constant $g \equiv G\Lambda^2/(2\pi)^2$. In fact, for the whole range of subcritical values, $g < 1$, the only solution to this gap equation is $m = 0$. The nontrivial solution appears only in the case of sufficiently strong coupling, $g > 1$.

2.2.4 Symmetry Breaking as Bound State Problem

In this subsection, we consider an alternative approach to the problem of chiral symmetry breaking in the NJL model in a constant magnetic field. As we will see, this approach is particularly beneficial for illuminating the role of the dimensional reduction in magnetic catalysis.

Instead of solving the gap equation, we consider the problem of bound states with the quantum numbers of the Nambu-Goldstone bosons, using the method of a homogeneous Bethe-Salpeter equation (for a review, see Ref. [144]). The underlying idea for this framework is motivated by the Goldstone theorem [82, 83, 150]. The theorem states that spontaneous breaking of a continuous global symmetry leads to the appearance of new massless scalar particles (i.e., Nambu-Goldstone bosons) in the low-energy spectrum of the theory. The total number of Nambu-Goldstone bosons and their quantum numbers are determined by the broken symmetry generators.

The homogeneous Bethe-Salpeter equation for a pion-like state takes the form [103, 144]:

$$
\chi_{ab}(x, y; P) = -i \int d^4x' d^4y' d^4x'' d^4y'' S_{aa_1}(x, x') K_{a_1 b_1; a_2 b_2}(x'y', x''y'')
$$

$$
\times \chi_{a_2 b_2}(x'', y''; P) S_{b_1 b}(y', y),
\tag{2.28}
$$

where $\chi_{ab}(x, y; P) \equiv \langle 0|T\psi_a(x)\bar\psi_b(y)|P; \pi\rangle$ is the Bethe-Salpeter wave function of the bound state boson with four-momentum P, and $S_{ab}(x, y) = \langle 0|T\psi_a(x)\bar\psi_b(y)|0\rangle$ is the fermion propagator. Here and below, the sum over repeated Dirac indices (a_1, b_1, a_2, b_2) is assumed. The explicit form the Bethe-Salpeter kernel is [103, 144]:

$$
K_{a_1 b_1; a_2 b_2}(x'y', x''y'') = G\big[\delta_{a_1 b_1}\delta_{b_2 a_2} + (i\gamma_5)_{a_1 b_1}(i\gamma_5)_{b_2 a_2} - (i\gamma_5)_{a_1 a_2}(i\gamma_5)_{b_2 b_1}
$$

$$
- \delta_{a_1 a_2}\delta_{b_2 b_1}\big]\delta^4(x' - y')\delta^4(x' - x'')\delta^4(x' - y'').
\tag{2.29}
$$

It is convenient to rewrite the wave function in terms of the relative coordinate $z \equiv x - y$ and the center of mass coordinate $X \equiv (x + y)/2$,

$$
\chi_{ab}(X, z; P) = e^{is_\perp X^1 z^2/\ell^2} e^{-i P_\mu X^\mu} \tilde\chi_{ab}(X, z; P),
\tag{2.30}
$$

where we factorized the Schwinger phase factor, see (2.88), and introduced the notation: $s_\perp \equiv \text{sign}(eB)$ and $\ell = 1/\sqrt{|eB|}$. After substituting the wave function (2.30) and the kernel (2.29) into (2.28), we arrive at the following equation:

$$
\tilde\chi_{ab}(z; P) = -iG \int d^4X' \tilde S_{aa_1}\left(\frac{z}{2} - X'\right)\big[\delta_{a_1 b_1} \text{tr}[\tilde\chi(0; P)] - (\gamma_5)_{a_1 b_1}\text{tr}[\gamma_5\tilde\chi(0; P)]
$$

$$
- \tilde\chi_{a_1 b_1}(0; P) + (\gamma_5)_{a_1 a_2}\tilde\chi_{a_2 b_2}(0; P)(\gamma_5)_{b_2 b_1}\big]\tilde S_{b_1 b}\left(\frac{z}{2} + X'\right)
$$

$$
\times e^{i\frac{s_\perp}{2\ell^2}(z^1 X'^2 - X'^1 z^2)} e^{i P_\mu X'^\mu}.
\tag{2.31}
$$

Here we took into account that the equation admits a translation invariant solution and replaced $\tilde\chi_{ab}(X, z; P) \to \tilde\chi_{ab}(z, P)$. Note that, on the right-hand side of (2.31), the dependence on the center of mass coordinate X completely disappeared after a shift of the integration variable $X' \to X' - X$ was made.

In the lowest Landau level approximation, one can show that the Fourier transform of the Bethe-Salpeter wave function takes the form [103]:

$$
\tilde\chi_{ab}(p; P \to 0) = A(p_\parallel)e^{-p_\perp^2 \ell^2}\frac{\gamma^0\omega - \gamma^3 p^3 - m}{\omega^2 - (p^3)^2 - m^2}\gamma^5\mathscr{P}_+\frac{\gamma^0\omega - \gamma^3 p^3 - m}{\omega^2 - (p^3)^2 - m^2},
\tag{2.32}
$$

where $p_\parallel = (\omega, p^3)$, $p_\perp^2 = (p^1)^2 + (p^2)^2$, and $\mathscr{P}_\pm = \frac{1}{2}(1 \pm i s_\perp \gamma^1 \gamma^2)$. The new function $A(p_\parallel)$ satisfies the equation:

$$A(p_\parallel) = \frac{G|eB|}{4\pi^3} \int \frac{A(k_\parallel) d^2 k_\parallel}{k_\parallel^2 + m^2}, \tag{2.33}$$

where we made the Wick rotation ($\omega \to i\omega$). The solution to this equation is a constant: $A(p_\parallel) = C$. Dropping nonzero C and cutting off the integration at Λ, we finally arrive at the gap equation for the mass parameter m:

$$1 \simeq \frac{G|eB|}{4\pi^2} \int_0^{\Lambda^2} \frac{dk_\parallel^2}{k_\parallel^2 + m^2}. \tag{2.34}$$

The solution to this equation is

$$m \simeq \Lambda^2 \exp\left(-\frac{2\pi^2}{G|eB|}\right), \tag{2.35}$$

which, to leading order, agrees with the solution obtained in (2.25).

The Bethe-Salpeter equation (2.33) can be rewritten in the form of a two-dimensional Schrödinger equation with an attractive δ-function potential. In order to see this explicitly, let us introduce the following wave function

$$\psi(\mathbf{r}) = \int \frac{d^2 k_\parallel}{(2\pi)^2} \frac{e^{-i\mathbf{k}_\parallel \mathbf{r}}}{k_\parallel^2 + m^2} A(k_\parallel). \tag{2.36}$$

Taking into account that $A(p_\parallel)$ satisfies (2.33), it is straightforward to show that the wave function $\psi(\mathbf{r})$ satisfies the following Schrödinger type equation:

$$\left(-\frac{\partial^2}{\partial r_1^2} - \frac{\partial^2}{\partial r_2^2} + m^2 - \frac{G|eB|}{\pi} \delta_\Lambda^2(\mathbf{r})\right) \psi(\mathbf{r}) = 0, \tag{2.37}$$

in which $-m^2$ plays the role of the energy E. Since m^2 must be positive, the problem is reduced to finding the spectrum of bound states (with $E = -m^2 < 0$) in the Schrödinger problem. The potential energy in 3(2.37) is expressed in terms of

$$\delta_\Lambda^2(\mathbf{r}) = \int_\Lambda \frac{d^2 k_\parallel}{(2\pi)^2} e^{-i\mathbf{k}_\parallel \mathbf{r}}, \tag{2.38}$$

which is a regularized version of the δ-function that describes the local interaction in the NJL model.

Notice that, by using the same approach, one can show that the Bethe-Salpeter equation for a massless NG-boson state in the NJL model in $2 + 1$ dimensions can be reduced to the gap equation

$$A(p) = \frac{G|eB|}{2\pi^2} \int_{-\Lambda}^{\Lambda} \frac{A(k) dk}{k^2 + m^2}. \tag{2.39}$$

This has the same solution for the mass as in (2.12). Also, this integral equation is equivalent to the following one-dimensional Schrödinger equation:

$$\left(-\frac{d^2}{dx^2} + m^2 - \frac{G|eB|}{\pi^2}\delta_\Lambda(x)\right)\psi(x) = 0, \tag{2.40}$$

where the regularized version of the δ-function is given by $\delta_\Lambda(x) = \int_{-\Lambda}^{\Lambda} \frac{dk}{2\pi} e^{-ikx}$.

2.2.5 Analogy with Superconductivity

It is interesting to point that the dynamics described by the gap equation in the case of magnetic catalysis has a lot of conceptual similarities to the mechanism of super-conductivity in metals and alloys. This is despite the clear differences between the two phenomena that we discussed in the Introduction. (In order to avoid a possible confusion, let us emphasize that here we compare the nonrelativistic Cooper pairing dynamics in superconductivity in the *absence* of magnetic fields with the relativistic dynamical generation of a mass in the *presence* of a constant magnetic field.)

The corresponding gap equation in the Bardeen-Cooper-Schrieffer theory [17] of superconductivity can be written in the following form:

$$1 = GN(0) \int_0^{\hbar\omega_D} \frac{d\varepsilon}{\sqrt{\varepsilon^2 + \Delta^2}}, \tag{2.41}$$

where $N(0)$ is the density of electron states at the Fermi surface, ω_D is the Debye frequency, and Δ is the energy gap associated with superconductivity. The solution for the gap reads

$$\Delta \simeq \hbar\omega_D \exp\left(-\frac{1}{GN(0)}\right). \tag{2.42}$$

At weak coupling, this solution has the same essential singularity as the dynamical mass parameter in (2.35). We can argue that the similarity is not accidental. To see this clearly, let us rewrite the gap equation in the problem of magnetic catalysis in the lowest Landau level approximation, see (2.34), as follows:

$$1 = G\frac{|eB|}{2\pi} \int \frac{d\omega dp^3}{\omega^2 + (p^3)^2 + m^2} \simeq G\frac{|eB|}{2\pi} \int_0^\Lambda \frac{d\omega}{\sqrt{\omega^2 + m^2}}, \tag{2.43}$$

where the Wick rotation was performed ($\omega \to i\omega$). As we see, the structure of this gap equation is identical to its counterpart in the BCS theory after we identify the density of states $N(0)$ with the density of states in the lowest Landau level, $|eB|/(2\pi)$, and the Debye frequency ω_D with the cutoff parameter Λ.

The similarity between the BCS theory of superconductivity and magnetic catalysis goes deeper. In particular, the generation of a nonzero gap in superconductors can be also thought of as the result of a $3 + 1 \to 1 + 1$ dimensional reduction of the phase space around the Fermi surface. Also, just like in magnetic catalysis, it is essential that the density of states at the Fermi surface is nonzero.

2.2.6 Bound States in Lower Dimensions

As we saw in Sect. 2.2.4, the problem of spontaneous symmetry breaking and the associated dynamical generation of the Dirac mass can be reformulated as a problem of composite massless states with the quantum numbers of Nambu-Goldstone bosons.

In the presence of a constant magnetic field, in particular, we also found that the corresponding Bethe-Salpeter equation for the bound states can be recast in an equivalent form as a Schrödinger equation in a dimensionally reduced space. The dimensional reduction is $D \to D - 2$ and, therefore, the relevant problem of bound states is considered in spaces of lower dimensions.

In order to prove that the essence of magnetic catalysis is directly connected with this reduction, let us consider a simple quantum mechanical problem: the formation of bound states in a shallow potential well in spaces of various dimensions. As we will see, at least one bound state does exist in one- and two-dimensional cases [19, 136, 177, 178], but not always in three dimensions. We will also see that, while the result for the binding energy is perturbative in the coupling constant in one dimension, it has an essential singularity in two dimensions.

2.2.6.1 Bound States in a One-Dimensional Potential Well

Let us start from the simplest one-dimensional problem of a nonrelativistic particle of mass m_* confined to move on a line. Let the potential energy of the well be given by $U(x)$, which is negative and quickly approaches zero when $|x| \to \infty$. One can show that even a vanishingly small depth of the potential well is sufficient to produce a bound state (i.e., a quantum state with a negative energy). The corresponding binding energy is given by [136]

$$|E_{1D}| \simeq \frac{m_*}{2\hbar^2}\left(-\int_{-\infty}^{\infty} U(x)dx\right)^2. \tag{2.44}$$

If we rescale the potential energy $U(x)$ by a "coupling constant" factor g, i.e., $U(x) \to gU(x)$, we find that $|E_{1D}| \sim g^2$ as $g \to 0$. In other words, the binding energy has a power-law dependence as a function of the depth of the potential energy $U(x)$. This is a typical result that can be obtained by perturbative techniques, controlled by powers of the small parameter g [19].

The above conclusion remains valid basically for any attractive potential $U(x)$. For example, one can rigorously prove that, if $\int(1 + |x|)|U(x)|dx < \infty$, there is a bound state for all small positive g if and only if $\int U(x)dx \le 0$ (i.e., the potential is attractive at least on average) [126].

2.2.6.2 Bound States in a Two-Dimensional Potential Well

In the case of a two-dimensional system (i.e., a nonrelativistic particle of mass m_* confined to move on a plane), the general conclusion about the existence of a bound state around a potential well of a vanishingly small depth still remains valid. However, an important qualitative difference appears in the result. The binding energy reveals an essential singularity as a function of the depth of the potential well. In order to understand this better, let us consider a problem with a cylindrically symmetric potential energy $U(r)$, where r is the radial polar coordinate in the plane. If the potential energy is sufficiently shallow and localized (i.e., $|\int_0^\infty rU(r)dr| \ll m_*/\hbar^2$), one finds that the energy of the bound state is given by [136]

$$|E_{2D}| \simeq \frac{\hbar^2}{m_* a^2} \exp\left(-\frac{\hbar^2}{m_*}\left|\int_0^\infty rU(r)dr\right|^{-1}\right), \qquad (2.45)$$

where a is the characteristic size of the potential well. The fact that this energy is singular can be made explicit by rescaling the potential energy $U(r)$: $U(r) \to gU(r)$. Then, we find that $|E_{2D}| \sim \exp(-C/g)$ as $g \to 0$ (here C is a constant determined by the shape of the potential well). Unlike the g^2 power-law suppression of the binding energy in one dimension, this is a much stronger suppression indicating a much weaker binding. Moreover and perhaps more importantly, such an essential singularity cannot possibly be obtained by resuming any finite number of perturbative corrections, controlled by powers of the small parameter g. Therefore, the singular behavior of the binding energy in two dimensions is a sign of a truly nonperturbative (albeit weakly-interacting) physics.

Again, this result is very general. It can be rigorously proven that, in the case when $\int |U(x)|^{1+\varepsilon} d^2x < \infty$ (with some $\varepsilon > 0$) and $\int (1+x^2)^\varepsilon |U(x)| d^2x < \infty$, there is a bound state for all small positive g if and only if $\int U(x) d^2x \leq 0$ (i.e., the potential is attractive at least on average) [177, 178].

2.2.6.3 Bound States in a Three-Dimensional Potential Well

Now, in the three-dimensional case, there are no bound states if the potential well is too shallow in depth. This was first shown by Peierls in 1929 [160]. This can be demonstrated, for example, in a special case of a spherically symmetric potential well of a finite size,

$$U(r) = \begin{cases} -g\dfrac{\pi^2\hbar^2}{8m_* a^2} & \text{for } r \leq a, \\ 0 & \text{for } r > a. \end{cases} \qquad (2.46)$$

The condition to have at least one bound state is $g > 1$ [136]. In other words, the depths of the potential well (or the strength of the "coupling constant" g) should be larger than the critical value, given by $g_{cr} = 1$. In the supercritical regime, $g = 1 + \varepsilon$ with $0 < \varepsilon \ll 1$, the binding energy is given by [136]

$$|E_{3D}| = \frac{\pi^4 \hbar^2}{2^7 m_* a^2} \varepsilon^2.$$

(2.47)

In the subcritical regime $g < 1$, on the other hand, there are no bound states at all.

2.3 Magnetic Catalysis in Gauge Theories

Motivated by the fact that magnetic catalysis has a rather general underlying physics, explained by the dimensional reduction of the particle-antiparticle pairing, it is natural to ask how it is realized in gauge theories with long-range interactions, such as QED. This problem was discussed in numerous studies [5, 6, 11, 12, 51, 53, 101, 104–107, 110, 111, 137, 138, 158, 159, 165]. Here we will briefly review only the key results and refer the reader to the original papers for further details.

2.3.1 Magnetic Catalysis in QED

Using the same conceptual approach as outlined in Sect. 2.2.4 for the NJL model, one can show that, in Euclidean space, the equation describing a pion-like Nambu-Goldstone boson in QED in a magnetic field has the form of a two-dimensional Schrödinger equation [101]:

$$\left[-\frac{\partial^2}{\partial r_1^2} - \frac{\partial^2}{\partial r_2^2} + m^2 + V(\mathbf{r}) \right] \psi(\mathbf{r}) = 0.$$

(2.48)

The function $\psi(\mathbf{r})$ is defined in terms of the Bethe-Salpeter wave function $A(p)$ in exactly the same way as in the NJL model, see (2.36). This time, however, $A(p)$ satisfies a different integral equation,

$$A(p) = \frac{\alpha}{2\pi^2} \int \frac{d^2 k A(k)}{k^2 + m^2} \int\limits_0^\infty \frac{dx \exp(-x \ell^2 / 2)}{(\mathbf{k} - \mathbf{p})^2 + x},$$

(2.49)

where $\ell = 1/\sqrt{|eB|}$ is the magnetic length. Note that, in addition to using the lowest Landau level approximation, we assumed that the photon screening effects are negligible. As is easy to check, the explicit form of the potential $V(\mathbf{r})$ is given by [101]

$$V(\mathbf{r}) = \frac{\alpha}{\pi \ell^2} \exp\left(\frac{r^2}{2\ell^2} \right) \mathrm{Ei}\left(-\frac{r^2}{2\ell^2} \right),$$

(2.50)

where $r^2 = r_1^2 + r_2^2$ and $\mathrm{Ei}(x) = -\int_{-x}^{\infty} dt \exp(-t)/t$ is the integral exponential function [94]. Since $V(\mathbf{r})$ is negative, we have a Schrödinger equation with an attractive potential, in which the parameter $-m^2$ plays the role of the energy E. Therefore, the problem is again reduced to finding the spectrum of bound states with $E = -m^2 < 0$.

It is known that the energy of the lowest level $E(\alpha)$ for the two-dimensional Schrödinger equation is a nonanalytic function of the coupling constant α at $\alpha = 0$ [178]. If the potential $V(\mathbf{r})$ were short-range, the result would have the form $m^2 = -E(\alpha) \propto \exp[-1/(C\alpha)]$, where C is a positive constant [177, 178]. In our case, however, we have a long-range potential. Indeed, using the asymptotic expansion for $\mathrm{Ei}(x)$ [94], we get:

$$V(\mathbf{r}) \simeq -\frac{2\alpha}{\pi}\frac{1}{r^2}, \quad r \to \infty. \tag{2.51}$$

In order to find an approximate solution for m^2, one can use the integral equation (2.49) at $p = 0$. As $\alpha \to 0$, the dominant contribution in the integral on the right-hand side comes from the infrared region $k^2 \lesssim m^2$. Therefore,

$$A(0) \simeq \frac{\alpha}{2\pi^2}A(0)\int \frac{d^2k}{k^2+m^2}\int\limits_0^\infty \frac{dx\,\exp(-y/2)}{l^2k^2+y} \simeq \frac{\alpha}{4\pi}A(0)\left[\ln\left(\frac{m^2\ell^2}{2}\right)\right]^2,$$

$$\tag{2.52}$$

which implies that [101]

$$m \propto \sqrt{|eB|}\exp\left(-\sqrt{\frac{\pi}{\alpha}}\right). \tag{2.53}$$

A slightly more careful analysis of the integral equation (2.49) can be made by approximating the interaction kernel so that the exchange momentum $(\mathbf{k}-\mathbf{p})^2$ in the denominator is replaced by $\max(k^2, p^2)$. The problem then reduces to an ordinary differential equation with two (infrared and ultraviolet) boundary conditions. The approximate analytical solution reveals that the lowest energy bound state, which describes the stable vacuum solution in quantum field theory, corresponds to the following value of the dynamical mass [101]:

$$m \simeq C\sqrt{|eB|}\exp\left(-\frac{\pi}{2}\sqrt{\frac{\pi}{2\alpha}}\right). \tag{2.54}$$

Unfortunately, the approximation used in this analysis is not completely reliable. There are higher order diagrams that can substantially modify the interaction potential and, in turn, the result for the dynamical mass. For example, taking into account the vacuum polarization effects in the improved rainbow (ladder) approximation, in which the free photon propagator is replaced by a screened interaction with the one-loop photon self-energy, the result changes. The corrected expression for the mass has the same form as in (2.54), but with α replaced by $\alpha/2$ [101]. This is a clear indication that, despite weak coupling, there can exist other relevant contributions, coming from higher order diagrams.

A further study showed that, by using a similarity between the magnetic catalysis problem in QED and the exactly solvable Schwinger model [73, 169], one can find a special nonlocal gauge, in which the leading singularity of the dynamical mass can be extracted exactly [104],

$$m \simeq \tilde{C}\sqrt{|eB|}\, F(\alpha) \exp\!\left[-\frac{\pi}{\alpha \ln(C_1/N\alpha)}\right], \tag{2.55}$$

where N is the number of fermion flavors, $F(\alpha) \simeq (N\alpha)^{1/3}$, $C_1 \simeq 1.82 \pm 0.06$ and $\tilde{C} \sim O(1)$. Note that the leading singularity in the final expression for the mass is quite different from that in the rainbow approximation (2.54).

The magnetic catalysis of chiral symmetry breaking in QED yields a rare example of dynamical symmetry breaking in a $(3+1)$-dimensional gauge theory without fundamental scalar fields, in which there exists a consistent truncation of the Schwinger-Dyson equation.

2.3.2 Magnetic Catalysis in QCD

Recently there was an increased interest in studies of QCD in a strong magnetic field [2, 14–16, 23, 25, 31, 34, 67–72, 77, 78, 121, 146–149]. There are several reasons why such investigations may be of interest. Very strong magnetic fields are known to have existed in the Early Universe [18, 30, 44, 95, 183] and are expected to be generated in relativistic heavy ion collisions [123, 179]. Since the chiral symmetry plays a profound role in QCD, it is interesting to study also the role of magnetic catalysis in this theory [121, 146].

Because of the property of asymptotic freedom, one can argue that the dynamics underlying magnetic catalysis in QCD is, at least in principle, weakly coupled at sufficiently large magnetic fields [121]. This fact can be used to justify a consistent truncation of the Schwinger-Dyson equation, resembling that in QED, which we discussed in the preceding section.

Let us start by introducing a QCD like theory with N_u up flavors of quarks having electric charges $2e/3$ and N_d down flavors of quarks having electric charges $-e/3$. (The total number of flavors is $N_f = N_u + N_d$.) It is important to distinguish the up and down types of quarks because the chiral symmetry subgroup that mixes them is explicitly broken by the external magnetic field. Taking this into account, we find that the model is invariant under the $SU(N_u)_L \times SU(N_u)_R \times SU(N_d)_L \times SU(N_d)_R \times U^{(-)}(1)_A$ chiral symmetry. The anomaly free subgroup $U^{(-)}(1)_A$ is connected with the conserved current which is the difference of the $U^{(d)}(1)_A$ and $U^{(u)}(1)_A$ currents. [The $U^{(-)}(1)_A$ symmetry is of course absent when either N_d or N_u equals zero.] A dynamical generation of quark masses spontaneously breaks the chiral symmetry down to $SU(N_u)_V \times SU(N_d)_V$ and gives rise to $N_u^2 + N_d^2 - 1$ massless Nambu-Goldstone bosons in the low-energy spectrum.

Just like in QED, the vacuum polarization effects play a very important role in QCD in the presence of a strong magnetic field. By properly modifying the known result from the Abelian gauge theory [26, 36, 140] to the case of QCD, we find that the gluon polarization tensor has the following behavior:

$$\Pi^{AB,\mu\nu} \simeq \frac{\alpha_s}{6\pi}\delta^{AB}\left(k_\parallel^\mu k_\parallel^\nu - k_\parallel^2 g_\parallel^{\mu\nu}\right)\sum_{q=1}^{N_f}\frac{|e_q B|}{m_q^2}, \quad \text{for } |k_\parallel^2| \ll m_q^2, \tag{2.56}$$

$$\Pi^{AB,\mu\nu} \simeq -\frac{\alpha_s}{\pi}\delta^{AB}\left(k_\|^\mu k_\|^\nu - k_\|^2 g_\|^{\mu\nu}\right)\sum_{q=1}^{N_f}\frac{|e_q B|}{k_\|^2}, \quad \text{for } m_q^2 \ll |k_\|^2| \ll |eB|, \quad (2.57)$$

where $k_\|^\mu \equiv g_\|^{\mu\nu} k_\nu$ and $g_\|^{\mu\nu} \equiv \mathrm{diag}(1,0,0,-1)$ is the projector onto the longitudinal subspace. Notice that quarks in a strong magnetic field do not couple to the transverse subspace spanned by $g_\perp^{\mu\nu} \equiv g^{\mu\nu} - g_\|^{\mu\nu} = \mathrm{diag}(0,-1,-1,0)$ and $k_\perp^\mu \equiv g_\perp^{\mu\nu} k_\nu$. This is connected with the dominant role of the lowest Landau level, in which quarks are polarized along the magnetic field.

The expressions (2.56) and (2.57) coincide with those for the polarization operator in the massive Schwinger model [169] if the parameter $\alpha_s |e_q B|/2$ here is replaced by the dimensional coupling α_1 of $(1+1)$-dimensional QED. In particular, (2.57) implies that there is a massive gluon resonance with the mass given by

$$M_g^2 = \sum_{q=1}^{N_f}\frac{\alpha_s}{\pi}|e_q B| = (2N_u + N_d)\frac{\alpha_s}{3\pi}|eB|. \quad (2.58)$$

This is reminiscent of the pseudo-Higgs effect in the $(1+1)$-dimensional massive QED. It is not the genuine Higgs effect because there is no complete screening of gluons in the far infrared region with $|k_\|^2| \ll m_g^2$, see (2.56). Nevertheless, the pseudo-Higgs effect is manifested in creating a massive resonance and this resonance provides the dominant force leading to chiral symmetry breaking.

In the end, the dynamics in QCD in a strong magnetic field appears to be essentially the same as in QED, except for purely kinematic changes. After expressing the magnetic field in terms of the running coupling α_s at the scale $\sqrt{|eB|}$ using

$$\frac{1}{\alpha_s} \simeq b\ln\frac{|eB|}{\Lambda_{\mathrm{QCD}}^2}, \quad \text{where } b = \frac{11N_c - 2N_f}{12\pi}, \quad (2.59)$$

we obtain the result for the dynamical mass in the following form [146]:

$$m_q^2 \simeq 2C_1\left|\frac{e_q}{e}\right|\Lambda_{\mathrm{QCD}}^2(c_q\alpha_s)^{2/3}\exp\left[\frac{1}{b\alpha_s} - \frac{4N_c\pi}{\alpha_s(N_c^2-1)\ln(C_2/c_q\alpha_s)}\right], \quad (2.60)$$

where e_q is the electric charge of the q-th quark and N_c is the number of colors. The numerical factors C_1 and C_2 are of order 1, and the value of c_q is given by

$$c_q = \frac{1}{6\pi}(2N_u + N_d)\left|\frac{e}{e_q}\right|. \quad (2.61)$$

Because of the difference in electric charges, the dynamical mass of the up-type quarks is considerably larger than that of the down-type quarks.

It is interesting to point that the dynamical quark masses in a wide range of strong magnetic fields, $\Lambda_{\mathrm{QCD}}^2 \ll |eB| \lesssim (10\,\mathrm{TeV})^2$, remain much smaller than the dynamical (constituent) masses of quarks $m_q^{(0)} \simeq 300\,\mathrm{MeV}$ in vacuum QCD *without* a magnetic field. This may suggest that QCD can have an intermediate regime, in which the magnetic field is strong enough to provide a gluon screening to interfere with the vacuum pairing dynamics [76, 146], but not sufficiently strong to produce large dynamical masses through magnetic catalysis. In this intermediate regime, the dynamical mass and the associated chiral condensate could be *decreasing* with the magnetic field. The corresponding regime may start already at magnetic fields as low as 10^{19} G, when the gluon mass M_g, given by (2.58), becomes comparable to Λ_{QCD}. (For the estimate, we assumed that the value of the coupling constant is of order 1 at the QCD energy scale.)

2.3.3 Magnetic Catalysis in Graphene

In this section, we briefly discuss the application of the magnetic catalysis ideas to graphene in the regime of the quantum Hall effect.

Graphene is a single atomic layer of graphite [156] that has many interesting properties and promises widespread applications (for reviews, see Refs. [1, 27, 109]). The uniqueness of graphene is largely due to its unusual band structure with two Dirac points at the corners of the Brillouin zone. Its low-energy excitations are described by massless Dirac fermions [170]. Because of a relatively small Fermi velocity of quasiparticles, $v_F \approx c/300$, the effecting coupling constant for the Coulomb interaction in graphene, $\alpha \equiv e^2/(\varepsilon_0 v_F)$, is about 300 times larger than the fine structure constant in QED, $e^2/(\varepsilon_0 c) \approx 1/137$.

When graphene is placed in a perpendicular magnetic field, it reveals an anomalous quantum Hall effect [155, 189], exactly as predicted in theory [108, 161, 190]. The anomalous plateaus in the Hall conductivity are observed at the filling factors $\nu = \pm 4(n + 1/2)$, where $n = 0, 1, 2, \ldots$ is the Landau level index. The factor 4 in the filling factor is due to a fourfold (spin and valley) degeneracy of each Landau level. As for the half-integer shift in the filling factor, it is directly connected with the Dirac nature of quasiparticles [90, 112, 124, 125, 170].

It was observed experimentally [22, 37, 120, 188] that there appear additional plateaus in the Hall conductivity when graphene is placed in a very strong magnetic field. The new plateaus can be interpreted as the result of lifting the fourfold degeneracy of the Landau levels. In the case of the lowest Landau level, in particular, some of the degeneracy, i.e., between the particle and hole states, can be removed when there is a dynamical generation of a Dirac mass. Considering the possibility of magnetic catalysis, such an outcome seems almost unavoidable [90, 91, 93, 98, 113, 124, 125, 174].

The low-energy quasiparticle excitations in graphene are conveniently described in terms of four-component Dirac spinors $\Psi_s^T = (\psi_{KAs}, \psi_{KBs}, \psi_{K'Bs}, \psi_{K'As})$, introduced for each spin state $s = \uparrow, \downarrow$. Note that the components of Ψ_s are the Bloch states from two sublattices (A, B) of the graphene hexagonal lattice and two valleys (K, K') at the opposite corners of the Brillouin zone. The approximate low-energy Hamiltonian, including the kinetic and Coulomb interaction terms, is given by

$$H = v_F \sum_s \int d^2 r \bar{\Psi}_s \left(\gamma^1 \pi_x + \gamma^2 \pi_y \right) \Psi_s$$

$$+ \frac{1}{2} \sum_{s,s'} \int d^2 r d^2 r' \Psi_s^\dagger(\mathbf{r}) \Psi_s(\mathbf{r}) U_C(\mathbf{r} - \mathbf{r}') \Psi_{s'}^\dagger(\mathbf{r}') \Psi_{s'}(\mathbf{r}'), \qquad (2.62)$$

where $U_C(\mathbf{r})$ is the Coulomb potential, which takes into account the polarization effects in a magnetic field [90, 93]. Note that the two electron spins ($s = \uparrow, \downarrow$) in graphene give rise to two independent species of Dirac fermions. As a result, the Hamiltonian possesses an approximate $U(4)$ symmetry [90], which is a generalization of the $U(2)$ flavor symmetry discussed in the case of the one-species model in Sect. 2.2.2. The 16 generators of the extended $U(4)$ flavor symmetry are obtained by a direct product of the 4 generators of the $U(2)$ group acting in the valley space (K, K'), and the 4 generators of the $U(2)$ spin symmetry.

The $U(4)$ symmetry is preserved even when the electron chemical potential term, $-\mu \Psi^\dagger \Psi$, is added. The inclusion of the Zeeman term, which distinguishes the electron states with opposite spins, breaks the symmetry down to the $U_\uparrow(2) \times U_\downarrow(2)$ subgroup. The explicit form of the Zeeman term is given by $\mu_B B \Psi^\dagger \sigma_3 \Psi$, where B is the magnetic field, $\mu_B = e\hbar/(2mc)$ is the Bohr magneton, and σ_3 is the third Pauli matrix in spin space. An interesting thing is that this explicit symmetry breaking is a small effect even in very strong magnetic fields. To see this, we can compare the Zeeman energy ε_Z with the Landau energy ε_ℓ,

$$\varepsilon_Z = \mu_B B = 5.8 \times 10^{-2} B \, [\text{T}] \, \text{meV}, \qquad (2.63)$$

$$\varepsilon_\ell = \sqrt{\hbar v_F^2 |eB|/c} = 26\sqrt{B \, [\text{T}]} \, \text{meV}. \qquad (2.64)$$

Therefore, the Zeeman energy is less then a few percent of the Landau energy even for the largest (continuous) magnetic fields created in a laboratory, $B \lesssim 50$ T.

Because of the large flavor symmetry, there are many potential ways how it can be broken [91, 93]. Here we mention only the possibilities that are connected to the magnetic catalysis scenario at zero filling $\nu = 0$ (i.e., the lowest Landau level is half-filled).

We will allow independent symmetry breaking condensates for fermions with opposite spins. Also, in addition to the usual $\langle \bar{\Psi}_s \Psi_s \rangle$ condensates (no sum over the repeated spin indices here), we introduce the time reversal odd ones, $\langle \bar{\Psi}_s \gamma^3 \gamma^5 \Psi_s \rangle$ [91, 93]. While the former will give rise to Dirac masses m_s ($s = \uparrow, \downarrow$) in the low-energy theory, the latter will result in the Haldane masses Δ_s ($s = \uparrow, \downarrow$) [112].

In the ground state, one can also have additional condensates, $\langle \Psi^\dagger \sigma^3 \Psi \rangle$ and $\langle \Psi^\dagger \gamma^3 \gamma^5 P_s \Psi \rangle$, associated with nonzero spin and pseudo-spin (valley) densities. To capture this possibility in the variational ansatz, one needs to include a spin chemical potential μ_3 and two pseudo-spin chemical potentials $\tilde{\mu}_s$ ($s = \uparrow, \downarrow$). Thus, the general structure of the (inverse) full fermion propagator for quasiparticles of a fixed spin has the following form:

$$S_s^{-1}(\omega; \mathbf{r}, \mathbf{r}') = -i \left[\gamma^0 \omega - v_F (\pi \cdot \gamma) + \hat{\Sigma}_s^+ \right] \delta^2(\mathbf{r} - \mathbf{r}'), \qquad (2.65)$$

where the generalized self-energy operator $\hat{\Sigma}^+$ is given by

$$\hat{\Sigma}^+ = -m_s + \gamma^0 \mu_s + is_\perp \gamma^1 \gamma^2 \tilde{\mu}_s + is_\perp \gamma^0 \gamma^1 \gamma^2 \Delta_s. \quad (2.66)$$

Functions m_s, μ_s, $\tilde{\mu}_s$, and Δ_s on the right-hand side depend on the operator valued argument $(\boldsymbol{\pi} \cdot \boldsymbol{\gamma})^2 \ell^2$, whose eigenvalues are nonpositive even integers: $-2n$, where $n = 0, 1, 2, \ldots$. Therefore, in the Landau level representation, m_s, μ_s, $\tilde{\mu}_s$, and Δ_s will get an additional Landau index n dependence: $m_{n,s}$, $\mu_{n,s}$, $\tilde{\mu}_{n,s}$, and $\Delta_{n,s}$.

The Schwinger–Dyson equation for the full fermion propagator takes the form

$$S^{-1}(t - t'; \mathbf{r}, \mathbf{r}') = S_0^{-1}(t - t'; \mathbf{r}, \mathbf{r}') + e^2 \gamma^0 S(t - t'; \mathbf{r}, \mathbf{r}') \gamma^0 D(t' - t; \mathbf{r}' - \mathbf{r}), \quad (2.67)$$

where $D(t; \mathbf{r})$ is the photon propagator mediating the Coulomb interaction. The latter is approximately instantaneous because the quasiparticle velocities are much smaller than the speed of light. In momentum space, the photon propagator takes the following form:

$$D(\omega, k) \approx D(0, k) = \frac{i}{\varepsilon_0 [k + \Pi(0, k)]}, \quad (2.68)$$

where $\Pi(0, k)$ is the static polarization function and ε_0 is a dielectric constant.

It should be noted that, in the coordinate-space representation, both the fermion propagator and its inverse contain exactly the same Schwinger phase, see (2.88). After omitting such a (nonzero) phase on both sides of (2.67) and performing the Fourier transform with respect to the time variable, we will arrive at the following equation for the translationally invariant part of the fermion propagator [93]:

$$\tilde{S}^{-1}(\omega; \mathbf{r}) = \tilde{S}_0^{-1}(\omega; \mathbf{r}) + i \frac{e^2}{\varepsilon_0} \int_{-\infty}^{\infty} \frac{d\Omega}{2\pi} \int_0^\infty \frac{dk}{2\pi} \frac{k J_0(kr)}{k + \Pi(0, k)} \gamma^0 \tilde{S}(\Omega; \mathbf{r}) \gamma^0. \quad (2.69)$$

In the Landau level representation, this equation is equivalent to a coupled set of $4 \times 2 \times n_{\max}$ equations, where we counted 4 parameters (m, μ, $\tilde{\mu}$, and Δ), 2 spins ($s = \uparrow, \downarrow$), and $n_{\max} \simeq [\Lambda^2/(2|eB|)]$ Landau levels below the ultraviolet energy cutoff Λ, where the low-energy theory is valid.

The explicit form of the gap equations can be found elsewhere [93]. The corresponding set of equations can be solved by making use of numerical methods. Here, instead, we will discuss only some general features of the solutions in the lowest Landau level approximation, which can be obtained with analytical methods.

Let us start by considering the solutions to the gap equations for quasiparticles of a fixed spin. In the lowest Landau level approximation, there are two independent gap equations, i.e.,

$$\mu_{\text{eff}} - \mu = \frac{\alpha \varepsilon_\ell}{2} \mathcal{K}_0 \left[n_F(m_{\text{eff}} - \mu_{\text{eff}}) - n_F(m_{\text{eff}} + \mu_{\text{eff}}) \right], \quad (2.70)$$

$$m_{\text{eff}} = \frac{\alpha \varepsilon_\ell}{2} \mathcal{K}_0 \left[1 - n_F(m_{\text{eff}} - \mu_{\text{eff}}) - n_F(m_{\text{eff}} + \mu_{\text{eff}}) \right], \quad (2.71)$$

where $\alpha \equiv e^2/(\varepsilon_0 v_F) \approx 2.2/\varepsilon_0$ is the coupling constant, $n_F(x) \equiv 1/(e^{x/T} + 1)$ is the Fermi distribution function, and $\mathscr{K}_0$ is the interaction kernel due to the Coulomb interaction in the lowest Landau level approximation. In the above equations, we used the shorthand notation $\mu_{\text{eff}} = \mu - \Delta$ and $m_{\text{eff}} = m - \tilde{\mu}$ for the two independent combination of parameters that determine the spectrum of the lowest Landau level quasiparticles,

$$\omega_- = -\mu_{\text{eff}} - m_{\text{eff}}, \quad \text{and} \quad \omega_+ = -\mu_{\text{eff}} + m_{\text{eff}}. \tag{2.72}$$

At zero temperature, the gap equations reduce down to

$$\mu_{\text{eff}} = \mu + \frac{\alpha \varepsilon_\ell}{4\sqrt{2\pi}} \text{sign}(\mu_{\text{eff}}) \theta\left(|\mu_{\text{eff}}| - |m_{\text{eff}}|\right), \tag{2.73}$$

$$|m_{\text{eff}}| = \frac{\alpha \varepsilon_\ell}{4\sqrt{2\pi}} \theta\left(|m_{\text{eff}}| - |\mu_{\text{eff}}|\right). \tag{2.74}$$

Here we used the value for the interaction kernel $\mathscr{K}_0 = 1/(2\sqrt{2\pi})$, which is obtained in the approximation with screening effects neglected [93]. One of the solutions to this set of equations has a nonzero dynamical Dirac mass ($m \propto \alpha \varepsilon_\ell$), i.e.,

$$|m_{\text{eff}}| = \frac{\alpha \varepsilon_\ell}{4\sqrt{2\pi}}, \quad \Delta = 0, \quad -\frac{\alpha \varepsilon_\ell}{4\sqrt{2\pi}} < \mu < \frac{\alpha \varepsilon_\ell}{4\sqrt{2\pi}}. \tag{2.75}$$

The other two solutions have nonzero Haldane masses ($\Delta \propto \alpha \varepsilon_\ell$), i.e.,

$$m_{\text{eff}} = 0, \quad \Delta = \frac{\alpha \varepsilon_\ell}{4\sqrt{2\pi}}, \quad -\infty < \mu < \frac{\alpha \varepsilon_\ell}{4\sqrt{2\pi}}, \tag{2.76}$$

$$m_{\text{eff}} = 0, \quad \Delta = -\frac{\alpha \varepsilon_\ell}{4\sqrt{2\pi}}, \quad -\frac{\alpha \varepsilon_\ell}{4\sqrt{2\pi}} < \mu < \infty. \tag{2.77}$$

In both types of solutions, the values of the masses are proportional to a power of the coupling constant α, as expected from the dimensional reduction [136, 177, 178].

In order to determine the ground state in graphene when both spin states are accounted for, one has to find among many possible solutions the one with the lowest free energy. In the approximation used here, the ground state solution at $\nu = 0$ filling (i.e., an analog of the vacuum state in particle physics) corresponds to a spin-singlet state with equal in magnitude, but opposite in sign Haldane masses for the two spin states [93]: $\Delta_\uparrow = -\Delta_\downarrow$, i.e., a mixture of the two solutions in (2.76) and (2.77).

The symmetry of the corresponding ground state is $U_\uparrow(2) \times U_\downarrow(2)$, but with the Zeeman energy splitting dynamically enhanced by the nonzero Haldane masses. The quasiparticle energies of the dynamically modified lowest Landau level are [93]

$$\omega_\uparrow = -\mu + \varepsilon_Z + |\Delta_\uparrow| > 0, \quad (\times 2), \tag{2.78}$$

$$\omega_\downarrow = -\mu - \varepsilon_Z - |\Delta_\downarrow| < 0, \quad (\times 2), \tag{2.79}$$

which show that the original fourfold degeneracy is indeed partially lifted.

2.4 Concluding Remarks

We hope that this review of magnetic catalysis is sufficient to convey the main idea of the phenomenon in terms of simple and rather general physics concepts. From the outset, this review was never meant to be comprehensive. Here we concentrated only on the bare minimum needed to understand the phenomenon as a consequence of the underlying dimensional reduction of the fermion-antifermion pairing in a magnetic field [99–102]. For further reading and for deeper insights into various aspects of the magnetic catalysis, it is suggested that the reader refers to the original literature on the topic.

Over nearly 20 years of research, there has been a lot of progress made in our understanding of magnetic catalysis. A rather long list of research papers at the end of this review is a pretty objective proof of that. At present, it is evident that the key features of the underlying physics are well established and understood. At the same time, it is also evident that there are still many theoretical questions about the applications of magnetic catalysis under various conditions, where factors other than the magnetic field may also play a substantial role.

One prominent example is the dynamics of chiral symmetry breaking in QCD in a magnetic field. Because of a poorly understood interplay between the dynamics responsible for the quark (de-)confinement on the one hand and the magnetic catalysis on the other, there are a lot of uncertainties about the precise role of the magnetic field in this case [2, 14–16, 23, 25, 31, 34, 67–72, 77, 78, 121, 146–149]. One can even suggest that there exists an intermediate regime in QCD, starting at magnetic fields of order $B \simeq 10^{19}$ G or so, in which the magnetic field is sufficiently strong to provide a gluon screening [76] and, thus, suppress the vacuum chiral condensate, but still is not strong enough to produce equally large quark masses through magnetic catalysis [146]. At finite temperature, further complications could appear because of the interplay of the magnetic field and the temperature in gluon screening [16]. All in all, it is obvious that there are many research directions remaining to be pursued in the future.

As we argued in Sect. 2.3.3, magnetic catalysis may play a profound role in the quantum Hall effect in monolayer graphene. It appears, however, that an interesting variation of magnetic catalysis can be also realized in bilayer graphene [86, 88, 89, 92]. In essence, it is a *nonrelativistic* analog of the magnetic catalysis. This fact alone is of interest because of a large diversity of solid state physics systems and the relative ease of their studies in table-top experiments.

Finally, one should keep in mind that the fundamental studies of gauge field theories, which are known to have an extremely rich and complicated dynamics, is of general interest even in the regimes that are not readily accessible in current experiments. Such studies usually provide invaluable information about the complicated theories in the regimes that are under theoretical control. This often allows one to understand better the structure of the theory and even predict its testable limitations. In the case of QCD in a magnetic field, e.g., we may gain not only a better understand of the fundamental properties, but also get an insight into the physics in the Early Universe and in heavy ion collisions.

Acknowledgements   The author thanks E.V. Gorbar, V.P. Gusynin and V.A. Miransky for reading the early version of the review and offering many useful comments. This work was supported in part by the U.S. National Science Foundation under Grant No. PHY-0969844.

Appendix: Fermion Propagator in a Magnetic Field

Let us start from the discussion of the Dirac fermion propagator in a magnetic field in $3 + 1$ dimensions. It is formally defined by the following expression:

$$S(x, x') = i\left[i\gamma^0\partial_t - (\boldsymbol{\pi}_\perp \cdot \boldsymbol{\gamma}_\perp) - \pi^3\gamma^3 - m\right]^{-1}\delta^4(x - x'), \qquad (2.80)$$

where $x \equiv (x^0, x^1, x^2, x^3) = (t, \mathbf{r})$. By definition, the spatial components of the canonical momenta are $\pi^i \equiv -i\partial_i - eA^i$, where $i = 1, 2, 3$. (The perpendicular components are $i = 1, 2$.) Here we assume that e is the fermion electric charge (i.e., one should take $e < 0$ in the case of the electron) and use the Landau gauge $\mathbf{A} = (0, Bx^1, 0)$, where B is the magnetic field pointing in the x^3-direction. By definition, the components of the usual three-dimensional vectors $\mathbf{A}$ (vector potential) and $\mathbf{r}$ (position vector) are identified with the *contravariant* components A^i and x^i, respectively.

In the Landau gauge used, it is convenient to perform a Fourier transform in the time $(t - t')$ and the longitudinal $(x^3 - x'^3)$ coordinates. Then, we obtain

$$S(\omega, p^3; \mathbf{r}_\perp, \mathbf{r}_\perp') = i\left[\gamma^0\omega - (\boldsymbol{\pi}_\perp \cdot \boldsymbol{\gamma}_\perp) - \gamma^3p^3 - m\right]^{-1}\delta^2(\mathbf{r}_\perp - \mathbf{r}_\perp')$$

$$= i\left[\gamma^0\omega - (\boldsymbol{\pi}_\perp \cdot \boldsymbol{\gamma}_\perp) - \gamma^3p^3 + m\right]$$

$$\times \left[\omega^2 - \pi_\perp^2 + ieB\gamma^1\gamma^2 - (p^3)^2 - m^2\right]^{-1}\delta^2(\mathbf{r}_\perp - \mathbf{r}_\perp'),$$

$$(2.81)$$

where $\mathbf{r}_\perp$ is the position vector in the plane perpendicular to the magnetic field.

In order to obtain a Landau level representation for the propagator (2.81), it is convenient to utilize the complete set of eigenstates of the operator $\pi_\perp^2$. This operator has the eigenvalues $(2k + 1)|eB|$, where $k = 0, 1, 2, \ldots$ is the quantum number associated with the orbital motion in the perpendicular plane. The corresponding normalized wave functions read

$$\psi_{kp_2}(\mathbf{r}_\perp) = \frac{1}{\sqrt{2\pi}\ell}\frac{1}{\sqrt{2^k k!\sqrt{\pi}}}H_k\left(\frac{x^1}{\ell} + p_2\ell\right)e^{-\frac{1}{2\ell^2}(x^1 + p_2\ell^2)^2}e^{-is_\perp x^2 p_2}, \quad (2.82)$$

where $H_k(z)$ are the Hermite polynomials [94], $\ell = 1/\sqrt{|eB|}$ is the magnetic length, and $s_\perp \equiv \mathrm{sign}(eB)$. The wave functions satisfy the conditions of normalizability and completeness,

$$\int d^2\mathbf{r}_\perp \psi_{kp_2}^*(\mathbf{r}_\perp)\psi_{k'p_2'}(\mathbf{r}_\perp) = \delta_{kk'}\delta(p_2 - p_2'), \qquad (2.83)$$

$$\int_{-\infty}^{\infty} dp_2 \sum_{k=0}^{\infty} \psi_{kp_2}(\mathbf{r}_\perp)\psi^*_{kp_2}(\mathbf{r}'_\perp) = \delta^2(\mathbf{r}_\perp - \mathbf{r}'_\perp), \qquad (2.84)$$

respectively.

By making use of the spectral expansion of the δ-function in (2.84), as well as the following identities:

$$(\boldsymbol{\pi}_\perp \cdot \boldsymbol{\gamma}_\perp)\psi_{kp_2} = \frac{i}{\ell}\gamma^1\left[\sqrt{2(k+1)}\psi_{k+1,p_2}\mathscr{P}_- - \sqrt{2k}\psi_{k-1,p_2}\mathscr{P}_+\right], \quad (2.85)$$

$$\boldsymbol{\pi}_\perp^2\psi_{kp_2} = \frac{2k+1}{\ell^2}\psi_{kp_2}, \qquad (2.86)$$

with $\mathscr{P}_\pm = \frac{1}{2}(1 \pm is_\perp\gamma^1\gamma^2)$ being the spin projectors onto the direction of the magnetic field, we can rewrite the propagator in (2.81) as follows:

$$S(\omega, p^3; \mathbf{r}_\perp, \mathbf{r}'_\perp) = \int_{-\infty}^{\infty} dp_2 \sum_{k=0}^{\infty} i\left[\gamma^0\omega - (\boldsymbol{\pi}_\perp \cdot \boldsymbol{\gamma}_\perp) - \gamma^3 p^3 + m\right]\left[\omega^2 - (p^3)^2\right.$$

$$\left. - (2k+1)|eB| + ieB\gamma^1\gamma^2 - m^2\right]^{-1}\psi_{kp_2}(\mathbf{r}_\perp)\psi^*_{kp_2}(\mathbf{r}'_\perp)$$

$$= e^{i\Phi(\mathbf{r}_\perp, \mathbf{r}'_\perp)}\tilde{S}(\omega, p^3; \mathbf{r}_\perp - \mathbf{r}'_\perp). \qquad (2.87)$$

The Schwinger phase is given by

$$\Phi(\mathbf{r}_\perp, \mathbf{r}'_\perp) = s_\perp \frac{(x^1 + x'^1)(x^2 - x'^2)}{2\ell^2}, \qquad (2.88)$$

and the translationary invariant part of the propagator reads

$$\tilde{S}(\omega, p^3; \mathbf{r}_\perp - \mathbf{r}'_\perp) = i\frac{e^{-\xi/2}}{2\pi\ell^2}\sum_{n=0}^{\infty} \frac{F_n(\omega, p^3; \mathbf{r}_\perp - \mathbf{r}'_\perp)}{\omega^2 - 2n|eB| - (p^3)^2 - m^2}, \qquad (2.89)$$

$$F_n(\omega, p^3; \mathbf{r}_\perp - \mathbf{r}'_\perp) = (\gamma^0\omega - \gamma^3 p^3 + m)\left[L_n(\xi)\mathscr{P}_+ + L_{n-1}(\xi)\mathscr{P}_-\right]$$

$$- \frac{i}{\ell^2}\boldsymbol{\gamma}_\perp \cdot (\mathbf{r}_\perp - \mathbf{r}'_\perp)L^1_{n-1}(\xi), \qquad (2.90)$$

where we used the short-hand notation

$$\xi = \frac{(\mathbf{r}_\perp - \mathbf{r}'_\perp)^2}{2\ell^2}. \qquad (2.91)$$

In order to integrate over the quantum number p_2 in (2.87), we took into account the following table integral [94]:

$$\int_{-\infty}^{\infty} e^{-x^2} H_m(x+y)H_n(x+z)dx = 2^n\pi^{1/2}m!z^{n-m}L_m^{n-m}(-2yz), \qquad (2.92)$$

which is valid when $m \leq n$. Here L_n^α are the generalized Laguerre polynomials, and $L_n \equiv L_n^0$.

Here a short remark is in order regarding the general structure of the Dirac propagator in a magnetic field. It is not a translationally invariant function, but has the form of a product of the Schwinger phase factor $e^{i\Phi(\mathbf{r}_\perp, \mathbf{r}'_\perp)}$ and a translationally invariant part. The Schwinger phase spoils the translational invariance. From a physics viewpoint, this reflects a simple fact that the fermion momenta in the two spatial directions perpendicular to the field are not conserved quantum numbers.

The Fourier transform of the translationary invariant part of the propagator (2.89) reads

$$\tilde{S}(\omega, p^3; \mathbf{p}_\perp) = 2ie^{-p_\perp^2 \ell^2} \sum_{n=0}^{\infty} \frac{(-1)^n D_n(\omega, p^3; \mathbf{p}_\perp)}{\omega^2 - 2n|eB| - (p^3)^2 - m^2}, \qquad (2.93)$$

where

$$D_n(\omega, p^3; \mathbf{p}_\perp) = (\gamma^0 \omega - \gamma^3 p^3 + m)\left[L_n(2p_\perp^2 \ell^2)\mathscr{P}_+ - L_{n-1}(2p_\perp^2 \ell^2)\mathscr{P}_-\right]$$
$$+ 2(\boldsymbol{\gamma}_\perp \cdot \mathbf{p}_\perp)L_{n-1}^1(2p_\perp^2 \ell^2). \qquad (2.94)$$

Taking into account the earlier comment that the perpendicular momenta of charged particles are not conserved quantum numbers, this representation may appear surprising. However, one should keep in mind that the result in (2.93) is not a usual momentum representation of the propagator, but the Fourier transform of its translationary invariant part only.

In some applications, it is convenient to make use of the so-called proper-time representation [168], in which the sum over Landau levels is traded for a proper-time integration. This is easily derived from (2.93) by making the following substitution:

$$\frac{i}{\omega^2 - 2n|eB| - (p^3)^2 - m^2 + i0} = \int_0^\infty ds e^{is[\omega^2 - 2n|eB| - (p^3)^2 - m^2 + i0]}. \qquad (2.95)$$

Then, the sum over Landau levels can be easily performed with the help of the summation formula for Laguerre polynomials [94],

$$\sum_{n=0}^{\infty} L_n^\alpha(x)z^n = (1-z)^{-(\alpha+1)} \exp\left(\frac{xz}{z-1}\right). \qquad (2.96)$$

The final expression for the propagator in the proper-time representation reads

$$\tilde{S}(\omega, p^3; \mathbf{p}_\perp) = \int_0^\infty ds e^{is[\omega^2 - m^2 - (p^3)^2] - i(p_\perp^2 \ell^2)\tan(s|eB|)}\left[\gamma^0 \omega - (\boldsymbol{\gamma} \cdot \mathbf{p}) + m\right.$$
$$\left. + (p^1 \gamma^2 - p^2 \gamma^1)\tan(seB)\right]\left[1 - \gamma^1 \gamma^2 \tan(seB)\right], \qquad (2.97)$$

where $(\boldsymbol{\gamma} \cdot \mathbf{p}) \equiv (\boldsymbol{\gamma}_\perp \cdot \mathbf{p}_\perp) + \gamma^3 p^3$.

Using the same method, one can also derive the Dirac fermion propagator in a magnetic field in $2+1$ dimensions. It has the same structure as the propagator in Eqs. (2.87), (2.88), (2.89), and (2.90), but with $p^3 = 0$, i.e.,

$$S_{2+1}(\omega; \mathbf{r}, \mathbf{r}') = e^{i\Phi(\mathbf{r}, \mathbf{r}')}\tilde{S}_{2+1}(\omega; \mathbf{r} - \mathbf{r}'), \qquad (2.98)$$

where

$$\tilde{S}_{2+1}(\omega; \mathbf{r} - \mathbf{r}') = i\frac{e^{-\xi/2}}{2\pi \ell^2}\sum_{n=0}^{\infty}\left[\frac{\gamma^0\omega + m}{\omega^2 - 2n|eB| - m^2}[L_n(\xi)\mathscr{P}_- + L_{n-1}(\xi)\mathscr{P}_+]\right.$$
$$\left. -\frac{i}{\ell^2}\frac{\boldsymbol{\gamma}\cdot(\mathbf{r}-\mathbf{r}')}{\omega^2 - 2n|eB| - m^2}L^1_{n-1}(\xi)\right]. \qquad (2.99)$$

The Fourier transform of the translationally invariant part is

$$\tilde{S}_{2+1}(\omega; \mathbf{p}) = 2ie^{-p^2\ell^2}\sum_{n=0}^{\infty}(-1)^n\left[\frac{(\gamma^0\omega + m)[L_n(2p^2\ell^2)\mathscr{P}_+ - L_{n-1}(2p^2\ell^2)\mathscr{P}_-]}{\omega^2 - 2n|eB| - m^2}\right.$$
$$\left. +2\frac{(\boldsymbol{\gamma}\cdot\mathbf{p})}{\omega^2 - 2n|eB| - m^2}L^1_{n-1}(2p^2\ell^2)\right]. \qquad (2.100)$$

Finally, the proper-time representation reads

$$\tilde{S}_{2+1}(\omega; \mathbf{p}) = \int_0^{\infty} ds\, e^{is[\omega^2 - m^2] - i(p^2\ell^2)\tan(s|eB|)}\left[\gamma^0\omega - (\boldsymbol{\gamma}\cdot\mathbf{p}) + m\right.$$
$$\left. + (p^1\gamma^2 - p^2\gamma^1)\tan(seB)\right]\left[1 - \gamma^1\gamma^2\tan(seB)\right]. \qquad (2.101)$$

References

1. D. Abergel, V. Apalkov, J. Berashevich, K. Ziegler, T. Chakraborty, Properties of graphene: a theoretical perspective. Adv. Phys. **59**, 261 (2010)
2. N. Agasian, Nonperturbative phenomena in QCD at finite temperature in a magnetic field. Phys. At. Nucl. **71**, 1967 (2008)
3. A. Akhiezer, V. Berestetsky, *Quantum Electrodynamics* (Interscience, New York, 1965)
4. M.S. Alam, V.S. Kaplunovsky, A. Kundu, Chiral symmetry breaking and external fields in the Kuperstein-Sonnenschein model. J. High Energy Phys. **1204**, 111 (2012). arXiv: 1202.3488
5. J. Alexandre, K. Farakos, G. Koutsoumbas, Magnetic catalysis in QED(3) at finite temperature: beyond the constant mass approximation. Phys. Rev. D **63**, 065015 (2001)
6. J. Alexandre, K. Farakos, G. Koutsoumbas, Remark on the momentum dependence of the magnetic catalysis in QED. Phys. Rev. D **64**, 067702 (2001)
7. M.G. Alford, J. Berges, K. Rajagopal, Magnetic fields within color superconducting neutron star cores. Nucl. Phys. B **571**, 269 (2000)

8. J.O. Andersen, R. Khan, Chiral transition in a magnetic field and at finite baryon density. Phys. Rev. D **85**, 065026 (2012)
9. J.O. Andersen, A. Tranberg, The chiral transition in a magnetic background: finite density effects and the functional renormalization group. J. High Energy Phys. **1208**, 002 (2012). arXiv:1204.3360
10. S.S. Avancini, D.P. Menezes, M.B. Pinto, C. Providencia, The QCD critical end point under strong magnetic fields. Phys. Rev. D **85**, 091901 (2012)
11. A. Ayala, A. Bashir, E. Gutierrez, A. Raya, A. Sanchez, Chiral and parity symmetry breaking for planar fermions: effects of a heat bath and uniform external magnetic field. Phys. Rev. D **82**, 056011 (2010)
12. A. Ayala, A. Bashir, A. Raya, A. Sanchez, Impact of a uniform magnetic field and nonzero temperature on explicit chiral symmetry breaking in QED: arbitrary hierarchy of energy scales. J. Phys. G **37**, 015001 (2010)
13. A.Y. Babansky, E. Gorbar, G. Shchepanyuk, Chiral symmetry breaking in the Nambu-Jona-Lasinio model in external constant electromagnetic field. Phys. Lett. B **419**, 272 (1998)
14. G. Bali, F. Bruckmann, G. Endrodi, Z. Fodor, S. Katz et al., The finite temperature QCD transition in external magnetic fields. PoS LATTICE2011, 192 (2011)
15. G. Bali, F. Bruckmann, G. Endrodi, Z. Fodor, S. Katz et al., QCD quark condensate in external magnetic fields (2012). arXiv:1206.4205
16. G. Bali, F. Bruckmann, G. Endrodi, Z. Fodor, S. Katz et al., The QCD phase diagram for external magnetic fields. J. High Energy Phys. **1202**, 044 (2012)
17. J. Bardeen, L. Cooper, J. Schrieffer, Theory of superconductivity. Phys. Rev. **108**, 1175 (1957)
18. G. Baym, D. Bodeker, L.D. McLerran, Magnetic fields produced by phase transition bubbles in the electroweak phase transition. Phys. Rev. D **53**, 662 (1996)
19. R. Blankenbecler, M.L. Goldberger, B. Simon, The bound states of weakly coupled long-range one-dimensional quantum hamiltonians. Ann. Phys. **108**, 69 (1977)
20. S. Bolognesi, J.N. Laia, D. Tong, K. Wong, A gapless hard wall: magnetic catalysis in bulk and boundary (2012). arXiv:1204.6029
21. S. Bolognesi, D. Tong, Magnetic catalysis in AdS4. Class. Quantum Gravity **29**, 194003 (2012). arXiv:1110.5902
22. K.I. Bolotin, F. Ghahari, M.D. Shulman, H.L. Stormer, P. Kim, Observation of the fractional quantum Hall effect in graphene. Nature **462**, 196 (2009)
23. F. Bruckmann, G. Endrodi, Dressed Wilson loops as dual condensates in response to magnetic and electric fields. Phys. Rev. D **84**, 074506 (2011)
24. P. Buividovich, M. Chernodub, D. Kharzeev, T. Kalaydzhyan, E. Luschevskaya et al., Magnetic-field-induced insulator-conductor transition in SU(2) quenched lattice gauge theory. Phys. Rev. Lett. **105**, 132001 (2010)
25. P. Buividovich, M. Chernodub, E. Luschevskaya, M. Polikarpov, Lattice QCD in strong magnetic fields (2009). arXiv:0909.1808
26. G. Calucci, R. Ragazzon, Nonlogarithmic terms in the strong field dependence of the photon propagator. J. Phys. A **27**, 2161 (1994)
27. A.H. Castro Neto, F. Guinea, N.M.R. Peres, K.S. Novoselov, A.K. Geim, The electronic properties of graphene. Rev. Mod. Phys. **81**, 109 (2009)
28. P. Cea, L. Cosmai, P. Giudice, A. Papa, Lattice Planar QED in external magnetic field. PoS **LATTICE2011**, 307 (2011)
29. B. Chatterjee, H. Mishra, A. Mishra, Chiral symmetry breaking in 3-flavor Nambu-Jona Lasinio model in magnetic background. Nucl. Phys. A **862–863**, 312 (2011)
30. B. Cheng, A.V. Olinto, Primordial magnetic fields generated in the quark–hadron transition. Phys. Rev. D **50**, 2421 (1994)
31. M. Chernodub, Superconductivity of QCD vacuum in strong magnetic field. Phys. Rev. D **82**, 085011 (2010)
32. T.D. Cohen, D.A. McGady, E.S. Werbos, The chiral condensate in a constant electromagnetic field. Phys. Rev. C **76**, 055201 (2007)

33. T.D. Cohen, E.S. Werbos, Magnetization of the QCD vacuum at large fields. Phys. Rev. C **80**, 015203 (2009)
34. M. D'Elia, S. Mukherjee, F. Sanfilippo, QCD phase transition in a strong magnetic background. Phys. Rev. D **82**, 051501 (2010)
35. M. D'Elia, F. Negro, Chiral properties of strong interactions in a magnetic background. Phys. Rev. D **83**, 114028 (2011)
36. W. Dittrich, M. Reuter, in *Effective Lagrangians in Quantum Electrodynamics*. Lecture Notes in Physics, vol. 220 (Springer, Berlin, 1985)
37. X. Du, I. Skachko, F. Duerr, A. Luican, E.Y. Andrei, Fractional quantum Hall effect and insulating phase of Dirac electrons in graphene. Nature **462**, 192 (2009)
38. R.C. Duncan, C. Thompson, Formation of very strongly magnetized neutron stars—implications for gamma-ray bursts. Astrophys. J. **392**, L9 (1992)
39. D. Ebert, K. Klimenko, H. Toki, V. Zhukovsky, Chromomagnetic catalysis of color superconductivity and dimensional reduction. Prog. Theor. Phys. **106**, 835 (2001)
40. D. Ebert, K. Klimenko, M. Vdovichenko, A. Vshivtsev, Magnetic oscillations in dense cold quark matter with four fermion interactions. Phys. Rev. D **61**, 025005 (2000)
41. D. Ebert, V.C. Zhukovsky, Chiral phase transitions in strong chromomagnetic fields at finite temperature and dimensional reduction. Mod. Phys. Lett. A **12**, 2567 (1997)
42. V. Elias, D. McKeon, V. Miransky, I. Shovkovy, The Gross-Neveu model and the supersymmetric and nonsupersymmetric Nambu-Jona-Lasinio model in a magnetic field. Phys. Rev. D **54**, 7884 (1996)
43. E. Elizalde, E. Ferrer, V. de la Incera, Beyond constant mass approximation magnetic catalysis in the gauge Higgs-Yukawa model. Phys. Rev. D **68**, 096004 (2003)
44. K. Enqvist, P. Olesen, On primordial magnetic fields of electroweak origin. Phys. Lett. B **319**, 178 (1993)
45. J. Erdmenger, V.G. Filev, D. Zoakos, Magnetic catalysis with massive dynamical flavours. J. High Energy Phys. **1208**, 004 (2012). arXiv:1112.4807
46. N. Evans, A. Gebauer, K.Y. Kim, M. Magou, Holographic description of the phase diagram of a chiral symmetry breaking gauge theory. J. High Energy Phys. **1003**, 132 (2010)
47. N. Evans, A. Gebauer, K.Y. Kim, M. Magou, Phase diagram of the D3/D5 system in a magnetic field and a BKT transition. Phys. Lett. B **698**, 91 (2011)
48. N. Evans, A. Gebauer, K.Y. Kim, E, B, μ, T phase structure of the D3/D7 holographic dual. J. High Energy Phys. **1105**, 067 (2011)
49. N. Evans, T. Kalaydzhyan, K.Y. Kim, I. Kirsch, Non-equilibrium physics at a holographic chiral phase transition. J. High Energy Phys. **1101**, 050 (2011)
50. N. Evans, K.Y. Kim, J.P. Shock, Chiral phase transitions and quantum critical points of the D3/D7(D5) system with mutually perpendicular E and B fields at finite temperature and density. J. High Energy Phys. **1109**, 021 (2011)
51. K. Farakos, G. Koutsoumbas, N. Mavromatos, Dynamical flavor symmetry breaking by a magnetic field in lattice QED in three-dimensions. Phys. Lett. B **431**, 147 (1998)
52. K. Farakos, G. Koutsoumbas, N. Mavromatos, A. Momen, Catalysis of chiral symmetry breaking by external magnetic fields in three-dimensional lattice QED (1998). hep-lat/9902017
53. K. Farakos, G. Koutsoumbas, N. Mavromatos, A. Momen, On magnetic catalysis in even flavor QED(3). Phys. Rev. D **61**, 045005 (2000)
54. S. Fayazbakhsh, N. Sadooghi, Color neutral 2SC phase of cold and dense quark matter in the presence of constant magnetic fields. Phys. Rev. D **82**, 045010 (2010)
55. S. Fayazbakhsh, N. Sadooghi, Phase diagram of hot magnetized two-flavor color superconducting quark matter. Phys. Rev. D **83**, 025026 (2011)
56. G.N. Ferrari, A.F. Garcia, M.B. Pinto, Chiral transition within effective quark models under magnetic fields. Phys. Rev. D **86**, 096005 (2012)
57. E. Ferrer, V. Gusynin, V. de la Incera, Magnetic field induced gap and kink behavior of thermal conductivity in cuprates. Mod. Phys. Lett. B **16**, 107 (2002)

58. E. Ferrer, V. Gusynin, V. de la Incera, Thermal conductivity in 3-D NJL model under external magnetic field. Eur. Phys. J. B **33**, 397 (2003)
59. E.J. Ferrer, V. de la Incera, Yukawa interactions and dynamical generation of mass in an external magnetic field. AIP Conf. Proc. **444**, 452 (1998)
60. E.J. Ferrer, V. de la Incera, Yukawa coupling contribution to magnetic field induced dynamical mass. Int. J. Mod. Phys. **14**, 3963 (1999)
61. E. Ferrer, V. de la Incera, Magnetic catalysis in the presence of scalar fields. Phys. Lett. B **481**, 287 (2000)
62. E.J. Ferrer, V. de la Incera, C. Manuel, Color-superconducting gap in the presence of a magnetic field. Nucl. Phys. B **747**, 88 (2006)
63. E.J. Ferrer, V. de la Incera, C. Manuel, Colour superconductivity in a strong magnetic field. J. Phys. A **39**, 6349 (2006)
64. V.G. Filev, C.V. Johnson, J.P. Shock, Universal holographic chiral dynamics in an external magnetic field. J. High Energy Phys. **0908**, 013 (2009)
65. V.G. Filev, R.C. Raskov, Magnetic catalysis of chiral symmetry breaking. A holographic prospective. Adv. High Energy Phys. **2010**, 473206 (2010)
66. V.G. Filev, D. Zoakos, Towards unquenched holographic magnetic catalysis. J. High Energy Phys. **1108**, 022 (2011)
67. E.S. Fraga, A.J. Mizher, Chiral transition in a strong magnetic background. Phys. Rev. D **78**, 025016 (2008)
68. E.S. Fraga, A.J. Mizher, Can a strong magnetic background modify the nature of the chiral transition in QCD? Nucl. Phys. A **820**, 103C (2009)
69. E.S. Fraga, A.J. Mizher, Chiral symmetry restoration and strong CP violation in a strong magnetic background. PoS **CPOD2009**, 037 (2009)
70. E.S. Fraga, L.F. Palhares, Deconfinement in the presence of a strong magnetic background: an exercise within the MIT bag model. Phys. Rev. D **86**, 016008 (2012)
71. E.S. Fraga, A.J. Mizher, M. Chernodub, Possible splitting of deconfinement and chiral transitions in strong magnetic fields in QCD. PoS **ICHEP2010**, 340 (2010)
72. M. Frasca, M. Ruggieri, Magnetic susceptibility of the quark condensate and polarization from chiral models. Phys. Rev. D **83**, 094024 (2011)
73. Y. Frishman, *Particles, Quantum Fields and Statistical Particles, Quantum Fields and Statistical Mechanics*. Lecture Notes in Physics, vol. 32 (Springer, Berlin, 1975)
74. K. Fukushima, J.M. Pawlowski, Magnetic catalysis in hot and dense quark matter and quantum fluctuations. Phys. Rev. D **86**, 076013 (2012). arXiv:1203.4330
75. K. Fukushima, H.J. Warringa, Color superconducting matter in a magnetic field. Phys. Rev. Lett. **100**, 032007 (2008)
76. B.V. Galilo, S.N. Nedelko, Impact of the strong electromagnetic field on the QCD effective potential for homogeneous Abelian gluon field configurations. Phys. Rev. D **84**, 094017 (2011)
77. R. Gatto, M. Ruggieri, Dressed Polyakov loop and phase diagram of hot quark matter under magnetic field. Phys. Rev. D **82**, 054027 (2010)
78. R. Gatto, M. Ruggieri, Deconfinement and chiral symmetry restoration in a strong magnetic background. Phys. Rev. D **83**, 034016 (2011)
79. B. Geyer, L. Granda, S. Odintsov, Nambu-Jona-Lasinio model in curved space-time with magnetic field. Mod. Phys. Lett. A **11**, 2053 (1996)
80. S. Ghosh, S. Mandal, S. Chakrabarty, Chiral properties of QCD vacuum in magnetars—a Nambu-Jona-Lasinio model with semi-classical approximation. Phys. Rev. C **75**, 015805 (2007)
81. D. Gitman, S. Odintsov, Y. Shilnov, Chiral symmetry breaking in d = 3 NJL model in external gravitational and magnetic fields. Phys. Rev. D **54**, 2968 (1996)
82. J. Goldstone, Field theories with superconductor solutions. Nuovo Cimento **19**, 154 (1961)
83. J. Goldstone, A. Salam, S. Weinberg, Broken symmetries. Phys. Rev. **127**, 965 (1962)
84. E. Gorbar, On chiral symmetry breaking in a constant magnetic field in higher dimension. Phys. Lett. B **491**, 305 (2000)

85. E. Gorbar, On color superconductivity in external magnetic field. Phys. Rev. D **62**, 014007 (2000)

86. E. Gorbar, V. Gusynin, J. Jia, V. Miransky, Broken-symmetry states and phase diagram of the lowest Landau level in bilayer graphene. Phys. Rev. B **84**, 235449 (2011)

87. E. Gorbar, V. Gusynin, V. Miransky, Toward theory of quantum Hall effect in graphene. J. Low Temp. Phys. **34**, 790 (2008)

88. E.V. Gorbar, V.P. Gusynin, V.A. Miransky, Dynamics and phase diagram of the $\nu = 0$ quantum Hall state in bilayer graphene. Phys. Rev. B **81**, 155451 (2010)

89. E.V. Gorbar, V.P. Gusynin, V.A. Miransky, Energy gaps at neutrality point in bilayer graphene in a magnetic field. JETP Lett. **91**, 314 (2010)

90. E. Gorbar, V. Gusynin, V. Miransky, I. Shovkovy, Magnetic field driven metal insulator phase transition in planar systems. Phys. Rev. B **66**, 045108 (2002)

91. E. Gorbar, V. Gusynin, V. Miransky, I. Shovkovy, Dynamics in the quantum Hall effect and the phase diagram of graphene. Phys. Rev. B **78**, 085437 (2008)

92. E. Gorbar, V. Gusynin, V. Miransky, I. Shovkovy, Broken-symmetry $\nu = 0$ quantum Hall states in bilayer graphene: Landau level mixing and dynamical screening. Phys. Rev. B **85**, 235460 (2012)

93. E. Gorbar, V. Gusynin, V. Miransky, I. Shovkovy, Coulomb interaction and magnetic catalysis in the quantum Hall effect in graphene. Phys. Scr. T **146**, 014018 (2012)

94. I.S. Gradshteyn, I.M. Ryzhik, *Table of Integrals, Series and Products* (Academic Press, Orlando, 1980)

95. D. Grasso, H.R. Rubinstein, Magnetic fields in the early universe. Phys. Rept. **348**, 163 (2001)

96. V. Gusynin, Magnetic catalysis of chiral symmetry breaking in gauge theories. Ukr. J. Phys. **45**, 603 (2000)

97. V. Gusynin, D. Hong, I. Shovkovy, Chiral symmetry breaking by a nonAbelian external field in $(2 + 1)$-dimensions. Phys. Rev. D **57**, 5230 (1998)

98. V. Gusynin, V. Miransky, S. Sharapov, I. Shovkovy, Excitonic gap, phase transition, and quantum Hall effect in graphene. Phys. Rev. B **74**, 195429 (2006)

99. V. Gusynin, V. Miransky, I. Shovkovy, Catalysis of dynamical flavor symmetry breaking by a magnetic field in $(2 + 1)$-dimensions. Phys. Rev. Lett. **73**, 3499 (1994)

100. V. Gusynin, V. Miransky, I. Shovkovy, Dimensional reduction and dynamical chiral symmetry breaking by a magnetic field in $(3 + 1)$-dimensions. Phys. Lett. B **349**, 477 (1995)

101. V. Gusynin, V. Miransky, I. Shovkovy, Dynamical chiral symmetry breaking by a magnetic field in QED. Phys. Rev. D **52**, 4747 (1995)

102. V. Gusynin, V. Miransky, I. Shovkovy, Dynamical flavor symmetry breaking by a magnetic field in $(2 + 1)$-dimensions. Phys. Rev. D **52**, 4718 (1995)

103. V. Gusynin, V. Miransky, I. Shovkovy, Dimensional reduction and catalysis of dynamical symmetry breaking by a magnetic field. Nucl. Phys. B **462**, 249 (1996)

104. V. Gusynin, V. Miransky, I. Shovkovy, Dynamical chiral symmetry breaking in QED in a magnetic field: toward exact results. Phys. Rev. Lett. **83**, 1291 (1999)

105. V. Gusynin, V. Miransky, I. Shovkovy, Theory of the magnetic catalysis of chiral symmetry breaking in QED. Nucl. Phys. B **563**, 361 (1999)

106. V. Gusynin, V. Miransky, I. Shovkovy, Physical gauge in the problem of dynamical chiral symmetry breaking in QED in a magnetic field. Found. Phys. **30**, 349 (2000)

107. V. Gusynin, V. Miransky, I. Shovkovy, Large N dynamics in QED in a magnetic field. Phys. Rev. D **67**, 107703 (2003)

108. V. Gusynin, S. Sharapov, Unconventional integer quantum Hall effect in graphene. Phys. Rev. Lett. **95**, 146801 (2005)

109. V. Gusynin, S. Sharapov, J. Carbotte, AC conductivity of graphene: from tight-binding model to $2 + 1$-dimensional quantum electrodynamics. Int. J. Mod. Phys. B **21**, 4611 (2007)

110. V. Gusynin, I. Shovkovy, Chiral symmetry breaking in QED in a magnetic field at finite temperature. Phys. Rev. D **56**, 5251 (1997)

111. V. Gusynin, A.V. Smilga, Electron selfenergy in strong magnetic field: summation of double logarithmic terms. Phys. Lett. B **450**, 267 (1999)

112. F. Haldane, Model for a quantum Hall effect without Landau levels: condensed-matter realization of the 'parity anomaly'. Phys. Rev. Lett. **61**, 2015 (1988)

113. I.F. Herbut, B. Roy, Quantum critical scaling in magnetic field near the Dirac point in graphene. Phys. Rev. B **77**, 245438 (2008)

114. B. Hiller, A.A. Osipov, A.H. Blin, J. da Providencia, Effects of quark interactions on dynamical chiral symmetry breaking by a magnetic field. SIGMA **4**, 024 (2008)

115. J. Hubbard, Calculation of partition functions. Phys. Rev. Lett. **3**, 77 (1959)

116. T. Inagaki, D. Kimura, T. Murata, Four fermion interaction model in a constant magnetic field at finite temperature and chemical potential. Prog. Theor. Phys. **111**, 371 (2004)

117. T. Inagaki, D. Kimura, T. Murata, NJL model at finite chemical potential in a constant magnetic field. Prog. Theor. Phys. Suppl. **153**, 321 (2004)

118. T. Inagaki, S. Odintsov, Y. Shil'nov, Dynamical symmetry breaking in the external gravitational and constant magnetic fields. Int. J. Mod. Phys. A **14**, 481 (1999)

119. M. Ishi-i, T. Kashiwa, N. Tanimura, Effect of dynamical SU(2) gluons to the gap equation of Nambu-Jona-Lasinio model in constant background magnetic field. Phys. Rev. D **65**, 065025 (2002)

120. Z. Jiang, Y. Zhang, H.L. Stormer, P. Kim, Quantum Hall states near the charge-neutral Dirac point in graphene. Phys. Rev. Lett. **99**, 106802 (2007)

121. D.N. Kabat, K.M. Lee, E.J. Weinberg, QCD vacuum structure in strong magnetic fields. Phys. Rev. D **66**, 014004 (2002)

122. S. Kawati, G. Konisi, H. Miyata, Symmetry behavior in an external field. Phys. Rev. D **28**, 1537 (1983)

123. D.E. Kharzeev, L.D. McLerran, H.J. Warringa, The effects of topological charge change in heavy ion collisions: 'event by event P and CP violation'. Nucl. Phys. A **803**, 227 (2008)

124. D. Khveshchenko, Ghost excitonic insulator transition in layered graphite. Phys. Rev. Lett. **87**, 246802 (2001)

125. D. Khveshchenko, Magnetic-field-induced insulating behavior in highly oriented pyrolitic graphite. Phys. Rev. Lett. **87**, 206401 (2001)

126. M. Klaus, On the bound state of Schrodinger operators in one-dimension. Ann. Phys. **108**, 288 (1977)

127. S. Klevansky, The Nambu-Jona-Lasinio model of quantum chromodynamics. Rev. Mod. Phys. **64**, 649 (1992)

128. S. Klevansky, R.H. Lemmer, Chiral symmetry restoration in the Nambu-Jona-Lasinio model with a constant electromagnetic field. Phys. Rev. D **39**, 3478 (1989)

129. K. Klimenko, Three-dimensional Gross-Neveu model at nonzero temperature and in an external magnetic field. Z. Phys. C **54**, 323 (1992)

130. K. Klimenko, Three-dimensional Gross-Neveu model in an external magnetic field. Theor. Math. Phys. **89**, 1161 (1992)

131. K.G. Klimenko, Three-dimensional Gross-Neveu model at nonzero temperature and in an external magnetic field. Theor. Math. Phys. **90**, 1 (1992)

132. K. Klimenko, Magnetic catalysis and oscillating effects in Nambu-Jona-Lasinio model at nonzero chemical potential (1998). hep-ph/9809218

133. K. Klimenko, B. Magnitsky, A. Vshivtsev, Three-dimensional $(\psi\bar{\psi})^2$ model with an external nonAbelian field, temperature and chemical potential. Nuovo Cimento A **107**, 439 (1994)

134. K. Klimenko, V. Zhukovsky, Does there arise a significant enhancement of the dynamical quark mass in a strong magnetic field? Phys. Lett. B **665**, 352 (2008)

135. I. Krive, S. Naftulin, Dynamical symmetry breaking and phase transitions in a three-dimensional Gross-Neveu model in a strong magnetic field. Phys. Rev. D **46**, 2737 (1992)

136. L.D. Landau, E.M. Lifshitz, *Quantum Mechanics: Non-Relativistic Theory*, 2nd edn. (Pergamon, London, 1965)

137. D. Lee, P. McGraw, Y. Ng, I. Shovkovy, The effective potential of composite fields in weakly coupled QED in a uniform external magnetic field. Phys. Rev. D **59**, 085008 (1999)

138. C.N. Leung, Y. Ng, A. Ackley, Chiral symmetry breaking by a magnetic field in weak coupling QED (1995). hep-th/9512114
139. W.V. Liu, Parity breaking and phase transition induced by a magnetic field in high T(c) superconductors. Nucl. Phys. B **556**, 563 (1999)
140. Y.M. Loskutov, V.V. Skobelev, Nonlinear electrodynamics in a superstrong magnetic field. Phys. Lett. A **56**, 151 (1976)
141. C. Manuel, Color superconductivity in a strong external magnetic field. Nucl. Phys. A **785**, 114 (2007)
142. D.P. Menezes, M. Benghi Pinto, S.S. Avancini, A. Perez Martinez, C. Providencia, Quark matter under strong magnetic fields in the Nambu-Jona-Lasinio model. Phys. Rev. C **79**, 035807 (2009)
143. D.P. Menezes, M. Benghi Pinto, S.S. Avancini, C. Providencia, Quark matter under strong magnetic fields in the su(3) Nambu-Jona-Lasinio model. Phys. Rev. C **80**, 065805 (2009)
144. V. Miransky, *Dynamical Symmetry Breaking in Quantum Field Theories* (World Scientific, Singapore, 1994)
145. V. Miransky, Catalysis of dynamical symmetry breaking by a magnetic field. Prog. Theor. Phys. Suppl. **123**, 49 (1996)
146. V. Miransky, I. Shovkovy, Magnetic catalysis and anisotropic confinement in QCD. Phys. Rev. D **66**, 045006 (2002)
147. A.J. Mizher, M. Chernodub, E.S. Fraga, Phase diagram of hot QCD in an external magnetic field: possible splitting of deconfinement and chiral transitions. Phys. Rev. D **82**, 105016 (2010)
148. A.J. Mizher, E.S. Fraga, CP violation and chiral symmetry restoration in the hot linear sigma model in a strong magnetic background. Nucl. Phys. A **831**, 91 (2009)
149. A.J. Mizher, E.S. Fraga, M. Chernodub, Phase diagram of strong interactions in an external magnetic field. PoS **FACESQCD**, 020 (2010)
150. Y. Nambu, Axial vector current conservation in weak interactions. Phys. Rev. Lett. **4**, 380 (1960)
151. Y. Nambu, G. Jona-Lasinio, Dynamical model of elementary particles based on an analogy with superconductivity. I. Phys. Rev. **122**, 345 (1961)
152. Y. Nambu, G. Jona-Lasinio, Dynamical model of elementary particles based on an analogy with superconductivity. II. Phys. Rev. **124**, 246 (1961)
153. A. Niemi, G. Semenoff, Axial anomaly induced fermion fractionization and effective gauge theory actions in odd dimensional space-times. Phys. Rev. Lett. **51**, 2077 (1983)
154. J.L. Noronha, I.A. Shovkovy, Color-flavor locked superconductor in a magnetic field. Phys. Rev. D **76**, 105030 (2007)
155. K. Novoselov, A. Geim, S. Morozov, D. Jiang, M. Katsnelson et al., Two-dimensional gas of massless Dirac fermions in graphene. Nature **438**, 197 (2005)
156. K.S. Novoselov, A.K. Geim, S.V. Morozov, D. Jiang, Y. Zhang et al., Electric field effect in atomically thin carbon films. Science **306**, 666 (2004)
157. A. Osipov, B. Hiller, A. Blin, J. da Providencia, Dynamical chiral symmetry breaking by a magnetic field and multi-quark interactions. Phys. Lett. B **650**, 262 (2007)
158. R.R. Parwani, On chiral symmetry breaking by external magnetic fields in QED in three-dimensions. Phys. Lett. B **358**, 101 (1995)
159. R.R. Parwani, Spin polarization by external magnetic fields, Aharonov-Bohm flux strings, and chiral symmetry breaking in QED in three-dimensions. Int. J. Mod. Phys. A **11**, 1715 (1996)
160. R. Peierls, Über die existenz stationärer zustände. Z. Phys. **58**, 59 (1929)
161. N. Peres, F. Guinea, A. Castro Neto, Electronic properties of disordered two-dimensional carbon. Phys. Rev. B **73**, 125411 (2006)
162. F. Preis, A. Rebhan, A. Schmitt, Inverse magnetic catalysis in dense holographic matter. J. High Energy Phys. **1103**, 033 (2011)
163. F. Preis, A. Rebhan, A. Schmitt, Holographic baryonic matter in a background magnetic field. J. Phys. G **39**, 054006 (2012)

164. A. Redlich, Gauge noninvariance and parity violation of three-dimensional fermions. Phys. Rev. Lett. **52**, 18 (1984)

165. N. Sadooghi, A. Sodeiri Jalili, New look at the modified Coulomb potential in a strong magnetic field. Phys. Rev. D **76**, 065013 (2007)

166. D.D. Scherer, H. Gies, Renormalization group study of magnetic catalysis in the 3d Gross-Neveu model. Phys. Rev. B **85**, 195417 (2012). arXiv:1201.3746

167. S. Schramm, B. Muller, A.J. Schramm, Quark–anti-quark condensates in strong magnetic fields. Mod. Phys. Lett. A **7**, 973 (1992)

168. J.S. Schwinger, On gauge invariance and vacuum polarization. Phys. Rev. **82**, 664 (1951)

169. J.S. Schwinger, Gauge invariance and mass. 2. Phys. Rev. **128**, 2425 (1962)

170. G.W. Semenoff, Condensed matter simulation of a three-dimensional anomaly. Phys. Rev. Lett. **53**, 2449 (1984)

171. G.W. Semenoff, Electronic zero modes of vortices in Hall states of gapped graphene. Phys. Rev. B **83**, 115450 (2011). arXiv:1005.0572

172. G. Semenoff, I. Shovkovy, L. Wijewardhana, Phase transition induced by a magnetic field. Mod. Phys. Lett. A **13**, 1143 (1998)

173. G. Semenoff, I. Shovkovy, L. Wijewardhana, Universality and the magnetic catalysis of chiral symmetry breaking. Phys. Rev. D **60**, 105024 (1999)

174. G.W. Semenoff, F. Zhou, Magnetic catalysis and quantum Hall ferromagnetism in weakly coupled graphene. J. High Energy Phys. **1107**, 037 (2011)

175. I. Shovkovy, V. Turkowski, Dimensional reduction in Nambu-Jona-Lasinio model in external chromomagnetic field. Phys. Lett. B **367**, 213 (1996)

176. I. Shushpanov, A.V. Smilga, Quark condensate in a magnetic field. Phys. Lett. B **402**, 351 (1997)

177. B. Simon, *On the Number of Bound States of Two-Body Schrodinger Operators—A Review* (Princeton University Press, Princeton, 1976)

178. B. Simon, The bound state of weakly coupled Schrodinger operators in one and two-dimensions. Ann. Phys. **97**, 279 (1976)

179. V. Skokov, A.Y. Illarionov, V. Toneev, Estimate of the magnetic field strength in heavy-ion collisions. Int. J. Mod. Phys. A **24**, 5925 (2009)

180. R.L. Stratonovich, On a method of calculating quantum distribution functions. Sov. Phys. Dokl. **2**, 416 (1957)

181. H. Suganuma, T. Tatsumi, On the behavior of symmetry and phase transitions in a strong electromagnetic field. Ann. Phys. **208**, 470 (1991)

182. C. Thompson, R.C. Duncan, Neutron star dynamos and the origins of pulsar magnetism. Astrophys. J. **408**, 194 (1993)

183. T. Vachaspati, Magnetic fields from cosmological phase transitions. Phys. Lett. B **265**, 258 (1991)

184. M. Vdovichenko, A. Vshivtsev, K. Klimenko, Magnetic catalysis and magnetic oscillations in the Nambu-Jona-Lasinio model. Phys. At. Nucl. **63**, 470 (2000)

185. A. Vshivtsev, K. Klimenko, B. Magnitsky, Three-dimensional Gross-Neveu model in the external chromomagnetic fields at finite temperature. Theor. Math. Phys. **101**, 1436 (1994)

186. L. Yu, I.A. Shovkovy, Directional dependence of color superconducting gap in two-flavor QCD in a magnetic field. Phys. Rev. D **85**, 085022 (2012)

187. A. Zayakin, QCD vacuum properties in a magnetic field from AdS/CFT: chiral condensate and Goldstone mass. J. High Energy Phys. **0807**, 116 (2008)

188. Y. Zhang, Z. Jiang, J.P. Small, M.S. Purewal, Y.W. Tan et al., Landau-level splitting in graphene in high magnetic fields. Phys. Rev. Lett. **96**, 136806 (2006)

189. Y. Zhang, Y.W. Tan, H.L. Stormer, P. Kim, Experimental observation of the quantum Hall effect and Berry's phase in graphene. Nature **438**, 201 (2005)

190. Y. Zheng, T. Ando, Hall conductivity of a two-dimensional graphite system. Phys. Rev. B **65**, 245420 (2002)

191. V. Zhukovsky, V. Khudyakov, K. Klimenko, D. Ebert, Chromomagnetic catalysis of color superconductivity. JETP Lett. **74**, 523 (2001)

192. V. Zhukovsky, K. Klimenko, Magnetic catalysis of the P-parity-breaking phase transition of the first order and high-temperature superconductivity. Theor. Math. Phys. **134**, 254 (2003)
193. V.C. Zhukovsky, K. Klimenko, V. Khudyakov, Magnetic catalysis in a P-even, chiral invariant three-dimensional model with four-fermion interaction. Theor. Math. Phys. **124**, 1132 (2000)
194. V.C. Zhukovsky, K. Klimenko, V. Khudyakov, D. Ebert, Magnetic catalysis of parity breaking in a massive Gross-Neveu model and high temperature superconductivity. JETP Lett. **73**, 121 (2001)

Chapter 3
Inverse Magnetic Catalysis in Field Theory and Gauge-Gravity Duality

Florian Preis, Anton Rebhan, and Andreas Schmitt

3.1 Introduction

Two of the most important laboratories for the theory of strong interactions exhibit large magnetic fields: firstly, in non-central relativistic heavy-ion collisions the magnetic field perpendicular to the collision plane can be as high as 10^{18} G [1], and, secondly, in certain compact stars called magnetars the surface magnetic field is of the order of 10^{15} G [2], while the application of the virial theorem suggests that in the interior the magnitude of the magnetic field might reach 10^{18} G [3]. Since this is comparable to the QCD scale $\Lambda_{QCD} \simeq 200$ MeV [in natural Heaviside–Lorentz units, 10^{18} G $\simeq (140 \text{ MeV})^2$], the magnetic field in these laboratories might have a significant influence on properties governed by the strong interaction. For example, in the case of heavy-ion collisions, the magnetic field might be responsible for an observed charge separation, which has been attributed to the so-called chiral magnetic effect [4–6]. On the other hand, in compact stars the structural composition of the star's interior, i.e., the equation of state, and transport properties could be affected.

Also from a theoretical point of view these two physical systems present a great challenge. Both of them cover regions of the QCD phase diagram that are very difficult to study from first principles, since the large coupling strength prevents the application of perturbative methods. Relativistic heavy-ion collisions explore the phase diagram at low chemical potential and intermediate temperature (of the order of the QCD scale). This region is best tackled by lattice QCD, which has in the

F. Preis (✉) · A. Rebhan · A. Schmitt
Institut für Theoretische Physik, Technische Universität Wien, 1040 Vienna, Austria
e-mail: fpreis@hep.itp.tuwien.ac.at

A. Rebhan
e-mail: rebhana@hep.itp.tuwien.ac.at

A. Schmitt
e-mail: aschmitt@hep.itp.tuwien.ac.at

D. Kharzeev et al. (eds.), *Strongly Interacting Matter in Magnetic Fields*,
Lecture Notes in Physics 871, DOI 10.1007/978-3-642-37305-3_3,
© Springer-Verlag Berlin Heidelberg 2013

recent years been able to quantify the equilibrium properties in this regime [7, 8]. For transport phenomena, however, lattice simulations are not well suited. Here, the application of the AdS/CFT correspondence [9], a method developed in string theory, has contributed the celebrated result for the ratio of shear viscosity η over entropy density s [10]. The result, $\eta/s = 1/4\pi$, is currently unrivalled by other methods and appears to agree very well with experimental data. Furthermore, this value has been conjectured to be a lower bound for all isotropic fluids [11]. (This bound has been lowered in higher-derivative gravity duals [12], while in anisotropic fluids there appears to be no lower bound [13].) Unfortunately, the region of the phase diagram relevant for compact stars, where the temperature is low and the quark chemical potential is of the order of the QCD scale, is inaccessible for lattice simulations due to the so-called sign problem. Here one has to rely on extrapolations (down in density) from perturbative calculations or extrapolations (up in density) from nuclear physics or on suitable models, two of which will be of relevance for this article: the Nambu–Jona-Lasinio (NJL) model [14, 15] and the Sakai–Sugimoto model [16, 17].

In this review we focus on the effect of a background magnetic field on the chiral phase transition of QCD. The Lagrangian of QCD with N_f flavors exhibits an approximate global $U(N_f)_L \times U(N_f)_R$ symmetry group, which is broken down to a global $SU(N_f)_L \times SU(N_f)_R \times U(1)_{L+R}$ by the axial anomaly. At small temperatures and chemical potentials, i.e., in the hadronic phase, the chiral symmetry $SU(N_f)_L \times SU(N_f)_R$ is spontaneously broken to $SU(N_f)_{L+R}$ through the formation of a quark–anti-quark condensate. In this scheme the light mesons are understood as the (pseudo-)Goldstone modes corresponding to the broken generators of the symmetry group. By turning on a chemical potential one introduces an asymmetry between quarks and anti-quarks and thus exerts a stress on their pairing. As a consequence, one expects to eventually restore chiral symmetry.[1]

The two models under consideration realize the implementation and breaking of chiral symmetry quite differently. In its original formulation, the NJL model was supposed to explain the mass of nucleons via chiral symmetry breaking. With the advent of QCD it was reformulated as a model of quarks [21, 22]. It is a non-renormalizable field theory since it approximates the interaction of quarks by a four-point fermion interaction, and therefore the results of the model depend on the regularization scheme and on the UV cut-off that is used. Furthermore, the NJL model in its standard form lacks confinement. In the chiral limit, the Lagrangian of the NJL model is invariant under the same symmetry group as the QCD Lagrangian with massless quarks—the *global* $SU(N_c) \times U(N_f)_L \times U(N_f)_R$, where N_c denotes the number of colors. The chiral symmetry is broken explicitly by a bare quark mass,

---

[1]At asymptotically large chemical potentials it is known from first principles that chiral symmetry is also broken, however via a different mechanism, namely by the formation of a diquark condensate in the color-flavor-locked phase [18, 19]. Whether the hadronic phase is superseded by normal quark matter or by CFL or by some other color-superconducting phase is a matter of debate. Here we shall ignore color superconductivity. For the inclusion of color superconductivity in the context of the chiral phase transition in a magnetic field see Ref. [20].

which has to be sufficiently small compared to the momentum cut-off. Spontaneous breaking of chiral symmetry is then realized very similarly to the BCS theory of superconductivity [23], which actually served as a guideline in the development of the model. Nevertheless, there are important differences between the condensation of Cooper pairs in a superconductor and the condensation of fermion–anti-fermion pairs. For instance, the presence of a Fermi surface in the former case implies the instability with respect to Cooper pairing for arbitrarily weak attractive interactions, while, as we shall see in the NJL model, there is a finite critical coupling that is needed to form a chiral condensate. We shall also see that the analogy becomes better for chiral condensation *in a magnetic field*.

Our second model, the Sakai–Sugimoto model, is a top–down string-theoretical approach to a holographic dual of large-N_c QCD. It exploits a non-supersymmetric variation of the original gauge-gravity duality conjectured in [9] known as the Witten model [24]. Sakai and Sugimoto introduced fundamental quarks in the chiral limit by placing a stack of N_f probe D-branes for the left-handed and anti-D-branes for the right-handed sector into the supergravity limit of the Witten model. According to the holographic dictionary, the local gauge symmetry on these "flavor branes" translates into a global symmetry on the boundary, i.e., into the chiral symmetry of the dual field theory. The question of whether the symmetry is intact or broken amounts to asking whether one can perform gauge transformations on D-branes and anti-D-branes independently or not, i.e., whether the D-branes connect with the anti-D-branes in the bulk. Thus the symmetry breaking mechanism in the Sakai–Sugimoto model is of geometrical nature.

In order to understand what effects a magnetic field might have on the formation of the chiral condensate, let us recapitulate the general discussion found in [25]. Calculating the chiral condensate in field theory amounts to calculating a fermion loop. Let the bare fermion mass be finite for the moment and regularize the UV divergence via a cut-off in some suitable scheme, e.g., Schwinger's proper time method. In the presence of a magnetic field one has to take Landau quantization of the transverse momentum of the charged fermions into account. It turns out that if one performs the chiral limit on the result, an IR singularity appears, which can be shown to originate from the lowest Landau level. As a consequence, a mass gap is dynamically generated in order to avoid this IR singularity. The precise form of the gap is of course dictated by the form of the interactions in the model under consideration. This effect—termed magnetic catalysis—was first found in the Gross–Neveu model [26, 27] and later on in several NJL model calculations [25, 28–30] and in QED [31] as well as in holographic approaches [32–38]. It also plays an important role in the context of graphene [39, 40]. For QCD, it was found in a lattice calculation (however, with unphysical quark masses) that the critical temperature increases with the magnetic field [41], in accordance with magnetic catalysis. However, recently the Budapest–Wuppertal collaboration found (with physical quark masses) that the maximum of the quark susceptibility drops significantly at temperatures about 140 MeV under the influence of a magnetic field [42], i.e., the opposite effect was observed. It remains an open and interesting question what prevents magnetic catalysis to persist for larger temperatures in QCD.

This article is mostly, but not exclusively, a review of existing work in the field-theoretical NJL and the holographic Sakai–Sugimoto model. In Sect. 3.2, we review the effect of a magnetic field on chiral symmetry breaking in the NJL model in a pedagogical way, starting from the simplest case without magnetic field. This section also contains several new, so far unpublished, aspects, for instance the analytic approximations and related discussions regarding inverse magnetic catalysis in Sect. 3.2.2.3. After a pedagogical introduction to the Sakai–Sugimoto model in Sect. 3.3.1, we discuss its limit of a small asymptotic separation of the flavor branes and map out the critical surface of chiral symmetry breaking in the parameter space of temperature, chemical potential and magnetic field. This part is mostly a review of our own works [43] and [44], with emphasis on the analytic approximations of the results and their comparison to the field-theoretical analogues.

3.2 Chiral Phase Transition in the Nambu–Jona-Lasinio Model

We start from the standard Lagrangian of the NJL model (for an overview over the various NJL-type models see [45]),

$$\mathscr{L} = \overline{\psi}\left(i\gamma^{\mu}D_{\mu} - m + \mu\gamma_0\right)\psi + G\left[(\overline{\psi}\psi)^2 + (\overline{\psi}\gamma_5\psi)^2\right]. \tag{3.1}$$

We restrict ourselves to $N_f = 1$; μ denotes the quark chemical potential, m the bare quark mass, $D_{\mu} = \partial_{\mu} + iqA_{\mu}$ the covariant derivative with q the electric charge of the quarks and the electromagnetic gauge potential $A_{\mu} = (\phi, -\mathbf{A})$. We shall work with imaginary (Euclidean) time $\tau = -ix^0$ compactified on a circle, the circumference of which is identified with the inverse of the temperature T.

As a next step, we assume the pseudoscalar condensate to vanish, $\langle\overline{\psi}\gamma_5\psi\rangle \equiv 0$, and apply the mean-field approximation,

$$(\overline{\psi}\psi)^2 \simeq -\langle\overline{\psi}\psi\rangle^2 + 2\langle\overline{\psi}\psi\rangle\overline{\psi}\psi. \tag{3.2}$$

We assume the quark–anti-quark condensate $\langle\overline{\psi}\psi\rangle$ to be homogeneous and isotropic. For more general ansätze see [46, 47], where a dual chiral density wave (a.k.a. chiral spiral) has been discussed, and [48] for more general inhomogeneous phases. We define the constituent quark mass as

$$M = m - 2G\langle\overline{\psi}\psi\rangle. \tag{3.3}$$

In the following we assume stationarity and apply the temporal gauge fixing condition. Therefore, the temporal dependence of the eigenfunctions of the Dirac operator is simply an exponential, $e^{i\omega_n\tau}$, with the fermionic Matsubara frequencies $\omega_n = (2n + 1)\pi T$. Then, the thermodynamic potential becomes

$$\Omega = -\frac{T}{V}\ln Z = \frac{(M - m)^2}{4G} - \frac{T}{V}\operatorname{Tr}\ln\frac{-i\omega_n + \mu - \varepsilon}{T}, \tag{3.4}$$

where Z is the partition function. The trace includes summation over the fermionic Matsubara frequencies and some suitable spectral decomposition ε of the Dirac Hamiltonian,

$$H_D = \gamma^0 \boldsymbol{\gamma} \cdot (-i\nabla - q\mathbf{A}) + \gamma^0 M. \tag{3.5}$$

In analogy to BCS theory, the equation for minimizing the effective potential with respect to M is called gap equation, which reads

$$\langle \bar{\psi}\psi \rangle = -\frac{T}{V} \operatorname{Tr} \frac{\gamma^0}{i\omega_n + \mu - \varepsilon}. \tag{3.6}$$

In the context of a background magnetic field, we shall also discuss the axial current. Its expectation value is given by

$$\langle j_5^\mu \rangle = -\frac{T}{V} \operatorname{Tr} \frac{\gamma^0 \gamma^\mu \gamma_5}{i\omega_n + \mu - \varepsilon}. \tag{3.7}$$

3.2.1 Chiral Symmetry Breaking Without External Fields

Without external fields, the normalized eigenfunctions of the Dirac Hamiltonian are given (in a Weyl basis) by the momentum eigenfunctions

$$\psi_{e,s,k}(\mathbf{x}) = \frac{1}{\sqrt{V}} e^{i\mathbf{k}\cdot\mathbf{x}} \frac{1}{\sqrt{2\varepsilon_k}} \left(\xi^s \sqrt{\varepsilon_k - esk}, \xi^s e \sqrt{\varepsilon_k + esk} \right)^{\mathrm{T}}, \tag{3.8}$$

where ξ^s are two-vectors defined by the eigenvalue equation $\hat{\mathbf{k}} \cdot \boldsymbol{\sigma} \xi^s = s\xi^s$ with the usual Pauli matrices σ^i, and where $e\varepsilon_k = e\sqrt{k^2 + M^2}$ with $e = \pm$ is the eigenvalue of the Dirac Hamiltonian, which in turn is two-fold degenerate with respect to $s = \pm$. For the diagonal matrix elements of the gamma matrices in this basis we find

$$\gamma^0_{e,s,k} = e\frac{M}{\varepsilon_k}, \qquad \left(\gamma^5 \gamma^0 \hat{\mathbf{k}} \cdot \boldsymbol{\gamma} \right)_{e,s,k} = s. \tag{3.9}$$

From the second relation and (3.7) we conclude that the axial current vanishes, $\langle j_5^i \rangle = 0$. The reason is the spin degeneracy in each state. This will no longer be true in the presence of a magnetic field, as we shall discuss in Sect. 3.2.2.3. We can compute the thermodynamic potential and, by inserting the first relation into (3.6), the gap equation in the thermodynamic limit (at vanishing magnetic field B),

$$\Omega_{B=0} = \frac{(M-m)^2}{4G} - 2 \sum_{e=\pm} \int \frac{d^3k}{(2\pi)^3} \left[\frac{\varepsilon_k}{2} + T \ln\left(1 + e^{-\frac{\varepsilon_k - e\mu}{T}}\right) \right], \tag{3.10}$$

$$M - m = 4G \int \frac{d^3k}{(2\pi)^3} \frac{M}{\varepsilon_k} \left[1 - f(\varepsilon_k - \mu) - f(\varepsilon_k + \mu) \right], \tag{3.11}$$

where $f(x) \equiv 1/(e^{x/T} + 1)$ is the Fermi–Dirac distribution function. The (vacuum parts of the) momentum integrals are UV divergent and have to be regularized. Since the NJL model is non-renormalizable, all results, e.g., the magnitude of the gap and the order of phase transitions, will depend on the regulator as well as on the regularization scheme. We use the proper time regularization scheme [49]. In this procedure, the integrand of divergent expressions is recast into so-called proper time integrals,

$$
\left(k^2 + b^2\right)^{-a} = \frac{1}{\Gamma(a)} \int_0^\infty d\tau \, \tau^{a-1} e^{-\tau(k^2 + b^2)}, \tag{3.12}
$$

and one then performs the momentum integral before the proper time integral. The UV divergence of the momentum integral reappears at the lower bound of the proper time integral, which therefore has to be regularized. We set the lower bound to $1/\Lambda^2$.

This yields the thermodynamic potential at zero temperature

$$
16\pi^2 \Omega_{B=T=0} = \frac{2\Lambda^2 (M - m)^2}{g} + \Lambda^2 (\Lambda^2 - M^2) e^{-M^2/\Lambda^2} + M^4 \Gamma\left(0, \frac{M^2}{\Lambda^2}\right)
$$
$$
- 2\theta(\mu - M) \left[\frac{\mu k_F}{3} (2\mu^2 - 5M^2) + M^4 \ln \frac{\mu + k_F}{M} \right], \tag{3.13}
$$

where $\Gamma(a, x)$ is the incomplete gamma function, and the gap equation

$$
M - m = Mg\left[e^{-M^2/\Lambda^2} - \frac{M^2}{\Lambda^2} \Gamma\left(0, \frac{M^2}{\Lambda^2}\right) \right]
$$
$$
- 2Mg\theta(\mu - M)\left(\frac{\mu k_F}{\Lambda^2} - \frac{M^2}{\Lambda^2} \ln \frac{\mu + k_F}{M} \right), \tag{3.14}
$$

where we have defined the Fermi momentum $k_F = \sqrt{\mu^2 - M^2}$ and the dimensionless coupling constant

$$
g \equiv \frac{G\Lambda^2}{2\pi^2}. \tag{3.15}
$$

For simplicity we shall discuss the chiral limit $m = 0$ in the rest of the paper. In this case, $M = 0$ is always a solution to the gap equation.

For $\mu = 0$, the gap equation further simplifies since the term $\propto \theta(\mu - M)$ does not contribute. Then, after dividing (3.14) by M and g, its right-hand side is always smaller than 1. Therefore, a nontrivial solution for M only exists if the dimensionless coupling constant g is larger than 1. When it exists, this solution is preferred over the trivial solution, as one can verify with the help of the thermodynamic potential (3.13).

In Fig. 3.1 we show the numerical solution for the gap equation as a function of μ for three different coupling constants larger than 1 (i.e., they all admit a nontrivial solution for $\mu = 0$). For all couplings, there is a certain critical μ where M goes to

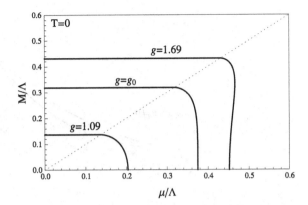

Fig. 3.1 The zero-temperature solution to the gap
equation for three different
values of the coupling g. The
thin dotted line is the line
$\mu = M$. The solution
becomes multi-valued in the
region $\mu > M$ for all
couplings larger than g_0 with
g_0 given in (3.19)

zero. By first dividing the gap equation by M and then setting $M = 0$, it is easy to
show that this critical μ is given by

$$\frac{\mu_0(g)}{\Lambda} = \frac{1}{\sqrt{2}}\sqrt{1 - \frac{1}{g}}. \tag{3.16}$$

If and only if the solution is single-valued, this is the critical μ at which the (then
second-order) phase transition to the chirally restored phase occurs.

Above a certain coupling, the solution becomes multi-valued. The coupling
where this qualitative change occurs can be computed as follows. By differentiating
the gap equation with respect to μ we find

$$\frac{\partial M}{\partial \mu} = -\frac{2k_F}{M[\Gamma(0, \frac{M^2}{\Lambda^2}) - 2\ln\frac{\mu + k_F}{M}]}. \tag{3.17}$$

In accordance to the numerical plot, this derivative is infinite for $M = 0$. For all
couplings for which the solution is multi-valued, there is another point where the
derivative is infinite, which is given by the second pole of the denominator,

$$\mu = M \cosh\frac{\Gamma(0, M^2/\Lambda^2)}{2}. \tag{3.18}$$

We can now ask for the value of g at which this point coincides with $\mu_0(g)$ for
$M \to 0$. The resulting equation then yields the coupling where the multi-valuedness
sets in. We find

$$g_0 = \frac{1}{1 - \frac{e^{-\gamma_E}}{2}} \simeq 1.390, \tag{3.19}$$

where γ_E is the Euler–Mascheroni constant. In the regime $1 < g < g_0$ the chiral
phase transition is second order and takes place at $\mu_0(g)$.

For couplings larger than g_0 the transition is first order and has to be determined
numerically. It turns out that the branch with a positive slope is always energetically disfavored. Therefore, in terms of Fig. 3.1, the preferred solution follows the

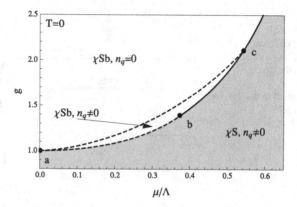

Fig. 3.2 The phase diagram at $T = 0$ in the μ–g-plane. *Dashed lines* indicate second-order, *solid lines* first-order phase transitions. In the *shaded region* chiral symmetry is restored (χS). The *points* a, b and ç correspond to $(\mu/\Lambda, g) = (0, 1)$, $(e^{-\gamma_E/2}/2, g_0)$, and $(0.542, 2.106)$, respectively, with g_0 given in (3.19). Between *points* a and b the transition line is given by $\mu_0(g)$ from (3.16). The *dashed line* between a and c indicates the onset of a finite quark number density n_q within the chirally broken phase (χSb)

horizontal line $M(\mu = 0)$ and, for all multi-valued cases, jumps to zero at a certain chemical potential. Whether (and how far) the preferred solution follows the curve into the region $\mu > M$ depends on the coupling. We find numerically that for couplings below (above) $g \simeq 2.106$ it does (doesn't). This is a first example of the nontrivial effect of μ on the preferred phase: it is not always the phase with the largest dynamical mass that is favored. In more physical terms, for couplings above $g \simeq 2.106$ the chirally broken phase with vanishing quark density is directly superseded by the quark matter phase, while for smaller couplings there is a region of finite density between these two phases. Since for $g > 2.106$ there are no complicated effects of the quark density, we can write down a very simple expression for the free energy difference between the broken phase and the restored phase, evaluated at the solution of the gap equation (and using $M \ll \Lambda$),

$$\Delta\Omega = -\frac{M_0^2 \Lambda^2}{16\pi^2}\left(1 - \frac{1}{g}\right) + \frac{\mu^4}{12\pi^2}, \qquad (3.20)$$

with M_0 being the (non-analytical) solution to the gap equation for $\mu = 0$. This result is very intuitive: the first, negative, term is the condensation energy, i.e., the energy gain from the chiral condensate, while the second, positive, term corresponds to the energy costs for pairing which must be paid because the chemical potential has separated fermions from anti-fermions. When the costs exceed the gain, chiral symmetry is restored. This determines the phase transition line. Below we shall derive the analogue of this strong-coupling free energy difference in the presence of a background magnetic field, see (3.40). We summarize our discussion of the chiral phase transition at $B = T = 0$ in Fig. 3.2.

For nonzero temperatures, we need to solve the gap equation (3.11) [with the regularization of the vacuum part shown in (3.14)] numerically. The result for various

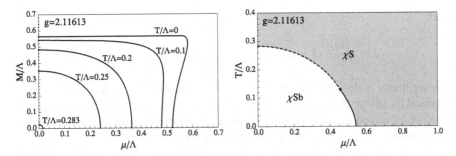

Fig. 3.3 Finite-temperature effects on the chiral phase transition in the NJL model. *Left panel*: the gap as a function of the chemical potential for a given coupling strength and different values of temperature. *Right panel*: the phase diagram in the μ–T-plane for the same coupling. The (*dashed*) second-order phase transition line is given by the analytic expression (3.21)

temperatures and a large coupling (larger than that of point c in Fig. 3.2) is shown in the left panel of Fig. 3.3. In general, the temperature decreases the gap. Moreover, the temperature can also change the order of the chiral phase transition by removing the multi-valuedness of the solution to the gap equation. The critical temperature of the chiral phase transition in the T–μ phase diagram is shown in the right panel of Fig. 3.3. The critical point moves towards higher temperatures with increasing coupling. If the phase transition is second order it is possible to find a closed form for the critical temperature. To this end, one divides (3.11) (with $m = 0$) by M and then sets $M = 0$ in the remaining equation. Then, solving for T yields the critical temperature

$$\frac{T_c(\mu)}{\Lambda} = \sqrt{\frac{3}{2\pi^2}} \sqrt{1 - \frac{1}{g} - 2\frac{\mu^2}{\Lambda^2}}. \tag{3.21}$$

3.2.2 Chiral Symmetry Breaking in the Presence of a Magnetic Field

3.2.2.1 Structure of the Fermion States in a Background Magnetic Field

Let us consider a homogeneous background magnetic field $\mathbf{B} = (0, 0, B)$ by choosing the Landau gauge fixing condition with $\mathbf{A} = (-yB, 0, 0)$. Within this ansatz, the eigenfunctions of the Hamiltonian are proportional to $\exp[i(\omega_n\tau + k_x x + k_z z)]$. Using this, we split the Dirac Hamiltonian in a longitudinal and a transverse part with respect to the direction of the magnetic field, $H_D = H_L + H_T$, where

$$H_L = \gamma^0\gamma^3 k_z + \gamma^0 M, \qquad H_T = \text{sgn}(q)\sqrt{2|q|B}\begin{pmatrix} -1 & 0 \\ 0 & 1 \end{pmatrix} \otimes \begin{pmatrix} 0 & a^\dagger \\ a & 0 \end{pmatrix}, \tag{3.22}$$

with

$$a \equiv \sqrt{\frac{|q|B}{2}}\,\xi + \text{sgn}(q)\mathrm{i}\frac{1}{\sqrt{2|q|B}}(-\mathrm{i}\partial_\xi), \quad \xi \equiv y + \frac{k_x}{qB}. \qquad (3.23)$$

We see that a is the annihilation (creation for $q < 0$) operator of the quantum mechanical oscillator, which gives rise to the Landau quantization of the energy spectrum of a charged fermion moving in a background magnetic field. For $q > 0$, the orthogonalized eigenfunctions of the Hamiltonian are given by

$$\psi^{e,s}_{k_x,k_z,\ell}(\mathbf{x}) = \frac{e^{\mathrm{i}(k_z z + k_x x)}}{\sqrt{L_x L_z}}\frac{1}{2\sqrt{\kappa_{k_z,\ell}\varepsilon_{k_z,\ell}}}\begin{pmatrix} s\sqrt{\kappa_{k_z,\ell} + sk_z}\sqrt{\varepsilon_{k_z,\ell} - s\kappa_{k_z,\ell}}\,\langle\xi|\ell\rangle \\ \sqrt{\kappa_{k_z,\ell} - sk_z}\sqrt{\varepsilon_{k_z,\ell} - s\kappa_{k_z,\ell}}\,\langle\xi|\ell - 1\rangle \\ es\sqrt{\kappa_{k_z,\ell} + sk_z}\sqrt{\varepsilon_{k_z,\ell} + s\kappa_{k_z,\ell}}\,\langle\xi|\ell\rangle \\ e\sqrt{\kappa_{k_z,\ell} - sk_z}\sqrt{\varepsilon_{k_z,\ell} + s\kappa_{k_z,\ell}}\,\langle\xi|\ell - 1\rangle \end{pmatrix}$$

$$(3.24)$$

where $\ell = 0, 1, 2, 3, \ldots$ denotes the Landau level, where

$$\langle\xi|\ell\rangle = \frac{1}{\sqrt{2^\ell \ell!}}\left(\frac{|q|B}{\pi}\right)^{1/4} e^{-|q|B\xi^2/2} H_\ell\left(\sqrt{|q|B}\,\xi\right), \qquad (3.25)$$

$\langle\xi|-1\rangle \equiv 0$, and

$$\varepsilon_{k_z,\ell} = \sqrt{k_z^2 + M^2 + 2|q|B\ell}, \qquad \kappa_{k_z,\ell} = \sqrt{k_z^2 + 2|q|B\ell}. \qquad (3.26)$$

Here, H_ℓ is the ℓth Hermite polynomial and L_i the length of a box with volume V in the ith direction. In order to obtain the eigenfunctions for the case $q < 0$, one simply replaces $\langle\xi|\ell\rangle$ with $\langle\xi|\ell - 1\rangle$ and vice versa. For the diagonal matrix elements of γ^0 and $\gamma^0\gamma^3\gamma_5$ we find

$$\gamma^0_{e,s,k_z,\ell} = e\frac{M}{\varepsilon_{k_z,\ell}}, \qquad \left(\gamma^0\gamma^3\gamma_5\right)_{e,s,k_z,\ell} = \text{sgn}(q)\frac{sk_z}{\kappa_{k_z,\ell}}. \qquad (3.27)$$

From (3.24) we see that in the lowest Landau level (LLL) $\ell = 0$ only the $\text{sgn}(q)\,s = 1$-states survive, which are also eigenstates of the spin operator $\Sigma_3 = \gamma^0\gamma^3\gamma_5$ as well as zero-eigenmodes of H_T. Therefore, the dynamics of the LLL becomes effectively $(1 + 1)$-dimensional. Moreover, in the limit $M \to 0$ for $\text{sgn}(q)e\, k_z > 0\ (< 0)$ these states are right- (left-)handed only. This is an indication that the magnetic field induces an axial current [50]. More precisely, due to the sum over s in the axial current (3.7), the relation (3.27) shows that only the LLL level contributes. Due to the sum over e there can only be a finite contribution if $\mu \neq 0$. Since we have put the fermions into a box with volume $V = L_x L_y L_z$, the range of y is restricted to $[-L_y/2, L_y/2]$ and therefore $k_{x,\max} - k_{x,\min} = L_y|q|B$ since we have absorbed k_x into the new coordinate ξ. Hence, because of $\Delta k_x = 2\pi/L_x$, each energy level for given e, k_z, s and ℓ has a degeneracy of $L_x L_y|q|B/(2\pi)$. In two cases the result for the axial current along the magnetic field can be given in closed form,

$$M = 0, \ \forall T: \quad \langle j_5^3 \rangle = \frac{q B \mu}{2\pi^2}, \tag{3.28}$$

$$T = 0, \ \forall M < \mu: \quad \langle j_5^3 \rangle = \frac{q B \sqrt{\mu^2 - M^2}}{2\pi^2}. \tag{3.29}$$

The prefactor $|q|B/(2\pi)$ found by phase space considerations has a very special role here. It is the difference of the number of zero-eigenmodes of H_T with $s = 1$ and $s = -1$ respectively. This is a topological result since it is given by the index of each 2×2 block of H_T, which in turn is linked to the Euclidean chiral anomaly in two dimensions via the index theorem. Furthermore, the first result is independent of T which is a special feature of massless $(1+1)$-dimensional fermions and hence reflects the effective dimensional reduction.

3.2.2.2 Magnetic Catalysis

Let us return to chiral symmetry breaking, now in the presence of a magnetic field. The thermodynamic potential and the gap equation read

$$\Omega = \frac{M^2}{4G} - \frac{|q|B}{2\pi} \sum_{e=\pm} \sum_{\ell=0}^{\infty} \alpha_\ell \int_{-\infty}^{\infty} \frac{dk_z}{2\pi} \left[\frac{\varepsilon_{k_z,\ell}}{2} + T \ln\left(1 + e^{-\frac{\varepsilon_{k_z,\ell} - e\mu}{T}}\right) \right], \tag{3.30}$$

$$M = 2G \frac{|q|B}{2\pi} \sum_{\ell=0}^{\infty} \alpha_\ell \int_{-\infty}^{\infty} \frac{dk_z}{2\pi} \frac{M}{\varepsilon_{k_z,\ell}} \left[1 - f(\varepsilon_{k_z,\ell} - \mu) - f(\varepsilon_{k_z,\ell} + \mu) \right], \tag{3.31}$$

where $\alpha_\ell \equiv 2 - \delta_{0\ell}$. Comparing with the corresponding $B = 0$ expressions in (3.10) and (3.11), we see that the effect of the magnetic field is to replace $\varepsilon_k \to \varepsilon_{k_z,\ell}$ and

$$2 \int \frac{d^3 k}{(2\pi)^3} \to \frac{|q|B}{2\pi} \sum_{\ell=0}^{\infty} \alpha_\ell \int_{-\infty}^{\infty} \frac{dk_z}{2\pi}. \tag{3.32}$$

Using again proper time regularization, the thermodynamic potential at vanishing temperature becomes

$$\Omega_{T=0} = \Omega_{\mu=T=B=0} - \frac{(qB)^2}{2\pi^2} \left[\frac{x^4}{4} (3 - 2\ln x) + \frac{x}{2} \left(\ln \frac{x}{2\pi} - 1 \right) + \psi^{(-2)}(x) \right]$$
$$- \frac{|q|B}{4\pi^2} \theta(\mu - M) \sum_{\ell=0}^{\ell_{max}} \alpha_\ell \left(\mu k_{F,\ell} - M_\ell^2 \ln \frac{\mu + k_{F,\ell}}{M_\ell} \right). \tag{3.33}$$

Here, $\Omega_{\mu=T=B=0}$ is the vacuum part from (3.13), $\psi^{(n)}$ the nth polygamma function (analytically continued to negative values of n), we have abbreviated $x \equiv M^2/(2|q|B)$, and

$$M_\ell \equiv \sqrt{M^2 + 2|q|B\ell}, \qquad k_{F,\ell} \equiv \sqrt{\mu^2 - M_\ell^2}, \qquad \ell_{max} \equiv \left\lfloor \frac{\mu^2 - M^2}{2|q|B} \right\rfloor. \tag{3.34}$$

Different regularization schemes—compare for instance with [51], where dimensional regularization is used—only differ in the $B = 0$ result and in (divergent) terms that depend on B but are constant in M, which are omitted. The latter can be viewed as renormalizing the energy content coming solely from the magnetic field.

The corresponding gap equation is

$$M = Mg \left[e^{-M^2/\Lambda^2} - \frac{M^2}{\Lambda^2} \Gamma \left(0, \frac{M^2}{\Lambda^2} \right) \right]$$

$$+ 2Mg \frac{|q|B}{\Lambda^2} \left[\left(\frac{1}{2} - x \right) \ln x + x + \ln \Gamma(x) - \frac{1}{2} \ln 2\pi \right]$$

$$- 2Mg \frac{|q|B}{\Lambda^2} \sum_{\ell=0}^{\ell_{max}} \alpha_\ell \ln \frac{\mu + k_{F,\ell}}{M_\ell} \theta(\mu - M). \tag{3.35}$$

Let us first consider the case $\mu = 0$, i.e., we can ignore the terms $\propto \theta(\mu - M)$ in (3.33) and (3.35). For small coupling $g \ll 1$, the dynamical mass squared will be much smaller than the magnetic field, $M^2 \ll |q|B$. Then, with $M \ll \Lambda$, the gap equation becomes

$$\frac{1}{g} \simeq \frac{2|q|B}{\Lambda^2} \ln \sqrt{\frac{|q|B}{\pi M^2}}. \tag{3.36}$$

Now, there is a nontrivial solution for arbitrarily small g. This is in contrast to the case $B = 0$ where chiral symmetry can be broken only for $g > 1$. The solution is obviously

$$M \simeq \sqrt{\frac{|q|B}{\pi}} e^{-\frac{\pi^2}{|q|BG}}. \tag{3.37}$$

This qualitative effect of the magnetic field on chiral symmetry breaking was termed "magnetic catalysis" in [28] and was since observed in numerous different models. Interestingly, as already mentioned in the introduction, this effect stems mainly from the physics in the LLL. In order to show that, one omits all contributions from $\ell > 0$ in (3.31) and cuts off the momentum integral at $\sqrt{|q|B/4\pi}$, since below that cut-off the LLL dominates. Then, one obtains exactly the result (3.37). Furthermore, the logarithmic IR singularity in (3.36) regulated by the dynamically generated mass is precisely due to the LLL contribution and its $(1 + 1)$-dimensional nature. The form of the gap in the weak coupling limit is reminiscent of the BCS gap in a superconductor [52]. In both expressions for the respective gap the relevant density of states appears in the denominator of the exponent. Here it is the density of states of the massless fermions at $\varepsilon_{k_z, \ell=0} = 0$, whereas in the BCS gap it is the density of states at the Fermi surface. In both cases the dynamics is essentially $(1 + 1)$-dimensional. While in BCS theory this effective dimensional reduction is a consequence of the Fermi surface, here it is provided by the magnetic field. Note that the dimensional reduction is not in conflict with the Mermin–Wagner–Coleman theorem that states

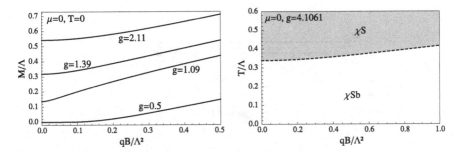

Fig. 3.4 Effects of magnetic catalysis on the dynamical mass M and the critical temperature. *Left*: the gap at $T = \mu = 0$ for different couplings. The lowest coupling shown corresponds to a subcritical coupling at $B = 0$, i.e., its nonzero value is solely induced by B. Its behavior at small B is given by the exponential in (3.37). *Right*: the critical temperature for chiral symmetry restoration as a function of B

that no spontaneous symmetry breaking can occur in $(1 + 1)$ dimensions. The reason is that the Nambu–Goldstone modes are neutral, and hence their motion is not restricted by the magnetic field. At extremely large magnetic fields the internal structure of these modes can be resolved which might invalidate this argument [53].

We show the numerical solution of the gap equation for various coupling strengths for $T = \mu = 0$ in the left panel of Fig. 3.4. Magnetic catalysis also manifests itself in the critical temperature for chiral symmetry restoration, which, at $\mu = 0$, monotonically increases with increasing magnetic field, see right panel of Fig. 3.4.

3.2.2.3 Inverse Magnetic Catalysis

We now include the contributions from a nonvanishing chemical potential μ. First we discuss the case of weak coupling which corresponds to $M^2 \ll |q|B$. Since the chiral phase transition can be expected to occur at chemical potentials of the order of the mass gap, we may thus also assume $\mu^2 \ll |q|B$ (we are not interested in the physics far beyond the phase transition). As a consequence, we can employ the lowest Landau level approximation, i.e., drop the contribution of all higher Landau levels. Then, from (3.33) we conclude that the difference of the thermodynamical potentials of the chirally broken phase and the quark matter phase is

$$\Delta\Omega \simeq \frac{|q|B}{4\pi^2}\left(\mu^2 - \frac{M^2}{2}\right) - \frac{|q|B}{4\pi^2}\mu k_{F,0}\theta(\mu - M)$$

$$+ \frac{\Lambda^2 M^2}{8\pi^2}\underbrace{\left(\frac{1}{g} - \frac{2|q|B}{\Lambda^2}\ln\sqrt{\frac{|q|B}{\pi M^2}} + \frac{2|q|B}{\Lambda^2}\theta(\mu - M)\ln\frac{\mu + k_{F,0}}{M}\right)}_{=0 \text{ via gap equation}}. \quad (3.38)$$

Again, we find a very interesting analogy to superconductivity: the resulting expression is exactly the same as for a BCS superconductor with mismatched Fermi

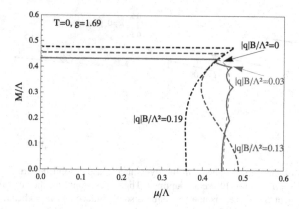

Fig. 3.5 The zero-temperature dynamical mass as a function of the chemical potential for different values of the magnetic field. For the lowest nonzero value of $|q|B$ shown (*solid line*), Landau level oscillations can be seen. The magnetic field for the two other curves (*dashed* and *dashed–dotted lines*) is sufficiently large to suppress all Landau levels except for the lowest

momenta—first discussed by Clogston [54] and Chandrasekhar [55]—after M is replaced with the superconducting gap Δ, $|q|B$ with the average Fermi momentum (squared) of the constituents of a Cooper pair, and μ with the difference of the respective Fermi momenta. (Note that again the degeneracy factor of the LLL emulates the role of the Fermi surface.)

To discuss the meaning of $\Delta\Omega$ for the chiral phase transition, let us first consider the case of a fixed magnetic field B and start from $\mu = 0$, i.e., in the chirally broken phase. Upon increasing μ, we will reach the point $\mu = M/\sqrt{2}$ where $\Delta\Omega$ changes its sign and thus the phase transition to the chirally restored phase occurs. This point is, in the context of superconductivity, called the Clogston limit. It occurs before the second term has a chance to contribute since still $\mu < M$. Now, more importantly for our purpose, let us again start in the chirally broken phase, i.e., from $\Delta\Omega < 0$, but now we increase the magnetic field at fixed μ (as we have just seen, for the discussion of the phase transition we may assume $\mu < M$ and thus ignore the term $\propto \theta(\mu - M)$). Since we have started from a negative $\mu^2 - M^2/2$, increasing the magnetic field can only make $\Delta\Omega$ more negative because the dynamical mass increases with B. Consequently, the magnetic field only brings us "deeper" into the chirally broken phase. This is what we have expected from magnetic catalysis.

However, as we will now explain, for $g > 1$ and finite chemical potential this expectation is incorrect. We shall rather find that, for intermediate values of the magnetic field, an increasing magnetic field *does* restore chiral symmetry. Let us, to this end, first discuss the numerical solution of the gap equation, see Fig. 3.5. Due to the sum over the Landau levels, the gap exhibits the well-known de Haas–van Alphen oscillations. Similar to the behavior found for $B = 0$, only the branches with a negative slope of $M(\mu)$ can be energetically preferred. Depending on the specific value of g there might be several phase transitions within the gapped phase into regions with $\mu > M$, i.e., with a finite quark number density, before entering the restored phase $M = 0$. In general it is also possible that the order of the phase transition into

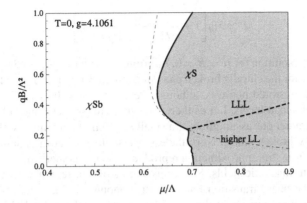

Fig. 3.6 Zero-temperature chiral phase transition in the plane of magnetic field and quark chemical potential at a rather large value of the coupling constant such that the phase transition is first order for all magnetic fields. (For smaller values the shape of the transition line is similar, but the order can vary between first and second.) Apart from oscillations at small B due to higher Landau levels in the chirally restored phase, the critical chemical potential *decreases* up to $qB/\Lambda^2 \simeq 0.5$, see explanation in the text. The *dashed–dotted line* is the approximation to the phase transition line from (3.40)

the restored phase oscillates between first and second order upon varying B: in the example shown in the plot, at vanishing magnetic field the phase transition is first order, while at $|q|B/\Lambda^2 = 0.13$ it is second order and at $|q|B/\Lambda^2 = 0.19$ again first order. We also see that the dashed (blue) curve for the *lower* magnetic field reaches farther in the μ direction than the dashed–dotted (black) curve for the *larger* magnetic field. This is a surprise from the point of view of magnetic catalysis: it seems to indicate that the critical chemical potential for chiral symmetry breaking can decrease with increasing magnetic field. We discuss this "inverse magnetic catalysis" in more detail now.

To this end, let us consider the "cleaner" case of sufficiently large couplings where symmetry restoration happens in the region $\mu < M$ for all magnetic fields. In this case, oscillations of the critical line in the phase diagram originate solely from the restored phase (not from the solution of the gap equation), and the phase transition is always first order. The numerically obtained phase diagram for such a case is shown in Fig. 3.6. From the arguments in the previous subsection, one might have expected that magnetic catalysis leads to a monotonically increasing critical chemical potential as a function of B (just like the critical temperature in the right panel of Fig. 3.4). However, this is not the case: there is a region in the phase diagram where, upon increasing B at fixed μ, chiral symmetry is *restored*, in contrast to the weak-coupling case discussed below (3.38).

In order to understand this phenomenon, let us derive an analytic expression for $\Delta\Omega$, analogous to the weak-coupling case. As discussed, for the given large coupling, the solution to the gap equation is simply given by the $\mu = 0$ solution. For small magnetic fields, $|q|B \ll M^2$, we can expand the solution up to second order in the magnetic field,

$$M \simeq M_0 \left[1 + \frac{(qB)^2}{6M_0^4 \Gamma(0, M_0^2/\Lambda^2)} \right], \tag{3.39}$$

where M_0 is the solution for $\mu = B = 0$. Inserting this solution into (3.33), we obtain the free energy for the chirally broken phase up to second order in B. The free energy for the chirally restored phase is, although we can set $M = 0$, complicated due to the sum over Landau levels. Let us therefore ignore the higher Landau levels. This seems to contradict our assumption of a small magnetic field which we have made for the chirally broken phase. Nevertheless, we shall see that the phase transition line obtained from this approximation reproduces the full numerical line in a region of intermediate magnetic fields. Since this is exactly the region where the "back bending" of the phase transition line is most pronounced, this serves our purpose to capture the main physics of the inverse magnetic catalysis. With $M_0 \ll \Lambda$, the resulting free energy difference is

$$\Delta\Omega \simeq -\frac{M_0^2 \Lambda^2}{16\pi^2} \left(1 - \frac{1}{g} \right) + \frac{|q|B}{4\pi^2} \mu^2 - \frac{(qB)^2}{24\pi^2} \left[1 - 12\zeta'(-1) + \ln \frac{M_0^2}{2|q|B} \right]. \tag{3.40}$$

(This is the generalization of (3.20) to nonzero (but small) magnetic fields.) This expression allows for a qualitatively different phase transition line compared to the weak-coupling limit (3.38) for the following reason. The term linear in B corresponds to the cost in free energy to form a fermion–anti-fermion condensate at finite μ. Importantly, this cost depends not only on μ, but also on the magnetic field. This is also true at weak coupling. However, in that case, the gain from the condensation energy was also linear in B. This is different here: now, if we start from the chirally broken phase, i.e., from $\Delta\Omega < 0$, increasing the magnetic field *can* lead to a change of sign for $\Delta\Omega$ and thus restore chiral symmetry. This is what we have termed inverse magnetic catalysis in [43]. In this reference, we have also explained that the physical picture can be understood once again in analogy to superconductivity, where, in the presence of a mismatch in Fermi momenta, it is useful to think of a fictitious state where both fermion species are filled up to a common Fermi momentum. Creating such a state costs free energy which may or may not be compensated by condensation. The point of inverse magnetic catalysis is that creating such a fictitious state (where fermions and anti-fermions are not separated by μ) becomes more costly with increasing B, while B still enhances the dynamical gap due to magnetic catalysis. The magnetic field thus plays an ambivalent role by counteracting its own catalysing effect.

This effect was first observed in the NJL model in [56] at $T = 0$ and in [57] for the full three-dimensional T–μ–B parameter space, and has been confirmed in various other calculations [20, 58–63]. Only for sufficiently strong magnetic fields the system enters the regime where magnetic catalysis is dominant. Typical fits of the model-parameters yield a cut-off of the order of a few hundred MeV [45]. Translating this into a scale for the magnetic field shows that the regime of magnetic catalysis is beyond the magnetic field strength expected in compact stars, and thus, if there is any observable effect of the magnetic field for the phase transition between hadronic and quark matter, it is inverse magnetic catalysis.

3.3 Chiral Phase Transition in the Sakai–Sugimoto Model

3.3.1 Introducing the Model

The model discussed in this section is based on the conjecture that particular strongly coupled quantum gauge theories are equivalent to certain classical gravitational theories in higher dimensions. In the context of string theory, the first realization of this holographic principle known as AdS/CFT correspondence was proposed by Maldacena [9]. In a nutshell, it utilizes two different limits of describing so-called D-branes, which are dynamical objects in string theory that impose Dirichlet boundary conditions on the endpoints of open strings. On the one hand, a stack of N_c D-branes hosts a maximally supersymmetric $U(N_c)$ gauge theory coming from the massless excitations of open superstrings; on the other hand, the stack of D-branes is a massive object that curves space-time by coupling to gravitons—coming from the closed strings—with the strength $\lambda \propto g_s N_c$, where g_s denotes the string coupling. Now, let $N_c \to \infty$ and keep λ fixed. In the limit $\lambda \ll 1$, gravity decouples from the open strings, whose low-energy effective theory is given by the mentioned $U(N_c)$ super Yang–Mills theory. In the case of D3-branes, this gauge theory is four-dimensional. In the opposite limit, $\lambda \gg 1$, the stack of D-branes back-reacts strongly on the background. Gravity far in the asymptotic region also decouples from the system due to the gravitational red shift. Therefore, one can zoom in to the near-horizon region of the space-time, which in the case of D3-branes is given by $AdS_5 \times S^5$. The idea behind the AdS/CFT duality is that the classical (super-)gravitational description is fully equivalent to the quantum theory of the large-N_c, large λ limit of the super-Yang–Mills theory. This particular gauge/gravity duality, which has passed many nontrivial tests, has since been greatly generalized and also been used in the form of phenomenological (bottom–up) models.

    The Sakai–Sugimoto model [16, 17] is a string-theoretical top–down approach to large-N_c QCD. It is based on a proposal for a holographic dual of a non-supersymmetric large-N_c Yang–Mills theory in four effective dimensions by Witten [24]. In contrast to the original AdS/CFT correspondence, the background is provided by the gravitational field of a stack of D4-branes. The dual field theory now is $(4 + 1)$-dimensional since this is the dimension of the world volume of the D4-branes. The extra dimension is compactified on an S^1 and thus breaks supersymmetry on the field theory side: by imposing anti-periodic boundary conditions on the adjoint fermions, they obtain a mass of the order of the inverse radius of the S^1, called Kaluza–Klein mass M_{KK}. At one loop level, also the adjoint scalars become massive. Hence, by choosing the radius of the extra dimension small enough and by restricting to low energies, one effectively breaks supersymmetry and effectively reduces the number of dimensions to $3 + 1$. However, there is a price to pay for introducing the extra dimension: in order to justify the supergravity approximation for the D4-brane background, the five-dimensional (dimensionful) 't Hooft coupling λ_5 has to be large compared to M_{KK}^{-1}. This corresponds to a large four-dimensional (dimensionless) 't Hooft coupling $\lambda = \lambda_5/(2\pi M_{KK}^{-1})$. In this case, however, the mass gap of the field theory is of the same order as M_{KK} and thus the Kaluza–Klein modes do not decouple. Only in the opposite limit $\lambda \ll 1$, where string corrections are important and which thus is inaccessible, the Kaluza–Klein modes do decouple and the theory becomes dual to large-N_c QCD in $3+1$ dimensions (at small energies below the Kaluza–Klein scale). It has nevertheless turned out that the classical gravity limit of the D4-brane background is a remarkably useful tool for understanding certain nonperturbative properties of (large-N_c) QCD.

An important property of the Sakai–Sugimoto model is the existence of a Hawking–Page transition between a soft-wall and a black hole background, which encodes a confinement–deconfinement transition. This feature can be understood either from power counting in N_c of the corresponding thermodynamic potentials of the gravity backgrounds or by studying the dual to the Wilson line. Confined and deconfined phases correspond to two different geometric backgrounds which are, in coordinates made dimensionless by dividing by the curvature radius R, given by

$$\frac{\mathrm{d}s^2}{R^2} = u^{3/2}\left[-h_d(u)\,\mathrm{d}t^2 + \delta_{ij}\,\mathrm{d}x^i\,\mathrm{d}x^j + h_c(u)\,\mathrm{d}x_4^2\right] + \frac{\mathrm{d}u^2}{f(u)u^{3/2}} + u^{1/2}\,\mathrm{d}\Omega_4^2, \quad (3.41)$$

where

$$f(u) = \begin{cases} 1 - \frac{u_{\mathrm{KK}}^3}{u^3}, \\ 1 - \frac{u_T^3}{u^3}, \end{cases} \qquad h_d(u) = \begin{cases} 1, \\ 1 - \frac{u_T^3}{u^3}, \end{cases} \qquad h_c(u) = \begin{cases} 1 - \frac{u_{\mathrm{KK}}^3}{u^3} & \text{conf.} \\ 1 & \text{deconf.} \end{cases}$$

$$(3.42)$$

and

$$u_{\mathrm{KK}} = \left(\frac{4\pi}{3}\right)^2 \frac{R^2}{\beta_{x_4}^2} = \frac{4}{9}R^2 M_{\mathrm{KK}}^2, \qquad u_T = \left(\frac{4\pi}{3}\right)^2 \frac{R^2}{\beta_\tau^2}. \quad (3.43)$$

Here, β_{x_4} is the period of x_4—the coordinate of the additional S^1—necessary to prevent a conical singularity at $u = u_{\mathrm{KK}}$ in the confined phase. The curvature radius is related to the Yang–Mills coupling g_{YM} by

$$R^3 = \pi g_s N_c \ell_s^3 = \frac{g_{\mathrm{YM}}^2 N_c \alpha'}{2M_{\mathrm{KK}}}, \quad (3.44)$$

where $\ell_s^2 = \alpha'$ is the squared string length. In the analytic continuation to Euclidean signature, time is also compactified to a circle with circumference $\beta_\tau = T^{-1}$, analogously to finite temperature field theory. Increasing the temperature shrinks the Euclidean time circle. At the point where the circumference of the time circle and the extra dimensional circle match, the Hawking–Page transition takes place. Apart from the metric field the Witten model also contains a nontrivial dilaton and Ramond–Ramond (RR) flux background given by

$$e^\Phi = u^{3/4}g_s, \qquad F_4 = \frac{(2\pi)^3 \ell_s^3 N_c}{\Omega_4}\,\mathrm{d}\Omega_4, \quad (3.45)$$

where Ω_4 is the volume of the 4-sphere.

Sakai and Sugimoto introduced fundamental quarks by placing two stacks of N_f D8-branes with opposite orientation into the background in the so-called probe limit $N_f \ll N_c$, i.e., back-reactions on the geometry are neglected. In the asymptotic region $u \to \infty$ the two stacks of D-branes are separated on the Kaluza–Klein circle. In the original model they reside at antipodal points. In the bulk, the D-branes are space filling in the field theory directions, x^μ, as well as in the S^4 and are specified by an embedding function in the u–x_4 subspace. Before going to the gravity description of the D4-branes one can interpret the underlying string picture as follows: strings connecting the D4 with the D8-branes carry one flavor and one color index, hence representing (massless) quarks in the fundamental representation, whereas strings stretching between D8-branes represent mesons. The local symmetry of the $U(N_f) \times U(N_f)$ gauge theory supported on the world volume of the stacks of D8-branes translates into a global symmetry via the holographic dictionary, which is interpreted as the chiral symmetry of the field theory. In the confined background, the two stacks of D8-branes are forced to join at u_{KK} where the additional S^1 degenerates and therefore form a single stack with gauge symmetry $U(N_f)$. On the field theory side, this reflects the chiral symmetry breaking mechanism. One can use a diagonal subgroup of the full symmetry group to introduce chemical potentials and electromagnetic quantities such as an external, non-dynamical magnetic field. Usually the gauge is chosen such that for example the asymptotic value of the zeroth component of the Abelian gauge field is identified with the quark chemical potential. Due to the probe limit, the deconfinement transition is not affected by a finite chemical potential, trivially leading to a phase diagram in the plane T–μ similar to the one discussed for large-N_c QCD in [64].

The low-energy effective theory describing the open string fluctuations is a non-Abelian Dirac–Born–Infeld (DBI) theory on the probe branes; calculating the fluctuations of the gauge field corresponds to calculating the meson spectrum. Indeed, after fitting the value of the 't Hooft coupling λ and M_{KK} to the rho meson mass and the pion decay constant, the spectrum matches experimental data nicely. The mode expansion used in the calculation of the meson spectrum can also be used to link the Sakai–Sugimoto model to the Skyrme model. Apart from the DBI action, the dynamics of D8-branes in a background with nontrivial RR-flux is governed by a Chern–Simons (CS) action, since the D8-brane is magnetically charged under that flux. This contribution allows for introducing baryons in the model and is related to chiral solitons in the Skyrme model. Therefore, the full action reads

$$S = S_{DBI} + S_{CS}$$

$$= T_8 \int_{D8} d\tau \, d^8x \, e^{-\Phi} \, \mathrm{Tr} \sqrt{\left| \det\left(g_{mn} + 2\pi\alpha' \mathscr{F}_{mn}\right) \right|}$$

$$+ \frac{T_8}{6} \int_{D8} C_3 \, \mathrm{Tr}\left(2\pi\alpha' \mathscr{F}\right)^3, \tag{3.46}$$

where T_8 is the D8-brane tension, and $dC_3 = F_4$. Usually one integrates the last term by parts, omitting all boundary terms, to obtain a gauge-variant action where the RR-flux couples to the Chern–Simons 5-form.

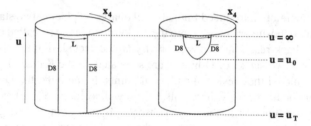

Fig. 3.7 The chirally restored (*left*) and chirally broken (*right*) phases of the non-antipodal Sakai–Sugimoto model in the deconfined background. The calculation reviewed here determines which of the two D8-brane embeddings is favored as a function of temperature, chemical potential, and magnetic field for a small asymptotic separation L. Only in that limit (in which the dual field theory resembles the NJL model) does the chiral phase transition in the probe brane approximation depend on chemical potential and magnetic field

The Sakai–Sugimoto model also has a connection to the NJL model. In the "decompactified" limit where the asymptotic coordinate distance between the D8- and anti-D8-branes is much smaller than the radius of the extra compactified dimension, $L \ll M_{KK}^{-1}$, the Sakai–Sugimoto model is dual to a non-local NJL model [65]. As a consequence, in the scenario with broken chiral symmetry, the D8-branes now in general join at $u_0 > u_{KK}$. The difference $u_0 - u_{KK}$ is commonly interpreted as the constituent quark mass within a meson, which is realized as a string with both end points attached to the tip of the joined D8-branes hanging down to the bottom of the geometry. With a sufficiently small asymptotic separation of the flavor branes, it is also possible to find an energetically preferred phase with broken chiral symmetry in the deconfined background [66], see Fig. 3.7. The resulting phase diagram at finite chemical potential was first discussed in [67]. By reducing L compared to M_{KK}^{-1}, the temperature range where the system is confined becomes small compared to the temperature range governed by the deconfined and chirally broken phase. Eventually, the resulting phase diagram resembles the NJL result (where no confined phase is present) shown in the right panel of Fig. 3.3. Consequently, the Sakai–Sugimoto model allows for interpolating between a non-local NJL model ($L \ll M_{KK}^{-1}$) and—modulo the above mentioned caveats—large-N_c QCD ($L = \pi M_{KK}^{-1}$). In the former limit, the flavor D8-branes do not probe deeply the background geometry produced by the color D4-branes (which corresponds to neglecting gluon dynamics), while in the latter the gluons dominate.

The effect of a homogeneous background magnetic field has first been considered in [68]. Shortly thereafter, the effect on the critical temperature for chiral symmetry restoration at vanishing chemical potential has been analyzed [69]. Like in the NJL result from Fig. 3.4, the critical temperature increases with the magnetic field, which shows that the Sakai–Sugimoto model exhibits magnetic catalysis. Finite chemical potentials have been introduced together with a magnetic field in [70] in the original Sakai–Sugimoto model. The deconfined, chirally symmetric phase was discussed in [71], where a magnetic phase transition within the symmetric phase was found that is reminiscent of a transition to the lowest Landau level. The full phase diagram in the parameter space T–μ–B in the deconfined phase was presented in our work [43]. In particular, the inverse magnetic catalysis effect was found and discussed in this reference.

The Sakai–Sugimoto model can be developed further to include homogeneous baryonic matter, made from point-like approximations to the solitonic baryons mentioned above [72]. Applications in the context of a background magnetic field have been studied in the confined [73] and deconfined [44] backgrounds, the latter study investigating the effect of baryonic matter on inverse magnetic catalysis. Here we shall mostly focus on the case without baryons, and only at the end of Sect. 3.3.6 briefly review their effect on the phase diagram.

3.3.2 Equations of Motion and Axial Current

In terms of an embedding function $x_4(u)$ for the D8-branes, the induced metric reads

$$\frac{ds_{D8}^2}{R^2} = u^{3/2} h_d \, dt^2 + u^{3/2} \delta_{ij} \, dx^i \, dx^j + u^{3/2} \left(x_4'^2 h_c + \frac{1}{f u^3} \right) du^2 + u^{1/2} \, d\Omega_4^2,$$

(3.47)

where the prime denotes derivative with respect to u. We work with one flavor, $N_f = 1$, and for the (dimensionless) $U(1)$ gauge field we choose the ansatz

$$a = \frac{2\pi \alpha'}{R} \mathscr{A}_\mu \, dx^\mu = a_0(u) \, dt + b x^1 \, dx^2 + a_3(u) \, dx^3,$$

(3.48)

where $b = 2\pi \alpha' B$ denotes the magnitude of the dimensionless magnetic field strength. Note that the necessity of introducing the third component of the gauge field, which is P-odd, is due to the coupling to a_0 and b via the (P-odd) CS-action. We denote the asymptotic values of the gauge field by

$$a_0(\infty) = \mu \equiv \mu_q \frac{2\pi \alpha'}{R}, \qquad a_3(\infty) = j, \qquad x_4(\infty) = \frac{\ell}{2},$$

(3.49)

where μ_q is the dimensionful quark chemical potential,[2] and $\ell \equiv L/R$ is the dimensionless asymptotic separation of the flavor branes. The boundary value of a_3 can be shown to correspond to a finite expectation value for the pion gradient in the direction of the magnetic field, hence it will only be nonvanishing when chiral symmetry is broken. In that case, one has to extremize the on-shell action with respect to j [70, 73, 74]. From the field theory perspective this means that, if $j \neq 0$, the chiral condensate is rotating between a scalar and a pseudoscalar condensate when moving along the z-direction, i.e., it forms a so-called chiral spiral [75]. Each full turn of the spiral raises the baryon number by one. Therefore, since j measures the rate of turns per unit length, it is related to the baryon density. Equivalently, one can regard

---

[2] Here we keep the notation of Refs. [43, 44]. Note that in the NJL section μ is the dimensionful quark chemical potential.

j as a supercurrent, in analogy to superfluidity, where the phase of the condensate gives the superfluid velocity.

Before continuing we put some restrictions on the gauge field and the embedding: in the joined configuration we assume that the fields are continuous at the junction point u_0 since for now we omit any point-like sources, hence $a_3(u_0) = 0$ and $x_4'(u_0) = \infty$. In the restored phase, due to the presence of the horizon, we have to satisfy the regularity constraint $a_0(u_T) = 0$ [67].

Within our ansatz, the action for the D8-brane describing left-handed fermions becomes

$$S' = \frac{\mathcal{N} V}{2T} \int_{u_0/u_T}^{\infty} du \left[\sqrt{u^5 + b^2 u^2} \sqrt{u^3 f x_4'^2 + \frac{h_d}{f} - a_0'^2 + a_3'^2 h_d} \right.$$
$$\left. + \frac{3b}{2} \left(a_3 a_0' - a_0 a_3' \right) \right], \tag{3.50}$$

where the lower bound of the integration has to be chosen according to the phase under consideration and where

$$\mathcal{N} \equiv 2 \frac{T_8 R^5 \Omega_4}{g_s} = \frac{N_c R^2}{6\pi^2 (2\pi\alpha')^3}. \tag{3.51}$$

Here we have modified the original action S and denoted the new action by S'. The reason is that proceeding with S results in an inconsistency: the conserved currents sourced by the boundary values of the gauge field turn out to be different from those currents calculated using the thermodynamic relations. In [73] this issue was related to the gauge variance of the CS action. The solution to this problem is to supplement the CS action with boundary terms residing at the holographic as well as at the spatial boundaries. After integration by parts this modification amounts to simply multiplying the CS contribution with a factor $3/2$.

The integrated equations of motion are

$$\frac{\sqrt{u^5 + b^2 u^2} a_0'}{\sqrt{u^3 f x_4'^2 + \frac{h_d}{f} - a_0'^2 + a_3'^2 h_d}} = 3ba_3 + c, \tag{3.52}$$

$$\frac{\sqrt{u^5 + b^2 u^2} h_d a_3'}{\sqrt{u^3 f x_4'^2 + \frac{h_d}{f} - a_0'^2 + a_3'^2 h_d}} = 3ba_0 + d, \tag{3.53}$$

$$\frac{\sqrt{u^5 + b^2 u^2} f u^3 x_4'}{\sqrt{u^3 f x_4'^2 + \frac{h_d}{f} - a_0'^2 + a_3'^2 h_d}} = k. \tag{3.54}$$

The left-hand side of (3.52) is the magnitude of the (bulk) electric field corresponding to the gradient of a_0 in a curved background on one D8-brane pointing towards larger values of u. When we move past the point u_0 in the joined D-brane configuration the direction of the electric field is flipped since we assume that a_0 is P-even.

Therefore, since $a_3(u_0) = 0$, the integration constant c corresponds to a point-like source at u_0. For now, we do not include any point-like baryons and thus set $c = 0$ in the broken phase. In the restored phase, on the other hand, $c \neq 0$, hence the horizon provides a charge that will be translated into the quark density at the boundary. (In the restored phase, $x_4'(u) \equiv 0$ and thus $k = 0$, i.e., only two nontrivial equations remain.) Furthermore, if the magnetic field is nonzero there is an additional contribution to the quark density from the gradient of a_3, which in general is distributed over the whole D8-brane world volume. Equation (3.53) evaluated at u_T enforces us to set $d = 0$ in the restored phase in order to maintain consistency since $h_d(u_T) = a_0(u_T) = 0$.

The nonvanishing components of the current densities sourced by the asymptotic gauge field components are given by

$$\mathscr{J}_V^0 = \mathscr{J}_R^0 + \mathscr{J}_L^0 = \frac{2\pi\alpha'\mathscr{N}}{R}\left(\frac{3b}{2}j + c\right), \tag{3.55}$$

$$\mathscr{J}_A^3 = \mathscr{J}_R^3 - \mathscr{J}_L^3 = \frac{2\pi\alpha'\mathscr{N}}{R}\left(\frac{3b}{2}\mu + d\right), \tag{3.56}$$

where we have used the equations of motion. The first line relates the baryon density with the *magnetic* chiral spiral and the point-like charges in the bulk. The second line is the axial current which we have already encountered in Sect. 3.2.2.2. Because we have to extremize the thermodynamic potential with respect to j, i.e., with respect to $a_3(\infty)$, we can immediately conclude that in the broken phase the axial current has to vanish, hence $d = -3/2\,b\mu$. In the chirally symmetric phase the axial current at *any temperature*—reinstating dimensionful quantities—reads

$$\mathscr{J}_A^3 = \frac{N_c}{4\pi^2}B\mu_q. \tag{3.57}$$

This result differs from the corresponding expression (3.28) obtained in the NJL model by a factor 2, which is related to the modification of the CS term in the action in order to obtain a consistent thermodynamic description of the currents. For a thorough discussion of the effect of this modification on the chiral anomaly see Ref. [76].

3.3.3 Semianalytic Solution to the Equations of Motion

In general, from this point on one has to rely on numerical methods. However, using $f(u) \simeq 1$ we can go a little further. This approximation is valid either in the deconfining background if $T = 0$ or in the decompactified limit of the confined background. Moreover, as will be justified a posteriori, for $L \ll M_{KK}^{-1}$ and in the chirally broken phase we have $u_0 \gg u_{KK}$ (confined) or $u_0 \gg u_T$ for sufficiently small T (deconfined). We will later work in the deconfined geometry and apply this approximation for the chirally broken phase at any T (i.e., our approximation becomes less accurate for large T). If chiral symmetry is restored this is of course not allowed, since the D8-branes extend from the holographic boundary down to the horizon at u_T. Hence, when computing the phase diagram we will compare the grand canonical potential of the broken phase using the $f(u) \simeq 1$ approximation with the full numerical result obtained for the restored phase. Note that in the special case $b = 0$ or $\mu = 0$ the temperature can be easily introduced in the symmetric phase since the "blackening" function $f(u)$ does not appear explicitly in the equations of motion. There temperature enters only in the lower bound u_T of the integrals over the holographic coordinate.

With $f(u) \simeq 1$, we can simplify (3.52) and (3.53) considerably. We define the new coordinate field $y(u)$ via the differential equation

$$y' = \frac{3bu^{3/2}}{\sqrt{u^8 + u^5 b^2 - k^2 + (3b)^2 u^3 [(\partial_y a_0)^2 - (\partial_y a_3)^2]}}, \tag{3.58}$$

for which we have the freedom to choose $y(u_0) = 0$ or $y(0) = 0$ in the broken and symmetric phase respectively. Its value at the holographic boundary will be denoted by y_∞ in the following. In the joined D8-brane configuration the boundary condition $x_4'(u_0) \to \infty$ implies that $y'(u_0) \to \infty$. After algebraically rearranging (3.52)–(3.54) such that all derivatives with respect to u are placed on the left-hand side, the equations of motion for the gauge fields as a function of the new coordinate y are

$$\partial_y a_0 = a_3 + \frac{c}{3b}, \qquad \partial_y a_3 = a_0 + \frac{d}{3b}, \tag{3.59}$$

for which we can easily find the solutions

$$a_0 = c_1 \cosh y + c_2 \sinh y - \frac{d}{3b}, \qquad a_3 = c_1 \sinh y + c_2 \cosh y - \frac{c}{3b}. \tag{3.60}$$

This allows us to write the grand canonical potential, i.e., the on-shell action, as

$$\Omega = \mathcal{N} \left[\int_{u_0/u_T}^{\infty} \frac{3b}{y'} du + \frac{kL}{2} - \frac{3b}{2} y_\infty (c_2^2 - c_1^2) - \frac{c}{2}(\mu - c_1) + \frac{d}{2}(j - c_2) \right]. \tag{3.61}$$

This expression is divergent. In order to obtain finite expressions we renormalize the grand canonical potential by the chirally symmetric vacuum contribution

$$\Omega(\mu = T = 0) = \mathcal{N} \int_0^{\Lambda} du \sqrt{u^5 + b^2 u^2}. \tag{3.62}$$

The integration constants found by imposing the boundary conditions discussed below (3.48) and the supercurrent $j = a_3(\infty)$ are summarized in Table 3.1.

Table 3.1 The integration constants d, c_1, c_2, c, k and the supercurrent j for the chirally broken and restored phases

	d	c_1	c_2	c	k	j
Broken	$-\frac{3}{2}\mu$	$\frac{\mu}{2\cosh y_\infty}$	0	0	$\sqrt{u_0^8 + b^2 u_0^5 - (\frac{3b\mu}{2\cosh y_\infty})^2 u_0^3}$	$\frac{\mu}{2}\tanh y_\infty$
Restored	0	0	$\frac{\mu}{\sinh y_\infty}$	$3b\mu\coth y_\infty$	0	0

3.3.4 Broken Chiral Symmetry

Inserting the supercurrent j and the constant c from Table 3.1 into (3.55) yields the quark number density

$$n_q \equiv \mathscr{J}_V^0 = \frac{N_c}{8\pi^2} B\mu_q \tanh y_\infty. \tag{3.63}$$

The only equations that remain and in general have to be solved numerically for the variables u_0 and y_∞ are

$$\frac{\ell}{2} = \sqrt{u_0^8 + b^2 u_0^5 - \left(\frac{3b\mu}{2\cosh y_\infty}\right)^2} \int_{u_0}^\infty \frac{du}{u^{3/2} g(u)}, \quad y_\infty = 3b\int_{u_0}^\infty \frac{u^{3/2}\,du}{g(u)}, \tag{3.64}$$

where we have abbreviated

$$g(u) \equiv \sqrt{u^8 + b^2 u^5 - \left(\frac{3b\mu}{2\cosh y_\infty}\right)^2 u^3 - u_0^8 - b^2 u_0^5 - \left(\frac{3b\mu}{2\cosh y_\infty}\right)^2 u_0^3}. \tag{3.65}$$

Note that the explicit dependence on the asymptotic separation ℓ can be eliminated by the rescaling $u \to \ell^2 u$, $\mu \to \ell^2\mu$, $b \to \ell^3 b$ and $\Omega \to \ell^7\Omega$. Therefore, in all plots shown below, the axes are measured in appropriate units of the D8-brane separation.

Before coming to the full numerical results, let us first discuss the two limits of small and large magnetic fields b. For a detailed derivation of the approximations consult Appendix D in Ref. [43].

For small magnetic fields, y_∞ and thus the supercurrent j rise linearly with b, and therefore the lowest order contribution to the quark number density induced by the chiral spiral is quadratic in b. The location of the tip of the connected flavor branes is $u_0 \simeq u_0^{(0)} + \eta_1(\mu)b^2$ with the value of u_0 at $b = 0$,

$$u_0^{(0)} = \left[\frac{4\sqrt{\pi}\,\Gamma(\frac{9}{16})}{\ell\,\Gamma(\frac{1}{16})}\right]^2 \simeq 0.5249\,\ell^{-2}. \tag{3.66}$$

Interestingly, the μ-dependent coefficient η_1 possesses a zero at $\mu \simeq 0.2905/\ell^2$, above which it becomes negative. This shows that the constituent quark mass (which is given by u_0) can *decrease* with the magnetic field for sufficiently large chemical potentials. This behavior can be traced back to the incorporation of the chiral spiral.

The grand canonical potential (renormalized by the vacuum contribution (3.62)) is approximated for small b by

$$\Omega_{\mathrm{ren}} \simeq -\mathcal{N}\left[\frac{2}{7}(u_0^{(0)})^{7/2}\frac{\sqrt{\pi}\,\Gamma(\frac{9}{16})}{\Gamma(\frac{1}{16})} + \eta_2(\mu)b^2\right], \qquad (3.67)$$

where

$$\eta_2(\mu) \equiv \frac{\sqrt{\pi}\,\Gamma(\frac{9}{16})}{\Gamma(\frac{1}{16})}\sqrt{u_0^{(0)}}\left[\cot\frac{\pi}{16} + \left(\frac{3\mu}{2u_0^{(0)}}\right)^2\frac{\Gamma(\frac{3}{16})\Gamma(\frac{17}{16})}{\Gamma(\frac{9}{16})\Gamma(\frac{11}{16})}\right]. \qquad (3.68)$$

(As explained in [43], there exists a second solution in the region of small b, where u_0 is small and y_∞ is large, which is separated from the solution discussed here by a first order phase transition. However, this first-order phase transition occurs in a region of large μ where the chirally restored phase is preferred. Therefore we will not discuss this second solution here.)

At asymptotically large magnetic field, y_∞ diverges faster than linearly, thus $j \simeq \mu/2$, while u_0 saturates at the value

$$u_0^{(\infty)} = \left[\frac{4\sqrt{\pi}\,\Gamma(\frac{3}{5})}{\ell\Gamma(\frac{1}{10})}\right]^2 \simeq 1.2317\ell^{-2}. \qquad (3.69)$$

We see that $u_0^{(\infty)} > u_0^{(0)}$, i.e., for any μ the constituent quark mass at asymptotically large b is larger than that at $b=0$. This can be interpreted as magnetic catalysis and is similar to the NJL model. However, as we have shown in the left panel of Fig. 3.4, in the NJL model the constituent quark mass does not saturate for asymptotically large magnetic fields.

Plugging these results into Ω and n_q yields

$$\Omega_{\mathrm{ren}} \simeq -\mathcal{N}b\left[\frac{\sqrt{\pi}\,\Gamma(\frac{3}{5})}{2\Gamma(\frac{1}{10})}(u_0^{(\infty)})^2 + \frac{3\mu^2}{8}\right], \qquad n_q \simeq \frac{N_c}{8\pi^2}B\mu_q. \qquad (3.70)$$

Remarkably, all model parameters have dropped out of the quark number density, which thus is solely expressed in terms of the dimensionful quantities B and μ_q.

3.3.5 Symmetric Phase

The following analytical expressions are all valid in the zero-temperature limit. Only in the plots at the end of this subsection we include numerical finite-temperature results. Now only one equation remains to be solved numerically for y_∞,

$$y_\infty = \int_0^\infty \frac{3bu^{3/2}}{\sqrt{u^8 + b^2u^5 + (\frac{3b\mu}{\sinh y_\infty})^2u^3}}\,du. \qquad (3.71)$$

For $b > 0$, this equation has in general three solutions: $y_\infty = \infty$, which is always a solution, and two finite solutions, the larger of which turns out to be unstable. At sufficiently large values of b for a given μ only the divergent solution survives. For the quark density we find

$$n_q = \frac{N_c}{2\pi^2} B\mu_q \coth y_\infty. \tag{3.72}$$

Let us first take the limit where b is small. In this case, y_∞ is linear in b, and we obtain for the (dimensionful) quark number density

$$n_q = \frac{\sqrt{N_c M_{KK}}}{3g_{YM}\pi^{3/2}} \mu_q^{5/2} \left[\frac{\sqrt{\pi}}{\Gamma(\frac{3}{10})\Gamma(\frac{6}{5})} \right]^{5/2} + \mathcal{O}(B^2). \tag{3.73}$$

The unusual exponent $5/2$ of μ_q can only occur due to the presence of the dimensionful model parameter M_{KK} (due to the extra dimension in the model), which provides the missing mass dimensions.

The grand canonical potential becomes for small b

$$\Omega_{\text{ren}} \simeq -\mathcal{N} \left\{ \frac{2}{7}\mu^{7/2} \left[\frac{\sqrt{\pi}}{\Gamma(\frac{3}{10})\Gamma(\frac{6}{5})} \right]^{5/2} + \eta_3 b^2 \sqrt{\mu} \right\}, \tag{3.74}$$

with

$$\eta_3 \equiv \frac{3}{2} \left[\frac{\Gamma(\frac{3}{10})\Gamma(\frac{6}{5})}{\sqrt{\pi}} \right]^{5/2} + \frac{\Gamma(\frac{9}{10})\Gamma(\frac{3}{5})}{\pi^{1/4}\sqrt{\Gamma(\frac{3}{10})\Gamma(\frac{6}{5})}}. \tag{3.75}$$

Taking the limit $b \to \infty$ allows only the solution $y_\infty = \infty$, as mentioned before. However, note that this is also a valid solution at finite b, hence the following results carry over to any value of b as long as this particular phase is considered. Interestingly, the density in this case is

$$n_q = \frac{N_c}{2\pi^2} B\mu_q, \tag{3.76}$$

which takes precisely the form of the density of gapless free fermions in the lowest Landau level. Therefore, we may speak of an LLL-like phase in the Sakai–Sugimoto model, although there are, because of the strong-coupling nature, no quasiparticles and thus no Landau levels in the actual sense. The grand canonical potential is

$$\Omega_{\text{ren}} = -\mathcal{N} \frac{3b\mu^2}{2}. \tag{3.77}$$

Using (3.74) together with (3.77) we can derive the critical magnetic field of the first-order transition within the chirally restored phase to the LLL-like phase as a function of the chemical potential,

$$b_c \simeq 0.095\mu^{3/2}. \tag{3.78}$$

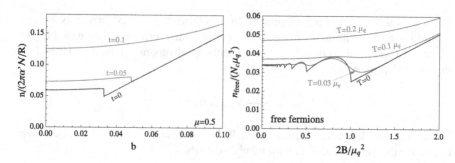

Fig. 3.8 Quark number density as a function of the background magnetic field for a given chemical potential at various (dimensionless) temperatures $t \equiv T R$ in the Sakai–Sugimoto model (*left*) and the NJL model (*right*)

In the left panel of Fig. 3.8 we plot the quark number density for different temperatures. As a comparison, we also plot the corresponding density for (massless) free fermions in a magnetic field, obtained by taking the derivative with respect to the chemical potential of the thermodynamic potential (3.30).

In the case of free fermions, the higher Landau levels cause oscillations in the density at small magnetic field. These oscillations are absent in the "higher Landau level phase" in the Sakai–Sugimoto model, given by the solution $y_\infty < \infty$. This might be a consequence of the strong coupling, in which case we do not expect a sharp Fermi surface, even at $T = 0$. Furthermore, in the NJL model, the transitions between the phases with differently filled Landau levels, in particular also the transition to the LLL phase, is second order, while in the Sakai–Sugimoto model it is first order. At finite temperature, the transitions become immediately smooth in the NJL model, while for given μ it remains first order in the Sakai–Sugimoto model until a critical temperature is reached, which increases with increasing μ. Above this temperature only one minimizing solution for y_∞ exists for all b and given μ. As a result, the transition line in the b–μ plane has a critical endpoint for a given temperature, resulting in a critical line in the three-dimensional phase diagram, see Fig. 3.9. Another important difference is the location of the LLL-transition in the μ–b diagram: the critical magnetic field at zero temperature is proportional to $\mu^{3/2}$, compared to $\mu^2/2$ for free fermions. Again this is due to the occurrence of $\sqrt{M_{\mathrm{KK}}}$.

3.3.6 Chiral Phase Transition

First we discuss the critical temperature for chiral symmetry restoration at vanishing chemical potential. In this case, in the restored phase the only temperature dependence enters via the lower bound of the integrals over the holographic coordinate, $u_T = (4\pi t/3)^2$, with $t = RT$. Therefore, one easily determines the renormalized grand canonical potential of the restored phase for the cases $b = 0$,

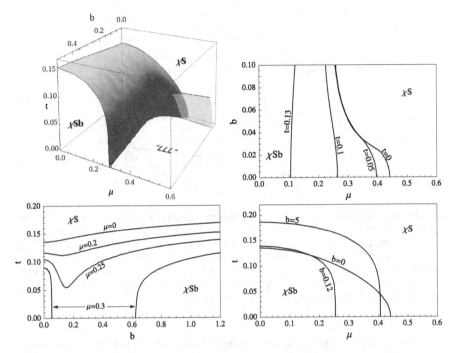

Fig. 3.9 *Upper left panel*: the surface of the chiral phase transition (*blue*) in the deconfined phase of the Sakai–Sugimoto model in the T–μ–B space. The small (*green*) surface shows the transition from the "higher LL" phase to the "LLL" phase, explained in Sect. 3.3.5. *Upper right, lower left* and *lower right panels*: two-dimensional cuts at various fixed temperatures, chemical potentials and magnetic fields, respectively, through the three-dimensional phase diagram. In the *lower left* plot, for instance, we see that the monotonically increasing critical temperature at $\mu = 0$ becomes a non-monotonic curve at finite μ and may even turn into two disconnected pieces, separating two chirally broken phases at small and large magnetic fields (Color figure online)

$\Omega_{\rm ren} = -2/7 \mathcal{N} u_T^{7/2}$, and $b \to \infty$, $\Omega_{\rm ren} = -\mathcal{N} b u_T^2 /2$. Then, together with the corresponding expressions for the broken phase from (3.67) and (3.70) we compute the critical temperatures

$$t_c(\mu = b = 0) = 0.1355/\ell, \tag{3.79}$$

$$t_c(\mu =, b \to \infty) = 0.1923/\ell. \tag{3.80}$$

(Remember that we have used the $f(u) \simeq 1$ approximation for the broken phase which, strictly speaking, is only valid for very small temperatures.) We see that the Sakai–Sugimoto model reproduces the usual magnetic catalysis effect at zero chemical potential because the critical temperature at asymptotically large b is larger than that at vanishing b. This is supported by the numerical solution which shows that the critical temperature increases monotonically with the magnetic field. In contrast to the NJL model, the critical temperature saturates at the value given in

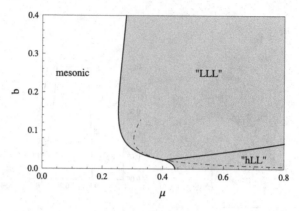

Fig. 3.10 The chiral phase transition at zero temperature from the Sakai–Sugimoto model (ignoring baryonic matter). The chirally broken phase (*white*) is separated by a first-order phase transition (*solid line*) from the chirally restored phase (*gray*). The *dashed–dotted line* is the approximation from (3.84). Translating the dimensionless quantities b and μ into physical units [43], one concludes that the magnetic field decreases the critical chemical potential from $\mu_q \simeq 400$ MeV at $|qB| = 0$ down to $\mu_q \simeq 230$ MeV at $|qB| \simeq 1.0 \times 10^{19}$ G where the critical line turns around and the critical chemical potential starts to increase with $|qB|$

(3.80), because the value for u_0, i.e., the holographic constituent quark mass, saturates.

At zero temperature, we use (3.67) and (3.70) for the broken phase and (3.74) and (3.77) for the restored phase to compute the critical chemical potentials

$$\mu_c(t = b = 0) = 0.4405/\ell^2, \tag{3.81}$$

$$\mu_c(t = 0, b \to \infty) = 0.4325/\ell^2. \tag{3.82}$$

This result already shows that inverse magnetic catalysis in the sense explained in Sect. 3.2.2.3 must be present in the Sakai–Sugimoto model. The full numerical solution of the surface of the chiral phase transition in the three-dimensional T–μ–B space, including cuts through the surface at fixed t, μ, and b, is shown in Fig. 3.9.

In order to discuss the inverse magnetic catalysis, we have plotted the zero-temperature phase diagram separately in Fig. 3.10. This phase diagram shows intriguing similarities with the corresponding NJL phase diagram in Fig. 3.6: inverse magnetic catalysis is present at small magnetic fields and is most pronounced when the restored phase has an LLL-like behavior. Even the manifestation of inverse magnetic catalysis in the analytical approximations is qualitatively the same as in the field-theoretical model as we now show. For large magnetic fields, (3.70) and (3.74) can be used to write the free energy difference between restored and broken phases as

$$\Delta\Omega = \frac{N_c B}{4\pi^2}\left[\mu_q^2 - M^2\frac{\sqrt{\pi}\,\Gamma(\frac{3}{5})}{3\Gamma(\frac{1}{10})}\right] - \frac{N_c B}{16\pi^2}\mu_q^2, \tag{3.83}$$

Fig. 3.11 As Fig. 3.10, but including baryonic matter (from Ref. [44]). The *dashed line* is the (second-order) onset of baryonic matter. The transition within the chirally restored phase between the "LLL" and "hLL" phases has disappeared because baryonic matter is preferred in this region of the phase diagram

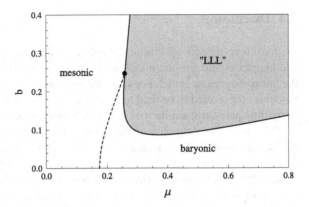

where we have identified $Ru_0/(2\pi\alpha')$ with the constituent quark mass M [66, 69]. This large-B expression for $\Delta\Omega$ is remarkably similar to the weak-coupling expression (3.38) in the NJL model. We can thus conclude, for the reasons explained below (3.38), that in the large-B regime the critical chemical potential must increase with B. This is confirmed by the chiral phase transition line of Fig. 3.10. Note the difference between the terms $\propto \theta(\mu - M)$ in the NJL expression and the last term in (3.83). Both terms come from a nonzero quark density which in our NJL calculation is only present if $\mu > M$, while in our Sakai–Sugimoto calculation there is a topological quark density at nonzero B for all μ due to the chiral spiral.

For small magnetic fields we may apply an approximation in the spirit of (3.40). We compare the free energy of the broken phase for small magnetic fields (3.67) with the free energy of the LLL phase (3.77). The result can be written as

$$\Delta\Omega \simeq -\frac{2N_c^{1/2}\Gamma(\frac{9}{16})}{21\pi g_{\text{YM}}\Gamma(\frac{1}{16})}M_{\text{KK}}^{1/2}M_0^{7/2} + \frac{N_c}{4\pi^2}B\mu_q^2 - \frac{N_c^2 g_{\text{YM}}^2 \eta_2(\mu)}{24\pi^3 M_{\text{KK}}R}B^2, \quad (3.84)$$

where $M_0 \propto u_0^{(0)}$ is the constituent quark mass at $B = 0$. Again we recover the form of the NJL result (3.40). The main conclusion is that the energy cost for condensation is linear in B, whereas the energy gain from condensation, i.e., the magnetic catalysis is only quadratic in B for small B. This allows for inverse magnetic catalysis. The dashed–dotted line in Fig. 3.10 is the approximate phase transition from (3.84). Comparison with the full numerical result shows that the approximation captures the physics of inverse magnetic catalysis where it is most pronounced and that the "hLL" phase counteracts inverse magnetic catalysis.

In Fig. 3.11 we show the phase diagram including baryonic matter discussed in [44]. The main observations are that (i) baryonic matter prevents chiral symmetry restoration for small magnetic field for any value of μ (as already found in Ref. [72] for $B = 0$) and that (ii) for sufficiently large magnetic fields, baryons become disfavored, i.e., the chirally broken, mesonic, phase is directly superseded by the quark matter phase. Interestingly, in the presence of baryonic matter, inverse magnetic catalysis becomes even more prominent in the phase diagram: now, the magnetic field restores chiral symmetry for any $\mu > 0.25$.

3.4 Discussion

We have investigated equilibrium phases at finite temperature, chemical potential, and magnetic field for one massless flavor in the Nambu–Jona-Lasinio model and the Sakai–Sugimoto model. For small flavor brane separations, the Sakai–Sugimoto model is conjectured to be dual to a (non-local) NJL model. Indeed, we have found intriguing qualitative similarities between both models.

There is an exact equality of the number density at zero temperature of the lowest Landau level in the restored phase of the NJL model and the large magnetic field phase with restored chiral symmetry in the Sakai–Sugimoto model. The higher Landau level phase in the NJL model, however, differs from the small magnetic field phase with restored chiral symmetry in the Sakai–Sugimoto model. For example, there occur no de Haas–van Alphen oscillations in the holographic model. One possible interpretation is that in the holographic model—dual to a strongly coupled gauge theory—there are no quasiparticles and no sharp Fermi surface. Furthermore, the axial current found on the field theory side is also reproduced in the holographic model. In the version of the model discussed here [73], the holographic current reproduces the field-theoretical current only up to a factor of 2. This discrepancy can be resolved by properly implementing the axial anomaly [76], however for the price of losing a consistent thermodynamic description.

Also the phase diagrams in both models share the same qualitative features. The main differences are the order of the phase transitions (first and second order in NJL vs. first order in Sakai–Sugimoto), the saturation of the critical temperature and the critical chemical potential at asymptotically large magnetic fields (which only occurs in Sakai–Sugimoto), and the absence of de Haas–van Alphen oscillations of the phase transition line in the Sakai–Sugimoto model. The main physical effect, first discussed in detail in the holographic context [43], is the nontrivial behavior of the chiral phase transition in a magnetic field at finite quark chemical potential. Somewhat unexpectedly, at sufficiently large chemical potentials and small temperatures and not too large magnetic fields, the effect of inverse magnetic catalysis dominates. We have explained inverse magnetic catalysis in both models by a free energy argument. This argument shows that, even if the magnetic field increases the constituent quark mass (due to the usual magnetic catalysis) and thus increases the condensation energy, it also increases the energy cost for forming a chiral condensate. In particular, in the LLL, where the effect is most pronounced, the cost for overcoming the separation of fermions and antifermions due to the chemical potential increases linearly in B, while the constituent quark mass rises quadratically. It is interesting that at asymptotically large magnetic fields the free energy difference in the Sakai–Sugimoto model resembles the corresponding expression in the weak-coupling limit of the NJL model. In this regime magnetic catalysis is dominant in both models, and the situation is analogous to weak-coupling superconductivity with mismatched Fermi surfaces.

By fitting the parameters of the holographic model with the help of the critical temperature at $\mu = B = 0$ from QCD lattice calculations [7, 8] and the (not very well known) critical chemical potential at $T = B = 0$ from model calculations [77, 78], we find that inverse magnetic catalysis persists up to $B \simeq 1.0 \times 10^{19}$ G, where the critical chemical potential has decreased from 400 MeV to about 230 MeV. It is not clear whether the magnetic field inside compact stars is large enough to have any effect on the chiral phase transition. Our results show, however, that if it is large enough then only *inverse* magnetic catalysis will play a role, i.e., the transition from hadronic to quark matter occurs at smaller densities than naively expected from the $B = 0$ case.

We have included an anisotropic chiral condensate in the Sakai–Sugimoto model, but not in the NJL model. For comparison, it is easy to show that in the holographic calculation the assumption of an isotropic chiral condensate does not change the qualitative features of the phase diagram. One finds that the effects of inverse magnetic catalysis are rather enhanced. On the other hand, including an anisotropic chiral condensate in the NJL model changes the phase diagram drastically [79]. Most notably, there exists a phase with anisotropic chiral condensate even at $B = 0$; in the Sakai–Sugimoto model, $B \neq 0$ is necessary for having such a phase. Moreover, this phase inevitably has a finite quark density. In order to realize this in the holographic model at $B = 0$ one needs solitonic baryon sources which are related to Skyrmions and thus rather different from "baryons" in the NJL model which consist of dislocated quarks. We have briefly discussed the effect of such baryonic matter in the Sakai–Sugimoto model, based on Ref. [44]. One of the most important changes is the non-existence of a chiral symmetry restoration at $B = 0$ for any value of the chemical potential.

Another phenomenon that was not included in our discussion is the so-called chiral shift [80, 81], a chiral asymmetry in the Fermi surfaces of right- and left-handed charged fermions induced by a magnetic field. It would be interesting to discuss its effect on the chiral phase transition and thus on inverse magnetic catalysis. However, the chiral shift is related to the Fock exchange terms, which are suppressed at large N_c. Therefore, this effect is difficult to study in a holographic model where $N_c \to \infty$ is necessary for the validity of the supergravity approximation.

Acknowledgements This work has been supported by the Austrian science foundation FWF under project No. P22114-N16.

References

1. V. Skokov, A.Yu. Illarionov, V. Toneev, Estimate of the magnetic field strength in heavy-ion collisions. Int. J. Mod. Phys. A **24**, 5925–5932 (2009)
2. R.C. Duncan, C. Thompson, Formation of very strongly magnetized neutron stars—implications for gamma-ray bursts. Astrophys. J. **392**, L9 (1992)
3. D. Lai, S.L. Shapiro, Cold equation of state in a strong magnetic field—effects of inverse beta-decay. Astrophys. J. **383**, 745–751 (1991)
4. D.E. Kharzeev, L.D. McLerran, H.J. Warringa, The effects of topological charge change in heavy ion collisions: 'Event by event P and CP violation'. Nucl. Phys. A **803**, 227–253 (2008)
5. K. Fukushima, D.E. Kharzeev, H.J. Warringa, The chiral magnetic effect. Phys. Rev. D **78**, 074033 (2008)

6. D.E. Kharzeev, H.J. Warringa, Chiral magnetic conductivity. Phys. Rev. D **80**, 034028 (2009). 10 pages, 4 figures

7. Y. Aoki, Z. Fodor, S.D. Katz, K.K. Szabó, The QCD transition temperature: results with physical masses in the continuum limit. Phys. Lett. B **643**, 46–54 (2006)

8. Y. Aoki, G. Endrődi, Z. Fodor, S.D. Katz, K.K. Szabó, The order of the quantum chromodynamics transition predicted by the standard model of particle physics. Nature **443**, 675–678 (2006)

9. J.M. Maldacena, The large N limit of superconformal field theories and supergravity. Adv. Theor. Math. Phys. **2**, 231–252 (1998)

10. G. Policastro, D.T. Son, A.O. Starinets, The shear viscosity of strongly coupled N = 4 supersymmetric Yang–Mills plasma. Phys. Rev. Lett. **87**, 081601 (2001)

11. P. Kovtun, D.T. Son, A.O. Starinets, Viscosity in strongly interacting quantum field theories from black hole physics. Phys. Rev. Lett. **94**, 111601 (2005)

12. M. Brigante, H. Liu, R.C. Myers, S. Shenker, S. Yaida, Viscosity bound violation in higher derivative gravity. Phys. Rev. D **77**, 126006 (2008)

13. A. Rebhan, D. Steineder, Violation of the holographic viscosity bound in a strongly coupled anisotropic plasma. Phys. Rev. Lett. **108**, 021601 (2012)

14. Y. Nambu, G. Jona-Lasinio, Dynamical model of elementary particles based on an analogy with superconductivity. 1. Phys. Rev. **122**, 345–358 (1961)

15. Y. Nambu, G. Jona-Lasinio, Dynamical model of elementary particles based on an analogy with superconductivity. II. Phys. Rev. **124**, 246–254 (1961)

16. T. Sakai, S. Sugimoto, Low energy hadron physics in holographic QCD. Prog. Theor. Phys. **113**, 843–882 (2005)

17. T. Sakai, S. Sugimoto, More on a holographic dual of QCD. Prog. Theor. Phys. **114**, 1083–1118 (2005)

18. M.G. Alford, K. Rajagopal, F. Wilczek, Color flavor locking and chiral symmetry breaking in high density QCD. Nucl. Phys. B **537**, 443–458 (1999)

19. M.G. Alford, A. Schmitt, K. Rajagopal, T. Schäfer, Color superconductivity in dense quark matter. Rev. Mod. Phys. **80**, 1455–1515 (2008)

20. Sh. Fayazbakhsh, N. Sadooghi, Phase diagram of hot magnetized two-flavor color superconducting quark matter. Phys. Rev. D **83**, 025026 (2011)

21. M.K. Volkov, Meson Lagrangians in a superconductor quark model. Ann. Phys. **157**, 282–303 (1984)

22. T. Hatsuda, T. Kunihiro, Possible critical phenomena associated with the chiral symmetry breaking. Phys. Lett. B **145**, 7–10 (1984)

23. J. Bardeen, L.N. Cooper, J.R. Schrieffer, Microscopic theory of superconductivity. Phys. Rev. **106**, 162 (1957)

24. E. Witten, Anti-de Sitter space, thermal phase transition, and confinement in gauge theories. Adv. Theor. Math. Phys. **2**, 505–532 (1998)

25. V.P. Gusynin, V.A. Miransky, I.A. Shovkovy, Dimensional reduction and dynamical chiral symmetry breaking by a magnetic field in $(3 + 1)$-dimensions. Phys. Lett. B **349**, 477–483 (1995)

26. K.G. Klimenko, Three-dimensional Gross–Neveu model in an external magnetic field. Theor. Math. Phys. **89**, 1161–1168 (1992)

27. K.G. Klimenko, Three-dimensional Gross–Neveu model at nonzero temperature and in an external magnetic field. Theor. Math. Phys. **90**, 1–6 (1992)

28. V.P. Gusynin, V.A. Miransky, I.A. Shovkovy, Catalysis of dynamical flavor symmetry breaking by a magnetic field in $(2 + 1)$-dimensions. Phys. Rev. Lett. **73**, 3499–3502 (1994)

29. V.P. Gusynin, V.A. Miransky, I.A. Shovkovy, Dynamical flavor symmetry breaking by a magnetic field in $(2 + 1)$-dimensions. Phys. Rev. D **52**, 4718–4735 (1995)

30. K. Fukushima, J.M. Pawlowski, Magnetic catalysis in hot and dense quark matter and quantum fluctuations. Phys. Rev. D **86**, 076013 (2012). arXiv:1203.4330 [hep-ph]

31. V.P. Gusynin, V.A. Miransky, I.A. Shovkovy, Dynamical chiral symmetry breaking by a magnetic field in QED. Phys. Rev. D **52**, 4747–4751 (1995)

32. V.G. Filev, C.V. Johnson, R.C. Rashkov, K.S. Viswanathan, Flavoured large N gauge theory in an external magnetic field. J. High Energy Phys. **0710**, 019 (2007)
33. J. Erdmenger, R. Meyer, J.P. Shock, AdS/CFT with flavour in electric and magnetic Kalb–Ramond fields. J. High Energy Phys. **0712**, 091 (2007)
34. V.G. Filev, C.V. Johnson, J.P. Shock, Universal holographic chiral dynamics in an external magnetic field. J. High Energy Phys. **0908**, 013 (2009)
35. V.G. Filev, R.C. Rashkov, Magnetic catalysis of chiral symmetry breaking. A holographic prospective. Adv. High Energy Phys. **2010**, 473206 (2010)
36. N. Callebaut, D. Dudal, H. Verschelde, Holographic rho mesons in an external magnetic field. J. High Energy Phys. **1303**, 033 (2013). arXiv:1105.2217 [hep-th]
37. S. Bolognesi, D. Tong, Magnetic catalysis in AdS4. Class. Quantum Gravity **29**, 194003 (2012). arXiv:1110.5902 [hep-th]
38. J. Erdmenger, V.G. Filev, D. Zoakos, Magnetic catalysis with massive dynamical flavours. J. High Energy Phys. **1208**, 004 (2012). arXiv:1112.4807 [hep-th]
39. V.P. Gusynin, V.A. Miransky, S.G. Sharapov, I.A. Shovkovy, Excitonic gap, phase transition, and quantum Hall effect in graphene. Phys. Rev. B **74**, 195429 (2006)
40. E.V. Gorbar, V.P. Gusynin, V.A. Miransky, I.A. Shovkovy, Dynamics in the quantum Hall effect and the phase diagram of graphene. Phys. Rev. B **78**, 085437 (2008)
41. M. D'Elia, S. Mukherjee, F. Sanfilippo, QCD phase transition in a strong magnetic background. Phys. Rev. D **82**, 051501 (2010)
42. G.S. Bali, F. Bruckmann, G. Endrődi, Z. Fodor, S.D. Katz et al., The QCD phase diagram for external magnetic fields. J. High Energy Phys. **1202**, 044 (2012)
43. F. Preis, A. Rebhan, A. Schmitt, Inverse magnetic catalysis in dense holographic matter. J. High Energy Phys. **1103**, 033 (2011)
44. F. Preis, A. Rebhan, A. Schmitt, Holographic baryonic matter in a background magnetic field. J. Phys. G **39**, 054006 (2012)
45. M. Buballa, NJL model analysis of quark matter at large density. Phys. Rep. **407**, 205–376 (2005)
46. T. Tatsumi, E. Nakano, Dual chiral density wave in quark matter (2004). arXiv:hep-ph/0408294
47. E. Nakano, T. Tatsumi, Chiral symmetry and density wave in quark matter. Phys. Rev. D **71**, 114006 (2005)
48. D. Nickel, Inhomogeneous phases in the Nambu–Jona-Lasino and quark–meson model. Phys. Rev. D **80**, 074025 (2009)
49. J.S. Schwinger, On gauge invariance and vacuum polarization. Phys. Rev. **82**, 664–679 (1951)
50. M.A. Metlitski, A.R. Zhitnitsky, Anomalous axion interactions and topological currents in dense matter. Phys. Rev. D **72**, 045011 (2005)
51. D.P. Menezes, M. Benghi Pinto, S.S. Avancini, A. Perez Martinez, C. Providencia, Quark matter under strong magnetic fields in the Nambu–Jona-Lasinio model. Phys. Rev. C **79**, 035807 (2009)
52. J. Bardeen, L.N. Cooper, J.R. Schrieffer, Theory of superconductivity. Phys. Rev. **108**, 1175–1204 (1957)
53. K. Fukushima, Y. Hidaka, Magnetic catalysis vs magnetic inhibition. Phys. Rev. Lett. **110**, 031601 (2013). arXiv:1209.1319 [hep-th]
54. A.M. Clogston, Upper limit for the critical field in hard superconductors. Phys. Rev. Lett. **9**, 266–267 (1962)
55. B.S. Chandrasekhar, A note on the maximum critical field of high-field superconductors. Appl. Phys. Lett. **1**, 7 (1962)
56. D. Ebert, K.G. Klimenko, M.A. Vdovichenko, A.S. Vshivtsev, Magnetic oscillations in dense cold quark matter with four fermion interactions. Phys. Rev. D **61**, 025005 (2000)
57. T. Inagaki, D. Kimura, T. Murata, Four fermion interaction model in a constant magnetic field at finite temperature and chemical potential. Prog. Theor. Phys. **111**, 371–386 (2004)
58. J.K. Boomsma, D. Boer, The influence of strong magnetic fields and instantons on the phase structure of the two-flavor NJL model. Phys. Rev. D **81**, 074005 (2010)

59. B. Chatterjee, H. Mishra, A. Mishra, Vacuum structure and chiral symmetry breaking in strong magnetic fields for hot and dense quark matter. Phys. Rev. D **84**, 014016 (2011)
60. S.S. Avancini, D.P. Menezes, M.B. Pinto, C. Providencia, The QCD critical end point under strong magnetic fields. Phys. Rev. D **85**, 091901 (2012)
61. J.O. Andersen, A. Tranberg, The chiral transition in a magnetic background: finite density effects and the functional renormalization group. J. High Energy Phys. **1208**, 002 (2012). arXiv:1204.3360 [hep-ph]
62. Sh. Fayazbakhsh, S. Sadeghian, N. Sadooghi, Properties of neutral mesons in a hot and magnetized quark matter. Phys. Rev. D **86**, 085042 (2012). arXiv:1206.6051 [hep-ph]
63. G.N. Ferrari, A.F. Garcia, M.B. Pinto, Chiral transition within effective quark models under magnetic fields. Phys. Rev. D **86**, 096005 (2012). arXiv:1207.3714 [hep-ph]
64. L. McLerran, R.D. Pisarski, Phases of cold, dense quarks at large N_c. Nucl. Phys. A **796**, 83–100 (2007)
65. E. Antonyan, J.A. Harvey, S. Jensen, D. Kutasov, NJL and QCD from string theory (2006). arXiv:hep-th/0604017
66. O. Aharony, J. Sonnenschein, S. Yankielowicz, A holographic model of deconfinement and chiral symmetry restoration. Ann. Phys. **322**, 1420–1443 (2007)
67. N. Horigome, Y. Tanii, Holographic chiral phase transition with chemical potential. J. High Energy Phys. **0701**, 072 (2007)
68. O. Bergman, G. Lifschytz, M. Lippert, Response of holographic QCD to electric and magnetic fields. J. High Energy Phys. **0805**, 007 (2008)
69. C.V. Johnson, A. Kundu, External fields and chiral symmetry breaking in the Sakai–Sugimoto model. J. High Energy Phys. **0812**, 053 (2008)
70. E.G. Thompson, D.T. Son, Magnetized baryonic matter in holographic QCD. Phys. Rev. D **78**, 066007 (2008)
71. G. Lifschytz, M. Lippert, Holographic magnetic phase transition. Phys. Rev. D **80**, 066007 (2009)
72. O. Bergman, G. Lifschytz, M. Lippert, Holographic nuclear physics. J. High Energy Phys. **0711**, 056 (2007)
73. O. Bergman, G. Lifschytz, M. Lippert, Magnetic properties of dense holographic QCD. Phys. Rev. D **79**, 105024 (2009)
74. A. Rebhan, A. Schmitt, S.A. Stricker, Meson supercurrents and the Meissner effect in the Sakai–Sugimoto model. J. High Energy Phys. **0905**, 084 (2009)
75. V. Schön, M. Thies, Emergence of Skyrme crystal in Gross–Neveu and 't Hooft models at finite density. Phys. Rev. D **62**, 096002 (2000)
76. A. Rebhan, A. Schmitt, S.A. Stricker, Anomalies and the chiral magnetic effect in the Sakai–Sugimoto model. J. High Energy Phys. **1001**, 026 (2010)
77. A. Rebhan, P. Romatschke, HTL quasiparticle models of deconfined QCD at finite chemical potential. Phys. Rev. D **68**, 025022 (2003)
78. A. Kurkela, P. Romatschke, A. Vuorinen, Cold quark matter. Phys. Rev. D **81**, 105021 (2010)
79. I.E. Frolov, V.Ch. Zhukovsky, K.G. Klimenko, Chiral density waves in quark matter within the Nambu–Jona-Lasinio model in an external magnetic field. Phys. Rev. D **82**, 076002 (2010)
80. E.V. Gorbar, V.A. Miransky, I.A. Shovkovy, Chiral asymmetry of the Fermi surface in dense relativistic matter in a magnetic field. Phys. Rev. C **80**, 032801 (2009)
81. E.V. Gorbar, V.A. Miransky, I.A. Shovkovy, Normal ground state of dense relativistic matter in a magnetic field. Phys. Rev. D **83**, 085003 (2011)

Chapter 4
Quark Matter in a Strong Magnetic Background

Raoul Gatto and Marco Ruggieri

4.1 Introduction

Quantum Chromodynamics (QCD) is the gauge theory of strong interactions. The understanding of its vacuum, and how it is modified by a large temperature and/or a baryon density, is one of the most intriguing aspects of modern physics. However, it is very hard to get a full understanding of its properties, because its most important characteristics, namely chiral symmetry breaking and color confinement, have a non-perturbative origin, and the use of perturbative methods is useless. One of the best strategies to overcome this problem is offered by Lattice QCD simulations at zero chemical potential (see [1–8] for several examples and see also references therein). At vanishing quark chemical potential, two crossovers take place in a broad range of temperatures; one for quark deconfinement, and another one for the (approximate) restoration of chiral symmetry.

An alternative approach to the physics of strong interactions, which is capable to capture some of the non-perturbative properties of the QCD vacuum, is the use of models. Among them, we will consider here the Nambu-Jona Lasinio (NJL) model [9, 10], see Refs. [11–14] for reviews. In this gluon-less model, which was inspired by the microscopic theory of superconductivity, the QCD interactions are replaced by effective interactions, which are built in order to respect the global symmetries of QCD. Since gluons are absent in the NJL model, it is not a gauge theory. However, it shares the global symmetries of the QCD action; moreover, the parameters of the NJL model are fixed to reproduce some phenomenological quantity of the QCD vacuum. Therefore, it is the main characteristics of its phase diagram should represent, at least qualitatively, those of QCD.

R. Gatto (✉)
Departement de Physique Theorique, Universite de Geneve, CH-1211 Geneve 4, Switzerland
e-mail: raoul.gatto@unige.ch

M. Ruggieri
Department of Astronomy and Physics, Catania University, Via S. Sofia 64, I-95125 Catania, Italy
e-mail: marco.ruggieri@lns.infn.it

D. Kharzeev et al. (eds.), *Strongly Interacting Matter in Magnetic Fields*,
Lecture Notes in Physics 871, DOI 10.1007/978-3-642-37305-3_4,
© Springer-Verlag Berlin Heidelberg 2013

The other side of the NJL model is that it lacks confinement. The latter, in the case of a pure gauge theory, can be described in terms of the center symmetry of the color gauge group and of the Polyakov loop [15–18], which is an order parameter for the center symmetry. Motivated by this property, the Polyakov extended Nambu-Jona Lasinio model (PNJL model) has been introduced [19, 20], in which the concept of statistical confinement replaces that of the true confinement of QCD, and an effective interaction among the chiral condensate and the Polyakov loop is achieved by a covariant coupling of quarks with a background temporal gluon field. In the literature there are several studies on the PNJL model, see Refs. [21–43] and references therein.

In this chapter, we make use of the PNJL model to study the interplay between chiral symmetry breaking and deconfinement in a *strong magnetic background*. Moreover, we compute several quantities which are relevant for the phenomenology of strong interactions physics in presence of a magnetic background. These topics are widely studied in the literature using many theoretical approaches. Lattice studies on the response to external magnetic (and chromo-magnetic) fields can be found in [44–51]. Previous studies of QCD in magnetic fields, and of QCD-like theories as well, can be found in Refs. [52–68]. Self-consistent model calculations of deconfinement pseudo-critical temperature in magnetic field have been performed [69–71].

An important motivation for these kind of studies is phenomenological. In fact, strong magnetic fields are produced in non-central heavy ion collisions [72–74]. For example, at the energy scale for RHIC it is found $eB_{max} \approx 5m_\pi^2$; for collisions at the LHC energy scale $eB_{max} \approx 15m_\pi^2$. In this case, it has been argued that the non-trivial topological structure of thermal QCD, namely the excitation of the strong sphalerons [75, 76], locally changes the chirality of quarks; this is reflected to event-by-event charge separation, a phenomenon which is dubbed Chiral Magnetic Effect (CME) [72, 77–79]. The possibility that the CME is observed in heavy ion collision experiments has motivated the study of strong interactions in presence of a chirality imbalance and a magnetic background, see [79–84] and references therein. An experimental measurement of observables connected to charge separation has been reported by the ALICE Collaboration in [85]. It is fair to say that realistic simulations of heavy ion collisions show that the magnetic fields have a very short lifetime, and decay before the local equilibrium is reached in the fireball. Moreover, the magnetic field is highly inhomogeneous. Furthermore, electric fields are produced beside the magnetic fields. Therefore, in order to describe realistically hot matter produced in the collisions, one should take care of the aforementioned details. However, for simplicity we neglect them, and leave the (much harder) complete problem to future studies.

We mainly base the present chapter on our previous works [86–88]. We firstly discuss chiral symmetry restoration in a strong magnetic background at finite temperature, using the PNJL model augmented with the eight-quark interaction [89–92]. In this case we also compute the dressed Polyakov loop in a magnetic field. The scenario which turns out from our calculations is compatible with that of the magnetic catalysis, in which the magnetic field acts as a catalyzer for chiral symmetry breaking. Moreover, we discuss on the role of the entanglement NJL vertex on the separation between deconfinement and chiral symmetry restoration in the background field. Finally, we summarize our computation of the magnetic susceptibility of the chiral condensate and of quark polarization in a strong magnetic background at zero temperature. We base the latter analysis on the Quark-Meson (QM) model, which offers the simplest renormalizable extension of the NJL model. Throughout this chapter we consider QCD in the vacuum, that is at zero baryon (as well as isospin) chemical potential. Computations at finite chemical potential are present in the literature, as we will mention in the main body of the chapter.

4.2 The PNJL Model with a Magnetic Background

In this section, we mainly summarize the results obtained in [86, 87]. We consider quark matter modeled by the following Lagrangian density:

$$\mathcal{L} = \bar{q}(i\gamma^\mu D_\mu - m_0)q + g_\sigma\left[(\bar{q}q)^2 + (\bar{q}i\gamma_5\tau q)^2\right]$$
$$+ g_8\left[(\bar{q}q)^2 + (\bar{q}i\gamma_5\tau q)^2\right]^2, \tag{4.1}$$

which corresponds to the NJL Lagrangian with multi-quark interactions [89]. The covariant derivative embeds the quark coupling to the external magnetic field and to the background gluon field as well, as we will see explicitly below. In (4.1), q represents a quark field in the fundamental representation of color and flavor (indices are suppressed for notational simplicity); m_0 is the bare quark mass, which is fixed to reproduce the pion mass in the vacuum, $m_\pi = 139$ MeV. Our interaction in (4.1) consists of a four-quark term, whose coupling g_σ has inverse mass dimension of two, and an eight-quark term, whose coupling constant g_8 has inverse mass dimension of eight.

We are considering the effect of a strong magnetic background on chiral symmetry restoration as well as deconfinement at finite temperature. We assume the magnetic field to be along the positive z-axis; we chose to work in the Landau gauge, specified by the vector potential $A = (0, Bx, 0)$.

To compute a temperature for the deconfinement crossover, we use the expectation value of the Polyakov loop, that we denote by L. In order to compute L we introduce a static, homogeneous and Euclidean background temporal gluon field, $A_0 = iA_4 = i\lambda_a A_4^a$, coupled minimally to the quarks via the QCD covariant derivative [20]. Then

$$L = \frac{1}{3}\text{Tr}_c \exp(i\beta\lambda_a A_4^a), \tag{4.2}$$

where $\beta = 1/T$. In the Polyakov gauge, which is convenient for this study, $A_0 = i\lambda_3\phi + i\lambda_8\phi^8$; moreover, we work at zero quark chemical potential, therefore we can take $L = L^\dagger$ from the beginning, which implies $A_4^8 = 0$. This choice is also motivated by the study of [71], where it is shown that the paramagnetic contribution of the quarks to the thermodynamic potential induces the breaking of the Z_3 symmetry, favoring the configurations with a real-valued Polyakov loop.

Besides the Polyakov loop, it is interesting to compute the dressed Polyakov loop [93]. In order to define this quantity, we work in a finite Euclidean volume with temperature extension $\beta = 1/T$. We take *twisted* fermion boundary conditions along the compact temporal direction,

$$q(x, \beta) = e^{-i\varphi} q(x, 0), \quad \varphi \in [0, 2\pi], \tag{4.3}$$

while for spatial directions the usual periodic boundary condition is taken. The canonical antiperiodic boundary condition for the quantization of fermions at finite temperature, is obtained by taking $\varphi = \pi$ in the previous equation. The dual quark condensate, $\tilde{\Sigma}_n$, is defined as

$$\tilde{\Sigma}_n(m, V) = \int_0^{2\pi} \frac{d\varphi}{2\pi} \frac{e^{-i\varphi n}}{V} \langle \bar{q} q \rangle_G, \tag{4.4}$$

where n is an integer. The expectation value $\langle \cdot \rangle_G$ denotes the path integral over gauge field configurations. An important point is that in the computation of the expectation value, the twisted boundary conditions acts only on the fermion determinant; the gauge fields are taken to be quantized with the canonical periodic boundary condition.

Using a lattice regularization, it has been shown in [93] that (4.4) can be expanded in terms of loops which wind n times along the compact time direction. In particular, the case $n = 1$ is called the *dressed Polyakov loop*; it corresponds to a sum of loops winding just once along the time direction. These correspond to the thin Polyakov loop (the loop with shortest length) plus higher order loops, the order being proportional to the length of the loop. Each higher order loop is weighed by an inverse power of the quark mass. Because of the weight, in the infinite quark mass limit only the thin Polyakov loop survives; for this reason, the dressed Polyakov loop can be viewed as a mathematical dressing of the thin loop, by virtue of longer loops, the latter being more and more important as the quark mass tends to smaller values.

If we denote by z an element of the center of the color gauge group, then $\tilde{\Sigma}_n \to z^n \tilde{\Sigma}_n$. It then follows that, under the center of the symmetry group Z_3, the dressed Polyakov loop ($n = 1$) is an order parameter for the center symmetry, with the same transformation rule of the thin Polyakov loop. Since the center symmetry is spontaneously broken in the deconfinement phase and restored in the confinement phase [15–18] (in presence of dynamical quarks, it is only approximately restored), the dressed Polyakov loop can be regarded as an order parameter for the confinement-deconfinement transition as well.

4.2.1 The One-Loop Quark Propagator

The evaluation of the bulk thermodynamic quantities requires we compute the quantum effective action of the model. This cannot be done exactly; hence we rely ourselves to the one-loop approximation for the partition function, which amounts to take the classical contribution plus the fermion determinant. At this level, the effect of the strong interactions is to modify the quark mass as follows:

$$M = m_0 - 2\sigma - 4\sigma^3 g_8/g_\sigma^3, \tag{4.5}$$

where $\sigma = g_\sigma \langle \bar{q}q \rangle$. As a further simplification, we neglect condensation in the pseudoscalar channel. We notice that the quark mass depends only on the sum of the u and d chiral condensates; therefore the mean field quark mass does not depend on the flavor.

To write the one-loop quark propagator in the background of the magnetic field we make use of the Leung-Ritus-Wang method [94, 95], which allows to expand the propagator on the complete and orthonormal set made of the eigenfunctions of a charged fermion in a homogeneous and static magnetic field. This is a well known procedure, discussed many times in the literature, see for example [96–102]; therefore it is enough to quote the final result:

$$S_f(x, y) = \sum_{k=0}^{\infty} \int \frac{dp_0 dp_2 dp_3}{(2\pi)^4} E_P(x) \Lambda_k \frac{1}{P \cdot \gamma - M} \bar{E}_P(y), \tag{4.6}$$

where $E_P(x)$ corresponds to the eigenfunction of a charged fermion in magnetic field, and $\bar{E}_P(x) \equiv \gamma_0 (E_P(x))^\dagger \gamma_0$. In the above equation,

$$P = \left(p_0 + iA_4, 0, \mathcal{Q}\sqrt{2k|Q_f eB|}, p_z\right), \tag{4.7}$$

where $k = 0, 1, 2, \ldots$ labels the kth Landau level, and $\mathcal{Q} \equiv \text{sign}(Q_f)$, with Q_f denoting the charge of the flavor f; Λ_k is a projector in Dirac space which keeps into account the degeneracy of the Landau levels; it is given by

$$\Lambda_k = \delta_{k0}[\mathcal{P}_+ \delta_{\mathcal{Q},+1} + \mathcal{P}_- \delta_{\mathcal{Q},-1}] + (1 - \delta_{k0})I, \tag{4.8}$$

where $\mathcal{P}_\pm$ are spin projectors and I is the identity matrix in Dirac spinor indices. The propagator in (4.6) has a non-trivial color structure, due to the coupling to the background gauge field, see (4.7).

It is useful to write down explicitly the expression of the chiral condensates for the flavor f with $f = u, d$. The chiral condensate is easily computed by taking minus the trace of the f-quark propagator. It is easy to show that the following equation holds:

$$\langle \bar{f}f \rangle = -N_c \frac{|Q_f eB|}{2\pi} \sum_{k=0}^{\infty} \beta_k \int \frac{dp_z}{2\pi} \frac{M_f}{\omega_f} \mathscr{C}(L, \bar{L}, T|p_z, k). \tag{4.9}$$

Here,

$$\mathscr{C}(L, \bar{L}, T|p_z, k) = U_\Lambda - 2\mathcal{N}(L, \bar{L}, T|p_z, k), \tag{4.10}$$

and $\mathcal{N}$ denotes the statistically confining Fermi distribution function,

$$\mathscr{C}(L,\bar{L},T\,|\,p_z,k) = \frac{1+2Le^{\beta\omega_f}+Le^{2\beta\omega_f}}{1+3Le^{\beta\omega_f}+3Le^{2\beta\omega_f}+e^{3\beta\omega_f}},$$

(4.11)

where

$$\omega_f^2 = p_z^2 + 2|Q_feB|k + M^2.$$

(4.12)

The first and the second addenda in the r.h.s. of (4.9) correspond to the vacuum and the thermal fluctuations contribution to the chiral condensate, respectively. The coefficient $\beta_k = 2 - \delta_{k0}$ keeps into account the degeneracy of the Landau levels. The vacuum contribution is ultraviolet divergent. In order to regularize it, we adopt a smooth regulator U_Λ, which is more suitable, from the numerical point of view, in our model calculation with respect to the hard-cutoff which is used in analogous calculations without magnetic field. We chose

$$U_\Lambda = \frac{\Lambda^{2N}}{\Lambda^{2N}+(p_z^2+2|Q_feB|k)^N}.$$

(4.13)

4.2.2 The One-Loop Thermodynamic Potential

The one-loop thermodynamic potential of quark matter in external fields has been discussed in [69, 70], in the case of canonical antiperiodic boundary conditions; following [36], it is easy to generalize it to the more general case of twisted boundary conditions:

$$\begin{aligned}
\Omega = \mathscr{U}(L,\bar{L},T) + \frac{\sigma^2}{g_\sigma} + \frac{3\sigma^4 g_8}{g_\sigma^4} \\
- \sum_{f=u,d} \frac{|Q_feB|}{2\pi} \sum_k \beta_k \int_{-\infty}^{+\infty} \frac{dp_z}{2\pi} g_\Lambda(p_z,k)\omega_k(p_z) \\
- T\sum_{f=u,d} \frac{|Q_feB|}{2\pi} \sum_k \beta_k \int_{-\infty}^{+\infty} \frac{dp_z}{2\pi} \log\bigl(1+3Le^{-\beta\mathscr{E}_-}+3\bar{L}e^{-2\beta\mathscr{E}_-}+e^{-3\beta\mathscr{E}_-}\bigr) \\
- T\sum_{f=u,d} \frac{|Q_feB|}{2\pi} \sum_k \beta_k \int_{-\infty}^{+\infty} \frac{dp_z}{2\pi} \log\bigl(1+3\bar{L}e^{-\beta\mathscr{E}_+}+3Le^{-2\beta\mathscr{E}_+}+e^{-3\beta\mathscr{E}_+}\bigr).
\end{aligned}$$

(4.14)

In the previous equation the arguments of the thermal exponentials are defined as

$$\mathscr{E}_\pm = \omega_f(p_z) \pm \frac{i(\varphi-\pi)}{\beta},$$

(4.15)

with φ defined in (4.3).

The potential term $\mathscr{U}[L, \bar{L}, T]$ in (4.14) is built by hand in order to reproduce the pure gluonic lattice data [21]. Among several different potential choices [103] we adopt the following logarithmic form [20, 21],

$$\frac{\mathscr{U}[L, \bar{L}, T]}{T^4} = -\frac{a(T)}{2}\bar{L}L + b(T)\ln\left[1 - 6\bar{L}L + 4(\bar{L}^3 + L^3) - 3(\bar{L}L)^2\right], \quad (4.16)$$

with three model parameters (one of four is constrained by the Stefan-Boltzmann limit),

$$a(T) = a_0 + a_1\left(\frac{T_0}{T}\right) + a_2\left(\frac{T_0}{T}\right)^2,$$

$$b(T) = b_3\left(\frac{T_0}{T}\right)^3. \quad (4.17)$$

The standard choice of the parameters reads [21];

$$a_0 = 3.51, \qquad a_1 = -2.47, \qquad a_2 = 15.2, \qquad b_3 = -1.75. \quad (4.18)$$

The parameter T_0 in (4.16) sets the deconfinement scale in the pure gauge theory, i.e. $T_c = 270$ MeV.

4.3 Numerical Results

In this section, we show our results. The main goal to achieve numerically is the solution of the gap equations,

$$\frac{\partial \Omega}{\partial \sigma} = 0, \qquad \frac{\partial \Omega}{\partial L} = 0. \quad (4.19)$$

This is done by using a globally convergent algorithm with backtrack [104]. From the very definition of the dressed Polyakov loop, (4.4), the twisted boundary condition, (4.3), must be imposed only in D_φ. Therefore, we firstly compute the expectation value of the Polyakov loop and to the chiral condensate, taking $\varphi = \pi$. Then, in order to compute the dressed Polyakov loop, we compute the φ-dependent chiral condensate using the first of (4.19), keeping the expectation value of the Polyakov loop fixed at its value at $\varphi = \pi$ [36].

Following [86, 87] we report results obtained using the UV-regulator with $N = 5$. As expected, in the other cases no different qualitative results are found; the parameter set is specified in Table 4.1. In the case $N = 5$, they are obtained by the requirements that the vacuum pion mass is $m_\pi = 139$ MeV, the pion decay constant $f_\pi = 92.4$ MeV and the vacuum chiral condensate $\langle \bar{u}u \rangle \approx (-241 \text{ MeV})^3$. In this case, the chiral and deconfinement pseudo-critical temperatures at zero magnetic field are $T_0^\chi = T_0^P = 175$ MeV.

Table 4.1 Parameters of the model for the two choices of the UV-regulator

	Λ (MeV)	m_0 (MeV)	g_σ (MeV)$^{-2}$	g_8 (MeV)$^{-8}$
$N = 5$	588.657	5.61	5×10^{-6}	6×10^{-22}

The main effect of the eight-quark interaction in (4.1) is to lower the pseudo-critical temperature of the crossovers. This has been already discussed several times in the literature [89, 90], in the context of both the NJL and the PNJL models. Therefore, it is not necessary to discuss it further here, while at the same time we prefer to stress the results that have not been discussed yet.

In order to identify the pseudo-critical temperatures, we have define the *effective susceptibilities* as

$$\chi_A = (m_\pi)^g \left| \frac{dA}{dT} \right|, \quad A = \sigma, P, \Sigma_1. \tag{4.20}$$

Strictly speaking, the quantities defined in the previous equation are not true susceptibilities. Nevertheless, they allow to represent faithfully the pseudo-critical region, that is, the range in temperature in which the various crossovers take place. Therefore, for our purposes it is enough to compute these quantities. In (4.20), the appropriate power of m_π is introduced just for a matter of convenience, in order to have a dimensionless quantity; therefore, $g = 0$ if $A = \sigma, \Sigma_1$, and $g = 1$ if $A = P$.

4.3.1 Condensates and Dressed Polyakov Loop

From now on, we fix $N = 5$ unless specified. The results for this case are collected in the form of surface plots in Fig. 4.1. In more detail, in the figure we plot the chiral condensate $S \equiv (\sigma/2)^{1/3}$, the expectation value of the Polyakov loop, and the dressed Polyakov loop Σ_1, as a function of temperature and magnetic field.

We slice the surface plots in Fig. 4.1 at fixed value of the magnetic field strength, and show the results in Fig. 4.2, where we plot the chiral condensate (upper panel), the Polyakov loop (middle panel) and Σ_1 (lower panel) as a function of temperature, for several values of the applied magnetic field strength, measured in units of m_π^2. In the right panel, we plot fits of the effective susceptibilities in the critical regions, as a function of temperature. The fits are obtained from the raw data, using Breit-Wigner-like fitting functions.

The qualitative behavior of the chiral condensate, and of the Polyakov loop as well, is similar to that found in a previous study within the PNJL model in the chiral limit [69]. Quantitatively, the main difference with the case of the chiral limit, is that in the latter the chiral restoration at large temperature is a true second order phase transition (in other model calculations it has been reported that the phase transition might become of the first order at very large magnetic field strengths [66, 67]). On the other hand, in the case under investigation, chiral symmetry is always broken explicitly because of the bare quark masses; as a consequence, the second order phase transition is replaced by a crossover.

Fig. 4.1 Chiral
condensate, Polyakov
loop and dressed Polyakov
loop as a function of
temperature and magnetic
field, for the case $N = 5$.
From Ref. [87]. Copyright
(2012) by the American
Physical Society

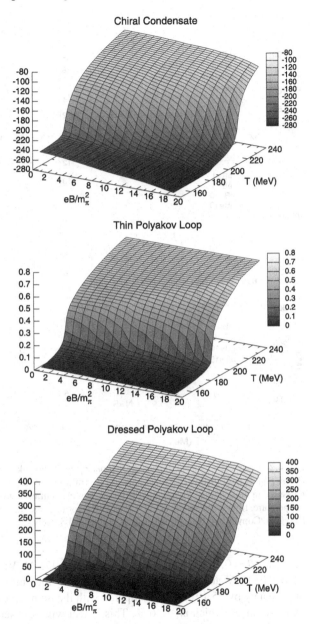

Another aspect is that the Polyakov loop crossover temperature is less sensitive
to the strength of the magnetic field than the same quantity computed for the chiral condensate. It is useful, for illustration purpose, to quantify the net shift of the
pseudo-critical temperatures, for the largest value of magnetic field we have studied,
$eB = 19m_\pi^2$. In this case, if we take $N = 5$, then the two crossovers occur simultane-

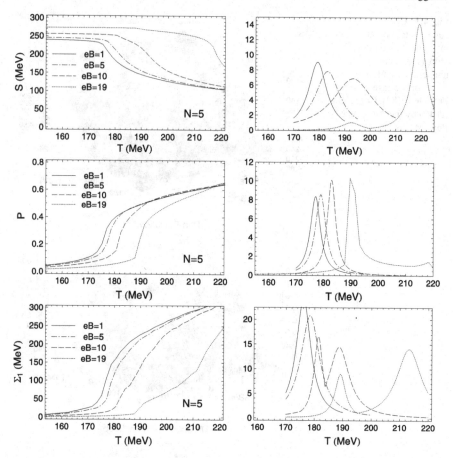

Fig. 4.2 *Left panel*: Chiral condensate S (*upper panel*), Polyakov loop (*middle panel*) and Σ_1 (*lower panel*) as a function of temperature, for several values of the applied magnetic field strength, measured in units of m_π^2. *Right panel*: Effective susceptibilities, defined in (4.20), as a function of temperature, for several values of eB. Conventions for lines are the same as in the *left panel*. From Ref. [87]. Copyright (2012) by the American Physical Society

ously at $eB = 0$, at the temperature $T_0^\chi = T_0^P = 175$ MeV; for $eB = 19m_\pi^2$, we find $T_\chi = 219$ MeV and $T_P = 190$ MeV. Therefore, the chiral crossover is shifted approximately by 25.1 %, to be compared with the more modest shift of the Polyakov loop crossover, which is ≈ 8.6 %. This aspect will be discussed further in the next section, in which we will comment on the possibility of entanglement between the NJL coupling at finite temperature and the Polyakov loop.

In the lower panels of Figs. 4.1 and 4.2, we plot the dressed Polyakov loop as a function of temperature, for several values of eB. We have normalized Σ_1 multiplying the one defined in [93] by the NJL coupling constant. For small values of eB/m_π^2, the behavior of Σ_1 as temperature is increased, is qualitatively similar to that at $eB = 0$, which has been discussed within effective models in [36, 105].

In particular, the dressed Polyakov loop is very small for temperatures below the pseudo-critical temperature of the simultaneous crossover. Then, it experiences a crossover in correspondence of the simultaneous Polyakov loop and chiral condensate crossovers. It eventually saturates at very large temperature (for example, in [36] the saturation occurs at a temperature of the order of 0.4 GeV, in agreement with the results of [105]). However, we do not push up our numerical calculation to such high temperature, because we expect that the effective model in that case is well beyond its range of validity.

As we increase the value of eB, as noticed previously, we observe a tiny splitting of the chiral and the Polyakov loop crossovers. Correspondingly, the qualitative behavior of the dressed Polyakov loop changes dramatically: the range of temperature in which the Σ_1 crossover takes place is enlarged, if compared to the thin temperature interval in which the crossover takes place at the lowest value of eB (compare the solid and the dotted lines in Fig. 4.2, as well as the lower panel of Fig. 4.1).

On passing, we discuss briefly the effective susceptibility, $d\Sigma_1/dT$, plotted in the lower right panel of Fig. 4.2, since its qualitative behavior is very interesting. We observe a double peak structure, which we interpret as the fact that the dressed Polyakov loop is capable to feel (and hence, describe) both the crossovers. If we were to interpret Σ_1 as the order parameter for deconfinement, and the temperature with the largest susceptibility as the crossover pseudo-critical temperature, then we obtain almost simultaneous crossover even for very large magnetic field.

4.3.2 Entanglement of NJL Coupling and Polyakov Loop

In [106] it has been shown that the NJL vertex can be deduced under some assumption from the QCD action; following this derivation a non-local structure of the interaction turns out. An analogous conclusion is achieved in [107–109]. More important for our study, the NJL vertex acquires a non-trivial dependence on the phase of the Polyakov loop. Therefore, in the model we consider here, it is important to keep into account this dependence. Here we follow the phenomenological ansatz introduced in [110], that is

$$G = g_\sigma \left[1 - \alpha_1 L\bar{L} - \alpha_2 \left(L^3 + \bar{L}^3 \right) \right], \qquad (4.21)$$

and we take $L = \bar{L}$. Moreover, we mainly discuss here the case without 8-quark interaction. The model with coupling constant specified in (4.21) is named Entangled-Polyakov improved-NJL model (EPNJL in the following) [110], since the vertex describes an entanglement between Polyakov loop and the interaction responsible for chiral symmetry breaking.

The functional form in the above equation is constrained by C and extended Z_3 symmetry. We refer to [110] for a more detailed discussion. The numerical values of α_1 and α_2 have been fixed in [110] by a best fit of the available Lattice data at zero and imaginary chemical potential of Ref. [111], which have been confirmed

Fig. 4.3 *Upper panel*: Chiral condensates of u and d quarks as functions of temperatures in the pseudo-critical region, at $eB = 15m_\pi^2$ and $eB = 30m_\pi^2$. Condensates are measured in units of their value at zero magnetic field and zero temperature, namely $\sigma_0 = (-253 \text{ MeV})^3$. *Lower panel*: Polyakov loop expectation value as a function of temperature, at $eB = 15m_\pi^2$ and $eB = 30m_\pi^2$. Data correspond to $\alpha = 0.2$. From Ref. [87]. Copyright (2012) by American Physical Society

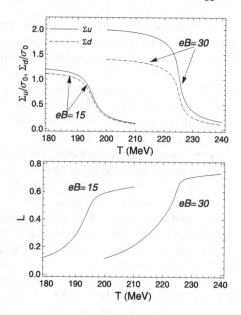

recently in [112]. In particular, the fitted data are the critical temperature at zero chemical potential, and the dependence of the Roberge-Weiss endpoint on the bare quark mass.

The values $\alpha_1 = \alpha_2 \equiv \alpha = 0.2 \pm 0.05$ have been obtained in [110] using a hard cutoff regularization scheme. We will focus mainly on the case $\alpha = 0.2$ as in [110]. In [86] we have verified that in the regularization scheme with the smooth cutoff, the results are in quantitative agreement with those of [110]. There, a detailed discussion of the role of α can be found as well (we will skip this discussion in this chapter).

We plot in Fig. 4.3 the chiral condensates of u and d quarks as a function of temperature, at $eB = 15m_\pi^2$ and $eB = 30m_\pi^2$. In the lower panel of the figure, we plot the expectation value of the Polyakov loop as a function of temperature. The condensates are measured in units of their value at zero magnetic field and zero temperature, namely $\sigma_0 \equiv \langle \bar{u}u \rangle = \langle \bar{d}d \rangle = (-253 \text{ MeV})^3$. They are computed by a two-step procedure: firstly we find the values of σ and L that minimize the thermodynamic potential; then, we make use of (4.9) to compute the expectation values of $\bar{u}u$ and $\bar{d}d$ in magnetic field. If we measure the strength of the crossover by the value of the peak of $|d\sigma/dT|$, it is obvious from the figure that the chiral crossover becomes stronger and stronger as the strength of the magnetic field is increased, in agreement with [44].

The results in Fig. 4.3 show that, identifying the deconfinement crossover with the temperature T_L at which dL/dT is maximum, and the chiral crossover with the temperature T_χ at which $|d\sigma/dT|$ is maximum, the two temperatures are very close also in a strong magnetic field. From the model point of view, it is easy to understand why deconfinement and chiral symmetry restoration are entangled also in strong magnetic field. As a matter of fact, using the data shown in Fig. 4.3, it is possible to compute the NJL coupling constant in the pseudo-critical region, which turns out to decrease of the 15 % as a consequence of the deconfinement crossover. Therefore, the strength of the interaction responsible for the spontaneous chiral symmetry breaking is strongly affected by the deconfinement, with the obvious consequence that the numerical value of the chiral condensate drops down and the chiral crossover takes place. We have verified that the picture remains qualitatively and quantitatively unchanged if we perform a calculation at $eB = 30m_\pi^2$. In this case, we find $T_L = 224$ MeV and $T_\chi = 225$ MeV.

This result can be compared with the previous calculations [69], described also in the previous section, in which the Polyakov loop dependence of the NJL coupling constant was not included. In [69] we worked in the chiral limit and we observed that the Polyakov loop crossover in the PNJL model is almost insensitive to the magnetic field; on the other hand, the chiral phase transition temperature was found to be very sensitive to the strength of the applied magnetic field, in agreement with the well known magnetic catalysis scenario [52–58]. This model prediction has been confirmed within the Polyakov extended quark-meson model in [71], when the contribution from the vacuum fermion fluctuations to the energy density is kept into account;[1] we then obtained a similar result in [87], in which we turned from the chiral to the physical limit at which $m_\pi = 139$ MeV, and introduced the 8-quark term as well (PNJL$_8$ model, according to the nomenclature of [110]). The comparison with the results of the PNJL$_8$ model of [87] is interesting because the model considered there, was tuned in order to reproduce the Lattice data at zero and imaginary chemical potential [35], like the model we use in this study. Therefore, they share the property of describing the QCD thermodynamics at zero and imaginary chemical potential; it is therefore instructive to compare their predictions at finite eB.

For concreteness, in [87] we found $T_P = 185$ MeV and $T_\chi = 208$ MeV at $eB = 19m_\pi^2$, corresponding to a split of ≈ 12 %. On the other hand, in the present calculation we measure a split of ≈ 1.5 % at the largest value of eB considered. Therefore, the results of the two models are in slight quantitative disagreement; this disagreement is then reflected in a slightly different phase diagram. We will draw the phase diagram of the two models in a next section; however, since now it is easy to understand what the main difference consists in: the PNJL$_8$ model predicts some window in the eB–T plane in which chiral symmetry is still broken by a chiral condensate, but deconfinement already took place. In the case of the EPNJL model, this window is shrunk to a very small one, because of the entanglement of the two crossovers at finite eB. On the other hand, it is worth to stress that the two models share an important qualitative feature: both chiral restoration and deconfinement temperatures are enhanced by a strong magnetic field; the latter conclusion is in qualitative agreement with the Lattice data of D'Elia et al. [44], but in disagreement with more recent data [50, 51]. We will come back to a comparison with Lattice data, as well as with other computations, in the next section.

---

[1]If the vacuum corrections are neglected, the deconfinement and chiral crossovers are found to be coincident even in very strong magnetic fields [71], but the critical temperature decreases as a function of eB; this scenario is very interesting theoretically, but it seems excluded from the recent Lattice simulations [44].

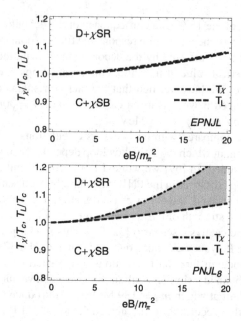

Fig. 4.4 *Upper panel*: Phase diagram in the $eB–T$ plane for the EPNJL model. Temperatures on the *vertical axis* are measured in units of the pseudo-critical temperature for deconfinement at $eB = 0$, namely $T_c = 185.5$ MeV. *Lower panel*: Phase diagram in the $eB–T$ plane for the PNJL$_8$ model. Temperatures on the *vertical axis* are measured in units of the pseudo-critical temperature for deconfinement at $eB = 0$, namely $T_c = 175$ MeV. In both the phase diagrams, T_χ, T_L correspond to the chiral and deconfinement pseudo-critical temperatures, respectively. The *grey shaded region* denotes the portion of phase diagram in which hot quark matter is deconfined and chiral symmetry is still broken spontaneously. From Ref. [87]. Copyright (2012) by American Physical Society

4.4 Phase Diagram in the $eB–T$ Plane

In Fig. 4.4 we collect our data on the pseudo-critical temperatures for deconfinement and chiral symmetry restoration, in the form of a phase diagram in the $eB–T$ plane. In the upper panel we show the results obtained within the EPNJL model; in the lower panel, we plot the results of the PNJL$_8$ model, that are obtained using the fitting functions computed in [87]. In the figure, the magnetic field is measured in units of m_π^2; temperature is measured in units of the deconfinement pseudo-critical temperature at zero magnetic field, namely $T_{B=0} = 185.5$ MeV for the EPNJL model, and $T_{B=0} = 175$ MeV for the PNJL$_8$ model. For any value of eB, we identify the pseudo-critical temperature with the peak of the effective susceptibility.

It should be kept in mind, however, that the definition of a pseudo-critical temperature in this case is not unique, because of the crossover nature of the phenomena that we describe. Other satisfactory definitions include the temperature at which the order parameter reaches one half of its asymptotic value (which corresponds to the $T \to 0$ limit for the chiral condensate, and to the $T \to +\infty$ for the Polyakov loop), and the position of the peak in the true susceptibilities. The expectation is that the critical temperatures computed in these different ways differ from each other only of few percent. This can be confirmed concretely using the data in Fig. 4.3 at $eB = 30m_\pi^2$. Using the peak of the effective susceptibility we find $T_\chi = 225$ MeV and $T_L = 224$ MeV; on the other hand, using the half-value criterion, we find $T_\chi = 227$ MeV and $T_L = 222$ MeV, in very good agreement with the previous estimate. Therefore, the qualitative picture that we derive within our simple calculational scheme, namely the entanglement of the two crossovers in a strong magnetic field, should not be affected by using different definitions of the critical temperatures.

Firstly we focus on the phase diagram of the EPNJL model. In the upper panel of Fig. 4.4, the dashed and dot-dashed lines correspond to the deconfinement and chiral symmetry restoration pseudo-critical temperatures, respectively. As a consequence of the entanglement, the two crossovers stay closed also in very strong magnetic field, as we have already discussed in the previous section. The grey region in the figure denotes a phase in which quark matter is (statistically) deconfined, but chiral symmetry is still broken. According to [113, 114], we can call this phase Constituent Quark Phase (CQP).

On the lower panel of Fig. 4.4 we have drawn the phase diagram for the PNJL8 model based on Ref. [87] and discussed in the previous section. The most astonishing feature of the phase diagram of the PNJL8 model is the entity of the split among the deconfinement and the chiral restoration crossover. The difference with the result of the EPNJL model is that in the former, the entanglement with the Polyakov loop is neglected in the NJL coupling constant. As we have already mentioned in the previous section, the maximum amount of split that we find within the EPNJL model, at the largest value of magnetic field considered here, is of the order of 2 %; this number has to be compared with the split at $eB = 20m_\pi^2$ in the PNJL8 model, namely ≈ 12 %. The larger split causes a considerable portion of the phase diagram to be occupied by the CQP.

4.4.1 Comparison with Other Computations

In this section we summarize the main results obtained in the literature, comparing them with the scenario depicted in our works. Before going ahead, it is useful to summarize the two main results obtained within our one-loop computations:

- The critical temperature for chiral symmetry restoration is *increased* by an external magnetic field;
- The split between deconfinement and chiral symmetry restoration temperatures in a strong magnetic background can *be reduced* if the entanglement vertex is considered.

The first conclusion is in agreement with most of the computations: calculations based on the quark-meson-model with and without quantum fluctuations [71, 115–120], on chiral perturbation theory at finite temperature [121], on the PNJL model [69, 122], on the holographic correspondence [68, 123–125].

Besides models predictions, lattice computations of the critical temperatures in a magnetic background have been performed [44, 50, 51]. The computations of [44] have been performed with a pion mass of the order of 400 MeV, hence a little bit far from the physical limit. In this case, both the chiral symmetry restoration and the deconfinement temperatures are measured, and they are found to increase slightly with the magnetic field strength; moreover, the two transitions seem to be entangled even at the largest value of the magnetic field considered in the study. On the other hand, in [50, 51] quark masses are chosen such that the lattice pion has its physical mass; in this case a non-trivial dependence of the critical temperature for chiral symmetry restoration on the magnetic field strength and the quark mass is found. For the up and down quark condensates at the physical limit, the critical temperature decreases with the magnetic field strength. The deconfinement temperature, measured either by the Polyakov loop and by the strange quark number susceptibility, is found to decrease as well. If the light quark masses are increased up to the value of the strange quark mass, see Fig. 5 of [51], then the critical temperature, measured by the peak of the u-quark chiral susceptibility, remains almost constant as a function of the magnetic field strength.

We also mention a recent work [126] in which the influence of a magnetic field on the finite temperature phase structure and the chiral properties of 2-colors QCD with four species of dynamical staggered fermions is investigated. In this case, at a fixed mass the critical temperature is seen to rise with the magnetic field strength.

Computations within the MIT bag model [127] do not have direct access to the chiral symmetry restoration, but to the deconfinement temperature. Within this model it is found that the critical temperature for deconfinement is a decreasing function of the external magnetic field strength. This conclusion is in agreement with a previous computation [66, 67]. Furthermore, a decreasing temperature is found also within the quark-meson model if the fermion vacuum contribution is neglected [71]. If deconfinement and chiral symmetry restoration are entangled at finite magnetic field, then the MIT model based study would give some hint on the mechanism which makes the temperature for chiral symmetry restoration in a magnetic background lower than that at zero field. However, if this is the case, then the role of the quark mass on the dependence of the critical temperature on the magnetic field strength [50, 51] should be transparent. In our opinion, more study is needed to understand the puzzling behavior of T_c as a function of the magnetic field found on the Lattice: beside model computations, independent Lattice simulations should be performed, in order to confirm the results of [50, 51].

4.5 Polarization of the Quark Condensate

It has been realized that external fields can induce QCD condensates that are absent otherwise [128]. Here we focus on the magnetic moment, $\langle \bar{f} \Sigma^{\mu\nu} f \rangle$ where f denotes the fermion field of the flavor fth, and $\Sigma^{\mu\nu} = -i(\gamma^\mu \gamma^\nu - \gamma^\nu \gamma^\mu)/2$. At small fields one can write, according to [128],

$$\langle \bar{f} \Sigma^{\mu\nu} f \rangle = \chi \langle \bar{f} f \rangle Q_f |eB|, \qquad (4.22)$$

and χ is a constant independent of flavor, which is dubbed magnetic susceptibility of the quark condensate. In [128] it is proved that the role of the condensate (4.22) to QCD sum rules in external fields is significant, and it cannot be ignored. The quantity χ has been computed by means of special sum rules [128–132], OPE combined with Pion Dominance [133], holography [134, 135], instanton vacuum model [136], analytically from the zero mode of the Dirac operator in the background of a $SU(2)$ instanton [137], and on the Lattice in two color quenched simulations at zero and finite temperature [138]. It has also been suggested that in the photoproduction of lepton pairs, the interference of the Drell-Yan amplitude with the amplitude of a process where the photon couples to quarks through its chiral-odd distribution amplitude, which is normalized to the magnetic susceptibility of the QCD vacuum, is possible [139]. This interference allows in principle to access the chiral odd transversity parton distribution in the proton. Therefore, this quantity is interesting both theoretically and phenomenologically. The several estimates, that we briefly review in Sect. 4.3, lead to the numerical value of χ as follows:

$$\chi\langle \bar{f} f \rangle = 40\text{--}70 \text{ MeV}. \tag{4.23}$$

A second quantity, which embeds non-linear effects at large fields, is the polarization, μ_f, defined as

$$\mu_f = \left| \frac{\Sigma_f}{\langle \bar{f} f \rangle} \right|, \qquad \Sigma_f = \langle \bar{f} \Sigma^{12} f \rangle, \tag{4.24}$$

which has been computed on the Lattice in [138] for a wide range of magnetic fields, in the framework of two-color QCD with quenched fermions. At small fields $\mu_f = |\chi Q_f eB|$ naturally; at large fields, non-linear effects dominate and an interesting saturation of μ_f to the asymptotic value $\mu_\infty = 1$ is measured. According to [138] the behavior of the polarization as a function of eB in the whole range examined, can be described by a simple inverse tangent function. Besides, magnetization of the QCD vacuum has been computed in the strong field limit in [140] using perturbative QCD, where it is found it grows as $B \log B$.

In [88] we compute the magnetic susceptibility of the quark condensate by means of the NJL and the QM models. This study is interesting because in the chiral models, it is possible to compute self-consistently the numerical values of the condensates as a function of eB, once the parameters are fixed to reproduce some characteristic of the QCD vacuum. We firstly perform a numerical study of the problem, which is then complemented by some analytic estimate of the same quantity within the renormalized QM model. Moreover, we compute the polarization of quarks at small as well as large fields, both numerically and analytically. In agreement with the Lattice results [138], we also measure a saturation of μ_f to one at large fields, in the case of the effective models. Our results push towards the interpretation of the saturation as a non-artifact of the Lattice. On the contrary, we can offer a simple physical understanding of this behavior, in terms of lowest Landau level dominance of the chiral condensate. As a matter of fact, using the simple equations of the models for the chiral condensate and for the magnetic moment, we can show that at large magnetic field μ_f has to saturate to one, because in this limit the higher Landau levels are expelled from the chiral condensate; as a consequence, the ratio of the two approaches one asymptotically.

We also obtain a saturation of the polarization within the renormalized QM model. There are some differences, however, in comparison with the results of the non-renormalized models. In the former case, the asymptotic value of μ_f is charge-dependent; moreover, the interpretation of the saturation as a lowest Landau level (LLL) dominance is not straightforward, because the renormalized contribution of the higher Landau levels is important in the chiral condensate, even in the limit of very strong fields. It is possible that the results obtained within the renormalized model are a little bit far from true QCD. As a matter of fact, in the renormalized model we assume that the quark self-energy is independent on momentum; thus, when we take the limit of infinite quark momentum in the gap equation, and absorb the ultraviolet divergences by means of counterterms and renormalization conditions, we implicitly assume that quark mass at large momenta is equal to its value at zero momentum. We know that this is not true, see for example [141–143]: even in the renormalized theory, the quark self-energy naturally cuts off the large momenta, leading to LLL dominance in the traces of quark propagator which are relevant for our study. Nevertheless, it is worth to study this problem within the renormalized QM model in its simplest version, because it helps to understand the structure of this theory under the influence of a strong magnetic field.

In our calculations we neglect, for simplicity, the possible condensation of ρ-mesons at strong fields [144, 145]. Vector meson dominance [144] and the Sakai-Sugimoto model [146] suggest for the condensation a critical value of $eB_c \approx m_\rho^2 \approx$ 0.57 GeV^2, where m_ρ is the ρ-meson mass in the vacuum. Beside these, a NJL-based calculation within the lowest Landau level (LLL) approximation [145] predicts ρ-meson condensation at strong fields as well, even if in the latter case it is hard to estimate exactly eB_c, mainly because of the uncertainty of the parameters of the model. It would certainly be interesting to address this problem within our calculations, in which not only the LLL but also the higher Landau levels are considered, and in which the spontaneous breaking of chiral symmetry is kept into account self-consistently. However, this would complicate significantly the calculational setup. Therefore, for simplicity we leave this issue to a future project.

In [88] the computation of the polarization and of the magnetic susceptibility has been performed both within the NJL and the Quark-Meson (QM) models; the qualitative picture does not depend on the model considered. Moreover, within the QM model an analytical computation of the aforementioned quantities within the renormalized quantum effective potential is feasible (the NJL model, at least with a contact interaction as considered in [88], is not renormalizable). Therefore in this review chapter we limit ourselves to summarize the results obtained within the QM model, both numerical and analytical, deferring to the original reference for further details.

4.5.1 Non-renormalized Quark-Meson Model Results

In the QM model, a meson sector described by the linear sigma model Lagrangian, is coupled to quarks via a Yukawa-type interaction. The model is renormalizable in $D = 3 + 1$ dimensions. However, since we adopt the point of view of it as an effective description of QCD, it is not necessary to use the renormalized version of the model itself. On the contrary, it is enough to fix an ultraviolet scale to cut-off the divergent expectation values; the UV scale is then chosen phenomenologically, by requiring that the numerical value of the chiral condensate in the vacuum obtained within the model, is consistent with the results obtained from the sum rules [147, 148]. This is a rough approximation of the QCD effective quark mass, which smoothly decays at large momenta [141–143]. In Sect. 4.4 we will use a renormalized version of the model, to derive semi-analytically some results in the two regimes of weak and strong fields.

The Lagrangian density of the model is given by

$$\mathcal{L} = \bar{q}\left[iD_\mu\gamma^\mu - g(\sigma + i\gamma_5\boldsymbol{\tau}\cdot\boldsymbol{\pi})\right]q$$
$$+ \frac{1}{2}(\partial_\mu\sigma)^2 + \frac{1}{2}(\partial_\mu\boldsymbol{\pi})^2 - U(\sigma, \boldsymbol{\pi}). \tag{4.25}$$

In the above equation, q corresponds to a quark field in the fundamental representation of color group $SU(3)$ and flavor group $SU(2)$; the covariant derivative, $D_\mu = \partial_\mu - Q_f e A_\mu$, describes the coupling to the background magnetic field, where Q_f denotes the charge of the flavor f. Besides, σ, $\boldsymbol{\pi}$ correspond to the scalar singlet and the pseudo-scalar iso-triplet fields, respectively. The potential U describes tree-level interactions among the meson fields. In this article, we take its analytic form as

$$U(\sigma, \boldsymbol{\pi}) = \frac{\lambda}{4}\left(\sigma^2 + \boldsymbol{\pi}^2 - v^2\right)^2 - h\sigma, \tag{4.26}$$

where the first addendum is chiral invariant; the second one describes a soft explicit breaking of chiral symmetry, and it is thus responsible for the non-zero value of the pion mass. For $h = 0$, the interaction terms of the model are invariant under $SU(2)_V \otimes SU(2)_A \otimes U(1)_V$. This group is broken explicitly to $U(1)_V^3 \otimes U(1)_A^3 \otimes U(1)_V$ if the magnetic field is coupled to the quarks, because of the different electric charge of u and d quarks. Here, the superscript 3 in the V and A groups denotes the transformations generated by τ_3, $\tau_3\gamma_5$ respectively. Therefore, the chiral group in presence of a magnetic field is $U(1)_V^3 \otimes U(1)_A^3$. This group is then explicitly broken by the h-term to $U(1)_V^3$.

The formalism which is used to compute the magnetic susceptibility and the polarization of the quark condensate is similar to the one described in the previous sections; therefore it is not necessary to give the details here. It is enough to write down the expressions for the chiral condensate at zero temperature,

$$\langle \bar{f}f \rangle = -N_c\frac{|Q_f eB|}{2\pi}\sum_{k=0}^{\infty}\beta_k\int\frac{dp_3}{2\pi}\frac{m_q}{\omega_k(p_3)}, \tag{4.27}$$

where the divergent integral on the r.h.s. of the above equation has to be understood regularized as in (4.13), and for the magnetic moment for the flavor f,

$$\langle \bar{f} \Sigma^{\mu\nu} f \rangle = -\text{Tr}[\Sigma^{\mu\nu} S_f(x,x)].$$ (4.28)

We take $\boldsymbol{B} = (0, 0, B)$; in this case, only $\Sigma^{12} \equiv \Sigma_f$ is non-vanishing. Using the properties of γ-matrices it is easy to show that only the Lowest Landau Level (LLL) gives a non-vanishing contribution to the trace:

$$\Sigma_f = N_c \frac{Q_f |eB|}{2\pi} \int \frac{dp_3}{2\pi} \frac{m_q}{\omega_0(p_3)},$$ (4.29)

where $\omega_0 = \omega_{k=0}$.

From (4.27) we notice that the prescription (4.13) is almost equivalent to the introduction of a running effective quark mass,

$$m_q = g\sigma \Theta \left(\Lambda^2 - p_3^2 - 2k|Q_f eB| \right),$$ (4.30)

that can be considered as a rough approximation to the effective running quark mass in QCD [141, 142] which decays at large quark momenta, see also the discussion in [143]. Once the scale Λ is fixed, the Landau levels with $n \geq 1$ are removed from the chiral condensate if $eB \gg \Lambda^2$.

In the upper panel of Fig. 4.5, we plot the chiral condensates for u and d quarks, as a function of eB, for the QM model. The magnetic field splits the two quantities because of the different charge for the two quarks. The small oscillations, which are more evident for the case of the u-quark, are an artifact of the regularization scheme, and disappear if smoother regulators are used, see the discussion in [70]. In the regime of weak fields, our data are consistent with the scaling $\langle \bar{f} f \rangle \propto |eB|^2/M$ where M denotes some mass scale; in the strong field limit we find instead $\langle \bar{f} f \rangle \propto |eB|^{3/2}$. The behavior of the quark condensate as a function of magnetic field is in agreement with the magnetic catalysis scenario.

In the middle panel of Fig. 4.5 we plot our data for the expectation value of the magnetic moment. At weak fields, $\Sigma_f \propto |eB|$ as expected from (4.28). In the strong field limit, non-linearity arises because of the scaling of quark mass (or chiral condensate); we find $\Sigma_f \propto |eB|^{3/2}$ in this limit.

In the lower panel of Fig. 4.5 we plot our results for the polarization. Data are obtained by the previous ones, using the definition (4.24). At small fields, the polarization clearly grows linearly with the magnetic field. This is a natural consequence of the linear behavior of the magnetic moment as a function of eB for small fields, see Fig. 4.5. On the other hand, within the chiral models we measure a saturation of μ_f at large values of eB, to an asymptotic value $\mu_\infty = 1$. This conclusion remains unchanged if we consider the NJL model, and it is in agreement with the recent Lattice findings [138]. It should be noticed that, at least for the u-quark, saturation is achieved before the expected threshold for ρ-meson condensation [144–146]. Therefore, our expectation is that our result is stable also if vector meson condensation is considered.

Fig. 4.5 *Upper panel*: Chiral
condensates of *u*-quarks (*red*)
and *d*-quarks (*blue*), in units
of the same quantities at zero
magnetic field, as a function
of the magnetic field. *Middle
panel*: Expectation value of
the magnetic moment
operator, in units of f_π^3, as a
function of eB. *Lower panel*:
Polarization of *u*-quarks (*red*)
and *d*-quarks (*blue*) as a
function the magnetic field
strength. From Ref. [88].
Copyright (2012) by
American Physical Society
(Color figure online)

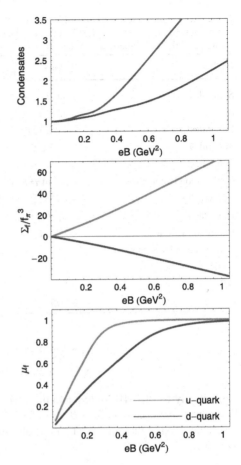

The saturation to the asymptotic value $\mu_\infty = 1$ of polarization is naturally understood within the models we investigate, as a LLL dominance in the chiral condensate (i.e., full polarization). As a matter of fact, Σ_f and $\langle \bar{f} f \rangle$ turn out to be proportional in the strong field limit, since only the LLL gives a contribution to the latter, comparing (4.27) and (4.29) which imply

$$\mu_f = 1 - \frac{\langle \bar{f} f \rangle_{\text{HLL}}}{\langle \bar{f} f \rangle}, \tag{4.31}$$

where $\langle \bar{f} f \rangle_{\text{HLL}}$ corresponds to the higher Landau levels contribution to the chiral condensate. In the strong field limit $\langle \bar{f} f \rangle_{\text{HLL}} \to 0$, see (4.27); hence, μ_f has to approach the asymptotic value $\mu_\infty = 1$. On the other hand, in the weak field limit $\langle \bar{f} f \rangle_{\text{HLL}} \to \langle \bar{f} f \rangle$ and the proportionality among Σ_f and $\langle \bar{f} f \rangle$ is lost.

At small fields $\mu_f = |\chi Q_f eB|$ from (4.22). Hence, we use the data on polarization at small fields, to obtain the numerical value of the magnetic susceptibility of the chiral condensate. Our results are as follows:

Table 4.2 Magnetic susceptibility of the quark condensate obtained within several theoretical approaches. In the table, $F_\pi = 130.7$ MeV. See the text for more details. Adapted from Ref. [88]. Copyright (2012) by American Physical Society

Method	χ (GeV^{-2})	Ren. point (GeV)	Ref.
Sum rules	-8.6 ± 0.24	1	[128]
Sum rules	-5.7	0.5	[129]
Sum rules	-4.4 ± 0.4	1	[130]
Sum rules	-3.15 ± 0.3	1	[131]
Sum rules	-2.85 ± 0.5	1	[132]
OPE + Pion Dominance	$-N_c/(4\pi^2 F_\pi^2)$	0.5	[133]
Holography	$-1.075 N_c/(4\pi^2 F_\pi^2)$	$\ll 1$	[134]
Holography	$-N_c/(4\pi^2 F_\pi^2)$	$\ll 1$	[135]
Instanton vacuum	-2.5 ± 0.15	1	[136]
Zero mode of Dirac Operator	-3.52	1	[137]
Lattice	$-1.547(3)$	2	[138]
NJL model	-4.3	0.63	This work
QM model	-5.25	0.56	This work

$$\chi \approx -4.3 \text{ GeV}^{-2}, \quad \text{NJL} \tag{4.32}$$

$$\chi \approx -5.25 \text{ GeV}^{-2}, \quad \text{QM} \tag{4.33}$$

respectively for the NJL model and the QM model. To obtain the numerical values above we have used data for eB up to $5m_\pi^2 \approx 0.1$ GeV^2, which are then fit using a linear law. Using the numerical values of chiral condensate in the two models, we obtain

$$\chi \langle \bar{f} f \rangle \approx 69 \text{ MeV}, \quad \text{NJL} \tag{4.34}$$

$$\chi \langle \bar{f} f \rangle \approx 65 \text{ MeV}, \quad \text{QM} \tag{4.35}$$

The numerical values of χ that we obtain within the effective models are in fair agreement with recent results, see Table 4.2. In our model calculations, the role of the renormalization scale is played approximately by the ultraviolet cutoff, which is equal to 0.560 GeV in the QM model, and 0.627 GeV in the NJL model.

To facilitate the comparison with previous estimates, we review briefly the frameworks in which the results in Table 4.2 are obtained. In [133] the following result is found, within OPE combined with Pion Dominance (we follow when possible the notation used in [138]):

$$\chi^{PD} = -c_\chi \frac{N_c}{8\pi^2 F_\pi^2}, \quad \text{Pion Dominance} \tag{4.36}$$

with $F_\pi = \sqrt{2} f_\pi = 130.7$ MeV and $c_\chi = 2$; the estimate of [133] is done at a renormalization point $M = 0.5$ GeV. It is remarkable that (4.36) has been reproduced recently within AdS/QCD approach in [135]. Probably, this is the result more comparable to our estimate, because the reference scales in [133] and in this article are very close. Within our model calculations we find $c_\chi^{NJL} = 1.93$ and $c_\chi^{QM} = 2.36$. Using the numerical value of F_π and c_χ we get $\chi^{PD} = -4.45$ GeV^{-2}, which agrees within 3 % with our NJL model result, and within 18 % with our QM model result.

In [134] the authors find $c_\chi = 2.15$ within hard-wall holographic approach, at the scale $M \ll 1$ GeV. The results of [134] are thus in very good parametric agreement with [133]; on the other hand, the numerical value of F_π in the holographic model is smaller than the one used in [133], pushing the holographic prediction for χ to slightly higher values than in [133]. However, the scale at which the result of [134] is valid should be much smaller than $M = 1$ GeV, thus some quantitative disagreement with [133] is expected. As the authors have explained, it might be possible to tune the parameters of the holographic model, mainly the chiral condensate, to reproduce the correct value of F_π; their numerical tests suggest that by changing the ratio $\langle \bar{f} f \rangle / m_\rho$ of a factor of 8, then the numerical value of c_χ is influenced only by a 5 %. It is therefore plausible that a best tuning makes the quantitative prediction of [134] much closer to the estimate of [133].

In [136] an estimate of χ within the instanton vacuum model has been performed beyond the chiral limit, both for light and for strange quarks (the result quoted in Table 4.1 corresponds to the light quarks; for the strange quark, $\chi_s / \chi_{u,d} \approx 0.15$ is found). Taking into account the numerical value of the chiral condensate in the instanton vacuum, the numerical estimate of [136] leads to $\chi = -2.5 \pm 0.15$ GeV^{-2} at the scale $M = 1$ GeV. An analytic estimate within a similar framework has been obtained in [137], in which the zero-mode of the Dirac operator in the background of a $SU(2)$ instanton is used to compute the relevant expectation values. The result of [128] gives $\chi = -3.52$ GeV^{-2} at $M \approx 1$ GeV.

In [138] the result $\chi = -1.547$ GeV^{-2} is achieved within a two-color simulation with quenched fermions. It is interesting that in [138] the same quantity has been computed also at finite temperature in the confinement phase, at $T = 0.82T_c$, and the result seems to be independent on temperature. The reference scale of [138], determined by the inverse lattice spacing, is $M \approx 2$ GeV. Therefore the lattice results are not quantitatively comparable with our model calculation. However, they share an important feature with the results presented here, namely the saturation of the polarization at large values of the magnetic field. Finally, estimates of the magnetic susceptibility of the chiral condensate by means of several QCD sum rules exist [128–132]. The results are collected in Table 4.1.

4.5.2 Results Within the Renormalized QM Model

In this section, we make semi-analytic estimates of the polarization and the magnetic susceptibility of the quark condensate, as well as for the chiral condensate in magnetic background, within the renormalized QM model. This is done with the scope to compare the predictions of the renormalized model with those of the effective models, in which an ultraviolet cutoff is introduced to mimic the QCD effective quark mass.

In the renormalized model, we allow the effective quark mass to be a constant in the whole range of momenta, which is different from what happens in QCD [141, 142]. Thus, the higher Landau levels give a finite contribution to the vacuum chiral condensate even at very strong fields. This is easy to understand: the ultraviolet cutoff, Λ, in the renormalized model can be taken larger than any other mass scale, in particular $\Lambda \gg |eB|^{1/2}$; as a consequence, the condition $p_3^2 + 2n|eB| < \Lambda^2$ is satisfied taking into account many Landau levels even at very large eB. The contribution of the higher Landau levels, once renormalized, appears in the physical quantities to which we are interested here, in particular in the chiral condensate.

Since the computation is a little bit lengthy, it is useful to anticipate its several steps: firstly we perform regularization, and then renormalization, of the QEP at zero magnetic field (the corrections due to the magnetic field turn out to be free of ultraviolet divergences). Secondly, we solve analytically the gap equation for the σ condensate in the limit of weak fields, and semi-analytically in the opposite limit. The field-induced corrections to the QEP and to the solution of the gap equation are divergence-free in agreement with [149], and are therefore independent on the renormalization scheme adopted. Then, we compute the renormalized and self-consistent values of the chiral condensate and of the magnetic moment, as a function of eB, using the results for the gap equation. Within this theoretical framework, it is much more convenient to compute $\langle \bar{f} f \rangle$ and Σ_f by taking derivatives of the renormalized potential; in fact, the computation of the traces of the propagator in the renormalized model is much more involved if compared to the situation of the non-renormalized models, since in the former a non-perturbative (and non-trivial) renormalization procedure of composite local operators is required. Finally, we estimate χ, as well as the behavior of the polarization as a function of eB.

4.5.2.1 Renormalization of the QEP

To begin with, we need to regularize the one-loop fermion contribution, namely

$$V_{\text{1-loop}}^{\text{fermion}} = -N_c \sum_f \frac{|Q_f eB|}{2\pi} \sum_{n=0}^{\infty} \beta_n \int_{-\infty}^{+\infty} \frac{dk}{2\pi} \left(k^2 + 2n|Q_f eB| + m_q^2\right)^{1/2}. \quad (4.37)$$

To this end, we define the function, $\mathscr{V}(s)$, of a complex variable, s, as

$$\mathscr{V}(s) = -N_c \sum_f \frac{|Q_f eB|}{2\pi} \sum_{n=0}^{\infty} \beta_n \int_{-\infty}^{+\infty} \frac{dk}{2\pi} \left(k^2 + 2n|Q_f eB| + m_q^2\right)^{\frac{1-s}{2}}. \quad (4.38)$$

The function $\mathscr{V}(s)$ can be analytically continued to $s = 0$. We define then $V_{\text{1-loop}}^{\text{fermion}} = \lim_{s \to 0^+} \mathscr{V}(s)$. After elementary integration over k, summation over n and taking the limit $s \to 0^+$, we obtain the result

$$
V_{\text{1-loop}}^{\text{fermion}} = N_c \sum_f \frac{(Q_f eB)^2}{4\pi^2} \left(\frac{2}{s} - \log(2|Q_f eB|) + a \right) B_2(q)
$$

$$
- N_c \sum_f \frac{(Q_f eB)^2}{2\pi^2} \zeta'(-1, q)
$$

$$
- N_c \sum_f \frac{|Q_f eB| m_q^2}{8\pi^2} \left(\frac{2}{s} - \log(m_q^2) + a \right), \tag{4.39}
$$

where we have subtracted terms which do not depend explicitly on the condensate. In the above equation, $\zeta(t, q)$ is the Hurwitz zeta function; for $\text{Re}(t) > 1$ and $\text{Re}(q) > 0$, it is defined by the series $\zeta(t, q) = \sum_{n=0}^{\infty} (n + q)^{-t}$; the series can be analytically continued to a meromorphic function defined in the complex plane $t \neq 1$. Moreover we have defined $q = (m_q^2 + 2|Q_f eB|)/2|Q_f eB|$; furthermore, $a = 1 - \gamma_E - \psi(-1/2)$, where γ_E is the Eulero-Mascheroni number and ψ is the digamma function. The derivative $\zeta'(-1, q) = d\zeta(t, q)/dt$ is understood to be computed at $t = -1$.

The first two addenda in (4.39) arise from the higher Landau levels; on the other hand, the last addendum is the contribution of the LLL. The function B_2 is the second Bernoulli polynomial; using its explicit form, it is easy to show that the divergence in the LLL term in (4.39) is canceled by the analogous divergence in the first addendum of the same equation. It is interesting that the LLL contribution, which is in principle divergent, combines with a part of the contribution of the higher Landau levels, leading to a finite result. This can be interpreted as a renormalization of the LLL contribution. On the other hand, the remaining part arising from the higher Landau levels is still divergent; this divergence survives in the $B \to 0$ limit, and is due to the usual divergence of the vacuum contribution. We then have

$$
V_{\text{1-loop}}^{\text{fermion}} = N_c \sum_f \frac{m_q^4}{16\pi^2} \left(\frac{2}{s} - \log(2|Q_f eB|) + a \right)
$$

$$
+ N_c \sum_f \frac{|Q_f eB| m_q^2}{8\pi^2} \log \frac{m_q^2}{2|Q_f eB|}
$$

$$
- N_c \sum_f \frac{(Q_f eB)^2}{2\pi^2} \zeta'(-1, q). \tag{4.40}
$$

The renormalization procedure of the quantum effective potential is discussed in some detail in [88]. Here it is not necessary to discuss this procedure, and we just focus on the results.

4.5.2.2 Approximate Solutions of the Gap Equation

Weak Fields In the weak field limit ($eB \ll m_q^2$) the correction due to the magnetic field to the quantum effective potential can be computed:

$$V_1 \approx -N_c \sum_f \frac{(Q_f eB)^2}{24\pi^2} \log \frac{m_q^2}{2|Q_f eB|} = -N_c \sum_f \frac{(Q_f eB)^2}{24\pi^2} \log \frac{m_q^2}{\mu^2}, \quad (4.41)$$

which is in agreement with the result of [149]. In the above equation we have followed the notation of [140] introducing an infrared scale μ, isolating and then subtracting the term which does not depend on the condensate. The scale μ is arbitrary, and we cannot determine it from first principles; on the other hand, it is irrelevant for the determination of the σ-condensate. We expect $\mu \approx f_\pi$ since this is the typical scale of chiral symmetry breaking in the model for the σ field.

In this limit, it is easy to obtain analytically the behavior of the constituent quark mass as a function of eB. As a matter of fact, we can expand the derivative of the QEP with respect to σ, around the solution at $B = 0$, writing $\langle \sigma \rangle = f_\pi + \delta\sigma$. Then, a straightforward evaluation leads to

$$m_q = g f_\pi \left(1 + \frac{5}{9} \frac{N_c}{12\pi^2 f_\pi^2 m_\sigma^2} (eB)^2 \right). \quad (4.42)$$

As anticipated, the scale μ is absent in the solution of the gap equation.

Strong Fields In the limit $eB \gg m_q^2$, we can find an asymptotic representation of V_1 by using the expansion $\zeta'(-1, q) = c_0 + c_1(q - 1)$ valid for $q \approx 1$, with $c_0 = -0.17$ and $c_1 = -0.42$. Then we find

$$V_1 \approx -N_c \sum_f \frac{m_q^2}{8\pi^2} \left(\frac{m_q^2}{2} + |Q_f eB|\right) \log \frac{2|Q_f eB|}{m_q^2} - N_c \sum_f \frac{|Q_f eB| m_q^2}{2\pi^2} c_1, \quad (4.43)$$

where we have subtracted condensate-independent terms.

In the strong field limit it is not easy to find analytically an asymptotic representation for the sigma condensate as a function of eB; therefore we solve the gap equation numerically, and then fit data with a convenient analytic form as follows:

$$m_q = b|eB|^{1/2} + \frac{c f_\pi^3}{|eB|}, \quad (4.44)$$

where $b = 0.32$ and $c = 32.78$. At large fields the quark mass grows as $|eB|^{1/2}$ as expected by dimensional analysis; this is a check of the equations that we use.

4.5.2.3 Evaluation of Chiral Condensate and Magnetic Moment

Chiral Condensate To compute the chiral condensate we follow a standard procedure: we introduce source term for $\bar{f} f$, namely a bare quark mass m_f, then take derivative of the effective potential with respect to m_f evaluated at $m_f = 0$. For the weak field case we obtain

$$\langle \bar{f} f \rangle = \langle \bar{f} f \rangle_0 - \frac{N_c}{12\pi^2} \frac{|Q_f eB|^2}{m_q}.$$
(4.45)

On the other hand, in the strong field limit we have

$$\langle \bar{f} f \rangle = -\frac{N_c m_q}{4\pi^2} \left(|Q_f eB| + m_q^2 \right) \log \frac{2|Q_f eB|}{m_q^2}.$$
(4.46)

Using (4.42) and (4.44), we show that the chiral condensate scales as $a + b(eB)^2$ for small fields, and as $|eB|^{3/2}$ for large fields.

Magnetic Moment   Next we turn to the computation of the expectation value of the magnetic moment. The expression in terms of Landau levels is given by (4.29), which clearly shows that this quantity has a log-type divergence. In order to avoid a complicated renormalization procedure of a local composite operator, we notice that it is enough to take the minus derivative of V_1 with respect to $\boldsymbol{B}$ to get magnetization, $\mathcal{M}$ [140], then multiply by $2m/Q_f$ to get the magnetic moment. This procedure is very cheap, since the $\boldsymbol{B}$-dependent contributions to the effective potential are finite, and the resulting expectation value will turn out to be finite as well (that is, already renormalized).

In the case of weak fields, from (4.41) we find

$$\Sigma_f = N_c \frac{Q_f |eB| m_q}{6\pi^2} \log \frac{m_q^2}{\mu^2}.$$
(4.47)

On the other hand, in the strong field limit we get from (4.43)

$$\Sigma_f = N_c \frac{m_q^3}{4\pi^2} \log \frac{2|Q_f eB|}{m_q^2}.$$
(4.48)

The above result is in parametric agreement with the estimate of magnetization in [140]. In fact, $m_q^2 \approx |eB|$ in the strong field limit, which leads to a magnetization $\mathcal{M} \approx B \log B$.

Using the expansions for the sigma condensate at small and large values of the magnetic field strength, we argue that $\Sigma_f \approx |eB|$ in a weak field, and $\Sigma_f \approx |eB|^{3/2}$ in a strong field.

4.5.2.4   Computation of Chiral Magnetization and Polarization

We can now estimate the magnetic susceptibility of the quark condensate and the polarization as a function of eB. For the former, we need to know the behavior of the magnetic moment for weak fields. From (4.47) and from the definition (4.22) we read

$$\chi \langle \bar{f} f \rangle = \frac{N_c m_q}{6\pi^2} \log \frac{m_q^2}{\mu^2} \equiv f(\mu). \qquad (4.49)$$

The presence of the infrared scale μ makes the numerical estimate of χ uncertain; however, taking for it a value $\mu \approx f_\pi$, which is the typical scale of chiral symmetry breaking, we have $\chi \langle \bar{f} f \rangle \approx 44$ MeV, which is in agreement with the expected value, see (4.23).

Next we turn to the polarization. For weak fields we find trivially a linear dependence of μ_f on $|Q_f eB|$, with slope given by the absolute value of χ in (4.49). On the other hand, in the strong field limit we find, according to (4.48),

$$\mu_f \approx \frac{m_q^2}{m_q^2 + |Q_f eB|} \approx 1 - \frac{|Q_f|}{b + |Q_f|}, \qquad (4.50)$$

where we have used (4.44). This result shows that the polarization saturates at large values of eB, but the asymptotic value depends on the flavor charge.

It is interesting to compare the result of the renormalized model with that of the effective models considered in the previous section. In the former, the asymptotic value of μ_f is flavor-dependent; in the latter, $\mu_f \to 1$ independently on the value of the electric charge. Our interpretation of this difference is as follows: comparing (4.50) with the general model expectation, (4.31), we recognize in the factor $|Q_f|/(b + |Q_f|)$ the contribution of the higher Landau levels at zero temperature, which turns out to be finite and non-zero after the renormalization procedure. This contribution is then transmitted to the physical quantities that we have computed. The trace of the higher Landau levels is implicit in the solution of the gap equation in the strong field limit, namely the factor b in (4.44), and explicit in the additional $|Q_f|$ dependence in (4.50). A posteriori, this conclusion seems quite natural, because in the renormalization procedure we assume that the effective quark mass is independent of quark momentum, thus there is no cut of the large momenta in the gap equation (and in the equation for polarization as well). In the effective models considered in the first part of this article, on the other hand, the cutoff procedure is equivalent to have a momentum-dependent effective quark mass, $m_q = g\sigma \Theta(\Lambda^2 - p_3^2 - 2n|Q_f eB|)$, which naturally cuts off higher Landau levels when $eB \gg \Lambda^2$. At the end of the days, the expulsion of the higher Landau levels from the chiral condensate makes $\mu_f \to 1$ in the strong field limit. Our expectation is that if we allow the quark mass to run with momentum and decay rapidly at large momenta, mimicking the effective quark mass of QCD, higher Landau levels would be suppressed in the strong field limit, and the result (4.50) would tend to the result in Fig. 4.5.

4.6 Conclusions

In this chapter we have summarized our results for the phase structure of quark matter in a strong magnetic background. Our theoretical investigative tools are chiral quark models improved with the Polyakov loop, which allow to study simultaneously chiral symmetry breaking and deconfinement.

The main motivation of this series of studies is of a phenomenological nature. In fact, it has been shown that huge magnetic fields are produced in non-central heavy ion collisions. These fields might trigger the P- and CP-odd process dubbed Chiral Magnetic Effect (CME). Therefore, in order to make quantitative estimates of the observables which are sensitive to the CME, it is extremely important to understand how hot quark matter behaves under the influence of a strong magnetic background. The other side of the coin is that simulations show these huge fields have a very short lifetime: therefore the present studies should take into account this time dependence. Moreover electric fields, which we have neglected so far, are also produced in the collisions. Finally, the electromagnetic fields considerably depend on space coordinates on the scale of the volume of the expanding fireball. This dependence has been ignored in our studies, since the magnetic background is taken to be homogeneous in space and constant in time.

Our results support the scenario of magnetic catalysis, which manifests itself in both an increase of the chiral condensate at zero temperature, and an increase of the critical temperature for chiral symmetry restoration. Moreover, depending on the interaction used, deconfinement may occur either together with chiral symmetry restoration, or anticipate it. The latter possibility, even if more fascinating than the former since it opens a window for the Constituent Quark Phase, seems to be excluded by lattice simulations.

Recent lattice simulations show that the critical temperature for chiral symmetry restoration, T_c, is strongly affected by the quark mass. In particular, for small quark masses (hence, for the u and d quarks) the critical temperature *decreases* with the magnetic field strength; on the other hand, T_c increases with the magnetic field strength for the s quark. As we have discussed in the main body of this chapter, it seems that self-consistent computations within chiral quark models are not able to reproduce this feature, even when quantum fluctuations are taken into account. Thus, it remains an open problem to understand this unexpected behavior of T_c. Certainly independent simulations performed by other groups are necessary to confirm the present results.

We have also briefly summarized a computation of the magnetic susceptibility of the chiral condensate, χ, and of quark polarization, μ_f at zero temperature, based on the quark-meson model. The computed value of χ is in agreement with most of the previous estimates, and with experimental data. Moreover, this model gives a simple interpretation of the saturation of μ_f observed on the lattice: at very large magnetic field strength, the quarks occupy the lowest Landau level, expelling higher levels from the chiral condensate; hence, chiral condensate turns out to be proportional to the quark magnetic moment, making the ratio (that is, polarization) just a constant. In the case of the non-renormalized model, this constant turns out to be equal to one and flavor independent; on the other hand, in the case of the renormalized model, the constant is flavor dependent. The latter result is easily understood: the renormalization procedure of the momentum independent interaction of the quark-meson model brings all the Landau levels into the renormalized chiral condensate. We expect that the replacement of the simple interaction discussed here with a non-local one, which should mimic the quark self-energy measured on the lattice, will make the expulsion of the higher Landau levels active also in the renormalized model, hence reproducing the results of the non-renormalized model.

Acknowledgements We are pleased to acknowledge the editors of this volume of *Lecture Notes in Physics* for their interest in our work and their kind invitation to contribute to the book. We also acknowledge K. Fukushima and M. Frasca for scientific collaborations which led to some of the results presented here. Finally we also acknowledge many clarifying discussions with M. D'Elia and G. Endrodi about the topics discussed in this chapter, with particular reference to the ones related to Lattice simulations.

References

1. P. de Forcrand, O. Philipsen, J. High Energy Phys. **0701**, 077 (2007). arXiv:hep-lat/0607017
2. P. de Forcrand, O. Philipsen, J. High Energy Phys. **0811**, 012 (2008). arXiv:0808.1096 [hep-lat]
3. P. de Forcrand, O. Philipsen, PoS **LATTICE2008**, 208 (2008). arXiv:0811.3858 [hep-lat]
4. Y. Aoki, Z. Fodor, S.D. Katz, K.K. Szabo, Phys. Lett. B **643**, 46 (2006). arXiv:hep-lat/0609068
5. Y. Aoki, S. Borsanyi, S. Durr, Z. Fodor, S.D. Katz, S. Krieg, K.K. Szabo, J. High Energy Phys. **0906**, 088 (2009). arXiv:0903.4155 [hep-lat]
6. Y. Aoki, G. Endrodi, Z. Fodor, S.D. Katz, K.K. Szabo, Nature **443**, 675 (2006). hep-lat/0611014
7. A. Bazavov et al., Phys. Rev. D **80**, 014504 (2009). arXiv:0903.4379 [hep-lat]
8. M. Cheng et al., Phys. Rev. D **81**, 054510 (2010). arXiv:0911.3450 [hep-lat]
9. Y. Nambu, G. Jona-Lasinio, Phys. Rev. **122**, 345 (1961)
10. Y. Nambu, G. Jona-Lasinio, Phys. Rev. **124**, 246 (1961)
11. U. Vogl, W. Weise, Prog. Part. Nucl. Phys. **27**, 195 (1991)
12. S.P. Klevansky, Rev. Mod. Phys. **64**, 649 (1992)
13. T. Hatsuda, T. Kunihiro, Phys. Rept. **247**, 221 (1994)
14. M. Buballa, Phys. Rept. **407**, 205 (2005)
15. A.M. Polyakov, Phys. Lett. B **72**, 477 (1978)
16. L. Susskind, Phys. Rev. D **20**, 2610 (1979)
17. B. Svetitsky, L.G. Yaffe, Nucl. Phys. B **210**, 423 (1982)
18. B. Svetitsky, Phys. Rept. **132**, 1 (1986)
19. P.N. Meisinger, M.C. Ogilvie, Phys. Lett. B **379**, 163 (1996). arXiv:hep-lat/9512011
20. K. Fukushima, Phys. Lett. B **591**, 277 (2004). arXiv:hep-ph/0310121
21. C. Ratti, M.A. Thaler, W. Weise, Phys. Rev. D **73**, 014019 (2006)
22. S. Roessner, C. Ratti, W. Weise, Phys. Rev. D **75**, 034007 (2007)
23. E. Megias, E. Ruiz Arriola, L.L. Salcedo, Phys. Rev. D **74**, 114014 (2006)
24. E. Megias, E. Ruiz Arriola, L.L. Salcedo, Eur. Phys. J. A **31**, 553 (2007). arXiv:hep-ph/0610163
25. C. Sasaki, B. Friman, K. Redlich, Phys. Rev. D **75**, 074013 (2007)
26. S.K. Ghosh, T.K. Mukherjee, M.G. Mustafa, R. Ray, Phys. Rev. D **77**, 094024 (2008)
27. K. Fukushima, Phys. Rev. D **77**, 114028 (2008) [Erratum. Phys. Rev. D **78**, 039902 (2008)]
28. M. Ciminale, R. Gatto, N.D. Ippolito, G. Nardulli, M. Ruggieri, Phys. Rev. D **77**, 054023 (2008)
29. W.j. Fu, Z. Zhang, Y.x. Liu, Phys. Rev. D **77**, 014006 (2008)
30. T. Hell, S. Rossner, M. Cristoforetti, W. Weise, Phys. Rev. D **81**, 074034 (2010)
31. H. Abuki, R. Anglani, R. Gatto, G. Nardulli, M. Ruggieri, Phys. Rev. D **78**, 034034 (2008)
32. T. Kahara, K. Tuominen, Phys. Rev. D **82**, 114026 (2010). arXiv:1006.3931 [hep-ph]

33. Y. Sakai, K. Kashiwa, H. Kouno, M. Yahiro, Phys. Rev. D **77**, 051901 (2008)
34. Y. Sakai, K. Kashiwa, H. Kouno, M. Yahiro, Phys. Rev. D **78**, 036001 (2008)
35. Y. Sakai, K. Kashiwa, H. Kouno, M. Matsuzaki, M. Yahiro, Phys. Rev. D **79**, 096001 (2003). arXiv:0902.0487 [hep-ph]
36. K. Kashiwa, H. Kouno, M. Yahiro, Phys. Rev. D **80**, 117901 (2009)
37. H. Abuki, M. Ciminale, R. Gatto, N.D. Ippolito, G. Nardulli, M. Ruggieri, Phys. Rev. D **78**, 014002 (2008)
38. H. Abuki, M. Ciminale, R. Gatto, M. Ruggieri, Phys. Rev. D **79**, 034021 (2009). arXiv:0811. 1512 [hep-ph]
39. T. Sasaki, Y. Sakai, H. Kouno, M. Yahiro, Phys. Rev. D **82**, 116004 (2010). arXiv:1005.0910 [hep-ph]
40. T. Hell, S. Roessner, M. Cristoforetti, W. Weise, Phys. Rev. D **79**, 014022 (2009)
41. K. Kashiwa, H. Kouno, M. Matsuzaki, M. Yahiro, Phys. Lett. B **662**, 26 (2008). arXiv:0710. 2180 [hep-ph]
42. O. Lourenco, M. Dutra, T. Frederico, A. Delfino, M. Malheiro, Phys. Rev. D **85**, 097504 (2012)
43. O. Lourenco, M. Dutra, A. Delfino, M. Malheiro, Phys. Rev. D **84**, 125034 (2011)
44. M. D'Elia, S. Mukherjee, F. Sanfilippo, Phys. Rev. D **82**, 051501 (2010). arXiv:1005.5365 [hep-lat]
45. P.V. Buividovich, M.N. Chernodub, E.V. Luschevskaya, M.I. Polikarpov, Phys. Rev. D **81**, 036007 (2010). arXiv:0909.2350 [hep-ph]
46. P.V. Buividovich, M.N. Chernodub, E.V. Luschevskaya, M.I. Polikarpov, Phys. Lett. B **682**, 484 (2010). arXiv:0812.1740 [hep-lat]
47. P. Cea, L. Cosmai, J. High Energy Phys. **0302**, 031 (2003). arXiv:hep-lat/0204023
48. P. Cea, L. Cosmai, J. High Energy Phys. **0508**, 079 (2005). arXiv:hep-lat/0505007
49. P. Cea, L. Cosmai, M. D'Elia, J. High Energy Phys. **0712**, 097 (2007). arXiv:0707.1149 [hep-lat]
50. G.S. Bali, F. Bruckmann, G. Endrodi, Z. Fodor, S.D. Katz, A. Schafer, Phys. Rev. D **86**, 071502 (2012). arXiv:1206.4205 [hep-lat]
51. G.S. Bali, F. Bruckmann, G. Endrodi, Z. Fodor, S.D. Katz, S. Krieg, A. Schafer, K.K. Szabo, J. High Energy Phys. **1202**, 044 (2012)
52. S.P. Klevansky, R.H. Lemmer, Phys. Rev. D **39**, 3478 (1989)
53. I.A. Shushpanov, A.V. Smilga, Phys. Lett. B **402**, 351 (1997). arXiv:hep-ph/9703201
54. D.N. Kabat, K.M. Lee, E.J. Weinberg, Phys. Rev. D **66**, 014004 (2002). arXiv:hep-ph/0204120
55. T. Inagaki, D. Kimura, T. Murata, Prog. Theor. Phys. **111**, 371 (2004). arXiv:hep-ph/0312005
56. T.D. Cohen, D.A. McGady, E.S. Werbos, Phys. Rev. C **76**, 055201 (2007). arXiv:0706.3208 [hep-ph]
57. K. Fukushima, H.J. Warringa, Phys. Rev. Lett. **100**, 032007 (2008). arXiv:0707.3785 [hep-ph]
58. J.L. Noronha, I.A. Shovkovy, Phys. Rev. D **76**, 105030 (2007). arXiv:0708.0307 [hep-ph]
59. V.P. Gusynin, V.A. Miransky, I.A. Shovkovy, Nucl. Phys. B **462**, 249 (1996). arXiv:hep-ph/9509320
60. V.P. Gusynin, V.A. Miransky, I.A. Shovkovy, Nucl. Phys. B **563**, 361 (1999). arXiv:hep-ph/9908320
61. G.W. Semenoff, I.A. Shovkovy, L.C.R. Wijewardhana, Phys. Rev. D **60**, 105024 (1999). arXiv:hep-th/9905116
62. V.A. Miransky, I.A. Shovkovy, Phys. Rev. D **66**, 045006 (2002). arXiv:hep-ph/0205348
63. K.G. Klimenko, Theor. Math. Phys. **89**, 1161 (1992) [Teor. Mat. Fiz. **89**, 211 (1991)]
64. K.G. Klimenko, Z. Phys. C **54**, 323 (1992)
65. K.G. Klimenko, Theor. Math. Phys. **90**, 1 (1992) [Teor. Mat. Fiz. **90**, 3 (1992)]
66. N.O. Agasian, S.M. Fedorov, Phys. Lett. B **663**, 445 (2008). arXiv:0803.3156 [hep-ph]
67. E.S. Fraga, A.J. Mizher, Phys. Rev. D **78**, 025016 (2008). arXiv:0804.1452 [hep-ph]

68. F. Preis, A. Rebhan, A. Schmitt, J. High Energy Phys. **1103**, 033 (2011)
69. K. Fukushima, M. Ruggieri, R. Gatto, Phys. Rev. D **81**, 114031 (2010). arXiv:1003.0047 [hep-ph]
70. L. Campanelli, M. Ruggieri, Phys. Rev. D **80**, 034014 (2009). arXiv:0905.0853 [hep-ph]
71. A.J. Mizher, M.N. Chernodub, E.S. Fraga, Phys. Rev. D **82**, 105016 (2010). arXiv:1004.2712 [hep-ph]
72. D.E. Kharzeev, L.D. McLerran, H.J. Warringa, Nucl. Phys. A **803**, 227 (2008). arXiv: 0711.0950 [hep-ph]
73. V. Skokov, A.Y. Illarionov, V. Toneev, Int. J. Mod. Phys. A **24**, 5925 (2009). arXiv:0907.1396 [nucl-th]
74. V. Voronyuk, V.D. Toneev, W. Cassing, E.L. Bratkovskaya, V.P. Konchakovski, S.A. Voloshin, Phys. Rev. C **83**, 054911 (2011)
75. G.D. Moore, M. Tassler, J. High Energy Phys. **1102**, 105 (2011)
76. G.D. Moore, hep-ph/0009161
77. P.V. Buividovich, M.N. Chernodub, E.V. Luschevskaya, M.I. Polikarpov, Phys. Rev. D **80**, 054503 (2009). arXiv:0907.0494 [hep-lat]
78. M. Abramczyk, T. Blum, G. Petropoulos, R. Zhou, PoS **LATTICE2009**, 181 (2009). arXiv: 0911.1348 [hep-lat]
79. K. Fukushima, D.E. Kharzeev, H.J. Warringa, Phys. Rev. D **78**, 074033 (2008). arXiv:0808. 3382 [hep-ph]
80. K. Fukushima, M. Ruggieri, Phys. Rev. D **82**, 054001 (2010)
81. R. Gatto, M. Ruggieri, Phys. Rev. D **85**, 054013 (2012)
82. M. Ruggieri, Phys. Rev. D **84**, 014011 (2011)
83. M.N. Chernodub, A.S. Nedelin, Phys. Rev. D **83**, 105008 (2011)
84. C.A.B. Bayona, K. Peeters, M. Zamaklar, J. High Energy Phys. **1106**, 092 (2011)
85. B. Abelev et al. (ALICE Collaboration), Phys. Rev. Lett. **110**, 012301 (2013). arXiv: 1207.0900 [nucl-ex]
86. R. Gatto, M. Ruggieri, Phys. Rev. D **83**, 034016 (2011)
87. R. Gatto, M. Ruggieri, Phys. Rev. D **82**, 054027 (2010)
88. M. Frasca, M. Ruggieri, Phys. Rev. D **83**, 094024 (2011)
89. A.A. Osipov, B. Hiller, J. Moreira, A.H. Blin, J. da Providencia, Phys. Lett. B **646**, 91 (2007). arXiv:hep-ph/0612082
90. K. Kashiwa, H. Kouno, T. Sakaguchi, M. Matsuzaki, M. Yahiro, Phys. Lett. B **647**, 446 (2007). arXiv:nucl-th/0608078
91. A.A. Osipov, B. Hiller, J. da Providencia, Phys. Lett. B **634**, 48 (2006). arXiv:hep-ph/ 0508058
92. A.A. Osipov, B. Hiller, A.H. Blin, J. da Providencia, Phys. Lett. B **650**, 262 (2007). arXiv: hep-ph/0701090
93. E. Bilgici, F. Bruckmann, C. Gattringer, C. Hagen, Phys. Rev. D **77**, 094007 (2008). arXiv: 0801.4051 [hep-lat]
94. V.I. Ritus, Ann. Phys. **69**, 555 (1972)
95. C.N. Leung, S.Y. Wang, Nucl. Phys. B **747**, 266 (2006)
96. E. Elizalde, E.J. Ferrer, V. de la Incera, Ann. Phys. **295**, 33 (2002)
97. E. Elizalde, E.J. Ferrer, V. de la Incera, Phys. Rev. D **70**, 043012 (2004)
98. E.J. Ferrer, V. de la Incera, C. Manuel, Phys. Rev. Lett. **95**, 152002 (2005)
99. E.J. Ferrer, V. de la Incera, C. Manuel, Nucl. Phys. B **747**, 88 (2006)
100. K. Fukushima, D.E. Kharzeev, H.J. Warringa, Nucl. Phys. A **836**, 311 (2010)
101. K. Fukushima, H.J. Warringa, Phys. Rev. Lett. **100**, 032007 (2008)
102. J.L. Noronha, I.A. Shovkovy, Phys. Rev. D **76**, 105030 (2007)
103. B.J. Schaefer, M. Wagner, J. Wambach, Phys. Rev. D **81**, 074013 (2010). arXiv:0910.5628 [hep-ph]
104. W. Press, S.A. Teukolsky, W.T. Vetterling, B.P. Flannery, *Numerical Recipes: The Art of Scientific Computing*, 3rd edn. (Cambridge University Press, Cambridge, 2007)

105. T.K. Mukherjee, H. Chen, M. Huang, Phys. Rev. D **82**, 034015 (2010). arXiv:1005.2482 [hep-ph]
106. K.I. Kondo, Phys. Rev. D **82**, 065024 (2010)
107. M. Frasca, Int. J. Mod. Phys. E **18**, 693 (2009)
108. M. Frasca, arXiv:1002.4600 [hep-ph]
109. M. Frasca, Phys. Rev. C **84**, 055208 (2011)
110. Y. Sakai, T. Sasaki, H. Kouno, M. Yahiro, Phys. Rev. D **82**, 076003 (2010). arXiv:1006.3648 [hep-ph]
111. M. D'Elia, F. Sanfilippo, Phys. Rev. D **80**, 111501 (2009)
112. C. Bonati, G. Cossu, M. D'Elia, F. Sanfilippo, Phys. Rev. D **83**, 054505 (2011). arXiv: 1011.4515 [hep-lat]
113. J. Cleymans, K. Redlich, H. Satz, E. Suhonen, Z. Phys. C **33**, 151 (1986)
114. H. Kouno, F. Takagi, Z. Phys. C **42**, 209 (1989)
115. S.S. Avancini, D.P. Menezes, M.B. Pinto, C. Providencia, Phys. Rev. D **85**, 091901 (2012)
116. D.P. Menezes, M. Benghi Pinto, S.S. Avancini, C. Providencia, Phys. Rev. C **80**, 065805 (2009)
117. J.O. Andersen, A. Tranberg, J. High Energy Phys. **1208**, 002 (2012). arXiv:1204.3360 [hep-ph]
118. J.O. Andersen, R. Khan, Phys. Rev. D **85**, 065026 (2012)
119. V. Skokov, Phys. Rev. D **85**, 034026 (2012)
120. K. Fukushima, J.M. Pawlowski, Phys. Rev. D **86**, 076013 (2012). arXiv:1203.4330 [hep-ph]
121. J.O. Andersen, J. High Energy Phys. **1210**, 005 (2012). arXiv:1205.6978 [hep-ph]
122. K. Kashiwa, Phys. Rev. D **83**, 117901 (2011)
123. O. Bergman, G. Lifschytz, M. Lippert, J. High Energy Phys. **0805**, 007 (2008)
124. C.V. Johnson, A. Kundu, J. High Energy Phys. **0812**, 053 (2008)
125. A.V. Zayakin, J. High Energy Phys. **0807**, 116 (2008)
126. E.-M. Ilgenfritz, M. Kalinowski, M. Muller-Preussker, B. Petersson, A. Schreiber, Phys. Rev. D **85**, 114504 (2012)
127. E.S. Fraga, L.F. Palhares, Phys. Rev. D **86**, 016008 (2012). arXiv:1201.5881 [hep-ph]
128. B.L. Ioffe, A.V. Smilga, Nucl. Phys. B **232**, 109 (1984)
129. V.M. Belyaev, Y.I. Kogan, Yad. Fiz. **40**, 1035 (1984)
130. I.I. Balitsky, A.V. Kolesnichenko, A.V. Yung, Sov. J. Nucl. Phys. **41**, 178 (1985) [Yad. Fiz. **41**, 282 (1985)]
131. P. Ball, V.M. Braun, N. Kivel, Nucl. Phys. B **649**, 263 (2003)
132. J. Rohrwild, J. High Energy Phys. **0709**, 073 (2007)
133. A. Vainshtein, Phys. Lett. B **569**, 187 (2003)
134. A. Gorsky, A. Krikun, Phys. Rev. D **79**, 086015 (2009)
135. D.T. Son, N. Yamamoto, arXiv:1010.0718 [hep-ph]
136. H.C. Kim, M. Musakhanov, M. Siddikov, Phys. Lett. B **608**, 95 (2005)
137. B.L. Ioffe, Phys. Lett. B **678**, 512 (2009)
138. P.V. Buividovich, M.N. Chernodub, E.V. Luschevskaya, M.I. Polikarpov, Nucl. Phys. B **826**, 313 (2010)
139. B. Pire, L. Szymanowski, Phys. Rev. Lett. **103**, 072002 (2009)
140. T.D. Cohen, E.S. Werbos, Phys. Rev. C **80**, 015203 (2009)
141. H.D. Politzer, Nucl. Phys. B **117**, 397 (1976)
142. H.D. Politzer, Phys. Lett. B **116**, 171 (1982)
143. K. Langfeld, C. Kettner, H. Reinhardt, Nucl. Phys. A **608**, 331 (1996)
144. M.N. Chernodub, Phys. Rev. D **82**, 085011 (2010)
145. M.N. Chernodub, Phys. Rev. Lett. **106**, 142003 (2011). arXiv:1101.0117 [hep-ph]
146. N. Callebaut, D. Dudal, H. Verschelde, PoS **FacesQCD**, 046 (2010). arXiv:1102.3103 [hep-ph]
147. H.G. Dosch, S. Narison, Phys. Lett. B **417**, 173 (1998)
148. S. Narison, Phys. Rev. D **74**, 034013 (2006)
149. H. Suganuma, T. Tatsumi, Ann. Phys. **208**, 470 (1991)

Chapter 5
Thermal Chiral and Deconfining Transitions in the Presence of a Magnetic Background

Eduardo S. Fraga

5.1 Introduction

The thermodynamics of strong interactions under a strong magnetic background has proven to be a very rich and subtle subject. Recent developments were initially motivated by the utility of magnetic fields in separating charge in space, which would render the possible formation of sphaleron-induced CP-odd domains in the plasma created in high-energy heavy ion collisions, in the so-called chiral magnetic effect [1–7], measurable. In fact, the magnetic fields created in non-central collisions in heavy ion experiments at RHIC-BNL and the LHC-CERN are possibly the highest since the epoch of the electroweak phase transition, reaching values such as $B \sim 10^{19}$ Gauss ($eB \sim 6m_\pi^2$) for peripheral collisions at RHIC [8, 9] and even much higher at the LHC due to the fluctuations in the distribution of protons inside the nuclei [10, 11].

From the theoretical point of view, the non-trivial role played by magnetic fields in the nature of phase transitions has been known for a long time [12]. Modifications in the vacuum of QED and QCD have also been investigated within different frameworks, mainly using effective models [13–27], especially the NJL model [28], and chiral perturbation theory [29–31], but also resorting to the quark model [32] and certain limits of QCD [33]. Interesting phases in dense systems [34–41], as well as effects on the dynamical quark mass [42] were also considered. Nevertheless, the mapping of the new $T-eB$ phase diagram is still an open problem. There are clear indications that sufficiently large magnetic fields could significantly modify the behavior of the chiral and the deconfinement phase transition lines [43–69], or even transform the vacuum into a superconducting medium via ρ-meson condensation [70–73]. Although most of the analyses so far relied on effective models, lattice

E.S. Fraga (✉)
Instituto de Física, Universidade Federal do Rio de Janeiro, Caixa Postal 68528, Rio de Janeiro, RJ 21941-972, Brazil
e-mail: fraga@if.ufrj.br

D. Kharzeev et al. (eds.), *Strongly Interacting Matter in Magnetic Fields*,
Lecture Notes in Physics 871, DOI 10.1007/978-3-642-37305-3_5,
© Springer-Verlag Berlin Heidelberg 2013

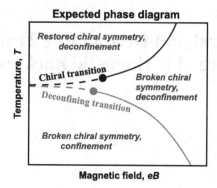

Fig. 5.1 Originally expected magnetic field–temperature phase diagram of strong interactions. The *thick lines* indicate first-order transitions, the *filled circles* are the (second-order) endpoints of these lines, and the *thin dashed lines* stand for the corresponding crossovers. A new phase with broken chiral symmetry and deconfinement appears at high magnetic fields. Extracted from Ref. [48]

QCD has definitely entered the field and has been producing its first results for the phase diagram [74–81].

From the first results obtained within effective models for the deconfining [43] and chiral [44, 45] transition lines, one would expect the phase diagram structure illustrated in Fig. 5.1, as discussed in Ref. [48]. Indeed, after the prediction of a splitting between the chiral and deconfining transition lines, with the appearance of a new phase, in Ref. [48], several model descriptions produced the same effect [52–58, 67]. However, until 2011, all model studies have yielded either a monotonically increasing or an essentially flat functional form for the deconfinement critical line as B increases to very large values.[1] Pioneering lattice simulations [77, 78] also found an essentially flat behavior for both critical lines, that seemed to increase together at a very low rate. Nevertheless, since the pion mass used in these simulations was still very high, this could be an indication that one would probably need huge magnetic fields in the simulations in order to be able to compare to effective model predictions.

This was the scenario, rather coherent in terms of expectations for the behavior of the critical lines for the chiral and deconfining transitions in the presence of a magnetic background, until lattice simulations of magnetic QCD with physical masses and fine grids were performed [80] and showed that both critical temperatures actually go down for increasing B, saturating for very large fields, very differently from what has been predicted by all previous effective model calculations and found in previous lattice simulations.

---

[1]Contrastingly, a significant decrease in the critical temperature as a function of B, vanishing at $eB_c \sim 25m_\pi^2$, was found in Ref. [43], featuring the disappearance of the confined phase at large magnetic fields. This phenomenon that was not reproduced by any other effective model nor observed on the lattice (even for much larger fields).

Soon after the appearance of the new lattice results, the behavior of the critical temperature for deconfinement in the presence of a very large magnetic field was addressed within the MIT bag model [68], a very economic model in terms of parameters to be fixed (essentially one) and other ingredients usually hard to control in more sophisticated effective theories. The model is, of course, crude in numerical precision and misses the correct nature of the (crossover) transition. Nevertheless, it provides a simple setup for the discussion of some subtleties of vacuum and thermal contributions in each phase. It was shown in Ref. [68] that the influence of the magnetic field on the thermodynamics of both extreme energy domains is captured, so that the model furnishes a reasonable qualitative description of the behavior of the critical temperature in the presence of B, decreasing and saturating.

The fact that chiral models, even when coupled with the (static) Polyakov loop sector, seem to fail in the description of the behavior of $T_c \times eB$, whereas the (assumedly simple) MIT bag approach finds a good qualitative agreement, suggests that the critical temperature in QCD is a confinement-driven observable. This was also hinted by a previous successful description of the behavior of the critical temperature as a function of the pion mass and isospin chemical potential, as compared to lattice data, where chiral models also failed even qualitatively [82–84]. If confinement dynamics plays a central role in guiding the functional behavior of T_c, a the large-N_c limit of QCD should provide an adequate and powerful framework to study associated magnetic thermodynamics. In fact, it was shown in Ref. [85] on very general grounds that the fact that the deconfining temperature decreases and tends to saturate for large B, although this last point cannot be proven in a model-independent way, depends solely on quarks behaving paramagnetically.

In the sequel we summarize results for the chiral and deconfining transitions obtained in the framework of the linear sigma model coupled to quarks and to the Polyakov loop, especially the prediction of a splitting of the two critical lines, and how they compare to other effective model approaches as well as to lattice QCD. Then we discuss the outcome of the magnetic MIT bag model that yields a behavior for the critical deconfining temperature compatible with the most recent lattice simulations and magnetic catalysis. We continue with a discussion of very recent results, starting with the rather general findings within the large-N_c limit of magnetic QCD. Finally, we present our conclusions.

5.2 Modified Dispersion Relations and Integral Measures

In the presence of a classical, constant and uniform (Abelian) magnetic field, dispersion relations and momentum integrals will be modified. In order to compute vacuum and thermal determinants and Feynman diagrams, it is necessary to express these quantities in a convenient fashion. Lorentz invariance is broken by the preferred direction established by the external field, and Landau orbits redefine the new counting of quantum states [12].

For definiteness, let us take the direction of the magnetic field as the z-direction, $\mathbf{B} = B\hat{\mathbf{z}}$. One can compute, for instance, the modified effective potential or the modified pressure to lowest order by redefining the dispersion relations of charged scalar and spinor fields in the presence of $\mathbf{B}$, using the minimal coupling shift in the gradient and the field equations of motion.[2] For this purpose, it is convenient to choose the gauge such that $A^{\mu} = (A^0, \mathbf{A}) = (0, -By, 0, 0)$.

For scalar fields with electric charge q, such as pions, one has

$$\left(\partial^2 + m^2\right)\phi = 0, \tag{5.1}$$

$$\partial_{\mu} \to \partial_{\mu} + iqA_{\mu}. \tag{5.2}$$

After decomposing ϕ into Fourier modes, except for the dependence in the coordinate y, one obtains

$$\varphi''(y) + 2m\left[\left(\frac{p_0^2 - p_z^2 - m^2}{2m}\right) - \frac{q^2B^2}{2m}\left(y + \frac{p_x}{qB}\right)^2\right]\varphi(y) = 0, \tag{5.3}$$

which has the form of a Schrödinger equation for a harmonic oscillator. Its eigenmodes correspond to the well-known Landau levels

$$\varepsilon_n \equiv \left(\frac{p_{0n}^2 - p_z^2 - m^2}{2m}\right) = \left(\ell + \frac{1}{2}\right)\omega_B, \tag{5.4}$$

where $\omega_B = |q|B/m$ and ℓ is a positive ($\ell \geq 0$) integer, and provide the new dispersion relation:

$$p_{0n}^2 = p_z^2 + m^2 + (2\ell + 1)|q|B. \tag{5.5}$$

One can proceed in an analogous way for fermions with charge q. From the free Dirac equation $(i\gamma^{\mu}\partial_{\mu} - m)\psi = 0$, and the shift in ∂_{μ}, one arrives at the following Schrödinger equation

$$u_s''(y) + 2m\left[\left(\frac{p_0^2 - p_z^2 - m^2 + |q|Bs}{2m}\right) - \frac{q^2B^2}{2m}\left(y + \frac{p_x}{qB}\right)^2\right]u_s(y) = 0, \tag{5.6}$$

which yields the new dispersion relation for quarks:

$$p_{0n}^2 = p_z^2 + m^2 + (2\ell + 1 - s)|q|B, \tag{5.7}$$

where $s = \pm 1$ is the spin projection in the $\hat{\mathbf{z}}$ direction.

It is also straightforward to show that integrals over four momenta and thermal sum-integrals acquire the following forms, respectively [44, 45, 86]:

$$\int \frac{d^4k}{(2\pi)^4} \mapsto \frac{|q|B}{2\pi} \sum_{\ell=0}^{\infty} \int \frac{dk_0\,dk_z}{2\pi\,2\pi}, \tag{5.8}$$

---

[2]Higher-order (loop) corrections need the full propagator, not only its poles.

$$T \sum_n \int \frac{d^3k}{(2\pi)^3} \mapsto \frac{|q|BT}{2\pi} \sum_n \sum_{\ell=0}^{\infty} \int \frac{dk_z}{2\pi}, \qquad (5.9)$$

where ℓ represents the different Landau levels and n stands for the Matsubara frequency indices [87].

5.3 PLSM$_q$ Effective Model and the Splitting of the Chiral and Deconfining Transition Lines

Let us consider the two-flavor linear sigma model coupled to quarks and to the Polyakov loop, the PLSM$_q$ effective model, in the presence of an external magnetic field [48].

The confining properties of QCD are encoded in the complex-valued Polyakov loop variable L. As a matter of fact, the Polyakov loop sector only provides a description of the behavior of the approximate order parameter for the $Z(3)$ symmetry, which is explicitly broken by the presence of quarks. It is convenient for modeling the deconfining transition and has a good agreement with lattice results for most thermodynamic quantities such as the pressure and energy density, especially for the pure glue theory, but it does not provide a dynamical description of confinement.[3]

The expectation value of the Polyakov loop L is an *exact* order parameter for color confinement in the limit of infinitely massive quarks:

Confinement: $\begin{cases} \langle L \rangle = 0, & \text{low } T, \\ \langle L \rangle \neq 0, & \text{high } T, \end{cases}$ $\quad L(x) = \frac{1}{3} \text{Tr} \, \mathscr{P} \exp\left[i \int_0^{1/T} d\tau A_4(\mathbf{x}, \tau) \right],$

$$(5.10)$$

where $A_4 = i A_0$ is the matrix-valued temporal component of the Euclidean gauge field A_μ and the symbol $\mathscr{P}$ denotes path ordering. The integration takes place over compactified imaginary time τ, with periodic boundary conditions.

The chiral features of the model are encoded in the dynamics of the $O(4)$ chiral field, which is an exact order parameter in the chiral limit, in which quarks and pions are massless degrees of freedom:

Chiral symmetry: $\begin{cases} \langle \sigma \rangle \neq 0, & \text{low } T, \\ \langle \sigma \rangle = 0, & \text{high } T, \end{cases}$ $\quad \begin{array}{l} \phi = (\sigma, \boldsymbol{\pi}), \\ \boldsymbol{\pi} = (\pi^+, \pi^0, \pi^-). \end{array}$ $\quad (5.11)$

Here $\boldsymbol{\pi}$ is the isotriplet of the pseudoscalar pion fields and σ is the chiral scalar field which plays the role of an approximate order parameter of the chiral transition in QCD, since chiral symmetry is explicitly broken by the nonzero quark masses.

---

[3]This will be a key feature in the discussion of recent results for the critical temperature, since T_c seems to be a confinement-driven observable for both QCD transitions.

Within this effective model, the quark field ψ connects the Polyakov loop L and the chiral field ϕ, making a bridge between confining and chiral properties. Quarks are also coupled to the external magnetic field since the u and d quarks are electrically charged. Thus, it is clear that the external magnetic field will affect the chiral dynamics as well as the confining properties of the model, as much as the latter can be captured by the Polyakov loop sector.

This represents a natural generalization of the linear sigma model coupled to quarks [88], an effective theory that has been widely used to describe different aspects of the chiral transition, such as thermodynamic properties [89–102] and the nonequilibrium phase conversion process [103]. This generalization differs from previous ones [104–107] by the inclusion of a bridge via the covariant derivative and, of course, because of the modifications brought about by the magnetic field.

The Lagrangian of PLSM$_q$ describes the constituent quarks ψ, which interact with the meson fields σ, $\pi^\pm = (\pi^1 \pm i\pi^2)/\sqrt{2}$ and $\pi^0 = \pi^3$, the Abelian gauge field $a_\mu = (a^0, \mathbf{a}) = (0, -By, 0, 0)$, and the $SU(3)$ gauge field A_μ via the covariant derivative $D_\mu^{(q)} = (\partial_\mu - iQa_\mu - iA_\mu)$ with the charge matrix $Q = \mathrm{diag}(+2e/3, -e/3)$. Its explicit form is given by

$$\mathcal{L} = \overline{\psi}\left[i\gamma^\mu D_\mu^{(q)} - g(\sigma + i\gamma_5\boldsymbol{\tau} \cdot \boldsymbol{\pi})\right]\psi + \frac{1}{2}\left[(\partial_\mu\sigma)^2 + (\partial_\mu\pi^0)^2\right]$$
$$+ \left|D_\mu^{(\pi)}\right|^2 - V_\phi(\sigma, \boldsymbol{\pi}) - V_L(L, T), \qquad (5.12)$$

where $D_\mu^{(\pi)} = \partial_\mu + iea_\mu$ is the covariant derivative acting on colorless pions.

The chiral potential has the form

$$V_\phi(\sigma, \boldsymbol{\pi}) = \frac{\lambda}{4}\left(\sigma^2 + \pi^2 - v^2\right)^2 - h\sigma, \qquad (5.13)$$

where $h = f_\pi m_\pi^2$, $v^2 = f_\pi^2 - m_\pi^2/\lambda$, $\lambda = 20$, $f_\pi \approx 93$ MeV and $m_\pi \approx 138$ MeV. The constituent quark mass is given by $m_q \equiv m_q(\langle\sigma\rangle) = g\langle\sigma\rangle$, and, choosing $g = 3.3$ at $T = 0$, one obtains for the constituent quarks in the vacuum $m_q \approx 310$ MeV. At low temperatures quarks are not excited, and the model reproduces results from the usual linear σ-model without quarks.

The Polyakov potential adopted is given by [108–110]

$$\frac{V_L(L, T)}{T^4} = -\frac{L^*L}{2}\sum_{l=0}^{2} a_l\left(\frac{T_0}{T}\right)^l$$
$$+ b_3\left(\frac{T_0}{T}\right)^3 \log\left[1 - 6L^*L + 4(L^{*3} + L^3) - 3(L^*L)^2\right], \quad (5.14)$$

where $T_0 \equiv T_{SU(3)} = 270$ MeV is the critical temperature in the pure gauge case and $a_0 = 16\pi^2/45 \approx 3.51$, $a_1 = -2.47$, $a_2 = 15.2$, and $b_3 = -1.75$. Below we follow a mean-field analysis in which the mesonic sector is treated classically whereas quarks represent fast degrees of freedom.

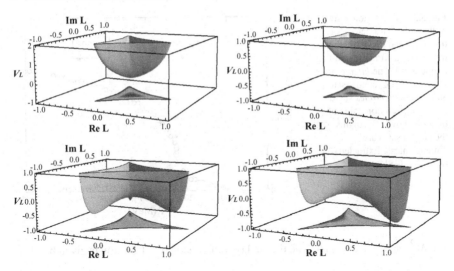

Fig. 5.2 Effects of temperature and magnetic field on quark confinement: The Polyakov loop potential at $T = 0.8T_0$ (*top*) and $T = 1.2T_0$ (*bottom*) and at zero magnetic field (*left*) and at $eB = 9T^2$ (*right*). Extracted from Ref. [48]

The one-loop corrections to the free energy Ω coming from quarks can be written as:

$$e^{iV_{3d}\Omega_q/T} = \left[\frac{\det(i\gamma^\mu D_\mu^{(q)} - m_q)}{\det(i\gamma^\mu \partial_\mu - m_q)}\right] \cdot \left[\frac{\det_T(i\gamma^\mu D_\mu^{(q)} - m_q)}{\det(i\gamma^\mu D_\mu^{(q)} - m_q)}\right], \quad (5.15)$$

so that the expectation values of the condensates can be obtained by minimizing the free energy

$$\Omega(\sigma, L; T, B) = V_\phi(\sigma, \pi) + V_L(L, T) + \Omega_q(\sigma, L, T), \quad (5.16)$$

at fixed values of temperature and magnetic field. The interaction piece $\Omega_q(\sigma, L, T)$ can be split into a vacuum (temperature-independent but still magnetic-field dependent) contribution and a thermal correction. The vacuum term has the form

$$\Omega_q^{\text{vac}}(B) = -\frac{N_c}{\pi} \sum_{f=u,d} |q_f| B \left[\left(\sum_{n=\ell}^\infty I_B^{(1)}(M_{\ell f}^2)\right) - \frac{I_B^{(1)}(m_f)}{2}\right]$$

minus the standard vacuum correction in the absence of the magnetic field,

$$\Omega_q^{(0)} = 2N_c \sum_{f=u,d} I_B^{(3)}(m_f^2), \quad (5.17)$$

where we have defined the integral

$$I_B^{(d)}(M^2) = \int \frac{d^d p}{(2\pi)^d} \sqrt{p^2 + M^2} \quad (5.18)$$

Fig. 5.3 Expectation values of the order parameters for the chiral and deconfinement transitions as functions of the temperature. The *filled circles* represent the σ-condensate, and the *empty circles* stand for the expectation value of the Polyakov loop. In this plot the condensates are dimensionless. Extracted from Ref. [48]

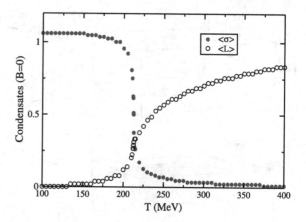

and $M_{\ell f}^2 = m_f^2 + 2\ell |q_f| B$. The thermal (paramagnetic) contribution is given by [48]

$$\Omega_q^{\text{para}} = \frac{|q_f| BT}{\pi^2} \sum_{s=\pm\frac{1}{2}} \sum_{\ell=0}^{\infty} \sum_{k=1}^{\infty} \frac{(-1)^k}{k} \text{Re}\big[\text{Tr}\,\Phi^k\big] \mu_{s\ell}(\sigma) K_1\left[\frac{k}{T}\mu_{s\ell}(\sigma)\right], \quad (5.19)$$

where $\mu_{s\ell}$ is the energy of the ℓth Landau level at zero longitudinal momentum, $\mu_{s\ell}(\sigma) = [g^2\sigma^2 + (2\ell + 1 - 2s)|q|B]^{1/2}$, and the untraced Polyakov loop is such that $\text{Re}[\text{Tr}\,\Phi^k] = \sum_{i=1}^{3} \cos(k\varphi_i)$, the integer k corresponding to the winding number of the Polyakov loops [48].

For finite temperature and $B = 0$ this model produces a crossover for both transitions. Figure 5.3 displays the condensates as functions of the temperature, and the critical temperature is defined by the change curvature in the curves. This occurs simultaneously for the chiral and deconfinement transitions within this model.

At zero temperature, the Polyakov loop variable does not play a role. The presence of a magnetic field enhances the chiral symmetry breaking, increasing the value of the chiral condensate, in line with the phenomenon of magnetic catalysis [14, 15, 17, 18, 33, 111]. This is shown in Fig. 5.4. It also deepens the minimum of the potential as B is increased, as illustrated in the same figure for several values of the magnetic field

The dependence of the chiral condensate on the magnetic field is approximately linear, as shown in Fig. 5.4. This is in line with results from chiral perturbation theory [29–31]. Recent lattice results [81] seem to deviate from a linear behavior for large B, growing faster, in better qualitative agreement with results from PNJL [52–55]. However, for larger values of B, all model calculations seem to deviate from the lattice data, whereas for very small B they all quantitatively agree [81].

Turning on the temperature, one can investigate the effects of the magnetic field on the thermodynamics and phase structure of strong interactions as captured by this model description. In the confining sector, the strong magnetic field affects the potential for the expectation value of the Polyakov loop via the intermediation of

Fig. 5.4 *Upper*: the expectation value of the (dimensionless, $\sigma = \xi v$) condensate as a function of the magnetic field. *Black dots* are obtained from the PLSMq and the *orange line* is the linear fit. *Lower*: effective potential for the condensate at zero temperature for several values of the magnetic field B. Extracted from Ref. [48] (Color figure online)

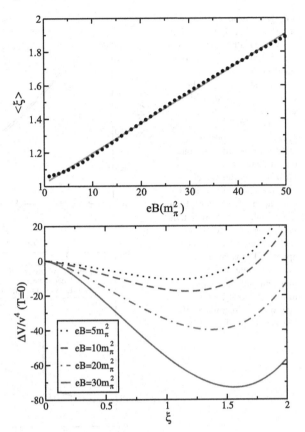

the quarks in three ways [48]: (i) the presence of the magnetic field intensifies the breaking of the global $\mathbb{Z}_3$ symmetry and makes the Polyakov loop real-valued, as shown in Fig. 5.2; (ii) the thermal contribution from quarks tends to destroy the confinement phase by increasing the expectation value of the Polyakov loop; (iii) on the contrary, the vacuum quark contribution tends to restore the confining phase by lowering the expectation value of the Polyakov loop.

In fact, the vacuum correction from quarks has a crucial impact on the phase structure. If one disregards the vacuum contribution from the quarks, as was done in Refs. [44, 45], one finds that the confinement and chiral phase transition lines coincide. Moreover, in this case an increasing magnetic field lowers the equivalent chiral-confinement transition temperature. On the other hand, the inclusion of the vacuum contribution from quark loops in a magnetic field modifies completely the picture: confinement and chiral transition lines split, and both chiral and deconfining critical temperatures become increasing functions of the magnetic field. Both scenarios are shown in Fig. 5.5, which exhibit the full calculation of the phase diagram from the effective potential within the PLSM$_q$ effective model. The vacuum contribution from the quarks affects drastically the chiral sector as well. Our calculations also show that the vacuum contribution seems to soften the order of the phase tran-

Fig. 5.5 Phase diagram in the *B-T* plane. *Upper*: without vacuum corrections: the critical temperatures of the deconfinement (*the dash-dotted line*) and chiral (*the dashed line*) transition coincide all the way, and decrease with *B*. *Lower*: with vacuum corrections: the critical temperatures of the deconfinement (*the dash-dotted line*) and chiral (*the dashed line*) transition coincide at $B = 0$ and split at higher values of the magnetic field. A deconfined phase with broken chiral symmetry appears. The *vertical line* is the magnitude of the magnetic field that expected to be realized at LHC heavy-ion collisions [8, 9]. Extracted from Ref. [48]

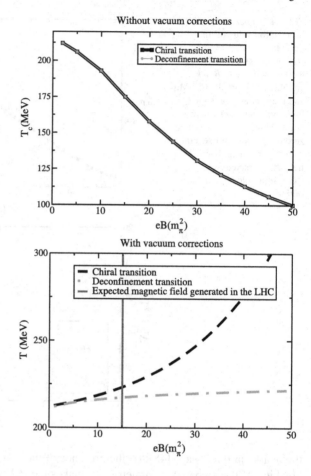

sition: the first-order phase transition—which would be realized in the absence of the vacuum contribution—becomes a smooth crossover in the system with vacuum quark loops included.

The modifications produced by strong magnetic fields over strong interactions seem very exciting, bringing new possibilities for the phase diagram: affecting the nature of the transitions, splitting different coexistence lines, possibly exhibiting new phases, increasing the breaking of $\mathbb{Z}_3$, and so on. As discussed in the Introduction, the second scenario has been also found in other effective models containing a chiral and a Polyakov loop sector [52–55, 67], as well as in preliminary lattice simulations [77, 78]. However, lattice simulations of magnetic QCD with physical masses and fine grids have shown that both critical temperatures actually go down for increasing B, saturating for very large fields [80], an unexpected behavior that is very different from the scenario depicted above.

This leads us to consider of a much simpler model that, yet, seems to contain the essential ingredients to describe the behavior of the deconfining line, and produces results that are in qualitative agreement with the lattice: the magnetic

MIT bag model [68]. As will be clear in the discussion, the subtraction procedure in renormalization is subtle (which can be seen as the choice of the renormalization scale) but can be guided by known physical phenomena and lattice results.

5.4 Magbag—The Thermal MIT Bag Model in the Presence of a Magnetic Background

In the MIT bag model framework for the pressure of strong interactions, one needs the free quark pressure. As seen previously, the presence of a magnetic field in the $\hat{z}$ direction affects this computation by modifying the dispersion relation to

$$\omega_{\ell s f}(k_z) = k_z^2 + m_f^2 + q_f B(2\ell + s + 1) \equiv k_z^2 + M_{\ell s f}^2, \qquad (5.20)$$

$\ell = 0, 1, 2, \ldots$ being the Landau level index, $s = \pm 1$ the spin projection, f the flavor index, and q_f the absolute value of the electric charge. Loop integrals are also affected as presented previously [44, 45, 86].

Since it has been shown that only very large magnetic fields do affect significantly the structure of the phase diagram for strong interactions [43–45, 51–55, 77, 78, 80], we can restrict the free quark pressure to the limit of very high magnetic fields, where it is possible to simplify some analytic expressions.

It is crucial to realize, however, that the lowest Landau level (LLL) approximation for the free gas pressure is not equivalent to the leading order of a large magnetic field expansion. For the zero-temperature, finite-B contribution to the pressure, the LLL is the energy level which less contributes in the limit of large B; the result being dominated by high values of ℓ. Nevertheless, the equivalence between the LLL approximation and the large B limit remains valid for the temperature-dependent part of the free pressure (as well as for the propagator), simplifying the numerical evaluation of thermal integrals [84].

The free magnetic contribution to the quark pressure has been considered in different contexts (usually, in effective field theories [44, 45, 49–51, 63, 64, 112]) and computed from the direct knowledge of the energy levels of the system, (5.20). The exact result, including all Landau levels, has to be computed from

$$P_q = 2N_c \sum_{\ell, s, f} \frac{q_f B}{2\pi} \int \frac{dk_z}{2\pi} \left\{ \frac{\omega_{\ell s f}(k_z)}{2} + T \ln\left[1 + e^{-\omega_{\ell s f}(k_z)/T}\right] \right\}, \qquad (5.21)$$

where the first term is a clearly divergent zero-point energy and the other one is the finite-temperature contribution for vanishing chemical potential. Since $\omega_{\ell s f}$ grows with B, the largest the ℓ labeling the Landau level considered the larger the zero-point energy term becomes, being minimal for the LLL, corroborating the previous discussion. Thus, in the limit of large B, the LLL approximation is inadequate here.

The decaying exponential dependence of the finite-temperature term on $\omega_{\ell s f}$, on the other hand, guarantees that the LLL dominates indeed this result for intense magnetic fields.

To obtain a good approximation for the large B limit of the free pressure, we choose to treat the full exact result and take the leading order of a $x_f \equiv m_f^2/2q_f B$ expansion in the final renormalized expression. Let us then discuss the treatment of the divergent zero-point term. Despite being a zero-temperature contribution, the first term in (5.21) cannot be fully subtracted because it carries the modification to the pressure brought about by the magnetic dressing of the quarks. Using dimensional regularization and the zeta-function representation, which is also a type of regularization, for the sums over Landau levels and subtracting the pure vacuum term in $(3+1)$ dimensions, one arrives at:

$$
P_q^V = \frac{N_c}{2\pi^2} \sum_f (q_f B)^2 \left[\zeta'(-1, x_f) + \frac{1}{2}(x_f - x_f^2) \ln x_f + \frac{x_f^2}{4} \right.
$$
$$
\left. - \frac{1}{12}\left(2/\varepsilon + \log(\Lambda^2/2q_f B) + 1\right) \right],
\tag{5.22}
$$

where a pole $\sim (q_f B)^2 [2/\varepsilon]$ still remains. This infinite contribution that survives the vacuum subtraction can be interpreted as a pure magnetic pressure coming from the artificial scenario adopted, with a constant and uniform B field covering the whole universe (analogous to the case of a cosmological constant). In this vein, one may neglect all terms $\sim (q_f B)^2$ and independent of masses and other couplings (as done, e.g. in Refs. [44, 45, 49, 50, 63, 64]), concentrating on the modification of the pressure of the quark matter under investigation. This can be seen as a choice for the renormalization scale after the renormalization of a $\sim F_{\mu\nu} F^{\mu\nu}$ term representing the magnetic field, as discussed, e.g. in Refs. [63, 64]. We will come back to this point in the sequel.

The final exact result for the free pressure of magnetically dressed quarks is therefore

$$
\frac{P_q}{N_c} = \sum_f \frac{(q_f B)^2}{2\pi^2} \left[\zeta'(-1, x_f) - \zeta'(-1, 0) + \frac{1}{2}(x_f - x_f^2) \ln x_f + \frac{x_f^2}{4} \right]
$$
$$
+ T \sum_{\ell, s, f} \frac{q_f B}{2\pi^2} \int dk_z \ln\left[1 + e^{-\omega_{\ell s f}(k_z)/T} \right].
\tag{5.23}
$$

In Refs. [44, 45, 49, 50, 63, 64], the constant $\zeta'(-1, 0) = -0.165421\ldots$ was not subtracted. In the case of pions, however, the full subtraction ensures that magnetic catalysis, i.e. an enhancement of chiral symmetry breaking, at zero temperature [14, 15, 17, 18, 33, 111], is realized. On the other hand, if this term is left, the pion contribution to the effective potential for the chiral condensate at large magnetic fields will eventually raise the minimum instead of lowering it.

In the limit of large magnetic field (i.e. $x_f = m_f^2/(2q_f B) \to 0$), we obtain

$$\frac{P_q}{N_c} \overset{\text{large } B}{=} \sum_f \frac{(q_f B)^2}{2\pi^2}[x_f \ln \sqrt{x_f}] + T \sum_f \frac{q_f B}{2\pi^2} \int dk_z \ln[1 + e^{-\sqrt{k_z^2 + m_f^2}/T}].$$

$$(5.24)$$

Adding the free piece of the gluonic contribution and the bag constant $\mathscr{B}$, the pressure of the QGP sector in the presence of an intense magnetic field reads:

$$P_{\text{QGP}}^B = 2(N_c^2 - 1)\frac{\pi^2 T^4}{90} + P_q - \mathscr{B}.$$

$$(5.25)$$

It is clear that, for $\sqrt{eB}$ much larger than all other energy scales, the pressure in the QGP phase increases with the magnetic field, which seems to favor a steady drop in the critical temperature with increasing B that would lead to a crossing of the critical line with the $T = 0$ axis at some critical value for the magnetic field. However, the behavior of $T_c(B)$ also depends on how the pions react to B, so that the outcome is not obvious.

In the confined sector, which we describe by a free pion gas, one may follow analogous steps in order to compute the contribution from the charged pions, which couple to the magnetic field, arriving at

$$P_{\pi^+} + P_{\pi^-} = -\frac{(eB)^2}{4\pi^2}\left[\zeta'\left(-1, \frac{1}{2} + x_\pi\right) - \zeta'\left(-1, \frac{1}{2}\right) + \frac{x_\pi^2}{4} - x_\pi^2 \ln \sqrt{x_\pi}\right]$$

$$-2\frac{eB}{4\pi^2}T\sum_\ell \int dk_z \ln[1 - e^{-\sqrt{k_z^2 + M_{\pi\ell}^2}/T}],$$

$$(5.26)$$

where $M_{\pi\ell}^2 \equiv m_\pi^2 + (2\ell + 1)eB$ and $x_\pi \equiv m_\pi^2/(2eB)$. In this final expression all terms $\sim(q_f B)^2$ and independent of masses and other couplings were subtracted, as discussed before. Notice that the spin-zero nature of the pions guarantees that *all* charged pion modes in a magnetic field, differently from what happens with the quark modes, are B-dependent. So, in the large magnetic field limit the thermal integral associated with π^+ and π^- is exponentially suppressed by an effective mass $\gtrsim(m_\pi^2 + eB)$, as was also noticed in Refs. [44, 45], and can be dropped. In this limit, we have

$$P_{\pi^+} + P_{\pi^-} \overset{\text{large } B}{=} -\frac{(eB)^2}{4\pi^2}\zeta^{(1,1)}(-1, 1/2)\, x_\pi,$$

$$(5.27)$$

where $\zeta^{(1,1)}(-1, 1/2) = -\ln(2)/2 = -0.346574\ldots$. Neutral pions do not couple to the magnetic field and contribute only with the usual thermal integral [87].

As before, for $\sqrt{eB}$ much larger than all other scales, the pion pressure rises with the magnetic field, as a consequence of the subtraction of all terms that are independent of temperature, masses and other couplings in the renormalization process, which renders the pressure positive. Differently from the quark pressure, however,

Fig. 5.6 Crossing pion gas and QGP pressures as functions of the temperature for different values of the magnetic field: $eB = 0$ (*black, solid, right-most*), $20m_\pi^2$, $40m_\pi^2$, $60m_\pi^2$ (*magenta, dash-dotted*) and $eB = 100m_\pi^2$ (*gray, solid, left-most*), where $m_\pi = 138$ MeV is the vacuum pion mass. Extracted from Ref. [68] (Color figure online)

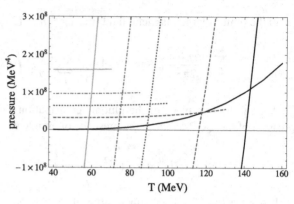

Fig. 5.7 Phase diagram in the presence of a strong magnetic field. We also keep the $T_c(B = 0)$ point. The *blue square* represents a very conservative estimate for the maximum value of eB expected to be achieved in non-central collisions at the LHC with the formation of deconfined matter. The *arrow* marks the critical temperature for $eB \approx 210m_\pi^2$ [113], expected to be found at the early universe. Extracted from Ref. [68] (Color figure online)

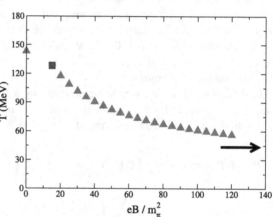

the $B = 0$ pion pressure takes over for temperatures of the order of the pion mass, which is not small and always enlarged by the presence of a magnetic field (given its scalar nature). Moreover, for large T, the magnetic pion pressures converge to $(1/3)$ of the $B = 0$ pressure, since π^0 is the only degree of freedom that contributes thermally for large B.

Each equilibrium phase should maximize the pressure, so that the critical line in the phase diagram can be constructed by directly extracting $T_c(B)$ from the equality of pressures. It is instructive, nevertheless, to consider a plot of the crossing pressures, as shown in Fig. 5.6. The figure shows, as expected, a decrease in the critical temperature (crossing points) as B is increased due to the corresponding positive shift of the QGP pressure. However, T_c seems to be saturating at a constant value. One can see that the critical pressure (crossing point) goes down, but then it bends up again due to the increase in the pion pressure with B. This combination avoids a steady and rapid decrease of the critical temperature, as becomes clear in the phase diagram shown in Fig. 5.7. In fact, inspection of the zero-temperature limit of (5.24) and (5.27) shows that there is no value of magnetic field that allows for a vanishing critical temperature.

The phase diagram in the plane $T-eB$ shows that the critical temperature for deconfinement falls as we increase the magnetic field. However, instead of falling with a rate that will bring it to zero at a given critical value of eB, it falls less and less rapidly, tending to saturate at large values of B. Remarkably, this qualitative behavior agrees quite well with the most recent lattice results with physical masses [80].[4] As discussed in the Introduction, previous models [44, 45, 48, 52–57, 67], have predicted either an increase or an essentially flat behavior for the deconfinement critical line as B is increased to very large values. The same was true for previous lattice simulations [77, 78], which could be reproduced by the authors of Ref. [80] by increasing the quark masses to unphysical values.

The renormalization procedure in the presence of a constant and uniform magnetic field seems to be very subtle and crucial for the phenomenological outcome for the phase structure. B-dependent, mass-independent terms survive pure vacuum ($B = 0$) subtraction and have to be subtracted either in an *ad hoc* fashion [68] or by including a background field counterterm associated with a term $\sim F_{\mu\nu} F^{\mu\nu}$ representing the magnetic field [63, 64]. The latter brings a renormalization scale and, upon an appropriate choice, reproduces the former. Subtracting all purely magnetic terms in the pressures seems to be the appropriate choice since: (i) one guarantees that the pion pressure grows with increasing magnetic field at zero temperature, which is consistent with the well-known phenomenon of magnetic catalysis; (ii) lattice simulations usually measure derivatives of the pressure with respect to temperature and quark mass, and do not access derivatives with respect to B, so that purely B-dependent terms are not included in their results; and (iii) the effect of a purely magnetic contribution to the pressure would only shift the effective potential as a whole. In particular, there would be no modification on relative positions and heights of different minima that represent different phases of matter.

The qualitative success of the description of the deconfinement transition in the presence of an external magnetic field in terms of the MIT bag model suggests that confinement dynamics plays a central role in guiding the functional behavior of T_c. In this case, a large N_c investigation of the associated magnetic thermodynamics seems appropriate.

5.5 Large N_c

Lattice QCD calculations [114] show that the deconfinement phase transition of pure glue $SU(N_C)$ gauge theory becomes first order when $N_c \geq 3$ [115–122] with a critical temperature given by [123]

$$\lim_{N_c \to \infty} \frac{T_c}{\sqrt{\sigma}} = 0.5949(17) + \frac{0.458(18)}{N_c^2}, \tag{5.28}$$

---

[4]Of course, our description necessarily predicts a first-order transition, as usual with the MIT bag model, and our numbers should be taken as rough estimates, as is always the case in effective models.

where $\sigma \sim (440\,\mathrm{MeV})^2$ is the string tension. The thermodynamic properties of pure glue do not seem to change appreciably when $N_c \geq 3$ [124–126], which suggests that large N_c arguments may indeed capture the main physical mechanism behind the deconfinement phase transition of QCD.

It has been shown in Ref. [85] that the deconfinement critical temperature must decrease in the presence of an external magnetic field in the large N_c limit of QCD, provided that quarks exhibit a paramagnetic behavior. Assuming that $N_f/N_c \ll 1$ and $m_q = 0$, the only contribution to the pressure of the confined phase that enters at $\mathcal{O}(N_c^2)$ is given by the vacuum ($B = 0$) gluon condensate $c_0^4 N_c^2 \sigma^2$. The gluon and quark condensates change in the presence of a magnetic field [30, 31, 127] but these modifications are negligible in the large N_c limit. Besides, the gluon contribution to the deconfined pressure is blind to the magnetic field.

On the other hand, the quark contribution is affected by the magnetic field and has the form

$$P_{quark}(T, eB) \sim N_c N_{pairs}(N_f) T^4 \tilde{f}_{quark}(T/\sqrt{\sigma}, eB/T^2), \qquad (5.29)$$

with $N_{pairs}(N_f)/N_c \ll 1$ being the number of pairs of quark flavors with electric charges $\{(N_c - 1)/N_c, -1/N_c\}$ in units of the fundamental charge. Only the largest ($\sim N_c^0$) charge in each pair contributes to leading order in N_f/N_c. Notice that the function $\tilde{f}_{quark}$ is positive definite and must increase monotonically with T for a fixed value of eB until it goes to 1 in the high temperature limit $T \gg \sqrt{\sigma}$, eB. Thus, one should expect that the critical temperature as a function of the magnetic field, $T_c(eB)$, must decrease with respect to the pure glue critical temperature, $T_c^{(0)}$, by an amount of $\mathcal{O}(N_f/N_c)$. This can be seen directly by equating the pressures at T_c, which yields [85]

$$\frac{T_c(eB)}{\sqrt{\sigma}} f_{glue}^{1/4}\left(\frac{T_c(eB)}{\sqrt{\sigma}}\right) = \frac{c_2(N_{pairs}, eB)}{c_{SB}}, \qquad (5.30)$$

where we defined

$$c_2(N_{pairs}, eB) \equiv c_0\left[1 - \frac{1}{4}\frac{N_{pairs}(N_f)}{N_c}\frac{c_{qSB}^4 \tilde{f}_{quark}\left(\frac{T_c^{(0)}}{\sqrt{\sigma}}, \frac{eB}{T_c^{(0)2}}\right)}{c_{SB}^4 f_{glue}\left(\frac{T_c^{(0)}}{\sqrt{\sigma}}\right)}\right]. \qquad (5.31)$$

Since $c_2(N_{pairs}, eB) < c_0$, one finds that $T_c(eB)/T_c^{(0)} < 1$ by an amount $\sim N_f/N_c$ [85]. Assuming that quarks behave paramagnetically for all values of B, then $c_2(N_{pairs}, eB) < c_1(N_f)$, its equivalent in the case with $B = 0$ and $N_f > 0$, and $T_c(eB)$ is also lower than the critical temperature in the presence of N_f/N_c flavors of massless quarks at $B = 0$ [85].

In a free gas implementation of the deconfined phase $f_{glue} = 1$ and, for very strong magnetic fields, $\tilde{f}_{quark} \sim eB/T_c^2$ [68], so that the magnetic suppression of the deconfinement critical temperature goes like $eBN_{pairs}/(N_c\sigma)$. This simple implementation in the limits of low and high magnetic fields provides a scenario in which

Fig. 5.8 Cartoon of the
$T_c \times eB$ phase diagram in the
large N_c limit, using the
approximation of free
deconfined quarks and
gluons. The numerical
value 0.59 shown in the plot
was extracted from
Ref. [123]. Extracted from
Ref. [85]

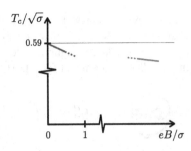

the slope in $T_c(eB)$ decreases for large fields, as illustrated in Fig. 5.8, which hints
for a saturation of T_c as a function of eB, as observed on the lattice [80] and in the
magnetic MIT bag model [68], but cannot be obtained in a model-independent way.

5.6 Conclusions and Perspectives

The investigation of the effects brought about by the presence of a magnetic background on the thermal chiral and deconfining transitions is in its infancy yet. Nevertheless, the promise of the outcome of a rich phenomenology in mapping this new
phase diagram of strong interactions is concrete.

First model calculations have revealed the possibility of modifications in the nature of the QCD phase transitions, and also the appearance of a new phase of strong
interactions in the case of a splitting of the critical (chiral and deconfining) lines.
Even if recent, more physical lattice simulations have drastically modified the initial
picture, they have also shown that the magnetic background has a very non-trivial
influence on strong interactions. For instance, the behavior of quark condensates at
finite temperature is non-monotonic [81], rendering well-established vacuum phenomena such as magnetic catalysis more subtle at finite temperature.

The functional behavior of the critical temperatures still has to be understood
more deeply. Although no model foresaw the fact that both, chiral and deconfining
temperatures, decrease then saturate at a nonzero value according to the lattice [80],
a posteriori the magnetic MIT bag model was successful to describe this behavior
for deconfinement qualitatively [68] and seems to capture some essential ingredients. A model-independent analysis in the large-N_c limit of QCD also points to this
behavior [85], which is reassuring from the theoretical standpoint.

Another key ingredient in building an understanding of the physics of the quark-
gluon plasma under these new conditions, which can be relevant for high-energy
heavy-ion collision experiments, the primordial quark-hadron transition and magnetars is the standard perturbative investigation of magnetic QCD. The calculation
of the pressure in thermal QCD to two loops in the strong sector using the full QED
propagator in the lowest Landau level approximation is subtle but possible, as done
originally in Ref. [84].

The computation makes use of the full magnetic propagator that was obtained by
Schwinger [128], but can be cast in a more convenient form in terms of a sum over

Landau levels as derived in Ref. [129] (see also Refs. [84, 130]). In particular, it has been shown in Ref. [84] that the chiral limit for the exchange diagram seems to be trivial for very large magnetic fields. Concretely, it can be written diagrammatically in the following compact form [84]:

$$
\text{(diagram)}^{\text{LLL}} = \left(\frac{q_f B}{2\pi}\right) \int \frac{dk_1 dk_2}{(2\pi)^2}\, e^{-\frac{k_1^2+k_2^2}{2q_f B}}\ \text{(diagram)}^{\bar{d}=2;\ m_k^2=k_1^2+k_2^2}, \tag{5.32}
$$

which realizes the intuitive expectation that the nontrivial dynamics in an extremely intense magnetic field should be one-dimensional. Since gluons do not couple directly to the magnetic field, their dispersion relation maintains its three-dimensional character, which effectively results in a "massive" gluon in the dimensionally-reduced diagram. In the end the exchange contribution to the pressure is essentially an average over the effective gluon transverse mass $m_k^2 = k_1^2 + k_2^2$ of the exchange diagram in $(1+1)$-dimensions with the Gaussian weight $(q_f B/2\pi)\exp[-m_k^2/2q_f B]$. Since the trace in the reduced diagram is proportional to m_f^2, the chiral limit seems trivial [84]. A detailed analysis of the dependence of the pressure on the mass and temperature and a semiclassical interpretation of this result will be reported soon [131].

The nature of the phase diagram of strong interactions in the presence of a magnetic background is still open. Recent lattice data, especially when compared to effective model predictions, seem to indicate that confinement dynamics plays an important role in the phase structure that emerges and should be incorporated in any effective description. Comparison between lattice data with very different quark masses [77, 78, 80, 81] also show that the dependence of the critical temperatures on this parameter is non-trivial: T_c increases at the percent level for large masses [77, 78] whereas it decreases appreciably for physical masses [80]. This competition between the effects from the magnetic field and quark masses on T_c was also found in the large-N_c QCD analysis of Ref. [85]. A more systematic analysis of this phenomenon on the lattice would be very helpful for the building of effective models.

Acknowledgements It is a pleasure to thank the editors of this volume of *Lecture Notes in Physics* for the invitation to contribute to this special edition. The discussion presented here in part summarizes work done in collaboration with M.N. Chernodub, A.J. Mizher, J. Noronha and L.F. Palhares to whom I am deeply grateful. I am especially indebted to my former students A.J. Mizher and L.F. Palhares, from whom I have learnt so much over several years. I also thank J.-P. Blaizot, M. D'Elia and G. Endrodi for fruitful discussions.

References

1. D.E. Kharzeev, L.D. McLerran, H.J. Warringa, Nucl. Phys. A **803**, 227 (2008)
2. K. Fukushima, D.E. Kharzeev, H.J. Warringa, Phys. Rev. D **78**, 074033 (2008)
3. D.E. Kharzeev, H.J. Warringa, Phys. Rev. D **80**, 034028 (2009)

4. D.E. Kharzeev, Nucl. Phys. A **830**, 543C (2009)
5. D.E. Kharzeev, Ann. Phys. **325**, 205 (2010)
6. K. Fukushima, D.E. Kharzeev, H.J. Warringa, Nucl. Phys. A **836**, 311 (2010)
7. K. Fukushima, D.E. Kharzeev, H.J. Warringa, Phys. Rev. Lett. **104**, 212001 (2010)
8. V. Skokov, A.Y. Illarionov, V. Toneev, Int. J. Mod. Phys. A **24**, 5925 (2009)
9. V. Voronyuk, V.D. Toneev, W. Cassing, E.L. Bratkovskaya, V.P. Konchakovski, S.A. Voloshin, Phys. Rev. C **83**, 054911 (2011)
10. A. Bzdak, V. Skokov, Phys. Lett. B **710**, 171 (2012)
11. W.-T. Deng, X.-G. Huang, Phys. Rev. C **85**, 044907 (2012)
12. L.D. Landau, E.M. Lifshitz, *Statistical Physics – Course of Theoretical Physics*, vol. 5 (Butterworth, Stoneham, 1984)
13. S.P. Klevansky, R.H. Lemmer, Phys. Rev. D **39**, 3478 (1989)
14. V.P. Gusynin, V.A. Miransky, I.A. Shovkovy, Phys. Lett. B **349**, 477 (1995)
15. V.P. Gusynin, V.A. Miransky, I.A. Shovkovy, Nucl. Phys. B **462**, 249 (1996)
16. A.Y. Babansky, E.V. Gorbar, G.V. Shchepanyuk, Phys. Lett. B **419**, 272 (1998)
17. K.G. Klimenko, arXiv:hep-ph/9809218
18. G.W. Semenoff, I.A. Shovkovy, L.C.R. Wijewardhana, Phys. Rev. D **60**, 105024 (1999)
19. A. Goyal, M. Dahiya, Phys. Rev. D **62**, 025022 (2000)
20. B. Hiller, A.A. Osipov, A.H. Blin, J. da Providencia, SIGMA **4**, 024 (2008)
21. A. Ayala, A. Sanchez, G. Piccinelli, S. Sahu, Phys. Rev. D **71**, 023004 (2005)
22. A. Ayala, A. Bashir, A. Raya, E. Rojas, Phys. Rev. D **73**, 105009 (2006)
23. E. Rojas, A. Ayala, A. Bashir, A. Raya, Phys. Rev. D **77**, 093004 (2008)
24. A. Ayala, A. Bashir, A. Raya, A. Sanchez, Phys. Rev. D **80**, 036005 (2009)
25. A. Ayala, A. Bashir, A. Raya, A. Sanchez, J. Phys. G **37**, 015001 (2010)
26. A. Ayala, A. Bashir, E. Gutierrez, A. Raya, A. Sanchez, Phys. Rev. D **82**, 056011 (2010)
27. J. Navarro, A. Sanchez, M.E. Tejeda-Yeomans, A. Ayala, G. Piccinelli, Phys. Rev. D **82**, 123007 (2010)
28. S.P. Klevansky, Rev. Mod. Phys. **64**, 649 (1992)
29. I.A. Shushpanov, A.V. Smilga, Phys. Lett. B **402**, 351 (1997)
30. N.O. Agasian, I.A. Shushpanov, Phys. Lett. B **472**, 143 (2000)
31. T.D. Cohen, D.A. McGady, E.S. Werbos, Phys. Rev. C **76**, 055201 (2007)
32. D. Kabat, K.M. Lee, E. Weinberg, Phys. Rev. D **66**, 014004 (2002)
33. V.A. Miransky, I.A. Shovkovy, Phys. Rev. D **66**, 045006 (2002)
34. E.J. Ferrer, V. de la Incera, C. Manuel, Phys. Rev. Lett. **95**, 152002 (2005)
35. E.J. Ferrer, V. de la Incera, C. Manuel, Nucl. Phys. B **747**, 88 (2006)
36. E.J. Ferrer, V. de la Incera, Phys. Rev. Lett. **97**, 122301 (2006)
37. E.J. Ferrer, V. de la Incera, Phys. Rev. D **76**, 045011 (2007)
38. E.J. Ferrer, V. de la Incera, Phys. Rev. D **76**, 114012 (2007)
39. K. Fukushima, H.J. Warringa, Phys. Rev. Lett. **100**, 032007 (2008)
40. J.L. Noronha, I.A. Shovkovy, Phys. Rev. D **76**, 105030 (2007)
41. D.T. Son, M.A. Stephanov, Phys. Rev. D **77**, 014021 (2008)
42. K.G. Klimenko, V.C. Zhukovsky, Phys. Lett. B **665**, 352 (2008)
43. N.O. Agasian, S.M. Fedorov, Phys. Lett. B **663**, 445 (2008)
44. E.S. Fraga, A.J. Mizher, Phys. Rev. D **78**, 025016 (2008)
45. E.S. Fraga, A.J. Mizher, Nucl. Phys. A **820**, 103C (2009)
46. A.J. Mizher, E.S. Fraga, Nucl. Phys. A **831**, 91 (2009)
47. A.J. Mizher, E.S. Fraga, M.N. Chernodub, PoS **FACESQCD**, 020 (2010)
48. A.J. Mizher, M.N. Chernodub, E.S. Fraga, Phys. Rev. D **82**, 105016 (2010)
49. D.P. Menezes, M. Benghi Pinto, S.S. Avancini, A. Perez Martinez, C. Providencia, Phys. Rev. C **79**, 035807 (2009)
50. G.N. Ferrari, A.F. Garcia, M.B. Pinto, Phys. Rev. D **86**, 096005 (2012)
51. J.K. Boomsma, D. Boer, Phys. Rev. D **81**, 074005 (2010)
52. K. Fukushima, M. Ruggieri, R. Gatto, Phys. Rev. D **81**, 114031 (2010)
53. R. Gatto, M. Ruggieri, Phys. Rev. D **82**, 054027 (2010)

54. R. Gatto, M. Ruggieri, Phys. Rev. D **83**, 034016 (2011)
55. R. Gatto, M. Ruggieri, arXiv:1207.3190 [hep-ph]
56. C.V. Johnson, A. Kundu, J. High Energy Phys. **0812**, 053 (2008)
57. F. Preis, A. Rebhan, A. Schmitt, J. High Energy Phys. **1103**, 033 (2011)
58. N. Callebaut, D. Dudal, H. Verschelde PoS **FACESQCD**, 046 (2010)
59. S.S. Avancini, D.P. Menezes, C. Providencia, Phys. Rev. C **83**, 065805 (2011)
60. S.S. Avancini, D.P. Menezes, M.B. Pinto, C. Providencia, Phys. Rev. D **85**, 091901 (2012)
61. K. Kashiwa, Phys. Rev. D **83**, 117901 (2011)
62. B. Chatterjee, H. Mishra, A. Mishra, Phys. Rev. D **84**, 014016 (2011)
63. J.O. Andersen, R. Khan, Phys. Rev. D **85**, 065026 (2012)
64. J.O. Andersen, A. Tranberg, J. High Energy Phys. **1208**, 002 (2012)
65. J.O. Andersen, Phys. Rev. D **86**, 025020 (2012)
66. J.O. Andersen, J. High Energy Phys. **1210**, 005 (2012)
67. V. Skokov, Phys. Rev. D **85**, 034026 (2012)
68. E.S. Fraga, L.F. Palhares, Phys. Rev. D **86**, 016008 (2012)
69. K. Fukushima, J.M. Pawlowski, Phys. Rev. D **86**, 076013 (2012)
70. M.N. Chernodub, Phys. Rev. D **82**, 085011 (2010)
71. M.N. Chernodub, Phys. Rev. Lett. **106**, 142003 (2011)
72. M.N. Chernodub, J. Van Doorsselaere, H. Verschelde, Phys. Rev. D **85**, 045002 (2012)
73. M.N. Chernodub, Int. J. Mod. Phys. A **27**, 1260003 (2012)
74. P.V. Buividovich, M.N. Chernodub, E.V. Luschevskaya, M.I. Polikarpov, Phys. Lett. B **682**, 484 (2010)
75. P.V. Buividovich, M.N. Chernodub, E.V. Luschevskaya, M.I. Polikarpov, Nucl. Phys. B **826**, 313 (2010)
76. P.V. Buividovich, M.N. Chernodub, E.V. Luschevskaya, M.I. Polikarpov, Phys. Rev. D **80**, 054503 (2009)
77. M. D'Elia, S. Mukherjee, F. Sanfilippo, Phys. Rev. D **82**, 051501 (2010)
78. M. D'Elia, F. Negro, Phys. Rev. D **83**, 114028 (2011)
79. V.V. Braguta, P.V. Buividovich, M.N. Chernodub, M.I. Polikarpov, Phys. Lett. B **718**, 667 (2012)
80. G.S. Bali, F. Bruckmann, G. Endrodi, Z. Fodor, S.D. Katz, S. Krieg, A. Schafer, K.K. Szabo, J. High Energy Phys. **1202**, 044 (2012)
81. G.S. Bali, F. Bruckmann, G. Endrodi, Z. Fodor, S.D. Katz, A. Schafer, Phys. Rev. D **86**, 071502 (2012)
82. E.S. Fraga, L.F. Palhares, C. Villavicencio, Phys. Rev. D **79**, 014021 (2009)
83. L.F. Palhares, E.S. Fraga, C. Villavicencio, Nucl. Phys. A **820**, 287C (2009)
84. L.F. Palhares, Exploring the different phase diagrams of strong interactions. PhD Thesis, Federal University of Rio de Janeiro, 2012. arXiv:1208.0574v1 [hep-ph]
85. E.S. Fraga, J. Noronha, L.F. Palhares, arXiv:1207.7094 [hep-ph]
86. S. Chakrabarty, Phys. Rev. D **54**, 1306 (1996)
87. J.I. Kapusta, C. Gale, *Finite-Temperature Field Theory: Principles and Applications* (Cambridge University Press, Cambridge, 2006)
88. M. Gell-Mann, M. Levy, Nuovo Cim. **16**, 705 (1960)
89. L.P. Csernai, I.N. Mishustin, Phys. Rev. Lett. **74**, 5005 (1995)
90. A. Abada, J. Aichelin, Phys. Rev. Lett. **74**, 3130 (1995)
91. A. Abada, M.C. Birse, Phys. Rev. D **55**, 6887 (1997)
92. I.N. Mishustin, O. Scavenius, Phys. Rev. Lett. **83**, 3134 (1999)
93. O. Scavenius, A. Dumitru, Phys. Rev. Lett. **83**, 4697 (1999)
94. H.C.G. Caldas, A.L. Mota, M.C. Nemes, Phys. Rev. D **63**, 056011 (2001)
95. O. Scavenius, A. Mocsy, I.N. Mishustin, D.H. Rischke, Phys. Rev. C **64**, 045202 (2001)
96. O. Scavenius, A. Dumitru, E.S. Fraga, J.T. Lenaghan, A.D. Jackson, Phys. Rev. D **63**, 116003 (2001)
97. K. Paech, H. Stoecker, A. Dumitru, Phys. Rev. C **68**, 044907 (2003)
98. K. Paech, A. Dumitru, Phys. Lett. B **623**, 200 (2005)

99. A. Mocsy, I.N. Mishustin, P.J. Ellis, Phys. Rev. C **70**, 015204 (2004)
100. C.E. Aguiar, E.S. Fraga, T. Kodama, J. Phys. G **32**, 179 (2006)
101. B.J. Schaefer, J. Wambach, Phys. Rev. D **75**, 085015 (2007)
102. B.G. Taketani, E.S. Fraga, Phys. Rev. D **74**, 085013 (2006)
103. E.S. Fraga, G. Krein, Phys. Lett. B **614**, 181 (2005)
104. R.D. Pisarski, Phys. Rev. D **62**, 111501 (2000)
105. A. Dumitru, R.D. Pisarski, Phys. Lett. B **504**, 282 (2001)
106. A. Dumitru, R.D. Pisarski, Nucl. Phys. A **698**, 444 (2002)
107. O. Scavenius, A. Dumitru, A.D. Jackson, Phys. Rev. Lett. **87**, 182302 (2001)
108. K. Fukushima, Phys. Lett. B **591**, 277 (2004)
109. S. Roessner, C. Ratti, W. Weise, Phys. Rev. D **75**, 034007 (2007)
110. S. Roessner, T. Hell, C. Ratti, W. Weise, Nucl. Phys. A **814**, 118 (2008)
111. I.A. Shovkovy, arXiv:1207.5081 [hep-ph]
112. D. Ebert, K.G. Klimenko, Nucl. Phys. A **728**, 203 (2003)
113. T. Vachaspati, Phys. Lett. B **265**, 258 (1991)
114. M. Teper, PoS **LATTICE2008**, 022 (2008)
115. G. Boyd, J. Engels, F. Karsch, E. Laermann, C. Legeland, M. Lutgemeier, B. Petersson, Nucl. Phys. B **469**, 419 (1996)
116. B. Lucini, M. Teper, U. Wenger, Phys. Lett. B **545**, 197 (2002)
117. B. Lucini, M. Teper, U. Wenger, J. High Energy Phys. **01**, 061 (2004)
118. K. Holland, M. Pepe, U.-J. Wiese, Nucl. Phys. B **694**, 35 (2004)
119. B. Lucini, M. Teper, U. Wenger, Nucl. Phys. B, Proc. Suppl. **129–130**, 712 (2004)
120. M. Pepe, Nucl. Phys. B, Proc. Suppl. **141**, 238 (2005)
121. B. Lucini, M. Teper, U. Wenger, J. High Energy Phys. **0502**, 033 (2005)
122. S. Borsanyi, G. Endrodi, Z. Fodor, S.D. Katz, K.K. Szabo, J. High Energy Phys. **1207**, 056 (2012)
123. B. Lucini, A. Rago, E. Rinaldi, Phys. Lett. B **712**, 279 (2012)
124. B. Bringoltz, M. Teper, Phys. Lett. B **628**, 113 (2005)
125. S. Datta, S. Gupta, Nucl. Phys. A **830**, 749c (2009)
126. M. Panero, Phys. Rev. Lett. **103**, 232001 (2009)
127. N.O. Agasian, Phys. Lett. B **488**, 39 (2000)
128. J.S. Schwinger, Phys. Rev. **82**, 664 (1951)
129. A. Chodos, K. Everding, D.A. Owen, Phys. Rev. D **42**, 2881 (1990)
130. K. Fukushima, Phys. Rev. D **83**, 111501 (2011)
131. J.-P. Blaizot, E.S. Fraga, L.F. Palhares, Phys. Lett. B **722**, 167 (2013)

Chapter 6
Electromagnetic Superconductivity of Vacuum Induced by Strong Magnetic Field

M.N. Chernodub

6.1 Introduction

Quantum Chromodynamics (QCD) exhibits many remarkable properties in the presence of a very strong magnetic field. The external magnetic field affects dynamics of quarks because the quarks are electrically charged particles. As a result, the external magnetic field enhances the chiral symmetry breaking by increasing the value of the (quark) chiral condensate [1–5]. The change in the dynamics of quarks is also felt by the gluon sector of QCD because the quarks are coupled to the gluons. Therefore, the magnetic field may affect the whole strongly interacting sector and influence very intrinsic properties of QCD such as, for example, the confinement of color [6–14].

In order to make a noticeable influence on the strongly interacting sector, the strength of the magnetic field should be of the order of a typical QCD mass scale, $eB \sim m_\pi^2$, where $m_\pi \approx 140$ MeV is a pion mass. The corresponding magnetic field strength, $B \sim 3 \times 10^{14}$ T, is enormous from a human perspective (1 T $\equiv 10^4$ G). However, such strong magnetic field can be achieved in noncentral heavy-ion collisions at Relativistic Heavy-Ion Collider (RHIC) [15]. At higher energies of Large Hadron Collider, noncentral heavy-ion collisions may generate even higher magnetic field of $eB \sim 15m_\pi^2$ ($B \sim 5 \times 10^{15}$ T) [15]. And in ultraperipheral collisions—when two nuclei pass near each other without a real collision—the

This work was supported by Grant No. ANR-10-JCJC-0408 HYPERMAG.

M.N. Chernodub (✉)
CNRS, Laboratoire de Mathématiques et Physique Théorique, Fédération Denis Poisson, Parc de Grandmont, Université François-Rabelais, 37200 Tours, France
e-mail: maxim.chernodub@lmpt.univ-tours.fr

M.N. Chernodub
Department of Physics and Astronomy, University of Gent, Krijgslaan 281, S9, 9000 Gent, Belgium

M.N. Chernodub
ITEP, B. Cheremushkinskaya 25, 117218 Moscow, Russia

D. Kharzeev et al. (eds.), *Strongly Interacting Matter in Magnetic Fields*,
Lecture Notes in Physics 871, DOI 10.1007/978-3-642-37305-3_6,
© Springer-Verlag Berlin Heidelberg 2013

magnetic field strength may reach $eB \sim (60 \dots 100)m_\pi^2$ or, in conventional units, $B \sim (2 \dots 3) \times 10^{16}$ T [16, 17].

Despite the magnetic field is generated in a heavy-ion collision for a very short time, it may have observable consequences. In noncentral collisions, the magnetic field is generated together with a hot expanding fireball of quark–gluon plasma. Topological QCD transitions may lead to a chiral imbalance of the plasma, and the chirally-imbalanced matter may produce electric current along the axis of the magnetic field [18, 19] driven by the chiral magnetic effect [20–23].

In a finite-density (quark) matter the magnetic catalysis [1–5] may be reversed [24, 25] and the phase diagram may be modified [24–26] substantially. In the absence of matter (i.e., in the vacuum), the external magnetic field affects the finite-temperature phase structure of the theory by shifting the critical temperatures and affecting the strength of the confinement–deconfinement and chiral transitions [6–14].

The vacuum may also spontaneously become an electromagnetic superconductor if the magnetic field strength exceeds the critical value [27, 28]

$$B_c \simeq 10^{16} \text{ Tesla} \quad \text{or} \quad eB_c \simeq 0.6 \text{ GeV}^2. \tag{6.1}$$

This counterintuitive effect should be realized in the absence of matter in a cold vacuum. Moreover, the superconductivity of, basically, empty space, should always be accompanied by a superfluid component [29–31]. We discuss these effects below.

In Sect. 6.2 we describe the mechanism and the basic features of the vacuum superconductivity in a very qualitative way. We compare in details the vacuum superconductivity with an ordinary superconductivity. We also highlight certain similarities of this exotic vacuum phase with a magnetic-field-assisted "reentrant superconductivity" in condensed matter and the electric-field-induced Schwinger electron-positron pair production in the vacuum of Quantum Electrodynamics.

In Sect. 6.3 the emergence of the superconducting phase is demonstrated both in a bosonic ρ-meson electrodynamics [32] and in an extended Nambu–Jona-Lasinio model [33, 34]. Various properties of the superconducting state are summarized in the last Section.

6.2 Conventional Superconductivity, Vacuum Superconductivity and Schwinger Pair Creation: Differences and Similarities

6.2.1 Conventional Superconductivity via Formation of Cooper Pairs

Before going into details of the magnetic-field-induced vacuum superconductivity let us discuss basic qualitative features of a conventional superconductivity. Why certain compounds are superconductors?

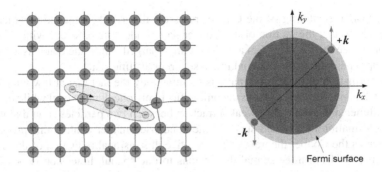

Fig. 6.1 (*Left*) Formation of the Cooper pair (the *yellowish oval*) of electrons (the *small green circles*) in an ionic lattice (the *large red circles*) due to phonon interaction. (*Right*) Two interacting electrons (the *small green circles*) and the Fermi sphere (the *large blue circle*) in the momentum space. The electrons in the Cooper pair have mutually opposite spins (the *green arrows*) and momenta (Color figure online)

In a simplified picture, electrons in a metal can be considered as (almost free) negatively charged particles which move through a periodically structured background of a lattice of positively charged ions (e.g., of a metal). The individual electrons scatter inelastically off the ions leading to a dissipation of an electric current and, consequently, to emergence of a nonvanishing electrical resistance in the metal.

As an electron moves through the ionic lattice it attracts neighboring ions via a Coulomb interaction. The attraction leads to a local deformation of the ionic lattice, and, simultaneously, to an excess of the positive electric charge in a vicinity of the electron. The excess of the positive charge, in turn, attracts another electron nearby, so that in a background of the positively charged ion lattice the like-charged electrons may experience a mutual attractive force, Fig. 6.1(left). The deformation of the ionic lattice can be viewed as a superposition of collective excitations of the ion lattice (phonons), so that the process of the electron–electron interaction via lattice deformations can be described by a phonon exchange.

The attractive force between the electrons may, in principle, lead to formation of electron–electron bound states. However, this attraction is extremely weak and therefore thermal fluctuations at room temperature easily destroy the two-electron bound states. On the other hand, at low temperature the attractive interaction between the electrons prevails the thermal disorder and, consequently, the bound states may indeed be formed. These bound state are called the Cooper pairs. The electrons in the Cooper pairs have mutually opposite spins and opposite momenta thus making the Cooper pair a (composite) spin-zero bosonic state. Bosons have a tendency to condense at low temperature so that the Cooper pair condensate may emerge.

In the condensed state all Cooper pairs behave as one collective entity. The Cooper-pair condensate can move frictionlessly through the ion lattice similarly to a motion of a superfluid. The motion without dissipation is guaranteed due to an energy gap, which separates the energy of the condensed ground state and an excited state with a lowest possible energy. Since the intermediate states are absent, at certain conditions (low enough temperature, weak enough electric current, etc.)

the dissipative scattering of the Cooper pairs off the ions becomes kinematically impossible so that the motion of the Cooper pair condensate proceeds without dissipation. Since the Cooper pairs are electrically charged states, the condensation of the Cooper pairs turns the material into a superconducting state.

The formation of the Cooper pairs is facilitated by the fact that at low temperature the dynamics of the electrons becomes effectively one dimensional while in one spatial dimension even a very weak attraction between two particles should always lead to formation of a bound state (in the condensed matter context this property is known as the Cooper theorem). The effective dimensional reduction of the electron dynamics from three spatial dimensions to one spatial dimension is possible because at low temperature the interaction between the electrons occurs if and only if the electrons are sufficiently close to the Fermi surface. One momentum coordinate of the Fermi surface counts the degeneracy of the electron states while another coordinate is dynamical. The Cooper pair is formed by two electrons with mutually opposite momenta and mutually opposite spins, Fig. 6.1(right).

Summarizing, in order to exhibit the conventional superconductivity a system should satisfy the following basic requirements:

(A) electric charge carriers should be present in the system (otherwise the system cannot support the electric current);
(B) dynamics of the electric charge carriers should effectively be one-dimensional (otherwise the Cooper pairs cannot be formed);
(C) the like-charged carriers should experience mutual (pairwise) attraction (otherwise the Cooper pairs cannot be formed).

Surprisingly, the same requirements are satisfied by vacuum in a background of a sufficiently strong magnetic field. In the next section we compare basic features of conventional and "vacuum" superconductors.

6.2.2 Vacuum Superconductivity

6.2.2.1 Condition A: Presence of Electric Charges

In order for the vacuum to behave as an electromagnetic superconductor one needs, at least, the presence of electrically charged particles in the superconducting phase of the vacuum (**condition A** on page 146). From a first sight, it is impossible to satisfy this requirement because under the usual conditions the vacuum is characterized by the absence of the free electric charges. Nevertheless, the quantum vacuum may be considered as an excellent "reservoir" of various particles including the electrically charged ones while under certain conditions the virtual particles may become real.

This "virtual-to-real" scenario does not sound unlikely. For example, there are at least two well-known cases of external conditions when a vacuum becomes an electrically conducting media: the vacuum may conduct electricity if it is either subjected under a strong electric field or if it is sufficiently hot.

Table 6.1 Conventional superconductivity vs. vacuum superconductivity: very general features (from Ref. [36])

Property	Conventional superconductivity	Vacuum superconductivity
Environment	a material (metal, alloy etc.)	vacuum (empty space)
Reservoir of carriers	real particles	virtual particles
Normal state	a conductor	an insulator
Basic carriers of electric charge	electrons (e)	light quarks (u, d) and light antiquarks $(\bar{u}, \bar{d})$
Electric charges of basic carries	$q_e = -e\ (e \equiv |e|)$	$q_u = +2e/3, q_d = -e/3$ $q_{\bar{u}} = -2e/3, q_{\bar{d}} = +e/3$

The first example of the "virtual-to-real" transition is the Schwinger effect in Quantum Electrodynamics (QED): a sufficiently strong external electric field generates electron-positron pairs out of the vacuum [35]. The created positrons and electrons move in opposite directions thus creating an electric current.[1] The critical strength of the electric field required for this process is $E_c = m_e^2/e \approx 10^{18}$ V/m.

The second "virtual-to-real" example is a simple thermal ionization of electron–positron pairs: the vacuum turns into an electron–positron plasma at temperatures $T \sim 0.1\ T^{\mathrm{QED}}$ where $T^{\mathrm{QED}} \approx 2m_e \approx 1$ MeV $\approx 10^{10}$ K is a typical QED temperature.

Thus, the quantum vacuum may be turned into a conductor if it is subjected to sufficiently strong electric field ($E \sim 10^{18}$ V/m) or to sufficiently high temperature ($T \sim 10^9$ K). Below we show that a sufficiently strong magnetic field ($B \sim 10^{16}$ T) may turn the vacuum into a superconducting state. The magnetic-field-induced vacuum superconductivity works at the QCD scale: the key role here is played by virtual quarks and antiquarks which have fractional electric charges. As we discuss below, the strong magnetic field catalyses the formation of the electrically charged condensates made from the quarks and antiquarks. Very general features of a conventional superconductor and the magnetic-field-induced vacuum superconductivity are summarized in Table 6.1.

6.2.2.2 Conditions B and C: Formation of Superconducting Carriers

In order for the superconducting carriers to be formed, the fermion dynamics should be reduced from three spatial dimensions to one spatial dimension (**condition B** of Sect. 6.2.2.1, page 146). In conventional superconductivity the dimensional reduction proceeds via formation of the Fermi surface at sufficiently low temperatures. This mechanism cannot work in our case because the Fermi surface, obviously, does not exist in the vacuum due the absence of matter. However, the dimensional reduction may be achieved with the help of a magnetic field background since electrically

---

[1] We briefly discuss an analogy between the magnetic-field-induced vacuum superconductivity and the Schwinger effect in Sect. 6.2.2.5, page 152.

charged particles with low-energy can move only along the axis of the magnetic field. This effect leads to the required dimensional reduction of the charge's dynamics from three to one spatial dimensions.

The described dimensional reduction effect in the background of the external magnetic field works for all electrically charged elementary particles, including electrons, positrons, quarks, antiquarks etc. However, the superconducting bound state may only be formed from a particular combination of these particles which should satisfy the following conditions:

(i) the superconducting bound state should be a boson;
(ii) the bound state should be electrically charged;
(iii) the interaction between the constituents of the bound state should be attractive.

Condition (i) implies that the superconducting bound state should contain even number of constituents because the known carriers of the electric charge are fermions (quarks, electron and positron, etc.). We consider a simplest two-fermion states.

Condition (ii) implies that the bound state cannot be composed of a particle and its antiparticle. In combination with condition (iii) it means that the vacuum superconductivity cannot—unlike the Schwinger's pair creation—emerge in the pure QED vacuum sector which describes electrons, positrons and photons. Indeed, the electron–electron interaction is mediated by a repulsive photon exchange [condition (iii) is not satisfied]. The interaction between electron and positron is attractive, but the electron–positron bound state is electrically neutral [condition (ii) is not satisfied]. Thus, the superconductivity cannot emerge in the pure QED.

Therefore, the candidates for the superconducting charged bound states should be outside of the purely electrodynamics sector. Below we concentrate on the next (in terms of energy scale), strongly interacting sector which describes the dynamics of quarks and gluons.

The QCD sector of the vacuum contains the gluon particle which is a carrier of the strong force. From our perspective the gluon is an analogue of the phonon of conversional superconductivity since the gluon may provide an attractive interaction between quarks and antiquarks regardless of their electric charges. In particular, the gluon may bind a quark and an anti-quark into an electrically charged meson. The attractive nature of the gluon interaction allows us to satisfy **condition C** of superconductivity on page 146.

Thus, the suggested mechanism of the vacuum superconductivity may indeed work at the interface of the QED and QCD sectors. The simplest example of the superconducting carrier may be given by a bound state of a u quark with the electric charge $q_u = +2e/3$ and a $\bar{d}$ antiquark with the electric charge $q_{\bar{d}} \equiv -q_d = +e/3$. The attractive nature of the gluon-mediated interaction between the quark and antiquark of different flavors is only possible if these constituents reside in a triplet state, so that the $u\bar{d}$ bound state should be a spin-1 state (the ρ meson).

Therefore, the vacuum analogue of the Cooper pair are the charged $\rho^{\pm}$ meson states. And in next sections we show that the ρ-meson condensates do indeed appear in the vacuum in background of the strong magnetic field, and we argue that the emergent state is indeed an electromagnetic superconductor.

Table 6.2 Conventional superconductivity vs. vacuum superconductivity: superconducting carriers

Property of carrier	Conventional superconductivity	Vacuum superconductivity
Type	Cooper pair	ρ-meson excitations, ρ^+ and ρ^-
Composition	electron–electron state (ee)	quark–antiquark states ($\rho^+ = u\bar{d}$ and $\rho^- = d\bar{u}$)
Electric charge	$-2e$	$+e$ and $-e$, respectively
Spin	typically spin-zero state (scalar)	one-one state (vector)
The carriers are formed due to	1) reduction of dynamics of basic electric charges from three spatial dimensions to one dimension, $3d \rightarrow 1d$	
	2) attraction force between two electrons	2) attraction force between a quark and an antiquark
1) a reason for the reduction $3d \rightarrow 1d$	at very low temperatures electrons interact with each other near the Fermi surface	in strong magnetic field the motion of electrically charged particles is one dimensional
2) attraction is due to	phonons (lattice vibrations)	gluons (strong force, QCD)
Isotropy of superconducting properties	Yes: superconducting in all spatial directions	No: superconducting along the axis of the magnetic field, insulator in other directions

Notice that the formation of the bound state is facilitated by the dimensional reduction of the quark's dynamics in the background of the magnetic field (**condition B**). The dimensional reduction implies automatically a strong anisotropy of the suggested superconductivity since the electric charges (the quarks u and d and their antiquarks) may move only along the axis of the magnetic field. As a result, the superconducting charge carriers (the ρ mesons in our case) may also flow along the axis of the magnetic field only. Thus, the vacuum exhibit a superconducting property in the longitudinal direction (along the magnetic field) while in the two transverse directions the superconductivity of the vacuum should be absent.

It is interesting to note that due to the anisotropic superconducting properties the vacuum in the strong magnetic field acquires a very unusual optical property: the vacuum becomes as (hyperbolic) metamaterial which behaves as diffractionless "perfect lenses" [37].

In Table 6.2 we compare of certain basic features of the superconducting carriers in a conventional (low-temperature) superconductivity and in the vacuum (high-magnetic-field) superconductivity.

6.2.2.3 Counterintuitive Coexistence of Magnetic Field and Superconductivity due to Strong Anisotropy of Magnetic-Field-Induced Superconductivity

So far we have ignored a well-known property of all known superconductors:

• A very magnetic field and conventional superconductivity cannot coexist!

Table 6.3 A comparison of the effects of magnetic field and thermal effects on conventional superconductivities and electromagnetic superconductivity of vacuum (from Ref. [36])

Property	Conventional superconductivity	Vacuum superconductivity
Magnetic field	destroys superconductivity	induces superconductivity
The Meissner effect	present	absent
Thermal fluctuations	destroy superconductivity	destroy superconductivity

Thus, we can ask ourselves: why do we believe that the superconducting phase of the vacuum can exist in (and, moreover, be induced by) the strong magnetic field? In fact, this single question contains two puzzles (Table 6.3):

- Why the Meissner effect is absent in the superconducting phase of vacuum?
- Why strong magnetic field does not destroy the superconductivity of vacuum?

A short "technical" answer to these questions is that in the background of the magnetic field the superconducting state of the vacuum has lower energy compared to the energy of the normal (insulator) state (Sect. 6.3.1, page 153). A physical argument is that the strong magnetic field may coexist with the vacuum superconductivity because the latter is highly anisotropic. Let us consider this point in detail.

Qualitative arguments against the Meissner effect in the vacuum superconductor are as follows. The Meissner effect is the screening of weak external magnetic field by a superconducting state so that a magnetic field cannot penetrate deep into a superconductor. Qualitatively, the Meissner effect is caused by superconducting currents which are induced by the external magnetic field in the bulk of a superconductor. The circulation of these currents in the transversal (with respect to the magnetic field axis) plane generates a backreacting magnetic field, which screens the external magnetic field in the bulk of the superconducting material. The backreacting currents are geometrically large, so that the corresponding magnetic length (i.e., the radius of the lowest Landau level), $1/\sqrt{|eB|}$, is much larger than the correlation length ξ of the superconductor. Since the vacuum superconductivity is realized only along the axis of the magnetic field, the large transversal currents are absent and the Meissner effect cannot be realized.

If the axis of the external magnetic field is oriented along the normal to a boundary of an ordinary superconductor, then the backreacting magnetic field squeezes the external magnetic field into thin Abrikosov vortices[2] which form a sparse vortex lattice in a background of weak magnetic field. As we show below (Sect. 6.3.4.4, page 166), the superconducting ground state of the vacuum is the dense lattice of the Abrikosov-type vortices for which the magnetic length is of the order of (or even smaller than) the correlating length. This is a quantum regime of the Abrikosov lattice, in which the geometrically short transverse currents in the cannot screen geometrically large external magnetic field. In this case the physical situation is similar

---

[2]Since the magnetic flux coming through the superconductor's boundary is a conserved quantity, the superconductor expels it from the superconductor's bulk into thin vortexlike structures.

to a "reentrant superconductivity" of extreme type-II superconductor in a high magnetic field [38–40] (Sect. 6.2.2.6, page 152).

If the axis of the weak external magnetic field is oriented tangentially to a boundary of an ordinary superconductor then the external field is expelled from the superconductor's bulk without formation of the Abrikosov vortices. In our case, the superconducting ground state is created by a strong magnetic field, therefore an imposition of a weak tangential magnetic field is impossible from simple geometrical reasons: a sole result of the superposition of the weak "testing" magnetic field onto the strong "creating" magnetic field is a slight turn of the stronger field. In other words, the weak testing field should slightly reorientate the anisotropy axis of the superconductor without destroying it.

The ordinary superconductivity is destroyed by sufficiently strong magnetic field. Qualitatively, one can understand this effect as follows: in strong enough external magnetic field the (positive) excess in energy of the induced transverse superconducting currents prevails the (negative) condensation energy of the superconducting carriers. As a result, at certain critical field the conventional superconductivity becomes energetically unfavorable and the material turns from the superconducting state back to the normal (nonsuperconducting) state. On the contrary, in the vacuum superconductivity the large superconducting currents are absent due to strong anisotropy of the superconducting currents, so that the mentioned argument should not work. Moreover, as we discuss below, the energy of the short transverse currents is diminished as the magnetic field becomes stronger.

Thus, the electromagnetic superconductivity of the vacuum coexists with high magnetic field due to the anisotropy of the magnetic-field-induced superconducting properties. It is the anisotropy which makes the vacuum superconductivity to be different from the conventional one.

6.2.2.4 Magnetic-Field-Induced Vacuum Superconductivity: Temperature Effects

The common key element of the ordinary and vacuum superconductivities is the dimensional reduction of the dynamics of the charge carriers (condition B on page 146). Thermal fluctuations should destroy this property regardless of the mechanism of the dimensional reduction. In ordinary superconductivity, if the energy of the thermal fluctuations becomes of the order of the Fermi energy then the Fermi surface broadens and the dimensional reduction no more works.

The thermal fluctuations should destroy the vacuum superconductivity because of the same reason as in the ordinary superconductivity: the loss of the dimensional reduction (Table 6.3). Indeed, the one-dimensional motion of electric charges in strong magnetic field is due to the fact that the electric charges occupy the lowest Landau level which is localized in the transverse plane. The one-dimensional motion can only be spoiled by transitions of the particles to higher Landau levels which are less localized. Generally, for a typical gap between the Landau energy

Fig. 6.2 Schematic plot of
the QCD phase diagram in
the presence of magnetic field
in the low temperature region

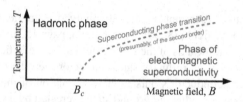

levels is expected to be of the QCD scale, $\delta E \sim \Lambda_{\mathrm{QCD}} \approx 100$ MeV, so that thermal fluctuations of a typical QCD scale, $T \sim \Lambda_{\mathrm{QCD}} \approx 100$ MeV should destroy the dimensional reduction property.

Therefore, we conclude that the superconductivity should be lost at certain critical temperature, $T_c \equiv T_c(B)$. At the critical magnetic field, $B = B_c$, the critical temperature is zero, $T_c(B_c) = 0$. The corresponding phase diagram is schematically shown in Fig. 6.2: the superconducting and hadronic phases are separated by a phase transition of (presumably) second order [28].

6.2.2.5 Electric-Field-Induced Pair Production (the Schwinger Effect) and Magnetic-Field-Induced Superconductivity: A Comparison

The Schwinger effect is a generation of the electron–positron pairs from the vacuum in a background of a strong enough *electric* field [35]. The created particles form a momentary electric current which tend to screen the external electric field which has created them. The electron–positron pair production is a process which is described entirely by the QED sector of the vacuum.

The vacuum superconductivity is associated with the emergence of the electrically charged quark–antiquark condensates out of vacuum provided the vacuum is subjected to the strong enough *magnetic* field [27, 28]. Contrary to the Schwinger effect, these electrically charged condensates do not screen the external magnetic field which has created them.

Following Ref. [36], in Table 6.4 we compare the very basic features of the Schwinger effect and the vacuum superconductivity.

6.2.2.6 Electromagnetic Superconductivity of Vacuum and "Reentrant Superconductivity" in Strong Magnetic Field

Unexpectedly, the magnetic-field-induced electromagnetic superconductivity of vacuum may have counterparts in certain condensed matter systems. It was suggested in Refs. [38–40] that in a very strong magnetic field the Abrikosov flux lattice of a type-II superconductor may enter a quantum limit of the "low Landau level dominance", characterized by a spin-triplet pairing, absence of the Meissner effect, and a superconducting flow along the magnetic field axis. The mentioned quantum limit is reached when the magnetic length $1/\sqrt{|eB|}$ becomes of the order of (or smaller than) the correlation length ξ.

Table 6.4 Basic features of the Schwinger effect and the electromagnetic vacuum superconductivity

Property	Schwinger effect	Vacuum superconductivity
Environment	vacuum	vacuum
Background of	strong electric field, E	strong magnetic field, B
Interactions involved	electromagnetic (QED) only	electromagnetic (QED) and strong (QCD)
Typical energy scales	megaelectronvolts (10^6 eV)	gigaelectronvolts (10^9 eV)
Critical value	$E_c = m_e^2/e \approx 10^{18}$ V/m ($m_e = 0.511$ MeV is electron mass)	$B_c = m_\rho^2/e \approx 10^{16}$ T ($m_\rho = 0.775$ GeV is ρ-meson mass)
Nature of the effect	virtual electron-positron $(e^- e^+)$ pop up from the vacuum and become real $e^- e^+$ pairs	virtual quark–antiquark pairs ($u\bar{u}$ and $d\bar{d}$) pop up and form real $u\bar{d}$ and $d\bar{u}$ condensates
Backreaction	created $e^- e^+$ pairs tend to screen the external field	created $u\bar{d}$ and $d\bar{u}$ condensates do not screen the external field
Stability	a process (unstable)	a ground state (stable)
Transport property	an electromagnetic conductor: electric current is generated	a superconducting state

In condensed matter, the magnetic-field-induced anisotropic superconductivity is sometimes called the "reentrant" superconductivity because the system should normally "exit" a superconducting state as an increasing external magnetic field suppresses superconductivity via diamagnetic and Pauli pair breaking effects. Although it is unclear whether this particular mechanism of the reentrant superconductivity works in real superconductors, the restoration of the superconducting properties was experimentally observed in certain materials like an uranium superconductor URhGe [41–43].

Our proposal [27, 28] of the vacuum superconductivity has basically the same features as the reentrant superconductivity [38–40]: the electrically charged condensates correspond to a spin-one quark–antiquark states (ρ mesons), the vacuum superconductor exhibits no Meissner effect while the vacuum superconductivity is highly anisotropic.

6.3 Ground State of Vacuum Superconductor

6.3.1 Energetic Favorability of the Superconducting State

As we have argued in the previous section, a quark–antiquark pair of different flavors may condense in sufficiently strong magnetic field. How strong should the relevant magnetic field be? Following Ref. [27], let us make a simple estimation of the critical magnetic field B_c using very general arguments.

An electromagnetic superconductivity emerges when an electrically charged field starts to condense. Assume that we have a free relativistic particle with mass m, electric charge e and spin s. In a background of a constant uniform magnetic field B the relativistic energy levels ε of this particle are given by the following formula:

$$\varepsilon_{n,s_z}^2(p_z) = p_z^2 + (2n - gs_z + 1)|eB| + m^2, \tag{6.2}$$

where $n \geqslant 0$ is the nonnegative integer, $s_z = -s, \ldots, s$ is the projection of the spin s on the field's axis, p_z is the particle's momentum along the field's axis and g is the gyromagnetic ratio (or, "g-factor") of the particle.

Let us consider quark–antiquark bound states made of lightest, u and d quarks. The corresponding simplest spin-zero and spin-one bound states are called π and ρ mesons, respectively. In our simplest estimation we ignore the internal structure (formfactors) of the mesons treating them as pointlike bound states.

The ground state energy (or, "mass") of the $\pi^\pm$ mesons in the background of the magnetic field corresponds to the quantum numbers $p_z = 0$ and $n_z = 0$ (notice that $s_z \equiv 0$ since the π meson is a spineless particle):

$$m_{\pi^\pm}^2(B) = m_\pi^2 + |eB|, \tag{6.3}$$

where $m_\pi = 139.6\,\text{MeV}$ is the mass of the π meson in the absence of the magnetic field. The mass of the neutral π^0 meson is insensitive to the external magnetic field in our approximation, $m_{\pi^0}(B) = m_\pi$ (here we ignore a small splitting between masses of charged and neutral π mesons at $B = 0$).

Analogously, the ground state energy ("mass") of the charged $\rho^\pm$ mesons correspond to the quantum numbers $p_z = 0$, $n_z = 0$ and $s_z = 1$:

$$m_{\rho^\pm}^2(B) = m_\rho^2 - |eB|, \tag{6.4}$$

where $m_\rho = 775.5\,\text{MeV}$ is the mass of the charged ρ meson in the absence of the magnetic field. The mass of the neutral ρ^0 meson is a B-independent quantity in this approximation, $m_{\rho^0}(B) = m_\rho$ (a small difference in masses of charged and neutral ρ mesons at $B = 0$ is again ignored).

It is important to mention that in (6.4) the gyromagnetic ratio of the ρ meson is taken to be $g = 2$. This anomalously large value was independently obtained in the framework of the QCD sum rules [44–46] and in the Dyson–Schwinger approach to QCD [47]. It was also conformed by the first-principle numerical simulations of lattice QCD [48, 49] providing us with a value $g \approx 2$. The condensation of the charged ρ mesons in the vacuum of QCD is very similar to the Nielsen–Olesen instability of the pure gluonic vacuum in Yang–Mills theory [50] and to the magnetic-field-induced Ambjørn–Olesen condensation of the W-bosons in the vacuum of standard electroweak model [51–53]. Both the ρ mesons in QCD, the gluons in Yang–Mills theory, and the W bosons in the electroweak model have the anomalously large g-factor, $g \approx 2$. Notice, that the phase diagrams of QCD at finite density (at finite chemical and/or isospin potential) contain certain phases characterized by the presence of exotic vector condensates [26, 54–59]. Some of these phases exhibit superconducting/superfluid behavior [56–59].

In the absence of the magnetic field background the ρ meson is a very unstable particle. One can notice, however, that the ground-state mass of the charged π meson is an increasing function of the magnetic field strength B while the mass of the charged ρ mesons is a decreasing function of B. As all known modes of the $\rho^{\pm}$-meson decays proceed via emission of $\pi^{\pm}$ mesons, $\rho^{\pm} \to \pi^{\pm} X$ [60], it is clear that at certain strength of magnetic field the fast hadronic decays of the ρ meson become forbidden due to simple kinematical arguments (the mass of the would-be decay products exceed the mass of the ρ meson itself). Thus, in a background of sufficiently strong magnetic field the charged ρ meson should be stable against all known hadronic decay modes [27].

The presence of the superconducting ground state at high magnetic field can be seen as follows. The square of mass of the charged ρ meson should decrease as the magnetic field B increases, (6.4). When the magnetic field reaches the critical value (6.1), the ground state energy of the $\rho^{\pm}$ mesons becomes zero. As the magnetic field becomes even stronger, the ground state energy becomes a purely imaginary quantity indicating the presence of a tachyonic instability of vacuum. In other words, the trivial ground state, $\langle \rho \rangle = 0$, is no more stable at $B > B_c$, and the vacuum should slide towards a new state with a nonzero ρ-meson condensate, $\langle \rho \rangle \neq 0$. Since the condensation of the electrically charged field indicates the presence of electromagnetic superconductivity, the vacuum should become a superconductor at $B > B_c$.

6.3.2 Approaches: Ginzburg–Landau vs Bardeen–Cooper–Schrieffer

The condensation of the ρ mesons in QCD in the background of the strong magnetic field may be treated in the same way as the condensation of the Cooper pairs in the conventional superconductivity. The conventional superconductivity may be described in the framework of both microscopic fermionic models and macroscopic bosonic theories.

The fermionic models of the Bardeen–Cooper–Schrieffer (BCS) type describe basic carriers of electric charge (electrons and/or holes), and these models are, generally, nonrenormalizable because their Lagrangians include a four-fermion interaction term. The bosonic models of the Ginzburg–Landau (GL) type are usually based on renormalizable effective Lagrangians which describe superconducting excitations (the Cooper pairs) [61].

Despite of the fact that fermionic and bosonic approaches are formulated in a very different way, they both can describe the superconductivity at a good quantitative level. Moreover, the bosonic and fermionic approaches are mathematically equivalent near the superconducting phase transition [62]. In Table 6.5 we outline a correspondence between the traditional (BCS and GL) models of conventional superconductivity and their vacuum counterparts which are used to describe the magnetic-field-induced vacuum superconductor.

Table 6.5 Simplest models which can be chosen to study physical properties of conventional and vacuum superconductivities

Basic field describes...	Conventional	Vacuum
Bosonic, condensed superconducting carriers	Ginzburg–Landau model [63]	ρ-meson electrodynamics [32] based on vector dominance model [64]
Fermionic constituents of superconducting carriers	Bardeen–Cooper–Schrieffer model [65]	Nambu–Jona-Lasinio model [66] extended with vector interactions [33, 34]

It is important to notice that in the effective GL approach the Cooper pairs are treated as pointlike particles. However, physically the Cooper pairs are rather non-local objects because their size is much larger than the average distance between electrons in metals. Nevertheless, the GL model describes superconductivity very well especially near the second-order phase transition where the symmetries of the system dominate its dynamics according to the universality argument.

Following our experience in the conventional superconductivity, in next sections we describe a GL like approach to the vacuum superconductivity using a model which describes a ρ-meson sector of vacuum [32]. Then, we briefly outline a Bardeen–Cooper–Schrieffer approach to the vacuum superconductor using a well-known extension [33, 34] of the Nambu–Jona-Lasinio model [66]. We would like to mention that signatures of the vacuum superconductor were also found in holographic effective theories [67, 68] and in numerical ("lattice") approaches to QCD [69].

6.3.3 Example: Ginzburg–Landau Model

Before going into the details of the ρ condensation in QCD, it is very useful to outline a few basic properties of conventional superconductivity in the GL model which provides us with a simplest phenomenological description of the conventional superconductivity.

6.3.3.1 The Relativistic Version of the Ginzburg–Landau Lagrangian

Basic properties of a conventional superconductor can be described by the following GL Lagrangian:

$$\mathscr{L}_{GL} = -\frac{1}{4}F_{\mu\nu}F^{\mu\nu} + (\mathfrak{D}_{\mu}\Phi)^*\mathfrak{D}^{\mu}\Phi - \lambda\big(|\Phi|^2 - \eta^2\big)^2, \qquad (6.5)$$

where $\mathfrak{D}_\mu = \partial_\mu - ieA_\mu$ is the covariant derivative and Φ is the complex scalar field carrying the electric charge[3] e.

The superconducting ground state of the GL model, $\eta^2 > 0$, is characterized by the homogeneous condensate of the scalar field, $\langle\Phi\rangle$ with $|\langle\Phi\rangle| = \eta$. It is the condensate $\langle\Phi\rangle$ which is responsible for the superconductivity.

In the superconducting phase the mass of the scalar excitation, $\delta\Phi = \Phi - \langle\Phi\rangle$, and the mass of the photon field A_μ are, respectively, as follows:

$$m_\Phi = \sqrt{4\lambda}\eta, \qquad m_A = \sqrt{2}e\eta. \tag{6.6}$$

The classical equations of motion of the GL Lagrangian (6.5) are as follows:

$$\mathfrak{D}_\mu \mathfrak{D}^\mu \Phi + 2\lambda(|\Phi|^2 - \eta^2)\Phi = 0, \tag{6.7}$$

$$\partial_\nu F^{\nu\mu} + J^\mu_{GL} = 0, \tag{6.8}$$

where the electric current is

$$J^\mu_{GL} = -ie[\Phi^* \mathfrak{D}^\mu \Phi - (\mathfrak{D}^\mu \Phi)^* \Phi]. \tag{6.9}$$

Thermal fluctuations make the condensate $\langle\Phi\rangle$ smaller, eventually destroying the superconductivity at certain critical temperature, $T = T_c$. In order to describe this effect in the GL approach, one usually assumes that the quadratic coefficient of the potential term $V_{GL} = \lambda(|\Phi|^2 - \eta^2)^2$ in the GL Lagrangian (6.5) exhibits a linear dependence on temperature T:

$$V_{GL}(\Phi) = \alpha_0(T - T_c)|\Phi|^2 + \lambda|\Phi|^4 + \text{const}. \tag{6.10}$$

The superconducting condensate is present at $T < T_c$, while $\langle\Phi\rangle = 0$ at $T \geqslant T_c$.

6.3.3.2 Magnetic Field Destroys Conventional Superconductivity

In a background of a sufficiently strong magnetic field, $B > B_c^{GL}$, the superconducting condensate disappears. The corresponding critical value of the magnetic field,

$$B_c^{GL} = \frac{m_\Phi^2}{2e} \equiv \frac{2\lambda}{e}\eta^2, \tag{6.11}$$

can be obtained as follows. Consider a near-critical case when the uniform time-independent magnetic field $B \equiv F_{12}$ is slightly weaker than the critical value (6.11), $0 < B_c^{GL} - B \ll B_c^{GL}$. As the magnetic field approaches the critical value (6.11), the superconducting condensate becomes very small,

$$|\langle\Phi\rangle(B)| \ll \eta, \tag{6.12}$$

---

[3]Without loss of generality we consider the singly-charged bosons Φ instead of the usual doubly-charged Cooper pairs and we use a relativistic description of superconductivity.

and, consequently, the classical equation of motion (6.7) can be linearized:

$$\left\{(\mathfrak{D}_1 - i\mathfrak{D}_2)(\mathfrak{D}_1 + i\mathfrak{D}_2) + e\left[B_c^{GL} - B(x)\right]\right\}\Phi = 0. \tag{6.13}$$

where B_c^{GL} is given in (6.11) and $B(x)$ is the magnetic field inside the superconductor (it appears due to a rearrangement of the derivatives of the first term of this equation).

In the vicinity of the critical magnetic field, $B \simeq B_c$, so that the second term in (6.13) can be neglected and (6.13) is satisfied if

$$\mathfrak{D}\Phi \simeq 0 \quad \text{with } \mathfrak{D} = \mathfrak{D}_1 + i\mathfrak{D}_2. \tag{6.14}$$

6.3.3.3 Lattice of Abrikosov Vortices in Background of Magnetic Field

If the strength of the external magnetic field is smaller then the critical value (6.11) then the superconductor may squeeze the magnetic field into the vortexlike structures which are known as the Abrikosov vortices [70]. The Abrikosov vortex is a topological stringlike solution to the classical equations of motion (6.7) and (6.8). A single Abrikosov vortex carries a quantized flux of the magnetic field,

$$\int d^2 x^\perp B(x^\perp) = \frac{2\pi}{e}, \tag{6.15}$$

where the integral of the vortex magnetic field B is taken over the two-dimensional coordinates $x^\perp = (x^1, x^2)$ of the plane which is transverse to the infinitely-long, strait and static vortex [notice, that the flux (6.15) is twice larger than a conventional value since we consider the condensed bosons Φ with the electric charge e and not $2e$].

The Abrikosov vortex has a well-defined center where the condensate Φ vanishes and the normal (nonsuperconducting) phase is restored. The phase of scalar field is singular at the vortex center. The behavior of the scalar field in the vicinity of the an elementary vortex, situating at the origin and carrying the flux (6.15), is as follows

$$\Phi(x^\perp) \propto |x^\perp| e^{i\varphi} \equiv x_1 + i x_2, \tag{6.16}$$

where φ is the azimuthal angle in the transverse plane, and $|x^\perp|$ is the distance from the vortex center. Equation (6.16) is valid provided $m_\Phi|x^\perp| \ll 1$ and $m_A|x^\perp| \ll 1$, where the mass parameters m_A and m_Φ are given in (6.6).

If the external magnetic field is strong enough but still weaker than the critical value (6.11), then multiple elementary Abrikosov vortices may be created. Parallel Abrikosov vortices repel each other in a type-II superconductor, for which the mass of the scalar excitation is larger then the photon mass, $m_\Phi > m_A$. Due to the mutual repulsion, these vortices arrange themselves in a regular periodic structure known as the Abrikosov lattice [61]. The Abrikosov lattice corresponds to the so called "mixed state" of the conventional superconductor, in which both normal phase (inside the vortex cores) and superconducting phase (outside the vortex cores) coexist.

There are various arrangements of vortices corresponding to different types of Abrikosov lattices [61]. The simplest Abrikosov lattice is given by the square lattice solution of (6.14),

$$\Phi = \Phi_0 K(z/L_B), \quad K(z) = e^{-\frac{\pi}{2}(|z|^2 + z^2)} \sum_{n=-\infty}^{+\infty} e^{-\pi n^2 + 2\pi n z}, \tag{6.17}$$

where Φ_0 is a dimensional complex parameter, $z = x^1 + ix^2$, and

$$L_B = \sqrt{2\pi}\ell_B, \quad \ell_B = \frac{1}{\sqrt{eB}}, \tag{6.18}$$

is the inter-vortex distance L_B which is expressed via the magnetic length ℓ_B. The area of an elementary lattice cell (i.e. of a square cell which contains one Abrikosov vortex) is $L_B^2 \equiv 2\pi\ell_B$. In the solution (6.17) the vortices are located at the sites of the square lattice,

$$\frac{x_i}{L_B} = n_i + \frac{1}{2}, \quad n_i \in \mathbb{Z}, \ i = 1, 2, \tag{6.19}$$

at which the condensate $\Phi(x_1, x_2)$ vanishes exactly. In the vicinity of these sites the scalar field (6.17) is described by (6.16).

The ground state of the system in the mixed phase is characterized, by definition, by a minimal energy of the vortex lattice. One can show that a global minimum of the energy density of the GL model (6.5) corresponds to a global minimum a convenient dimensionless quantity which is called the Abrikosov ratio [61]:

$$\beta_A = \frac{\langle |\phi|^4 \rangle}{\langle |\phi|^2 \rangle^2}. \tag{6.20}$$

It turns out that the ground state of the system corresponds an equilateral triangular lattice (which is sometimes also called "hexagonal" lattice) with the Abrikosov ratio β_A(Triang) ≈ 1.1596, Fig. 6.3(right). For the square Abrikosov lattice the Abrikosov ratio (6.20) is slightly higher, β_A(Square) ≈ 1.180, Fig. 6.3(left). Notice that despite very different visual appearances of these two lattices, the difference in their energies (and in the corresponding Abrikosov ratios, β_A) is of the order of a few percent. A Ginzburg–Landau description of the type-II superconductors in the background of magnetic field can be found in the nice review [71].

6.3.3.4 London Equations and Complex Electric Conductivity

The GL model (6.5) in the condensed phase describes a superconductor. Indeed, taking into account that in the ground state the condensate is a uniform time-independent quantity, one finds from the definition of the electric current (6.9):

$$\partial^\mu J_{GL}^\nu - \partial^\nu J_{GL}^\mu = -m_A^2 F^{\mu\nu}, \tag{6.21}$$

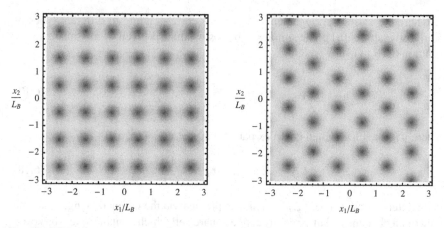

Fig. 6.3 Minimal-energy vortex lattices for square (*left*) and equilateral triangular (*right*) lattices in the Ginzburg–Landau model. The *dark dots* correspond to the positions of the Abrikosov vortices. The equilateral triangular lattice corresponds to the global minimum of the energy. From Refs. [30, 31]

where m_A is given in (6.6). Setting $\mu = 0$ and $\nu = 1, 2, 3$, in (6.21) we recover the first London relation for a locally neutral $[J_0(x) = 0]$ superconductor:

$$\frac{\partial \mathbf{J}_{\mathrm{GL}}}{\partial t} = m_A^2 \mathbf{E}, \tag{6.22}$$

where $\mathbf{E}$ or $E^i \equiv -F^{0i}$ is a weak (test) external electric field. Equation (6.22) describes a linear growth of the electric current in the constant electric field, thus indicating a vanishing electric resistance of the system. The superconducting properties described by (6.22) are homogeneous (independent of the spatial coordinates) and isotropic (independent of the direction of the electric field).

The London equation (6.22) corresponds to a singular part of the complex conductivity tensor $\sigma_{kl} = \mathrm{Re}\, \sigma_{kl} + i\, \mathrm{Im}\, \sigma_{kl}$. The conductivity tensor is defined as follows:

$$J_k(\mathbf{x}, t; \omega) = \sum_{k=1}^{3} \sigma_{kl}(\omega) E_l(\mathbf{x}, t), \tag{6.23}$$

where $\mathbf{E}(\mathbf{x}, t) = \mathbf{E}_0 e^{i(\mathbf{x} \cdot \mathbf{q} - \omega t)}$ is the alternating external current in the long-wavelength limit, $|\mathbf{q}| \to 0$. The London equation (6.22) indicates that $\sigma_{kl}(\omega) = \sigma_{kl}^{\mathrm{sing}}(\omega) + \sigma_{kl}^{\mathrm{reg}}(\omega)$, where the singular part comes from paired (superconducting) electrons,

$$\sigma_{kl}^{\mathrm{sing}}(\omega) = \frac{\pi m_A^2}{2}\left[\delta(\omega) + \frac{2i}{\pi \omega}\right]\delta_{kl}, \tag{6.24}$$

while regular part σ^{reg} accounts for other contributions to the conductivity.

6.3.3.5 Meissner Effect

The spatial components of (6.21) provide us with the second London relation:

$$\partial \times \mathbf{J}_{GL} = -m_A^2 \mathbf{B}.$$ (6.25)

In the absence of a background electric field ($\mathbf{E} = 0$), (6.8) implies $\mathbf{J}_{GL} = \partial \times \mathbf{B}$, so that (6.25) can now be reformulated as follows:

$$(-\Delta + m_A^2)\mathbf{B} = 0.$$ (6.26)

Equation (6.26) describes the Meissner effect: inside a superconductor the photon becomes massive so that an external magnetic field $B < B_c$ is expelled. Physically, the Meissner effect appears due to the fact that the external magnetic field induces circulating superconducting currents (6.25) which, in turn, generate their own magnetic field. As the generated field is directed in the opposite direction with respect to the external field, the magnetic field is eventually screened inside superconductor.

If the external magnetic field is directed tangentially with respect the superconductor's boundary then this field is always screened inside the bulk of the superconductor. However, if the external magnetic field is directed along a normal of the boundary of a type-II superconductor, then the magnetic flux—which is a conserved quantity—may penetrate the superconductor and create a mixed phase of the Abrikosov vortices (Sect. 6.3.3.3, page 158).

6.3.4 Superconductivity of Vacuum in Strong Magnetic Field

6.3.4.1 Electrodynamics of ρ Mesons

The conventional superconductivity is driven by the condensation of the Cooper pairs which are described by the local scalar field Φ in the Ginzburg–Landau approach. The superconductivity of vacuum in a sufficiently strong magnetic field is caused by emergence of quark–antiquark condensates which carry quantum numbers of (charged) ρ mesons [27]. Below we consider the electrodynamics of ρ mesons in strong magnetic field using the following effective Lagrangian [32]:

$$\mathcal{L} = -\frac{1}{4}F_{\mu\nu}F^{\mu\nu} - \frac{1}{2}(D_{[\mu},\rho_{\nu]})^{\dagger}D^{[\mu},\rho^{\nu]} + m_\rho^2 \rho_\mu^{\dagger}\rho^\mu$$
$$-\frac{1}{4}\rho_{\mu\nu}^{(0)}\rho^{(0)\mu\nu} + \frac{m_\rho^2}{2}\rho_\mu^{(0)}\rho^{(0)\mu} + \frac{e}{2g_s}F^{\mu\nu}\rho_{\mu\nu}^{(0)},$$ (6.27)

where the complex vector field $\rho_\mu = (\rho_\mu^{(1)} - i\rho_\mu^{(2)})/\sqrt{2}$ and the real-valued vector field $\rho_\mu^{(0)} \equiv \rho_\mu^{(3)}$, correspond, respectively, to the charged and neutral vector mesons with the bare mass m_ρ. The charged and neutral fields are made of components of the triplet of the ρ field:

$$\rho_\mu = \begin{pmatrix} \rho_\mu^{(1)} \\ \rho_\mu^{(2)} \\ \rho_\mu^{(3)} \end{pmatrix}. \tag{6.28}$$

The last term in (6.27) describes a nonminimal coupling of the ρ mesons to the electromagnetism via the field strength $F_{\mu\nu} = \partial_{[\mu,} A_{\nu]}$ of the photon field A_μ. The presence of the nonminimal coupling implies, in particular, the anomalously large value of the gyromagnetic ratio of the ρ meson, $g = 2$ (discussed already in Sect. 6.3.1, page 154). Both the covariant derivative $D_\mu = \partial_\mu + i g_s \rho_\mu^{(0)} - i e A_\mu$ and the strength tensor $\rho_{\mu\nu}^{(0)} = \partial_{[\mu,} \rho_{\nu]}^{(0)} - i g_s \rho_{[\mu,}^\dagger \rho_{\nu]}$ involve the $\rho\pi\pi$ coupling g_s which has the known phenomenological value of $g_s \approx 5.88$.

The model (6.27) enjoys the electromagnetic $U(1)$ gauge invariance:

$$\rho_\mu^{(0)}(x) \to \rho_\mu^{(0)}(x),$$

$$\rho_\mu(x) \to e^{i e \omega(x)} \rho_\mu(x), \tag{6.29}$$

$$A_\mu(x) \to A_\mu(x) + \partial_\mu \omega(x).$$

The ρ-meson Lagrangian (6.27) is an analogue of the Ginzburg–Landau Lagrangian (6.5), while the ρ-meson field (6.28) plays the role of the GL scalar field Φ. The electric current of ρ mesons is given by the following analogue of (6.9):

$$J_\mu = i e [\rho^{\nu\dagger} \rho_{\nu\mu} - \rho^\nu \rho_{\nu\mu}^\dagger + \partial^\nu (\rho_\nu^\dagger \rho_\mu - \rho_\mu^\dagger \rho_\nu)] - \frac{e}{g_s} \partial^\nu f_{\nu\mu}^{(0)}. \tag{6.30}$$

6.3.4.2 Instability of Vacuum: Potential Energy in Strong Magnetic Field

The energy density $\mathscr{E}$ of the ρ-meson ground state is given by the T_{00} component,

$$\mathscr{E} \equiv T_{00} = \frac{1}{2} F_{0i}^2 + \frac{1}{4} F_{ij}^2 + \frac{1}{2} (\rho_{0i}^{(0)})^2 + \frac{1}{4} (\rho_{ij}^{(0)})^2 + \frac{m_\rho^2}{2} [(\rho_0^{(0)})^2 + (\rho_i^{(0)})^2]$$

$$+ \rho_{0i}^\dagger \rho_{0i} + \frac{1}{2} \rho_{ij}^\dagger \rho_{ij} + m_\rho^2 (\rho_0^\dagger \rho_0 + \rho_i^\dagger \rho_i) - \frac{e}{g_s} F_{0i} \rho_{0i}^{(0)} - \frac{e}{2 g_s} F_{ij} \rho_{ij}^{(0)}, \tag{6.31}$$

of the energy–momentum tensor of the ρ-meson electrodynamics (6.27):

$$T_{\mu\nu} = 2 \frac{\partial \mathscr{L}}{\partial g^{\mu\nu}} - \mathscr{L} g_{\mu\nu}. \tag{6.32}$$

It is useful to consider a "homogeneous" approximation and ignore for a moment all derivatives and covariant derivatives in the energy density (6.31). This procedure corresponds, roughly speaking, to selection of a potential part of the energy density:

$$V\left(\rho_\mu, \rho_\mu^{(0)}\right) = \frac{1}{2}B^2 + \frac{g_s^2}{4} \sum_{\mu,\nu=0}^{4} \left[i\left(\rho_\mu^\dagger\rho_\nu - \rho_\nu^\dagger\rho_\mu\right)\right]^2 + ieB\left(\rho_1^\dagger\rho_2 - \rho_2^\dagger\rho_1\right)$$

$$+ \frac{m_\rho^2}{2} \sum_{\mu=0}^{4} \left(\rho_\mu^{(0)}\right)^2 + m_\rho^2 \sum_{\mu=0}^{4} \rho_\mu^\dagger\rho_\mu. \tag{6.33}$$

In the homogeneous approximation the ground state can be found via the minimization of the potential energy (6.33) with respect to the fields ρ_μ and $\rho_\mu^{(0)}$. It turns out that in this approximation the vacuum expectation value of the neutral ρ-meson field is zero, $\rho_\mu^{(0)} = 0$. The quadratic part of the charged field is the following:

$$V^{(2)}(\rho_\mu) = \sum_{a,b=1}^{2} \rho_a^\dagger \mathcal{M}_{ab}\rho_b + m_\rho^2\left(\rho_0^\dagger\rho_0 + \rho_3^\dagger\rho_3\right), \quad \mathcal{M} = \begin{pmatrix} m_\rho^2 & ieB \\ -ieB & m_\rho^2 \end{pmatrix}, \tag{6.34}$$

where the mass matrix $\mathcal{M}$ for the Lorentz components ρ_1 and ρ_2 is non-diagonal.

The eigenvalues $\mu_\pm$ and the corresponding eigenvectors $\rho_\pm$ of the mass matrix (6.34) are, respectively, as follows:

$$\mu_\pm^2 = m_\rho^2 \pm eB, \qquad \rho_\pm = \frac{1}{\sqrt{2}}(\rho_1 \mp i\rho_2). \tag{6.35}$$

It is clearly seen that one of mass states, either μ_- or μ_+ depending on the sign of eB, is getting smaller as the magnetic field increases. Taking for definiteness $eB > 0$, we chose the ground state of the system in the following form:

$$\rho_1 = \rho, \qquad \rho_2 = -i\rho, \qquad \rho_0 = 0, \qquad \rho_3 = 0. \tag{6.36}$$

The longitudinal components ρ_0 and ρ_3 are always zero because for any value of the magnetic field the corresponding terms in (6.34) are positive and diagonal.

The potential energy part of the ρ meson system can be calculated with the help of (6.33) and (6.36):

$$V(\rho) = \frac{1}{2}B^2 + 2e(B_c - B)|\rho|^2 + 2g_s^2|\rho|^4, \tag{6.37}$$

where

$$B_c = \frac{m_\rho^2}{e}, \tag{6.38}$$

is the critical magnetic field (6.1).

Thus, we get the familiar Mexican-hat potential (6.37) for the ρ-meson condensate ρ. In particular, the very same form of the potential appears in the GL model of superconductivity (6.10), with one very important exception: in the conventional superconductivity the condensation of the Cooper pairs emerges at low temperatures, $T < T_c$, while the ρ mesons start to condense in the vacuum in the presence of sufficiently strong magnetic fields, $B > B_c$:

$$|\langle \rho \rangle|_V = \begin{cases} \sqrt{\frac{e(B - B_c)}{2g_s^2}}, & B \geqslant B_c, \\ 0, & B < B_c. \end{cases} \tag{6.39}$$

Here the subscript V indicates that we consider the potential part $V(\Phi)$ only and we ignore all kinetic terms. If $B \geqslant B_c$ then the condensate (6.39) breaks spontaneously the electromagnetic symmetry group (6.29), similarly to the spontaneous symmetry breaking in the superconducting phase of the GL model.

6.3.4.3 Negative Condensation Energy due to ρ-Meson Condensate

The homogeneous nature of the ρ-meson condensate (6.39) and, consequently, of the ground state (6.36) is an artifact of the potential approximation. This simple approximation was useful for us to find the very presence of the tachyonic instability of the noncondensed state at $B > B_c$, while the detailed local structure of the ground state can only be revealed beyond the potential approximation. To this end we notice that a wavefunction of the lowest energy state of a free particle in a uniform static magnetic field is independent on the coordinate $z \equiv x^3$ along the magnetic field axis. Moreover, for the ground state, the dependence on the time coordinate $t \equiv x^0$ should appear only in a form of a trivial phase factor. These facts suggest us to concentrate on x^0- and x^3-independent solutions to the classical equation of motions for the ρ mesons, similarly to the case of the Abrikosov lattice solutions in the type-II superconductors. Technically, it is convenient to choose the complex coordinate $z = x^1 + ix^2$ where $x^\perp = (x^1, x^2)$ and define the complex variables $\mathcal{O} = \mathcal{O}_1 + i\mathcal{O}_2$ and their conjugates $\overline{\mathcal{O}} = \mathcal{O}_1 - i\mathcal{O}_2$ for vector quantities $\mathcal{O}_\mu$.

The classical equations of motion of the ρ-meson model (6.27) can be written in the following "complexified" form:

$$g_s \partial B + iem_0^2 \rho^{(0)} = 0, \tag{6.40}$$

$$(-\bar{\partial}\partial + m_0^2 + 2g_s^2|\rho|^2)\rho^{(0)} - 2ig_s\partial|\rho|^2 = 0, \tag{6.41}$$

$$[-\bar{D}D + 2(g_s C - eB + 2g_s^2|\rho|^2 + m_\rho^2)]\rho = 0, \tag{6.42}$$

where the z-components of the magnetic field and its analogue for the neutral ρ mesons are, respectively, as follows

$$B(z) \equiv \partial_1 A_2 - \partial_2 A_1 = \mathrm{Im}(\bar{\partial}A), \qquad C(z) \equiv \partial_1 \rho_2^{(0)} - \partial_2 \rho_1^{(0)} = \mathrm{Im}(\bar{\partial}\rho^{(0)}). \tag{6.43}$$

We have also introduced two covariant derivatives:

$$D \equiv D_1 + iD_2 = \mathcal{D} + ig_s\rho^{(0)}, \quad \mathcal{D} = \partial - ieA, \quad A = \frac{B_{\mathrm{ext}}}{2i}z, \tag{6.44}$$

where the external (background) magnetic field B_{ext} should be distinguished from the full magnetic field in the superconducting state, $B(x_1, x_2) \equiv B(x_1 + ix_2)$: in the latter a backreaction from the superconducting ground state is included.

The transverse components of the electric current (6.30) are as follows:

$$J^\perp \equiv J_1 + iJ_2 = 2ie\left(\rho^\dagger D\rho + \partial|\rho|^2\right) + i\frac{e}{g_s}\partial C. \qquad (6.45)$$

In the vicinity of the phase transition, $B_{ext} > B_c$ with $|B_{ext} - B_c| \ll B_c$, the equations of motion (6.40), (6.41) and (6.42) can be linearized. It turns out that the equation for the ρ-meson condensate in the overcritical magnetic field ($B_{ext} \gtrsim B_c$) in the vacuum is identical to the equation for the Cooper pair condensate (6.14) in the subcritical magnetic field ($B_{ext} \lesssim B_c$) in the GL model of conventional superconductivity [27]:

$$\mathcal{D}\rho \equiv \left(\partial - \frac{e}{2}B_{ext}z\right)\rho = 0. \qquad (6.46)$$

Among infinite number of solutions to (6.46), the ground state solution corresponds to the global minimum of the mean energy density (6.31). In the chosen approximation, the latter can be expressed via the ρ-meson condensate [30, 31]

$$\langle \mathscr{E} \rangle \equiv \langle T_{00} \rangle = \frac{1}{2}B_{ext}^2 + 2e(B_c - B_{ext})\langle|\rho|^2\rangle + 2e^2\langle|\rho|^2\rangle^2$$

$$+ 2(g_s^2 - e^2)\left\langle|\rho|^2\frac{m_0^2}{-\Delta + m_0^2}|\rho|^2\right\rangle, \qquad (6.47)$$

where $\partial_\perp^2 \equiv \partial_1^2 + \partial_2^2$ is the two-dimensional Laplace operator,

$$\frac{1}{-\partial_\perp^2 + m_0^2}(x^\perp) = \frac{1}{2\pi}K_0(m_0|x^\perp|) \qquad (6.48)$$

is a two-dimensional Euclidean propagator of a scalar massive particle with the mass of the neutral $\rho^{(0)}$ meson,

$$m_0 \equiv m_{\rho^{(0)}} = m_\rho\left(1 - \frac{e^2}{g_s^2}\right)^{-\frac{1}{2}}, \qquad (6.49)$$

and K_0 is a modified Bessel function. Contrary to the potential part of the energy density (6.37), the full expression (6.47) depends nonlocally on the condensate ρ.

A general solution of (6.46) is the following generalization to the square Abrikosov lattice (6.17):

$$\rho(z) = \sum_{n\in\mathbb{Z}}C_n\exp\left\{-\frac{\pi}{2}\left(|z|^2 + \bar{z}^2\right) - \pi v^2 n^2 + 2\pi vn\bar{z}\right\}, \quad z = x_1 + ix_2, \quad (6.50)$$

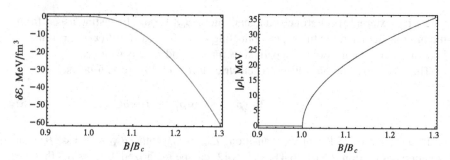

Fig. 6.4 At $B > B_c$ the superconducting state—corresponding to the equilateral triangular lattice Fig. 6.3(right)—is more energetically favorable compared to the trivial vacuum state: at the critical magnetic field (6.1) $B = B_c$, the condensation energy (6.52), the *left panel*, becomes negative due to emergence of the superconducting condensate (the *right panel*). From Refs. [30, 31]

where ν is an arbitrary real parameter, L_B is the magnetic length (6.18) and C_n are arbitrary complex coefficients.

The ground state solution is given by an equilateral triangular lattice, Fig. 6.3 (right). The corresponding coefficients C_n obey the two-fold symmetry $C_{n+2} = C_n$ with $C_1 = iC_0$, while the independent parameters ν and C_0 cannot be calculated analyticity so that they should be found by a numerical minimization of the energy density (6.47). Equivalently, one can also minimize an analogue the Abrikosov ratio (6.20):

$$\beta_\rho = \left\langle \frac{|\rho|^2}{\langle|\rho|^2\rangle} \frac{m_0^2}{-\Delta + m_0^2} \frac{|\rho|^2}{\langle|\rho|^2\rangle} \right\rangle. \tag{6.51}$$

The left and right panels of Fig. 6.4 show, respectively, the condensation energy

$$\delta\mathscr{E} = \langle\mathscr{E}\rangle - \frac{1}{2}B_{\text{ext}}^2, \tag{6.52}$$

and the superconducting condensate $|\rho| \equiv \sqrt{\langle|\rho|^2\rangle}$ in the ground state. The rise of the condensate at $B > B_c$ makes the ground state energy smaller compared to the normal, noncondensed state.

6.3.4.4 Periodic Pattern of Ground State: Superfluid and Superconductor Vortices

Solutions of (6.50) are inhomogeneous functions in the transversal (x_1, x_2) plane. The inhomogeneities in the charged ρ condensate induce an unexpected condensation of the *neutral* ρ mesons:

$$\rho^{(0)}(x^\perp) = \frac{2ig_s}{-\partial_\perp^2 + m_0^2}\partial|\rho|^2 \equiv \frac{ig_s}{\pi}\partial \int d^2y^\perp K_0(m_0|x^\perp - y^\perp|)|\rho(y^\perp)|^2. \tag{6.53}$$

The longitudinal components of the neutral condensate are zero, $\rho_0^{(0)} = \rho_3^{(0)} = 0$.

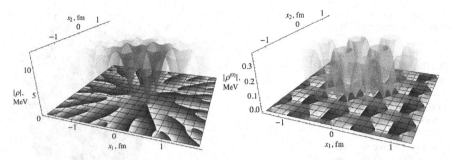

Fig. 6.5 The charged superconducting (*left*) and neutral superfluid (*right*) condensates in the transversal (x_1, x_2) plane at $B = 1.01B_c$. The 3D plots illustrate the absolute values of the condensates (in MeV) while the corresponding projections on the (x_1, x_2) planes of these figures are the density plots of the phases of the corresponding condensates. The *white lines* of the projections are gauge-dependent singularities (the Dirac sheets) attached to the superconductor vortices (the *left panel*) and stretched between the superfluid vortices and antivortices (the *right panel*)

The external magnetic field B_{ext} leads to a backreaction from the charged condensate (6.50), which creates a transverse electric current (6.45),

$$J^{\perp}(x^{\perp}) = \left(\frac{2iem_0^2 \partial}{-\partial_{\perp}^2 + m_0^2} |\rho|^2 \right)(x^{\perp}) \equiv \frac{iem_0^2}{\pi} \partial \int d^2y^{\perp} K_0(m_0|x^{\perp} - y^{\perp}|)|\rho(y^{\perp})|^2,$$

$$(6.54)$$

which affects the magnetic field inside the superconductor:

$$B(x^{\perp}) = B_{\text{ext}} + \frac{2em_0^2}{-\partial_{\perp}^2 + m_0^2} \left[|\rho(x^{\perp})|^2 - \langle |\rho|^2 \rangle \right].$$

$$(6.55)$$

In Fig. 6.5 the charged and neutral condensates are plotted as functions of the transverse coordinates x_1 and x_2 for the magnetic field $B = 1.01B_c$. The periodic equilateral-triangle structure of the absolute value of the charged ρ-meson condensate is identical—apart from the magnitude of the physical scales—to the one of the GL model, Fig. 6.3(right). The absolute value of the neutral condensate also exhibits a lattice pattern which has, however, a bit more involved appearance: hexagonally-shaped structures are arranged into the equilateral triangular lattice, Fig. 6.6.

The charged and the neutral ρ-meson condensates coexist together. Since the magnetic field cannot directly induce the neutral condensate, the mechanism of its appearance is as follows: the background magnetic field induces the charged condensate ρ, (6.50), while the charged condensate gives rise to the neutral one, $\rho^{(0)}$, (6.53). As a result, the neutral condensate is an order of magnitude smaller than the charged condensate, Fig. 6.5. These condensates form a nested structure, Fig. 6.6.

The projection of Fig. 6.5 (left) shows the density plot of the phase of the charged condensate, $\arg \rho$. The phase—which is not a periodic function of the transverse coordinates x_1 and x_2—exhibits discontinuities across which the phase is changed

Fig. 6.6 A visualization of the nested structure of the electrically charged, superconducting condensate (6.50) and the electrically neutral, superfluid condensate (6.53), plotted in *dark magenta* and *light green*, respectively. Shown are the regions where these condensates take maximal values. The transverse plane corresponds to 2 fm × 2 fm region at the magnetic field $B = 1.01 B_c$ (Color figure online)

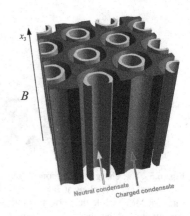

by 2π. These discontinuities correspond the Dirac sheets, shown as the white lines in the same projection of Fig. 6.5 (left). The Dirac sheets are attached to the new class of vortices, "the superconductor ρ vortices". The positions of these vortices correspond to the endpoints of the Dirac sheets. According to the 3D plot of the same figure, the absolute value of the ρ-meson condensate is vanishing at the centers of the superconductor vortices. Locally, these superconductor vortices have the structure which is similar (up to a gauge-dependent phase) to the Abrikosov vortices in the conventional superconductors (6.16): $\rho(x^\perp) \equiv \rho_1(x^\perp) \equiv i\rho_2(x^\perp) \propto |x^\perp| e^{i\varphi} \equiv x_1 + ix_2$.

Contrary to the phase of the charged condensate (6.50), the phase of the neutral condensate (6.53) is a periodic function of the x_1 and x_2 coordinates. The neutral phase exhibits the 2π-discontinuities as well. These discontinuities are visualized as white lines in the projection on the bottom-left panel of Fig. 6.5. The end-points of these discontinuities mark positions of a new type of vortices called "superfluid ρ vortices" connected by the corresponding 2π-discontinuity to the superfluid antivortices. Locally, the superfluid vortices have, up to a phase, the familiar structure: $\rho^{(0)}(x^\perp) \equiv \rho_1^{(0)}(x^\perp) + i\rho_2^{(0)}(x^\perp) \propto |x^\perp| e^{i\varphi} \equiv x_1 + ix_2$.

The vacuum ground state has a rich "kaleidoscopic" structure in terms of the vortex content: the equilateral triangular lattice of the superconductor vortices is superimposed on the hexagonal lattice of the superfluid vortices and antivortices. An elementary lattice cell of this superposition contains one superconductor vortex in the electrically charged ρ condensate as well as three superfluid vortices and three superfluid antivortices in the neutral $\rho^{(0)}$ condensate, Fig. 6.7(left).

6.3.4.5 Superconductivity and Superfluidity in the Ground State

In this section we show that the ground state of the ρ meson condensates is a superconductor and a superfluid simultaneously.

Let us first consider the conducting properties of the new vacuum state. Electric transport properties of a material (such as, for example, the electrical conductivity)

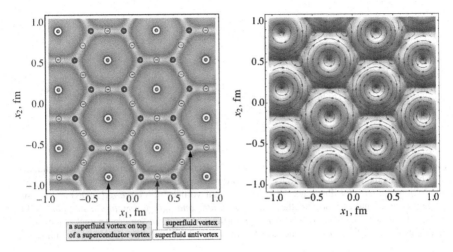

a superfluid vortex on top | superfluid vortex
of a superconductor vortex | superfluid antivortex

Fig. 6.7 (*Left*) The kaleidoscopic vortex structure of the charged and neutral condensates induced by the magnetic field $B = 1.01 B_c$: the superconductor vortices (the *large circles*) are always superimposed on the superfluid vortices (the *small dark blue disks* marked by the plus signs) forming an equilateral triangular lattice in the (x_1, x_2) plane. The isolated superfluid vortices and antivortices (the *small light yellow disks* with the minus signs) are arranged in the hexagonal lattice. The *shades of green* illustrate the absolute value of the neutral ρ-meson condensate (6.53) (from Refs. [30, 31]). (*Right*) The density and the vector flow of the superconducting currents in the (x_1, x_2) plane at $B = 1.01 B_c$ (Color figure online)

are usually determined in a linear response approximation in which one studies an electric current generated inside the material by a weak (test) external electric field background. The electric field should be weak enough in order to preserve, in a leading order, the ground state of the studied material.

In our ground state the transverse (with respect to the strong magnetic field) electric currents (6.54) of charged condensates are confined to elementary cells of the periodic ground state, Fig. 6.7(right). The size of the elementary cell is of the order of the size of the wavefunction of lowest Landau level (in physical units the size of the cell is approximately 0.5 fm for the near-critical magnetic field $B \sim B_c$). In order for a net electric current to be induced in the transverse directions, the quarks need to be excited to the next Landau level which is, however, separated from the lowest Landau level by a large energy gap of the order of $\delta E \sim \sqrt{|eB|}$. It is the energy gap which makes the vacuum state to behave as an insulator in the transverse directions because a weak ($|\mathbf{E}| \ll |\mathbf{B}|$) transverse ($\mathbf{E} \perp \mathbf{B}$) electric field $\mathbf{E}$ cannot create large enough excitation of overcome the gap. The presence of the gap is the very reason why the Meissner effect is absent in the superconducting ground state [27] so that the emerging superconductivity does not screen the external magnetic field (Sect. 6.2.2.3, page 149).

Contrary to the transverse electric currents, the longitudinal currents are not restricted by the external magnetic field. Let us apply a weak electric field $\mathbf{E} = (0, 0, E_z)$ along the axis of the strong magnetic field $\mathbf{B} \equiv (0, 0, B)$. According to the equations of motion of the ρ-meson model (6.27), the induced electric cur-

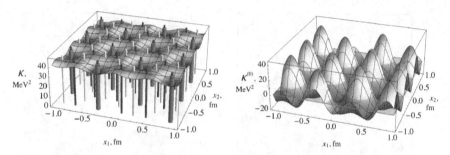

Fig. 6.8 The strength of the vacuum (*left*) superconductivity κ, (6.57), and (*right*) superfluidity $\kappa^{(0)}$, (6.59), as the function of the transverse coordinates x_1 and x_2 at magnetic field $B = 1.01 B_c$. In addition, the *left plot* illustrates the superconductor vortices (the *large red tubes*) and the superfluid vortices and antivortices (the *smaller dark blue and light yellow tubes*, simultaneously) in accordance with notations of Fig. 6.7(*left*). In the *right plot* the semitransparent plane shows highlights the line $\kappa^{(0)} = 0$ where the superfluid strength changes its sign (Color figure online)

rents (6.30) satisfy the following equations [27]:

$$\frac{\partial J_3(x)}{\partial x^0} - \frac{\partial J_0(x)}{\partial x^3} = -\kappa(x^\perp)E_z, \qquad \frac{\partial J_k(x)}{\partial x^\mu} - \frac{\partial J_\mu(x)}{\partial x^k} \equiv 0, \qquad (6.56)$$

where $\mu = 0, \dots, 3$ and $k = 1, 2$.

The set of equations (6.56) is nothing but an anisotropic "vacuum" analogue of the London equation (6.21) of superconductivity. Equations (6.56) show that the electric current—induced by a weak electric "test" field—flows without resistance along the magnetic field axis while in the transverse directions the superconductivity is absent. The strength of the vacuum superconductivity is characterized by the quantity κ, which is a nonlocal function of the superconducting ρ-meson condensate:

$$\kappa(x^\perp) = \left(\frac{4e^2 m_0^2}{-\partial_\perp^2 + m_0^2}|\rho|^2\right)(x^\perp) \equiv \frac{2e^2 m_0^2}{\pi}\int d^2 y^\perp K_0(m_0|x^\perp - y^\perp|)|\rho(y^\perp)|^2.$$
$$(6.57)$$

According to Fig. 6.8(left), the strength of the superconductivity (6.57) is a weakly dependent function of the transverse coordinates x_1 and x_2. In a response to a weak electric current, the superconducting currents are generated outside of the superconductor vortex cores while the maxima of the induced electric currents are located at the centers of the superfluid vortices. Contrary to the ordinary superconductivity, the vacuum superconductivity is not completely suppressed inside the vortices due to the nonlocal nature of the relation between the transport coefficient κ and the superconducting condensate (6.57).

Unexpectedly, the condensate of the neutral ρ mesons is also sensitive to the presence of the external electric current. It turns out that the electrically neutral current of $\rho^{(0)}$ mesons, defined via the relation $J_\mu^{(0)} = -\frac{e}{g_s}\partial^\nu f_{\nu\mu}^{(0)}$, satisfies a London-like equation as well [72]:

$$\frac{\partial J_3^{(0)}(x)}{\partial x^0} - \frac{\partial J_0^{(0)}(x)}{\partial x^3} = -\kappa^{(0)}(x_1, x_2) E_z(x), \tag{6.58}$$

$$\kappa^{(0)}(x_1, x_2) = \left(\frac{4e^2 \partial_\perp^2}{-\partial_\perp^2 + m_0^2} |\rho|^2\right)(x_1, x_2) \equiv \frac{1}{m_0^2} \partial_\perp^2 \kappa(x^\perp), \tag{6.59}$$

where the superfluid coefficient $\kappa^{(0)}$ is visualized in Fig. 6.8(right).

Equations (6.56) and (6.58) indicate that, respectively, the charged and neutral currents should flow frictionlessly (i.e., accelerating ballistically) along the magnetic field axis if an weak external electric field is applied along same direction. Then, if even at some moment of time the electric field is set back to zero, both the superconducting current and the superfluid flow would continue to flow permanently because of the absence of the dissipation forces for the corresponding condensates.

Notice that the electric-field-induced superfluid flow is a locally nonvanishing quantity, while the total superfluid flow of each elementary lattice cell is zero,

$$\frac{\partial \langle J_3^{(0)} \rangle_\perp(x)}{\partial x^0} - \frac{\partial \langle J_0^{(0)} \rangle_\perp(x)}{\partial x^3} = 0, \qquad \langle \mathcal{O} \rangle_\perp \equiv \int d^2 x^\perp \mathcal{O}(x), \tag{6.60}$$

because the average of the superfluid coefficient (6.59) is zero, $\langle \kappa^{(0)}(x^\perp) \rangle_\perp = 0$. A comparison of the left and right plots of Fig. 6.8 reveals that the external electric field generates the positive superfluid flow at the positions of the superconductor vortices which are always accompanied by a superfluid vortices according to Fig. 6.7(left). The negative superfluid flow is generated at the positions of other, unaccompanied superfluid vortices and antivortices.

Finally, we would like to stress that the global quantum numbers of the new exotic superconducting (and, simultaneously, superfluid) phase correspond to those of the vacuum. For example, all chemical potentials in the superconducting phase are vanishing. The vacuum is an electrically neutral object: the presence of the positively charged condensate ρ implies an automatic appearance of a negatively charged condensate ρ^* of the equal magnitude, $\rho \equiv |\rho^*|$. As a result, in strong magnetic field the energy of the vacuum is lowered due to the emergence of the charged condensates, while the net electric charge of the vacuum stays always zero [27, 28]. Despite of the net electric neutrality, the vacuum should superconduct since a weak external electric field pushes the positively and negatively charged condensates in opposite directions along the magnetic field axis, thus creating a net electric current of a double magnitude.

6.3.4.6 Abelian Gauge Symmetry Breaking and Gauge-Lorentz Locking

What is the symmetry breaking pattern in the new superconducting phase of the vacuum? The vacuum superconductivity appears due to the emergence of the magnetic-field-induced ρ-meson condensate (6.36). In the presence of the background magnetic field **B**, the group of global rotations $SO(3)_{\text{rot}}$ of the coordinate space is explicitly broken to its $O(2)_{\text{rot}}$ subgroup which is generated by rotations around the magnetic field axis. The scalar condensate ρ, which describes the vector condensation (6.36) in ground state of the system, transforms under the residual rotational group $O(2)_{\text{rot}}$ as follows:

$$O(2)_{\text{rot}}: \quad \rho(x) \to e^{i\varphi}\rho(x), \tag{6.61}$$

where φ is the azimuthal angle of the rotation in the transverse plane about the x_3 axis. In addition, the ρ meson field transforms under the electromagnetic $U(1)_{\text{e.m.}}$ gauge group (6.29) as follows:

$$U(1)_{\text{e.m.}}: \quad \rho(x) \to e^{ie\omega(x)}\rho(x). \tag{6.62}$$

Thus, if the condensate ρ were a homogeneous (i.e., coordinate independent) quantity then the ground state would "lock" the residual rotational symmetry with the electromagnetic gauge symmetry,[4] $U(1)_{\text{e.m.}} \times O(2)_{\text{rot}} \to U(1)_{\text{locked}}$, since a rotation of the coordinate space at the angle φ about the axis x_3 and a simultaneous gauge transformation with a constant gauge-angle $\omega = -\varphi/e$ leave the homogeneous condensate ρ intact. The inhomogeneities in the ρ condensate break the locked subgroup further from the global $U(1)$ group down to a discrete subgroup of the lattice rotations $G_{\text{locked}}^{\text{lat}}$ of the kaleidoscopic lattice state, Fig. 6.7(left), Ref. [27]:

$$U(1)_{\text{e.m.}} \times O(2)_{\text{rot}} \to G_{\text{locked}}^{\text{lat}}. \tag{6.63}$$

Thus, the superconducting condensate locks the electromagnetic gauge group with the group of the coordinate space rotations.

6.3.5 Superconductivity of Vacuum in Nambu–Jona-Lasinio Model

Basic properties of ordinary superconductivity can equally be described either by the Ginzburg–Landau (GL) phenomenological approach which describes the scalar field of the superconducting carriers or by the Bardeen–Cooper–Schrieffer (BCS) model which accounts for dynamics of electrons in superconductors (Sect. 6.3.2, page 155). Both approaches are mathematically equal if the temperature is sufficiently close to the superconducting phase transition [62].

So far we have discussed the basic features of the vacuum superconductivity in the effective ρ meson electrodynamics [32], which serves as a "vacuum" analogue of the GL approach to the ordinary superconductivity [63]. In this section we

---

[4]A philosophically similar phenomenon, a color-flavor locking, is realized in a different context of the color superconductivity in a dense quark matter [73, 74].

briefly consider, following Ref. [28], the ρ-meson condensation in strong magnetic field in the BCS-like approach [65], which is based on the Nambu–Jona-Lasinio model [33, 34, 66].

6.3.5.1 Effective Action in Strong Magnetic Field

We consider an extended two-flavor ($N_f = 2$) three-color ($N_c = 3$) Nambu–Jona-Lasinio model [33, 34]:

$$\mathcal{L}(\psi, \bar{\psi}) = \bar{\psi}\left(i\partial\!\!\!/ + \hat{Q}\,A\!\!\!/ - \hat{M}^0\right)\psi + \frac{G_S^{(0)}}{2}\left[(\bar{\psi}\psi)^2 + (\bar{\psi}i\gamma^5\boldsymbol{\tau}\psi)^2\right]$$

$$- \frac{G_V^{(0)}}{2}\sum_{i=0}^{3}\left[(\bar{\psi}\gamma_\mu\tau^i\psi)^2 + (\bar{\psi}\gamma_\mu\gamma_5\tau^i\psi)^2\right], \qquad (6.64)$$

where the light quarks are represented by the doublet $\psi = (u, d)^T$, and $G_S^{(0)}$ and $G_V^{(0)}$ are the corresponding bare couplings of scalar and vector quarks' interactions. The masses m_u and m_d, and electric charges ($q_u = +2e/3$ and $q_d = -e/3$) of the quarks are combined into the bare mass matrix $\hat{M}^0 = \mathrm{diag}(m_u^0, m_d^0)$ and the charge matrix $\hat{Q} = \mathrm{diag}(q_u, q_d)$, respectively. The 2×2 matrices in the flavor space are denoted by hats over the corresponding symbols and $\boldsymbol{\tau}$ is a vector made of the Pauli matrices.

The partition function of the NJL model can be represented as an integral,

$$\mathscr{Z} = \int D\bar{\psi}D\psi\, e^{i\int d^4x \mathcal{L}} = \int D\sigma\, D\pi\, DV\, DA\, e^{iS[\sigma,\pi,V,A]}, \qquad (6.65)$$

over bosonic fields [33, 34]. The bosonic fields are given by one scalar field $\sigma \sim \bar{\psi}\psi$, the triplet of three pseudoscalar fields $\pi \sim \bar{\psi}\gamma^5\boldsymbol{\tau}\psi$ [made of the electrically neutral, $\pi^0 \equiv \pi^3$, and electrically charged, $\pi^\pm = (\pi^1 \mp i\pi^2)/\sqrt{2}$, pions], four vector fields,

$$\hat{V}_\mu \equiv \sum_{i=0}^{3}\tau^i V_\mu^i = \begin{pmatrix} \omega_\mu + \rho_\mu^0 & \sqrt{2}\rho_\mu^+ \\ \sqrt{2}\rho_\mu^- & \omega_\mu - \rho_\mu^0 \end{pmatrix}, \quad V_\mu^i \sim \bar{\psi}\gamma_\mu\tau^i\psi, \quad (6.66)$$

[composed of the flavor-singlet coordinate-vector ω-meson field ω_μ, and of the electrically neutral, $\rho_\mu^0 \equiv \rho_\mu^3$, and charged, $\rho_\mu^\pm = (\rho_\mu^1 \mp i\rho_\mu^2)/\sqrt{2}$, components of the ρ-meson triplet], and four pseudovector (axial) fields,

$$\hat{A}_\mu \equiv \sum_{i=0}^{3}\tau^i A_\mu^i = \begin{pmatrix} f_\mu + a_\mu^0 & \sqrt{2}a_\mu^+ \\ \sqrt{2}a_\mu^- & f_\mu - a_\mu^0 \end{pmatrix}, \quad A_\mu^i \sim \bar{\psi}\gamma^5\gamma_\mu\tau^i\psi. \quad (6.67)$$

where the fields f_μ and $(a_\mu^0, a_\mu^\pm)$ represent the singlet axial f_1 meson and the $\mathbf{a}_1$ triplet of the axial mesons, respectively.

The effective bosonic action in (6.65) is as follows

$$S[\sigma, \boldsymbol{\pi}, V, A] = S_\psi - \int d^4x \left[\frac{1}{2G_S^{(0)}} (\sigma^2 + \boldsymbol{\pi}^2) - \frac{1}{2G_V^{(0)}} \left(V_\mu^k V^{k\mu} + A_\mu^k A^{k\mu} \right) \right],$$

(6.68)

$$S_\psi[\sigma, \boldsymbol{\pi}, V, A] = -i N_c \, \text{Tr} \, \text{Ln}(i \mathfrak{D}),$$

(6.69)

$$i\mathfrak{D} = i\slashed{\partial} + \hat{Q}\slashed{A} - \hat{M}^0 + \hat{V}_\mu + \gamma^5 \hat{A} - (\sigma + i\gamma^5 \boldsymbol{\pi} \boldsymbol{\tau}),$$

(6.70)

where we have used simplified notations for the expectation values of the fields, $\langle \sigma \rangle = \sigma$ etc. In the absence of a magnetic field background the expectation value of σ plays a role of the constituent quark mass, $m_q = \sigma \sim 300$ MeV while the expectation values of the fields $\boldsymbol{\pi}$, V, and A are zero [33, 34].

The effective action (6.68) in the strong magnetic field background was calculated in Ref. [28] in the lowest Landau level (LLL) approach using a mean-field technique inspired by calculations of the magnetic catalysis phenomenon [1–5]. In the regime of the LLL dominance the propagator of a f's quark

$$S_f^{\text{LLL}}(x, y) = P_f^\perp \left(x^\perp, y^\perp \right) S_f^\parallel \left(x^\parallel - y^\parallel \right),$$

(6.71)

factorizes into the B-transverse projector onto the LLL states

$$P_f^\perp \left(x^\perp, y^\perp \right) = \frac{|q_f B|}{2\pi} e^{\frac{i}{2} q_f B \varepsilon_{ab} x^a x^b - \frac{1}{4} |q_f B|(x^\perp - y^\perp)^2},$$

(6.72)

and B-longitudinal fermion propagator in the $(1+1)$-dimensions,

$$S_f^\parallel(k_\parallel) \equiv S_{\text{sign}}^\parallel(k_\parallel) = \frac{i}{\gamma^\parallel k_\parallel - m} P_f^\parallel, \qquad P_f^\parallel = \frac{1 - i f \gamma^1 \gamma^2}{2},$$

(6.73)

which depend, respectively, on the B-transverse, $x^\perp = (x^1, x^2)$, and B-longitudinal, $x^\parallel = (x^0, x^3)$, coordinates [1–5]. Here q_f is the electric charge of the f th quark and $eB > 0$ is taken for definiteness.

In (6.73) the matrix $P_f^\parallel$ is the spin projector operator onto the fermion states with the spin polarized along (for u quarks) or opposite (for d quarks) to the magnetic field (we use $f = \pm 1$ for, respectively, $f = u, d$). The operator $P_f^\parallel$ projects the original four $3 + 1$ fermionic states onto two $(1 + 1)$-dimensional fermionic states, so that in the LLL approximation the quarks can move only along the axis of the magnetic field (the latter fact is a natural sequence of the LLL dominance [1–5]).

The operator (6.72) satisfies the projector relation,

$$P_f^\perp \circ P_f^\perp = P_f^\perp, \qquad A \circ B \equiv \int d^2 y^\perp A\left(\ldots, y^\perp \right) B\left(y^\perp, \ldots \right),$$

(6.74)

where "∘" is the convolution operator in the B-transverse space.

In the one-loop order the effective action (6.69) contains a scalar part and vector parts, respectively:

$$S = S_S(\sigma, \pi) + S_V(A, V). \tag{6.75}$$

In terms of the nontrivial condensates,[5] the potential term in the scalar part of the action has the following (renormalized) form:

$$S_S = -\int d^4x \left[\frac{1}{2G_S}\sigma^2 + \frac{|eB|N_c}{8\pi^2}\left(\ln\frac{\sigma^2}{\mu^2} - 1\right)\sigma^2 \right], \tag{6.76}$$

which reflects one of the most important features of the magnetic catalysis [1–5]: an enhancement of quarks' masses by the magnetic field background,

$$m_q(B) = \sigma_{\min}(B) = \mu\exp\{-2\pi^2/(G_S N_c|eB|)\}, \tag{6.77}$$

given by the minimum $\sigma_{\min}$ of potential (6.76). The mass scale, $\mu \propto \sqrt{|eB|}$ is to be fixed beyond the LLL approximation because it is determined, in particular, by the $(1+1)$-dimensional motion of the quarks along the magnetic field [1–5]. As noticed in Refs. [1–5], the renormalization of the scalar NJL coupling G_S in the $\overline{\text{MS}}$ scheme,

$$\frac{1}{G_S} = \frac{1}{G_S^{(0)}} - \frac{N_c|eB|}{4\pi^2\bar{\varepsilon}}, \quad G_S \equiv \frac{2\pi G_{\text{GN}}}{N_c|eB|}, \tag{6.78}$$

is very similar to the renormalization of the coupling constant G_{GN} in the $(1+1)$-dimensional Gross–Neveu model [75]. The divergencies of the $(1+1)$-dimensional fermions are treated in the dimensional regularization in $d = 2 - 2\varepsilon$ dimensions, $1/\bar{\varepsilon} = 1/\varepsilon - \gamma_E + \log 4\pi$ and $\gamma_E \approx 0.57722$ is Euler's constant.

A potentially nontrivial part of the (non-renormalized yet) effective vector action,

$$\begin{aligned}
S_V^{(0)} &\equiv \frac{iN_c}{2}\text{Tr}\left[\frac{1}{i\mathfrak{D}_0}(\hat{V}_\mu + \gamma^5\hat{A})\frac{1}{i\mathfrak{D}_0}(\hat{V}_\mu + \gamma^5\hat{A})\right] \\
&= \frac{4N_c|eB|}{9\pi^2}\cdot\int d^2x^\| \left[\left(\frac{1}{\bar{\varepsilon}} - \ln\frac{\sigma^2}{\mu^2}\right)(\phi^* \circ P_e \circ \phi)\right. \\
&\quad \left. + \left(\frac{1}{\bar{\varepsilon}} - \ln\frac{\sigma^2}{\mu^2} - 2\right)(\xi^* \circ P_e \circ \xi)\right],
\end{aligned} \tag{6.79}$$

involves only the B-transverse combinations of the vector and axial mesons, $\phi = (\rho_1^+ + i\rho_2^+)/2$ and $\xi = (a_1^+ + ia_2^+)/2$. In (6.79) the B-transverse projector for the unit charged particle P_e is given by (6.72) with the replacement $q_f \to e$:

$$P_e^\perp(x^\perp, y^\perp) = \frac{9\pi}{|eB|}P_u^\perp(x^\perp, y^\perp)P_d^\perp(y^\perp, x^\perp). \tag{6.80}$$

---

[5]Here we omit all terms with vanishing condensates as well as all kinetic terms.

The potential (6.79) has an unstable tachyonic mode which is determined by an inhomogeneous eigenstate of the charge-1 projection operator (6.80):

$$(P_e \circ \phi)(x^\perp) = \phi(x^\perp). \tag{6.81}$$

The solution to this equation is a general periodic Abrikosov-like configuration [61] which is given, up to a proportionality coefficient, by (6.50): $\phi(x^\perp) \propto \rho(x^\perp)$. One can also show [28] that there are no unstable modes for the axial mesons and, in accordance with (6.36), for B-longitudinal components of the ρ mesons. Thus, we set the corresponding expectation values to zero.

For the sake of simplicity, we set in (6.50) all coefficients C_n equal, $C_n = \phi_0$, and evaluate certain basic quantities for the simplest square lattice (6.17). As we have mentioned, despite different visual appearances of the square and equilateral triangular lattices, Fig. 6.3, the corresponding bulk quantities (as, for example, the energy density) evaluated at these condensate solutions are almost the same as the difference between them lies within (sometimes, a fraction of) percents.

The leading, quadratic and quartic terms in the effective potential for the square lattice solution (6.17) of the eigenvalue equation (6.81) are given by

$$V = \sqrt{2}\left(\frac{1}{G_B} - \frac{2N_c|eB|}{9\pi^2}\right)|\phi_0|^2 + C_0 \frac{|eB|N_c}{2\pi^2 m^2}|\phi_0|^4, \quad \frac{1}{G_B} = \frac{1}{G_V} - \frac{8}{9G_S}, \tag{6.82}$$

where $C_0 \approx 1.2$ is a numerical (geometrical) factor. If the magnetic field exceeds certain critical strength,[6] $eB_c^{\mathrm{NJL}} = 9\pi^2/(2N_c G_B) \sim 1$ GeV^2, then the potential (6.82) becomes unstable towards a spontaneous creation of the B-transverse ρ-meson condensates with the tachyonic mode $\rho_1^-(x^\perp) = -i\rho_2^-(x^\perp) = \phi(x^\perp)$ [and, respectively, $\rho_1^+(x^\perp) = i\rho_2^+(x^\perp) = \phi^*(x^\perp)$], where $\phi(x^\perp)$ is a solution of (6.81).

6.3.5.2 Electromagnetically Superconducting Ground State in the NJL Model

In the magnetic field background, the effective ρ-meson potential in the NJL model (6.82) has the same Ginzburg–Landau form as the potential (6.37) for the ρ-meson field in the ρ-meson electrodynamics (6.27). If the magnetic field exceeds the critical value, $B \geqslant B_c^{\mathrm{NJL}}$, then the ρ-meson condensate emerge. In terms of the quark fields the new vector quark–antiquark condensates have the following form:

$$\langle \bar{u}\gamma_1 d \rangle = -i\langle \bar{u}\gamma_2 d \rangle = \frac{\phi_0(B)}{G_V} K\left(\frac{x_1 + ix_2}{L_B}\right), \tag{6.83}$$

---

[6]We have estimated the critical field only approximately since the phenomenological values of the NJL parameters $G_{S,V}$ are not known precisely [76]. Moreover, subtleties of the renormalization of the effective dimensionally reduced $(1+1)$-dimensional theory embedded in $3+1$ dimensions provide us with an additional uncertainty.

where the function $K(z)$ is given in (6.17). The magnitude and global phase θ_0 of the condensate (6.83) are determined by the following formula:

$$\phi_0(B) = e^{i\theta_0} C_\phi m_q(B) \left(1 - \frac{B_c^{NJL}}{B}\right)^{1/2} \quad \text{for } B > B_c^{NJL}. \quad (6.84)$$

Here $C_\phi \approx 0.51$ is a numerical prefactor and the B-dependent quark mass m_q is given in (6.77). At $B < B_c^{NJL}$ the ρ-meson condensate (6.84) is zero. The superconducting phase transition at $B = B_c$ is of the second order with the critical exponent $1/2$, similarly to the phase transition in the ρ-meson electrodynamics (6.27).

The quark condensates (6.83) have the quantum numbers of the ρ mesons. They form an inhomogeneous ground state identical to the one found in the ρ-meson electrodynamics (6.50). It is very interesting to notice that the ground state in the NJL model (6.64), determined by the integral equation (6.81), and the ground state in the ρ-meson electrodynamics (6.27), determined by the differential equation of motion (6.46), have the same functional form (6.50).

The vacuum state (6.83) of the NJL model is superconducting. The validity of the anisotropic London equation (6.56) for the quarks' electric current,

$$J^\mu(x) = \sum_{f=u,d} q_f \langle \bar{\psi}_f \gamma^\mu \psi_f \rangle \equiv -\operatorname{tr}[\gamma^\mu \hat{Q} S(x, x)], \quad (6.85)$$

can be shown in a linear-response approach using retarded Green functions [28]:

$$\frac{\partial \langle J_3 \rangle (x^{\parallel})}{\partial x^0} - \frac{\partial \langle J_0 \rangle (x^{\parallel})}{\partial x^3} = -\frac{2C_q}{(2\pi)^3} e^3 \left(B - B_c^{NJL}\right) E_3 \quad \text{for } B > B_c^{NJL}, \quad (6.86)$$

where $C_q \approx 1$ is a numerical prefactor and at $B < B_c$ the right hand side of (6.86) is zero. For the sake of simplicity, in (6.86) we have averaged the electric charge density J^0 and the electric current $J^z \equiv J^0$ over the transverse plane (6.60).

Apart from the prefactors, the London equations in the NJL model (6.86) and in the ρ-meson electrodynamics (6.56) are identical. In a linear-response approximation these laws can be generalized to a completely Lorentz-covariant form [28],

$$\partial_\mu J_\nu - \partial_\nu J_\mu = \gamma \cdot (F, \tilde{F}) \tilde{F}_{\mu\nu}, \quad (6.87)$$

via the Lorentz invariants $(F, \tilde{F}) = 4(\mathbf{B}, \mathbf{E})$ and $(F, F) = 2(\mathbf{B}^2 - \mathbf{E}^2)$. Here γ is a function of the scalar invariant (F, F) and $\tilde{F}_{\mu\nu} = \varepsilon_{\mu\nu\alpha\beta} F^{\alpha\beta}/2$. The local form of the London laws for the superconductor (6.56) and superfluid (6.58) components can be rewritten in a similar way.

6.4 Conclusion

We have shown that in sufficiently strong magnetic field the empty space becomes an electromagnetic superconductor. The new state of the vacuum has many unusual features [27, 28, 30, 31, 72]:

- The magnetic field induces the superconductivity instead of destroying it.
- The Meissner effect is absent.
- The superconductivity has a strong anisotropy: the electric currents may flow without resistance only along the axis of the magnetic field.
- The superconductivity appears in the empty space as a result of the restructuring of the quantum fluctuations due to the presence magnetic field. The overcritical magnetic field ($B > B_c \approx 10^{16}$ T) induces the quark–antiquark condensates which have the quantum numbers of the electrically charged ρ-mesons.
- The electromagnetic superconductivity is always accompanied by the superfluid component caused by emergence of a neutral ρ-meson condensate.
- The tandem superconductor-superfluid ground state is inhomogeneous, it resembles an Abrikosov lattice in a mixed state of an ordinary type-II superconductor.
- The charged and neutral vector quark–antiquark condensates have stringlike topological singularities: superconductor and superfluid ρ vortices, respectively.
- The ground state is characterized by a "kaleidoscopic" lattice made an equilateral triangular lattice of the superconductor vortices superimposed on the hexagonal lattice of the superfluid vortices.

The vacuum superconductivity may be considered as a "magnetic" analogue of the Schwinger effect. Indeed, the Schwinger effect (*the vacuum superconductivity*) is the electron-positron pair production (*the emergence of the quark–antiquark condensates*) due to strong electric (*magnetic*) field background in the vacuum.

The sufficiently strong magnetic fields, of the strength from two to three times higher than the required critical value (6.1) may emerge in the ultraperipheral heavy-ion collisions at the Large Hadron Collider (LHC) at CERN [16, 17]. Thus, signatures of the magnetic-field-induced superconductivity have a chance to be found in laboratory conditions.

References

1. I.A. Shovkovy, arXiv:1207.5081 [hep-ph]
2. K.G. Klimenko, Z. Phys. C **54**, 323 (1992)
3. V.P. Gusynin, V.A. Miransky, I.A. Shovkovy, Phys. Rev. Lett. **73**, 3499 (1994)
4. V.P. Gusynin, V.A. Miransky, I.A. Shovkovy, Phys. Lett. B **349**, 477 (1995)
5. V.P. Gusynin, V.A. Miransky, I.A. Shovkovy, Nucl. Phys. B **462**, 249 (1996)
6. R. Gatto, M. Ruggieri, arXiv:1207.3190 [hep-ph]
7. R. Gatto, M. Ruggieri, Phys. Rev. D **83**, 034016 (2011)
8. R. Gatto, M. Ruggieri, Phys. Rev. D **82**, 054027 (2010)
9. K. Fukushima, M. Ruggieri, R. Gatto, Phys. Rev. D **81**, 114031 (2010)
10. E.S. Fraga, arXiv:1208.0917 [hep-ph]
11. E.S. Fraga, A.J. Mizher, Phys. Rev. D **78**, 025016 (2008)
12. A.J. Mizher, M.N. Chernodub, E.S. Fraga, Phys. Rev. D **82**, 105016 (2010)
13. M. D'Elia, S. Mukherjee, F. Sanfilippo, Phys. Rev. D **82**, 051501 (2010)
14. G.S. Bali, F. Bruckmann, G. Endrodi, Z. Fodor, S.D. Katz, S. Krieg, A. Schafer, K.K. Szabo, J. High Energy Phys. **1202**, 044 (2012)
15. V. Skokov, A.Y. Illarionov, V. Toneev, Int. J. Mod. Phys. A **24**, 5925 (2009)
16. W.-T. Deng, X.-G. Huang, Phys. Rev. C **85**, 044907 (2012)

17. A. Bzdak, V. Skokov, Phys. Lett. B **710**, 171 (2012)
18. A. Vilenkin, Phys. Rev. D **22**, 3080 (1980)
19. M.A. Metlitski, A.R. Zhitnitsky, Phys. Rev. D **72**, 045011 (2005)
20. D. Kharzeev, Phys. Lett. B **633**, 260 (2006)
21. K. Fukushima, D.E. Kharzeev, H.J. Warringa, Phys. Rev. D **78**, 074033 (2008)
22. D.E. Kharzeev, L.D. McLerran, H.J. Warringa, Nucl. Phys. A **803**, 227 (2008)
23. G. Basar, G.V. Dunne, arXiv:1207.4199 [hep-th]
24. F. Preis, A. Rebhan, A. Schmitt, arXiv:1208.0536 [hep-ph]
25. F. Preis, A. Rebhan, A. Schmitt, J. Phys. G **39**, 054006 (2012)
26. O. Bergman, J. Erdmenger, G. Lifschytz, arXiv:1207.5953 [hep-th]
27. M.N. Chernodub, Phys. Rev. D **82**, 085011 (2010)
28. M.N. Chernodub, Phys. Rev. Lett. **106**, 142003 (2011)
29. M.N. Chernodub, PoS **FACESQCD**, 021 (2010)
30. M.N. Chernodub, J. Van Doorsselaere, H. Verschelde, Phys. Rev. D **85**, 045002 (2012)
31. J. Van Doorsselaere, arXiv:1201.0909 [hep-ph]
32. D. Djukanovic, M.R. Schindler, J. Gegelia, S. Scherer, Phys. Rev. Lett. **95**, 012001 (2005)
33. D. Ebert, M.K. Volkov, Z. Phys. C **16**, 205 (1983)
34. D. Ebert, H. Reinhardt, Nucl. Phys. B **271**, 188 (1986)
35. J.S. Schwinger, Phys. Rev. **82**, 664 (1951)
36. M.N. Chernodub, Int. J. Mod. Phys. A **27**, 1260003 (2012)
37. I.I. Smolyaninov, Phys. Rev. Lett. **107**, 253903 (2011)
38. M. Rasolt, Phys. Rev. Lett. **58**, 1482 (1987)
39. Z. Tešanović, M. Rasolt, L. Xing, Phys. Rev. Lett. **63**, 2425 (1989)
40. M. Rasolt, Z. Tešanović, Rev. Mod. Phys. **64**, 709 (1992)
41. F. Lévy, I. Sheikin, B. Grenier, A.D. Huxley, Science **309**, 1343 (2005)
42. D. Aoki et al., J. Phys. Soc. Jpn. **78**, 113709 (2009)
43. D. Aoki et al., J. Phys. Soc. Jpn. **80**, SA008 (2011). arXiv:1012.1987
44. A. Samsonov, J. High Energy Phys. **0312**, 061 (2003)
45. V.V. Braguta, A.I. Onishchenko, Phys. Rev. D **70**, 033001 (2004)
46. T.M. Aliev, M. Savci, Phys. Rev. D **70**, 094007 (2004)
47. M.S. Bhagwat, P. Maris, Phys. Rev. C **77**, 025203 (2008)
48. J.N. Hedditch et al., Phys. Rev. D **75**, 094504 (2007)
49. F.X. Lee, S. Moerschbacher, W. Wilcox, Phys. Rev. D **78**, 094502 (2008)
50. N.K. Nielsen, P. Olesen, Nucl. Phys. B **144**, 376 (1978)
51. J. Ambjorn, P. Olesen, Nucl. Phys. B **315**, 606 (1989)
52. J. Ambjorn, P. Olesen, Int. J. Mod. Phys. A **5**, 4525 (1990)
53. J. Ambjorn, P. Olesen, Phys. Lett. B **218**, 67 (1989)
54. E.V. Gorbar, M. Hashimoto, V.A. Miransky, Phys. Rev. D **75**, 085012 (2007)
55. V.A. Miransky, Prog. Theor. Phys. Suppl. **168**, 405 (2007)
56. M. Ammon, J. Erdmenger, M. Kaminski, P. Kerner, Phys. Lett. B **680**, 516 (2009)
57. M. Ammon, J. Erdmenger, M. Kaminski, P. Kerner, J. High Energy Phys. **0910**, 067 (2009)
58. D.N. Voskresensky, Phys. Lett. B **392**, 262 (1997)
59. O. Aharony, K. Peeters, J. Sonnenschein, M. Zamaklar, J. High Energy Phys. **0802**, 071 (2008). arXiv:0709.3948 [hep-th]
60. J. Beringer et al. (Particle Data Group), Phys. Rev. D **86**, 010001 (2012)
61. A.A. Abrikosov, *Fundamentals of the Theory of Metals* (North-Holland, Amsterdam, 1988)
62. L.P. Gor'kov, Sov. Phys. JETP **9**, 1364 (1959)
63. V.L. Ginzburg, L.D. Landau, Zh. Eksp. Teor. Fiz. **20**, 1064 (1950)
64. J.J. Sakurai, Ann. Phys. **11**, 1 (1960)
65. J. Bardeen, L.N. Cooper, J.R. Schrieffer, Phys. Rev. **106**, 162 (1957)
66. Y. Nambu, G. Jona-Lasinio, Phys. Rev. **122**, 345 (1961)
67. M. Ammon, J. Erdmenger, P. Kerner, M. Strydom, Phys. Lett. B **706**, 94 (2011)
68. N. Callebaut, D. Dudal, H. Verschelde, arXiv:1105.2217
69. V.V. Braguta et al., Phys. Lett. B **718**, 667 (2012). arXiv:1104.3767

70. A.A. Abrikosov, Sov. Phys. JETP **5**, 1174 (1957)
71. B. Rosenstein, D. Li, Rev. Mod. Phys. **82**, 109 (2010)
72. M.N. Chernodub, J. Van Doorsselaere, H. Verschelde, Proc. Sci. **QNP2012**, 109 (2012). arXiv:1206.2845 [hep-ph]
73. M.G. Alford, Annu. Rev. Nucl. Part. Sci. **51**, 131 (2001)
74. K. Rajagopal, F. Wilczek, hep-ph/0011333
75. D.J. Gross, A. Neveu, Phys. Rev. D **10**, 3235 (1974)
76. V. Bernard, A.H. Blin, B. Hiller, Y.P. Ivanov, A.A. Osipov, U.G. Meissner, Ann. Phys. **249**, 499 (1996)

Chapter 7
Lattice QCD Simulations in External Background Fields

Massimo D'Elia

7.1 Introduction

The properties of strong interactions in presence of strong external fields are of great phenomenological and theoretical importance. Large magnetic or chromomagnetic background fields, of the order of 10^{16} Tesla, i.e. $\sqrt{|e|B} \sim 1.5$ GeV, may have been produced at the time of the cosmological electroweak phase transition [111] and may have influenced the subsequent evolution of the Universe, in particular the transition from deconfined to confined strongly interacting matter. Slightly lower magnetic fields are expected to be produced in non-central heavy ion collisions (up to $\sim 10^{15}$ Tesla at LHC [78, 104]), where they may give rise, in presence of non-trivial topological vacuum fluctuations, to CP-odd effects consisting in the separation of electric charge along the direction of the background magnetic field, the so-called *chiral magnetic effect* [63, 78, 112]. Finally, large magnetic (or even chromomagnetic [29]) fields are expected in some compact astrophysical objects, such as magnetars [50] (see [89] for a recent review on the subject).

Apart from the phenomenological issues above, there is great interest also at a purely theoretical level, since background fields can be useful probes to get insight into the properties of strong interactions and the non-perturbative structure of the QCD vacuum. A typical example is chiral symmetry breaking, which is predicted to be enhanced by the presence of a magnetic background, a phenomenon which is known as magnetic catalysis [3–5, 14, 19, 42, 51, 52, 58, 64, 70–72, 75, 77, 79–84, 90, 94, 95, 97–99, 101, 102, 106, 115].

Such interest justifies the large efforts which have been dedicated in the recent past to this subject by a variety of different approaches, most of which are reviewed in the present volume. Lattice QCD represents, in general, the ideal tool for a first

M. D'Elia (✉)
Dipartimento di Fisica dell'Università di Pisa and INFN—Sezione di Pisa, Largo Pontecorvo 3, I-56127 Pisa, Italy
e-mail: delia@df.unipi.it

D. Kharzeev et al. (eds.), *Strongly Interacting Matter in Magnetic Fields*, Lecture Notes in Physics 871, DOI 10.1007/978-3-642-37305-3_7,
© Springer-Verlag Berlin Heidelberg 2013

principle investigation of the non-perturbative properties of strong interactions. This is particularly true for the study of QCD in presence of a strong external magnetic or chromomagnetic field, since no kind of technical problem, such as a sign problem, appears, preventing standard Monte-Carlo simulations, as it happens instead at finite baryon chemical potential.

Similarly to the continuum theory, the magnetic field is introduced on the lattice in terms of additional $U(1)$ degrees of freedom (see Sect. 7.2.1), which are not directly coupled to the original $SU(3)$ link variables and affect quark propagation by a modification of the covariant derivative (i.e. the fermion matrix). The fermion determinant, contrary to the case of a baryon chemical potential μ_B or of an electric background field, is still real and positive, allowing for a probabilistic interpretation of the path integral measure. Lattice studies including the presence of electromagnetic fields have been done since long, originally with the purpose of studying the electromagnetic properties of hadrons [6, 7, 20, 47–49, 88, 108–110]. The introduction of a chromomagnetic field requires a different approach, since in this case the background variables are strictly related to the quantum gluon degrees of freedom: a standard procedure is that defined in the framework of the lattice Schrödinger functional [11, 30–37, 85, 86] (see Sect. 7.2.2).

The recent interest about external field effects in the QCD vacuum and around the deconfining transition has stimulated lattice investigations by many groups [1, 15–17, 22–28, 44, 45, 74, 87, 114]. Several studies have followed a quenched or partially quenched approach, considering only the effect of the magnetic background on physical observables, i.e. on quarks propagating in the background of the given magnetic field and of non-Abelian gauge configurations, the latter being sampled without taking into account the magnetic field.

However, other studies have shown that the electromagnetic background can have, via quark loop effects, a strong influence also on the distribution of non-Abelian fields, thus requiring an unquenched approach. A considerable contribution to magnetic catalysis appears to be related to gluon field modifications [45]; the magnetic field clearly affects also gluonic observables, like the plaquette [74] or those typically related to confinement/deconfinement, like the Polyakov loop [15, 44, 74]. That claims for a more systematic investigation about the effects of electromagnetic backgrounds on the gluonic sector of QCD and about the possible interplay between magnetic and chromomagnetic backgrounds, which could have many implications both at a theoretical and at a phenomenological level (e.g. for cosmology or heavy ion collisions).

The review is organized as follows. In Sect. 7.2 we discuss the formulation of QCD in external fields, considering both the case of an electromagnetic and of a chromomagnetic field. In Sect. 7.3 we review studies regarding vacuum properties in external fields, in particular concerning magnetic catalysis. Section 7.4 is devoted to a discussion about the influence of magnetic or chromomagnetic background fields on deconfinement and chiral symmetry restoration. In Sect. 7.5 we extend the discussion about the influence that electromagnetic fields may have on the gluonic sector, with a particular emphasis on symmetries and on suggestions for future studies on the subject.

7.2 Background Fields on the Lattice

In this section we will briefly review the methods which are commonly adopted for a numerical study of QCD in presence of background fields. There is a difference between the case of electromagnetic and chromomagnetic backgrounds: in the former, the external field simply adds new $U(1)$ degrees of freedom which contribute to quark propagation but are decoupled from the $SU(3)$ degrees of freedom and from their updating during the Monte-Carlo simulation; in the latter, the background field is made up of the same degrees of freedom which are dynamically updated during the simulation, i.e. the $SU(3)$ gauge variables, and one has to follow a Schrödinger functional approach.

7.2.1 Electromagnetic Fields

A quark field q propagating in the background of a non-Abelian $SU(3)$ gauge field plus an electromagnetic field is described, in the Euclidean space-time, by a Lagrangian $\bar{\psi}(\slashed{D} + m)\psi$, where the covariant derivative

$$D_\mu = \partial_\mu + i g A_\mu^a T^a + i q a_\mu \tag{7.1}$$

contains contributions from the non-Abelian gauge field $A_\mu = A_\mu^a T^a$ and from the Abelian gauge field a_μ. Here T^a are the $SU(3)$ generators, g is the $SU(3)$ gauge coupling and q is the quark electric charge.

Going to a discrete lattice formulation, $SU(3)$ gauge invariance is naturally preserved [113] by requiring that quarks pick up an appropriate non-Abelian phase when hopping from one site of the lattice to the other, so that the gauge field A_μ is substituted by the elementary parallel transports $U_\mu(n)$, which reduce to $U_\mu(n) \simeq 1 + i g a A_\mu(n)$ as the lattice spacing a vanishes in the continuum limit.

A gauge invariant quantity involving an antiquark at site n and a quark at site $n + \hat{\mu}$ is therefore $\bar{\psi}(n)U_\mu(n)\psi(n + \hat{\mu})$ ($\hat{\mu}$ is a unit vector on the lattice). The same approach can be taken for the Abelian field, so that quarks going from site n to site $n + \mu$ pick up also the Abelian phase $u_\mu(n) = \exp(i q \int_{an}^{a(n+\hat{\mu})} dx_\mu a_\mu) \simeq \exp(i a q a_\mu(n))$. A possible discretization of the covariant derivative is then

$$D_\mu\psi \rightarrow \frac{1}{2a}\left(U_\mu(n)u_\mu(n)\psi(n + \hat{\mu}) - U_\mu^\dagger(n - \hat{\mu})u_\mu^*(n - \hat{\mu})\psi(n - \hat{\mu})\right). \tag{7.2}$$

Of course, in presence of different quark species, carrying different electric charges q, the $U(1)$ part of the covariant derivative, $u_\mu(n)$, will change from quark to quark. Therefore, as usual, the discretized version of the fermion action is a bilinear form in the fermion fields, $\bar{\psi}(i)M_{i,j}\psi(j)$, however the elements of M now belong to $U(3)$ instead of $SU(3)$.

7.2.1.1 Magnetic Fields

To take an explicit example, which is directly linked to most of the results that we discuss in the following, let us consider the finite temperature partition function of two degenerate quarks, e.g. the u and d quarks, in presence of a constant and uniform magnetic background field. We shall consider a rooted staggered fermion discretized version of the theory, in which each quark is described by the fourth root of the determinant of a staggered fermion matrix:

$$Z(T, B) \equiv \int \mathscr{D}U e^{-S_G} \det M^{\frac{1}{4}}[B, q_u] \det M^{\frac{1}{4}}[B, q_d] \tag{7.3}$$

where in the standard formulation

$$M_{i,j}[B, q] = am\delta_{i,j} + \frac{1}{2} \sum_{\nu=1}^{4} \eta_{\nu}(i)\big(u_{\nu}(B, q)(i)U_{\nu}(i)\delta_{i, j-\hat{\nu}}$$

$$- u_{\nu}^{*}(B, q)(i - \hat{\nu})U_{\nu}^{\dagger}(i - \hat{\nu})\delta_{i, j+\hat{\nu}}\big). \tag{7.4}$$

$\mathscr{D}U$ is the functional integration over the non-Abelian gauge link variables, S_G is the discretized pure gauge action and $u(B, q)_{i,\nu}$ are the Abelian links corresponding to the background field; i and j refer to lattice sites and $\eta_{i,\nu}$ are the staggered phases. We shall consider two different charges for the two flavors, $q_u = 2|e|/3$ and $q_d = -|e|/3$. Periodic or antiperiodic boundary conditions along the Euclidean time direction must be taken, respectively for gauge or fermion fields, in order to define a thermal theory: the temperature T corresponds to the inverse temporal extension.

A special discussion must be devoted to the issue of spatial boundary conditions. It is well known that, for a given lattice size (which is usually constrained by the computational power available) periodic boundary conditions in space (for all fields) are the one which best approximate the infinite volume limit, and are therefore the standard choice in lattice simulations. On the other hand, they place some constraints on the possible magnetic fields [8, 43, 105, 107].

Let us consider a magnetic field directed along the $\hat{z}$ axis of a three-dimensional torus, $\mathbf{B} = B\hat{z}$, and let l_x and l_y be the torus extensions in the orthogonal directions. The circulation of a_{μ} along any closed path in the $x-y$ plane, enclosing a region of area A, is given by Stokes theorem

$$\oint a_{\mu}dx_{\mu} = AB. \tag{7.5}$$

On the other hand it is ambiguous, on a closed surface like a torus, to state which is the region enclosed by a given path: we can choose the complementary region of area $l_x l_y - A$ and equally state

$$\oint a_{\mu}dx_{\mu} = (A - l_x l_y)B. \tag{7.6}$$

The ambiguity can be resolved either by admitting discontinuities for a_μ somewhere on the torus, or by covering the torus with various patches corresponding to different gauge choices. However it is essential to guarantee that the ambiguity be not visible by any charged particle moving on the torus, i.e. the phase factor picked up by a particle of charge q moving along the given path must be a well defined quantity. Such requirement leads to magnetic field quantization, indeed it is satisfied only if

$$\exp(iqBA) = \exp\big(iqB(A - l_x l_y)\big) \tag{7.7}$$

i.e. if

$$qB = \frac{2\pi b}{l_x l_y} \tag{7.8}$$

where b is an integer. It is interesting to notice that this is exactly the same argument leading to Dirac quantization of the magnetic monopole charge (in that case the closed surface is any sphere centered around the monopole). The quantization rule depends on the electric charges of the particles moving on the torus,[1] in our case it is set by the d quark, which brings the smallest charge unit $q_d = -|e|/3$, hence

$$|e|B = \frac{6\pi b}{l_x l_y} = \frac{6\pi b}{a^2 L_x L_y} \tag{7.9}$$

where L_x and L_y are the dimensionless lattice extensions in the x, y directions.

A possible choice for the gauge links, corresponding to the continuum gauge field

$$a_y = Bx; \qquad a_\mu = 0 \quad \text{for } \mu = x, z, t \tag{7.10}$$

is the following:

$$u_y(B, q)(n) = e^{ia^2 q B n_x}; \qquad u_\mu(B, q)(n) = 1 \quad \text{for } \mu = x, z, t. \tag{7.11}$$

The smoothness of the background field across the boundaries and the gauge invariance of the fermion action are guaranteed if appropriate boundary conditions are taken for the fermion fields along the x direction [8], that corresponds to modifying the $U(1)$ gauge links in the x direction as follows:

$$u_x(B, q)(n)\big|_{n_x = L_x} = e^{-ia^2 q L_x B n_y}. \tag{7.12}$$

A different condition on the possible magnetic field values explorable on the lattice is placed by the ultraviolet (UV) cutoff, i.e. by discretization itself. All the information about the presence of the magnetic field is contained in the phase factors

---

[1] Of course we assume all particles living on the torus to carry integer multiples of some elementary electric charge, otherwise a consistent quantization of the magnetic field would not be possible.

picked up by particles moving on the lattice. There is a minimum such path on a cubic lattice, the plaquette; in particular, a particle moving around a $x-y$ plaquette takes a phase factor

$$\exp(\mathrm{i}\, qa^2 B) = \exp\left(\mathrm{i}\, \frac{2\pi b}{L_x L_y}\right) \tag{7.13}$$

and all other possible phase factors, corresponding to different paths, are positive integer powers of that. It is evident that the phase factor in (7.13) cannot distinguish magnetic fields such that $qa^2 B$ differs by multiples of 2π: one would need smaller paths to do that, which are unavailable because of the UV cutoff. Hence it is possible to define a sort of first Brillouin zone for the magnetic field

$$-\frac{\pi}{a^2} < qB < \frac{\pi}{a^2} \tag{7.14}$$

and we expect that physical quantities must be periodic in qB with a period $2\pi/a^2$; such periodicity is unphysical and one should always be cautious when exploring magnetic field values which get close to the limiting values in (7.14).

Another issue which is relevant to the lattice implementation of background fields regards translational invariance. The magnetic field B breaks the translational invariance of the continuum torus in the $x-y$ directions explicitly [8]. That is clearly seen by looking at the $U(1)$ Wilson lines (holonomies) W_x and W_y, i.e. the $U(1)$ parallel transports along straight paths in the x or y direction and closed around the torus by periodic boundary conditions, which are equal to

$$W_x = \exp(-\mathrm{i}\, qB\, l_x\, y); \qquad W_y = \exp(\mathrm{i}\, qB\, l_y\, x) \tag{7.15}$$

as can be verified explicitly by means of the elementary parallel transports given in (7.11) and (7.12). The Wilson lines therefore leave only a residual symmetry corresponding to discrete translations by multiples of

$$\tilde{a}_x = \frac{2\pi}{qB l_y} = \frac{l_x}{b}; \qquad \tilde{a}_y = \frac{2\pi}{qB l_x} = \frac{l_y}{b}; \tag{7.16}$$

note that Wilson lines are gauge invariant quantities, hence the breaking of translational invariance is not a matter of gauge choice.

However, on a discrete cubic lattice, translational invariance is already broken to a discrete residual group, corresponding to multiples of the lattice spacing a, hence for a lattice theory in presence of a constant magnetic field, translational invariance is further reduced to discrete steps which are multiples of both $\tilde{a}_{x/y}$ and a. Since $a/\tilde{a}_{x/y} = b/L_{x/y}$, lattice translational symmetry is preserved for b multiple of $L_{x/y}$, but is strongly reduced or completely lost for different values, despite the fact that the magnetic background field is uniform.

We have discussed so far about one possible lattice implementation of the fermionic action in presence of a magnetic field, (7.2) and (7.4), which is based on the simplest symmetric discretization of the covariant derivative, which takes quarks from nearest neighbours sites, rotating them by a non-Abelian and an Abelian phase, both corresponding to the simplest straight path, i.e. the elementary parallel transport. Such kind of discretization has been adopted in many lattice studies, see e.g. [44, 45, 74].

It is well known, however, that different, improved discretizations can lead to an improved convergence towards the continuum limit. Improvement can proceed in different ways: one can consider less simple discretizations of the derivative, involving fermion fields which are not nearest neighbours; one can also consider and average over paths different from the simplest straight one, so as to smear out UV fluctuations: this is idea adopted, for instance, in [15–17], where stout smearing has been used. If one considers lattice derivatives involving non-nearest neighbours lattice sites, one must include appropriate composite $U(1)$ parallel transports, so as to preserve the $U(1)$ gauge invariance of the fermion action. Which paths, however, and how the chosen paths are related to the paths considered for the non-Abelian phases, is a question which leaves much freedom.

Let us consider, for simplicity, a gauge invariant bilinear term involving nearest neighbour quarks, $\bar{\psi}(n)U_\mu(n)\psi(n+\hat{\mu})$, which reduces to $\bar{\psi}(x)(1+aD_\mu)\psi(x)+O(a^2)$ in the naïve continuum limit. We can choose a more general, improved such term, involving different paths, i.e. we can modify

$$\bar{\psi}(n)U_\mu(n)\psi(n+\hat{\mu}) \to \bar{\psi}(n)\left(\sum_{C(n,n+\hat{\mu})} \alpha_C U_C\right)\psi(n+\hat{\mu}) \qquad (7.17)$$

where $C(n, n+\hat{\mu})$ are a set of paths connecting n to $n+\hat{\mu}$ (e.g. for smearing one includes staples), U_C are the parallel transports taken along those paths and the coefficients α_C are chosen in order to keep the naïve continuum limit unchanged, up to $O(a^2)$. Let us now consider the inclusion of a $U(1)$ external field: a possible prescription is that, as quarks explore all considered paths C, taking the associated non-Abelian phases, they also take the corresponding $U(1)$ phases along the same paths, i.e.

$$\bar{\psi}(n)U_\mu(n)u_\mu(n)\psi(n+\hat{\mu}) \to \bar{\psi}(n)\left(\sum_{C(n,n+\hat{\mu})} \alpha_C U_C u_C\right)\psi(n+\hat{\mu}). \qquad (7.18)$$

However, different prescriptions can be taken and, in general, the sum over $U(1)$ phases can be decoupled from that on $SU(3)$ phases, without affecting gauge invariance. We can consider for example

$$\bar{\psi}(n)\left(\sum_{C(n,n+\hat{\mu})} \alpha_C U_C\right)\left(\sum_{C'(n,n+\hat{\mu})} \alpha'_{C'} u'_C\right)\psi(n+\hat{\mu}) \qquad (7.19)$$

where C' is a different set of paths. This is the choice made, for instance, in [15–17], where stout smearing is applied to $SU(3)$ links, while the $U(1)$ phases are left unchanged (i.e. C' runs over the elementary straight path only); that seems a reasonable choice, since the $U(1)$ field is already a classical smooth field.

It can be easily shown that for a uniform background and for free fermions, i.e. in absence of non-Abelian fields, all possible different choices of paths for the $U(1)$ term are equivalent, apart from a global phase stemming from interference (Aharonov-Bohm like) effects among the different paths. In presence of non-Abelian fields, however, this is in general not true and it would be interesting to explore the systematics connected to different choices.

A final comment concerns the issue of possible renormalizations related to the introduction of the external magnetic field. The quantity entering the covariant derivative and quark dynamics is the combination qa_μ, which does not renormalize. The situation is clearer in the lattice formulation, where the objects carrying information about the external field are the parallel transports of the $U(1)$ gauge field along closed loops (think of the loop expansion of the fermionic determinant). Such phase factors are gauge inviariant quantities and cannot renormalize. The only remaining possibility is that the lattice spacing a itself depends on the magnetic field B, because of external field effects at the scale of the UV cutoff, so that the physical area enclosed by a given loop changes, leading to an effective external field renormalization. Such possibility, however, has been excluded by a detailed study presented in [15], showing that a does not depend on B within errors.

7.2.1.2 Electric Fields

After the inclusion of a background magnetic field, the $\slashed{D}$ operator is still anti-Hermitean and it still anticommutes with γ_5, hence its spectrum is purely imaginary with non-zero eigenvalues coming in conjugate pairs, so that $\det(\slashed{D} + m) > 0$ and Monte-Carlo simulations are feasible, i.e. the path integral measure can be interpreted as a probability distribution function over gauge configurations.

The situation is different in presence of background electric fields [6, 7, 100, 108–110]. Let us consider a constant electric field $\mathbf{E} = E\hat{z}$, there are various possible choices for a corresponding vector potential in Minkowski space, like $(A_t = -Ez, A_i = 0)$ or $(A_z = -Et, A_x = A_y = A_t = 0)$. In every case, after continuation to Euclidean space time, $t \to -i\tau$ and $A_t \to iA_0$, the vector potential becomes purely imaginary, thus destroying the anti-Hermitean properties of the Dirac matrix.

That means that the fermion determinant is complex and numerical simulations are not feasible. This "sign problem" has a very strict relation with the usual sign problem which is encountered in the study of QCD at finite baryon chemical potential, indeed also a baryon chemical potential can be viewed as a constant background potential $A_0 \neq 0$. As for finite density QCD, also in this case a possible way out is to consider a purely imaginary electric field: the lattice formulation is then completely analogous to that of a magnetic field. For example, a purely imaginary electric field along the $\hat{z}$ direction can be formulated on the lattice following (7.11) and (7.12), by just replacing $(x, y) \to (z, t)$ and $B \to \text{Im}(E)$. A similar approach, followed by analytic continuation to real electric fields, is that usually adopted for the study of hadron electric polarizabilities.

7.2.2 Chromomagnetic Background Fields

As we have discussed above, the implementation of an electromagnetic background field amounts to adding extra $U(1)$ degrees of freedom to the Dirac matrix: the added $U(1)$ field is a classic, static field which is decoupled from the $SU(3)$ gauge field appearing in the functional integration. The situation is different for a color background field, since in this case one would like to consider quantum fluctuations of the non-Abelian field around a given background, e.g. a static ad uniform chromomagnetic field, however the functional integration over gauge variables can destroy information about the background field completely.

A common way to approach the problem is based on the Schrödinger functional approach [85, 86]: one considers functional integration over Euclidean space-time with fixed temporal boundaries, τ_1 and τ_2, and over gauge configurations which are frozen to particular assigned values, $A_i^{ext1}(\mathbf{x}, \tau_1)$ and $A_i^{ext2}(\mathbf{x}, \tau_2)$, at the initial and final times. This is related to the quantum amplitude of passing from the field eigenstate $|A_i^{ext1}\rangle$ to the field eigenstate $|A_i^{ext2}\rangle$ in the time $(\tau_2 - \tau_1)$. The amplitude is dominated, in the classical limit, by the gauge configuration which has the minimal action among those with the given boundary conditions; functional integration can then be viewed as an integration over quantum fluctuations around this classical background field.

In Refs. [30–35, 37] such formalism has been implemented on the lattice by considering equal initial and final fields, $A_i^{ext1}(\mathbf{x}, \tau_1) = A_i^{ext2}(\mathbf{x}, \tau_2) = A_i^{ext}(\mathbf{x})$. In this way one can define a lattice gauge invariant effective action $\Gamma[\mathbf{A}^{ext}]$ for the external background field $\mathbf{A}^{ext}$:

$$\Gamma[\mathbf{A}^{ext}] = -\frac{1}{L_t} \ln\left\{\frac{\mathscr{Z}[\mathbf{A}^{ext}]}{\mathscr{Z}[0]}\right\} \tag{7.20}$$

where L_t is the lattice temporal extension and $\mathscr{Z}[\mathbf{A}^{ext}]$ is the lattice functional integral

$$\mathscr{Z}_T[\mathbf{A}^{ext}] = \int_{U_k(L_t,\mathbf{x})=U_k(0,\mathbf{x})=U_k^{ext}(\mathbf{x})} \mathscr{D}U\,\mathscr{D}\psi\,\mathscr{D}\bar{\psi}\,e^{-(S_W+S_F)}$$

$$= \int_{U_k(L_t,\mathbf{x})=U_k(0,\mathbf{x})=U_k^{ext}(\mathbf{x})} \mathscr{D}U e^{-S_W} \det M, \tag{7.21}$$

where S_G and S_F are the pure gauge and fermion action respectively, M is the fermionic matrix. The functional integration is performed over the lattice links, with the constraint

$$U_k(\mathbf{x}, x_t = 0) = U_k(\mathbf{x}, x_t = L_t) = U_k^{ext}(\mathbf{x}), \quad (k = 1, 2, 3), \tag{7.22}$$

where $U_k^{ext}(\mathbf{x})$ are the elementary parallel transports corresponding to the continuum gauge potential $\mathbf{A}^{ext}(x) = \mathbf{A}_a^{ext}(x)\lambda_a/2$; fermion fields as well as temporal links are left unconstrained. $\mathscr{Z}[0]$ is defined as in (7.21), but adopting a zero external field.

In the following we shall consider the case of an Abelian static and uniform chromomagnetic background field: in this case one can choose λ_a belonging to the Cartan subalgebra of the gauge group, e.g. $\lambda_a = \lambda_3$, while the explicit form of $\mathbf{A}_a^{\text{ext}}(x)$, hence of the lattice links, can be chosen in the same way as for a standard magnetic field, see (7.10), (7.11) and (7.12). In this case, as well as in other cases in which the static background field does not vanish at spatial infinity, one usually imposes also that spatial links exiting from sites belonging to the spatial boundaries are fixed, at all times, according to (7.22): that corresponds to the requirement that fluctuations over the background field vanish at infinity.

Notice that a colored background field influences directly, and through the same gauge coupling g, the dynamics of both quark and gluon fields. Indeed numerical simulations in presence of colored background fields have considered originally just the case of pure Yang-Mills theories.

Differently from the usual formulation of the lattice Schrödinger functional [85, 86], where a cylindrical geometry is adopted, the effective action defined by (7.20) assumes an hypertoroidal geometry, i.e. the first and the last time slice are identified and periodic boundary conditions are assumed in the time direction for gluon fields. For finite values of L_t, having adopted also the prescription of antiperiodic boundary conditions in time direction for quark fields, (7.21) can be interpreted as the thermal partition function $\mathscr{Z}_T[\mathbf{A}^{\text{ext}}]$ in presence of the background field $\mathbf{A}^{\text{ext}}$, with the temperature given by $T = 1/(aL_t)$. The gauge invariant effective action, (7.20), is then replaced by the free energy functional:

$$\mathscr{F}[\mathbf{A}^{\text{ext}}] = -\frac{1}{L_t} \ln \left\{ \frac{\mathscr{Z}_T[\mathbf{A}^{\text{ext}}]}{\mathscr{Z}_T[0]} \right\}. \tag{7.23}$$

7.3 Vacuum Properties in Background Fields: Magnetic Catalysis

One of the most significant effects that a magnetic background field can induce on the QCD vacuum, as well as on other systems characterized by the chiral properties of fermion fields, is known as magnetic catalysis. It consists in an enhancement of chiral symmetry breaking, or spontaneous mass generation, which can be thought as related to the dimensional reduction taking place in the dynamics of particles moving in a strong external magnetic field (see [101] for a recent comprehensive review).

The enhancement of chiral symmetry breaking reveals itself in a dependence of the chiral condensate $\langle \bar{\psi}\psi \rangle$ on the magnetic field B; there are several predictions for the actual functional form $\langle \bar{\psi}\psi \rangle(B)$, depending on the adopted model, however some general features can be summarized as follows. By charge conjugation symmetry, $\langle \bar{\psi}\psi \rangle$ must be an even function of B, hence, if the theory is analytic around $B = 0$, one expects that $\langle \bar{\psi}\psi \rangle$ depends quadratically on B in a regime of small enough fields. The analyticity assumption is violated in presence of charged massless fermions, indeed chiral perturbation theory [102] predicts a linear behavior in

B if $m_\pi = 0$; such linear dependence is recovered in computations at $m_\pi \neq 0$, in the limit $eB \gg m_\pi^2$ [42]. Different power behaviors in B can be found in other model computations. In the following we shall make use of the relative increment of the chiral condensate, defined as

$$r(B) = \frac{\langle \bar{\psi}\psi \rangle(B) - \langle \bar{\psi}\psi \rangle(B=0)}{\langle \bar{\psi}\psi \rangle(B=0)}. \tag{7.24}$$

Early lattice investigations of magnetic catalysis have been done in the quenched approximation, both for $SU(2)$ [25, 26] and for $SU(3)$ [23] gauge theories. In this case, the chiral condensate is computed by inverting a Dirac operator which contains the contribution from the external field, but on gauge configurations sampled in absence of dynamical fermion contributions. The outcome is that $r(B) \propto B$ for $SU(2)$ [25, 26], while $r(B) \propto B^\nu$, with $\nu \sim 1.6$, for $SU(3)$ [23].

On the other hand, the inclusion of contributions from dynamical fermions moving in the background field may produce significant effects. The chiral condensate is related, via the Banks—Casher relation [18], to the density $\rho(\lambda)$ of eigenvalues of the Dirac operator around $\lambda = 0$: $\langle \bar{\psi}\psi \rangle = \pi\rho(0)$. Such density can change, as a function of B, both because the Dirac operator definition itself changes, and because the distribution of gauge configurations, over which the operator is computed, is modified by dynamical fermion contributions. Studies of magnetic catalysis involving full dynamical simulations have been reported in [16, 45], for QCD with 3 colors, and in [74] for QCD with 2 colors.

In Ref. [45], where a standard staggered fermion and plaquette gauge discretization has been adopted, corresponding to a pseudo-Goldstone pion mass $m_\pi \sim 200$ MeV, an attempt has been made to separate the contributions to magnetic catalysis coming from the modification of the Dirac operator from those coming from the modified gauge field distribution: such contributions have been named "valence" and "dynamical", respectively. One can show that such separation can be done consistently in the limit of small external fields. Indeed, let us define the quantities

$$\langle \bar{\psi}\psi \rangle^{val}_{u/d}(B) \equiv \int \mathcal{D}U\, \mathcal{P}[m, U, 0]\, \text{Tr}\big(M^{-1}[m, B, q_{u/d}]\big) \tag{7.25}$$

$$\langle \bar{\psi}\psi \rangle^{dyn}_{u/d}(B) \equiv \int \mathcal{D}U\, \mathcal{P}[m, U, B]\, \text{Tr}\big(M^{-1}[m, 0, q_{u/d}]\big) \tag{7.26}$$

and, from those, the average (over flavor) quantities $\langle \bar{\psi}\psi \rangle^{val}$ and $\langle \bar{\psi}\psi \rangle^{dyn}$; $\mathcal{P}$ is the probability distribution for gauge fields (including quark loop effects) and M is the fermion matrix. In the first case, one looks at the spectrum of the fermion matrix which includes the magnetic field explicitly, but is defined on non-Abelian configurations sampled with a measure $\mathcal{D}U\, \mathcal{P}[m, U, 0]$ which, even including dynamical fermion contributions, is taken at $B = 0$ (partial quenching). In the second case the measure term takes into account the external field, which is however neglected in the definition of the fermion matrix. From $\langle \bar{\psi}\psi \rangle^{val}(B)$ and $\langle \bar{\psi}\psi \rangle^{dyn}(B)$ we can define the corresponding quantities $r^{val/dyn}$.

Fig. 7.1 We show the relative increment $r(B)$ (see (7.24)) as a function of $|e|B$ for $N_f = 2$ QCD with $m_\pi \sim 200$ MeV. Data are reported separately for the u and d quark condensates, as well as for their average, together with a best fit according to the functional dependence inspired by chiral perturbation theory [42]. Figure taken from [45]

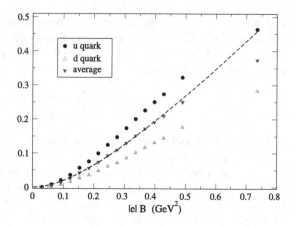

In the limit of small fields, B acts as a perturbation for both the measure term $\mathscr{P}[m, U, B]$ and the observable $\mathrm{Tr}(M^{-1}[m, B, q_{u/d}])$: assuming quadratic corrections in B (however this is not an essential assumption), one can write, configuration by configuration:

$$\mathscr{P}[m, U, B] = \mathscr{P}[m, U, 0] + CB^2 + O(B^4) \qquad (7.27)$$

and

$$\mathrm{Tr}(M^{-1}[B]) = \mathrm{Tr}(M^{-1}[0]) + C'B^2 + O(B^4) \qquad (7.28)$$

where C and C' depend in general on the quark mass and on the chosen configuration. Putting together the two expansions, one obtains [45]

$$r(B) = r^{val}(B) + r^{dyn}(B) + O(B^4), \qquad (7.29)$$

therefore, at least in the limit of small fields, the separation of magnetic catalysis in a valence and a dynamical contribution is a well defined concept.

In Fig. 7.1 we report data obtained in [45] for the relative increment of the u and d quark condensates and for their average, as a function of $|e|B$. Magnetic catalysis is quite larger for the u quark with respect to the d quark: this is expected on the basis of the larger electric charge of the u quark. Regarding their average, one observes that a functional dependence inspired by chiral perturbation theory [42] fits well data, when taking into account the unphysical quark mass spectrum considered in [45]. A simple quadratic dependence describes well data in the regime of small background fields, approaching a linear behavior for larger fields, while saturation effects starts to be visible as $\sqrt{|e|B}$ reaches the scale of the UV cutoff, which is quite low for the lattice setup used in [45], $a^{-1} \sim 0.7$ GeV. Regarding the separation into valence and dynamical contributions, Fig. 7.2, taken again from [45], shows that it is indeed well defined in a significant range of magnetic fields, where the two contributions are roughly additive, in the sense that their sum gives back the full signal, as expected at least in the limit of small fields. Moreover, from Fig. 7.2 one learns

Fig. 7.2 Relative increment of the average of the u and d quark condensates as a function of the magnetic field. $r(B)$, $r^{val}(B)$, $r^{dyn}(B)$ and $r^{val}(B) + r^{dyn}(B)$ are reported separately. From [45]

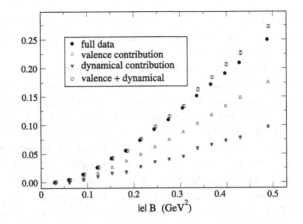

that the dynamical contribution is roughly 40 % of the total signal, at least for the discretization and quark mass spectrum adopted in [45]: this is an important contribution, which is larger than other usual unquenching effects, which are typically of the order of 20 %, and reflects a significant modification in the distribution of gauge fields, induced by the magnetic background field, which should be further investigated by future studies.

A new investigation of magnetic catalysis for the QCD vacuum has been done recently [16], making use of an improved gauge and (stout) rooted staggered discretization of the theory: results for the average increment of the u and d quark condensates have been reported for $N_f = 2 + 1$ flavors (i.e. including also strange quark contributions), adopting physical quark masses and after extrapolation to the continuum limit. Also in this case one observes, for the relative increment $r(B)$ (see Fig. 1 of Ref. [16]), a quadratic dependence on $|e|B$ for small external fields, followed by an almost linear dependence as $|e|B \gg m_\pi^2$; a nice agreement is found with predictions from chiral perturbation theory and from PNJL models [69], at least for not too large fields. It is interesting to notice that, in the regime of small fields, the results of [45] and those of [16] are compatible with each other if they are rescaled by a factor m_π^2 (which is different in the two studies):[2] the prediction from chiral perturbation theory is indeed, in the limit of small fields [42, 45]:

$$r(B) \simeq \frac{(|e|B)^2}{96\pi^2 F_\pi^2 m_\pi^2}. \tag{7.30}$$

The authors of [74], instead, have presented an investigation of magnetic catalysis for QCD with two colors and 4 staggered flavors, which are degenerate both in mass and electric charge: that permits, contrary to Refs. [16, 45], to avoid rooting and the possible systematic effects related to it. The lattice discretization is standard, as in [45], with also a similar range of pseudo-Goldstone pion masses. Results

---

[2]The author thanks G. Endrodi for giving him access to the continuum extrapolated data of [16].

are qualitatively similar to those obtained in [16, 45], however the authors try also an extrapolation to the chiral limit, thus verifying the prediction [102] for a linear dependence on $|e|B$ in such limit.

Till now we have discussed about magnetic catalysis in the QCD vacuum, i.e. in the low temperature, deconfined phase, where a consistent picture emerges, both from model and lattice computations. The situation is less clear as one approaches the high temperature, deconfined regime: results from [15, 16] have shown that the growth of the chiral condensate with the magnetic field may stop and even turn into an inverse magnetic catalysis at high enough temperatures; such effect may have a possible interpretation based on the effects of a strong magnetic field on gluody-namics (see [66, 90, 101] and the discussion in Sect. 7.5). Inverse magnetic cataly-sis has been predicted by some model computations [94, 95], however in a regime of low temperature and high baryon density, which is different from that explored in [15, 16]; recently a possible explanation has been proposed according to which inverse catalysis derives from dimensional reduction induced by the magnetic field on neutral pions [62]. On the other hand, the lattice studies reported in [44, 74] do not show such effect, reporting instead for standard magnetic catalysis at all ex-plored temperatures. We will comment on the possible origin of such discrepancies in the following section, where we discuss about the effects of strong external fields on the QCD phase diagram.

Finally, among the studies aimed at clarifying the effects of strong magnetic fields on chiral dynamics, we mention those addressing the determination of the magnetic susceptibility of the chiral condensate, both in quenched [23, 27] and in unquenched QCD [17], which is part of the total contribution to the magnetic sus-ceptibility of the QCD vacuum. In particular, recent unquenched results [17] show that such contribution is of diamagnetic nature.

7.4 QCD Phase Diagram in External Fields

Similarly to what happens with other external parameters, like a finite baryon chemi-cal potential, the introduction of a background field, either magnetic of chromomag-netic, can modify the phase structure of QCD. The interest in such issue is theoret-ical, on one side, since a background field can be viewed as yet another parameter of the QCD phase diagram, which can help in getting a deeper understanding of the dynamics underlying deconfinement and chiral symmetry restoration. There is however also great phenomenological interest, since strong background fields may be relevant to the cosmological QCD transition, to heavy ion collisions and to the physics of some compact astrophysical objects. The main questions regarding the QCD transition that one would like to answer can be summarized as follows:

(1) do deconfinement and chiral symmetry restoration remain entangled, or is a strong enough magnetic field capable of splitting the two transitions?
(2) does the (pseudo)critical temperature depend on the background field strength, and how?

Fig. 7.3 Behavior of the chiral condensate and of the Polyakov loop, as a function of the inverse gauge coupling β and for different magnetic field quanta b, on a $16^3 \times 4$ lattice and for a pion mass $m_\pi \sim 200$ MeV. Figure taken from [44]

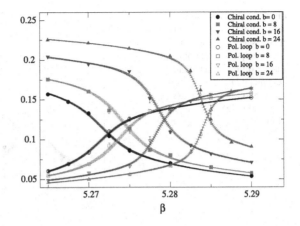

(3) does the nature of the transition depend on the background field strength?
(4) does any new, unexpected phase of strongly interacting matter emerge for strong enough background fields?

Many model computations exist, which try to answer those questions [2, 9, 10, 13, 21, 38, 55–57, 59–61, 65, 67, 68, 76, 91, 93, 96, 103], in the following we will discuss results based on lattice simulations. The focus will be on aspects regarding deconfinement and chiral symmetry restoration, leaving aside other issues, like the possible emergence of a superconductive phase for strong enough magnetic fields [22, 39–41, 73].

7.4.1 Deconfinement Transition in a Strong Magnetic Background

Investigating the effects of a background magnetic field on thermodynamics and on the phase diagram of QCD necessarily requires an approach in which the presence of the magnetic field is taken into account at the level of dynamical fermions.

A first study along these lines has been presented in [44], where finite temperature QCD with two degenerate flavors has been simulated in presence of a constant, uniform magnetic background. A standard rooted staggered discretization (see (7.3)) and a plaquette gauge action have been adopted, with a lattice spacing of the order of 0.3 fm, and different quark masses, corresponding to a pseudo-Goldstone pion mass ranging from 200 MeV to 480 MeV, with magnetic fields going up to $|e|B \sim 0.75$ GeV^2.

Some results from [44] are reported in Figs. 7.3, 7.4, 7.5 and 7.6. In particular, Fig. 7.3 shows the behavior of the chiral condensate and of the Polyakov loop as a function of the inverse gauge coupling β, for the lowest pion mass explored ($m_\pi \sim 200$ MeV) and for various magnetic fields (expressed in units b of the minimum quantum allowed by the periodic boundary conditions). Remember that the

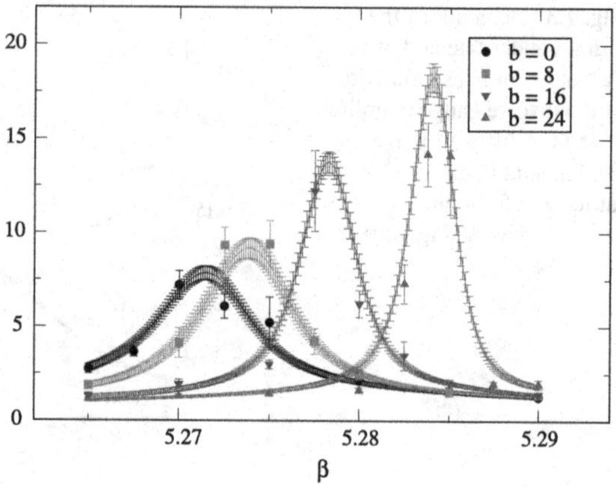

Fig. 7.4 Disconnected susceptibility of the chiral condensate as a function of the inverse gauge coupling β and for different magnetic field quanta b, on a $16^3 \times 4$ lattice and for a pion mass $m_\pi \sim 200$ MeV. Figure taken from [44]

Fig. 7.5 Polyakov loop susceptibility as a function of the inverse gauge coupling β and for different magnetic field quanta b, on a $16^3 \times 4$ lattice and for a pion mass $m_\pi \sim 200$ MeV. Figure taken from [44]

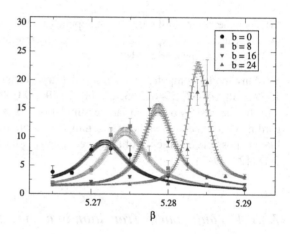

physical temperature, $T = 1/L_t a(\beta)$, is a monotonic, increasing function of β, and that, on the $16^3 \times 4$ lattice explored in [44], the magnetic field is given by $|e|B = (3\pi T^2/8)\, b \sim (0.03 \text{ GeV}^2)\, b$ around the transition.

Three facts are evident from Fig. 7.3. The chiral condensate increases, as a function of B, for all explored temperatures. The inflection points of the chiral condensate and of the Polyakov loop, signalling the location of the pseudo-critical temperature, move together towards higher temperatures as the magnetic field increases, meaning that deconfinement and chiral symmetry restoration do not disentangle, at least in the explored range of external fields. The drop (rise) of the chiral condensate (Polyakov loop) at the transition seems sharper and sharper as the magnetic field increases, meaning that the (pseudo)transition is strengthening.

Such facts are confirmed by Figs. 7.4 and 7.5, where the susceptibilities of the chiral condensate and of the Polyakov loop are reported, for different values of the magnetic field: the peaks move to higher β (hence to higher T) and become sharper and sharper as B increases, i.e. their height increases and their width decreases. The increased strength of the transition is also appreciable from the plaque-

Fig. 7.6 Plaquette
distribution at the
pseudocritical temperature for
$m_\pi \sim 200$ MeV and for
various different values of the
magnetic field. Figure taken
from [44]

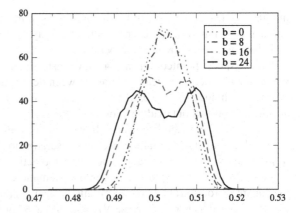

tte (pure gauge action) distribution at the pseudocritical coupling, which is reported
in Fig. 7.6 and which seems to evolve towards a double peak distribution, typical
of a first order transition, as B increases. Such hints for a change in the nature of
the transition, however, have not yet been confirmed by simulations on larger lattices. Regarding the increase in pseudo-critical temperature, we notice that it is quite
modest in magnitude and of the order of 2 % at $|e|B \sim 1$ GeV^2.

After the exploratory study of Ref. [44], new studies have appeared in the literature, which have added essential information and also opened new interesting
questions. Let us first consider the investigation reported in [15]. There are three
essential differences, with respect to [44], in the lattice discretization and approximation of QCD:

(1) still within a rooted staggered fermion formulation, improved gauge (tree level
    Symanzik improved) and fermionic (stout link) discretizations have been imple-
    mented, and different lattice spacings have been explored, in order to get control
    over the continuum limit;
(2) the authors have explored $N_f = 2 + 1$ QCD, i.e. they have considered strange
    quark effects;
(3) a physical quark mass spectrum has been adopted.

Concerning results, instead, the most striking difference with respect to [44]
is that the pseudocritical temperature decreases, instead of increasing, in presence of the magnetic background, in particular it becomes 10–20 % lower for
$|e|B \sim 1$ GeV^2. The authors of [15] have put the decrease of the critical temperature in connection with another unexpected phenomenon that they observe, namely
inverse magnetic catalysis, i.e. the fact that, at high enough temperatures, the chiral
condensate starts decreasing, instead of increasing, as a function of B. A possibility
pointed out in [16] is that inverse catalysis may originate from the gluonic sector,
i.e. the distribution of gluon fields may change, as an indirect effect of the magnetic field mediated by quark loops, so as to destroy the chiral condensate. We point
out that the study of [45] has shown instead that, at least at zero temperature, the

modified distribution of gluon fields contributes to increase magnetic catalysis (dynamical contribution), however the situation may indeed be quite different around and above the deconfinement temperature.

Other aspects pointed out in [44] have instead been confirmed by [15]: in particular the absence of a clear splitting between deconfinement and chiral symmetry restoration, induced by the external field, and an increased strength of the transition at $B \neq 0$, even without evidence for a change of its nature.

A different approach has been followed by the authors of Ref. [74]: they have explored 2 color QCD, making use of a discretization very close to that adopted in [44], i.e. with the same gauge and staggered action and a similar range of pion masses. However, to avoid the possible systematic effects connected with taking the square or fourth root of the fermionic determinant, they have considered a theory with four degenerate flavors, $N_f = 4$, all carrying the same mass and electric charge. Their results are in quite good agreement with those of [44]: deconfinement and chiral symmetry restoration do not split; the transition gets sharper; the pseudo-critical temperature increases as a function of B and no inverse catalysis is observed, i.e. the chiral condensate is an increasing function of B for all explored temperatures. The increase in T_c is also larger than what observed in [44], being of the order of 10 % at $|e|B \sim \text{GeV}^2$: that can possibly be explained by the fact that the authors of Ref. [74] make use of 4 flavors, instead of 2, moreover all carrying the charge of the u quark, hence the influence of the magnetic field on the system can be larger.

Another interesting aspect, explored in [74], regards the actual fate of chiral symmetry above the transition and in presence of the background field. The authors have explicitly verified, by performing an extrapolation to the chiral limit, that the chiral condensate vanishes in the high temperature phase, i.e. that chiral symmetry is exact above T_c, also in presence of the magnetic background. It is interesting to notice that, in the system explored in [74], the magnetic field does not break any of the flavor symmetries of the theory, since all quarks carry the same electric charge. However also in the standard case, in which quarks carry different charges, there are still diagonal chiral flavor generators which commute with the electric charge matrix and are not broken by the introduction of an external electromagnetic background field, so that an investigation similar to that performed in [74] should be done, to enquire about the actual realization of the unbroken symmetries above T_c.

To summarize, present lattice investigations about the influence of a strong magnetic field on the QCD phase diagram have given consistent indications about two facts: a magnetic field increases the strength of the QCD transition, however it does not change its nature, at least for $|e|B$ up to 1 GeV^2; in the same range of external fields, no significant splitting of deconfinement and chiral symmetry restoration has been detected.

There is a controversial issue, instead, regarding the location of the pseudocritical temperature, T_c, which increases as a function of $|e|B$ according to the results of [44, 74], while it decreases according to the results of [15]. It is perfectly possible that various improvements regarding the lattice discretization may turn the very slowly increasing function $T_c(B)$ determined in [44] into a decreasing function. The main point is to understand which aspect is more directly related to the change of behavior and to the appearance of inverse catalysis. Is it just a problem of lattice discretization and approach to the continuum limit, or can the difference be traced back to the larger pion mass used in [44]? Could instead the introduction of the strange quark explain the differences? We believe that a clear answer to those questions can be given by considering each single aspect separately, and that the answer will be important by itself, since it will help clarifying the origin of unexpected behaviors such as inverse magnetic catalysis.

7.4.2 Deconfinement Transition in a Chromomagnetic Background

A chromomagnetic background field influences directly gluodynamics, hence it makes sense to investigate its effects also in pure gauge simulations. A standard approach to introduce colored background fields has been described in Sect. 7.2.2 and is based on the formalism of the lattice Schrödinger functional [85, 86]: such approach has been adopted in the literature to study the influence of background fields both in pure Yang-Mills theories and in full QCD, and for various kinds of background fields, going from those corresponding to a uniform magnetic field to those produced by magnetic monopoles [11, 30–37]. In the following we shall consider the case of a constant background field [31–35].

Like a magnetic field, also a chromomagnetic background field leads to a shift of both chiral symmetry restoration, signalled by the drop of the chiral condensate, and of deconfinement, signalled by the rise of the Polyakov loop. The two transitions move together, moreover it is interesting to notice that the transition point coincides with the temperature at which the free energy (effective potential) of the background field shows a sudden change, the background field being screened (not screened) in the confined (deconfined) phase.

As an example, in Figs. 7.7 and 7.8, taken from [33–35], we show the behavior of the chiral condensate and of the Polyakov loop versus the inverse gauge coupling, β, together with the derivative of the free energy with respect to β, $f'(\beta)$, which shows a sharp peak at the same point at which chiral symmetry restoration and deconfinement take place. Results make reference to simulations of $N_f = 2$ QCD on a $32^3 \times 8$ lattice, with a standard pure gauge and staggered fermion formulation, a lattice spacing $a \simeq 0.15$ fm and a bare quark mass $am = 0.075$; the value of the magnetic background field is $gB = \pi/(16a^2) \sim 0.35$ GeV^2.

Present lattice results indicate that, both for pure gauge and full QCD, the transition temperature decreases in presence of a constant background chromomagnetic field. This is shown in Fig. 7.9 [33–35], where the critical temperature T_c (expressed in units of the parameter $\Lambda \sim 6$ MeV), is shown as a function of gB, both for the pure gauge theory and $N_f = 2$ QCD. The change in the critical temperature is, in general, much larger than what happens in presence of an electromagnetic back-

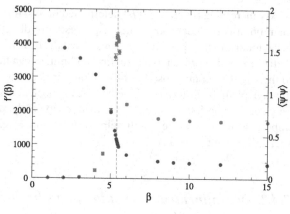

Fig. 7.7 Peak of the derivative of the free energy of the background field with respect to the inverse gauge coupling, together with the chiral condensate, for $N_f = 2$ QCD and for a background chromomagnetic field $gB = \pi/(16a^2) \sim 0.35\,\text{GeV}^2$. Figure taken from [33–35]

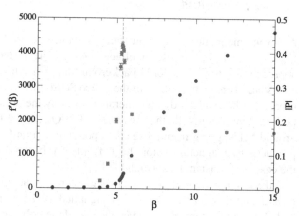

Fig. 7.8 Peak of the derivative of the free energy of the background field with respect to the inverse gauge coupling, the Polyakov loop, for $N_f = 2$ QCD and for a background chromomagnetic field $gB = \pi/(16a^2) \sim 0.35\,\text{GeV}^2$. Figure taken from [33–35]

ground: this can be interpreted in terms of the fact that a colored background field directly affects gluodynamics. An extrapolation of lattice results would even hint, as shown in Fig. 7.9, at the presence of a zero temperature deconfining transition, for $\sqrt{gB}$ of the order of 1 GeV. It would be interesting to investigate this possibility further, in the future, for the possible cosmological and astrophysical implications it could have. One should repeat the study of [33–35] in presence of more physical quark masses (with the values adopted in [33–35], m_π is of the order of 400–500 MeV) and closer to the continuum limit, also in order to reach higher values of the external field.

As a final comment, we notice that [33–35] also reports evidence about magnetic catalysis induced by the chromomagnetic field. Figure 7.10, in particular, shows the behavior of the chiral condensate for a few values of gB: it is clear that, at least around the transition, where data for all external fields are available, the chiral condensate grows as a function of gB. It is interesting to notice that in this case T_c decreases anyway, even in presence of normal catalysis, i.e. inverse catalysis does not seem to be a necessary condition for a decreasing critical temperature, at least in presence of finite quark masses.

Fig. 7.9 Dependence of the critical temperature T_c on the external chromomagnetic field for the pure gauge theory and for $N_f = 2$ QCD. Physical quantities are expressed in terms of the parameter $\Lambda \sim 6$ MeV

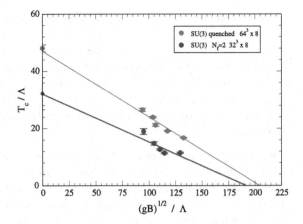

Fig. 7.10 Chiral condensate as a function of the inverse gauge coupling β for different values of the external chromomagnetic field $gB = \pi n_{ext}/(16a^2)$ and for $N_f = 2$ QCD. Figure taken from [33–35]

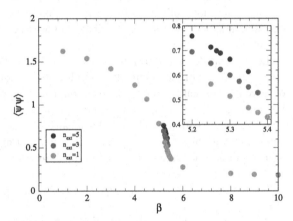

7.5 More on Gauge Field Modifications in External Electromagnetic Fields

The lattice results that we have discussed in the previous sections show that an electromagnetic background field can have a strong influence not only on quark dynamics, but also on gluodynamics, even if the interaction is not direct but mediated by quark loop effects. That does not come completely unexpected, at least in the strongly interacting, non-perturbative regime. Let us summarize a few facts:

(1) we have shown that large part of magnetic catalysis is due to the modification of the gauge field distribution (dynamical contribution), i.e. that the increment of chiral symmetry breaking is lower by about 40 % if gauge field configurations are sampled without taking into account the background field [45];

(2) it is known, from model computations, that a strong electromagnetic background can modify gluon screening properties and influence confinement dynamics, reducing the confinement scale [66, 90, 101]: such effects on gluodynamics have been proposed in [16] as a possible explanation for inverse catalysis;

(3) lattice computations have also explicitly shown that quantities related to quark confinement, like the Polyakov loop, are influenced by a background electromagnetic field, not only around the transition but also well inside the confined and the deconfined phases (see as an example the data reported in Fig. 7.2);

(4) the results of [74] have shown that the magnetic field can also induce an asymmetry in the non-Abelian plaquette values, especially around and above the transition, with a possible significant effect on the equation of state of the Quark-Gluon Plasma.

It is surely of great importance to study and understand these effects in a deeper and more systematic way. Is it possible, for instance, that electromagnetic background fields may induce chromoelectric or chromomagnetic background fields? That has been discussed in some recent model studies (see e.g. [66]) and could have important phenomenological consequences, both for cosmology and astrophysics and for heavy ion collisions. In the following we would like to discuss how lattice simulations could further contribute to the issue, with an accent on symmetry aspects.

Let us consider a uniform and constant magnetic background field. In order to understand some of the gluonic field modifications induced by it, it is interesting to observe that it breaks explicitly charge conjugation symmetry, and to ask how such breaking propagates to the gluon sector.

Charge conjugation on gluon fields means $A_\mu \to -A_\mu$ or, at the level of lattice gauge link variables, $U_\mu \to U_\mu^*$. Symmetry under it means that a gauge configuration $U_\mu(x)$ has the same path integral probability of its complex conjugate $U_\mu^*(x)$:

$$\mathscr{D}U\,\mathscr{P}[U] = \mathscr{D}U^*\,\mathscr{P}[U^*]. \tag{7.31}$$

As a consequence, all gluonic gauge invariant quantities, i.e. traces over closed parallel transports $W_C[U]$, including plaquettes and Wilson loops, must be real:

$$\langle W_C[U] \rangle^* = \langle W_C^*[U] \rangle = \langle W_C[U^*] \rangle = \int \mathscr{D}U\,\mathscr{P}[U]W_C[U^*]$$

$$= \int \mathscr{D}U^*\,\mathscr{P}[U^*]W_C[U] = \int \mathscr{D}U\,\mathscr{P}[U]W_C[U] = \langle W_C[U] \rangle \tag{7.32}$$

where the first equality holds if the path integral measure is real, the second express the fact that W_C is a trace over a product of links, while the fourth is simply a change of integration variables. Equation (7.32) may be violated by possible spontaneous symmetry breaking effects, which may affect some particular loops, as it happens for the Polyakov loop in the deconfined phase.

In presence of an electromagnetic background field, such property must be lost, since the breaking of charge conjugation symmetry will propagate from the quark to the gluonic sector: $\mathscr{P}[U] \neq \mathscr{P}[U^*]$ because of the contribution from the fermion determinant. Let us see that more explicitly, considering the particular case of the trace over a plaquette.

The most direct way to look for effective quark contributions to the gauge field distribution is to consider the loop expansion of the fermionic determinant. If we write the fermion matrix as $M = m_f \mathrm{Id} + D$, where D is the discretization of the Dirac operator and m_f is the bare quark mass, then the following formal expansion holds:

$$\det M = \exp\left(\mathrm{Tr}\log(m_f \mathrm{Id} + D)\right) \propto \exp\left(-\sum_k \frac{(-1)^k}{k\, m_f^k} \mathrm{Tr}\, D^k\right). \qquad (7.33)$$

D^k is made up of parallel transports connecting lattice sites; the trace operator implies that only closed parallel transports will contribute. Moreover, for each contributing loop, there is an equal contribution from its hermitian conjugate. If D is a standard, nearest neighbour discretization, the first non-trivial term comes for $k = 4$ and contains a coupling to the plaquette operator, which can be expressed as:

$$\frac{\Delta\beta(m_f)}{N_c} \frac{1}{2}\left(\mathrm{Tr}\, \Pi_{\mu\nu}(x) + \mathrm{Tr}\, \Pi_{\mu\nu}^\dagger(x)\right) = \frac{\Delta\beta(m_f)}{N_c} \mathrm{Re}\,\mathrm{Tr}\, \Pi_{\mu\nu}(x) \qquad (7.34)$$

where $\Delta\beta(m_f)$ is a coefficient which depends on m_f and on the particular fermion discretization adopted. The total effect can be viewed as a simple renormalization of the inverse bare gauge coupling, $\beta \to \beta + \sum_f \Delta\beta(m_f)$, where the sum runs over flavors.

Let us now consider the effect of an external electromagnetic field. That will change the definition of D, see e.g. (7.2), so that each loop contribution to the determinant will get a $U(1)$ phase from the external field. In particular, the expression in (7.34) for plaquettes will be modified into

$$\frac{\Delta\beta(m_f)}{N_c} \frac{1}{2}\left(e^{i\phi_{\mu\nu}(x)} \mathrm{Tr}\, \Pi_{\mu\nu}(x) + e^{-i\phi_{\mu\nu}(x)} \mathrm{Tr}\, \Pi_{\mu\nu}^\dagger(x)\right)$$

$$= \frac{\Delta\beta(m_f)}{N_c}\left(\cos(\phi_{\mu\nu}(x))\,\mathrm{Re}\,\mathrm{Tr}\, \Pi_{\mu\nu}(x) - \sin(\phi_{\mu\nu}(x))\,\mathrm{Im}\,\mathrm{Tr}\, \Pi_{\mu\nu}(x)\right). \qquad (7.35)$$

The phases $\phi_{\mu\nu}$ are in general non-trivial. For the particular case of a uniform magnetic field B in the $\hat{z}$ direction we have $\phi_{xy} = q_f B a^2$, where q_f is the quark charge. This implies that, apart from the case in which there is exact cancellation among quark flavors carrying equal masses and opposite electric charges (but this does not happen in real QCD), there will be a non-trivial coupling to the imaginary part of some plaquette traces, which will then develop a non-zero expectation value. Notice that this is exactly what happens also for the Polyakov loop in presence of an imaginary baryon chemical potential.

It is interesting to understand the meaning of a non-zero expectation value for the imaginary part of the trace of the plaquette, in terms of continuum quantities. Remember that, in the formal continuum limit, the plaquette operator is linked to the gauge field strength $G_{\mu\nu} = G_{\mu\nu}^a T^a$ as follows:

$$\Pi_{\mu\nu} \simeq \exp(i\, G_{\mu\nu} a^2). \qquad (7.36)$$

Considering the expansion of the exponential on the right hand side, the first non-trivial contribution to the imaginary part of its trace comes from the third order, which indeed is proportional to

$$-i \operatorname{Tr} G_{\mu\nu}^3 = -i \, G_{\mu\nu}^a G_{\mu\nu}^b G_{\mu\nu}^c \operatorname{Tr}\left(T^a T^b T^c\right) = \frac{1}{4}\left(f^{abc} - i \, d^{abc}\right) G_{\mu\nu}^a G_{\mu\nu}^b G_{\mu\nu}^c$$

$$(7.37)$$

where no sum over μ, ν is understood. A non-zero expectation value for the imaginary part of the plaquette, therefore, corresponds to a non-zero charge odd, three gluon condensate, constructed in terms of the symmetric tensor d^{abc}:

$$\left\langle d^{abc} G_{\mu\nu}^a G_{\mu\nu}^b G_{\mu\nu}^c\right\rangle \neq 0.$$

$$(7.38)$$

Such condensate, which is zero by charge conjugation symmetry in the normal QCD vacuum, will develop a non-zero value along some directions, e.g. for $(\mu, \nu) = (x, y)$ in case of a magnetic field in the $\hat{z}$ direction, and can be considered as a non-trivial, charge-odd gluon background induced by the external field. Of course the effect will be present only for $N_c \geq 3$ colors, since for $SU(2)$ all traces are real and indeed $d^{abc} = 0$.

The loop expansion of the determinant, examined above, is just a way to show explicitly that such effects can be present, but cannot be quantitative, since the expansion is purely formal (apart from the limit of large quark masses). A systematic investigation can and must be performed in the future by means of lattice simulations.

Similarly to charge conjugation symmetry, one can investigate electromagnetic backgrounds which break explicitly the symmetry under CP, i.e. such that $\mathbf{E} \cdot \mathbf{B} \neq 0$, and ask how CP violation propagates to the gluon sector, giving rise to an effective θ parameter. Such phenomenon is in some sense complementary to the chiral magnetic effect and is related in general to the effective QED-QCD interactions in the pseudoscalar channel [12, 53, 54, 92]. It has been studied recently by a first exploratory lattice investigation [46], where numerical simulations at imaginary electric fields plus analytic continuation have been exploited to determine the susceptibility χ_{CP} of the QCD vacuum to CP-odd electromagnetic fields, defined by $\theta_{\text{eff}} \simeq \chi_{CP} e^2 \mathbf{E} \cdot \mathbf{B}$, obtaining $\chi_{CP} \sim 7 \text{ GeV}^{-4}$ for QCD with two staggered flavors and $m_\pi \sim 480$ MeV.

Finally, let us discuss a few other lines along which the effects of electromagnetic background fields on the gluon sector could be further investigated. The first consists in the determination of the effective action of given gluonic backgrounds, using the techniques described in Sect. 7.2.2, but in combination with non-trivial electromagnetic external fields: that would give the possibility of studying if the latter can give rise to instabilities in the gluon sector, leading to the generation of non-trivial gluonic backgrounds (as proposed e.g. in [66]). The second regards the measurement of glueball masses and screening masses in presence of magnetic backgrounds, which could give the possibility, for instance, to test the proposal given in [90] about the lowering of the confinement scale in presence of a strong magnetic field. Finally, it would be important to perform a careful investigation about the influence of magnetic backgrounds on the QCD equation of state, as proposed in [74].

7.6 Conclusions

Numerical simulations in presence of external fields have been considered since the early stages of Lattice QCD. The last few years, however, have seen the development of considerable activity on the subject, which has been driven by the recent interest in theoretical and phenomenological issues regarding the behavior of strongly interacting matter in magnetic fields.

In this review, after a general overview about the formulation of lattice QCD in presence of magnetic or chromomagnetic background fields, we have discussed the present status of lattice studies regarding magnetic catalysis and the QCD phase diagram. In the low temperature phase, consistent results have been obtained, by different groups, which confirm magnetic catalysis and the predictions coming from chiral perturbation theory and some effective models. Good part of the enhancement of chiral symmetry breaking seems to be associated with modifications of gluon fields induced by the magnetic field via quark loops. A chromomagnetic background induces magnetic catalysis as well.

In the high temperature phase, consistent results have been obtained regarding the fact that a magnetic or chromomagnetic field shifts both chiral symmetry restoration and deconfinement and that the two transitions do not split, at least for fields up to $O(1)$ GeV^2. A magnetic field also leads to a strengthening of the transition, which however does not seem to turn into a strong first order, at least for fields up to $O(1)$ GeV^2; it would be interesting however, to further investigate the issue also in combination with other external parameters, like a baryon chemical potential, since that could be relevant to the search for a critical endpoint in the QCD phase diagram.

Regarding the shift in the pseudocritical temperature induced by a background magnetic field, existing studies have shown that a modification of the lattice implementation of QCD can change the outcome, going from a slight increase (of the order a few % for $|e|B$ of $O(1)$ GeV^2) to a decrease (of the order of 10 % for $|e|B$ of $O(1)$ GeV^2) when an improved discretization and a physical quark mass spectrum are used; in the latter case one also observes inverse magnetic catalysis, i.e. a decrease of the chiral condensate as a function of B, in the high temperature, deconfined phase. A decrease of the pseudocritical temperature is also observed in presence of a chromomagnetic background field. A remaining question is which effect is more relevant to explain the discrepancy among different lattice studies, i.e. whether lattice artifacts, or the unphysical quark spectrum, or both are at its origin.

One of the aspects that emerges more clearly from present studies is the fact that gluon fields are strongly modified by external electromagnetic fields, even though indirectly, by means of dynamical quark loops. We believe that this aspect can and should be investigated more systematically by lattice simulations: in Sect. 7.5 we have discussed and suggested a few possible ways to do that by future studies.

Acknowledgements The author is grateful to P. Cea, L. Cosmai, M. Mariti, S. Mukherjee, F. Negro and F. Sanfilippo for collaboration on some of topics discussed in this review. He also acknowledges M. Chernodub, G. Endrodi, E. Fraga, K. Fukushima, V. Miransky and M. Ruggieri for many useful discussions.

References

1. M. Abramczyk, T. Blum, G. Petropoulos, R. Zhou, PoS **LAT2009**, 181 (2009)
2. N.O. Agasian, S.M. Fedorov, Phys. Lett. B **663**, 445 (2008)
3. N.O. Agasian, I.A. Shushpanov, Phys. Lett. B **472**, 143 (2000)
4. M.S. Alam, V.S. Kaplunovsky, A. Kundu, J. High Energy Phys. **1204**, 111 (2012)
5. T. Albash, V.G. Filev, C.V. Johnson, A. Kundu, J. High Energy Phys. **0807**, 080 (2008)
6. A. Alexandru, F.X. Lee, PoS **LAT2008**, 145 (2008). arXiv:0810.2833 [hep-lat]
7. A. Alexandru, F.X. Lee, PoS **LAT2009**, 144 (2009). arXiv:0911.2520 [hep-lat]
8. M.H. Al-Hashimi, U.J. Wiese, Ann. Phys. **324**, 343 (2009)
9. J.O. Andersen, A. Tranberg, arXiv:1204.3360 [hep-ph]
10. J.O. Andersen, R. Khan, Phys. Rev. D **85**, 065026 (2012)
11. S. Antropov, M. Bordag, V. Demchik, V. Skalozub, Int. J. Mod. Phys. A **26**, 4831 (2011)
12. M. Asakawa, A. Majumder, B. Muller, Phys. Rev. C **81**, 064912 (2010)
13. S.S. Avancini, D.P. Menezes, M.B. Pinto, C. Providencia, Phys. Rev. D **85**, 091901 (2012)
14. A.Y. Babansky, E.V. Gorbar, G.V. Shchepanyuk, Phys. Lett. B **419**, 272–278 (1998)
15. G.S. Bali, F. Bruckmann, G. Endrodi, Z. Fodor, S.D. Katz, S. Krieg, A. Schafer, K.K. Szabo, J. High Energy Phys. **1202**, 044 (2012)
16. G.S. Bali, F. Bruckmann, G. Endrodi, Z. Fodor, S.D. Katz, A. Schafer, arXiv:1206.4205 [hep-lat]
17. G.S. Bali, F. Bruckmann, M. Constantinou, M. Costa, G. Endrodi, S.D. Katz, H. Panagopoulos, A. Schaefer, arXiv:1209.6015 [hep-lat]
18. T. Banks, A. Casher, Nucl. Phys. B **169**, 103 (1980)
19. O. Bergman, G. Lifschytz, M. Lippert, Phys. Rev. D **79**, 105024 (2009)
20. C.W. Bernard, T. Draper, K. Olynyk, M. Rushton, Phys. Rev. Lett. **49**, 1076 (1982)
21. J.K. Boomsma, D. Boer, Phys. Rev. D **81**, 074005 (2010)
22. V.V. Braguta, P.V. Buividovich, M.N. Chernodub, M.I. Polikarpov, Phys. Lett. B **718**, 667 (2012). arXiv:1104.3767 [hep-lat]
23. V.V. Braguta, P.V. Buividovich, T. Kalaydzhyan, S.V. Kuznetsov, M.I. Polikarpov, PoS **LAT2010**, 190 (2010). arXiv:1011.3795 [hep-lat]
24. P.V. Buividovich, M.N. Chernodub, D.E. Kharzeev, T. Kalaydzhyan, E.V. Luschevskaya, M.I. Polikarpov, Phys. Rev. Lett. **105**, 132001 (2010)
25. P.V. Buividovich, M.N. Chernodub, E.V. Luschevskaya, M.I. Polikarpov, Phys. Rev. D **80**, 054503 (2009)
26. P.V. Buividovich, M.N. Chernodub, E.V. Luschevskaya, M.I. Polikarpov, Phys. Lett. B **682**, 484 (2010)
27. P.V. Buividovich, M.N. Chernodub, E.V. Luschevskaya, M.I. Polikarpov, Nucl. Phys. B **826**, 313 (2010)
28. P.V. Buividovich, M.N. Chernodub, E.V. Luschevskaya, M.I. Polikarpov, Phys. Rev. D **81**, 036007 (2010)
29. P. Cea, Int. J. Mod. Phys. D **13**, 1917 (2004)
30. P. Cea, L. Cosmai, Phys. Lett. B **264**, 415 (1991)
31. P. Cea, L. Cosmai, J. High Energy Phys. **0302**, 031 (2003)
32. P. Cea, L. Cosmai, J. High Energy Phys. **0508**, 079 (2005)
33. P. Cea, L. Cosmai, M. D'Elia, J. High Energy Phys. **0712**, 097 (2007)
34. P. Cea, L. Cosmai, M. D'Elia, PoS **LAT2006**, 062 (2006). hep-lat/0610014

35. P. Cea, L. Cosmai, M. D'Elia, PoS **LAT2007**, 295 (2007). arXiv:0710.1449 [hep-lat]
36. P. Cea, L. Cosmai, M. D'Elia, J. High Energy Phys. **0402**, 018 (2004)
37. P. Cea, L. Cosmai, A.D. Polosa, Phys. Lett. B **392**, 177 (1997)
38. B. Chatterjee, H. Mishra, A. Mishra, Phys. Rev. D **84**, 014016 (2011)
39. M.N. Chernodub, Phys. Rev. D **82**, 085011 (2010)
40. M.N. Chernodub, Phys. Rev. Lett. **106**, 142003 (2011)
41. M.N. Chernodub, Phys. Rev. D **86**, 107703 (2012). arXiv:1209.3587 [hep-ph]
42. T.D. Cohen, D.A. McGady, E.S. Werbos, Phys. Rev. C **76**, 055201 (2007)
43. P.H. Damgaard, U.M. Heller, Nucl. Phys. B **309**, 625 (1988)
44. M. D'Elia, S. Mukherjee, F. Sanfilippo, Phys. Rev. D **82**, 051501 (2010)
45. M. D'Elia, F. Negro, Phys. Rev. D **83**, 114028 (2011)
46. M. D'Elia, M. Mariti, F. Negro, Phys. Rev. Lett. **110**, 082002 (2013). arXiv:1209.0722 [hep-lat]
47. W. Detmold, B.C. Tiburzi, A. Walker-Loud, Phys. Rev. D **73**, 114505 (2006). hep-lat/0603026
48. W. Detmold, B.C. Tiburzi, A. Walker-Loud, Phys. Rev. D **79**, 094505 (2009). arXiv:0904.1586 [hep-lat]
49. W. Detmold, B.C. Tiburzi, A. Walker-Loud, Phys. Rev. D **81**, 054502 (2010). arXiv:1001.1131 [hep-lat]
50. R.C. Duncan, C. Thompson, Astrophys. J. **392**, L9 (1992)
51. D. Ebert, V.V. Khudyakov, V.C. Zhukovsky, K.G. Klimenko, Phys. Rev. D **65**, 054024 (2002)
52. D. Ebert, K.G. Klimenko, M.A. Vdovichenko, A.S. Vshivtsev, Phys. Rev. D **61**, 025005 (2000)
53. H.T. Elze, B. Muller, J. Rafelski, hep-ph/9811372
54. H.T. Elze, J. Rafelski, in *Sandansky 1998, Frontier tests of QED and physics of the vacuum*, pp. 425–439. hep-ph/9806389
55. N. Evans, T. Kalaydzhyan, K.-y. Kim, I. Kirsch, J. High Energy Phys. **1101**, 050 (2011)
56. S. Fayazbakhsh, N. Sadooghi, Phys. Rev. D **83**, 025026 (2011)
57. G.N. Ferrari, A.F. Garcia, M.B. Pinto, arXiv:1207.3714 [hep-ph]
58. V.G. Filev, C.V. Johnson, R.C. Rashkov, K.S. Viswanathan, J. High Energy Phys. **0710**, 019 (2007)
59. E.S. Fraga, A.J. Mizher, Phys. Rev. D **78**, 025016 (2008)
60. E.S. Fraga, L.F. Palhares, Phys. Rev. D **86**, 016008 (2012). arXiv:1201.5881 [hep-ph]
61. E.S. Fraga, J. Noronha, L.F. Palhares, arXiv:1207.7094 [hep-ph]
62. K. Fukushima, Y. Hidaka, arXiv:1209.1319 [hep-ph]
63. K. Fukushima, D.E. Kharzeev, H.J. Warringa, Phys. Rev. D **78**, 074033 (2008)
64. K. Fukushima, J.M. Pawlowski, arXiv:1203.4330 [hep-ph]
65. K. Fukushima, M. Ruggieri, R. Gatto, Phys. Rev. D **81**, 114031 (2010)
66. B.V. Galilo, S.N. Nedelko, Phys. Rev. D **84**, 094017 (2011)
67. R. Gatto, M. Ruggieri, Phys. Rev. D **82**, 054027 (2010)
68. R. Gatto, M. Ruggieri, Phys. Rev. D **83**, 034016 (2011)
69. R. Gatto, M. Ruggieri, arXiv:1207.3190 [hep-ph]
70. A. Goyal, M. Dahiya, Phys. Rev. D **62**, 025022 (2000)
71. V.P. Gusynin, V.A. Miransky, I.A. Shovkovy, Phys. Rev. Lett. **73**, 3499 (1994). [Erratum. Phys. Rev. Lett. **76**, 1005 (1996)]
72. V.P. Gusynin, V.A. Miransky, I.A. Shovkovy, Phys. Lett. B **349**, 477 (1995)
73. Y. Hidaka, A. Yamamoto, arXiv:1209.0007 [hep-ph]
74. E.-M. Ilgenfritz, M. Kalinowski, M. Muller-Preussker, B. Petersson, A. Schreiber, Phys. Rev. D **85**, 114504 (2012)
75. D.N. Kabat, K.-M. Lee, E.J. Weinberg, Phys. Rev. D **66**, 014004 (2002)
76. K. Kashiwa, Phys. Rev. D **83**, 117901 (2011)
77. S. Kawati, G. Konisi, H. Miyata, Phys. Rev. D **28**, 1537–1541 (1983)
78. D.E. Kharzeev, L.D. McLerran, H.J. Warringa, Nucl. Phys. A **803**, 227 (2008)

79. S.P. Klevansky, R.H. Lemmer, Phys. Rev. D **39**, 3478 (1989)
80. K.G. Klimenko, Z. Phys. C **54**, 323 (1992)
81. K.G. Klimenko, Theor. Math. Phys. **94**, 393 (1993)
82. K.G. Klimenko, B.V. Magnitsky, A.S. Vshivtsev, Nuovo Cimento A **107**, 439–452 (1994)
83. C.V. Johnson, A. Kundu, J. High Energy Phys. **0812**, 053 (2008)
84. A.D. Linde, Phys. Lett. B **62**, 435 (1976)
85. M. Luscher, R. Narayanan, P. Weisz, U. Wolff, Nucl. Phys. B **384**, 168 (1992)
86. M. Luscher, P. Weisz, Nucl. Phys. B **452**, 213 (1995)
87. E.V. Luschevskaya, O.V. Larina, arXiv:1203.5699 [hep-lat]
88. G. Martinelli, G. Parisi, R. Petronzio, F. Rapuano, Phys. Lett. B **116**, 434 (1982)
89. S. Mereghetti, Astron. Astrophys. Rev. **15**, 225–287 (2008)
90. V.A. Miransky, I.A. Shovkovy, Phys. Rev. D **66**, 045006 (2002)
91. A.J. Mizher, M.N. Chernodub, E.S. Fraga, Phys. Rev. D **82**, 105016 (2010)
92. M.M. Musakhanov, F.C. Khanna, hep-ph/9605232
93. S.-i. Nam, C.-W. Kao, Phys. Rev. D **83**, 096009 (2011)
94. F. Preis, A. Rebhan, A. Schmitt, J. High Energy Phys. **1103**, 033 (2011)
95. F. Preis, A. Rebhan, A. Schmitt, J. Phys. G **39**, 054006 (2012). arXiv:1209.4468 [hep-ph]
96. A. Rabhi, C. Providencia, Phys. Rev. C **83**, 055801 (2011)
97. E. Rojas, A. Ayala, A. Bashir, A. Raya, Phys. Rev. D **77**, 093004 (2008)
98. A. Salam, J.A. Strathdee, Nucl. Phys. B **90**, 203 (1975)
99. S. Schramm, B. Muller, A.J. Schramm, Mod. Phys. Lett. A **7**, 973–982 (1992)
100. E. Shintani, S. Aoki, N. Ishizuka, K. Kanaya, Y. Kikukawa, Y. Kuramashi, M. Okawa, A. Ukawa et al., Phys. Rev. D **75**, 034507 (2007)
101. I.A. Shovkovy, arXiv:1207.5081 [hep-ph]
102. I.A. Shushpanov, A.V. Smilga, Phys. Lett. B **402**, 351 (1997)
103. V. Skokov, Phys. Rev. D **85**, 034026 (2012)
104. V. Skokov, A.Y. Illarionov, V. Toneev, Int. J. Mod. Phys. A **24**, 5925 (2009)
105. J. Smit, J.C. Vink, Nucl. Phys. B **286**, 485 (1987)
106. H. Suganuma, T. Tatsumi, Ann. Phys. **208**, 470 (1991)
107. G. 't Hooft, Nucl. Phys. B **153**, 141 (1979)
108. B.C. Tiburzi, PoS **LAT2011**, 020 (2011). arXiv:1110.6842 [hep-lat]
109. B.C. Tiburzi, Nucl. Phys. A **814**, 74 (2008)
110. B.C. Tiburzi, Phys. Lett. B **674**, 336 (2009)
111. T. Vachaspati, Phys. Lett. B **265**, 258 (1991)
112. A. Vilenkin, Phys. Rev. D **22**, 3080 (1980)
113. K.G. Wilson, Phys. Rev. D **10**, 2445 (1974)
114. A. Yamamoto, Phys. Rev. Lett. **107**, 031601 (2011)
115. A.V. Zayakin, J. High Energy Phys. **0807**, 116 (2008)

Chapter 8
𝒫-Odd Fluctuations in Heavy Ion Collisions. Deformed QCD as a Toy Model

Ariel R. Zhitnitsky

8.1 Introduction and Motivation

Recently it has become clear that quantum anomalies play very important role in the macroscopic dynamics of relativistic fluids. Much of this progress is motivated by very interesting ongoing experiments on local $\mathcal{P}$ and $\mathcal{CP}$ violation in QCD as studied at RHIC and ALICE at the LHC [1–12]. It is likely that the observed asymmetry is due to charge separation effect [13, 14] as a result of the chiral anomaly, see details below.

We start with review of the charge separation effect [13, 14] which can be explained in the following simple way. Let us assume that an effective $\theta(\mathbf{x}, t)_{\text{ind}} \neq 0$ is induced as a result of some non-equilibrium dynamics as suggested in Refs. [15–18]. The $\theta(\mathbf{x}, t)_{\text{ind}}$ parameter enters the effective Lagrangian as follows, $\mathscr{L}_\theta = -\theta_{\text{ind}}q$ where $q \equiv \frac{g^2}{64\pi^2}\varepsilon_{\mu\nu\rho\sigma}G^{a\mu\nu}G^{a\rho\sigma}$ such that local $\mathcal{P}$ and $\mathcal{CP}$ invariance of QCD is broken on the scales where correlated $\theta(\mathbf{x}, t)_{\text{ind}} \neq 0$ is induced. As a result of this violation, one should expect a number of $\mathcal{P}$ and $\mathcal{CP}$ violating effects taking place in the region where $\theta(\mathbf{x}, t)_{\text{ind}} \neq 0$.

This area of research became a very active field in recent years mainly due to very interesting ongoing experiments [1–12]. There is a number of different manifestations of this local $\mathcal{P}$ and $\mathcal{CP}$ violation, see [19–23] and many additional references therein. In particular, in the presence of an external magnetic field **B** or in case of the rotating system with angular velocity $\boldsymbol{\Omega}$ there will be induced electric current directed along **B** or $\boldsymbol{\Omega}$ correspondingly, resulting in separation of charges along those directions as mentioned above. One can interpret the same effects as a generation of induced electric field **E** directed along **B** or $\boldsymbol{\Omega}$ resulting in corresponding electric current flowing along $\mathbf{J} \sim \mathbf{B}$ or $\mathbf{J} \sim \boldsymbol{\Omega}$ directions. All these

A.R. Zhitnitsky (✉)
Department of Physics & Astronomy, University of British Columbia, Vancouver, BC V6T 1Z1, Canada
e-mail: arz@phas.ubc.ca

D. Kharzeev et al. (eds.), *Strongly Interacting Matter in Magnetic Fields*,
Lecture Notes in Physics 871, DOI 10.1007/978-3-642-37305-3_8,
© Springer-Verlag Berlin Heidelberg 2013

phenomena are obviously $\mathscr{P}$ and $\mathscr{CP}$ odd effects. Non-dissipating, induced vector current density has the form:

$$\mathbf{J} = \frac{e\mathbf{B}}{2\pi^2} \dot{\theta}(t)_{\text{ind}}, \tag{8.1}$$

where $\mathscr{P}$ odd effect is explicitly present in this expression as $\dot{\theta}(t)_{\text{ind}}$ after a corresponding $U(1)_A$ chiral time-dependent rotation can be interpreted as the combination $(\mu_L - \mu_R)$, which is the difference of chemical potentials of the right μ_R and left μ_L handed fermions, see also [21] for a physical interpretation of the relation $(\mu_L - \mu_R) = \dot{\theta}(t)_{\text{ind}}$. It is important to emphasize that the region where $\langle \theta(\mathbf{x}, t)_{\text{ind}} \rangle \neq 0$ should be much larger in size than the scale of conventional QCD fluctuations with correlation length $\sim \Lambda_{\text{QCD}}^{-1}$. Otherwise, the expression (8.1) derived under condition that $\dot{\theta}(t)_{\text{ind}} \ll \Lambda_{\text{QCD}}$ can not be trusted. In different words, if $\mathscr{P}$-odd fluctuations represented by $\dot{\theta}(t)_{\text{ind}}$ have typical wavelengths $\lambda_\theta \lesssim \Lambda_{\text{QCD}}^{-1}$ than the Effective Lagrangian approach can not be justified, formula (8.1) can not be trusted, and some other technique must be used instead.

Closely related phenomena have been previously discussed in physics of neutrinos [24–26] and quantum wires [27]. In QCD context formula (8.1) has been used in applications to neutron star physics where magnetic field is known to be large, and the corresponding $(\mu_L - \mu_R) \neq 0$ can be generated in neutron star environment as a result of continuos $\mathscr{P}$ violating processes happening in nuclear matter [28, 29]. It has been also applied to heavy ion collisions where an effective $(\mu_L - \mu_R) \neq 0$ is locally induced. The effect was estimated using the sphaleron transitions generating the topological charge density in the QCD plasma [19, 20]. The effect was coined as "chiral magnetic effect" (CME) [19, 20]. Formula (8.1) has been also derived a numerous number of times using varies techniques such as: effective Lagrangian approach developed in [30]; explicit mode's summation [31]; direct lattice computations [22, 23]. In addition, the effect has been studied in holographic models of QCD [32–38], see review [39] with large number of references on recent developments.

To conclude this introduction: on the theoretical side the effect is well established phenomenon. However, the crucial questions for the applications of the CME to heavy ion collisions is a correlation length of the induced $\langle \theta(\mathbf{x}, t)_{\text{ind}} \rangle \neq 0$: why the $\mathscr{P}$ odd domains are large, much larger than conventional $\Lambda_{\text{QCD}}^{-1}$ scale? Apparently, a relatively large correlation length is a required feature for interpretation of the observed asymmetry [1–12] in terms of CME as the conventional QCD vacuum processes are too small to explain the observed asymmetry [40]. This is in fact, the key element to be addressed in this review: why the correlation length of $\mathscr{P}$ odd fluctuations could be large in heavy ion collisions? We use a simplified version of QCD in order to answer this hard question. We want to see whether the long range order indeed emerges in "deformed QCD" model. We also want to understand the nature of this long range order which is apparently observed in the lattice simulations.

The review is organized as follows. In next Sect. 8.2 we review a standard derivation of the charge separation (CSE) and related effects such as Chiral Magnetic effect (CME), chiral Vortical Effects (CVE) using effective Lagrangian approach. Section 8.3 is a short detour where we review a number of independent lattice simulations, not related to CME/CVE effects, suggesting that long range order is indeed present in the system. In Sects. 8.4, 8.5, 8.6 we demonstrate that some kind of long range order indeed may emerge in "deformed QCD". It appears as a result of highly nontrivial topological features of this model which are known to be present in real strongly coupled QCD. Therefore, it is tempting to identify this long range order in "deformed QCD" with long range correlations observed in lattice simulations and reviewed in Sect. 8.3. We conjecture that a similar effect may also emerge in real strongly coupled QCD. In Sect. 8.7 we apply the results from previous sections to study CSE/CME/CVE in deformed QCD. We compute the magnitudes of these topological phenomena in deformed QCD. This toy model explicitly shows that the large observed intensity of the effect as studied at RHIC and ALICE at the LHC [1–12] might be due to a coherent phenomena when the large observed asymmetry is a result of accumulation of a small effect over large distances $\mathbb{L} \gg \Lambda_{QCD}^{-1}$. In conclusion we comment on a very deep relation between long range order observed in lattice simulations (which normally studied by measurement of the topological density distribution) and CSE/CME/CVE effects.

8.2 Quantum Anomalies. Effective Lagrangian Approach

Let us assume, following the proposals outlined above, that a dynamical local fluctuation of θ_{ind}-angle can be excited in QCD matter. From the viewpoint of the effective Lagrangian, $\theta_{\text{ind}}(x)$ is equivalent to a pseudo-scalar flavor-singlet quark–anti-quark field which couples to electromagnetism through the electric charges of its quark constituents as prescribed by axial anomaly:

$$L = \frac{1}{2}\mathbf{E}^2 - \frac{1}{2}\mathbf{B}^2 + N_c \sum_f \frac{e_f^2}{4\pi^2} \cdot \left(\frac{\theta_{\text{ind}}}{N_f} \right) (\mathbf{E} \cdot \mathbf{B}), \qquad (8.2)$$

where the sum runs over quark flavors f, and N is the number of colours. We treat (8.2) as an effective Lagrangian when all "fast" degrees of freedom are integrated out while "slow" degrees of freedom are explicitly present in expression (8.2). In particular, we assume that θ_{ind} is "slow" background field with typical wavelengths $\lambda_\theta \gg \Lambda_{QCD}^{-1}$, in which case the effective Lagrangian approach is justified.

8.2.1 Charge Separation Effect (CSE)

Let us now minimize the action density (8.2) with respect to the electric field E:

$$\frac{\delta L}{\delta E} = \mathbf{E} + N \sum_f \frac{e_f^2}{4\pi^2} \cdot \left(\frac{\theta_{\text{ind}}}{N_f} \right) \mathbf{B} = 0; \qquad (8.3)$$

we see that the magnetic field in the presence of $\theta_{\text{ind}} \neq 0$ generates an electric field $E \sim \theta_{\text{ind}} \cdot B$. Let us now consider the case of a uniform magnetic field B_z pointing in the z direction; we will assume that the field does not depend on x and y coordinates, so we are dealing with an effectively 2-dimensional theory. In this case, because of the quantization of the flux required by the single-valuedness of the particle wave function, we substitute $\int d^2x_\perp B_z = \Phi/e = 2\pi l/e$ into (8.3); l is an integer. We thus get [14]:

$$\mathbb{L}_x \mathbb{L}_y E_z = -\left(\frac{e\,\theta_{\text{ind}}}{2\pi}\right)l, \tag{8.4}$$

where $\mathbb{L}$ is the size of the system and we take $N_f = N = 1$, $e_a = e$ to simplify things. Therefore, the electric field along z will be induced in the presence of nonzero $\theta_{\text{ind}} \neq 0$ when magnetic field is applied along the z direction. For this simple geometry, the electric field E_z between two infinitely large charged plates with charge density σ_{xy} is exactly equal $E_z = \sigma_{xy}$. Therefore, (8.4) can be interpreted as CSE in the presence of the magnetic field if large θ_{ind} domain is formed,

$$Q \equiv \int dx\,dy\,\sigma_{xy} = (\mathbb{L}_x \mathbb{L}_y)E_z = -\left(\frac{e\,\theta_{\text{ind}}}{2\pi}\right)l. \tag{8.5}$$

8.2.2 Chiral Magnetic Effect (CME)

One can understand the same CSE also in a different way. Anomalous coupling (8.2) also implies that there will be induced current as a result of coordinate dependence of the induced $\theta_{\text{ind}}(x,t)$, which is convenient to represent in vector notations as follows

$$J_0 = \frac{e^2}{4\pi^2}\nabla\theta_{\text{ind}} \cdot \mathbf{B}, \qquad \mathbf{J} = \frac{e^2}{4\pi^2}\dot{\theta}_{\text{ind}}\mathbf{B}, \tag{8.6}$$

where we again take $N_f = N = 1$, $e_a = e$ to simplify things, and assume that the external magnetic field $\mathbf{B}$ is coordinate and time-independent. The expression similar to (8.6) for the anomalous current has been studied previously in literature in many fields, including particle physics, cosmology, condensed matter physics. In particular, in the context of the axion physics when $\theta_{\text{ind}}(x,t)$ is dynamical axion field, (8.6) was extensively discussed in [41].

In the present context relevant for the CME formula (8.6) reduces to our previous formula (8.5). Indeed, integrating $\int J_0 d^3x$ leads precisely to the known expression for the charge separation effect along z,

$$Q = \int d^3x\,J_0 = \frac{e^2}{4\pi^2}\int_0^{+\infty} dz\frac{d\theta_{\text{ind}}(z)}{dz}\int d^2x_\perp B^z = -\left(\frac{e\,\theta_{\text{ind}}}{2\pi}\right)l, \tag{8.7}$$

where we assume that $\theta_{\text{ind}}(z=0) = \theta_{\text{ind}}$ inside the cylinder and $\theta_{\text{ind}}(z=+\infty) = 0$ outside the domain with $\theta_{\text{ind}} \neq 0$. We also take into account the flux quantization,

$\int d^2 x_\perp B_{\text{ext}}^z = \Phi/e = 2\pi l/e$. In addition, the expression for **J** in (8.6) can be presented in much more familiar way if one replaces $\dot{\theta}_{\text{ind}} \to (\mu_L - \mu_R)$ to bring it to conventional form which is normally used in CME studies.

The electric field (8.4) aligned along the vector of angular momentum (and thus perpendicular to the reaction plane) will act on charged particles and cause the electric charge separation (8.5), (8.7) signalling the violation of $\mathscr{P}$ and $\mathscr{CP}$ invariances arising from $\theta_{\text{ind}} \neq 0$, which is precisely CSE in the presence of the external magnetic field [14].

We emphasize once again that formulae (8.4), (8.5), (8.6), (8.7) can only be trusted for slow-varying external background fields θ_{ind} and **B** when the effective Lagrangian approach is justified. In different words, it is assumed that the effective Lagrangian (8.2) describes slow degrees of freedom with typical wavelengths $\lambda_\theta \gg \Lambda_{\text{QCD}}^{-1}$, while fast degrees of freedom are already integrated out in formula (8.2). This condition can be easily satisfied for **B** field as it is truly external field which it is not originated from strongly coupled QCD dynamics. At the same time $\theta_{\text{ind}}(x) \neq 0$ is an internal QCD parameter. Why this parameter $\theta_{\text{ind}}(x)$ is so special that it may indeed demonstrate a long range order with typical $\lambda_\theta \gg \Lambda_{\text{QCD}}^{-1}$? This is the key question which will be addressed later in the text in Sects. 8.5, 8.6, 8.7.

Our last comment is about terminology. All critical elements on induced currents in presence of the background magnetic field **B** and $\theta_{\text{ind}}(x) \neq 0$ have been discussed in Ref. [14], though the term "chiral magnetic effect" (CME) itself was invented later in Ref. [20] when some specific applications to heavy ion collisions were considered.

8.2.3 Chiral Vortical Effect (CVE)

Now we consider a rotating system which is characterized by chemical potential μ and angular velocity $\mathbf{\Omega}$. We want to derive formulae similar to (8.4)–(8.7) where role of B is played by $\mathbf{\Omega}$. The corresponding technique with nonzero μ was developed in [30] and generalized for the axion field (nonzero $\theta(x)$) in [31]. The key idea is to introduce a fictitious vector gauge field V_μ which is nothing but 4-velocity of matter v_μ with chemical potential μ, see technical details in [30]. After that, one can derive an anomalous effective Lagrangian similar to (8.2) where auxiliary V_μ replaces the usual electromagnetic gauge field A_μ.

In this review we limit ourselves only to one term in the anomalous effective Lagrangian which plays the crucial role for the present study, see [30] and [31] for details:

$$L_{\theta\gamma V} = -N_c \sum_f \frac{e_f \mu_f}{4\pi^2 N_f} \cdot \varepsilon^{\mu\nu\lambda\sigma} \partial_\mu \theta_{\text{ind}} (\partial_\lambda V_\nu) A_\sigma, \tag{8.8}$$

where V_μ is a fictitious vector gauge field. The coupling of quarks to V_μ and to the usual electromagnetic gauge field A_μ are almost identical; the only difference is in the coupling constants. This new term (8.8) leads to a few very unusual phenomena. First, it leads to the separation of charges [14]:

$$J_0 = \frac{\delta L_{\theta \gamma V}}{\delta A_0} = N_c \sum_f \frac{e_f \mu_f}{4 N_f \pi^2} \cdot \varepsilon^{ijk} \partial_i \theta_{\text{ind}} (\partial_j V_k) = N_c \sum_f \frac{e_f \mu_f}{2 N_f \pi^2} \cdot (\nabla \theta_{\text{ind}} \cdot \boldsymbol{\Omega}),$$

(8.9)

where $\boldsymbol{\Omega}$ is defined as $2\varepsilon_{ijk} \Omega_k = (\partial_i V_j - \partial_j V_i)$ and is nothing but the angular velocity of the rotating system. Similarly, interaction (8.8) leads to the induced vector current along $\boldsymbol{\Omega}$ direction,

$$J = N_c \sum_f \frac{e_f \mu_f}{2 N_f \pi^2} \cdot (\dot{\theta}_{\text{ind}} \cdot \boldsymbol{\Omega}).$$

(8.10)

The results (8.9), (8.10) are very suggestive and imply that there is an induced current along $\boldsymbol{\Omega}$ which leads to the CSE in the presence of θ_{ind} in the rotating system. Equations (8.9) and (8.10) are very similar to the previously discussed case (8.6) where magnetic field B is to be replaced by angular velocity of the rotating system $\boldsymbol{\Omega}$.

This analogy is, in fact, very deep as advocated in Ref. [14]. In particular, one can rewrite (8.8) in the following way,

$$L = \frac{1}{2} \mathbf{E}^2 - \frac{1}{2} \mathbf{B}^2 + N_c \sum_f \frac{e_f \mu_f}{2 N_f \pi^2} \cdot \theta_{\text{ind}} (\mathbf{E} \cdot \boldsymbol{\Omega}),$$

(8.11)

where we added the Maxwell term $\frac{1}{2} \mathbf{E}^2 - \frac{1}{2} \mathbf{B}^2$ to (8.11). In this form the expression for the anomalous term is very similar to (8.2) discussed previously. Therefore, we apply the same procedure to compute the induced electric field in the background of external $\boldsymbol{\Omega}$ and θ_{ind} fields.

Without the anomalous term the ground state corresponds to $\langle \mathbf{E} \rangle = 0$. However, in the presence of the θ_{ind} term the minimum of the free energy corresponds to the state where $\langle \mathbf{E} \rangle \neq 0$. Indeed, minimization of (8.11) with respect to $\mathbf{E}$ gives,

$$\frac{\delta L}{\delta E} = \mathbf{E} + N_c \sum_f \frac{e_f \mu_f}{2 \pi^2} \cdot \left(\frac{\theta_{\text{ind}}}{N_f} \right) \cdot \boldsymbol{\Omega} = 0,$$

(8.12)

which is analogous to (8.3) derived previously.

Now we want to compute the charge separation effect as a result of rotation of the system with angular velocity $\boldsymbol{\Omega}$ in the background of $\theta_{\text{ind}} \neq 0$. In order to integrate over the surface $d^2 x_\perp$ in (8.9) we have to understand the way how the rotation $\boldsymbol{\Omega}$ is quantized. The answer lies in the definition of the fictitious field $V_\nu \equiv v_\nu$ which is the local 4-velocity of matter with chemical potential μ_f, see [30, 31]. Then the coupling of quarks to V_μ and to the usual electromagnetic gauge field A_μ takes the form:

$$L = \sum_f (\mu_f V_\nu - e_f A_\nu) \bar{\psi}_f \gamma_\nu \psi_f.$$

(8.13)

With this definition magnetic flux quantization as well as rotation quantization for a single species are very similar and take the form,

$$\int e\boldsymbol{B} \cdot d\boldsymbol{\Sigma} = 2\pi l,$$

$$\int \mu\boldsymbol{\Omega} \cdot d\boldsymbol{\Sigma} = 2\pi l. \tag{8.14}$$

Now we can derive formulae similar to (8.4), (8.5), (8.7) when the system is rotating rather than placed into the background magnetic field. In particular, formula, similar to (8.4) can be obtained from (8.12) and takes the form

$$\mathbb{L}_x \mathbb{L}_y E_z = -\left(\frac{e\,\theta_{\text{ind}}}{\pi}\right)l, \tag{8.15}$$

where $\mathbb{L}$ is the size of the system, $N_f = N = 1$, $e_a = e$, and we assume quantization (8.14) for $\boldsymbol{\Omega}$. As before, for this simple geometry, the electric field E_z between two infinitely large charged plates with charge density σ_{xy} is exactly equal $E_z = \sigma_{xy}$. Therefore, (8.15) can be interpreted as CSE if system is rotating and large θ_{ind} domain is formed,

$$Q \equiv \int dx dy \sigma_{xy} = (\mathbb{L}_x \mathbb{L}_y) E_z = -\left(\frac{e\,\theta_{\text{ind}}}{\pi}\right)l. \tag{8.16}$$

The same expression can be obtained by integrating expression (8.9) over the volume of the system. Indeed, integrating $\int J_0 d^3x$ leads precisely to expression (8.16) for the charge separation effect along z,

$$Q = \int d^3x J_0 = \frac{e\mu}{2\pi^2} \int_0^{+\infty} dz \frac{d\theta_{\text{ind}}(z)}{dz} \int d^2x_\perp \Omega^z = -\left(\frac{e\,\theta_{\text{ind}}}{\pi}\right)l. \tag{8.17}$$

where we assume that $\theta_{\text{ind}}(z = 0) = \theta_{\text{ind}}$ inside the cylinder and $\theta_{\text{ind}}(z = +\infty) = 0$ outside the domain with $\theta_{\text{ind}} \neq 0$. We also take into account the flux quantization according to (8.14) $\int d^2x_\perp \Omega_z = 2\pi l/\mu$.

One more remark on (8.12). An additional term in the Lagrangian (8.11) is a total divergence in the limit of slow varying θ_{ind}, and therefore does not change the equations of motion. However, in order to quantize the system (8.11) in canonical way, we have to express everything in terms of the generalized momentum $\pi_i = \frac{\delta L}{\delta \dot{A}_i}$ which is defined as

$$\pi_i = \frac{\delta L}{\delta \dot{A}_i} = E_i + N_c \sum_f \frac{e_f \mu_f}{2\pi^2} \cdot \left(\frac{\theta_{\text{ind}}}{N_f}\right) \cdot \Omega_i. \tag{8.18}$$

In the ground state we require that $\langle \pi_i \rangle = 0$ which is precisely the condition (8.12) derived earlier. In different words, though the θ_{ind} term is a total divergence and does not change the equations of motion, it does influence the physics (separation of charges, induced electric field, etc) in the presence of topological background field, $\Omega_k \sim \varepsilon_{ijk}(\partial_i V_j - \partial_j V_i) \neq 0$.

We emphasize once again that formulae (8.8), (8.9), (8.10), (8.11), (8.12), (8.15), (8.16), (8.17) can only be trusted for slow-varying external background fields θ_{ind} and Ω when the effective Lagrangian approach is justified. In different words, it is assumed that the effective Lagrangian (8.8) describes slow degrees of freedom with typical wavelengths $\lambda \gg \Lambda_{\text{QCD}}^{-1}$, while fast degrees of freedom are already integrated out in formula (8.8). This condition can be easily satisfied for Ω field as it is truly external field which is not originated from strongly coupled QCD dynamics. At the same time $\theta_{\text{ind}}(x) \neq 0$ is an internal QCD parameter. Why this parameter $\theta_{\text{ind}}(x)$ is so special that it may indeed demonstrate a long range order with typical $\lambda_\theta \gg \Lambda_{\text{QCD}}^{-1}$? This is the key question which will be addressed later in the text in Sects. 8.5, 8.6, 8.7.

Our last comment is about terminology. All formulae discussed above have been derived in Ref. [14], though the term "chiral vortical effect" (CVE) itself was invented much later in Ref. [42] when some specific applications to heavy ion collisions were considered.

8.3 Long Range Order as Seen on the Lattice

This section is a small detour from the main topic of this review. We do not discuss CSE, CME, CVE and related effects in this section. Nevertheless, we consider this section as an inherent part of this review. The crucial point is as follows. As we emphasized previously, the effective Lagrangian approach developed in Sect. 8.2 and represented by formulae (8.2) and (8.11) is only justified if parameter $\theta_{\text{ind}}(x)$ which enters these expressions can be considered as a "slow" background field. It is equivalent to requirement that the corresponding $\lambda_\theta \gg \Lambda_{\text{QCD}}^{-1}$. In different words, $\theta_{\text{ind}}(x)$ must be a long range field with large correlation length for the effective Lagrangian approach to be justified. Otherwise, there will be no region of validity for expressions (8.2) and (8.11) for applications to heavy ion collisions.[1]

The goal of this section is to argue that such kind of long range order apparently is indeed present in recent Monte Carlo simulations. Furthermore, this long range order is inherently related to the topological charge fluctuations, e.g. to the topological θ_{ind} parameter. We review these lattice results below.

The recent Monte Carlo studies of pure glue gauge theory have revealed some very unusual features. In particular, the gauge configurations display a laminar structure in the vacuum consisting of extended, thin, coherent, locally low-dimensional

---

[1]If the asymmetry $\dot{\theta}_{\text{ind}} \sim (\mu_L - \mu_R)$ is build in as a result of weak interactions, the parameter $\dot{\theta}_{\text{ind}} \ll \Lambda_{\text{QCD}}$ is obviously very small, and the effective Lagrangian approach represented by formulae (8.2) and (8.11) makes perfect sense. In particular, it has been argued in [28, 29] that the induced current (8.6) may have important applications for physics of neutron stars when $\dot{\theta}_{\text{ind}} \sim (\mu_L - \mu_R)$ is small, but does not vanish as a result of weak β decays.

sheets of topological charge embedded in 4d space, with opposite sign sheets interleaved, see original QCD lattice results [43–46]. A similar structure has been also observed in QCD by different groups [47–51] and also in two dimensional CP^{N-1} model [52]. Furthermore, the studies of localization properties of Dirac eigenmodes have also shown evidence for the delocalization of low-lying modes on effectively low-dimensional surfaces. Here is the list (not complete) of the key properties of these gauge configurations:

(a) The tension of the "low dimensional objects" vanishes below the critical temperature and these objects percolate through the vacuum, forming a kind of a vacuum condensate;
(b) These "objects" do not percolate through the whole 4d volume, but rather, lie on low dimensional surfaces $1 \le d < 4$ which organize a coherent double layer structure;
(c) The total area of the surfaces is dominated by a single percolating cluster of "low dimensional object";
(d) The opposite sign sheets interleaved such that the "low dimensional object" can be viewed as a coherent configuration in a form of "double layer structure";
(e) The width of the percolating objects apparently vanishes in the continuum limit.

It is very difficult to understand all those properties using conventional quantum field theory (QFT) analysis. Indeed, the QCD lattice results [43–50] essentially imply that the topological density distribution is spread out on the surface of low-dimensional sheets. Such a structure can not be immediately seen in gluodynamics, at least not at the semiclassical level. The most important element for the present studies is the observation that topological density distribution is not localized in any finite size configurations such as instantons; rather it demonstrates a long range structure in a form of some extended "low dimensional object". The dimensionality of these objects is difficult to extract from lattice simulations. In particular, in Ref. [51] it has been argued that these long range configurations actually might be characterized by Hausdorff dimension which gradually varies with cooling procedure. We interpret these percolating objects observed in lattice simulations as a manifestation of the long range structure which is a key required ingredient for the justification of the effective Lagrangian approach reviewed in Sect. 8.2.

We should also note that these Monte Carlo results, while very difficult to interpret within QFT, could be very nicely interpreted within holographic description. To be more specific, the observed long range structure can be identified with the extended D2 branes in holographic dual picture as conjectured in Refs. [53–55]. One of the key elements of this conjecture is assumption that the tension of the D2 branes vanishes below the QCD phase transition $T < T_c$ such that an arbitrary large number of these objects can be formed and they can percolate. Vanishing tension in the dual description in the confined phase is a result of the Hawking-Page phase transition [56] when the D2 brane shrinks to the tip of a cigar type geometry. This solution can be interpreted as a point of instability. It can be also interpreted as a formation of effectively tensionless objects when the entropy of an extended configuration may overcome its intrinsic tension. A similar interpretation, though in a different context, was also advocated in [57].

What is the nature of this long range structure observed in lattice simulations? We attempt to answer this hard question in next Sects. 8.4, 8.5, 8.6 using a simplified version of QCD, the so-called "deformed QCD" model. In Sect. 8.7 we also make a connection between CSE/CME/CVE reviewed in Sect. 8.2 and corresponding computations in "deformed QCD" model.

One should remark here that "deformed QCD" model has been previously successfully used to test some other highly nontrivial features of strongly coupled QCD such as emergence of non-dispersive contact term in topological susceptibility [58] and emergence of the topological Casimir behaviour in gauge theory with a gap [59]. In both cases the effect occurs as a result of highly nontrivial topological structure of the vacuum of this model. Furthermore, we interpret the topological Casimir behaviour [59] in a gapped "deformed QCD" model as another manifestation of the same long range order observed in lattice simulations. In different words, if the long range order did not exist in this model, it would be very difficult to imagine how the "deformed QCD" being a gapped theory exhibits a power like correction when no any physical massless degrees of freedom are present in the system.

8.4 Deformed QCD

Here we review the "center-stablized" deformed Yang-Mills developed in [60]. In the deformed theory an extra term is put into the Lagrangian in order to prevent the center symmetry breaking that characterizes the QCD phase transition between "confined" hadronic matter and "deconfined" quark-gluon plasma. Thus we have a theory which remains confined at high temperature in a weak coupling regime, and for which it is claimed [60] that there does not exist an order parameter to differentiate the low temperature (non-abelian) confined regime from the high temperature (abelian) confined regime. First, in Sect. 8.4.1 we review the relevant parts of the theory. After that, in Sect. 8.4.2 we review the low energy effective Lagrangian which plays a key role in this work as it explicitly shows the 2π periodic properties of the theory. This structure will play a crucial role in construction of the domain wall solutions.

8.4.1 Formulation of the Theory

We start with pure Yang-Mills (gluodynamics) with gauge group $SU(N)$ on the manifold $\mathbb{R}^3 \times S^1$ with the standard action

$$S^{\text{YM}} = \int_{\mathbb{R}^3 \times S^1} d^4x \frac{1}{2g^2} \text{tr}\big[F^2_{\mu\nu}(x)\big], \qquad (8.19)$$

and add to it a deformation action,

$$\Delta S \equiv \int_{\mathbb{R}^3} d^3x \frac{1}{L^3} P\big[\Omega(\mathbf{x})\big], \qquad (8.20)$$

built out of the Wilson loop (Polyakov loop) wrapping the compact dimension,

$$\Omega(\mathbf{x}) \equiv \mathscr{P}\left[e^{i \oint dx_4 A_4(\mathbf{x}, x_4)}\right]. \tag{8.21}$$

The "double-trace" deformation potential $P[\Omega]$ respects the symmetries of the original theory and is built to stabilize the phase with unbroken center symmetry. It is defined by

$$P[\Omega] \equiv \sum_{n=1}^{\lfloor N/2 \rfloor} a_n \left| \mathrm{tr}\left[\Omega^n\right] \right|^2. \tag{8.22}$$

Here $\lfloor N/2 \rfloor$ denotes the integer part of $N/2$ and $\{a_n\}$ is a set of suitably large positive coefficients.

In undeformed pure gluodynamics the effective potential for the Wilson loop is minimized for Ω an element of $\mathbb{Z}_N$. The deformation potential (8.22) with sufficiently large $\{a_n\}$ however changes the effective potential for the Wilson line so that it is minimized instead by configurations in which $\mathrm{tr}[\Omega^n] = 0$, which in turn implies that the eigenvalues of Ω are uniformly distributed around the unit circle. Thus, the set of eigenvalues is invariant under the $\mathbb{Z}_N$ transformations, which multiply each eigenvalue by $e^{2\pi i k/N}$ (rotate the unit circle by k/N). The center symmetry is then unbroken by construction. The coefficients, $\{a_n\}$, can be suitably chosen such that the deformation potential, $P[\Omega]$, forces unbroken symmetry at any compactification scales [60], but for our purposes we are only interested in small compactifications ($L \ll \Lambda^{-1}$ where L is the length of the compactified dimension and Λ is the QCD scale). At small compactification, the gauge coupling at the compactification scale is small so that the semiclassical computations are under complete theoretical control [60].

8.4.2 Infrared Description

As described in [60], the proper infrared description of the theory is a dilute gas of N types of monopoles, characterized by their magnetic charges, which are proportional to the simple roots and affine root of the Lie algebra for the gauge group $U(1)^N$. Although the symmetry breaking is $SU(N) \to U(1)^{N-1}$, it is simpler to work with $U(1)^N$. The extended root system is given by the simple roots,

$$\begin{aligned}
\alpha_1 &= (1, -1, 0, \ldots, 0) = \hat{e}_1 - \hat{e}_2, \\
\alpha_2 &= (0, 1, -1, \ldots, 0) = \hat{e}_2 - \hat{e}_3, \\
&\;\;\vdots \\
\alpha_{N-1} &= (0, \ldots, 0, 1, -1) = \hat{e}_{N-1} - \hat{e}_N,
\end{aligned} \tag{8.23}$$

and the affine root,

$$\alpha_N = (-1, 0, \ldots, 0, 1) = \hat{e}_N - \hat{e}_1.$$

We denote this root system by Δ_{aff} and note that the roots obey the inner product relation

$$\alpha_a \cdot \alpha_b = 2\delta_{a,b} - \delta_{a,b+1} - \delta_{a,b-1}. \tag{8.24}$$

For a fundamental monopole with magnetic charge $\alpha_a \in \Delta_{\text{aff}}$, the topological charge is given by

$$Q = \int_{\mathbb{R}^3 \times S^1} d^4x \frac{1}{16\pi^2} \text{tr}\left[F_{\mu\nu}\tilde{F}^{\mu\nu}\right] = \pm\frac{1}{N}, \tag{8.25}$$

and the Yang-Mills action is given by

$$S_{\text{YM}} = \int_{\mathbb{R}^3 \times S^1} d^4x \frac{1}{2g^2} \text{tr}\left[F_{\mu\nu}^2\right]$$

$$= \left|\int_{\mathbb{R}^3 \times S^1} d^4x \frac{1}{2g^2} \text{tr}\left[F_{\mu\nu}\tilde{F}^{\mu\nu}\right]\right| = \frac{8\pi^2}{g^2}|Q|. \tag{8.26}$$

The second equivalence hold because the classical monopole solutions are self dual, $F_{\mu\nu} = \pm\tilde{F}_{\mu\nu}$.

The θ-parameter in the Yang-Mills action can be included in conventional way,

$$S_{\text{YM}} \rightarrow S_{\text{YM}} + i\theta \int_{\mathbb{R}^3 \times S^1} d^4x \frac{1}{16\pi^2} \text{tr}\left[F_{\mu\nu}\tilde{F}^{\mu\nu}\right], \tag{8.27}$$

with $\tilde{F}^{\mu\nu} \equiv \varepsilon^{\mu\nu\rho\sigma} F_{\rho\sigma}$.

The system of interacting monopoles, including θ parameter, can be represented in the dual sine-Gordon form as follows [58, 60],

$$S_{\text{dual}} = \int_{\mathbb{R}^3} d^3x \frac{1}{2L}\left(\frac{g}{2\pi}\right)^2 (\nabla\sigma)^2 - \zeta \int_{\mathbb{R}^3} d^3x \sum_{a=1}^{N} \cos\left(\alpha_a \cdot \sigma + \frac{\theta}{N}\right), \tag{8.28}$$

where ζ is magnetic monopole fugacity which can be explicitly computed in this model using conventional semiclassical approximation. The θ parameter enters the effective Lagrangian (8.28) as θ/N which is the direct consequence of the fractional topological charges of the monopoles (8.25). Nevertheless, the theory is still 2π periodic. This 2π periodicity of the theory is restored not due to the 2π periodicity of Lagrangian (8.28). Rather, it is restored as a result of summation over all branches of the theory when the levels cross at $\theta = \pi \,(\text{mod}\, 2\pi)$ and one branch replaces another and becomes the lowest energy state as discussed in [58]. Indeed, the ground state energy density is determined by minimization of the effective potential (8.28) when summation over all branches is assumed in the definition of the canonical partition function. It is given by

$$E_{\min}(\theta) = -\lim_{V \to \infty} \frac{1}{VL} \ln\left[\sum_{l=0}^{N-1} e^{V\zeta N \cos(\frac{\theta+2\pi l}{N})}\right] \tag{8.29}$$

where V is 3d volume of the system. Equation (8.29) shows that in the limit $V \to \infty$ cusp singularities occur at the values at $\theta = \pi \pmod{2\pi}$ where the lowest energy vacuum state switches from one analytic branch to another one. Such a pattern is known to emerge in many four dimensional supersymmetric models, and also gluodynamics in the limit $N = \infty$ [61]. It has been further argued [62] that the same pattern also emerges in four dimensional gluodynamics at any finite N. The same pattern emerges in holographic description of QCD [63] at $N = \infty$ as well.

In what follows we need an explicit expression for the topological density and magnetic field in terms of scalar σ field,

$$q(x) = \frac{1}{16\pi^2} \mathrm{tr}\left[F_{\mu\nu}\tilde{F}^{\mu\nu}\right] = \frac{-1}{8\pi^2}\varepsilon^{ijk4}\sum_{a=1}^{N} F_{jk}^{(a)} F_{i4}^{(a)}$$

$$= \frac{g}{4\pi^2}\sum_{a=1}^{N}\langle A_4^{(a)}\rangle[\nabla \cdot \mathbf{B}^{(a)}(x)], \tag{8.30}$$

where the $U(1)^N$ magnetic field, $B^i = \varepsilon^{ijk4}F_{jk}/2g$ is expressed in terms of scalar magnetic potential as follows

$$F_{ij}^{(a)} = \frac{g^2}{2\pi L}\varepsilon_{ijk}\partial^k\sigma^{(a)}, \qquad \mathbf{B}^{(a)} = \frac{g}{2\pi L}\nabla\sigma^{(a)}. \tag{8.31}$$

The expression for the magnetic field in terms of scalar magnetic potential should not be surprising as our system is in fact magnetostatic and description in terms of $\sigma^{(a)}$ is quite appropriate to study the relevant dynamics.

In what follows we also need an explicit form for the creation operator for a monopole of type a at $\mathbf{x}$. It is given by [58]

$$\mathscr{M}_a(\mathbf{x}) = e^{i\alpha_a \cdot \sigma(\mathbf{x})}. \tag{8.32}$$

Likewise, the operator for an antimonopole is $\bar{\mathscr{M}}_a(\mathbf{x}) = e^{-i\alpha_a \cdot \sigma(\mathbf{x})}$. The expectation values of these operators $\langle \mathscr{M}_a(\mathbf{x})\rangle$ in fact determine the ground state of the theory. Formula (8.32) shows one more time that $\sigma(\mathbf{x})$ can be interpreted as a magnetic scalar potential. Finally, the dimensional parameter which governs the dynamics of the problem is mass of the σ field. It is given by

$$m_\sigma^2 \equiv L\zeta\left(\frac{4\pi}{g}\right)^2. \tag{8.33}$$

This parameter can be interpreted as Debye correlation length of the monopole's gas. The average number of monopoles in a "Debye volume" is given by

$$\mathcal{N} \equiv m_\sigma^{-3}\zeta = \left(\frac{g}{4\pi}\right)^3 \frac{1}{\sqrt{L^3\zeta}} \gg 1. \tag{8.34}$$

The last inequality holds since the monopole fugacity is exponentially suppressed, $\zeta \sim e^{-1/g^2}$, and in fact we can view (8.34) as a constraint on the validity of the approximation where semiclassical approximation is justified.

8.5 Domain Walls in Deformed QCD

This section of the review is mostly based on a recent preprint [64]. A discrete set of degenerate vacuum states as a result of the 2π periodicity of the effective Lagrangian (8.28) for σ field is a signal that the domain wall configurations interpolating between these states are present in the system. However, the corresponding configurations are not conventional domain walls similar to the well known ferromagnetic domain walls in condensed matter physics which interpolate between physically *distinct* vacuum states. In contrast, in present case a corresponding configuration interpolates between topologically different but physically equivalent winding states $|n\rangle$, which are connected to each other by large gauge transformation operator. Just because of that, the corresponding domain wall configurations in Euclidean space is interpreted as configurations describing the tunnelling processes in Minkowski space, similar to Euclidean monopoles which also interpolate between topologically different, but physically identical states. This interpretation should be contrasted with conventional interpretation of static domain walls defined in Minkowski space when corresponding solution interpolate between physically distinct states. One can view these "additional" vacuum states which are physically *identical* states and which have extra 2π phase in operator (8.32) as an analog to the Aharonov Bohm effect with integer magnetic fluxes where electrons do not distinguish integer fluxes from identically zero flux. Our domain wall solution (8.36) describes interpolation between these two physically identical states.

In fact, a similar domain wall which has analogous interpretation is known to exist in QCD at large temperature in weak coupling regime where it can be described in terms of classical equation of motion. These are so-called Z_N domain walls which separate domains characterized by a different value for the Polyakov loop at high temperature. As is known, see e.g. review papers [65, 66] and references therein, these Z_N domain walls interpolate between topologically different but physically identical states connected by large gauge transformations similar to our case. At high temperature these objects can be described in terms of classical equation of motion. In this regime they have finite tension $\sim T^3$ such that their contribution to path integral is strongly suppressed. While the corresponding topological sectors are still present in the system at low temperature (though they are realized in a different way) it is not known how to describe the fate of Z_N walls within QFT in strong coupling regime when semiclassical approximation breaks down.

The domain walls to be discussed below in deformed QCD are very much the same as Z_N domain walls at high temperature. In our case their contribution to path integral is also strongly suppressed as their tension is finite. Nevertheless, one can study the structure of these domain walls and its interaction with external fields, which is the main motivation for this study. Essentially, we will treat the domain walls as external sources rather than as fluctuating dynamical participants in the path integral computations. Furthermore, as we reviewed in Sect. 8.3 an extended structure, similar to the domain walls, is apparently observed in the lattice simulations, which imply that they may have effectively vanishing tension at low temperature. We conjecture that the domain walls we describe below in weak coupling regime in deformed QCD slowly become the objects which are observed in lattice simulations [43–50] in strong coupling regime, when we adiabatically increase the coupling constant without hitting the phase transition as argued in [60]. This portion of the theory can not be tested in our deformed QCD model using semiclassical approximation. Hopefully this portion of strongly coupled dynamics can be understood in future using a different technique such as dual holographic description as advocated in the present context in [55].

What could happen when we slow move to the strong coupling regime? Holographic picture suggests that the effective tension for the domain walls (DW) may vanish, and these objects can be easily formed in vacuum. It is difficult to trace how it happens in our weakly coupled theory. It is naturally to assume that the domain walls become very clumpy with large number of folders. Such fluctuations obviously increase the entropy of the DW which eventually may overcome the intrinsic tension. If this happens, the DWs would look like as very crumpled and wrinkled objects with large number of foldings. Even more that: the DWs may loose their natural dimensionality, and likely to be characterized by a Hausdorff dimension as recent lattice simulations suggest [51]. Furthermore, the DW described below are expressed in terms of pseudo-scalar long range field σ which effectively corresponds to spatially dependent $\theta_{\text{ind}}(x) \neq 0$. As it is well known a non-vanishing $\theta_{\text{ind}}(x)$ may change the electric and magnetic properties of constituents which are present in vicinity of the DW. We can not address, nor answer those hard questions in weakly coupled regime. The addressing of these questions is obviously the prerogative of numerical lattice simulations.

8.5.1 Domain Wall Solution

There are N different DW types in deformed QCD. However, there are only $(N-1)$ physical propagating scalars σ in the system as one singlet scalar field, though it remains massless, completely decouples from the system, and does not interact with other components at all [60]. Therefore, there are only $(N-1)$ independent DW solutions.

In what follows, without loosing any generalities, we consider $N = 2$ case. In this case there is only one physical field χ which corresponds to a single diagonal component from the original $SU(2)$ gauge group. The action (8.28) becomes,

$$S_\chi = \int_{\mathbb{R}^3} d^3x \frac{1}{4L} \left(\frac{g}{2\pi}\right)^2 (\nabla\chi)^2$$

$$- \zeta \int_{\mathbb{R}^3} d^3x \left[\cos\left(\chi + \frac{\theta}{2}\right) + \cos\left(-\chi + \frac{\theta}{2}\right)\right], \tag{8.35}$$

while the equation of motion and its solution take the form

$$\nabla^2\chi - m_\sigma^2 \sin\chi = 0,$$
$$\chi(z) = 4\arctan\left[\exp(m_\sigma z)\right] \tag{8.36}$$

where we take $\theta = 0$ to simplify things. The width of the domain wall is obviously determined by m_σ^{-1}, while the domain wall tension σ for profile (8.36) can be easily computed and it is given by

$$\sigma = 2 \cdot \int_{-\infty}^{+\infty} dz \frac{1}{4L^2} \left(\frac{g}{2\pi}\right)^2 (\nabla\chi)^2$$

$$= \frac{m_\sigma}{L^2} \left(\frac{g}{2\pi}\right)^2 \sim \sqrt{\frac{\zeta}{L^3}}. \tag{8.37}$$

The topological charge density for the profile (8.36) equals

$$q(z) = \frac{\zeta}{L} \sin\chi(z) = \frac{4\zeta}{L} \frac{(e^{m_\sigma z} - e^{-m_\sigma z})}{(e^{m_\sigma z} + e^{-m_\sigma z})^2}. \tag{8.38}$$

With explicit solution at hand (8.36) one can easily compute the magnetic field (8.31) distribution inside the domain wall,

$$B_z = \left(\frac{g}{4\pi L}\right) \frac{4m_\sigma}{(e^{m_\sigma z} + e^{-m_\sigma z})}. \tag{8.39}$$

We are now in position to explain the physical meaning of this solution. As we mentioned before, the domain wall (8.36) does not describe a physical DW similar to the DW in ferromagnetic system or in the Ising model, when solution interpolates between physically *distinct* vacuum states. In our case DW interpolates between topologically different but physically *identical* states, similar to Z_N walls mentioned previously. Such DW are obviously not stable objects, but will decay quantum mechanically as a result of tunnelling processes, see Appendix for corresponding estimates. One should remark that a similar construction has been considered previously in relation with the so-called $N = 1$ axion model [68, 69], and more recently in QCD context in [67] and in high density QCD in [70]. In all previously

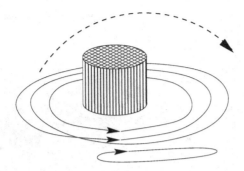

Fig. 8.1 This picture explains the transition corresponding to DW solution (8.36) which interpolates between one to the same vacuum state, in contrast with interpolation between physically distinct states. We can deform the paths by "lifting" them over the obstacle so that we can unwind them. If the paths were DW with some weight, then it would require some energy to "lift" the DW over the obstacle. If this energy was not available, then we would say that, classically, the configurations that wind around the peg are stable. Quantum mechanically, however, the DW could still tunnel through the peg, and so the configurations are unstable quantum mechanically, see estimate for this probability in Appendix. Picture adopted from [67]

considered cases [67–70] as well as in present case (8.36) there is a single physical unique vacuum state, and interpolation (8.36) corresponds to the transition from one to the same physical state. Nevertheless, if life time of configuration (8.36) is sufficiently large, it can be treated as stable classical background, and it can be used to study the structure of these extended configurations, which is one of the main objectives of this section, see Fig. 8.1 adopted from [67] with more explanations.

One can view these "additional" vacuum states which are physically *identical* states and which have extra 2π phase in operator (8.32) as analog of the Aharonov Bohm effect with integer magnetic fluxes when electrons do not distinguish integer fluxes from identically zero flux. Our domain wall solution (8.36) describes interpolation between these two physically identical states.

One should also comment that, formally, a similar soliton-like solution which follows from action (8.35) appears in computation of the string tension in 3d Polyakov's model [60, 71]. The solution considered there emerges as a result of insertion of the external sources in a course of computation of the vacuum expectation of the Wilson loop. In contrast, in our case, the solution (8.36) is internal part of the system without any external sources. Furthermore, the physical meaning of these solutions are fundamentally different: in our case the interpretation of solution (8.36) is similar to instanton describing the tunnelling processes in Minkowski space, while in computations [60, 71] it was an auxiliary object which appears in the course of computation of the string tension.

8.5.2 Double Layer Structure

From (8.38) one can explicitly see that the net topological charge $Q \sim \int_{-\infty}^{\infty} dz q(z)$ on the domain wall obviously vanishes. However, the charge density is distributed

Fig. 8.2 This picture shows a well organized double layer structure which emerges in deformed QCD model. The main assumption is that this structure persists even in strong coupling regime when semiclassical computations are not justified. Picture adopted from [64]

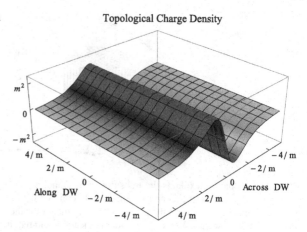

Topological Charge Density

not uniformly. Rather, it is organized in a double layer structure, see Fig. 8.2, which is precisely what apparently has been measured in the lattice simulations [43–46], see item (d) in Sect. 8.3.

The same double layer structure can be seen by computing the magnetic charge density ρ_M which is defined as

$$\rho_M \equiv \left[\nabla \cdot \mathbf{B}(\mathbf{x})\right] = \left(\frac{g}{4\pi L}\right) \frac{\partial^2 \chi}{\partial z^2} = 4\zeta \cdot \left(\frac{4\pi}{g}\right) \frac{(e^{m_\sigma z} - e^{-m_\sigma z})}{(e^{m_\sigma z} + e^{-m_\sigma z})^2}. \quad (8.40)$$

One can see that the relation between the topological charge density (8.38) and magnetic charge density (8.40) holds for the domain wall

$$q(z) = \left(\frac{g}{2\pi}\right) \cdot \left(\frac{1}{LN}\right) \cdot \rho_M(z) \quad (8.41)$$

in full agreement with general expression (8.30).

From (8.38), (8.40), we see that an average density of magnetic monopoles filling the interior of domain wall is expressed in terms of the same parameter ζ which characterizes the average monopole's density in the system (8.28). One can interpret this relation as a hint that the topological charge sources have a tendency to reside in vicinity of the domain walls rather than being uniformly distributed.

The most important lesson from this analysis is that the double layer structure naturally emerges in construction of the domain walls in weak coupling regime in deformed QCD. As claimed in [60] the transition from high temperature weak coupling regime to low temperature strong coupling regime should be smooth without any phase transitions on the way. Therefore, it would be tempting to identify the double layer structure described here and expressed by (8.38) with the double layer structure from lattice measurements [43–46] when one slowly moves along a smooth path from weak to strong coupling regime.

8.6 DW in the Presence of Matter Field

Next two sections of this review are mostly based on [72]. Our ultimate goal is to understand the long range structure described above in Sect. 8.5 in the presence of physical external $U(1)$ magnetic field, which is precisely the environment relevant for study of the CME. However, the gluons in pure glue theory represented in this model by low energy effective Lagrangian (8.28) do not couple to physical magnetic field. Therefore, we introduce a single massless quark field ψ into the model. The low energy description of the system in confined phase with a single quark is accomplished by introducing the η' colour singlet meson. As usual, the η' would be conventional massless Goldstone boson if the chiral anomaly is ignored. In the dual sine-Gordon theory the η' field appears exclusively in combination with the θ parameter as $\theta \to \theta - \eta'$ as a consequence of the Ward Identities. Indeed, the transformation properties of the path integral measure under the chiral transformations $\psi \to \exp(i\gamma_5 \frac{\eta'}{2})\psi$ dictate that η' appears only in the combination $\theta \to \theta - \eta'$. Therefore we have,

$$
\begin{aligned}
S_{\text{dual}} = & \int_{\mathbb{R}^3} d^3 x \frac{1}{2L} \left(\frac{g}{2\pi}\right)^2 \left[(\nabla\sigma)^2 + \frac{c}{2}(\nabla\eta')^2\right] \\
& - \zeta \int_{\mathbb{R}^3} d^3 x \sum_{a=1}^{N} \cos\left(\alpha_a \cdot \sigma + \frac{\theta - \eta'}{N}\right),
\end{aligned}
\tag{8.42}
$$

where dimensionless numerical coefficient $c \sim 1$ can be, in principle, computed in this model, though our results do not depend on its numerical value. One can explicitly compute the topological susceptibility χ in this model, and check that the Ward Identities are automatically satisfied when the η' field enters the Lagrangian precisely in form (8.42), see the detail computations in [58]. The η' mass computed from (8.42) has an extra $1/N$ suppression, as it should, in comparison with mass of the σ field

$$
m_{\eta'}^2 = \frac{2L\zeta}{cN}\left(\frac{2\pi}{g}\right)^2, \qquad \frac{m_{\eta'}^2}{m_\sigma^2} = \frac{1}{2cN}.
\tag{8.43}
$$

As we already mentioned, the system (8.42) is 2π periodic with respect $\theta \to \theta + 2\pi$. This 2π periodicity of the theory is not explicit in (8.42), but nevertheless is restored as a result of summation over all branches of the theory as (8.29) shows, see detail explanations in [58]. The fact that η' field enters the low energy effective Lagrangian precisely in combination $(\theta - \eta')$ implies that system (8.42) is also periodic with respect $\eta' \to \eta' + 2\pi$ when summation over all branches of the theory is properly implemented. In different words, the system (8.42) supports $\eta'(z)$ domain walls as a result of a very generic feature of the theory.[2] This argu-

---

[2]There are many different types of the domain walls which are supported by the Lagrangian (8.42). We leave this problem of classification of the DWs for a future study. In this work we concentrate on a simplest possible case with $N = 2$ to demonstrate few generic features of the system.

ment has been previously used to construct the η' domain walls in context of the axion physics [67]. One can derive the equation of motion for set of $\eta'(z)$ and $\sigma(z)$ fields determined by Lagrangian (8.42) and impose an appropriate boundary conditions $\eta'(z = +\infty) - \eta'(z = -\infty) = 2\pi$ to analyze this system numerically, in close analogy with procedure used, though in a different context, in Ref. [67].

However, for our present work it is sufficient to qualitatively describe the behaviour of the system in the limit when $m_{\eta'}^2/m_\sigma^2 \ll 1$, when the basic features can be easily understood even without numerical computations. The most important lesson of this qualitative analysis, as we shall see in a moment, is that the light η' field traces the $\sigma(z)$ field by exhibiting a similar double layer structure discussed for pure gauge theory in Sect. 8.5.

The low energy Lagrangian which describes the lightest degrees of freedom for $SU(2)$ gauge group is governed by the following action

$$
S_{\eta'} = \int_{\mathbb{R}^3} d^3x \frac{1}{4L} \left(\frac{g}{2\pi}\right)^2 \left[(\nabla\chi)^2 + c(\nabla\eta')^2\right]
$$
$$
- \zeta \int_{\mathbb{R}^3} d^3x \left[\cos\left(\chi + \frac{\theta - \eta'}{2}\right) + \cos\left(-\chi + \frac{\theta - \eta'}{2}\right)\right], \quad (8.44)
$$

where we inserted the η' field into (8.35) exactly in the form consistent with Ward Identities. Now, it is convenient to represent the action (8.44) of the system in the following way

$$
S_{\eta'} = \int_{\mathbb{R}^3} d^3x \frac{1}{4L} \left(\frac{g}{2\pi}\right)^2 \left[(\nabla\chi)^2 + c(\nabla\eta')^2\right]
$$
$$
- 2\zeta \int_{\mathbb{R}^3} d^3x \left[\cos\chi \cdot \cos\left(\frac{\eta'}{2}\right)\right], \quad (8.45)
$$

where we take $\theta = 0$ to simplify things. We are looking for a DW solution which satisfies the following boundary conditions:

$$
\left(\chi \to 0, \eta' \to 0\right) \quad \text{as } z \to -\infty,
$$
$$
\left(\chi \to \pi, \eta' \to 2\pi\right) \quad \text{as } z \to +\infty, \quad (8.46)
$$

One can explicitly see from (8.44) that the vacuum energy for $(\chi = \pi, \eta' = 2\pi)$ at $z = +\infty$ is identically coincide with vacuum energy when (χ, η') fields assume their trivial vacuum values: $(\chi = 0, \eta' = 0)$ at $z = -\infty$. As we already emphasized these states (with boundary conditions $(\chi = 0, \eta' = 0)$ and $(\chi = \pi, \eta' = 2\pi)$ correspondingly) must be interpreted as topologically different but physically equivalent states. Therefore, the corresponding domain wall configurations in Euclidean space should be interpreted as configurations describing the tunnelling processes rather than real DW in Minkowski space-time, as discussed in details in [55, 64]. The boundary conditions (8.46) correspond to the case when the changes in the χ field due to the large gauge transformation is compensated by the η' field which couples to the gluon density operator (8.27) in a unique and unambiguous way as η' enters the action in combination $(\theta - \eta')$ as explained above.

The corresponding equations of motion in this simple $N = 2$ case take the form

$$
\frac{1}{m_\sigma^2} \nabla^2 \chi = \sin \chi \cdot \cos\left(\frac{\eta'}{2}\right),
$$
$$
\frac{1}{2m_{\eta'}^2} \nabla^2 \eta' = \cos \chi \cdot \sin\left(\frac{\eta'}{2}\right).
$$

(8.47)

We are looking for a solution of the system (8.47) which satisfies the boundary conditions (8.46).

It is interesting to note that our system (8.47) with boundary conditions (8.46) formally is identical to the axion DW configuration analyzed in Ref. [41] in the limit when the isotopical symmetry is exact, i.e. $m_u = m_d$. In this case our χ field plays the role of the π^0 field denoted as γ in Ref. [41] while $\frac{\eta'}{2}$ plays the role of the axion field α from [41]. Precise relations are as follows:

$$
\chi \to \pi - \gamma, \qquad m_\sigma^2 \to m_\pi^2, \tag{8.48}
$$

$$
\frac{\eta'}{2} \to \alpha, \qquad m_{\eta'}^2 \to m_a^2 \tag{8.49}
$$

such that our equations (8.47) and boundary conditions (8.46) are identically coincide with the equations and the boundary conditions studied in ref [41]. Therefore, we simply formulate here the main points[3] which are relevant for our studies referring for the technical details to Ref. [41].

Most important result of analysis of Ref. [41] is that there is a unique solution of equations (8.47) which satisfies the boundary conditions (8.46). The simplest way to convince yourself that such a solution should exist is to use a mechanical analogy as suggested in [41]. While analytical formula of the solution is not known, its asymptotical behaviour at very large distances $|z| \gg m_\sigma^{-1}$ can be easily found as follows. It is clear that at large negative z a heavy $\chi(z)$ field already assumes its vacuum value $\chi = 0$ such that $\eta'(z)$ domain wall equation can be approximated in this region as

$$
\nabla^2 \eta' - 2m_{\eta'}^2 \sin\left(\frac{\eta'}{2}\right) \simeq 0, \quad z \ll -\frac{1}{m_\sigma}. \tag{8.50}
$$

---

[3] A simplest intuitive way to understand the qualitative behaviour of the system (8.47) is to use a mechanical analogy as suggested in [41] when variable z is replaced by time, while the fields (η', χ) can be thought as coordinates of two particles moving in one dimension with interaction determined by the potential term from (8.45).

Solution of this equation which vanishes at large distances $\eta'(z \to -\infty) \to 0$ can be approximated as

$$\eta'(z) \simeq 8 \arctan\left[\tan\left(\frac{\pi}{8}\right) e^{m_{\eta'} z}\right], \quad z \ll -\frac{1}{m_\sigma}. \tag{8.51}$$

Similarly, at large positive z a heavy $\chi(z)$ field already assumes its vacuum value $\chi = \pi$ such that $\eta'(z)$ domain wall equation can be approximated in this region as

$$\nabla^2 \eta' + 2m_{\eta'}^2 \sin\left(\frac{\eta'}{2}\right) \simeq 0, \quad z \gg \frac{1}{m_\sigma}. \tag{8.52}$$

Solution of this equation which approaches its vacuum value $\eta' = 2\pi$ at large distances $\eta'(z \to +\infty) \to 2\pi$ can be approximated as

$$\eta'(z) \simeq 2\pi - 8 \arctan\left[\tan\left(\frac{\pi}{8}\right) e^{-m_{\eta'} z}\right], \quad z \gg \frac{1}{m_\sigma}. \tag{8.53}$$

Formally, a similar construction when $\eta'(z)$ field interpolates between different branches representing the same physical vacuum state was considered previously, see [67] for the details and earlier references on the subject.[4]

What happens to the double layer structure (8.38), (8.40) for the topological charge distribution in the presence of the light dynamical quark? We anticipate, without any computations, that the light dynamical field suppresses the topological fluctuations similar to analysis of the topological susceptibility in this model [58]. Indeed, one can support this expectation by the following argument. First, we represent $q(z)$ in form similar to (8.38). The only difference in comparison with previous formula (8.38) is an emergence of the extra term due to the η' field on the right hand side of (8.54),

$$q(z) = \left(\frac{g}{4\pi}\right)^2 \frac{\nabla^2 \chi}{L^2} = \frac{\zeta}{L} \sin \chi(z) \cos\left(\frac{\eta'(z)}{2}\right). \tag{8.54}$$

In obtaining (8.54) we used (8.47) and (8.33) to simplify the expression for $q(z)$. As we already mentioned, the $\chi(z)$ field has a solitonic shape interpolating between $\chi = 0$ and $\chi = \pi$, see (8.46). Therefore, $q(z) \sim \partial^2 \chi / \partial z^2$ inevitably produces a double layer structure irrespectively to the details of the solitonic shape of χ, similar to our discussions of a pure gauge theory in Sect. 8.5. However, the magnitude of this structure is strongly suppressed as a result of dynamics of the light quark. Indeed, the $\eta'(z)$ field assumes its central value $\eta' \approx \pi$ in the region where $\chi(z)$ field

---

[4]The crucial difference with [67] is of course that the solutions for the system of the fields (η', χ) considered here are regular functions everywhere, while solution in Ref. [67] had a cusp singularity as a result of integrating out heavy fields played by the χ field in present "deformed QCD" model. Interpretations of these solutions in these two cases are also very different as we interpret the corresponding configurations as the transitions describing the tunnelling processes in Euclidean space-time rather than real static DW in Minkowski space-time as we mentioned above.

variation $\sim \partial^2 \chi / \partial z^2$ is sufficiently strong. This factor $\cos \frac{\eta'}{2}$ leads to a suppression of $q(z)$ in (8.54) as we had anticipated.

Few comments are in order. First of all, if we introduced the quark in "quenched approximation" rather than as a dynamical degree of freedom, we would return to our pure glue expression (8.38) as the "quenched approximation" corresponds to the $\eta' = 0$ in formula (8.54). Secondly, if we introduced a small quark's mass $m_\psi \neq 0$ into our Lagrangian it would not drastically change qualitative picture presented above as boundary conditions imposed on the system (8.46) can not depend on $m_\psi \neq 0$. Indeed, these boundary conditions are entirely based on exact symmetry of the system which requires that the energy of the system at $\theta = 0$ and $\theta = 2\pi$ is identically the same irrespectively to the quark's mass. Finally, the (χ, η') domain wall is a coherent configuration and can not easily decay into its constituents, even though the η' is a real physical asymptotic state of the system. Technically, it can be also explained using pure kinematical arguments: the η' constituents which are making the η' domain wall are off-shell, rather than on-shell, states. This DW can only decay through the tunnelling process, see (8.69) from Appendix.

To conclude this section: the main lesson of the present analysis is that the deformed QCD with matter field supports long range correlated configurations. In different words, the matter and glue fields accompany each other in their interpolations between topologically different, but physically identical states. This correlation is enforced by very generic features of the Lagrangian (8.42). First, it is a local enforcement as the Ward Identities require that θ parameter and η' field enter the effective Lagrangian in a specific way as (8.42) dictates. Secondly, it is a global enforcement as the 2π periodicity in θ implies there existence of interpolating (χ, η') configurations which inevitably present in the system.

We emphasize once again that the long range structure revealed in this section might be the trace of a similar structure measured on the lattices, as the transition from high temperature weak coupling regime to low temperature strong coupling regime should be smooth without any phase transitions on the way [60]. However, similar to our comment in Sect. 8.5, we expect that the (χ, η') domain walls determined by (8.47) with boundary conditions (8.46) and asymptotical behaviour (8.51), (8.53) become very crumpled and wrinkled objects with large number of foldings. Such fluctuations obviously increase the entropy of the DWs which eventually may overcome the intrinsic tension as holographic picture suggests, see [55] and references therein. In fact, it is quite likely that an appropriate description for this physics should be formulated in terms of holographic dual model, however we leave this subject for future studies.

8.7 CSE, CME, CVE and Related Topological Phenomena in Deformed QCD

As we already mentioned, the ultimate goal of our study here is to understand the infrared physics in the presence of physical external $U(1)$ magnetic field, which is precisely the environment relevant for study of the CME. We are now in position to couple our system (8.42) to external Maxwell $U(1)$ field A_μ as massless quark ψ carries the electric charge e of A_μ field. In this section we use conventional Minkowski metric in order to compare the obtained below formulae with known expressions written normally in Minkowski space-time. The η' field which appears in the low energy Lagrangian (8.42) does not couple to electromagnetic field directly as it is a neutral field. However, it does couple via triangle anomaly, similar to the textbook example describing $\pi^0 \to 2\gamma$ decay. The corresponding Maxwell term S_γ and anomalous term $S_{\eta'\gamma\gamma}$ have the form

$$S_\gamma = -\frac{1}{4} \int d^4x \left[F_{\mu\nu}^2 \right],$$

$$S_{\eta'\gamma\gamma} = \frac{e^2 N}{16\pi^2} \int d^4x \left[\eta' F_{\mu\nu} \tilde{F}^{\mu\nu} \right].$$

$$(8.55)$$

The structure of the $S_{\eta'\gamma\gamma}$ is unambiguously fixed by the anomaly. It describes the interaction of the Maxwell field[5] $F_{\mu\nu}$ with matter field.

The interaction (8.55) is normally used to describe $\eta' \to 2\gamma$ decay. However, in the context of the present work we treat $\eta'(x)$ as external background field describing the η' DW discussed in previous Sect. 8.6. Therefore, a number of new and unusual coherent effects will emerge as a result of long range structure represented by the η' domain wall described above. Let us emphasize again that this long range structure being represented by the η' component of the (η', χ) domain wall is the Euclidean long range configuration describing the tunnelling processes in Minkowski space-time, rather than a physical configuration in real Minkowski space-time. Nevertheless, this η' coherent component does interact with $F_{\mu\nu}$ field as (8.55) dictates.

We start by rewriting the action (8.55) using conventional vector notations for electric $\mathbf{E}$ and magnetic $\mathbf{B}$ fields,

$$S_\gamma + S_{\eta'\gamma\gamma} = \int d^4x \left[\frac{1}{2}\mathbf{E}^2 - \frac{1}{2}\mathbf{B}^2 - \frac{N e^2}{4\pi^2} \eta' \mathbf{E} \cdot \mathbf{B} \right].$$

$$(8.56)$$

One can immediately see that in the electric field will be induced in the presence of external magnetic field $\mathbf{B}_{ext}$ in the extended region where η' is not vanishing, i.e.

$$\mathbf{E} = \frac{N e^2}{4\pi^2} \eta' \mathbf{B}_{ext},$$

$$(8.57)$$

which is precisely the starting formula in [14] if we identify the coherent η' component from the (η', χ) domain wall with induced θ_{ind} parameter introduced in [14].

---

[5]Not to be confused with gluon field from (8.25), (8.27).

Furthermore, is we assume that the DW is extended along (x, y) directions, we get

$$(\mathbb{L}_x \mathbb{L}_y) E_z = \left(\frac{eNl}{2\pi}\right)\eta'(z) \tag{8.58}$$

where $\mathbb{L}$ is size of the system, and integer number l is the magnetic flux of the system, $\int d^2 x_\perp B_{\text{ext}}^z = \Phi/e = 2\pi l/e$. Formula (8.58) identically coincides with (8.4) from Sect. 8.2 and (7) from [14] if the induced θ_{ind} parameter from that work is identified with extended along (x, y) configuration represented by the $\eta'(z)$ domain wall described in Sect. 8.6. The difference in signs in (8.58) and (8.4) is due to our convention when we replace $\theta \to (\theta - \eta')$ in effective Lagrangian. The induced electric field along z obviously implies that the current will flow and charge will be separated along B_{ext}^z.

Anomalous coupling (8.56) also implies that there will be induced current as a result of coordinate dependence of the η' field,

$$J^\nu = -\frac{Ne^2}{8\pi^2}\partial_\mu\left(\eta' \tilde{F}^{\mu\nu}\right), \tag{8.59}$$

which is convenient to represent in vector notations as follows

$$J_0 = \frac{Ne^2}{4\pi^2}\boldsymbol{\nabla}\eta' \cdot \mathbf{B}_{\text{ext}}, \qquad \mathbf{J} = \frac{Ne^2}{4\pi^2}\dot{\eta}'\mathbf{B}_{\text{ext}}, \tag{8.60}$$

where we assume that the external magnetic field $\mathbf{B}_{\text{ext}}$ is coordinate independent. Formula (8.59) for the anomalous current has been studied previously in literature in many fields, including particle physics, cosmology, condensed matter physics. In particular, in the context of the axion domain wall it was extensively discussed in [41].

In the present context relevant for the CME formula (8.60) reduces to the well known result when the induced $\theta_{\text{ind}}(x, t)$ parameter is identified with extended η' domain wall from Sect. 8.6. Indeed, integrating $\int J_0 d^3 x$ leads precisely to the known expression for the charge separation effect along z,

$$Q = \int d^3 x\, J_0 = \frac{Ne^2}{4\pi^2}\int dz \frac{d\eta'}{dz}\int d^2 x_\perp B_{\text{ext}}^z = Nle, \tag{8.61}$$

where we took into account the boundary conditions for the η' field (8.46) and replaced $\int d^2 x_\perp B_{\text{ext}}^z = \Phi/e = 2\pi l/e$. Formula (8.61) identically coincides with our expression (8.5) if one replaces $\theta_{\text{ind}} = 2\pi$ as the boundary conditions (8.46) for the η' domain wall dictate. Furthermore, the expression for $\mathbf{J}$ in (8.60) can be presented in much more familiar way if one replaces $\dot{\eta}' \to \dot{\theta}_{\text{ind}} \equiv 2\mu_5$ as our identification suggests. With these replacements the expressions (8.60) and (8.61) assume their conventional forms which are normally used in CME studies (8.6), (8.7). One should comment here that while our solution for the η' domain wall considered in Sect. 8.6 in "deformed QCD" is time independent, in fact it will actually describe the tunnelling effects in strong coupling regime. Therefore, it is naturally to expect that $\eta'(x, t)$ component from (χ, η') domain wall becomes a time dependent configuration with a typical time scale $\dot{\eta'} \sim \Lambda_{QCD}$ in a course of a smooth transition from weak coupling to strong coupling regime, as discussed at the end of Sect. 8.6. However, one can not use very large magnitude for $\dot{\eta'} \sim 1$ GeV (as many people do) for numerical estimates as the effective Lagrangian approach which leads to formulae (8.6), (8.7) can only be justified for small values $|\eta'| \ll \Lambda_{QCD}$, and marginally justified for $|\eta'| \simeq \Lambda_{QCD}$. For large $|\dot{\eta'}| \gg \Lambda_{QCD}$ the effective Lagrangian approach can not be justified, and computations should be based on a different technique when underlying QCD degrees of freedom, quarks and gluons (rather than effective η' field) play the dynamical role.

One should also say that there are many other interesting topological effects originated from similar anomalous terms as originally discussed in terms of hadronic fields in [14, 30], and re-derived in terms of microscopical quark fields in [20, 31]. In particular, if a system with chemical potential μ rotates with angular velocity $\mathbf{\Omega}$, there will be a current flowing along $\mathbf{\Omega}$. The charges will be also separated along the same direction. To get corresponding formulae one should replace $e\mathbf{B} \to 2\mu\mathbf{\Omega}$ in (8.60) as discussed in [14], i.e.

$$J_0 = \frac{Ne\mu}{2\pi^2} \nabla \eta' \cdot \mathbf{\Omega}_{\text{ext}}, \qquad \mathbf{J} = \frac{Ne\mu}{2\pi^2} \dot{\eta'} \mathbf{\Omega}_{\text{ext}}, \qquad (8.62)$$

which precisely coincides with (A2) from [14] and with (8.9), (8.10) from Sect. 8.2 if one identifies η' field with induced parameter θ_{ind} from [14] as we already discussed above. Specific consequences of effect (8.9), (8.10), (8.62) relevant for heavy ion collisions were discussed quite recently in [42] where the effect was coined as the chiral vortical effect (CVE).

The main point of this section is as follows. The long range structure discussed in Sect. 8.6 might be the trace of a similar extended structure measured on the lattices, as the transition from high temperature weak coupling regime to low temperature strong coupling regime should be smooth [60]. It is important that this long range structure describes the tunnelling effects and represented by (χ, η') fields in deformed QCD. These configurations are not real physical configurations in Minkowski space-time. Nevertheless, these long range configurations do interact with real physical $E\&M$ field as a result of anomaly (8.56).

Such an interaction transfers unphysical long range correlations (expressed in terms of the Euclidean configurations describing the tunnelling processes) to physical long range correlated $E\&M$ effects (8.57), (8.58), (8.60), (8.62). The corresponding coherent effects are accumulated on large scales $\sim \mathbb{L}$ where the boundary conditions for different topological sectors are imposed.

This toy model explicitly shows that the large observed intensity of the effect as studied at RHIC and ALICE at the LHC [1–12] might be due to a coherent phenomena when the large observable asymmetry is a result of accumulation of a small effect over large distances $\sim \mathbb{L} \gg \Lambda_{QCD}^{-1}$. It should be contrasted with some other numerical estimates when the large intensity of the effect is achieved by choosing a relatively large $|\mu_5| \sim 1$ GeV $\gg \Lambda_{QCD} \sim 0.1$ GeV in which case there is no region of validity for the effective Lagrangian framework. In different words, the CSE given by formulae (8.61), (8.5), (8.7), (8.16), (8.17) leads to a large magnitude for an asymmetry in spite of the fact that parameters $|\dot{\eta}'|$, $|\nabla \eta'|$ remain small during entire tunnelling transition.

8.8 Conclusion and Future Directions

The question which is addressed in the present work is as follows: what could be the physics behind of the long range order which is postulated in [14], and which is apparently a required element for CSE, CME and CVE to be operational. We attempt to answer this question using the "deformed QCD" as a toy model where all computations are under complete theoretical control as this model is a weakly coupled gauge theory. Still, this model has all the relevant crucial elements allowing us to study difficult and nontrivial questions which are known to be present in real strongly coupled QCD. The study of these effects in this toy model reveals that the long range structure may result from the tunnelling effects and represented by (χ, η') fields in deformed QCD.

Apparently, such kind of transitions with long range structure in strong coupling regime are happening all the time, as it is observed in the lattice simulations. One should expect that the corresponding configurations at strong coupling regime should be very crumpled and wrinkled objects with large number of foldings in contrast with our (χ, η') domain walls described in Sect. 8.6. Such local crumples are expected to occur as it provides a large entropy for the DW configurations to overcome their intrinsic tension. However, the crucial element of these DWs, the long range coherence, is not lost in transition from weak to strong coupling regime. Precisely this feature, we believe, is the key element why the observed asymmetries are sufficiently strong and not washed out (which would be the case if one considers some conventional short range QCD fluctuations with a typical size Λ_{QCD}). We suspect that all other conventional mechanisms based on e.g. instanton/sphaleron transitions can not provide sufficient intensities observed at RHIC and the LHC as the observed asymmetries must be accumulated on large scales of order $\sim L$ rather than on scales of order $\Lambda_{\text{QCD}}^{-1}$.

On phenomenological side, the very basic observed features, such as energy and charge independence, of measured asymmetries in heavy ion collisions are automatically and naturally satisfied within the framework based on long range order, see recent papers [73, 74]. The same framework based on the idea of a coherent accumulation of the effect also provides a natural explanation for a strong dependence on centrality as observed at RHIC and ALICE at the LHC [1–12].

Essentially, our study in a simplified version of QCD provides a precise and very specific realization of an old idea [17, 18] (see also [15, 16] where similar idea was formulated in different terms), that a macroscopically large domain with $\theta_{\text{ind}} \neq 0$ can be formed in heavy ion collisions. Now we can precisely identify this domain characterized by $\theta_{\text{ind}} \neq 0$ with interpolating long range η' field which traces a pure glue configuration describing the transition between different topological sectors. In different words, the domains with $\theta_{\mathrm{ind}} \neq 0$ should not be thought as real extended regions formed in Minkowski space-time as a result of collision. Rather it *should be viewed as long range Euclidean coherent configurations which saturate the tunnelling transitions* in path integral computations. Nevertheless, these long range structure formulated in terms of auxiliary Euclidean configurations can be translated into observable long range effects (8.57), (8.58), (8.60), (8.62) in Minkowski space as a result of anomalous coupling with physical $E\&M$ field.

We should also mention that CME has been extensively studied in the lattice simulations [22, 75–77]. Independently, very different lattice studies reveal that the crucial topological configurations saturating the path integral are represented by extended, locally low-dimensional sheets of topological charge embedded in 4d space [43–51]. Our analysis based on computations in weakly coupled "deformed QCD" suggests that the long range configurations which are responsible for CSE, CME and CVE effects are *precisely the same objects* which we identify with long range extended objects from Refs. [43–51]. We presented a number of arguments suggesting that this relation is in fact quite generic as it is based on topological features of the theory, rather than on a specific details of the model. Therefore, we conjecture that this relation continues to hold in strongly coupled QCD. This conjecture can be explicitly tested in the lattice simulations as essentially this conjecture suggests that the topological charge distribution as it is done in [43–51] and electric charge distribution in the presence of the background magnetic field are strongly correlated and follow each other. Such a correlation also provides a new, and much easier way to study the original topological charge distribution by putting the system into the background magnetic field and studying the electric charge distribution in "quenched approximation" as it is previously done in Refs. [22, 75–77].

Our final comment is as follows. The transition from weakly coupled "deformed QCD" to strongly coupled regime should be smooth. Still, this transition is beyond the analytical control within QFT framework. What would be an appropriate tool to study this physics in strongly coupled regime? It is very likely that the description in terms of the holographic dual model may provide the required tools and technique. In fact, the long range structure is obviously present in holographic model as one can see from computations of the so-called "Topological Casimir Effect" [55, 78] when no massless degrees of freedom are present in the system, but dependence of physical observables demonstrate a power like sensitivity on size of the system $\sim \mathbb{L}^{-p}$. This scaling is in huge contrast with $\sim \exp(-\mathbb{L})$ dependence which is normally expected if a mass-gap is present in the system. Some analogies presented in [55] are actually suggesting that the ground state of QCD behaves very similarly to some condensed matter systems which are known to lie in topological phases. In principle, one could try to compute the topological entanglement entropy in "deformed QCD" (as it has been done in some condensed matter systems) to see if the theory indeed lies in topological phase. If it does, one could argue that the same topological phase must be realized in strongly coupled QCD as well as the path from weakly coupled "deformed QCD" to strongly coupled QCD must be smooth with no any phase transition on the way along this path [60]. We leave this subject for a future study. The last word whether these analogies can be extended to the strongly coupled four dimensional QCD remains, of course, the prerogative of the direct lattice computations.

Acknowledgements   I am thankful to Gokce Basar, Dima Kharzeev, Ho-Ung Yee, Edward Shuryak and other participants of the workshop "P-and CP-odd effects in hot and dense matter", Brookhaven, June, 2012, for useful and stimulating discussions related to the subject of the present work. This research was supported in part by the Natural Sciences and Engineering Research Council of Canada.

Appendix

The goal here is to estimate the life time of the DW studied in Sect. 8.5. These DW should be viewed as configurations which describe tunnelling processes, similar to instantons. In addition, these objects may decay themselves as a result of internal dynamics, similar to static (in Minkowski space) configurations studied previously [67–70].

The decay mechanism is due to a tunnelling process which creates a hole in the domain wall which connects the $\chi = 0$ domain on one side of the wall to the $\chi = 2\pi$ domain on the other, see (8.36). Passing through the hole, the fields remain in the ground state. This lowers the energy of the configuration over that where the hole was filled by the domain wall transition by an amount proportional to R^2 where R is the radius of the hole. The hole, however, must be surrounded by a string-like field configuration. This string represents an excitation in the heavy degrees of freedom and thus costs energy, however, this energy scales linearly as R. Thus, if a large enough hole can form, then it will be stable and the hole will expand and consume the wall. This process is commonly called quantum nucleation and is similar to the decay of a metastable wall bounded by strings, and we use a similar technique to estimate the tunnelling probability. The idea of the calculation was suggested in [68, 69] to estimate the decay rate in the so-called $N = 1$ axion model. In QCD context similar estimations have been discussed for the η' domain wall in large N QCD in [67] and for the η' domain wall in high density QCD in [70].

If the radius of the nucleating hole is much greater than the wall thickness, we can use the thin-string and thin-wall approximation. (The critical radius R_c will be estimated later and this approximation justified.) In this case, the action for the string and for the wall are proportional to the corresponding worldsheet areas

$$S_0\left(\mathbb{R}^3 \times \mathbb{S}^1\right) = 2\pi RL\alpha - \pi R^2 L\sigma. \tag{8.63}$$

The first term is the energy cost of forming a string: α is the string tension and $2\pi RL$ is its worldsheet area. The second term is energy gain by the hole over the domain wall: σ is the wall tension and $\pi R^2 L$ is its worldsheet volume. We should note that formula (8.63) replaces following, more familiar expression for the classical action which was used in many previous similar computations, see e.g. [67, 70]

$$S_0\left(\mathbb{R}^4\right) = 4\pi R^2\alpha - \frac{4\pi}{3}R^3\sigma. \tag{8.64}$$

Minimizing (8.63) with respect to R we find the critical radius R_c and the action S_0

$$R_c = \frac{\alpha}{\sigma}, \qquad S_0(\mathbb{R}^3 \times \mathbb{S}^1) = \frac{\pi \alpha^2 L}{\sigma}, \tag{8.65}$$

which replace more familiar expressions for the critical radius $R_c = \frac{2\alpha}{\sigma}$ and classical action $S_0(\mathbb{R}^4) = \frac{16\pi\alpha^3}{3\sigma^2}$ from [67, 70].

Therefore, the semiclassical probability of this process is proportional to

$$\Gamma \sim \exp\left(-\frac{\pi \alpha^2 L}{\sigma}\right) \tag{8.66}$$

where σ is the DW tension determined by (8.37), while α is the tension of the vortex line in the limit when the interaction term $\sim\zeta$ due to the monopole's interaction in low energy description (8.35) is neglected and $U(1)$ symmetry is restored. In this case the vortex line is a global string with logarithmically divergent tension

$$\alpha \sim 2\pi \frac{1}{4L^2} \left(\frac{g}{2\pi}\right)^2 \ln \frac{R}{R_{core}} \tag{8.67}$$

where $R \sim m_\sigma^{-1}$ is a long-distance cutoff which is determined by the width of DW, while $R_{core} \sim L$ when low energy description breaks down. The vortex tension is dominated by the region outside the core, so our estimates for computing α to the logarithmic accuracy are justified. Furthermore, the critical radius can be estimated as

$$R_c = \frac{\alpha}{\sigma} \sim \frac{\pi}{2m_\sigma} \ln\left(\frac{1}{m_\sigma L}\right), \tag{8.68}$$

which shows that the nucleating hole $\sim R_c$ is marginally greater than the wall thickness $\sim m_\sigma^{-1}$ as logarithmic factor $\ln(\frac{1}{m_\sigma L}) \sim \ln \mathcal{N} \gg 1$ where $\mathcal{N} \gg 1$ is large parameter of the model, see (8.34). Therefore, our thin-string and thin-wall approximation is marginally justified.

As a result of our estimates (8.66), (8.37), (8.67) the final expression for the decay rate of the domain wall is proportional to

$$\Gamma \sim \exp\left(-\frac{\pi \alpha^2 L}{\sigma}\right) \sim \exp\left(-\pi^3 \left(\frac{g}{4\pi}\right)^3 \frac{\ln^2(\frac{1}{m_\sigma L})}{\sqrt{L^3 \zeta}}\right)$$
$$\sim \exp\left(-\gamma \cdot \mathcal{N} \ln^2 \mathcal{N}\right) \ll 1, \tag{8.69}$$

with γ being some numerical coefficient. The estimate (8.69) supports our claim that in deformed QCD model when weak coupling regime is enforced and $\mathcal{N} \gg 1$ the domain walls are stable objects and, therefore, our treatment of the DW in Sect. 8.5 is justified.

References

1. S.A. Voloshin, Phys. Rev. C **70**, 057901 (2004). arXiv:hep-ph/0406311
2. I.V. Selyuzhenkov (STAR Collaboration) Rom. Rep. Phys. **58**, 049 (2006). arXiv:nucl-ex/0510069
3. S.A. Voloshin (STAR Collaboration), arXiv:0806.0029 [nucl-ex]
4. S.A. Voloshin, arXiv:1006.1020 [nucl-th]
5. B.I. Abelev et al. (STAR Collaboration), Phys. Rev. Lett. **103**, 251601 (2009). arXiv:0909.1739 [nucl-ex]
6. N.N. Ajitanand, S. Esumi, R.A. Lacey (PHENIX Collaboration), P- and CP-odd effects in hot and dense matter, in *Proc. of the RBRC Workshops*, vol. 96 (2010)
7. B. Mohanty et al. (STAR Collaboration), arXiv:1106.5902 [nucl-ex]
8. B.I. Abelev et al. (STAR Collaboration), Phys. Rev. C **81**, 054908 (2010). arXiv:0909.1717 [nucl-ex]
9. P. Christakoglou (ALICE Collaboration), J. Phys. G **38**, 124165 (2011). arXiv:1106.2826 [nucl-ex]
10. I. Selyuzhenkov (ALICE Collaboration), arXiv:1111.1875 [nucl-ex]
11. I. Selyuzhenkov (ALICE Collaboration). arXiv:1203.5230 [nucl-ex]
12. B. Abelev et al. (ALICE Collaboration), arXiv:1207.0900 [nucl-ex]
13. D. Kharzeev, Phys. Lett. B **633**, 260 (2006). arXiv:hep-ph/0406125
14. D. Kharzeev, A. Zhitnitsky, Nucl. Phys. A **797**, 67 (2007). arXiv:0706.1026 [hep-ph]
15. D. Kharzeev, R.D. Pisarski, M.H.G. Tytgat, Phys. Rev. Lett. **81**, 512 (1998). arXiv:hep-ph/9804221
16. D. Kharzeev, R.D. Pisarski, Phys. Rev. D **61**, 111901 (2000). arXiv:hep-ph/9906401
17. K. Buckley, T. Fugleberg, A. Zhitnitsky, Phys. Rev. Lett. **84**, 4814 (2000). arXiv:hep-ph/9910229
18. K. Buckley, T. Fugleberg, A. Zhitnitsky, Phys. Rev. C **63**, 034602 (2001). arXiv:hep-ph/0006057
19. D.E. Kharzeev, L.D. McLerran, H.J. Warringa, Nucl. Phys. A **803**, 227 (2008). arXiv:0711.0950 [hep-ph]
20. K. Fukushima, D.E. Kharzeev, H.J. Warringa, Phys. Rev. D **78**, 074033 (2008). arXiv:0808.3382 [hep-ph]
21. D.E. Kharzeev, Ann. Phys. **325**, 205 (2010). arXiv:0911.3715 [hep-ph]
22. P.V. Buividovich, M.N. Chernodub, E.V. Luschevskaya, M.I. Polikarpov, Phys. Rev. D **80**, 054503 (2009). arXiv:0907.0494 [hep-lat]
23. M. Abramczyk, T. Blum, G. Petropoulos, R. Zhou, PoS **LAT2009**, 181 (2009). arXiv:0911.1348 [hep-lat]
24. A. Vilenkin, Phys. Rev. D **20**, 1807 (1979)
25. A. Vilenkin, Phys. Rev. D **22**, 3067 (1980)
26. A. Vilenkin, Phys. Rev. D **22**, 3080 (1980)
27. A.Yu. Alekseev, V.V. Cheianov, J. Fröhlich, Phys. Rev. Lett. **81**, 3503 (1998)
28. J. Charbonneau, A. Zhitnitsky, Phys. Rev. C **76**, 015801 (2007). arXiv:astro-ph/0701308
29. J. Charbonneau, A. Zhitnitsky, JCAP **1008**, 010 (2010). arXiv:0903.4450 [astro-ph.HE]
30. D.T. Son, A.R. Zhitnitsky, Phys. Rev. D **70**, 074018 (2004). arXiv:hep-ph/0405216
31. M.A. Metlitski, A.R. Zhitnitsky, Phys. Rev. D **72**, 045011 (2005). arXiv:hep-ph/0505072
32. G. Lifschytz, M. Lippert, Phys. Rev. D **80**, 066005 (2009). 0904.4772
33. H.U. Yee, J. High Energy Phys. **0911**, 085 (2009). arXiv:0908.4189 [hep-th]
34. A. Rebhan, A. Schmitt, S.A. Stricker, J. High Energy Phys. **1001**, 026 (2010). arXiv:0909.4782 [hep-th]
35. A. Gorsky, P.N.. Kopnin, A.V. ZayakinStricker, Phys. Rev. D **83**, 014023 (2011). arXiv:1003.2293 [hep-ph]
36. A. Gynther, K. Landsteiner, F. Pena-Benitez, A. Rebhan, J. High Energy Phys. **1102**, 110 (2011). arXiv:1005.2587 [hep-th]
37. V.A. Rubakov, arXiv:1005.1888 [hep-ph]
38. L. Brits, J. Charbonneau, Phys. Rev. D **83**, 126013 (2011). arXiv:1009.4230 [hep-th]

39. D.E. Kharzeev, arXiv:1107.4004 [hep-ph]
40. M. Asakawa, A. Majumder, B. Muller, Phys. Rev. C **81**, 064912 (2010). arXiv:1003.2436 [hep-ph]
41. M.C. Huang, P. Sikivie, Phys. Rev. D **32**, 1560 (1985)
42. D.E. Kharzeev, D.T. Son, Phys. Rev. Lett. **106**, 062301 (2011). arXiv:1010.0038 [hep-ph]
43. I. Horvath, S.J. Dong, T. Draper, F.X. Lee, K.F. Liu, N. Mathur, H.B. Thacker, J.B. Zhang, Phys. Rev. D **68**, 114505 (2003). hep-lat/0302009
44. I. Horvath, A. Alexandru, J.B. Zhang, Y. Chen, S.J. Dong, T. Draper, F.X. Lee, K.F. Liu et al., Phys. Lett. B **612**, 21–28 (2005). hep-lat/0501025
45. I. Horvath, A. Alexandru, J.B. Zhang, Y. Chen, S.J. Dong, T. Draper, K.F. Liu, N. Mathur et al., Phys. Lett. B **617**, 49–59 (2005). hep-lat/0504005
46. A. Alexandru, I. Horvath, J.-b. Zhang, Phys. Rev. D **72**, 034506 (2005). hep-lat/0506018
47. E.-M. Ilgenfritz, K. Koller, Y. Koma, G. Schierholz, T. Streuer, V. Weinberg, Phys. Rev. D **76**, 034506 (2007). arXiv:0705.0018 [hep-lat]
48. E.-M. Ilgenfritz, D. Leinweber, P. Moran, K. Koller, G. Schierholz, V. Weinberg, Phys. Rev. D **77**, 074502 (2008) [Erratum. Phys. Rev. D **77**, 099902 (2008)]. arXiv:0801.1725 [hep-lat]
49. F. Bruckmann, F. Gruber, N. Cundy, A. Schafer, T. Lippert, Phys. Lett. B **707**, 278 (2012). arXiv:1107.0897 [hep-lat]
50. A.V. Kovalenko, M.I. Polikarpov, S.N. Syritsyn, V.I. Zakharov, Phys. Lett. B **613**, 52 (2005). hep-lat/0408014
51. P.V. Buividovich, T. Kalaydzhyan, M.I. Polikarpov, arXiv:1111.6733 [hep-lat]
52. S. Ahmad, J.T. Lenaghan, H.B. Thacker, Phys. Rev. D **72**, 114511 (2005). hep-lat/0509066
53. A. Gorsky, V. Zakharov, Phys. Rev. D **77**, 045017 (2008). arXiv:0707.1284 [hep-th]
54. A.S. Gorsky, V.I. Zakharov, A.R. Zhitnitsky, Phys. Rev. D **79**, 106003 (2009). arXiv:0902.1842 [hep-ph]
55. A.R. Zhitnitsky, Phys. Rev. D **86**, 045026 (2012). arXiv:1112.3365 [hep-ph]
56. E. Witten, Adv. Theor. Math. Phys. **2**, 505 (1998). arXiv:hep-th/9803131
57. A. Parnachev, A.R. Zhitnitsky, Phys. Rev. D **78**, 125002 (2008). arXiv:0806.1736 [hep-ph]
58. E. Thomas, A.R. Zhitnitsky, Phys. Rev. D **85**, 044039 (2012). arXiv:1109.2608 [hep-th]
59. E. Thomas, A.R. Zhitnitsky, Phys. Rev. D **86**, 065029 (2012). arXiv:1203.6073 [hep-ph]
60. M. Ünsal, L.G. Yaffe, Phys. Rev. D **78**, 065035 (2008). arXiv:0803.0344 [hep-th]
61. E. Witten, Ann. Phys. **128**, 363 (1980)
62. I.E. Halperin, A. Zhitnitsky, Phys. Rev. D **58**, 054016 (1998). hep-ph/9711398
63. E. Witten, Phys. Rev. Lett. **81**, 2862 (1998). arXiv:hep-th/9807109
64. E. Thomas, A.R. Zhitnitsky, Phys. Rev. D **87**, 085027 (2013). arXiv:1208.2030 [hep-ph]
65. A.V. Smilga, Phys. Rep. **291**, 1–106 (1997)
66. K. Fukushima, J. Phys. G **39**, 013101 (2012). arXiv:1108.2939 [hep-ph]
67. M.M. Forbes, A.R. Zhitnitsky, J. High Energy Phys. **0110**, 013 (2001). hep-ph/0008315
68. A. Vilenkin, A.E. Everett, Phys. Rev. Lett. **48**, 1867 (1982)
69. C. Hagmann, S. Chang, P. Sikivie, Phys. Rev. D **63**, 125018 (2001). hep-ph/0012361
70. D.T. Son, M.A. Stephanov, A.R. Zhitnitsky, Phys. Rev. Lett. **86**, 3955 (2001). hep-ph/0012041
71. A.M. Polyakov, Nucl. Phys. B **120**, 429 (1977)
72. A.R. Zhitnitsky, Nucl. Phys. A **897**, 93–108 (2013). arXiv:1208.2697 [hep-ph]
73. A.R. Zhitnitsky, Nucl. Phys. A **853**, 135–163 (2011). arXiv:1008.3598 [nucl-th]
74. A.R. Zhitnitsky, Nucl. Phys. A **886**, 17 (2012). arXiv:1201.2665 [hep-ph]
75. P.V. Buividovich, M.N. Chernodub, E.V. Luschevskaya, M.I. Polikarpov, Phys. Rev. D **81**, 036007 (2010). arXiv:0909.2350 [hep-ph]
76. P.V. Buividovich, M.N. Chernodub, D.E. Kharzeev, T. Kalaydzhyan, E.V. Luschevskaya, M.I. Polikarpov, Phys. Rev. Lett. **105**, 132001 (2010). arXiv:1003.2180 [hep-lat]
77. T. Blum, J. Phys. Conf. Ser. **180**, 012066 (2009). arXiv:0908.0937 [hep-lat]
78. G. Basar, D. Kharzeev, H.-U. Yee, A. Zhitnitsky, Contact term and its holographic description in QCD (to appear)

Chapter 9
Views of the Chiral Magnetic Effect

Kenji Fukushima

9.1 Introduction—Discovery of the Chiral Magnetic Effect

The Chiral Magnetic Effect (CME) is concisely summarized in the following handy formula;

$$\mathbf{j} = N_c \sum_f \frac{q_f^2 \mu_5}{2\pi^2} \mathbf{B}, \qquad (9.1)$$

which represents an electric current associated with the non-zero chirality and the external magnetic field $\mathbf{B}$. Here N_c stands for the number of colors in quantum chromodynamics (QCD) and q_f represents the electric charge carried by the quark flavor f where f runs over *up*, *down*, *strange*, etc. Equation (9.1) looks simple, but the physical meaning of this CME current is far from simple. Let me begin with telling some historical remarks on the discovery of the CME-current formula, hoping that it may be instructive and even inspiring to some readers.

When we, Harmen Warringa, Dima Kharzeev, and I, started working on the computation of $\mathbf{j}$, we had no *a priori* idea about the final answer, hence we did not really expect that the final result should be such elegant and beautiful. For several years Harmen and Dima had been working on the implication of axion physics in the context of the relativistic heavy-ion collision. [I will come back to the relevance of the CME to axion physics later.] At that time, around the year of 2007, I was thinking of a different (but related) physics problem, i.e. color-superconducting states in a strong B inspired by a pioneering work [8]. Harmen and I just chatted in the corridor of the RIKEN BNL Research Center (RBRC) about B-effects on color superconductivity, which was soon promoted to intriguing discussions, and a fruitful collaboration after all. Nearly simultaneously with the successful completion of

K. Fukushima (✉)
Department of Physics, Keio University, 3-14-1 Hiyoshi, Kohoku-ku, Yokohama-shi,
Kanagawa 223-8522, Japan
e-mail: fuku@rk.phys.keio.ac.jp

D. Kharzeev et al. (eds.), *Strongly Interacting Matter in Magnetic Fields*,
Lecture Notes in Physics 871, DOI 10.1007/978-3-642-37305-3_9,
© Springer-Verlag Berlin Heidelberg 2013

our project on color superconductivity in B [16] (see also Ref. [34] for an accidental coincidence of the research interest with our Ref. [16]), a monumental paper
by Harmen, Dima, and Larry McLerran appeared [28]. While we were finalizing
the color-superconductivity paper (or struggling with referees, probably), Harmen
excitedly explained the idea of the Chiral Magnetic Effect to me. Also, I clearly remember that Larry came over mischievously (as always) to ask about the strength of
my B in the neutron-star environment ($eB \sim 10^{15}$ gauss at most on the magnetar surface). As compared to *their* B produced in the relativistic heavy-ion collision where
$eB \sim 10^{20}$ gauss could be reached, mine was only negligible... Indeed, historically
speaking, the recognition of such B as strong as the QCD energy scale Λ_{QCD} in
realistic circumstances was an important turning point to get the B-physics research
into gear. In other words, physics researches at $eB \sim \Lambda_{\text{QCD}}^2$ have come to make
pragmatic sense rather than purely academic one since this turning point in 2007.
There was really a tremendous change in the attitude of researchers.

One year later, Harmen invited me to his continued project with Dima on the
Chiral Magnetic Effect. In their first paper the formula was given in a different style
from what is known today, namely, it was not the current but the charge separation
Q expressed as [28]

$$Q = 2Q_w \sum_f |q_f| \gamma \left(2|q_f \Phi| \right). \tag{9.2}$$

Here Q_w is the topological charge (i.e. counter part of μ_5 in (9.1)) and $\gamma(x)$ is a
function dependent on the microscopic dynamics of quark matter. According to the
analysis in Ref. [28] one can approximate $\gamma(x)$ by a simple function; $\gamma(x \leq 1) = x$,
$\gamma(x \geq 1) = 1$. This means that, if the magnetic flux per unit topological domain,
Φ, is large enough, $Q \approx 2Q_w \sum_f |q_f|$. This result is naturally understood from the
index theorem, i.e. $2Q_w = N_5 = N_L - N_R$. Under such strong B, all the spin directions should completely align in parallel with **B**, and thus the momentum directions
are uniquely determined in accord with the chirality. All produced chirality should
contribute to the charge separation, leading to $Q \approx N_5 \sum_f |q_f|$ that is nothing but
$2Q_w \sum_f |q_f|$. In the weak field case, on the other hand, $Q \approx 4\Phi Q_w \sum_f q_f^2$ was
the theoretical estimate.

Equation (9.2) is as a meaningful formula as (9.1), but the determination of $\gamma(x)$
requires some assumptions. Besides, since the formula involves Q_w, it is unavoidable to think of topologically non-trivial gauge configurations. As a matter of fact,
Harmen and I once tried to compute Q concretely on top of the real-time topological configuration, namely, the Lüscher-Schechter classical solution [31, 40], which
turned out to be too complicated to be of any practical use. Then, Harmen hit on
a brilliant idea to deal with Q_w, or strictly speaking, an idea to skirt around Q_w.
[He invented another nice trick later to treat Q_w more directly. I will come to this
point later.] The crucial point is the following; it is not the topological charge Q_w
but the chirality N_5 that causes the charge separation. It is tough to think of Q_w,
then what about starting with N_5 not caring too much about its microscopic origin?
If one wants to fix a value of some number, one should introduce a chemical potential conjugate to the number. In this case of N_5, the necessary ingredient is the

chiral chemical potential μ_5 that couples the chiral-charge operator $\bar{\psi}\gamma^0\gamma^5\psi$. In my opinion the introduction of μ_5 was a simple and great step to make the CME transparent to everybody. In this way the CME has eventually gotten equipped with enough simplicity and clarity.

The remaining task was to answer the following question; what is $\mathbf{j}$ in a system with both μ_5 and $\mathbf{B}$? Harmen and I were first going to calculate the expectation value of the current operator $\bar{\psi}\gamma^\mu\psi$ directly (see the derivation A in Ref. [11]). To this end we had to solve the Dirac equation in the presence of μ_5 and $\mathbf{B}$ to construct the propagator. Now I am very familiar with the way how to do this explicitly, but when we started working on this project, we had not had enough expertise yet, apart from some straightforward calculations in color superconductivity. Some years later Harmen, Dima, and I wrote a paper in which we reported the diagrammatic method to derive (9.1) (see Appendix A in Ref. [12]). Let me briefly explain this derivation here; the electric current in the z-axis direction is written in terms of the propagator as

$$j_z = N_c \sum_f \frac{q_f |q_f B|}{2\pi} \sum_n \int^T \frac{dp_0}{2\pi} \int \frac{dp_z}{2\pi} \int \frac{dx}{L_x}$$

$$\times \mathrm{tr}\left[\gamma^z P_n(x) \frac{i}{\tilde{p}_\mu \gamma^\mu + \mu_5 \gamma^0 \gamma^5 - M_f} P_n(x)\right], \tag{9.3}$$

where the p_0-integration is either at $T = 0$ or the Matsubara sum at $T \neq 0$. If we choose the gauge as $A_0 = A_x = A_z = 0$ and $A_y = Bx$, the tilde momentum in the denominator is $\tilde{p} = (p_0, 0, -\mathrm{sgn}(qB)\sqrt{2|qB|n}, p_z)$. We do not need the explicit form of the Landau wave-functions $P_n(x)$ that take a 4×4 matrix structure in Dirac space. Because we are interested in $\mathbf{j} \parallel \mathbf{B}$ here, γ^z commutes with $P_n(x)$ and thus we need only $P_n(x)^2$ which equals 1 for $n > 0$ and $(1 + i\,\mathrm{sgn}(q_f B)\gamma^x \gamma^y)/2$ for $n = 0$. After some calculations one can confirm that (9.3) is reduced to (9.1) regardless of the temperature T and the flavor-dependent mass M_f. Let me make a comment on this rather naïve calculation. In most cases the proper-time method is the best way to proceed in theoretical calculations [22, 42] and the above form of the quark propagator is not widely known. For the purpose of calculating a finite quantity like the CME current, I would like to stress that the above quark propagator should be equally useful. Actually it is almost obvious in (9.3) that any contributions from the Landau non-zero modes are vanishing and the current arises from the Landau zero-mode only.

Coming back to the story of our first attempt to discover $\mathbf{j}$, I remember that Harmen and I came to the office and brought different answers every morning and had the hottest discussions all the day. It took us a few days until we eventually convinced ourselves to arrive at the right answer. Later on, Harmen had great efforts to dig out several independent derivations of (9.1) while preparing for our paper. Among various derivations we first found the one based on the thermodynamic potential (i.e. the derivation C in Ref. [11]). Because this calculation plays some role in later discussions on the physical interpretation of the CME current, let us take a closer look at the detailed derivation using the thermodynamic potential.

The most essential ingredient is the quasi-particle energy dispersion relation in the presence of B and μ_5. For $\mathbf{B}$ along the z-axis, one can solve the Dirac equation to find the dispersion relation,

$$\omega_{p,s}^2 = \left[\left(p_z^2 + 2|q_f B|n\right)^{1/2} + \mathrm{sgn}(p_z)\, s\mu_5\right]^2 + M_f^2, \qquad (9.4)$$

where s is the spin, q_f and M_f are the electric charge and the mass of quark flavor f. Once the one-particle energy is given, one can immediately write the thermodynamic potential down as

$$\Omega = N_c \sum_f \frac{|q_f B|}{2\pi} \sum_{s=\pm} \sum_{n=0}^\infty \alpha_{n,s} \int_{-\infty}^\infty \frac{dp_z}{2\pi} \left[\omega_{p,s} + T \sum_\pm \ln\left(1 + e^{-(\omega_{p,s}\pm\mu)/T}\right)\right]$$

$$(9.5)$$

at finite temperature T and quark chemical potential μ. The spin factor, $\alpha_{n,s}$, is defined as $\alpha_{n,s} = 1$ $(n > 0)$, δ_{s+} $(n = 0, q_f B > 0)$, δ_{s-} $(n = 0, q_f B < 0)$. This factor is necessary to take care of the fact that the Landau zero-mode $(n = 0)$ exists for one spin state only. The current j_z is obtained by differentiating Ω with respect to the vector potential A_z. Because the vector potential in the matter sector resides only through the covariant derivative, the following replacement is possible inside of the p_z-integration,

$$\frac{\partial}{\partial A_z} = q \frac{d}{dp_z}. \qquad (9.6)$$

The combination of this derivative and the p_z-integration ends up with the surface terms. It is the characteristic feature of the quantum anomaly that a finite answer results from the ultraviolet edges in the momentum integration. That is, the CME current reads,

$$j_z = N_c \sum_f \frac{q_f|q_f B|}{2\pi} \sum_{s=\pm} \sum_n \alpha_{n,s} \int_{-\Lambda}^\Lambda \frac{dp_z}{2\pi} \frac{d}{dp_z} \left[\omega_{p,s} + T \sum_\pm \ln\left(1 + e^{-(\omega_{p,s}\pm\mu)/T}\right)\right]$$

$$= N_c \sum_f \frac{q_f|q_f B|}{4\pi^2} \left[\omega_{p,\pm}(p_z = \Lambda) - \omega_{p,\pm}(p_z = -\Lambda)\right]$$

$$= N_c \sum_f \frac{q_f|q_f B|}{4\pi^2} \left[(\Lambda \pm \mu_5) - (\Lambda \mp \mu_5)\right] = N_c \sum_f \frac{q_f^2 \mu_5}{2\pi^2} B. \qquad (9.7)$$

Here, in the second and the third lines, $\pm$ appears from the Landau zero-mode allowed by $\alpha_{n,s}$, i.e. $\pm$ amounts to $\mathrm{sgn}(q_f B)$ which cancels the modulus of $|q_f B|$, and the matter part drops off for infinitely large $\omega_{p,s}(p_z = \pm\Lambda)$. It would be just a several-line calculation to make sure that (9.3) is equivalent with (9.7) and they are calculations at the one-loop level. It is also a common character of the quantum anomaly that the one-loop calculation would often give the full quantum answer. Although I do not know any explicit check of the higher-order loop effects, the above method at the one-loop level is my favorite derivation of (9.1); all the calculation procedures are so elementary and transparent.

9.2 Chiral Separation Effect

Soon later, Harmen and I found that a very similar topological current had been discovered in the neutron-star environment, that is, the axial current associated with the quark chemical potential μ and the magnetic field $\mathbf{B}$ [32],

$$\mathbf{j}_5 = N_c \sum_f \frac{q_f^2 \mu}{2\pi^2} \mathbf{B}. \tag{9.8}$$

This is a chiral dual version of (9.1). Nowadays people call (9.8) the Chiral Separation Effect (CSE) in contrast to (9.1) referred to as the Chiral Magnetic Effect. When we learned the fact that (9.8) had been known earlier, our excitement got cooled down a bit. Also, three years later, we came to know that the CME formula had been discovered further earlier. Now there is a consensus in the community that the CME formula (9.1) was first derived by Alex Vilenkin [47]. It was an embarrassment for me to have overlooked his work until he brought our attention to his old papers. In fact an equivalent of (9.1) has been rediscovered over and over again [1, 19, 20] and I would not be surprised even if (9.1) is still buried in further unknown works. [I am not talking about the recent activities to derive (9.1) from a deeper insight into physics such as Berry's curvature [45, 50], hydro or kinetic approaches [18, 24, 25, 46], and so on, which really deserve more investigations.]

The derivation of (9.8) is worth discussing here. The topological effects in quantum electrodynamics (QED) from $N_c \times N_f$ quarks add terms in the action as

$$\delta S = \int d^4x \, \theta(x) \left[\partial_\mu j_5^\mu(x) + N_c \sum_f \frac{q_f^2}{16\pi^2} \varepsilon^{\mu\nu\rho\sigma} F_{\mu\nu}(x) F_{\rho\sigma}(x) \right], \tag{9.9}$$

associated with an axial rotation by $\theta(x)$. In this way we see that the axial current is not conserved but anomalous. With the replacement of $A_0 = \mu$ and $\varepsilon^{0ijk} \partial_j A_k = B^i$, one can transform this expression using the integration by parts into

$$\delta S = \int d^4x \, \partial_i \theta(x) \left[-j_5^i(x) - N_c \sum_f \frac{q_f^2}{2\pi^2} \varepsilon^{0ijk} A_0(x) \partial_j A_k(x) \right]$$

$$= \int d^4x \, \partial_i \theta(x) \left[-j_5^i(x) + N_c \sum_f \frac{q_f^2}{2\pi^2} \mu B^i(x) \right], \tag{9.10}$$

from which (9.8) immediately follows. This derivation also tells us that the B-induced current in the right-hand side of (9.8) is nothing but a part of the Chern-

Simons current $\sim \varepsilon^{\mu\nu\rho\sigma} A_\nu \partial_\rho A_\sigma$ in QED. It should be mentioned that the derivation presented above is a little bit cooked up by me for the illustration purpose and one should refer to the original paper [32] for more careful treatments of the surface integral.

Before going on our discussions, let me point out that the above derivation implicitly assumes massless quarks. If quarks are massive, (9.9) should be modified with an additional term $2iM_f\langle\bar{\psi}_f\gamma^5\psi_f\rangle$. This modification would be harmless as long as the pseudo-scalar condensate is vanishing, but in principle, (9.8) could be dependent on M_f unlike (9.1) as argued explicitly in Ref. [32]. In fact it is quite subtle whether (9.8) is sensitive to M_f or not, and I will address this question in an explicit way soon later.

It would be an interesting question how to derive (9.8) microscopically just like the ways addressed in the previous section. In fact I have once tried to prove (9.8) based on the thermodynamic potential by inserting an axial gauge field. There must be a way along this line, but I could not solve it (or I would say that I did not have enough time to find it out...). Instead, here, let me introduce another derivation based on the propagator as in (9.3).

The axial current is expressed as

$$j_z^A = N_c \sum_f \frac{q_f|q_f B|}{2\pi} \sum_n \int^T \frac{dp_0}{2\pi} \int \frac{dp_z}{2\pi} \int \frac{dx}{L_x}$$

$$\times \mathrm{tr}\left[\gamma^z\gamma^5 P_n(x)\frac{i}{\tilde{p}_\mu\gamma^\mu + \mu\gamma^0 - M_f}P_n(x)\right] \qquad (9.11)$$

at finite quark chemical potential μ. It is easy to see that any contributions from $n \neq 0$ vanish due to the Dirac trace. Only the Landau zero-mode produces a term involving $\gamma^x\gamma^y$ which makes $\mathrm{tr}(\gamma^0\gamma^x\gamma^y\gamma^z\gamma^5) = -4i \neq 0$. Then, the above expression simplifies as

$$j_z^A = -N_c \sum_f \frac{q_f^2 B}{2\pi} \int^T \frac{dp_0}{2\pi} \int \frac{dp_z}{2\pi} \mathrm{tr}\left[\gamma^z\gamma^5 \frac{\tilde{p}_\mu\gamma^\mu + \mu\gamma^0 + M_f}{(p_0+\mu)^2 - p_z^2 - M_f^2}\gamma^x\gamma^y\right]$$

$$= 4iN_c \sum_f \frac{q_f^2 B}{2\pi} \int^T \frac{dp_0}{2\pi} \int \frac{dp_z}{2\pi} \frac{p_0+\mu}{(p_0+\mu)^2 - p_z^2 - M_f^2}$$

$$= N_c \sum_f \frac{q_f^2 B}{2\pi} \frac{\partial Z(\mu)}{\partial\mu}, \qquad (9.12)$$

where $Z(\mu)$ denotes the partition function at finite density in $(1+1)$-dimensional theory (as a result of the dimensional reduction with the Landau zero-mode), and thus the μ-derivative leads to the quark density n. In the second line we used $2(p_0 + \mu)/[(p_0 + \mu)^2 - p_z^2 - M_f^2] = (\partial/\partial\mu)\ln[(p_0 + \mu)^2 - p_z^2 - M_f^2]$. One might have thought that it is a simple exercise to evaluate $Z(\mu)$ with the $(1+1)$-dimensional integration. The fact is, however, that the finite-μ system in $(1 + 1)$ dimensions is by no means simple.

In Ref. [32] one can find exactly the same expression as above in a slightly different calculation and the density is written as (see (37) in Ref. [32]),

$$n_f(T, \mu) = \int \frac{dp_z}{2\pi} \left[\frac{1}{e^{(\omega_f - \mu)/T} + 1} - \frac{1}{e^{(\omega_f + \mu)/T} + 1} \right] \tag{9.13}$$

with $\omega_f = \sqrt{p_z^2 + M_f^2}$. This result is certainly M_f-dependent as suggested in the paragraph below (9.10), and this would make a sharp contrast to the CME current (9.1).

We know, however, that the density in the $(1 + 1)$-dimensional fermionic theory arises from the anomaly [41] and the density (9.13) is not the right answer. In fact, in view of the second line of (9.12), it seems at a glance that the μ-dependence could be absorbed in the p_0-integration, which already gives us an impression that something non-natural should be happening. To see this, let us take one-step back to the microscopic expression, i.e., the $(1 + 1)$-dimensional partition function reads,

$$Z = 2i \int^T \frac{dp_0}{2\pi} \int \frac{dp_z}{2\pi} \ln\left[(p_0 + \mu)^2 - p_z^2 - M_f^2 \right]$$

$$= i \int^T \frac{dp_0}{2\pi} \int \frac{dp_z}{2\pi} \operatorname{tr}\left[\gamma^0 (p_0 + \mu) - \gamma^z p_z - M_f \right], \tag{9.14}$$

from which the μ-dependence could be eliminated by the chiral rotation (for the zero-mode basis only),

$$\psi_0 = e^{i\gamma^z \gamma^0 \mu z} \psi_0', \tag{9.15}$$

leading to (here, we shall show results at $T = 0$ for simplicity, but nothing is changed even at finite T),

$$Z = i \int \frac{dp_0}{2\pi} \int \frac{dp_z}{2\pi} \operatorname{tr}\left[e^{i\gamma^z \gamma^0 \mu z} \left(\gamma^0 (i\partial_0 + \mu) - \gamma^z i\partial_z - M_f \right) e^{i\gamma^z \gamma^0 \mu z} \right]$$

$$= i \int \frac{dp_0}{2\pi} \int \frac{dp_z}{2\pi} \operatorname{tr}\left[\gamma^0 i\partial_0 - \gamma^z i\partial_z - \tilde{M}_f \right]$$

$$= \int_{-\Lambda + \mu}^{\Lambda - \mu} \frac{dp_z}{2\pi} \frac{1}{2} \tilde{\omega}_f + \int_{-\Lambda - \mu}^{\Lambda + \mu} \frac{dp_z}{2\pi} \frac{1}{2} \tilde{\omega}_f \tag{9.16}$$

with $\tilde{\omega}_f = \sqrt{p_z^2 + |\tilde{M}_f|^2}$, where $\tilde{M}_f = M_f e^{2i\gamma^z \gamma^0 \mu z}$ is the chirally tilted mass. The momentum integration is shifted according to the chiral rotation (9.15). The first (second) integral corresponds to the particle (anti-particle, respectively) contribution. Thus, one can extract the μ-dependent piece from the surface terms as follows;

$$Z = \int_{-\Lambda}^{\Lambda} \frac{dp_z}{2\pi} \tilde{\omega}_f + \left(\int_{\Lambda}^{\Lambda+\mu} + \int_{\Lambda}^{\Lambda-\mu} \right) \frac{dp_z}{2\pi} \tilde{\omega}_f$$

$$= \mu^2 \frac{d^2}{dx^2} \int_{\Lambda}^{\Lambda+x} \frac{dp_z}{2\pi} \tilde{\omega}_f + (\mu\text{-independent terms})$$

$$= \frac{\mu^2}{2\pi} + (\mu\text{-independent terms}), \tag{9.17}$$

which results in the density $n_f = \mu/\pi$ that is independent of M_f [41]. It is clear from the second line of the above calculation that the density originates from the ultraviolet edges, which is the characteristic feature of the anomaly. The full quantum answer is then given as

$$n_f = \frac{\partial Z(\mu)}{\partial \mu} = \frac{\mu}{\pi} \quad \Rightarrow \quad j_z^A = N_c \sum_f \frac{q_f^2 \mu}{2\pi^2} B. \tag{9.18}$$

Equivalently, if one is interested in deriving the same answer from (9.12) directly, one should split the composite operator as $\bar{\psi}(x)\gamma^z\gamma^5\psi(x) \rightarrow \bar{\psi}(x+\varepsilon)\gamma^z\gamma^5\psi(x)$ and insert the infinitesimal gauge connection from x to $x+\varepsilon$. Interestingly, contrary to Ref. [32], the Chiral Separation Effect (9.8) is presumably insensitive to the quark mass just like the Chiral Magnetic Effect (9.1). Whether (9.8) is robust or not regardless of M_f is an important question particularly in the context of the Chiral Magnetic Wave (CMW) [30]. The anomalous nature of the density (9.18) implies that the CMW can exist also in the chiral-symmetry breaking phase where quarks acquire substantial mass dynamically.

I would not insist that I could prove the non-renormalization of (9.8) since the above is just a one-loop perturbation and non-perturbative interactions may change the story; I would like to thank Igor Shovkovy for raising this unanswerable but unforgettable question. The interested readers may consult Refs. [15, 21] for some examples of non-non-renormalization. Anyway, I can at least say with confidence that, if B is super-strong, the reduction to the $(1+1)$-dimensional system should be strict, and then (9.13) must be altered, conceivably as $n_f = \mu/\pi$ [9].

Although the interpretation of (9.18) may swerve a bit from our main stream, I would emphasize that (9.18) is extremely interesting and it would be definitely worth revisiting its profound meaning. Actually, (9.15) has an impact on the structure of the QCD vacuum. Let us consider the hadronic phase with spontaneous breakdown of chiral symmetry. After the rotation (9.15), apart from the anomalous term $\mu^2/(2\pi)$, the system is reduced to that at zero density, which means that $\chi = \langle \bar{\psi}_0' \psi_0' \rangle$ should take a finite value. Therefore, in terms of the original fields ψ_0, the chiral condensates form a spiral structure,

$$\langle \bar{\psi}_0 \psi_0 \rangle = \chi \cos(2\mu z), \qquad \langle \bar{\psi}_0 \gamma^z \gamma^0 \psi_0 \rangle = \chi \sin(2\mu z), \tag{9.19}$$

which is called the chiral spiral or the dual chiral-density wave (if γ^5 is involved) [7, 33]. In particular, if the above type of the inhomogeneous ground state is caused by B, it is sometimes called the chiral magnetic spiral [5].

I should emphasize that (9.18) does not really require the dimensional reduction, while the chiral magnetic spiral needs the pseudo $(1+1)$-dimensional nature under sufficiently strong B. This point might be a bit puzzling. As long as j_z^A is concerned, only the Landau zero-mode remains non-vanishingly for any B, and the momentum integration is purely $(1+1)$-dimensional. The chiral condensate is, however, not spiral but homogeneous for small B because of contributions from all non-zero Landau levels. That is, the genuine chiral condensate is $\langle \bar{\psi}\psi \rangle = \sum_n \langle \bar{\psi}_n \psi_n \rangle$, among which only the Landau zero-mode has a special structure as in (9.19). I would conjecture, hence, that there is no sharp phase transition from the homogeneous chiral condensate at $B = 0$ to the chiral magnetic spiral at $B \neq 0$, but it may be possible that the inhomogeneous zero-mode contribution gradually develops, which exhibits a smooth crossover to the chiral spirals with increasing B.

9.3 What Is the Chiral Chemical Potential?

Equation (9.8) is very similar to the CME current (9.1), so that one might have thought at a first glance that (9.1) emerges trivially from the insertion of γ^5 in both sides of (9.8). The relation between (9.1) and (9.8) is not such simple, though. As a matter of fact, this point was a major source of confusions about the validity of (9.1). One can readily extend the field-theoretical derivation of (9.8) using (9.9) in order to obtain (9.1) by introducing the axial vector fields A_μ^5, or the chiral gauge fields, $A_R = (A_\mu + A_\mu^5)/2$ and $A_L = (A_\mu - A_\mu^5)/2$. Then, in the same manner as in the previous section, one can formulate the counterpart of (9.9) associated with a vector rotation by $\beta(x)$, that is,

$$\delta S = \int d^4x\, \beta(x) \left[\partial_\mu j^\mu(x) + N_c \sum_f \frac{q_f^2}{16\pi^2} \varepsilon^{\mu\nu\rho\sigma} F_{\mu\nu}^R(x)\, F_{\rho\sigma}^R(x) \right.$$

$$\left. - N_c \sum_f \frac{q_f^2}{16\pi^2} \varepsilon^{\mu\nu\rho\sigma} F_{\mu\nu}^L(x)\, F_{\rho\sigma}^L(x) \right]. \tag{9.20}$$

This leads to $-j^i - N_c \sum_f (q_f^2/2\pi^2)\varepsilon^{0ijk}(A_0^R - A_0^L)\partial_j A_k = 0$ just as in (9.10), and this is nothing but (9.1) after the identification of A_0^5 as μ_5 (see the derivation D in Ref. [11]). Although the derivation may look flawless, it triggered suspicious views of (9.1), which was first addressed by Toni Rebhan, Andreas Schmitt, and Stefan Stricker using the Sakai-Sugimoto model [38]. [It should be noted that the CME current had been exactly reproduced in the holographic models [49].]

Obviously, one has to deal with the chiral gauge theory with both A_R and A_L to introduce μ_5 in the above way, and it is well-known that the anomaly in the chiral gauge theory has a more complicated structure than that in the vector gauge theory. Roughly speaking, the anomaly is a consequence from the inconsistency between chiral invariance and gauge symmetry. In the vector gauge theory, usually, the vector current is strictly conserved due to adherence to gauge symmetry, and the anomaly is seen in the axial vector channel only (see (9.9)). In the case in the chiral gauge theory, however, there is no such strict demand from the theory and it should be prescription dependent how the anomaly may appear in the vector and the axial vector currents. Indeed we can clearly see from (9.20) that the vector current is also anomalous. There are two representative results known as the covariant anomaly and the consistent anomaly, and they can coincide only when the anomaly cancellation holds, as is the case in the Standard Model. The authors of Ref. [38] claimed that the vector current should be free from the anomaly and the theory should accommodate the Bardeen counter-terms to cancel the anomalous terms in (9.20). Then, needless to say, the CME current is vanishing!

This argument scared Harmen and me very much. In 2009 when Ref. [38] came out, Harmen was a postdoc in Frankfurt and I was also there as a visitor. Harmen's face is always very white, but he got even more whity, and we had a lot of discussions on Ref. [38] in Frankfurt with a fear that we might have made a big steaming mistake... At that time, neither Harmen nor I was 100 % confident in (9.1) (maybe Dima was?), and the necessity of the Bardeen counter-terms sounded plausible. This puzzle was one of the issues discussed in a RBRC workshop, "P- and CP-odd Effects in Hot and Dense Matter" in May, 2010. One of the invited participants, Valery Rubakov, wrote a note to clarify this issue based on the discussions in the workshop [39]. The essence in his argument is the following. If one introduces μ_5 as the zeroth component of the axial gauge field, the CME current is gone indeed. However, QCD and QED are not the chiral gauge theory. One should then introduce μ_5 in a different way as a conjugate to the Chern-Simons charge. Therefore, instead of adding a term $\mu_5 \bar{\psi} \gamma^0 \gamma^5 \psi$ in a form of the covariant derivative in the Lagrangian, one should think of the Chern-Simons current K^μ which is deduced from

$$N_c \sum_f \frac{q_f^2}{16\pi^2} \varepsilon^{\mu\nu\rho\sigma} F_{\mu\nu}(x) F_{\rho\sigma}(x) = \partial_\mu \left[N_c \sum_f \frac{q_f^2}{4\pi^2} \varepsilon^{\mu\nu\rho\sigma} A_\nu(x) \partial_\rho A_\sigma(x) \right]$$

$$= \partial_\mu K^\mu(x) \qquad (9.21)$$

in the QED sector. The term to be added in the Lagrangian is,

$$S_{cs} = -\int d^4x \, \mu_5 K^0(x) = -N_c \sum_f \frac{q_f^2 \mu_5}{4\pi^2} \int d^4x \, \varepsilon^{0ijk} A_i(x) \partial_j A_k(x), \qquad (9.22)$$

from which (9.1) immediately follows as a result of the derivative, $j^i = \delta S_{cs}/\delta A_i(x)$. One may worry about gauge invariance in the above prescription. It would be then more convenient to rewrite S_{cs} in the following way after the integration by parts, that is manifestly gauge invariant,

$$S_{cs} = \int d^4x \, \theta(t) N_c \sum_f \frac{q_f^2}{16\pi^2} \varepsilon^{\mu\nu\rho\sigma} F_{\mu\nu}(x) F_{\rho\sigma}(x), \qquad (9.23)$$

where $\partial_0 \theta(t) = \mu_5$. In other words, we can say that μ_5 is the time derivative of the θ angle in the QED sector, which was pointed out already in Ref. [11] and the idea of the charge separation driven by inhomogeneous θ can be traced back to Ref. [26]. A subsequent question naturally arises; what happens if $\theta(x)$ has not only temporal but also spatial dependence in general? The Chern-Simons-Maxwell theory with general $\theta(x)$ provides us with the following modified Maxwell equations;

$$\nabla \cdot \mathbf{E} = \rho + N_c \sum_f \frac{q_f^2}{2\pi^2} (\nabla \theta) \cdot \mathbf{B}, \tag{9.24}$$

$$\nabla \times \mathbf{B} - \partial_0 \mathbf{E} = \mathbf{j} + N_c \sum_f \frac{q_f^2}{2\pi^2} [(\partial_0 \theta) \mathbf{B} - (\nabla \theta) \times \mathbf{E}], \tag{9.25}$$

and Faraday's law and Gauß's law are not altered. We see that the CME current appears in the right-hand side of (9.25) as if it is a part of the external current. In this manner we can conclude from (9.24) that an electric-charge density is induced by spatially inhomogeneous $\theta(x)$ in the presence of $\mathbf{B}$. To the best of my knowledge (9.24) and (9.25) are the quickest derivation of the Chiral Magnetic Effect, as discussed first in Ref. [27].

[After I finished writing this article, I was informed by Toni, one of the authors of Ref. [38], that the confusion about the CME in the holographic context seems to continue. I am not able enough to make any judgment here, and those who want to dive into this confusion can consult the recent analysis in Ref. [3].]

9.4 What Really Flows?

To tell the truth, I have never gotten any satisfactory answer to the following question; what really flows? I have had various discussions with people who have various backgrounds, but those discussions ended up with more confusions than before. Thanks to useful conversations, nevertheless, my eyes have been open to various views of (9.1). People (including me) say that the CME current is an *electric current* induced by $\mathbf{B}$ just like Ohm's law with the electric field $\mathbf{E}$. Let me begin with suspecting this interpretation that people just take for granted.

In classical electrodynamics (9.25) is usually written in a slightly different way, i.e.,

$$\nabla \times \mathbf{B} = \mathbf{j} + \partial_0 \mathbf{E} + N_c \sum_f \frac{q_f^2}{2\pi^2} [(\partial_0 \theta) \mathbf{B} - (\nabla \theta) \times \mathbf{E}], \tag{9.26}$$

and $\partial_0 \mathbf{E}$ is called the displacement *current*. We see that the CME current should be a genuine current *if* $\partial_0 \mathbf{E}$ can be regarded as a real electric current, for they enter Ampère's law on equal footing. In other words, *if* the displacement current is not a real current, the CME current is not, either. Now, we know from our experience that

$\partial_0 \mathbf{E}$ is only the time derivative of the electric field and no electric charge flows associated with the displacement current. The displacement current certainly plays the equivalent role as $\mathbf{j}$ as a source to create $\mathbf{B}$, but it is clear that there is no movement of electric charge at all. It would be therefore a legitimate claim to insist that the charge separation from the CME current might be an illusion. I would emphasize the importance to distinguish the current and the charge in the argument here. For example, the most well-known example of the displacement current is the problem of the capacitor that is composed of two separate conductors. Let a capacitor be connected to the wire with finite electric current. More and more electric charge accumulates on the conductors and produces stronger and stronger electric field inside as the time goes. Then, even though two conductors are physically separate and no electric current flows between them, the displacement current flows as if the electric current flowed along the wire without the capacitor. The distribution of the electric charge stored on the conductors is, however, totally different depending on the situation with and without the capacitor. In this sense, thus, the charge itself may not flow and the charge separation may not occur with the CME current also.

A related criticism against the CME current is that the current computed in (9.7) for example is the expectation value of the current operator, $\bar{\psi}\gamma^\mu\psi$, and it is not necessarily the current. In fact, there are some studies on the Chiral Magnetic Effect in the lattice gauge theory; the correlation functions of the chirality and the current were measured in Ref. [6], and later (9.1) was checked directly on the lattice [48]. It is not so straightforward, however, to interpret these lattice results properly. A system with a finite electric current could be steady but is out of equilibrium. What one can calculate in the thermal system in equilibrium like the situation of the lattice simulation in Euclidean space-time is the electric-current conductivity according to the Kubo formula. It is a tricky question what $\langle\bar{\psi}\gamma^\mu\psi\rangle$ really represents in the lattice simulation. Let me take one example for concreteness. If the system has a condensate of the omega meson, ω^μ, the interpolation field of ω^μ is $\sim \bar{\psi}\gamma^\mu\psi$ and then $j^\mu = \langle\bar{\psi}\gamma^\mu\psi\rangle \neq 0$, but this does not necessarily mean that the system has a persistent current. To make this point clearer, the spin operator in terms of the Dirac matrices is $\hat{S}^z = \frac{i}{4}[\gamma^x, \gamma^y] = \frac{1}{2}\text{diag}[\sigma^3, \sigma^3]$, so that the spin expectation value is $S^z = \langle\bar{\psi}\hat{S}^z\psi\rangle = \frac{1}{2}\langle\phi_R^\dagger\sigma^3\phi_L\rangle + \frac{1}{2}\langle\phi_L^\dagger\sigma^3\phi_R\rangle$, while the current expectation value is $j^z = \langle\bar{\psi}\gamma^z\psi\rangle = \langle\phi_R^\dagger\sigma^3\phi_R\rangle - \langle\phi_L^\dagger\sigma^3\phi_L\rangle$, where ϕ_L and ϕ_R are two-component spinors in the left-handed and right-handed chirality, respectively. Here, the similarity between S^z and j^z implies that we can regard j^z as a static quantity like the spin S^z, which may well be the most appropriate interpretation of the lattice measurement.

From the point of view of the theoretical treatment of the electric current, the formulation based on the linear response theory must be a good starting point. I believe that the work along this line in Ref. [29] should be one of the most important literature to think of physics of the Chiral Magnetic Effect. They computed the one-loop diagram on top of the μ_5 background to find the chiral magnetic conductivity $\sigma_\chi(\omega, p)$. The result is consistent with (9.1) in a particular limit; $\sigma_\chi(\omega = 0, p \to 0) = \lim_{p\to 0}\sigma_\chi(0, p) = N_c\sum_f(q_f^2\mu_5/2\pi^2)$ which correctly reproduces the CME current. In view of the result of Ref. [29], on the other hand, it seems

$\sigma_\chi(\omega \to 0, p = 0) = 0$. [This latter limit is not manifestly addressed in Ref. [29], but it is pointed out that the conductivity drops to one third just away from $\omega = 0$. It seems to be vanishing from (38) and Fig. 1 of Ref. [29].] This is a problem because the latter limit rather than the former one is more relevant to the real-time dynamics. The fact that the former limit ($\omega = 0$ first and $p = 0$ next) gives the CME current (9.1) suggests that the CME current should be a static quantity just as measured in the lattice simulation and thus not a genuine electric current!?

One may still consider that the intuitive argument leading to (9.2) should work anyway. My impression is also that all above-mentioned problems are just on the conceptual level (though I have no idea how to reconcile them) and in practice the CME current flows according to (9.1) after all. Indeed if there are almost massless quarks in a quark-gluon plasma and a strong **B** is imposed on a topological domain, an electric current must be induced for sure. An example of the real-time calculation of the CME current with not μ_5 but a topological domain is quite instructive in this sense [13]. The central innovation in Harmen's idea (as discussed in Ref. [13]) was to mimic the topological domain by putting **E** and **B** parallel to each other, with which $\varepsilon^{\mu\nu\rho\sigma} F_{\mu\nu} F_{\rho\sigma} \neq 0$. Then, the particle production occurs via the Schwinger mechanism and the produced particles are accelerated by the fields, and the electric current is generated. The current is time dependent and the current-generation rate can be analytically written down. In this setup the physical origin of the CME current is crystal-clear! So, if anything is fishy in physics of the Chiral Magnetic Effect, it should have something to do with technical defects of μ_5 in equilibrium circumstances.

Supposing that physics of the Chiral Magnetic Effect should be robust, let us admit the CME current (9.1) as it is to proceed to the next question, that is actually the central question in this section; what really flows?

An intuitive explanation tells us that quarks simply flow in a quark-gluon plasma. It is, however, based on a classical picture, and such a picture misses quantum character that is indispensable for phenomena related to the quantum anomaly. Look at the derivation of the CME current in (9.7). If this derivation captures the underlying physics of the CME current, the origin of the current comes from quarks with infinitely large momenta. Where are such fast-moving quarks in the real quark-gluon plasma? They may spill out from the vacuum through quantum processes, but how is it possible to retrieve particles with infinite momenta? Usually the quantum anomaly involves ultraviolet regions of the momentum integration as a loop of virtual particles, meanwhile ultraviolet particles directly participate in the physical observable in the CME problem. It is very hard (at least for me) to imagine that the current generation in such a way really happens in a physical plasma. This deliberation brings me a further doubt about the static evaluation of the CME current.

A natural extension of this question about the origin of the CME current is whether it exists in the hadronic phase and, if it does, how the current appears in terms of hadronic degrees of freedom. Actually this question has been something in mind for a long time since when we published Ref. [11]. In the hadronic phase an electric current should be attributed to charged pions, but pions are insensitive to chirality and thus μ_5 or the strong θ angle. One possible answer would be that there is no CME in the hadronic phase, and if so, it would be fantastic; the CME current can be a signature for quark deconfinement, as implied in Ref. [11]. I had heard that Harmen wanted to analyze the CME using the chiral perturbation theory, though he never worked it out.

Recently I have clarified what would happen in the hadronic phase and wrote a paper with one of my students, Kazuya Mameda [14]. Our conclusion was a very natural one, and a very perplexing one at the same time.

The CME current is unchanged even in the hadronic phase, which is very natural since the CME current has the anomalous origin that arises from ultraviolet fluctuations. At low energies the anomaly should be saturated by infrared degrees of freedom, which is sometimes referred to as the anomaly matching. This idea is formulated as the Wess-Zumino-Witten action and the current should be given by the derivative of the total effective action with respect to the gauge field. In this way we found that the leading-order term in the chiral Lagrangian leads to the current,

$$j_\chi^\mu = -i \frac{e f_\pi^2}{4} \mathrm{tr}\left[\left(\Sigma^\mu - \tilde{\Sigma}^\mu\right)\tau^3 \right] \simeq e\left(\pi^- i \partial^\mu \pi^+ - \pi^+ i \partial^\mu \pi^-\right) + \cdots, \qquad (9.27)$$

where $\Sigma^\mu = U^\dagger \partial^\mu U$, $\tilde{\Sigma}^\mu = (\partial^\mu U)U^\dagger$, and $U = e^{i\pi^a \tau^a / f_\pi}$ are the standard notation in the chiral Lagrangian. The physical meaning of the above expression is plain as seen from the expansion in terms of the pion fields. It is a common form of the probability flow in Quantum Mechanism representing the electric current associated with the flow of the charged pions.

A more non-trivial contribution comes from the Wess-Zumino-Witten part, which leads to the current associated with the π^0 domain-wall [44], i.e.,

$$j_{\mathrm{WZW}}^\mu = N_c \sum_f \frac{q_f}{8\pi^2 f_\pi} \varepsilon^{\mu\nu\rho\sigma} \left(\partial_\nu \pi^0\right) F_{\rho\sigma}. \qquad (9.28)$$

This current is very similar to the CME current (9.1) and $\theta(x)$ is just replaced by $\pi^0(x)/(4\pi^2 f_\pi)$. Although (9.28) is not the Chiral Magnetic Effect, it would give us a clue to think about the physical meaning of the CME current. Finally, the CME current appears from the so-called contact part of the Wess-Zumino-Witten action [23];

$$S_{\mathrm{P}} = N_c \sum_f \frac{q_f^2}{8 N_f \pi^2} \varepsilon^{\mu\nu\rho\sigma} \int d^4x \, A_\mu(x)\left(\partial_\nu A_\rho(x)\right)\partial_\sigma \theta(x), \qquad (9.29)$$

which is just equivalent to the Chern-Simons action already discussed in (9.22). [θ in (9.29) has a different normalization by $2N_f$ by convention.] Naturally the current derived from (9.29) should reproduce the CME current (9.1). This is how one can get the CME current in the hadronic phase and my surprise lies in the fact that the pion dynamics is completely decoupled from the CME current.

Because the π^0 domain-wall looks a bit more intuitive than the mystical θ angle, we shall consider a possible interpretation of the current (9.28). This is certainly a current, but no charged pions, $\pi^{\pm}$, are involved in the formula. Then, it is as puzzling in (9.28) how the current can flow and what really flows.

To answer this question, let me emphasize a very useful analogue of the Josephson current in superconductivity. The Josephson junction consists of superconducting materials and a thin layer of insulator (S-I-S) or non-superconducting metal (S-N-S) sandwiched by them. There was a big debate about whether the supercurrent can flow or not through the insulating barrier. Of course, there is no Cooper pair inside of the insulator, and thereby there is nothing that takes care of the supercurrent. It should have been a natural attitude to get skeptical about such a current [4]. This situation, a current without current carriers, is quite reminiscent of our problem of the CME current or the current accompanied by the π^0 domain-wall. Everyone knows that the Josephson current is the experimental fact today [2]. For the Josephson current, the coherence is the most important; in superconductor the quantum state is characterized by a wave-function just like a problem in Quantum Mechanics. In the QCD case, also, such a coherent state is realized by the condensation of fields, namely, $\pi^0(x)$ in (9.28) is to be regarded as a macroscopic wave-function. One may then raise a question; the current of the π^0 domain-wall may be okay, but what about the CME current? There is no coherent field but only $\theta(x)$ that is not a dynamical field but just a space-time dependent parameter! This is perfectly a sensible question. To answer this, I would say that $\theta(x)$ could be promoted to the dynamical field without mentioning on a possibility of axion, at least in the hadronic phase. If the system has a pseudo-scalar (and iso-scalar) condensate such as the condensate of the η^0 meson (that forms η' with a mixture with η^3), it could be mapped to $\theta(x)$ in the chiral Lagrangian approach. Once this mapping is noticed, there is no longer a big conceptual difference between the current in (9.28) and the CME current in (9.1). The analogy to the Josephson current may support the reality of the CME current, but this argument does not tell us anything about the microscopic constituent of the current yet.

Equation (9.28) means that the current can exist just with the $\pi^0(x)$ profile and the magnetic field, and then the only possible carrier of the current should be the quark content inside of π^0. Therefore, even in the hadronic phase, I must think that charged quarks flow to produce the electric current. Contrary to the intuition, there is no inconsistency with the notion of quark confinement. Regardless of the presence of the flow of quarks, these flowing quarks can be still confined in a big wave-function of the $\pi^0(x)$ profile. In this way, confined quarks can flow without breaking confinement because of the coherent background of the meson fields.

In reality it is next to impossible to achieve such an environment with abundant π^0 that forms a condensate to test (9.28) because π^0 quickly decays into photons via the anomalous QED process. This implies that the CME current may be also diminished by the photon production. Indeed, (9.29) exactly describes such a process of $\theta(x)$ decaying into 2γ. It is interesting, besides, that one γ can be provided from B in the case with background fields. More specifically, one injected γ and another γ from B can produce a θ (or the η^0 meson), that is nothing but the Primakoff effect [36]. The Primakoff effect has an application as a tool to detect the axion [37, 43], which is understandable from the above argument once $\theta(x)$ is augmented as a dynamical axion field. Because physics of the Chiral Magnetic Effect has a connection to axion physics through $\theta(x)$ (that was actually the very beginning of the path toward Ref. [28], as I mentioned), it should be naturally motivated to think of some application of the Primakoff effect in the context of the Chiral Magnetic Effect too. Then, the reverse process of the Primakoff effect, namely, $\gamma(B) + \theta \to \gamma$, should be the most relevant to the experimental opportunity. In the relativistic heavy-ion collision, the profile of the magnetic field $\mathbf{B}(x)$ can be estimated by the simulation, and the precise measurement of γ with subtraction of the background from the π^0 decay is available nowadays. The unknown piece in the reverse Primakoff effect is the profile of the $\theta(x)$ distribution. Needless to say, nothing is more important and ambiguous than the concrete distribution of $\theta(x)$ for any attempt to perform serious computation of the CME-related phenomena. This is why most of works on the Chiral Magnetic Effect address only qualitative predictions.

I think that it must be a very interesting challenge to find a condensed-matter counterpart in which the Chiral Magnetic Effect may be visible and testable experimentally. This is not an unrealistic desire; axion physics can be discussed in the so-called topological magnetic insulator [35], and why not the Chiral Magnetic Effect? In fact, recently, there are appearing some works one after another along this line.

9.5 My Outlook

Many people (including me) are still working on the theoretical aspects of the Chiral Magnetic Effect and its relatives such as the Chiral Separation Effect, the Chiral Magnetic Wave, etc. It is highly demanded to make some firm theoretical estimation about the experimental observables affected by the CME and related phenomena. To this end, however, one needs know the early-time dynamics even before the formation of the quark-gluon plasma. In fact, the most interesting regime that has the greatest impact on the CME happens to be the most difficult regime to describe theoretically.

A missing theoretical link between the coherent wave-function right after the collision and the thermalized plasma is the last piece of the jigsaw puzzle. Particles should be produced from quantum fluctuations on top of coherent fields (i.e. the Color Glass Condensate; CGC), that translates into the entropy production. It is known that the coherent background fields accommodate topological flux-tubes that play a role of Q_w in (9.2). Produced particles inside of those flux-tubes under a strong $\mathbf{B}$ should have a characteristic momentum distribution and this would embody the Chiral Magnetic Effect in a quantitative way. In my opinion, thus, the early-thermalization problem must be resolved even before talking about the observational possibility of the Chiral Magnetic Effect, or they may have something to do with each other. This is because, as I stated in the previous section, the real-time dynamics of the Chiral Magnetic Effect should involve the Schwinger process of particle production, and the particle production as in the Lund string model should be responsible for the entropy production from fields, and thus thermalization ultimately.

One may claim that the complete isotropization and thermal equilibrium should be no longer required to account for the experimental observation. This is true indeed, and this is a good news for the CME, for the thermalization means that the system should lose any memory and have only one information, i.e. the temperature. If the thermalization is incomplete, it would enhance the chance that the observed distribution of particles may still remember the early-time environment like the presence of the strong **B** and/or the topological flux-tubes. For the purpose of testing the idea as compared to the experimental data, it is indispensable to perform some serious simulation of the early-time dynamics of the heavy-ion collisions.

Unfortunately, there is no successful simulation starting from the CGC initial condition to achieve some reasonable input for the hydrodynamics within a reasonable time scale (see Ref. [17] for a latest attempt). There are so many theoretical efforts in this direction including mine [10] and it should be definitely worth discussing them, but not here and on another occasion maybe. In this article I have discussed physics of the Chiral Magnetic Effect and presented my views on the physical interpretation. Now my story has become a bit too diverging, and I should stop here, with a hope that some readers may find my views useful for future investigations.

Acknowledgements The author would like to express his sincere thanks to his collaborators, Dima Kharzeev, Kazuya Mameda, and Harmen Warringa. The contents of this article are based on the fruitful collaboration with them. Especially, this article is dedicated to Harmen Warringa, who inspired me enough to initiate my working on the Chiral Magnetic Effect. Harmen's contribution to physics of the Chiral Magnetic Effect should be memorable as long as the CME-related physics is of our interest. The author is also grateful to Tigran Kalaydzhyan, Dima Kharzeev, Toni Rebhan, Mikhail Shaposhnikov, Igor Shovkovy for useful comments on this article.

References

1. A.Y. Alekseev, V.V. Cheianov, J. Frohlich, Universality of transport properties in equilibrium, Goldstone theorem and chiral anomaly. Phys. Rev. Lett. **81**, 3503–3506 (1998). doi:10.1103/PhysRevLett.81.3503
2. P.W. Anderson, J.M. Rowell, Probable observation of the Josephson superconducting tunneling effect. Phys. Rev. Lett. **10**, 230 (1963). doi:10.1103/PhysRevLett.10.230
3. A. Ballon-Bayona, K. Peeters, M. Zamaklar, A chiral magnetic spiral in the holographic Sakai-Sugimoto model. J. High Energy Phys. **11**, 164 (2012). doi:10.1007/JHEP11(2012)164
4. J. Bardeen, Tunneling into superconductors. Phys. Rev. Lett. **9**, 147 (1962). doi:10.1103/PhysRevLett.9.147
5. G. Basar, G.V. Dunne, D.E. Kharzeev, Chiral magnetic spiral. Phys. Rev. Lett. **104**, 232301 (2010). doi:10.1103/PhysRevLett.104.232301
6. P. Buividovich, M. Chernodub, E. Luschevskaya, M. Polikarpov, Numerical evidence of chiral magnetic effect in lattice gauge theory. Phys. Rev. D **80**, 054503 (2009). doi:10.1103/PhysRevD.80.054503
7. D. Deryagin, D.Y. Grigoriev, V. Rubakov, Standing wave ground state in high density, zero temperature QCD at large N_c. Int. J. Mod. Phys. A **7**, 659–681 (1992). doi:10.1142/S0217751X92000302
8. E.J. Ferrer, V. de la Incera, C. Manuel, Magnetic color flavor locking phase in high density QCD. Phys. Rev. Lett. **95**, 152002 (2005). doi:10.1103/PhysRevLett.95.152002

9. K. Fukushima, QCD matter in extreme environments. J. Phys. G **39**, 013101 (2012). doi: 10.1088/0954-3899/39/1/013101

10. K. Fukushima, F. Gelis, The evolving Glasma. Nucl. Phys. A **874**, 108–129 (2012). doi: 10.1016/j.nuclphysa.2011.11.003

11. K. Fukushima, D.E. Kharzeev, H.J. Warringa, The chiral magnetic effect. Phys. Rev. D **78**, 074033 (2008). doi:10.1103/PhysRevD.78.074033

12. K. Fukushima, D.E. Kharzeev, H.J. Warringa, Electric-current susceptibility and the chiral magnetic effect. Nucl. Phys. A **836**, 311–336 (2010). doi:10.1016/j.nuclphysa.2010.02.003

13. K. Fukushima, D.E. Kharzeev, H.J. Warringa, Real-time dynamics of the chiral magnetic effect. Phys. Rev. Lett. **104**, 212001 (2010). doi:10.1103/PhysRevLett.104.212001

14. K. Fukushima, K. Mameda, Wess-Zumino-Witten action and photons from the chiral magnetic effect. Phys. Rev. D **86**, 071501 (2012). doi:10.1103/PhysRevD.86.071501

15. K. Fukushima, M. Ruggieri, Dielectric correction to the chiral magnetic effect. Phys. Rev. D **82**, 054001 (2010). doi:10.1103/PhysRevD.82.054001

16. K. Fukushima, H.J. Warringa, Color superconducting matter in a magnetic field. Phys. Rev. Lett. **100**, 032007 (2008). doi:10.1103/PhysRevLett.100.032007

17. C. Gale, S. Jeon, B. Schenke, P. Tribedy, R. Venugopalan, Event-by-event anisotropic flow in heavy-ion collisions from combined Yang-Mills and viscous fluid dynamics. Phys. Rev. Lett. **100**, 012302 (2013). doi:10.1103/PhysRevLett.110.012302

18. J.H. Gao, Z.T. Liang, S. Pu, Q. Wang, X.N. Wang, Chiral anomaly and local polarization effect from quantum kinetic approach. Phys. Rev. Lett. **109**, 232301 (2012). doi:10.1103/PhysRevLett.109.232301

19. M. Giovannini, M. Shaposhnikov, Primordial hypermagnetic fields and triangle anomaly. Phys. Rev. D **57**, 2186–2206 (1998). doi:10.1103/PhysRevD.57.2186

20. M. Giovannini, M. Shaposhnikov, Primordial magnetic fields, anomalous isocurvature fluctuations and big bang nucleosynthesis. Phys. Rev. Lett. **80**, 22–25 (1998). doi:10.1103/PhysRevLett.80.22

21. E. Gorbar, V. Miransky, I. Shovkovy, Chiral asymmetry of the Fermi surface in dense relativistic matter in a magnetic field. Phys. Rev. C **80**, 032801 (2009). doi:10.1103/PhysRevC.80.032801

22. V. Gusynin, V. Miransky, I. Shovkovy, Dimensional reduction and dynamical chiral symmetry breaking by a magnetic field in $(3 + 1)$-dimensions. Phys. Lett. B **349**, 477–483 (1995). doi:10.1016/0370-2693(95)00232-A

23. R. Kaiser, Anomalies and WZW term of two flavor QCD. Phys. Rev. D **63**, 076010 (2001). doi:10.1103/PhysRevD.63.076010

24. T. Kalaydzhyan, Chiral superfluidity of the quark-gluon plasma (2012)

25. T. Kalaydzhyan, I. Kirsch, Fluid/gravity model for the chiral magnetic effect. Phys. Rev. Lett. **106**, 211601 (2011). doi:10.1103/PhysRevLett.106.211601

26. D. Kharzeev, Parity violation in hot QCD: why it can happen, and how to look for it. Phys. Lett. B **633**, 260–264 (2006). doi:10.1016/j.physletb.2005.11.075

27. D.E. Kharzeev, Topologically induced local P and CP violation in QCD × QED. Ann. Phys. **325**, 205–218 (2010). doi:10.1016/j.aop.2009.11.002

28. D.E. Kharzeev, L.D. McLerran, H.J. Warringa, The effects of topological charge change in heavy ion collisions: 'Event by event P and CP violation'. Nucl. Phys. A **803**, 227–253 (2008). doi:10.1016/j.nuclphysa.2008.02.298

29. D.E. Kharzeev, H.J. Warringa, Chiral magnetic conductivity. Phys. Rev. D **80**, 034028 (2009). doi:10.1103/PhysRevD.80.034028

30. D.E. Kharzeev, H.U. Yee, Chiral magnetic wave. Phys. Rev. D **83**, 085007 (2011). doi: 10.1103/PhysRevD.83.085007

31. M. Luscher, SO(4) symmetric solutions of Minkowskian Yang-Mills field equations. Phys. Lett. B **70**, 321 (1977). doi:10.1016/0370-2693(77)90668-2

32. M.A. Metlitski, A.R. Zhitnitsky, Anomalous axion interactions and topological currents in dense matter. Phys. Rev. D **72**, 045011 (2005). doi:10.1103/PhysRevD.72.045011

33. E. Nakano, T. Tatsumi, Chiral symmetry and density wave in quark matter. Phys. Rev. D **71**, 114006 (2005). doi:10.1103/PhysRevD.71.114006

34. J.L. Noronha, I.A. Shovkovy, Color-flavor locked superconductor in a magnetic field. Phys. Rev. D **76**, 105030 (2007). doi:10.1103/PhysRevD.76.105030. Erratum: doi:10.1103/PhysRevD.86.049901

35. H. Ooguri, M. Oshikawa, Instability in magnetic materials with dynamical axion field. Phys. Rev. Lett. **108**, 161803 (2012). doi:10.1103/PhysRevLett.108.161803

36. H. Primakoff, Photoproduction of neutral mesons in nuclear electric fields and the mean life of the neutral meson. Phys. Rev. **81**, 899 (1951). doi:10.1103/PhysRev.81.899

37. G. Raffelt, L. Stodolsky, Mixing of the photon with low mass particles. Phys. Rev. D **37**, 1237 (1988). doi:10.1103/PhysRevD.37.1237

38. A. Rebhan, A. Schmitt, S.A. Stricker, Anomalies and the chiral magnetic effect in the Sakai-Sugimoto model. J. High Energy Phys. **1001**, 026 (2010). doi:10.1007/JHEP01(2010)026

39. V. Rubakov, On chiral magnetic effect and holography (2010)

40. B.M. Schechter, Yang-Mills theory on the hypertorus. Phys. Rev. D **16**, 3015 (1977). doi:10.1103/PhysRevD.16.3015

41. V. Schon, M. Thies, 2-D model field theories at finite temperature and density (2000)

42. J.S. Schwinger, On gauge invariance and vacuum polarization. Phys. Rev. **82**, 664–679 (1951). doi:10.1103/PhysRev.82.664

43. P. Sikivie, Experimental tests of the invisible axion. Phys. Rev. Lett. **51**, 1415 (1983). doi:10.1103/PhysRevLett.51.1415

44. D. Son, M. Stephanov, Axial anomaly and magnetism of nuclear and quark matter. Phys. Rev. D **77**, 014021 (2008). doi:10.1103/PhysRevD.77.014021

45. D.T. Son, N. Yamamoto, Berry curvature, triangle anomalies, and chiral magnetic effect in Fermi liquids. Phys. Rev. Lett. **109**, 181602 (2012). doi:10.1103/PhysRevLett.109.181602

46. M. Stephanov, Y. Yin, Chiral kinetic theory. Phys. Rev. Lett. **109**, 162001 (2012). doi:10.1103/PhysRevLett.109.162001

47. A. Vilenkin, Equilibrium parity violating current in a magnetic field. Phys. Rev. D **22**, 3080–3084 (1980). doi:10.1103/PhysRevD.22.3080

48. A. Yamamoto, Chiral magnetic effect in lattice QCD with a chiral chemical potential. Phys. Rev. Lett. **107**, 031601 (2011). doi:10.1103/PhysRevLett.107.031601

49. H.U. Yee, Holographic chiral magnetic conductivity. J. High Energy Phys. **0911**, 085 (2009). doi:10.1088/1126-6708/2009/11/085

50. I. Zahed, Anomalous chiral Fermi surface. Phys. Rev. Lett. **109**, 091603 (2012). doi:10.1103/PhysRevLett.109.091603

Chapter 10
The Chiral Magnetic Effect and Axial Anomalies

Gökçe Başar and Gerald V. Dunne

10.1 Dirac Operators, Dimensional Reduction and Axial Anomalies

In this first section we present an elementary relation between the chiral magnetic effect [1–6] and the axial anomalies in four-dimensional and two-dimensional space-time [7, 8]. This follows from the basic structure of the Dirac operator, together with the lowest-Landau level (LLL) projection produced by a strong magnetic field, a basic feature of the phenomenon of magnetic catalysis [9]. We first review the sub-block structure of the Dirac operator and the associated Schur decomposition [10] of the propagator.

10.1.1 Lowest Landau Level Projection

We adopt the following conventions for Dirac matrices in four dimensional Minkowski space:

$$\gamma^0 = \begin{pmatrix} 0 & \mathbb{1} \\ \mathbb{1} & 0 \end{pmatrix}, \qquad \gamma^j = \begin{pmatrix} 0 & -\sigma^j \\ \sigma^j & 0 \end{pmatrix}, \qquad \gamma_5 = \begin{pmatrix} \mathbb{1} & 0 \\ 0 & -\mathbb{1} \end{pmatrix} \qquad (10.1)$$

where σ^j, $j = 1, 2, 3$, are the 2×2 Pauli matrices, and $\mathbb{1}$ is the 2×2 identity matrix. It is convenient to write these 4×4 Dirac matrices in 2×2 block form:

G. Başar (✉)
Department of Physics, Stony Brook University, Stony Brook, USA
e-mail: basar@tonic.physics.sunysb.edu

G.V. Dunne
Physics Department, University of Connecticut, Storrs, USA
e-mail: gerald.dunne@uconn.edu

D. Kharzeev et al. (eds.), *Strongly Interacting Matter in Magnetic Fields*,
Lecture Notes in Physics 871, DOI 10.1007/978-3-642-37305-3_10,
© Springer-Verlag Berlin Heidelberg 2013

$$\gamma^\mu = \begin{pmatrix} 0 & \alpha^\mu \\ \tilde{\alpha}^\mu & 0 \end{pmatrix} \tag{10.2}$$

where the 2×2 matrices α^μ and $\tilde{\alpha}^\mu$ are:

$$\alpha^\mu = \left(\mathbb{1}, -\sigma^j\right), \qquad \tilde{\alpha}^\mu = \left(\mathbb{1}, \sigma^j\right) \tag{10.3}$$

The Dirac matrices satisfy the anti-commutation relations, $\{\gamma^\mu, \gamma^\nu\} = 2\eta^{\mu\nu}$, with Minkowski metric $\eta^{\mu\nu} = \mathrm{diag}(1, -1, -1, -1)$.

The Dirac operator is defined to be $\slashed{D} = \gamma^\mu(\partial_\mu - ieA_\mu)$, where later we will take the gauge field A_μ to have both abelian and non-abelian parts, but for now we take it to be abelian. Thus, we have a natural decomposition of the 4×4 Dirac operator into 2×2 sub-blocks:

$$\mathscr{D} = -i\slashed{D} + m = \begin{pmatrix} m & -i\alpha^\mu D_\mu \\ -i\tilde{\alpha}^\mu D_\mu & m \end{pmatrix}$$

$$\equiv \begin{pmatrix} m & D \\ -\tilde{D} & m \end{pmatrix} \tag{10.4}$$

Since $\det(-i\slashed{D} + m) = \det(i\slashed{D} + m)$, we consider

$$\det(i\slashed{D} + m)\det(-i\slashed{D} + m) = \begin{pmatrix} m^2 + D\tilde{D} & 0 \\ 0 & m^2 + \tilde{D}D \end{pmatrix} \tag{10.5}$$

The operators $D\tilde{D}$ and $\tilde{D}D$ differ from one another, and from the Klein-Gordon operator $D_\mu D^\mu$, in their spin-projection terms:

$$D\tilde{D} = D_\mu D^\mu + \frac{e}{2}\tilde{\sigma}^{\mu\nu}F_{\mu\nu} \tag{10.6}$$

$$\tilde{D}D = D_\mu D^\mu + \frac{e}{2}\sigma^{\mu\nu}F_{\mu\nu} \tag{10.7}$$

where the 2×2 spin matrices,

$$\tilde{\sigma}^{\mu\nu} = \frac{1}{2i}\left(\alpha^\mu\tilde{\alpha}^\nu - \alpha^\nu\tilde{\alpha}^\mu\right), \qquad \sigma^{\mu\nu} = \frac{1}{2i}\left(\tilde{\alpha}^\mu\alpha^\nu - \tilde{\alpha}^\nu\alpha^\mu\right) \tag{10.8}$$

are the sub-block components of the usual 4×4 spin matrices

$$\Sigma^{\mu\nu} = \frac{1}{2i}[\gamma^\mu, \gamma^\nu] = \begin{pmatrix} \tilde{\sigma}^{\mu\nu} & 0 \\ 0 & \sigma^{\mu\nu} \end{pmatrix} \tag{10.9}$$

Now, suppose we have a background electromagnetic field consisting of a magnetic field $\mathbf{B} = (0, 0, B)$, and an electric field $\mathbf{E} = (0, 0, E)$, both of which are pointing in the x^3 direction. We do not need to assume that these fields are uniform, but we choose $B = B(x^1, x^2)$ and $E = E(x^0, x^3)$, choosing gauge field $A^\mu = (A^0(x^3), A^1(x^2), A^2(x^1), A^3(x^0))$, which satisfies $\partial_\mu A^\mu = 0$. Then, with only a third component of both the electric and magnetic field, the 2×2 operators in (10.6) and (10.7) reduce to

$$D\tilde{D} = D_\mu D^\mu - e(B + iE)\sigma^3 \tag{10.10}$$

$$\tilde{D}D = D_\mu D^\mu - e(B - iE)\sigma^3 \tag{10.11}$$

because $\sigma^{12} = \tilde{\sigma}^{12} = -\sigma^3$, while $\sigma^{03} = -\tilde{\sigma}^{03} = i\sigma^3$. The important point in (10.10) and (10.11) is that both 2×2 operators $D\tilde{D}$ and $\tilde{D}D$ are **diagonal** in terms of their Dirac matrix structure.

To understand the effect of a strong magnetic field, and in particular its projection to the lowest Landau level (LLL), consider the factorization of the Klein-Gordon operator, for this case parallel $\mathbf{E}$ and $\mathbf{B}$ in the x^3 direction:

$$D_\mu D^\mu = (D_0 \mp D_3)(D_0 \pm D_3) \pm ieE - (D_1 \mp iD_2)(D_1 \pm iD_2) \pm eB \tag{10.12}$$

If the magnetic field is constant, and for example in the symmetric gauge $A_1 = -\frac{B}{2}x^2$, and $A_2 = \frac{B}{2}x^1$, then we can adopt complex coordinates, $z = x^1 + ix^2$, $\bar{z} = x^1 - ix^2$, to write

$$D_1 \pm iD_2 = \begin{cases} 2(\partial_{\bar{z}} + \frac{eB}{4}z) \\ 2(\partial_z - \frac{eB}{4}\bar{z}) \end{cases} \tag{10.13}$$

If $eB > 0$ we choose the upper sign to obtain normalizable solutions, with the Gaussian factor $\exp(-\frac{eB}{4}|z|^2)$, in which case we see that the eB term in (10.12) cancels the spin term in (10.10) and (10.11), when $\sigma^3 = +1$. Thus the magnetic component of the Dirac operators leads to a zero mode of the 2×2 operators $D\tilde{D}$ and $\tilde{D}D$, when the spin is aligned along the direction of the magnetic field. In fact, this also applies to the situation of an inhomogeneous $B(x^1, x^2)$ field, and the degeneracy of this lowest-Landau-level (LLL) is given by the integer part of the net magnetic flux; this is the Aharonov-Casher theorem [11–13], the projection onto the LLL.

10.1.2 Schur Decomposition of Dirac Propagator

The Schur decomposition gives an elementary algebraic decomposition of the inverse of a matrix in terms of its sub-block structure [10]. This leads immediately to an associated sub-block decomposition of the Dirac propagator, and as we show, it also provides a simple description of the lowest Landau level projection in a strong magnetic field, which is the key to magnetic catalysis [9].

Consider a matrix M written as

$$M = \begin{pmatrix} a & b \\ c & d \end{pmatrix} \tag{10.14}$$

where a and d are square, but not necessarily of the same size [and so b and c need not be square matrices], then we can write the inverse of M in block form in two different ways:

$$M^{-1} = \begin{pmatrix} s^{-1} & -s^{-1}bd^{-1} \\ -d^{-1}cs^{-1} & d^{-1}cs^{-1}bd^{-1}+d^{-1} \end{pmatrix}, \quad s \equiv a - bd^{-1}c \quad (10.15)$$

$$= \begin{pmatrix} a^{-1}bt^{-1}ca^{-1}+a^{-1} & -a^{-1}bt^{-1} \\ -t^{-1}ca^{-1} & t^{-1} \end{pmatrix}, \quad t \equiv d - ca^{-1}b \quad (10.16)$$

where s and t are the two different Schur complements of M. Note that the first expression only requires d and s to be invertible, while the second expression only requires a and t to be invertible. Applying this to the Dirac operator in (10.4), we find

$$s = \frac{1}{m}(m^2 + D\tilde{D}), \qquad t = \frac{1}{m}(m^2 + \tilde{D}D) \qquad (10.17)$$

which gives then two different decompositions of the Dirac propagator:

$$\mathscr{D}^{-1} = \begin{pmatrix} \frac{m}{m^2+D\tilde{D}} & \frac{1}{m^2+D\tilde{D}}D \\ \tilde{D}\frac{1}{m^2+D\tilde{D}} & \frac{1}{m}(1 - \tilde{D}\frac{1}{m^2+D\tilde{D}}D) \end{pmatrix} \qquad (10.18)$$

$$= \begin{pmatrix} \frac{1}{m}(1 - \tilde{D}\frac{1}{m^2+D\tilde{D}}D) & D\frac{1}{m^2+\tilde{D}D} \\ \frac{1}{m^2+\tilde{D}D}\tilde{D} & \frac{m}{m^2+D\tilde{D}} \end{pmatrix} \qquad (10.19)$$

The Euclidean analogue of this decomposition corresponds precisely to the chiral decompositions used in [14–16]:

$$\mathscr{D}^{-1} = \mathscr{D}\mathscr{G}^{(\pm)}\left(\frac{1\pm\gamma_5}{2}\right) + \mathscr{G}^{(\pm)}\mathcal{D}\left(\frac{1\mp\gamma_5}{2}\right) + \frac{1}{m}(1 - \mathcal{D}\mathscr{G}^{(\pm)}\mathcal{D})\left(\frac{1\mp\gamma_5}{2}\right) \qquad (10.20)$$

where $\mathscr{G}^{(+)} = 1/(m^2 + D\tilde{D})$, and $\mathscr{G}^{(-)} = 1/(m^2 + \tilde{D}D)$. These chiral decompositions of the fermion propagator are particularly useful since they show clearly the projection onto chiral zero modes.

In order to use such Schur decompositions to characterize also the lowest Landau level projection, in addition to the chiral decomposition, it is convenient to write the propagators in (10.18) and (10.19) in factored form:

$$\mathscr{D}^{-1} = \frac{1}{m}\begin{pmatrix} 1 & 0 \\ \frac{1}{m}\tilde{D} & 1 \end{pmatrix}\begin{pmatrix} \frac{m^2}{m^2+D\tilde{D}} & 0 \\ 0 & 1 \end{pmatrix}\begin{pmatrix} 1 & -\frac{1}{m}D \\ 0 & 1 \end{pmatrix} \qquad (10.21)$$

$$= \frac{1}{m}\begin{pmatrix} 1 & \frac{1}{m}D \\ 0 & 1 \end{pmatrix}\begin{pmatrix} 1 & 0 \\ 0 & \frac{m^2}{m^2+\tilde{D}D} \end{pmatrix}\begin{pmatrix} 1 & 0 \\ -\frac{1}{m}\tilde{D} & 1 \end{pmatrix} \qquad (10.22)$$

We stress that at this point we have only used elementary algebra to express a general Dirac propagator in terms of its 2×2 sub-block structure.

From these Schur decompositions, we can reduce the expectation values of the charge and axial currents to much simpler forms. By straightforward manipulations we find for the charge current

$$\langle j^\mu \rangle = ie \, \mathrm{tr}_{4 \times 4} \left(\gamma^\mu \mathscr{D}^{-1} \right) \tag{10.23}$$

$$= ie \, \mathrm{tr}_{2 \times 2} \left(\left(D \tilde{\alpha}^\mu - \alpha^\mu \tilde{D} \right) \frac{1}{m^2 + D \tilde{D}} \right) \tag{10.24}$$

$$= 2e\eta^{\mu\nu} \, \mathrm{tr}_{2 \times 2} \left(D_\nu \frac{1}{m^2 + D \tilde{D}} \right) \tag{10.25}$$

and for the axial current

$$\langle j_5^\mu \rangle = ie \, \mathrm{tr} \left(\gamma^\mu \gamma_5 \mathscr{D}^{-1} \right) \tag{10.26}$$

$$= ie \, \mathrm{tr}_{2 \times 2} \left(\left(D \tilde{\alpha}^\mu + \alpha^\mu \tilde{D} \right) \frac{1}{m^2 + D \tilde{D}} \right) \tag{10.27}$$

$$= 2ie \, \mathrm{tr}_{2 \times 2} \left(\tilde{\sigma}^{\mu\nu} D_\nu \frac{1}{m^2 + D \tilde{D}} \right) \tag{10.28}$$

Here we have used the facts that $(\alpha^\nu \tilde{\alpha}^\mu + \alpha^\mu \tilde{\alpha}^\nu) = 2\eta^{\mu\nu} \mathbb{1}_{2 \times 2}$, while $(\alpha^\nu \tilde{\alpha}^\mu - \alpha^\mu \tilde{\alpha}^\nu) = 2i \tilde{\sigma}^{\mu\nu}$. In particular, note that since $\tilde{\sigma}^{03} = -i\sigma^3 = -\tilde{\sigma}^{30}$,

$$\langle j^0 \rangle = 2e\eta^{\mu\nu} \, \mathrm{tr}_{2 \times 2} \left(D_0 \frac{1}{m^2 + D \tilde{D}} \right) \tag{10.29}$$

$$\langle j^3 \rangle = -2e\eta^{\mu\nu} \, \mathrm{tr}_{2 \times 2} \left(D_3 \frac{1}{m^2 + D \tilde{D}} \right) \tag{10.30}$$

$$\langle j_5^0 \rangle = 2e\eta^{\mu\nu} \, \mathrm{tr}_{2 \times 2} \left(\sigma^3 D_3 \frac{1}{m^2 + D \tilde{D}} \right) \tag{10.31}$$

$$\langle j_5^3 \rangle = -2e\eta^{\mu\nu} \, \mathrm{tr}_{2 \times 2} \left(\sigma^3 D_0 \frac{1}{m^2 + D \tilde{D}} \right) \tag{10.32}$$

Again, thus far we have only used elementary algebra to reduce the expectation values of the 4×4 matrices and propagators to expressions involving just 2×2 matrices and propagators.

10.1.3 Currents and Anomalies in the Lowest Landau Level Projection

Now suppose the background field consists of a very strong magnetic field in the x^3 direction. Then we project onto the LLL, which means that the states contributing to the 2×2 expectation values have $\sigma^3 = +1$. Therefore, we see immediately that in this LLL projection limit, where we project onto motion along the magnetic field direction, we can write for the remaining currents:

$$\langle j_5^M \rangle = \varepsilon^{MN} \langle j_N \rangle, \quad M, N = 0, 3 \tag{10.33}$$

where the epsilon symbol is $\varepsilon^{03} = +1 = -\varepsilon^{30}$. This is exactly the relation between the charge and axial current in two dimensional space-time.

Furthermore, suppose the four dimensional background field consists of a strong magnetic field and also an electric field, both directed along the x^3 axis. Then, following the analysis of the first section, we choose a gauge field of the form $A^\mu = (A^0(x^3), A^1(x^2), A^2(x^1), A^3(x^0))$, satisfying $\partial_\mu A^\mu = 0$. Then a simple two dimensional computation yields current expectation values for Dirac indices 0 and 3 (we adopt the convention of using capital Roman indices M, N to denote the components of the dimensionally reduced (x^0, x^3) plane):

$$\langle j^M \rangle = \frac{eB}{2\pi} \frac{eA^M}{\pi}, \qquad \langle j_5^M \rangle = \frac{eB}{2\pi} \varepsilon^{MN} \frac{eA_N}{\pi} \tag{10.34}$$

where $\frac{eB}{2\pi}$ is the Landau degeneracy factor, in its local Aharonov-Casher form. Note that these expressions are consistent with charge current conservation and the two dimensional axial anomaly:

$$\partial_M \langle j^M \rangle = 0, \qquad \partial_M \langle j_5^M \rangle = \frac{eB}{2\pi} \frac{eE}{\pi} \tag{10.35}$$

In this LLL projection limit, we can alternatively express this result in four dimensional language as

$$\partial_\mu \langle j^\mu \rangle_{LLL} = 0, \qquad \partial_\mu \langle j_5^\mu \rangle_{LLL} = \frac{e^2}{2\pi^2} \mathbf{B} \cdot \mathbf{E} = \frac{e^2}{8\pi^2} F^{\mu\nu} \tilde{F}_{\mu\nu} \tag{10.36}$$

which expresses charge current conservation and the four dimensional axial anomaly. This makes it clear that the relevant anomaly is the "covariant" anomaly, rather than the "consistent" anomaly. For abelian theories these differ by a factor of $1/d$ in $(2d - 2)$ space-time dimensions, while for non-abelian theories the covariant and consistent anomalies have different field structure [17, 18].

To make the connection with the chiral magnetic effect, we note that it is natural to identify $A^0(x^3)$ with a spatially dependent chemical potential, and $A^3(x^0)$ with a time dependent chiral chemical potential:

$$A^0 \leftrightarrow \mu, \qquad A^3 \leftrightarrow \mu_5 \tag{10.37}$$

For A^0 and μ this is obvious, because the coupling is given by $\mu \bar{\psi} \gamma^0 \psi$. For A^3 and μ_5, this follows because the coupling is $\mu_5 \bar{\psi} \gamma_5 \gamma^0 \psi$, and in the LLL projection

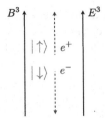

Fig. 10.1 Parallel electric and magnetic fields produce electron-positron pairs from vacuum, and because of the LLL projection caused by the strong magnetic field, correlated with spin and charge, this results in a net flow of chirality along the direction of the fields, in accordance with the Schwinger pair production rate and the chiral magnetic effect

$\gamma_5\gamma^0 \leftrightarrow \gamma^3$, since γ^3 has off-diagonal sub-blocks $\mp\sigma^3$, and $\sigma^3 \to +1$ in the LLL limit.

Therefore, we can understand the two dimensional currents in (10.34) as

$$\langle j^0 \rangle = \frac{\mu}{\pi} \frac{eB}{2\pi}, \qquad \langle j^3 \rangle = \frac{\mu_5}{\pi} \frac{eB}{2\pi} \qquad (10.38)$$

$$\langle j_5^0 \rangle = \frac{\mu_5}{\pi} \frac{eB}{2\pi}, \qquad \langle j_5^3 \rangle = \frac{\mu}{\pi} \frac{eB}{2\pi} \qquad (10.39)$$

These relations express the chiral magnetic effect, which we see is a direct consequence of the axial anomalies in two and four dimensional space-time, after the LLL projection caused by a strong magnetic field. The coefficients are fixed completely by the anomaly equations. For a complementary discussion of the relation between chiral asymmetry and the axial anomaly, see [19–21].

10.1.4 Chiral Magnetic Effect and the Schwinger Effect

The chiral magnetic effect can also be understood naturally in terms of the Schwinger effect [22, 23], particle production from vacuum, which occurs when there is a non-zero electric field background, as illustrated in Fig. 10.1. With approximately constant parallel electric and magnetic fields directed along the x^3 axis, the Schwinger pair production rate, per unit volume, is given by

$$\Gamma = e^2 \frac{EB}{4\pi^2} \coth\left(\frac{B}{E}\pi\right) e^{-m^2\pi/|eE|} \qquad (10.40)$$

When $B \to 0$ we recover the usual result for a pure electric field, and for nonzero magnetic field we find an enhancement of the rate which is linear in B in the strong magnetic field limit. Positrons are accelerated along the direction of the electric field, and electrons in the opposite direction. In the massless fermion limit this corresponds to a net productionand flow of chirality, because the lowest-Landau-level projection projects spin according to the charge and the direction of the magnetic field. Thus we find the rate of change of chirality

$$\frac{dj_5^0}{dt} = 2\Gamma = \frac{EB}{2\pi^2} \tag{10.41}$$

in agreement with the axial anomaly (10.36) and the chiral magnetic effect (10.39). The electric field produces the acceleration while the strong magnetic field provides the LLL projection that correlates spin with the direction of flow of charge, and hence also of chirality. Physically, a spatially dependent $A^0(x^3)$ produces charge separation, as for a local chemical potential, while a time dependent $A^3(x^0)$ drives a current along the direction of the electric field [24], which in the LLL projection corresponds to a flow of chirality.

10.1.5 Maxwell-Chern-Simons Theory and the Schwinger Model

The interpretation of the chiral magnetic effect in terms of the effect of an electric field directed along the same direction as the strong magnetic field is also very natural in terms of an effective Maxwell-Chern-Simons theory resulting from an adiabatic space- or time-dependent theta parameter [5]. Express the theta term in the Lagrangian as (up to a total derivative)

$$-\frac{e^2}{8\pi^2}\theta F^{\mu\nu}\tilde{F}^{\mu\nu} = P_\mu J_{CS}^\mu \tag{10.42}$$

where

$$P_\mu = \partial_\mu\theta, \qquad J_{CS}^\mu = \frac{e^2}{8\pi^2}\varepsilon^{\mu\nu\rho\sigma}A_\nu F_{\rho\sigma} \tag{10.43}$$

The pseudo vector P_μ encodes the anomalous terms from the chiral magnetic effect, modifying the usual inhomogeneous Maxwell equations to read

$$\partial_\mu F^{\mu\nu} = J^\nu - \frac{e^2}{2\pi^2}P_\mu\tilde{F}^{\mu\nu} \tag{10.44}$$

From (10.39), in the strong magnetic field limit we project to the 2d (x^0, x^3) plane with the identifications

$$P_0 = A_3, \qquad P_3 = A_0 \quad\Rightarrow\quad P^M = \varepsilon^{MN}A_N \tag{10.45}$$

Fig. 10.2 The effect of a spatially inhomogeneous theta parameter, with the interpretation of its gradient being a spatially inhomogeneous chemical potential $\mu \sim A^0$. The resulting spatially inhomogeneous electric field produces a build up of negative charge at the left-hand inhomogeneity and positive charge at the right-hand inhomogeneity, producing the electric charge separation of the chiral magnetic effect

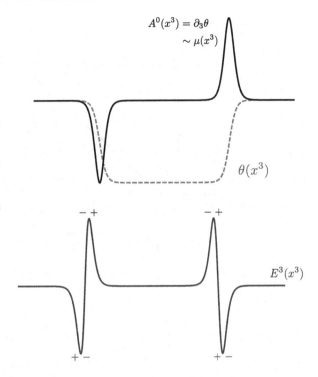

Furthermore, in the strong magnetic field limit, the Chern-Simons current projects to the 2d (x^0, x^3) plane as

$$J_{CS}^M = -\frac{e^2 B}{4\pi^2} \varepsilon^{MN} A_N \tag{10.46}$$

Therefore, in the strong magnetic field limit, the theta term in (10.42) reduces to a mass term of the 2d gauge field, providing an explicit realization of the 2d Schwinger model [7, 25, 26], with the effective Maxwell equations in the reduced 2d (x^0, x^3) plane being written as

$$J^M = \left(\Box + \frac{e^2 B}{2\pi^2}\right) A^M \tag{10.47}$$

In physical terms, we see from (10.45) that a spatial inhomogeneity in the theta parameter corresponds to a non-zero $A^0(x^3)$, which is a spatially inhomogeneous chemical potential. We expect this to produce charge separation. In terms of the Schwinger effect, this is illustrated in Fig. 10.2, depicting a spatial region of non-zero θ. At the edges, the gradient is non-zero, which produces a spatially inhomogeneous electric field, as shown in the figure. This leads to an accumulation of opposite charges at the edges of the region of non-zero theta. On the other hand, a time-dependence in the theta parameter corresponds to a non-zero $A^3(x^0)$, which acts as a chiral chemical potential in the LLL projection limit. For example, consider

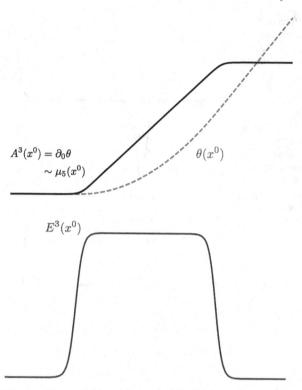

Fig. 10.3 The effect of a time dependent theta parameter, with the interpretation of its gradient being a time-dependent chiral chemical potential $\mu_5 \sim A^3$ in the LLL projection. Via the Schwinger effect, the resulting time dependent electric field produces a flow of electric current linear in $A^3(t)$, and in the LLL projection this corresponds to a net flow of chirality

an electric field turning on smoothly at some early time, and turning off smoothly at some later time, as shown in Fig. 10.3. Then the $A^3(x^0)$ field is approximately constant over the time of nonzero electric field, and this is known to drive a current $\langle J^3 \rangle$ that is linear in $A^3(x^0)$ [24], in agreement with the chiral magnetic effect relation (10.39).

10.2 Chiral Magnetic Spiral

In this section we elaborate more on the dimensional reduction to two-dimensions due to strong magnetic field. In particular, we will show that the expectation values of vector and axial currents, $\langle J^\mu \rangle$ and $\langle J_5^\mu \rangle$, can be expressed in terms of various fermionic bilinears whose dynamics are governed by a two-dimensional Lagrangian. The two dimensional axial anomaly reproduces the well known chiral magnetic effect. Furthermore, in addition to the chiral magnetic effect, the universal dynamics of two-dimensional chiral fermions implies the existence of additional currents which are **transverse** to the magnetic field, and have a spiral modulation along the direction of the field: the "chiral magnetic spiral" [27].

10.2.1 Basic Setup and Dimensional Reduction

We decompose the 4-component spinor in terms of eigenstates of the chiral projectors, $P_{R,L} = \frac{1}{2}(1 \pm \gamma^5)$, the spin projectors $P_{\uparrow,\downarrow} = \frac{1}{2}(1 \pm \Sigma^3)$, and the momentum direction projectors $P_{+,-} = \frac{1}{2}(1 \pm \gamma^0\gamma^3)$. The longitudinal spin operator is $\Sigma^3 = \gamma^0\gamma^3\gamma^5 = \mathrm{diag}(\sigma^3, \sigma^3)$, and $\gamma^0\gamma^3 = \mathrm{diag}(\sigma^3, -\sigma^3)$. We can write the 4-component spinor field as

$$\Psi = \begin{pmatrix} R_+ \\ R_- \\ L_- \\ L_+ \end{pmatrix} \tag{10.48}$$

The four-component spinor can be decomposed into two-component sub spinors in various ways. The chirality and spin decompositions respectively are

$$\psi_R = \begin{pmatrix} R_+ \\ R_- \end{pmatrix}, \qquad \psi_L = \begin{pmatrix} L_+ \\ L_- \end{pmatrix}$$

$$\phi_\uparrow = \begin{pmatrix} R_+ \\ L_- \end{pmatrix}, \qquad \phi_\downarrow = \begin{pmatrix} L_+ \\ R_- \end{pmatrix} \tag{10.49}$$

where $\pm$ denotes the direction of motion along x_3, the direction of the magnetic field. The corresponding four-dimensional currents can as well be decomposed in terms of chirality and spin sub-spinors. The vector current $\bar{\Psi}\gamma^\mu\Psi$ has the decomposition

$$J^0 = \psi_R^\dagger\psi_R + \psi_L^\dagger\psi_L = \phi_\uparrow^\dagger\phi_\uparrow + \phi_\downarrow^\dagger\phi_\downarrow$$

$$J^1 = \bar{\psi}_R\psi_R - \bar{\psi}_L\psi_L = -\bar{\phi}_\uparrow\Gamma^5\phi_\downarrow + \bar{\phi}_\downarrow\Gamma^5\phi_\uparrow$$

$$J^2 = i\bar{\psi}_R\Gamma^5\psi_R + i\bar{\psi}_L\Gamma^5\psi_L = i\bar{\phi}_\uparrow\Gamma^5\phi_\downarrow + i\bar{\phi}_\downarrow\Gamma^5\phi_\uparrow \tag{10.50}$$

$$J^3 = \bar{\psi}_R\Gamma^z\psi_R + \bar{\psi}_L\Gamma^z\psi_L = \bar{\phi}_\uparrow\Gamma^z\phi_\uparrow + \bar{\phi}_\downarrow\Gamma^z\phi_\downarrow$$

The axial current $J_5^\mu = \bar{\Psi}\gamma^\mu\gamma^5\Psi$ has a similar form:

$$J_5^0 = \psi_R^\dagger\psi_R - \psi_L^\dagger\psi_L = -i\bar{\phi}_\uparrow\Gamma^z\phi_\uparrow + i\bar{\phi}_\downarrow\Gamma^z\phi_\downarrow$$

$$J_5^1 = \bar{\psi}_R\psi_R + \bar{\psi}_L\psi_L = \bar{\phi}_\uparrow\phi_\downarrow + \bar{\phi}_\downarrow\phi_\uparrow$$

$$J_5^2 = i\bar{\psi}_R\Gamma^5\psi_R - i\bar{\psi}_L\Gamma^5\psi_L = -i\bar{\phi}_\uparrow\phi_\downarrow + i\bar{\phi}_\downarrow\phi_\uparrow \tag{10.51}$$

$$J_5^3 = \bar{\psi}_R\Gamma^z\psi_R - \bar{\psi}_L\Gamma^z\psi_L = \phi_\uparrow^\dagger\phi_\uparrow - \phi_\downarrow^\dagger\phi_\downarrow$$

Here we have defined the two-dimensional gamma matrices as $\Gamma^0 = \sigma^1$, $\Gamma^z = -i\sigma^2$, $\Gamma^5 = \sigma^3$.

Let us consider a generic chiral Lagrangian

$$\mathscr{L}_{4d} = \bar{\Psi}\left(i\gamma^\mu \partial_\mu\right)\Psi + \sum_{f=R,L} \mathscr{L}_{int,f}[\psi_f] \tag{10.52}$$

where the interaction term does not couple left and right sectors. Since $\gamma^0\gamma^\mu$ is also block-diagonal in the chiral basis, the left and right sectors are completely decoupled and can be treated as independent "flavors" in the Lagrangian (10.52). Each chiral sector has its own associated current $J_{R,L} = 1/2(J^\mu \pm J_5^\mu)$ which are given by

$$\begin{aligned}
J_f^0 &= \psi_f^\dagger \psi_f \\
J_f^1 &= \alpha_f \bar{\psi}_f \psi_f \\
J_f^2 &= i\bar{\psi}_f \Gamma^5 \psi_f \\
J_f^3 &= \bar{\psi}_f \Gamma^z \psi_f
\end{aligned} \tag{10.53}$$

Here $f = L, R$, and $\alpha_R = -\alpha_L = 1$. Even though the left and right currents seem to be completely independent, the conservation of vector current $\partial_\mu J^\mu = 0$ still holds.

Now let us consider the lowest Landau level projection in the presence of a very strong magnetic field B directed along x_3 direction as in Sect. 10.1.1. If the magnitude of the magnetic field is the largest scale in the problem, then the transition from the lowest Landau orbit to an excited orbit will be exponentially suppressed and motion along the transverse $x_1 x_2$ plane will be frozen. The kinetic term in the Lagrangian (10.52) then becomes:

$$\begin{aligned}
\bar{\Psi}\gamma^\mu \partial_\mu \Psi \to \bar{\Psi}\left(\gamma^0 \partial_0 + \gamma^3 \partial_3\right)\Psi &= \sum_{f=R,L} \psi_f^\dagger \partial_0 \psi_f + \psi_f^\dagger \sigma^3 \partial_3 \psi_f \\
&= \sum_{f=R,L} \bar{\psi}_f \Gamma^M \partial_M \psi_f
\end{aligned} \tag{10.54}$$

by using the definition of two-dimensional gamma matrices above. This is the kinetic term of a two-dimensional Lagrangian. Let us further assume that the dimensionally reduced interaction term $\mathscr{L}_{int,f}[\psi_f(z,t)]$ is invariant under the two-dimensional chiral rotation:

$$\psi_f(z,t) \to e^{i\Gamma^5 \zeta_f(z,t)} \psi_f(z,t) \tag{10.55}$$

for an arbitrary function $\zeta_f(z,t)$. This two-dimensional notion of chirality generated by Γ^5 should not be confused with the four-dimensional chirality generated by γ^5. The former acts on each right and left subspinor separately. Also, the four-dimensional chiral Lagrangian (10.52) does not have a term that couples right and left sectors so the system never develops a condensate such as $\langle \bar{\psi}_R \psi_L \rangle$, and the four-dimensional chiral symmetry is never broken in our consideration. However, the dimensionally reduced system may exhibit dynamical breaking of the two-dimensional chiral symmetry (10.55).[1] As an example let us consider the decomposition of the four-dimensional current-current interaction $\mathscr{L}_{int} = J^\mu J_\mu + J_5^\mu J_{5\mu}$. The $(0, 3)$ components of the interaction become

$$J_f^0 J_{f0} + J_f^3 J_{f3} \rightarrow (\bar{\psi}_f \Gamma^M \psi_f)(\bar{\psi}_f \Gamma_M \psi_f) \tag{10.56}$$

In two-dimensions this corresponds to a Thirring interaction. Similarly, the transverse components

$$J_f^1 J_{f1} + J_f^2 J_{f2} \rightarrow (\bar{\psi}_f \psi_f)^2 + (\bar{\psi}_f i \Gamma_5 \psi_f)^2 \tag{10.57}$$

generate a chiral Gross-Neveu or Nambu-Jona-Lasinio (NJL) interaction in two-dimensions [28]. Note that both (10.56) and (10.57) are invariant under (10.55) and exhibit chiral symmetry breaking in two-dimensions. The emergence of the Schwinger model from dimensional reduction of four-dimensional Maxwell-Chern-Simons theory investigated in Sect. 10.1.5 is another example. To sum up, the four-dimensional currents (10.53) are governed by a two-dimensional chiral Lagrangian

$$\mathscr{L}_{2d} = \sum_{f=R,L} \bar{\psi}_f \Gamma^M \partial_M \psi_f + \mathscr{L}_{int,f}[\psi_f(x^M)] \tag{10.58}$$

after dimensional reduction due to the strong magnetic field.

We now show that the Lagrangian (10.58) has certain model-independent properties concerning the expectation values of the fermion bilinears $\langle \bar{\psi}_f \Gamma \psi_f \rangle$. These are the building blocks of the expectation values of the four-dimensional currents (10.50, 10.51). Note from (10.50) that $\langle J_f^0 \rangle$ and $\langle J_f^3 \rangle$ are expressed in terms of two-dimensional densities and currents of ψ_f, while the perpendicular components $\langle J_f^1 \rangle, \langle J_f^2 \rangle$ are expressed in terms of two-dimensional scalar and pseudoscalar condensates of ψ_f.

10.2.2 Life in Two-Dimensions

Besides the magnetic field, the other necessary ingredient for chiral magnetic and/or separation effects is ambient charge density. In particular a nonzero baryon chemical potential μ leads to the chiral separation effect and a chiral (four-dimensional) chemical potential μ_5 leads to the chiral magnetic effect. To keep the discussion general, let us consider nonzero chemical potential for both right and left sectors separately

---

[1] Spontaneous symmetry breaking in two dimensions is a delicate subject whose details are beyond the scope of our topic. It suffices to mention that in the limit of large number of flavors it can be realized, for example in the Gross-Neveu model [28].

$$\mu_R = \mu + \mu_5 \neq 0, \qquad \mu_L = \mu - \mu_5 \neq 0 \tag{10.59}$$

which can be realized by adding the term $\sum_f \mu_f \psi_f^\dagger \psi_f$ to the Lagrangian. In two dimensions the same term can be generated by a special local chiral transformation with a *linear* dependence on the spatial coordinate z in the exponent:

$$\psi_f' = e^{-i\Gamma^5 \mu_f z} \psi_f \tag{10.60}$$

To see this it is sufficient to observe

$$\bar{\psi}_f'(i\Gamma^z \partial_z)\psi_f' = \bar{\psi}_f(i\Gamma^z \partial_z)\psi_f + \mu_f \psi_f^\dagger \psi_f \tag{10.61}$$

where we used the gamma matrix identity $\Gamma^z \Gamma^5 = \Gamma^0$. Since we assumed that the interaction term is invariant under any local chiral rotation of the form (10.55), the transformation (10.60) does not generate any other term than the coupling to the chemical potential. Note that this feature is special to two dimensions and does not generalize directly to higher dimensions. It is also crucial to have a chiral Lagrangian in four-dimensions. This would not hold for a massive fermion for instance.

Let us start with the density $\langle \psi_f^\dagger \psi_f \rangle$ and the reduced current along the z direction $\langle \bar{\psi}_f \Gamma^z \psi_f \rangle$ which constitute the four-dimensional density $(\langle J_f^0 \rangle)$ and current along the magnetic field $(\langle J_f^3 \rangle)$. In Sect. 10.1.3 we have seen that the axial anomaly leads to an anomalous density in the presence of a chemical potential. In two-dimensional language, the axial anomaly

$$\partial_M \langle j_{f5}^M \rangle = \frac{e}{2\pi} \varepsilon^{MN} \partial_M A_{fN}$$
$$\langle j_{f5}^M \rangle = \langle \bar{\psi}_f \Gamma^M \Gamma^5 \psi_f \rangle = \varepsilon^{MN} \langle \bar{\psi}_f \Gamma^N \psi_f \rangle \tag{10.62}$$

immediately reproduces $\langle \psi_f^\dagger \psi_f \rangle = \mu_f/\pi$ once eA_f^0 is identified with the chemical potential μ_f. Alternatively the same result can be obtained directly from path integral by observing the transformation (10.60) applied to the renormalized charge density $\langle \bar{\psi}_f'^\dagger \psi_f' \rangle = 0$ creates the anomalous term μ_f/π [29].

Once the projection to the lowest Landau level is implemented, only the spin up component $\phi_\uparrow$ survives. Therefore for a right (left) handed spinor, only the positive (negative) momentum component contributes to the anomalous density, reducing its value by a half:

$$\langle \psi_R^\dagger \psi_R \rangle \rightarrow \langle R_+^* R_+ \rangle = \frac{eB}{2\pi} \frac{\mu_R}{2\pi} \quad \text{(LLL projection)}$$
$$\langle \psi_L^\dagger \psi_L \rangle \rightarrow \langle L_-^* L_- \rangle = \frac{eB}{2\pi} \frac{\mu_L}{2\pi} \quad \text{(LLL projection)} \tag{10.63}$$

The overall factor $\frac{eB}{2\pi}$ is the density of the lowest Landau level in the transverse plane. There is no effect of the two-dimensional anomaly on the current $\bar{\psi}_f \Gamma^z \psi_f$ in general. However, the lowest Landau level projection leads to a nonzero anomalous contribution to the current as well:

$$\langle \bar{\psi}_R \Gamma^z \psi_R \rangle = \langle \psi_R^\dagger \sigma_3 \psi_R \rangle \to \langle R_+^* R_+ \rangle = \frac{eB}{2\pi} \frac{\mu_R}{2\pi}$$

$$\langle \bar{\psi}_L \Gamma^z \psi_L \rangle = \langle \psi_L^\dagger \sigma_3 \psi_L \rangle \to -\langle L_-^* L_- \rangle = -\frac{eB}{2\pi} \frac{\mu_L}{2\pi} \tag{10.64}$$

The strong magnetic field selects a particular spin. Provided that there is some excess charge in the medium, the magnetic field induces a current for each chiral sector along its direction. When we transform the right and left currents back to vector and axial currents, we see the chiral magnetic and chiral separation effects in their conventional form:

$$\langle J^3 \rangle = \langle J_5^0 \rangle = \frac{eB\mu_5}{2\pi^2}, \qquad \langle J_5^3 \rangle = \langle J^0 \rangle = \frac{eB\mu}{2\pi^2} \tag{10.65}$$

Now consider another set of bilinears, those forming the scalar $\langle \bar{\psi}_f \psi_f \rangle$ and pseudo-scalar $\langle \bar{\psi}_f i \Gamma^5 \psi_f \rangle$ condensates. They constitute the transverse components $\langle J_f^\perp \rangle$ of the four-dimensional currents. In the semiclassical limit where the number of fermion flavors is large, the two-dimensional systems generically exhibit dynamical breaking of the chiral symmetry (10.55) and the system typically acquires a nonzero scalar condensate $\langle \bar{\psi}' \psi' \rangle = m \neq 0$ at zero chemical potential. Since this is also a mass term, it means there is a gap in the energy spectrum. The existence of the gap in the energy spectrum in one spatial dimension lowers the free energy at low temperatures as it "pushes" the Dirac sea further down in the energy spectrum. Once a finite chemical potential μ is turned on, all the states with energy lower than μ will be occupied and the optimal configuration would be to open a gap right around μ to push the occupied states down and lower the free energy. This is the celebrated Peierls instability [30, 31] and has broad consequences in condensed matter physics. Our assumption of the invariance of the interaction under the chiral transformation (10.60) is sufficient to see that this scenario is indeed realized. The effect of (10.60) on the associated Dirac Hamiltonian is

$$H\psi' = -i\Gamma^5 \partial_z \psi' + H_{int} \psi' = (H - \mu)\psi \tag{10.66}$$

Therefore it is always possible to shift the energy spectrum, and hence the gap, by μ with a chiral rotation. This is the relativistic version of the Peierls instability and it is proven for instance explicitly in the 2d Nambu-Jona-Lasinio model [32, 33], and is expected to be ubiquitous in two-dimensional systems with continuous chiral symmetry.

The chiral transformation (10.60) mixes the scalar and pseudo-scalar condensates in the following way

$$\bar{\psi}'_f \psi'_f = \cos(2\mu_f z)\bar{\psi}_f \psi_f - \sin(2\mu z)\bar{\psi}_f i \Gamma^5 \psi_f \tag{10.67}$$

so the existence of a nonzero condensate $\langle \bar{\psi}'_f \psi'_f \rangle = m_f \neq 0$ at zero chemical potential generalizes into finite chemical potential as

$$\langle \bar{\psi}\psi \rangle = m\cos(2\mu z), \qquad \langle \bar{\psi} i \Gamma^5 \psi \rangle = -m\sin(2\mu z) \tag{10.68}$$

since (10.67) must hold for any z. This modulated scalar/pseudo-scalar condensate is referred to as the "chiral spiral" [34–36]. In the 2d Nambu-Jona-Lasinio model the chiral spiral is indeed the thermodynamically preferred phase at low temperatures [32, 33]. The chiral spiral translates into transverse components of the four-dimensional currents modulated in z:

$$\langle J_R^1 \rangle = c_R \cos(2\mu_R z + \phi_R), \qquad \langle J_L^1 \rangle = -c_L \cos(2\mu_L z + \phi_L)$$
$$\langle J_R^2 \rangle = c_R \sin(2\mu_R z + \phi_R), \qquad \langle J_L^2 \rangle = c_R \sin(2\mu_L z + \phi_L)$$
$$(10.69)$$

which we call the "chiral magnetic spiral" [27]. Here the amplitudes $c_{R,L}$ depend on the particular two-dimensional Lagrangian and are functions of the temperature, magnetic field and possibly other parameters in the model. However the chemical potential and space dependence of the currents is universal. $\phi_{R,L}$ are relative phases of the left and right chiral condensates.

It should be emphasized that, as opposed to the longitudinal currents, the chiral magnetic spiral mixes up and down spin components. This can be seen in the spin decomposition of the currents (10.50), (10.51). In the lowest Landau projection, these pairings between spin up and down spinors correspond to excitations which can be described as a pairing of a particle with momentum μ_f and a hole with momentum $-\mu_f$. This is because the particle and hole have opposite charges and therefore opposite spins in the lowest Landau level projection. Thus the excitation itself has momentum $\pm 2\mu_f$. This explains why the currents have the sinusoidal modulation in the z direction.

In heavy ion collisions, the chiral magnetic spiral can induce both out-of-plane and in-plane fluctuating charge asymmetries (the explicit separation of out-of-plane and in-plane fluctuations has been performed [37] on the basis of STAR data [38, 39]). In the absence of topological fluctuations ($\mu_5 = 0$), at finite baryon density ($\mu \neq 0$), and in the chirally broken phase, the charge current has only transverse components, and the charge asymmetry will fluctuate only in-plane. It should be kept in mind that the presence of magnetic field increases the chiral transition temperature [9]. If topological fluctuations are present in the chirally broken phase (e.g. due to the presence of meta-stable η' domains [40]), the CME current can be carried by the chiral magnetic spiral. The chiral magnetic spiral has also been seen in a holographic study [41] in the framework of Sakai-Sugimoto model of holographic QCD [42, 43].

10.3 Fermions in an Instanton and Magnetic Field Background

The previous discussion, along with many other papers, has presented the chiral magnetic effect as the flow of electrical charge as the result of some externally produced chirality imbalance, represented by a non-zero μ_5. In this section we consider the situation in which this chirality imbalance is produced not by an explicit μ_5, or by a time-dependent A_3, but instead as a result of a topologically non-trivial gauge background like an instanton. Since quarks carry both electric and color charge, they couple to both electromagnetic and gluonic gauge fields. In this section we discuss some features of the spectral problem for fermions in the combined background field of a strong magnetic field and an instanton [44]. To illustrate the effect most clearly we take a single instanton in $SU(2)$. We are motivated by situations in which quarks experience both types of fields, such as in dense astrophysical objects such as neutron stars and magnetars, and in heavy ion collisions such as those at RHIC and at CERN [3, 45, 46].

We are also motivated by recent lattice QCD analyses [47–54], which provide important numerical information about the Dirac spectrum in both QCD and magnetic field backgrounds. Analytically, while the effect of each individual background is very well known, their combined effect turns out to be quite intricate. In these lattice studies, certain matrix elements associated with chiral effects receive dominant contributions from zero-modes and near-zero-modes, so we pay particular attention to the low end of the spectrum, and show that certain generic features have a very simple analytic explanation.

As discussed already, a magnetic field introduces a Landau level structure to the fermion spectrum, in which the zero modes of the associated two-dimensional Euclidean Dirac operator have definite spin, aligned along the magnetic field [11–13]. For a constant magnetic field on a torus, as appropriate for lattice QCD analysis, this has been studied recently in [55–57]. The appropriate formalism is that of the magnetic translation group [58]. In a gluonic field with nontrivial topological charge (for example, an instanton), the fermion spectrum of the four-dimensional Euclidean Dirac operator also has zero modes, with chiralities determined locally by the local topological charge of the gauge field [59–63]. For a single instanton the fermion spectral problem has a conformal symmetry [64, 65], and the zero modes are localized on the instanton, falling off as a power law with Euclidean distance. The conformal symmetry is broken by the introduction of a magnetic field, and now the zero modes develop an asymmetry, falling off in Gaussian form in the plane transverse to the B field, but as a power law in the other two directions. This basic asymmetry is an important feature of the phenomena of magnetic catalysis [9] and the chiral magnetic effect [1–5], as sketched in Fig. 10.4.

10.3.1 Euclidean Dirac Operator

To discuss a combined background of an instanton and a static magnetic field we use Euclidean Dirac matrices, instead of the Minkowski ones used in the previous sections. Our conventions follow those of [62], expressing the 4×4 Dirac matrices, γ_μ, for $\mu = 1, 2, 3, 4$, in terms of the 2×2 matrices $\alpha_\mu = (\mathbb{1}, -i\sigma)$ and $\bar{\alpha}_\mu = (\mathbb{1}, i\sigma) = \alpha_\mu^\dagger$, [here σ are the usual 2×2 Pauli matrices]:

$$\gamma_\mu = \begin{pmatrix} 0 & \alpha_\mu \\ \bar{\alpha}_\mu & 0 \end{pmatrix}, \qquad \gamma_5 = \begin{pmatrix} \mathbb{1} & 0 \\ 0 & -\mathbb{1} \end{pmatrix} \qquad (10.70)$$

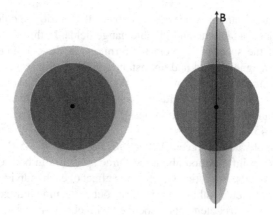

Fig. 10.4 A sketch of the topological charge density, $q \propto \text{tr} \, \mathcal{F}_{\mu\nu} \cdot \tilde{\mathcal{F}}_{\mu\nu}$, for a single instanton [*dark grey*], and the density of the quark zero mode [*light grey*]. On the *left*, there is a single instanton, and both densities fall off as power laws, with q falling off faster. On the *right*, with the introduction of a magnetic field, the topological charge density is unchanged but the zero mode density is distorted into an asymmetric shape, localized along the direction of the strong magnetic field

Thus, the Euclidean Dirac operator can be expressed as

$$
\not{D} = \begin{pmatrix} 0 & \alpha_\mu D_\mu \\ \bar{\alpha}_\mu D_\mu & 0 \end{pmatrix} \equiv \begin{pmatrix} 0 & D \\ -D^\dagger & 0 \end{pmatrix} \tag{10.71}
$$

where the covariant derivative, $D_\mu = \partial_\mu - i \mathcal{A}_\mu$, is written with a hermitean gauge field, $\mathcal{A}_\mu$, and x_4 is the Euclidean time coordinate. We write the gauge field $\mathcal{A}_\mu$ as a sum of a non-abelian part, A_μ, and an abelian part, a_μ:

$$
\mathcal{A}_\mu = A_\mu + a_\mu \tag{10.72}
$$

with the respective coupling constants absorbed into the gauge fields. The Dirac operator is anti-hermitean, so we write (with λ real)

$$
i\not{D}\psi_\lambda = \lambda\psi_\lambda \tag{10.73}
$$

Since $\{\gamma_5, \not{D}\} = 0$, we can take λ in (10.73) to be non-negative, with the negative eigenvalue solutions simply given by $\psi_{-\lambda} = \gamma_5 \psi_\lambda$. This means that we can effectively discuss the zero modes ($\lambda = 0$) separately, and for the nonzero modes ($\lambda \neq 0$) we consider the squared operator:

$$
(i\not{D})^2 \psi_\lambda = \begin{pmatrix} DD^\dagger & 0 \\ 0 & D^\dagger D \end{pmatrix} \psi_\lambda = \lambda^2 \psi_\lambda \tag{10.74}
$$

The positive chirality sector, $\chi = +1$, is described by the operator $DD^\dagger$, while the negative chirality sector, $\chi = -1$, is described by the operator $D^\dagger D$. We can write these operators as

$$\chi = +1: \quad DD^\dagger = -D_\mu^2 - \frac{1}{2}\mathscr{F}_{\mu\nu}\bar{\sigma}_{\mu\nu} \tag{10.75}$$

$$\chi = -1: \quad D^\dagger D = -D_\mu^2 - \frac{1}{2}\mathscr{F}_{\mu\nu}\sigma_{\mu\nu} \tag{10.76}$$

We have used $[D_\mu, D_\nu] = -i\mathscr{F}_{\mu\nu}$, where $\mathscr{F}_{\mu\nu}$ is the field strength associated with the gauge field $\mathscr{A}_\mu$, and the spin matrices $\bar{\sigma}_{\mu\nu}$ and $\sigma_{\mu\nu}$ are defined as

$$\bar{\sigma}_{\mu\nu} = \frac{1}{2i}(\alpha_\mu\bar{\alpha}_\nu - \alpha_\nu\bar{\alpha}_\mu), \qquad \sigma_{\mu\nu} = \frac{1}{2i}(\bar{\alpha}_\mu\alpha_\nu - \bar{\alpha}_\nu\alpha_\mu) \tag{10.77}$$

In (10.75), (10.76) we have used the properties [62]: $\bar{\alpha}_\mu\alpha_\nu = \delta_{\mu\nu} + i\sigma_{\mu\nu}$, and $\alpha_\mu\bar{\alpha}_\nu = \delta_{\mu\nu} + i\bar{\sigma}_{\mu\nu}$.

For non-zero modes [i.e., solutions to (10.74) with $\lambda \neq 0$], the operators $DD^\dagger$ and $D^\dagger D$ have identical spectra, for any background field. This is simply because we have an invertible map: suppose the 2-component spinor v satisfies $D^\dagger D v = \lambda^2 v$. Then $u = Dv$ is clearly an eigenfunction of the other operator, $DD^\dagger$, with precisely the same eigenvalue: $DD^\dagger u = DD^\dagger D v = \lambda^2 u$. Similarly, if u satisfies $DD^\dagger u = \lambda^2 u$, then $v = D^\dagger u$ is an eigenstate of $D^\dagger D$ with the same eigenvalue. Thus, when $\lambda \neq 0$, we can write the 4-component spinor solution in the form

$$\psi_\lambda = \begin{pmatrix} u_\lambda \\ -\frac{i}{\lambda}D^\dagger u_\lambda \end{pmatrix} \quad \text{where } DD^\dagger u_\lambda = \lambda^2 u_\lambda \tag{10.78}$$

or in the form

$$\psi_\lambda = \begin{pmatrix} \frac{i}{\lambda}Dv_\lambda \\ v_\lambda \end{pmatrix} \quad \text{where } D^\dagger D v_\lambda = \lambda^2 v_\lambda \tag{10.79}$$

This is true for any background field: non-abelian, abelian, or both.

10.3.2 Magnetic Field Background

For a constant (abelian) magnetic field, of strength B, pointing in the x_3 direction, we have an abelian field strength $f_{12} = B$, and so we find

$$D^\dagger D = DD^\dagger = -D_\mu^2 - B\sigma_3 \tag{10.80}$$

where we have used the fact that $\bar{\sigma}_{12} = \sigma_{12} = \sigma_3$. Due to the subtraction term, $-B\sigma_3$, it is possible to have zero modes, and since $DD^\dagger = D^\dagger D$ these zero modes occur in each chiral sector. More explicitly, we can make a Bogomolnyi-style factorization similar to (10.12) and write

$$-D_\mu^2 - B\sigma_3 = -\partial_3^2 - \partial_4^2 - (D_1 \mp iD_2)(D_1 \pm iD_2) \pm B - B\sigma_3$$

$$= -\partial_3^2 - \partial_4^2 - D_\mp D_\pm \pm B - B\sigma_3 \tag{10.81}$$

For zero modes, we take $\partial_3 = \partial_4 = 0$, and with $B > 0$ we choose the upper signs to ensure normalizable modes. For example, in the symmetric gauge where the abelian gauge field

$$a_\mu = \frac{B}{2}(-x_2, x_1, 0, 0) \tag{10.82}$$

the zero modes can be expressed in terms of the normalizable solutions to $(D_1 + iD_2)u = 0$:

$$\psi_0 = g(z_1)e^{-B|z_1|^2/4}\begin{pmatrix} 1 \\ 0 \\ 0 \\ 0 \end{pmatrix} \quad \text{or} \quad \psi_0 = g(z_1)e^{-B|z_1|^2/4}\begin{pmatrix} 0 \\ 0 \\ 1 \\ 0 \end{pmatrix} \tag{10.83}$$

Here $g(z_1)$ is a holomorphic function of the complex variable $z_1 = (x_1 + ix_2)$. Both sets of zero modes have spin up, aligned along the B field; this is just the familiar lowest Landau level projection onto spin up states. Note also that the zero modes have the characteristic Gaussian factor in the (x_1, x_2) plane, transverse to the direction of the magnetic field. This factor is the origin of the distortion sketched in the right frame of Fig. 10.4.

The number of zero modes per unit two-dimensional area [in the (x_1, x_2) plane] is given by the Landau degeneracy factor, the magnetic flux per unit area: $B/(2\pi)$. In fact, even for an inhomogeneous magnetic field $B(x_1, x_2)$, pointing in the x_3 direction, the number of zero modes [of each chirality] is determined by the integer part of the magnetic flux (this is the essence of the Aharonov-Casher theorem [11]). For example, on a torus [12, 13]:

$$N_+ = N_- = \frac{1}{2\pi}\int d^2x\, B \tag{10.84}$$

The higher Landau level states are the same for both spins, because $(-D_-D_+ + B)$ and $(-D_+D_- - B)$ have identical spectra, apart from the lowest level, which only has spin aligned along the magnetic field.

10.3.3 Instanton Background

For an instanton field, A_μ, the (non-abelian) field strength $F_{\mu\nu}$ is self-dual [that is: $F_{\mu\nu} = \tilde{F}_{\mu\nu}$, where the dual tensor is defined: $\tilde{F}_{\mu\nu} \equiv \frac{1}{2}\varepsilon_{\mu\nu\alpha\beta}F_{\alpha\beta}$]. Then the anti-self-duality property of $\bar{\sigma}_{\mu\nu}$ [that is: $\bar{\sigma}_{\mu\nu} = -\frac{1}{2}\varepsilon_{\mu\nu\rho\sigma}\bar{\sigma}_{\rho\sigma}$] implies:

$$\chi = +1: \quad DD^\dagger = -D_\mu^2 \tag{10.85}$$

$$\chi = -1: \quad D^\dagger D = -D_\mu^2 - \frac{1}{2}F_{\mu\nu}\sigma_{\mu\nu} \tag{10.86}$$

Since $-D_\mu^2$ is a positive operator, this means that for an instanton background there can be no zero mode in the positive chirality sector. On the other hand, due to the subtraction term, $-F_{\mu\nu}\sigma_{\mu\nu}$, in $D^\dagger D$, it is possible to have a zero eigenvalue solution in the negative chirality sector, and it has the form

$$\psi_0 = \begin{pmatrix} 0 \\ v \end{pmatrix}, \quad \text{where } Dv = 0 \tag{10.87}$$

[For an anti-instanton, an anti-self-dual field with $F_{\mu\nu} = -\tilde{F}_{\mu\nu}$, the zero mode lies in the positive chirality sector, because $\sigma_{\mu\nu}$ is self-dual: $\sigma_{\mu\nu} = \frac{1}{2}\varepsilon_{\mu\nu\rho\sigma}\sigma_{\rho\sigma}$.] For a general non-abelian gauge field A_μ, which is neither self-dual nor anti-self-dual, the Atiyah-Singer index theorem [63, 66] states that the difference between the number of positive and negative chirality zero modes is given by the topological charge of the gauge field:

$$N_+ - N_- = -\frac{1}{32\pi^2} \int d^4x\, F_{\mu\nu}^a \tilde{F}_{\mu\nu}^a \tag{10.88}$$

Here we have written $F_{\mu\nu} = F_{\mu\nu}^a T^a$, with generators normalized as $\text{tr}(T^a T^b) = \frac{1}{2}\delta^{ab}$. For gauge group $SU(2)$, with fermions in the defining representation, we take generators $T^a = \frac{1}{2}\tau^a$ in terms of the Pauli matrices τ, and we can write the single instanton gauge field [67], centered at the origin, in the regular gauge as

$$A_\mu^a = 2\frac{\eta_{\mu\nu}^a x_\nu}{x^2 + \rho^2} \tag{10.89}$$

where ρ is the instanton scale parameter, and $\eta_{\mu\nu}^a$ is the self-dual 't Hooft tensor [59, 62]. The topological charge density is

$$q(x) = \frac{1}{32\pi^2} F_{\mu\nu}^a \tilde{F}_{\mu\nu}^a = \frac{192\rho^4}{(x^2 + \rho^2)^4} \tag{10.90}$$

There is a single zero mode [14, 15, 59–61, 63], also localized at the origin, with density:

$$|\psi_0|^2 = \frac{64\rho^2}{(x^2 + \rho^2)^3} \tag{10.91}$$

These densities both fall off as power laws, with scale set by ρ, but the topological charge density is more localized, as indicated in the left-hand frame of Fig. 10.4. The nonzero modes are given by (10.78) or (10.79), and we note that the spectra are identical in each chiral sector, apart from the zero modes.

10.3.4 Combined Instanton and Magnetic Field Background

Physically, an instanton field projects the zero modes onto a definite chirality, while a constant magnetic field projects the zero modes onto definite spin, aligned along the direction of the magnetic field. When we combine the two background fields, both a non-abelian instanton field $F_{\mu\nu}$ and an abelian magnetic field $f_{12} = B$, there is a competition between the two projection mechanisms, and the outcome depends on their relative magnitude, as we show below. Technically speaking, the instanton zero mode has a specific ansatz form that unifies space-time and color indices, while the magnetic zero modes have a natural holomorphic structure, and these two different ansatz forms do not match one another. The competition between these two ansatz forms makes the combined problem nontrivial. For an instanton field, since the field falls off as a power law, all eigenmodes also fall off with power law behavior. On the other hand, once a constant magnetic field is introduced, for example in the gauge (10.82), all the eigenstates (even those in the higher Landau levels) have a Gaussian factor $\exp(-B|z_1|^2/4)$ that localizes the modes near the axis of the magnetic field. This is the reason for the distorted density in the right-hand frame of Fig. 10.1. In the extreme strong magnetic field limit this leads to a dimensional reduction to motion along the magnetic field, with interesting physical consequences such as magnetic catalysis [9] and the chiral magnetic effect [1, 4, 5].

Concerning zero modes, we begin with a simple but important comment: in the index theorem (10.88), the magnetic field makes no contribution, since with the field strength decomposed into its non-abelian and abelian parts, $\mathscr{F}_{\mu\nu} = F_{\mu\nu} + f_{\mu\nu}$, we have

$$\text{tr}(\mathscr{F}_{\mu\nu}\tilde{\mathscr{F}}_{\mu\nu}) = \text{tr}(F_{\mu\nu}\tilde{F}_{\mu\nu}) + (\dim) f_{\mu\nu}\tilde{f}_{\mu\nu} \tag{10.92}$$

$$= \text{tr}(F_{\mu\nu}\tilde{F}_{\mu\nu}) \tag{10.93}$$

where dim is the dimension of the Lie algebra representation of the non-abelian gauge fields. The cross terms vanish since the Lie algebra generators T^a are traceless, and the $f_{\mu\nu}\tilde{f}_{\mu\nu}$ term vanishes since there is no abelian electric field. For example, if there is no nonabelian field, just an abelian magnetic field, then the topological charge clearly vanishes, and the index theorem (10.88) is consistent with the fact that $DD^\dagger = D^\dagger D$ for an abelian magnetic background (recall (10.80)), so that there are the same number of zero modes in each chiral sector. Now, with both background fields present, we find

$$DD^\dagger = -D_\mu^2 - B\sigma_3 \tag{10.94}$$

$$D^\dagger D = -D_\mu^2 - \frac{1}{2}F_{\mu\nu}\sigma_{\mu\nu} - B\sigma_3 \tag{10.95}$$

Notice that the eigenvalues of $DD^\dagger$ are simply those of the scalar operator $-D_\mu^2$, with a spin term $\pm B$, as can be seen clearly in Fig. 10.5. The fact that there is a subtraction term from the positive operator $-D_\mu^2$ in both chirality sectors tells us that it is possible to have zero modes for each chirality, but their number will depend on the relative magnitude of F and B. In the next section we study a specific model where we can quantify this precisely. Another important implication is that we may also have some "near-zero-modes", where the F and B subtractions do not exactly

Fig. 10.5 Successive
approximations in the
derivative expansion of the
instanton background
interpolate between the
exponential form of the zero
mode [*dashed line*] to the
power-law form [*solid line*]
of the exact result

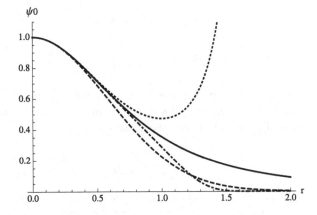

cancel the lowest eigenvalue of $-D_\mu^2$, but lower the eigenvalue of $DD^\dagger$ or $D^\dagger D$ to
near zero.

10.3.5 Large Instanton Limit: Covariantly Constant $SU(2)$ Instanton and Constant Abelian Magnetic Field

In the very strong magnetic field limit, where the magnetic length, $1/\sqrt{B}$, is small
compared to the instanton size ρ, we expect a significant distortion of instanton
modes and currents. In this limit we can make a simple approximation that reduces
the problem to a completely soluble system.

In the large instanton limit, we expand the instanton gauge field as:

$$A_\mu^a \approx \frac{2}{\rho^2}\eta_{\mu\nu}^a x_\nu + \cdots \tag{10.96}$$

To leading order in such a derivative expansion, the non-abelian gauge configura-
tion $A_\mu^a(x)$ is self-dual and has covariantly constant field strength: $F_{\mu\nu}^a = -\frac{4}{\rho^2}\eta_{\mu\nu}^a$.
In this limit we can make an $SU(2)$ "color" rotation, along with a choice of Lorentz
frame, to make the instanton field diagonal in the color space (we choose the τ^3 direction), so that the field is self-dual, covariantly constant and quasi-abelian. Defining the instanton scale $F = \frac{2}{\rho^2}$, the combined gauge field, including also the abelian
magnetic field as in (10.72), can be written as:

$$\mathscr{A}_\mu = -\frac{F}{2}(-x_2, x_1, -x_4, x_3)\tau^3 + \frac{B}{2}(-x_2, x_1, 0, 0)\mathbb{1}_{2\times2} \tag{10.97}$$

This gauge field is fully diagonal and moreover is linear in x_μ, so the problem is
analytically soluble (this is the basic premise of the derivative expansion). The only
nonzero entries of the field strength tensor are

$$\mathscr{F}_{12} = -F\tau^3 + B\mathbb{1} = \begin{pmatrix} B - F & 0 \\ 0 & B + F \end{pmatrix}$$

$$\mathscr{F}_{34} = -F\tau^3 = \begin{pmatrix} -F & 0 \\ 0 & +F \end{pmatrix} \tag{10.98}$$

In the absence of the magnetic field the field strength is self-dual, $\mathscr{F}_{12} = \mathscr{F}_{34}$, but a nonzero magnetic field breaks this symmetry. The topological charge density is (recall the normalization of the generators)

$$\frac{1}{32\pi^2}\mathscr{F}^a_{\mu\nu}\tilde{\mathscr{F}}^a_{\mu\nu} = \frac{4(2F)^2}{32\pi^2} = \frac{F^2}{2\pi^2} \tag{10.99}$$

A natural question to ask is: in such a constant field strength background, the wave functions have a *Gaussian* spatial dependence in the plane transverse to the direction of the field, characteristic of the Landau problem, so how can we recover the power-law dependence of the zero modes in an instanton background? This happens as follows. Recall [63] that with the appropriate ansatz for the zero mode, the zero mode equation reduces to a first-order radial equation

$$\psi_0' = -\frac{3r}{1+r^2}\psi_0 \quad \Rightarrow \quad \psi_0 = \frac{1}{(1+r^2)^{3/2}} \tag{10.100}$$

where r is the Euclidean distance. At short and long distances, this zero mode behaves as

$$\psi_0 \sim 1 - \frac{3}{2}r^2 + \cdots, \quad r \to 0 \tag{10.101}$$

$$\psi_0 \sim \frac{1}{r^3} + \cdots, \quad r \to \infty \tag{10.102}$$

On the other hand, if we make the above choice (10.97) of a constant field to represent a large instanton, we have instead the zero mode equation

$$\psi_0' = -3r\psi_0 \quad \Rightarrow \quad \psi_0 = e^{-3r^2/2} \tag{10.103}$$

which has the correct short-distance behavior but which is Gaussian rather than power law at large distances.

We can recover the correct power-law behavior by restoring the instanton size parameter, including the sub-leading terms in (10.96), and expanding the zero mode equation as

$$\psi_0' = -3r\left(1 - r^2 + r^4 - r^6 + r^8 - \cdots\right)\psi_0 \tag{10.104}$$

The leading term represents the leading term of the derivative expansion of the instanton field, and is solved by the Gaussian factor. Absorbing this Gaussian factor by writing $\psi_0 = e^{-3r^2/2}\chi_0(r)$, the resulting equation for $\chi_0(r)$ is

$$\chi_0' = 3r\left(r^2 - r^4 + r^6 - r^8 + \cdots\right)\chi_0 \tag{10.105}$$

which suggests writing $\chi_0 = e^{3r^4/4}\phi_0(r)$. Continuing this process order by order in the derivative expansion, we obtain

$$\psi_0(r) = \exp\left[-\frac{3}{2}r^2 + \frac{3}{4}r^4 - \frac{3}{6}r^6 + \frac{3}{8}r^8 - \cdots\right] = \exp\left[-\frac{3}{2}\ln(1+r^2)\right] \quad (10.106)$$

thereby recovering the correct power-law decay from the derivative expansion of the instanton background, as illustrated in Fig. 10.5.

10.3.6 Dirac Spectrum in the Strong Magnetic Field Limit

To study the Dirac spectrum with both a magnetic field and an instanton we consider the 2×2 operators $DD^\dagger$ and $D^\dagger D$ in (10.75, 10.76). Notice first that

$$\mathcal{F}_{\mu\nu}\bar{\sigma}_{\mu\nu} = 2(\mathcal{F}_{12} - \mathcal{F}_{34})\sigma_3 \quad (10.107)$$

$$\mathcal{F}_{\mu\nu}\sigma_{\mu\nu} = 2(\mathcal{F}_{12} + \mathcal{F}_{34})\sigma_3 \quad (10.108)$$

It is convenient to factor the 4-dimensional Euclidean space and consider separately the (x_1, x_2) plane and the (x_3, x_4) plane. Then in the (x_1, x_2) plane we have a (relativistic) Landau level problem with effective field strength $(B - F)$ in the $\tau^3 = +1$ sector, and with effective field strength $(B + F)$ in the $\tau^3 = -1$ sector. In the (x_3, x_4) plane we also have a (relativistic) Landau level problem, now with effective field strength $-F$ in the $\tau^3 = +1$ sector, and with effective field strength F in the $\tau^3 = -1$ sector. In the (x_1, x_2) plane the sign of the effective field strength depends on which of B or F is larger, and in the strong B field limit, both $B \pm F$ are positive.

When $B > F$, both $(B - F)$ and $(B + F)$ are positive. Thus, each color component of $\mathcal{F}_{12}$ is associated with a positive "magnetic" field. On the other hand, for $\mathcal{F}_{34}$, the $\tau^3 = +1$ sector has a negative field strength, while the $\tau^3 = -1$ sector has a positive field strength.

We first consider the $\tau^3 = +1$ case. Then $\mathcal{F}_{12} = (B - F)$, $\mathcal{F}_{34} = -F$, $\mathcal{F}_{\mu\nu}\bar{\sigma}_{\mu\nu} = 2B\sigma_3$, and $\mathcal{F}_{\mu\nu}\sigma_{\mu\nu} = 2(B - 2F)\sigma_3$. With a positive field strength the normalizable zero state is given by $(D_1 + iD_2)u = 0$. But since $\mathcal{F}_{34}$ is negative, we factorize the corresponding covariant derivatives in the opposite order, in order to obtain a normalizable state annihilated by $(D_3 - iD_4)$. Thus, we have, for chirality $\chi = \pm 1$, respectively:

$$DD^\dagger = -(D_1 - iD_2)(D_1 + iD_2) - (D_3 + iD_4)(D_3 - iD_4) + B - B\sigma_3$$

$$D^\dagger D = -(D_1 - iD_2)(D_1 + iD_2) - (D_3 + iD_4)(D_3 - iD_4) + B - (B - 2F)\sigma_3$$

$$(10.109)$$

This shows that there is a zero mode, when the spin term $B\sigma_3$ cancels the B term from the Bogomolnyi factorization of the covariant derivative term. This occurs in the positive chirality sector, $\chi = +1$, and with spin up: $\sigma_3 = +1$.

Now consider the $\tau^3 = -1$ case. Then $\mathscr{F}_{12} = (B + F)$, $\mathscr{F}_{34} = F$, $\mathscr{F}_{\mu\nu}\bar{\sigma}_{\mu\nu} = 2B\sigma_3$, and $\mathscr{F}_{\mu\nu}\sigma_{\mu\nu} = 2(B + 2F)\sigma_3$. All field strengths are positive, so we write, for chirality $\chi = \pm 1$, respectively:

$$
\begin{aligned}
DD^\dagger &= -(D_1 - iD_2)(D_1 + i\mathscr{D}_2) - (D_3 - iD_4)(D_3 + iD_4) \\
&\quad + (B + 2F) - B\sigma_3 \\
D^\dagger D &= -(D_1 - iD_2)(D_1 + iD_2) - (D_3 - iD_4)(D_3 + iD_4) \\
&\quad + (B + 2F) - (B + 2F)\sigma_3
\end{aligned}
\tag{10.110}
$$

This shows that there is a zero mode, but now in the opposite chirality sector, $\chi = -1$, and also with spin up: $\sigma_3 = +1$.

To summarize: when $B > F$, the $\tau_3 = +1$ color sector has spin up zero modes with positive chirality, while the $\tau_3 = -1$ color sector has spin up zero modes with negative chirality. We can count the number of zero modes in each chirality sector by simply taking the product of the Landau degeneracy factors for the (x_1, x_2) and (x_3, x_4) planes, with the corresponding effective magnetic field strengths. Therefore, the corresponding Landau degeneracy factors give the zero-mode number densities (i.e., the number per unit volume):

$$
\chi = +1: \quad n_+ = \frac{(B - F)}{2\pi} \frac{F}{2\pi} \quad (\tau_3 = +1, \, \sigma_3 = +1) \tag{10.111}
$$

$$
\chi = -1: \quad n_- = \frac{(B + F)}{2\pi} \frac{F}{2\pi} \quad (\tau_3 = -1, \, \sigma_3 = +1) \tag{10.112}
$$

The index (density) is the difference,

$$
n_+ - n_- = -\frac{F^2}{2\pi^2} \tag{10.113}
$$

in agreement with the general index theorem (10.88), in view of (10.99). We also note that the total number density of zero modes

$$
n_+ + n_- = \frac{BF}{2\pi^2} \tag{10.114}
$$

is linearly proportional to the magnetic field strength B. This is in agreement with numerical lattice gauge theory results [52].

It is worth emphasizing that if torus boundary conditions $x_\mu \sim x_\mu + L_\mu$ are imposed, then the fluxes are quantized as $BL^2 = 2\pi M$ and $FL^2 = 2\pi N$. Here M and N are positive integers. Consequently the index and total number of zero modes are given by also by integers

$$
\text{index}(\slashed{D}) \equiv N_+ - N_- = (N - M)M - (N + M)M = -2M^2
$$

$$
\text{total number of zero modes} = (N + M)M + (N - M) = 2NM
\tag{10.115}
$$

Moreover, the instanton solution with constant field strength (10.96) is shown to be an exact solution of Yang-Mills equations on a four torus [68–71]. Even with the inclusion of the magnetic field, (10.97) stays as a solution. Therefore on a on four-torus, there is no restriction on the magnitude of B and F (i.e. we do not have to consider a large instanton) and our counting of the zero modes (10.115) becomes exact for any integer M and N.

10.3.7 Physical Picture: Competition Between Spin and Chirality Projection

These results lead to the following simple physical picture. The instanton tries to generate a chirality imbalance but is neutral to the spin, whereas the magnetic field tries to generate a spin imbalance but does not affect the chirality. Depending on which is stronger, the zero modes have either a definite spin with a chirality imbalance ($B > F$), or a definite chirality with a spin imbalance ($F > B$). Also we see that in the former case, the *total* number of zero modes scales with B and is not equal to the index.

More explicitly, for the $B > F$ case, consider starting with just a strong magnetic field B, later turning on a weak instanton field. Without the instanton field, the zero modes and their degeneracy are given by the Aharonov-Casher theorem (10.84), so that the zero mode density is the Landau degeneracy factor $B/(2\pi)$ for each chirality sector. All the zero modes are spin up, as is familiar for the lowest Landau level (see Fig. 10.2). There is an equal number of positive and negative chirality zero-modes, which is consistent with the index theorem, since the topological charge vanishes for a constant B field. Now consider turning on an instanton field F, with $B > F > 0$. We see from (10.111, 10.112) that the effect of the instanton is to flip some of the chiralities: $(\frac{F}{2\pi})^2$ positive chirality modes become negative chirality modes, leading to a chirality imbalance of $\frac{F^2}{2\pi^2}$, in agreement with the index theorem (10.88). On the other hand, the total number of zero modes, $\frac{BF}{2\pi^2}$, grows linearly with the magnetic field when F is nonzero.

10.3.8 Matrix Elements and Dipole Moments

The computation of matrix elements is significantly simplified by using the Euclidean version of the Schur decomposition (10.18, 10.19) used earlier to relate the chiral magnetic effect to the two- and four-dimensional axial anomalies. Thus, introducing a small quark mass m, the propagator of the Dirac operator $\not{D} + m$ is given by its Schur decomposition:

$$\frac{1}{\not{D} + m} = \begin{pmatrix} \frac{m}{m^2 + DD^\dagger} & \frac{-1}{m^2 + DD^\dagger} D \\ \frac{1}{m^2 + D^\dagger D} D^\dagger & \frac{m}{m^2 + D^\dagger D} \end{pmatrix} \qquad (10.116)$$

Note that $DD^\dagger$ and $D^\dagger D$ have identical spectra, except for possible zero modes, so they can be viewed as square operators (matrices) of different dimension, as is clear when they are diagonalized in their respective eigenspaces. For a simple algebraic illustration, take D to be the 2-component column vector

$$D = \begin{pmatrix} 1 \\ 2 \end{pmatrix} \quad \Rightarrow \quad DD^\dagger = \begin{pmatrix} 1 & 2 \\ 2 & 4 \end{pmatrix}, \qquad D^\dagger D = 5 \qquad (10.117)$$

Thus, $DD^\dagger$ has eigenvalues 0 and 5, while $D^\dagger D$ clearly only has eigenvalue 5, the difference in rank being accounted for by the count of zero modes.

The zero mode contribution to the propagator can be separated by writing it in one of two ways, depending on which chirality supports zero modes,

$$\frac{1}{\not{D} + m} = \begin{pmatrix} \frac{m}{m^2 + DD^\dagger} & \frac{-1}{m^2 + DD^\dagger} D \\ D^\dagger \frac{1}{m^2 + DD^\dagger} & (\frac{1}{m} - \frac{1}{m} D^\dagger \frac{1}{m^2 + DD^\dagger} D) \end{pmatrix}$$

$$= \begin{pmatrix} (\frac{1}{m} - \frac{1}{m} D \frac{1}{m^2 + D^\dagger D} D^\dagger) & -D \frac{1}{m^2 + D^\dagger D} \\ \frac{1}{m^2 + D^\dagger D} D^\dagger & \frac{m}{m^2 + D^\dagger D} \end{pmatrix} \qquad (10.118)$$

An important set of quark bilinears involve the spin tensor $\Sigma_{\mu\nu}$:

$$\Sigma_{\mu\nu} = \frac{1}{2i}[\gamma_\mu, \gamma_\nu] = \begin{pmatrix} \bar{\sigma}_{\mu\nu} & 0 \\ 0 & \sigma_{\mu\nu} \end{pmatrix} \qquad (10.119)$$

This representation makes clear the natural decomposition of $\Sigma_{\mu\nu}$ into its self-dual part ($\sigma_{\mu\nu}$) and its anti-self-dual part ($\bar{\sigma}_{\mu\nu}$). The bilinears are

$$\langle \bar{\psi} \Sigma_{\mu\nu} \psi \rangle = \mathrm{tr}\left(\Sigma_{\mu\nu} \frac{1}{\not{D} + m} \right) \qquad (10.120)$$

using the 2×2 sub-block structure of the propagator, it is straightforward to derive the relations:

$$\langle \bar{\psi} \Sigma_{\mu\nu} \psi \rangle = m \, \mathrm{tr}_{2\times 2}\left(\sigma^{\mu\nu} \frac{1}{m^2 + DD^\dagger} + \bar{\sigma}^{\mu\nu} \frac{1}{m^2 + D^\dagger D} \right) \qquad (10.121)$$

$$\langle \bar{\psi} \psi \rangle = m \, \mathrm{tr}_{2\times 2}\left(\frac{1}{m^2 + DD^\dagger} + \frac{1}{m^2 + D^\dagger D} \right) \qquad (10.122)$$

$$\langle \bar{\psi} \Sigma_{\mu\nu} \gamma_5 \psi \rangle = m \, \mathrm{tr}_{2\times 2}\left(\sigma^{\mu\nu} \frac{1}{m^2 + DD^\dagger} - \bar{\sigma}^{\mu\nu} \frac{1}{m^2 + D^\dagger D} \right) \qquad (10.123)$$

$$\langle \bar{\psi} \gamma_5 \psi \rangle = m \, \mathrm{tr}_{2\times 2}\left(\frac{1}{m^2 + DD^\dagger} - \frac{1}{m^2 + D^\dagger D} \right) \qquad (10.124)$$

For applications to the chiral magnetic effect, we are interested in the magnetic and electric dipole moments:

$$\sigma_i^M = \frac{1}{2}\varepsilon_{ijk}\langle\bar\psi\,\Sigma_{jk}\psi\rangle \tag{10.125}$$

$$\sigma_i^E = \langle\bar\psi\,\Sigma_{i4}\psi\rangle \tag{10.126}$$

With a strong magnetic field in the x_3 direction, we concentrate on σ_3^M and σ_3^E, which require the spin tensors:

$$\Sigma_{12} = \begin{pmatrix} \sigma_3 & 0 \\ 0 & \sigma_3 \end{pmatrix}, \qquad \Sigma_{34} = \begin{pmatrix} -\sigma_3 & 0 \\ 0 & \sigma_3 \end{pmatrix} \tag{10.127}$$

Thus,

$$m\langle\bar\psi\,\Sigma_{12}\psi\rangle = \mathrm{tr}_{2\times2}\left(\sigma_3\frac{m^2}{m^2+DD^\dagger}\right) + \mathrm{tr}_{2\times2}\left(\sigma_3\frac{m^2}{m^2+D^\dagger D}\right) \tag{10.128}$$

$$m\langle\bar\psi\,\Sigma_{34}\psi\rangle = -\,\mathrm{tr}_{2\times2}\left(\sigma_3\frac{m^2}{m^2+DD^\dagger}\right) + \mathrm{tr}_{2\times2}\left(\sigma_3\frac{m^2}{m^2+D^\dagger D}\right) \tag{10.129}$$

The dominant contribution to the trace over the spectrum comes from the modes with low eigenvalues of $DD^\dagger$ and $D^\dagger D$. In the strong magnetic field limit, the zero modes and the near-zero-modes all have spin up, $\sigma_3 = +1$, as expected. The dominant contribution to the electric and magnetic moments are therefore:

$$m\langle\bar\psi\,\Sigma_{12}\psi\rangle \approx \mathrm{tr}_{2\times2}\left(\frac{m^2}{m^2+DD^\dagger}\right) + \mathrm{tr}_{2\times2}\left(\frac{m^2}{m^2+D^\dagger D}\right) \tag{10.130}$$

$$m\langle\bar\psi\,\Sigma_{34}\psi\rangle \approx -\,\mathrm{tr}_{2\times2}\left(\frac{m^2}{m^2+DD^\dagger}\right) + \mathrm{tr}_{2\times2}\left(\frac{m^2}{m^2+D^\dagger D}\right) \tag{10.131}$$

In particular, this means that in the strong magnetic field limit, we expect

$$\frac{\langle\bar\psi\,\Sigma_{12}\psi\rangle}{\langle\bar\psi\psi\rangle} \to 1, \qquad B\to\infty \tag{10.132}$$

as has been confirmed in a lattice study [47–49].

For the magnetic dipole moment, the main contribution comes from the zero modes, so we simply count the degeneracies in the various sectors:

$$m\langle\bar\psi\,\Sigma_{12}\psi\rangle \approx \left(\frac{B-F}{2\pi}\right)\left(\frac{F}{2\pi}\right) + \left(\frac{B+F}{2\pi}\right)\left(\frac{F}{2\pi}\right)$$

$$= \frac{BF}{2\pi^2} \tag{10.133}$$

which is linear in the magnetic field B. For the electric dipole moment, the near-zero-modes cancel, leaving just the zero mode contribution:

$$m\langle\bar{\psi}\Sigma_{34}\psi\rangle \approx -\left(\frac{B-F}{2\pi}\right)\left(\frac{F}{2\pi}\right) + \left(\frac{B+F}{2\pi}\right)\left(\frac{F}{2\pi}\right)$$

$$= \frac{F^2}{2\pi^2} \tag{10.134}$$

which is independent of B, and negligible compared to BF, for $B \gg F$. [Note that (10.134) does not imply that there is a residual electric dipole moment when B vanishes, because (10.134) applies only in the $B \gg F$ limit.] Thus, we see that the zero modes and near-zero-modes imply that

$$\langle\bar{\psi}\Sigma_{12}\psi\rangle \propto B, \qquad \langle\bar{\psi}\Sigma_{12}\psi\rangle \gg \langle\bar{\psi}\Sigma_{34}\psi\rangle \tag{10.135}$$

This is in agreement with the lattice results of [50]. We note that in a full QCD calculation with dynamical quarks there is an additional instanton measure factor that scales as m^{N_f}, which should be taken into account for these matrix elements.

If we now consider the fluctuations in the electric dipole moment, we find a dependence on B, because

$$\langle\bar{\psi}\Sigma_{34}\psi\bar{\psi}\Sigma_{34}\psi\rangle$$

$$= \text{tr}\left(\frac{1}{\slashed{D}+m}\Sigma_{34}\frac{1}{\slashed{D}+m}\Sigma_{34}\right)$$

$$= \text{tr}_{2\times2}\left(\frac{m^2}{(m^2+DD^\dagger)^2} + \frac{1}{(m^2+DD^\dagger)}D\sigma_3 D^\dagger\sigma_3\frac{1}{(m^2+DD^\dagger)}\right)$$

$$+ \text{tr}_{2\times2}\left(\frac{1}{(m^2+D^\dagger D)^2}D^\dagger\sigma_3 D\sigma_3 + \frac{m^2}{(m^2+D^\dagger D)}\sigma_3\frac{1}{(m^2+D^\dagger D)}\sigma_3\right)$$

$$\approx \text{tr}_{2\times2}\left(\frac{1}{(m^2+DD^\dagger)} + \frac{1}{(m^2+D^\dagger D)}\right) \tag{10.136}$$

where in the last step we have used the fact that the dominant contribution comes from zero modes and near-zero-modes, all of which have $\sigma_3 = +1$. Thus, comparing with (10.133) we see that the fluctuation is linear in B

$$\langle\bar{\psi}\Sigma_{34}\psi\bar{\psi}\Sigma_{34}\psi\rangle \approx \left(\frac{F}{2\pi^2 m^2 L^4}\right)B \tag{10.137}$$

again in agreement with the lattice results of [50].

10.4 Conclusions

In this article we have presented several different perspectives of the chiral magnetic effect. The unifying theme is that the effect arises due to the spin projection of the lowest-Landau-level projection that occurs in a very strong magnetic field, in conjunction with a chirality projection that relates to the axial anomaly, or to the effect of a topologically non-trivial background field such as an instanton. The effect could occur in an abelian theory with an electric field parallel to the strong magnetic field. In this case, the roles of the charge and chiral chemical potentials are played by A_0 and A_3, respectively, and the chiral magnetic effect is seen to be precisely equivalent to the 2d axial anomaly, which is itself the dimensionally reduced LLL projection of the 4d axial anomaly. Taking this dimensional reduction seriously, in a theory with a continuous chiral symmetry, we learn further from the 2d physics that there is a spiral condensate, an immediate consequence of the relativistic form of the Peierls instability. Finally, we considered the effect on light quarks of both a magnetic field and an instanton field, showing that the competition between the chiral projection in the instanton field and the spin projection in the magnetic field is responsible for the chiral magnetic effect. We demonstrated that this is consistent with the index theorem, and illustrated the mechanism with a soluble model by taking the leading derivative expansion form of the instanton field in which the magnetic length is much smaller than the instanton scale. It would be interesting to investigate the systematic corrections to this leading approximation using the Fock-Schwinger gauge [72–74] representation of the instanton field.

Acknowledgements   We thank T. Blum, D. Kharzeev and H-U. Yee for helpful discussions. This work was supported by the US Department of Energy under grants DE-FG02-92ER40716 (GD) and DE-AC02-98CH10886, DE-FG-88ER41723 (GB).

References

1. D.E. Kharzeev, Parity violation in hot QCD: why it can happen, and how to look for it. Phys. Lett. B **633**, 260 (2006). arXiv:hep-ph/0406125
2. D. Kharzeev, A. Zhitnitsky, Charge separation induced by P-odd bubbles in QCD matter. Nucl. Phys. A **797**, 67 (2007). arXiv:0706.1026 [hep-ph]
3. D.E. Kharzeev, L.D. McLerran, H.J. Warringa, The effects of topological charge change in heavy ion collisions: 'Event by event P and CP violation'. Nucl. Phys. A **803**, 227 (2008). arXiv:0711.0950 [hep-ph]
4. K. Fukushima, D.E. Kharzeev, H.J. Warringa, The chiral magnetic effect. Phys. Rev. D **78**, 074033 (2008). arXiv:0808.3382 [hep-ph]
5. D.E. Kharzeev, Topologically induced local P and CP violation in QCD × QED. Ann. Phys. **325**, 205 (2010). arXiv:0911.3715 [hep-ph]
6. H.J. Warringa, Dynamics of the chiral magnetic effect in a weak magnetic field. arXiv: 1205.5679 [hep-th]
7. R. Jackiw, Topological investigations of quantized gauge theories, in *Current Algebra and Anomalies*, ed. by S.B. Treiman, R. Jackiw, B. Zumino, E. Witten (Princeton University Press, Princeton, 1985)
8. M.A. Shifman, in *ITEP Lectures on Particle Physics and Field Theory, Vols. 1, 2*. World Sci. Lect. Notes Phys., vol. 62 (1999)
9. V.P. Gusynin, V.A. Miransky, I.A. Shovkovy, Dimensional reduction and catalysis of dynamical symmetry breaking by a magnetic field. Nucl. Phys. B **462**, 249 (1996). arXiv: hep-ph/9509320

10. G. Strang, *Linear Algebra and Its Applications* (Harcourt, San Diego, 1988)
11. Y. Aharonov, A. Casher, Ground state of a spin-1/2 charged particle in a two-dimensional magnetic field. Phys. Rev. A **19**, 2461 (1979)
12. S.P. Novikov, B.A. Dubrovin, Ground states of a two-dimensional electron in a periodic magnetic field. Zh. Èksp. Teor. Fiz. **79**, 1006 (1980). [Sov. Phys. JETP **52**, 511 (1980)]
13. S.P. Novikov, B.A. Dubrovin, Ground states in a periodic field. Magnetic Bloch functions and vector bundles. Dokl. Akad. Nauk SSSR **253**, 1293 (1980)
14. L.S. Brown, R.D. Carlitz, C. Lee, Massless excitations in instanton fields. Phys. Rev. D **16**, 417 (1977)
15. R.D. Carlitz, C. Lee, Physical processes in pseudoparticle fields: the role of fermionic zero modes. Phys. Rev. D **17**, 3238 (1978)
16. J. Hur, C. Lee, H. Min, Some chirality-related properties of the 4-D massive Dirac propagator and determinant in an arbitrary gauge field. Phys. Rev. D **82**, 085002 (2010). arXiv:1007.4616 [hep-th]
17. W.A. Bardeen, B. Zumino, Consistent and covariant anomalies in gauge and gravitational theories. Nucl. Phys. B **244**, 421 (1984)
18. G.V. Dunne, C.A. Trugenberger, Odd dimensional gauge theories and current algebra. Ann. Phys. **204**, 281 (1990)
19. E.V. Gorbar, V.A. Miransky, I.A. Shovkovy, Chiral asymmetry and axial anomaly in magnetized relativistic matter. Phys. Lett. B **695**, 354 (2011). arXiv:1009.1656 [hep-ph]
20. N. Sadooghi, A. Jafari Salim, Axial anomaly of QED in a strong magnetic field and noncommutative anomaly. Phys. Rev. D **74**, 085032 (2006). hep-th/0608112
21. J.H. Gao, Z.T. Liang, S. Pu, Q. Wang, X.N. Wang, Chiral anomaly and local polarization effect from quantum kinetic approach. Phys. Rev. Lett. **109**, 232301 (2012). arXiv:1203.0725 [hep-ph]
22. W. Heisenberg, H. Euler, Consequences of Dirac's theory of positrons. Z. Phys. **98**, 714 (1936)
23. J. Schwinger, On gauge invariance and vacuum polarization. Phys. Rev. **82**, 664 (1951)
24. Y. Kluger, E. Mottola, J.M. Eisenberg, The quantum Vlasov equation and its Markov limit. Phys. Rev. D **58**, 125015 (1998). hep-ph/9803372
25. J.S. Schwinger, Gauge invariance and mass. Phys. Rev. **125**, 397 (1962)
26. J.S. Schwinger, Gauge invariance and mass. 2. Phys. Rev. **128**, 2425 (1962)
27. G. Basar, G.V. Dunne, D.E. Kharzeev, Chiral magnetic spiral. Phys. Rev. Lett. **104**, 232301 (2010). arXiv:1003.3464 [hep-ph]
28. D.J. Gross, A. Neveu, Dynamical symmetry breaking in asymptotically free field theories. Phys. Rev. D **10**, 3235 (1974)
29. T. Kojo, Y. Hidaka, L. McLerran, R.D. Pisarski, Quarkyonic chiral spirals. Nucl. Phys. A **843**, 37–58 (2010). arXiv:0912.3800 [hep-ph]
30. R. Peierls, *The Quantum Theory of Solids* (Oxford University Press, Oxford, 1955)
31. R. Peierls, *More Surprises in Theoretical Physics* (Princeton University Press, Princeton, 1991)
32. G. Basar, G.V. Dunne, M. Thies, Inhomogeneous condensates in the thermodynamics of the chiral NJL$_2$ model. Phys. Rev. D **79**, 105012 (2009). arXiv:0903.1868 [hep-th]
33. G. Basar, G.V. Dunne, A twisted kink crystal in the chiral Gross-Neveu model. Phys. Rev. D **78**, 065022 (2008). arXiv:0806.2659 [hep-th]
34. V. Schon, M. Thies, Emergence of Skyrme crystal in Gross-Neveu and 't Hooft models at finite density. Phys. Rev. D **62**, 096002 (2000). arXiv:hep-th/0003195
35. V. Schon, M. Thies, 2D model field theories at finite temperature and density, in *At the Frontier of Particle Physics: Handbook of QCD*, vol. 3, ed. by M.A. Shifman (World Scientific, Singapore, 2000). arXiv:hep-th/0008175
36. M. Thies, From relativistic quantum fields to condensed matter and back again: updating the Gross-Neveu phase diagram. J. Phys. A **39**, 12707 (2006). hep-th/0601049
37. A. Bzdak, V. Koch, J. Liao, Remarks on possible local parity violation in heavy ion collisions. Phys. Rev. C **81**, 031901 (2010). arXiv:0912.5050 [nucl-th]

38. B.I. Abelev et al. (STAR Collaboration), Azimuthal charged-particle correlations and possible local strong parity violation. Phys. Rev. Lett. **103**, 251601 (2009). arXiv:0909.1739 [nucl-ex]
39. B.I. Abelev et al. (STAR Collaboration), Observation of charge-dependent azimuthal correlations and possible local strong parity violation in heavy ion collisions. Phys. Rev. C **81**, 054908 (2010). arXiv:0909.1717 [nucl-ex]
40. D. Kharzeev, R.D. Pisarski, M.H.G. Tytgat, Possibility of spontaneous parity violation in hot QCD. Phys. Rev. Lett. **81**, 512 (1998). arXiv:hep-ph/9804221
41. K.-Y. Kim, B. Sahoo, H.-U. Yee, Holographic chiral magnetic spiral. J. High Energy Phys. **1010**, 005 (2010). arXiv:1007.1985 [hep-th]
42. T. Sakai, S. Sugimoto, Low energy hadron physics in holographic QCD. Prog. Theor. Phys. **113**, 843 (2005). hep-th/0412141
43. T. Sakai, S. Sugimoto, More on a holographic dual of QCD. Prog. Theor. Phys. **114**, 1083 (2005). hep-th/0507073
44. G. Basar, G.V. Dunne, D.E. Kharzeev, Electric dipole moment induced by a QCD instanton in an external magnetic field. Phys. Rev. D **85**, 045026 (2012). arXiv:1112.0532 [hep-th]
45. V. Skokov, A.Y. Illarionov, V. Toneev, Estimate of the magnetic field strength in heavy-ion collisions. Int. J. Mod. Phys. A **24**, 5925 (2009). arXiv:0907.1396 [nucl-th]
46. A. Bzdak, V. Skokov, Event-by-event fluctuations of magnetic and electric fields in heavy ion collisions. Phys. Lett. B **710**, 171 (2012). arXiv:1111.1949 [hep-ph]
47. P.V. Buividovich, M.N. Chernodub, E.V. Luschevskaya, M.I. Polikarpov, Numerical study of chiral symmetry breaking in non-Abelian gauge theory with background magnetic field. Phys. Lett. B **682**, 484–489 (2010). arXiv:0812.1740 [hep-lat]
48. P.V. Buividovich, M.N. Chernodub, E.V. Luschevskaya, M.I. Polikarpov, Chiral magnetization of non-Abelian vacuum: a lattice study. Nucl. Phys. B **826**, 313–327 (2010). arXiv:0906.0488 [hep-lat]
49. P.V. Buividovich, M.N. Chernodub, E.V. Luschevskaya, M.I. Polikarpov, Numerical evidence of chiral magnetic effect in lattice gauge theory. Phys. Rev. D **80**, 054503 (2009). arXiv:0907.0494 [hep-lat]
50. P.V. Buividovich, M.N. Chernodub, E.V. Luschevskaya, M.I. Polikarpov, Quark electric dipole moment induced by magnetic field. Phys. Rev. D **81**, 036007 (2010). arXiv:0909.2350 [hep-ph]
51. M. Abramczyk, T. Blum, G. Petropoulos, R. Zhou, Chiral magnetic effect in $2 + 1$ flavor QCD + QED. PoS **LAT2009**, 181 (2009). arXiv:0911.1348 [hep-lat]
52. T. Blum, Talk at Workshop on P- and CP-odd Effects in Hot and Dense Matter, Brookhaven National Laboratory, April 2010
53. V.V. Braguta, P.V. Buividovich, T. Kalaydzhyan, S.V. Kuznetsov, M.I. Polikarpov, The chiral magnetic effect and chiral symmetry breaking in SU(3) quenched lattice gauge theory. PoS **LAT2010**, 190 (2010). arXiv:1011.3795 [hep-lat]
54. B.C. Tiburzi, Lattice QCD with classical and quantum electrodynamics. PoS **LAT2011**, 020 (2011). arXiv:1110.6842 [hep-lat]
55. L. Giusti, A. Gonzalez-Arroyo, C. Hoelbling, H. Neuberger, C. Rebbi, Fermions on tori in uniform Abelian fields. Phys. Rev. D **65**, 074506 (2002). arXiv:hep-lat/0112017
56. Y. Tenjinbayashi, H. Igarashi, T. Fujiwara, Dirac operator zero-modes on a torus. Ann. Phys. **322**, 460 (2007). arXiv:hep-th/0506259
57. M.H. Al-Hashimi, U.J. Wiese, Discrete accidental symmetry for a particle in a constant magnetic field on a torus. Ann. Phys. **324**, 343 (2009). arXiv:0807.0630 [quant-ph]
58. J. Zak, Magnetic translation group. Phys. Rev. **134**, A1602 (1964)
59. G. 't Hooft, Computation of the quantum effects due to a four-dimensional pseudoparticle. Phys. Rev. D **14**, 3432 (1976)
60. A.S. Schwarz, On regular solutions of Euclidean Yang-Mills equations. Phys. Lett. B **67**, 172 (1977)
61. J.E. Kiskis, Fermions in a pseudoparticle field. Phys. Rev. D **15**, 2329 (1977)
62. R. Jackiw, C. Rebbi, Spinor analysis of Yang-Mills theory. Phys. Rev. D **16**, 1052 (1977)

63. V.A. Rubakov, *Classical Theory of Gauge Fields* (Princeton University Press, Princeton, 2002)
64. R. Jackiw, C. Rebbi, Conformal properties of a Yang-Mills pseudoparticle. Phys. Rev. D **14**, 517 (1976)
65. S. Chadha, A. D'Adda, P. Di Vecchia, F. Nicodemi, Fermions in the background pseudoparticle field in an O(5) formulation. Phys. Lett. B **67**, 103 (1977)
66. M. Atiyah, V. Patodi, I. Singer, Spectral asymmetry and Riemannian geometry. Math. Proc. Camb. Philos. Soc. **77**, 43 (1975)
67. A.A. Belavin, A.M. Polyakov, A.S. Schwartz, Yu.S. Tyupkin, Pseudoparticle solutions of the Yang-Mills equations. Phys. Lett. B **59**, 85 (1975)
68. G. 't Hooft, Some twisted selfdual solutions for the Yang-Mills equations on a hypertorus. Commun. Math. Phys. **81**, 267 (1981)
69. P. van Baal, Some results for SU(N) gauge fields on the hypertorus. Commun. Math. Phys. **85**, 529 (1982)
70. P. van Baal, SU(N) Yang-Mills solutions with constant field strength on T^4. Commun. Math. Phys. **94**, 397 (1984)
71. P. van Baal, Instanton moduli for $T^3 \times \mathbb{R}$. Nucl. Phys. B, Proc. Suppl. **49**, 238 (1996). arXiv:hep-th/9512223
72. M.A. Shifman, Wilson loop in vacuum fields. Nucl. Phys. B **173**, 13 (1980)
73. M.S. Dubovikov, A.V. Smilga, Analytical properties of the quark polarization operator in an external selfdual field. Nucl. Phys. B **185**, 109–132 (1981)
74. B.L. Ioffe, A.V. Smilga, Nucleon magnetic moments and magnetic properties of vacuum in QCD. Nucl. Phys. B **232**, 109 (1984)

Chapter 11
Chiral Magnetic Effect in Hydrodynamic Approximation

Valentin I. Zakharov

11.1 Introduction

In this chapter[1] we will consider chiral liquids, that is liquids whose constituents are massless fermions. The motivation is an offspring from the discovery of the strongly interacting quark-gluon plasma (for review see, e.g., [1, 2]) which is, indeed, build on (nearly) massless quarks. The use of the (relativistic) hydrodynamic approximation is also suggested by the observations on the quark-gluon plasma. Moreover, the state of the chiral liquid is assumed to be asymmetric with respect to left- and right-fermions. In other words, we concentrate on the case of a non-vanishing chiral chemical potential μ_5.[2] The motivation to introduce $\mu_5 \neq 0$ is rather theoretical than experimental, however, and is rooted in the sphaleron-based picture which predicts that, event-by-event, the plasma is produced as chirally charged [3–5].

There are a few effects specific for the chiral liquids, the most famous one being the chiral magnetic effect (for review and further references see [6]). By the ChME

---

[1] The review is prepared for a volume of the Springer Lecture Notes in Physics "Strongly interacting matter in magnetic fields" edited by D. Kharzeev, K. Landsteiner, A. Schmitt, H.-U. Yee.

[2] In the realistic QCD case the singlet axial current is anomalous and is not conserved. Therefore, introduction of the chemical potential μ_5 is rather a subtle issue. In the bulk of the text we ignore this problem concentrating mostly on academic issues. One could have in mind, for example, that the chemical potential $\mu_5 \neq 0$ is associated in fact with the axial current with isospin $\Delta I = 1$ which is conserved in the limit of vanishing quark masses. Another possible line of reasoning is to invoke large-N_c limit of Yang-Mills theories. The contribution of the gluon anomaly is then suppressed by large N_c and the chemical potential μ_5 can be introduced consistently for the singlet current as well.

V.I. Zakharov (✉)
Institute of Theoretical and Experimental Physics, B. Cheremushkinskaya, 25, Moscow, Russia
e-mail: vzakharov@itep.ru

V.I. Zakharov
Max-Planck Institut fuer Physik, Werner-Heisenberg Institut, Foehringer Ring, 6, Munich, Germany

D. Kharzeev et al. (eds.), *Strongly Interacting Matter in Magnetic Fields*,
Lecture Notes in Physics 871, DOI 10.1007/978-3-642-37305-3_11,
© Springer-Verlag Berlin Heidelberg 2013

one understands the phenomenon of induction of electromagnetic current $\mathbf{j}_{el}$ by applying an external magnetic field $\mathbf{B}$ to a chiral medium with a non-vanishing μ_5:

$$\mathbf{j}_{el} = \frac{q^2 \mu_5}{2\pi^2} \mathbf{B},\tag{11.1}$$

where q is the electric charge of the constituents and $\mu_5 = (\mu_R - \mu_L)/2$ is the chiral chemical potential. The relation (11.1) plays a central role in our discussion and can be analyzed from various points of view. An exciting possibility is that the chiral magnetic effect (11.1) has already been observed in heavy-ion collisions, for a concise review and references see [7]. We will concentrate, however, on the underlying theory rather than on its experimental verification.

Qualitatively, (11.1) can be understood accounting only for the interaction of spin of quarks with an external magnetic field, $H_s \sim q(\sigma \cdot \mathbf{B})$. The overall coefficient in (11.1) can readily be found in case of free quarks [6, 8, 9]. The evaluation of the coefficient requires explicit counting of zero modes of the Dirac equation for chiral fermions interacting with an external magnetic field [9], see also Sect. 11.3.4. The number of chiral zero modes is controlled by the famous chiral anomaly [10, 11]. Thus, (11.1) is a manifestation of the chiral anomaly, as can actually be demonstrated in a number of ways, discussed later.

One of the central points is that the chiral magnetic effect can be derived not only for non-interacting chiral fermions but also in case of strong interactions between the constituents, provided that the hydrodynamic approximation is granted. There is, though, a change in (11.1) which is of pure kinematic nature. Namely, to describe liquid one introduces 4-velocity of an element of the liquid, $u_\mu(x)$ which is a function of the point x. The 4-velocity is normalized such that $-(u_0)^2 + u_i^2 = -1$, and in the non-relativistic limit $u_0 \approx 1$, $u_i \approx v_i$, where $i = 1, 2, 3$ and v_i is the 3-velocity entering the hydrodynamic equations in the non-relativistic limit. Equation (11.1) is valid now only if the whole of the liquid is at rest. If, on the other hand, u_μ is non-trivial (11.1) is generalized to

$$(j_\mu)_{el} = \frac{q^2 \mu_5}{2\pi^2} B_\mu,\tag{11.2}$$

where $B_\mu \equiv 1/2 \epsilon_{\mu\nu\rho\sigma} u_\nu (\partial_\rho A_\sigma - \partial_\sigma A_\rho)$, A_μ is the gauge potential of the external electromagnetic field and the chemical potential μ_5 can depend on the point x. In the rest frame of an element of the liquid $u_\mu \equiv (1, 0, 0, 0)$, and (11.2) coincides with (11.1).

Non-renormalization theorems in field theory are quite exceptional and attract a lot of attention. Essentially, there are two best known case studies of non-renormalizability. First, conserved charges are not renormalized, so that, for example, the absolute values of electric charges of electron and proton are the same. And the second example is the non-renormalizability of the chiral anomaly:

$$\partial_\mu j_\mu^5 = \frac{\alpha_{el}}{4\pi} \epsilon_{\alpha\beta\gamma\delta} F^{\alpha\beta} F^{\gamma\delta}.\tag{11.3}$$

The absence of higher-order corrections to this anomaly is guaranteed by the Adler-Bardeen theorem [12]. The proof of the non-renormalizability of the chiral magnetic effect utilizes both symmetry considerations and the miracle of the perturbative cancellations, revealed by the Adler-Bardeen theorem.

To put the consideration of the chiral magnetic effect into a field-theoretic framework one considers hydrodynamics as a kind of an effective field theory, see, e.g., [13, 14]. Viewed as an effective field theory, hydrodynamics reduces to an expansion in a number of derivatives from the velocity u_μ and thermodynamic quantities. On the microscopic level, hydrodynamics corresponds to the long-wave approximation,

$$l/a \gg 1,$$

where l is of order of wave length of the hydrodynamic excitations and a is of order of distance between the constituents.

The hydrodynamic equations reflect symmetries of a dynamical problem considered since they are nothing else but the conservation laws. In the absence of external fields

$$\partial_\mu T^{\mu\nu} = 0, \qquad \partial_\mu j_a^\mu = 0 \qquad (11.4)$$

where $T_{\mu\nu}$ is the energy-momentum tensor and j_a^μ are currents conserved in strong interactions, with the index a enumerating the currents. Consider liquid at rest and small fluctuations superimposed on it. Generically, fluctuations would be damped down on distances of order a and do not propagate far away. The exceptions are fluctuations of conserved quantities which cannot disappear and propagate far off. That is why the long-wave, or hydrodynamic approximation reduces to the conservation laws (11.4).

Explicit form of the hydrodynamic equations (11.4) depends on how many terms in the gradient expansion are kept in $T^{\mu\nu}$ and j_a^μ. In general,

$$T^{\mu\nu} = wu^\mu u^\nu + Pg^{\mu\nu} + \tau^{\mu\nu}, \qquad (11.5)$$

$$j_a^\mu = n_a u^\mu + v_a^\mu, \qquad (11.6)$$

where w, P, n_a are the standard thermodynamical variables, namely, enthalpy, $w \equiv \epsilon + P$, pressure and densities of charges. The quantities $\tau^{\mu\nu}$, v_a^μ satisfy conditions $u_\mu \tau^{\mu\nu} = u_\mu v_a^\mu = 0$. In the zeroth order in gradients $\tau^{\mu\nu} = 0$, $v_a^\mu = 0$.

The path from relativistic, or chiral hydrodynamics to the anomaly (11.3) was found first in Ref. [15]. One introduces external electric and magnetic fields so that the current conservation condition is changed into (11.3). The bridge between the fundamental-theory equation (11.3) and hydrodynamics is provided by considering the entropy current s_μ. In the presence of external fields the standard definition [16] of the current s_μ does not ensure growth of the entropy any longer. To avoid the contradiction with the second law of thermodynamics one includes terms proportional to external fields both into the newly defined entropy current s_μ and currents j_a^μ. The constraints imposed by the second law of thermodynamics involve the anomaly condition which is not renormalized by strong interactions and turn out to be strong enough to (almost uniquely) fix the currents in terms of the anomaly. We reproduce the basic point of this beautiful derivation in Sect. 11.2.1.

Most recently, it was observed [17–19] (see also [20]) that one can avoid considering the entropy current s_μ. Instead, one introduces not only external electromagnetic field but a static gravitational background as well. Equating the hydrodynamic stress tensor and currents (11.5) to the corresponding structures evaluated at the equilibrium in the gravitational background allows to fix the chiral current and stress tensor without considering the entropy current. This seems to be a very interesting extension of relativistic hydrodynamics. From the perspectives of the present review, derivation [17–19] reveals a novel feature of the chiral magnetic effect. Namely, the corresponding electromagnetic current appears to be non-dissipative since it persists in the equilibrium. We will come back to discuss this point later.

All these derivations of the ChME in fact uncover existence of some other effects as well. In particular, one predicts existence of the chiral vortical effect (ChVE), that is, flow of the axial current in the direction of the local angular velocity, $j^5 \sim \omega$. In relativistic covariant notations:

$$\delta j^5_\mu \approx \frac{\mu^2}{2\pi^2} \omega_\mu,$$
$$\omega_\mu = \frac{1}{2} \epsilon_{\mu\nu\alpha\beta} u^\nu \partial^\alpha u^\beta, \tag{11.7}$$

where ω_μ is the vorticity of the liquid and the chemical potential μ is considered to be small. The ChVE was derived first in a holographic set up [21] and is being actively discussed in the literature, along with the chiral magnetic effect. Another example is the axial-vector current $\mathbf{j}^5$ induced by a non-vanishing chemical potential μ_V [22, 23]:

$$\mathbf{j}^5 = \frac{q\mu_V}{2\pi^2} \mathbf{B}, \tag{11.8}$$

which is a kind of parity-reflected companion of (11.1).

It is worth emphasizing that all the chiral effects now considered were originally introduced quite long time ago basing on evaluation of loop graphs with non-interacting fermions [8, 24]. In particular, it was found in Ref. [24] that rotating a system of non-interacting massless fermions results in a vortical current:

$$\mathbf{j}^5 = \left(\frac{T^2}{12} + \frac{\mu^2}{4\pi^2} \right) \Omega, \tag{11.9}$$

where Ω is the angular velocity of the rotation. In the term proportional to μ^2 one readily recognizes (11.7) above.[3] It took many years, however, to prove that the result (11.9) is essentially not modified by strong interactions. As is mentioned above, the origin of the μ^2 in (11.9) term can be traced back to the chiral anomaly and it is not renormalizable.

---

[3] To compare (11.9) and (11.7) one should keep in mind that in the notations of Ref. [24] the current $\mathbf{j}$ in (11.9) is the current of right-handed fermions alone and, thus, constitutes one half of the chiral current entering (11.7).

The status of radiative corrections to the T^2 term in (11.9) has been clarified only very recently [25, 26]. First, one relates the chiral vortical effect to the static correlator of the axial current and momentum density. As far as only the fermionic part is kept in the operator of momentum density, all the higher-order contributions to the T^2 term cancel and (11.9) remains valid. The proof is based on an analysis of Feynman graphs in the effective 3d theories and echoes the proof [27] of non-renormalizability of the topological mass of a gauge field. There is, however, a gluonic part of the momentum density and it generates a calculable two-loop correction to (11.9). We reproduce the basic points of the proof [25] in Sect. 11.2.3.

Reference to anomalies of the fundamental theory which arise due to weak coupling to external fields (electromagnetic or gravitational) can be avoided by applying an effective field theory. This effective field theory elevates chemical potentials to interaction constants, see [28–30] and references therein. The corresponding vertices can be obtained from the standard electromagnetic interaction by substitution

$$q A_\mu \to \mu u_\mu, \tag{11.10}$$

where μ is the chemical potential associated with the conserved charge q. The effective theory is anomalous and does reproduce through these anomalies the chiral magnetic effect and the μ^2 term in the chiral vortical effect, see (11.9).[4] We will give further details in Sect. 11.2.4.

Non-renormalizability commonly implies topological nature of the corresponding term. Moreover, if the currents are topological they are non-dissipative. The best known example of such a type is provided by the integer quantum Hall effect (for the background and review see, e.g., [31]). Consider a two-dimensional system with an external electric field $\mathbf{E} = (E_1, 0)$ applied. Then there arises electric current j_i such that

$$j_i = \sigma_{ik} E_k,$$

where σ_{ik} are coefficients. The integer quantum Hall effect is characterized by a non-diagonal $\sigma_{12} \neq 0$ and $\sigma_{11} = 0$:

$$\sigma_{12} = \nu \frac{e^2}{h}, \tag{11.11}$$

where ν is integer. The work produced by the external electric field is equal to the product $W = j_i E_i$. The Hall current is clearly not associated with any work done and this observation suggests strongly that the Hall current is dissipation-less. For further examples of this type see, e.g., [32].

Since the chiral magnetic current (11.1) is not associated either with any work done by the external field it seems natural to assume that the chiral magnetic current is also dissipation-less. This suggestion is made first in Ref. [33] basing on somewhat different arguments, see Sect. 11.3.1.

---

[4]Note that in the underlying fundamental field theory there are no anomalies associated with a non-vanishing chemical potential μ. This observation is in no contradiction with the fact that such anomalies do arise in the language of the effective theory.

Now we come to a question, however, which has not been answered yet. Namely, topological, or dissipation-less currents usually manifest existence of a macroscopic quantum state. Well known examples are the superfluidity of weakly interacting Bose-liquid or the same Hall current (11.11). In these two cases the nature of the macroscopic quantum states is well understood. Also, in case of non-interacting fermions topological nature of the chiral magnetic effect has been demonstrated first long time ago [9]. Claiming the chiral magnetic current (11.1) to be topological in hydrodynamic approximation as well we imply quantum nature of the corresponding ground state.

This problem of constructing explicitly the quantum state can be addressed in some more detail within approach which starts with a microscopical picture and the central role is played then by low-dimensional defects. A well known example of this kind is provided by vortices in rotating superfluid. In more abstract language, this approach goes back also to papers [9, 34, 35]. It was demonstrated that defects in field theory are closely tied to the realization of anomaly. In particular, it was shown in [35] that anomaly in $2n+2$ dimensional theory is connected with $2n$ dimensional index density and can be understood in terms of fermion zero modes on strings and domain walls. In all the cases the chiral current is carried by fermionic zero modes living on the defects.

One can expect, therefore, that anomaly in effective, hydrodynamic theory is realized in chiral superfluid system on vortex-like defects. The continuum-medium results (11.7), (11.1) can arise then upon averaging over a large number of defects. In case of the chiral magnetic effect such a mechanism was considered, in particular, in Refs. [3–5, 23] and the final result (11.1) is reproduced on the microscopic level as well. The vortices considered in [3–5, 23] are simply the regions of space free of the medium substance. In case of superfluidity the vortices are better understood and the microscopical picture for the chiral effects can be clarified to some extent [36]. The outcome of the analysis in terms of defects, or vortices is that the chiral magnetic effect does survive without any change. As for the vortical chiral effect it is modified in the capillary picture by a factor of two:

$$\left(\delta J_\mu^5\right)_{capillary} = 2\frac{\mu^2}{2\pi^2}\omega_\mu. \tag{11.12}$$

We will give details in Sect. 11.3.4.

Upon introducing the reader to the topics to be discussed in this review, we would like to emphasize that there are many other interesting results which could have been included into the review but are actually not covered. The reason is mostly to avoid too much overlap with other chapters of this volume. A notable example of this kind is the holographic approach to the ChME which is reviewed, in particular, in Ref. [37]. The same remark applies to the phenomenological manifestations of the gravitational anomaly. Finally, there are very interesting applications of the technique used to condensed-matter systems. However, reviewing these applications goes beyond the scope of the present notes.

To summarize, we concentrate on two basic issues, non-renormalizability and dissipation-free nature of the chiral magnetic and chiral vortical effects. In Sect. 11.2

we consider non-renormalization theorems within various approaches outlined above (thermodynamic, geometric, diagrammatic, effective field theories). The derivations of the theorems make it also clear that the chiral effects considered are dissipation free. In Sect. 11.3 we review further arguments in favor of the dissipation-free nature of the chiral effects. In this section we also introduce a microscopic picture in terms of defects of lower dimensions. Sect. 11.4 is conclusions.

11.2 Non-renormalization Theorems

11.2.1 Non-renormalization Theorems in Thermodynamic Approach

In this subsection we reproduce the basic steps of the pioneering derivation [15] of chiral effects, the chiral vortical effect first of all, which utilizes only the chiral anomaly in external electromagnetic fields, thermodynamics and hydrodynamic approximation. To simplify the algebra, we consider first a single conserved current, chiral at that. Moreover, we consider the chiral symmetry not spontaneously broken (otherwise, we should have modified the hydrodynamic equations).

In presence of external electromagnetic fields both the energy-momentum tensor and chiral current are not conserved any longer. The current is not conserved because of the anomaly, while the energy is not conserved because external electric field executes work on the system. Thus, one starts with the equations:

$$\partial_\mu j^\mu = C E^\mu B_\mu, \tag{11.13}$$

$$\partial_\mu T^{\mu\nu} = F^{\nu\lambda} j_\lambda \tag{11.14}$$

where C is the coefficient determined by the anomaly (e.g., for QED $C = \frac{1}{2\pi^2}$). Turning to the thermodynamics, we have to introduce, following textbooks [16], an entropy current s_μ consistent with the second law of thermodynamics. In the ideal-liquid approximation the condition is $\partial_\mu s^\mu = 0$.

There are no general rules to construct s_μ. As a first guess, one can try $s_\mu = s u_\mu$ where s is the entropy density. Moreover, put the gradient terms $v^\mu, \tau^{\nu\mu} = 0$ for simplicity. However, using (11.13) and

$$dP = sdT + nd\mu$$

where μ is chemical potential, one readily derives

$$\partial_\mu(su^\mu) = -C\frac{\mu}{T}E \cdot B \quad (v^\mu, \tau^{\nu\mu} = 0). \tag{11.15}$$

The right-hand side of this equation does not have a definite sign and, therefore, one cannot accept su^μ as a definition of the entropy current in presence of the anomaly. Thus, we should continue with our guess-work to construct the entropy current. Note that it is quite a common situation. For example, consider non-ideal liquid with non-zero v^μ (and $\tau^{\mu\nu} = 0$). Then one has to modify the entropy current defining it as $s^\mu = s u^\mu - \frac{\mu}{T} v^\mu$ so that for the newly defined entropy current $\partial_\mu s^\mu = 0$ [16].

In presence of the chiral anomaly we can use the same idea and redefine the entropy current [15] by introducing terms proportional to the magnetic field and vorticity. To simplify equations we will not account for the dissipative terms, viscosities and electrical conductance. One can check that inclusion of these terms does not change the result [15]. Moreover, in the next subsection we will see that there are general reasons for the dissipative terms to be actually not relevant. Thus, expanding in the fields we look for solution for the matter current of the form:

$$j_\mu = n u_\mu + v_\mu$$
$$v_\mu = \xi_\omega \omega_\mu + \xi_B B_\mu, \tag{11.16}$$

where $\omega_\mu = \frac{1}{2} \epsilon_{\mu\nu\alpha\beta} u^\nu \partial^\alpha u^\beta$ is the vorticity, $B_\mu = \frac{1}{2} \epsilon_{\mu\nu\alpha\beta} u^\nu F^{\alpha\beta}$ is the magnetic field in the rest frame of liquid element (electric field $E_\mu = F_{\mu\nu} u^\nu$) and ξ_ω, ξ_B are unknown functions of the thermodynamic variables. For the entropy current, we assume:

$$s_\mu = s u_\mu - \frac{\mu}{T} v_\mu + D_\omega \omega_\mu + D_B B_\mu, \tag{11.17}$$

where D_ω, D_B are further unknown functions.

Conservation of the entropy current now reads:

$$\partial_\mu \left(D_\omega \omega^\mu \right) + \partial_\mu \left(D_B B^\mu \right) - v^\mu \left(\partial_\mu \frac{\mu}{T} - \frac{\mu}{T} \right) - C \frac{\mu}{T} E \cdot B = 0. \tag{11.18}$$

For the ideal liquid ($\tau^{\mu\nu} = 0$) the following identities hold:

$$\partial_\mu \omega^\mu = -\frac{2}{w} \omega^\mu (\partial_\mu P - n E_\mu), \tag{11.19}$$

$$\partial_\mu B^\mu = -2\omega \cdot E + \frac{1}{w}(-B \cdot \partial P + n E \cdot B). \tag{11.20}$$

Moreover, coefficients in front of the independent kinematical structures ω^μ, B^μ, $E \cdot \omega$, $E \cdot B$ in relation (11.18) should vanish:

$$\partial_\mu D_\omega - 2 \frac{\partial_\mu P}{\epsilon + p} D_\omega - \xi_\omega \partial_\mu \frac{\mu}{T} = 0,$$

$$\partial_\mu D_B - \frac{\partial_\mu P}{\epsilon + p} D_B - \xi_B \partial_\mu \frac{\mu}{T} = 0,$$

$$\frac{2n D_\omega}{\epsilon + p} - 2 D_B + \frac{\xi_\omega}{T} = 0, \tag{11.21}$$

$$\frac{n D_B}{\epsilon + p} + \frac{\xi_B}{T} - C \frac{\mu}{T} = 0.$$

To proceed further one has to choose a basis of thermodynamic variables and it is convenient to take this basis as $(P, \tilde{\mu} = \frac{\mu}{T})$. The thermodynamic derivatives in the basis look as $(\frac{\partial T}{\partial P})_{\tilde{\mu}} = \frac{T}{w}$, $(\frac{\partial T}{\partial \tilde{\mu}})_{\bar{P}} = -\frac{nT^2}{w}$. Since the thermodynamic gradients $\partial_\mu P$, $\partial_\mu \tilde{\mu}$ are independent the first two equations in (11.21) imply four conditions:

$$-\xi_\omega + \frac{\partial D_\omega}{\partial \tilde{\mu}} = 0, \qquad -\xi_B + \frac{\partial D_B}{\partial \tilde{\mu}} = 0,$$
$$\frac{\partial D_\omega}{\partial P} - \frac{2}{w} D_\omega = 0, \qquad \frac{\partial D_B}{\partial P} - \frac{1}{w} D_B = 0,$$

(11.22)

and the general solution for D_ω, D_B looks as

$$D_\omega = T^2 d_\omega(\tilde{\mu}), \qquad D_B = T d_B(\tilde{\mu}),$$

with functions d_ω, d_B being so far arbitrary. Then the last two equations in (11.21) reduce to simple differential equations which can be readily solved. As a result, the functions $d_\omega(\tilde{\mu})$, $d_B(\tilde{\mu})$ get fixed, up to the integration constants [29, 38] which are the values of the functions at $\tilde{\mu} = 0$.

For the chiral kinetic coefficients we finally obtain:

$$\xi_\omega = C\mu^2 \left(1 - \frac{2}{3}\frac{\mu \cdot n}{\epsilon + p}\right) + C_\omega T^2 \left(1 - \frac{\mu \cdot n}{\epsilon + p}\right)$$

(11.23)

and

$$\xi_B = C\left(\mu - \frac{1}{2}\frac{\mu^2 n}{\epsilon + p}\right),$$

(11.24)

where the constant C determines the anomaly and is fixed while the constant C_ω remains undetermined. We will come back to evaluate C_ω in Sect. 11.2.3 following the paper [25]. Note that we omitted a similar constant of integration from (11.24). The reason is that such a constant can appear in fact only due to parity-violating interactions [38] and, having in mind eventual applications to parity-conserving theories, we suppressed it right away.

The non-vanishing ξ_ω, ξ_B exhibit what we call chiral vortical and chiral magnetic effects, respectively. At temperature $T = 0$ the functions $\xi_\omega(\mu)$, $\xi_B(\mu)$ are fixed in terms of the coefficient C which can be read off from the chiral anomaly and is not renormalized by strong interactions. This is the content of the non-renormalization theorem of the chiral effects in hydrodynamic approximation. It is worth emphasizing that if parity is conserved the chiral magnetic and vortical effects are manifested in fact in different currents, axial and vector, respectively. On the other hand, according to (11.17), they appear in one and the same current. The reason is that through postulating conservation of a single chiral current we actually admitted for a parity violating "strong interaction". The case of a few currents, which allows for parity-conserving strong interactions was considered, in particular in Refs. [29, 38]. The outcome of the calculation is essentially the same: the chiral effects are fixed in all the currents, up to constants of integration.

11.2.2 Non-renormalization Theorems in Geometric Approach

Let us recall the reader one of simplest derivations of the chiral magnetic effect [3–5, 9]. The anomaly,

$$\partial^\mu j_\mu^5 = \frac{q^2}{2\pi^2} \mathbf{E} \cdot \mathbf{B} \tag{11.25}$$

can be rewritten as an equation for the production rate of chiral particles. Denoting the total chirality as N_5, where $N_5 \equiv N_R - N_L$ and $N_R(N_L)$ is the number of right-(left-)handed particles we have:

$$\frac{dN_5}{dt d^3 x} = \frac{q^2}{2\pi^2} \mathbf{E} \cdot \mathbf{B}. \tag{11.26}$$

Production of particles requires for energy to be deposit into the system. The source of this energy is the work done by the external electric field. Therefore:

$$\int d^3 x \mathbf{j}_{el} \cdot \mathbf{E} = \mu_5 \frac{dN_5}{dt} = \frac{q^2 \mu_5}{2\pi^2} \int d^3 x \mathbf{B} \cdot \mathbf{E} \tag{11.27}$$

where $\mathbf{j}_{el}$ is the electric current and μ_5 is the energy needed to produce a particle. Tending $\mathbf{E} \to 0$ we learn from (11.27) that there survives a non-vanishing current in this limit: $\mathbf{j}_{el} = (q^2 \mu_5 / 2\pi^2)\mathbf{B}$, and we come back to (11.1). As is mentioned above, the current is non-dissipative since magnetic field does not produce any work. One can also say that the current $\mathbf{j}_{el}$ exists in the equilibrium.

To summarize, one can calculate the non-dissipative current associated with the magnetic field by introducing electric field, taking the system out of the equilibrium in this way and then tending the electric field back to zero. A similar technique is commonly applied to study spontaneous symmetry breaking.

Recently it has been realized [17–19] that the procedure can be generalized in a rather unexpected way. Namely, one introduces not only external electromagnetic field but static gravitational field as well and studies equilibrium in this background. All the terms in the chiral currents and energy-momentum tensor are fixed in equilibrium and non-dissipative. Eventually one can go back to the flat space.

The basic object in the approach [17–19] is the generating functional W as a function of external electromagnetic and gravitational fields, or sources,

$$W = \int d^d x L(sources(x)),$$

where $W = \ln Z$ and Z is the partition function. Differentiating W with respect to the sources one evaluates in the standard way the energy-momentum tensor and currents, as well as their correlators in equilibrium. In particular,

$$\langle T_{\mu\nu} \rangle = \frac{2}{\sqrt{-g}} \frac{\delta W}{\delta g_{\mu\nu}}, \qquad \langle j_\mu \rangle = \frac{1}{\sqrt{-g}} \frac{\delta W}{\delta A_\mu}. \tag{11.28}$$

In the spirit of the hydrodynamic approximation, one expands W in the number of derivatives, both from the sources and thermodynamic variables and reiterates the procedure in each order in the expansion.

The medium is characterized by the time-like vector u^μ which in the zeroth order in the number of derivatives can be chosen as $u^\mu \sim (1, 0, 0, 0)$. Apart from u^μ the generating functional can depend on observables that are local in space but non-local in Euclidean time. In the zeroth order in derivatives, the invariant length L of the time circle is one of such observables. Also, there are Polyakov loops P_A of any U(1) gauge fields. Therefore, the temperature T and chemical potential μ are defined geometrically as

$$T = 1/L, \qquad \mu = \ln P_A/L. \tag{11.29}$$

These are simplest examples of diffeomorphic and gauge invariant scalars.

We pause here to emphasize that the outcome of the calculation are static correlators which are the same in the Euclidean and Minkowskian versions of the theory. Therefore the relations obtained are in fact thermodynamic in nature. Static correlators are to be distinguished from correlators which determine, through the Kubo formula, such transport coefficients as viscosity. In the latter case one considers correlator of certain components of the stress tensor at momentum transfer $\mathbf{q} \equiv 0$ and frequency $\omega \to 0$. On the other hand, correlators considered here correspond to $\omega \equiv 0$, $\mathbf{q} \to 0$. This, subtle at first sight, difference is crucial for continuation to the Euclidean space. It is straightforward to realize that the chiral magnetic effect for a time-independent magnetic field is indeed determined by a static correlator of components of the electromagnetic current, see, e.g., [39, 40]. To demonstrate this, it is convenient to begin with the standard Kubo relation for electric conductance:

$$\sigma_E = \lim_{\omega \to 0} \frac{i}{\omega} \langle j_i, j_i \rangle \big|_{\mathbf{q}=0} \tag{11.30}$$

where σ_E determines the electric current in terms of a time-independent electric field, $\mathbf{j}_{el} = \sigma_E \mathbf{E}$, and $\langle (j_{el})_i, (j_{el})_i \rangle$ is the retarded correlator of the components of the electromagnetic current (with no summation over the index i). Since both electric field $\mathbf{E}$ and magnetic field $\mathbf{B}$ are related to the same vector-potential $\mathbf{A}$ ($\mathbf{E} = -i\omega\mathbf{A}$, $\mathbf{B} = i\mathbf{q} \times \mathbf{A}$) one concludes:

$$\sigma_B = \lim_{q_n \to 0} \sum_{ij} \epsilon_{ijn} \frac{i}{2q_n} \langle (j_{el})_i, (j_{el})_j \rangle \big|_{\omega=0}, \tag{11.31}$$

where σ_B is defined as $\mathbf{j}_{el} = \sigma_B \mathbf{B}$. Thus, it is indeed a static correlator which we need to evaluate the ChME.

Probably, the best known example of the use of static correlators is the generation of photon screening mass through the Higgs mechanism. Namely, the correlator of components of electromagnetic current in superconducting case looks as:

$$\lim_{\mathbf{q} \to 0} \int d^3x \exp(i\mathbf{q} \cdot \mathbf{r}) \langle j_i(\mathbf{x}), j_k(0) \rangle \sim \left(\delta_{ik} - \mathbf{q}_i \mathbf{q}_k/\mathbf{q}^2 \right),$$

where presence of a pole signals superconductivity while the local term proportional to δ_{ik} signifies a non-vanishing photon mass. A similar role of a signature of superfluidity is played by a pole in the static correlator of components of momentum density:

$$\lim_{q \to 0} \int d^3x \exp(i\mathbf{q} \cdot \mathbf{r}) \langle T_{0i}(\mathbf{x}), T_{0k}(0) \rangle \sim \mathbf{q}_i \mathbf{q}_k / \mathbf{q}^2. \tag{11.32}$$

Note, however, absence of the local term proportional to δ_{ik}. This result is readily understood if we start from considering non-trivial gravitational background. The local terms are associated then with covariant derivatives, say, $D_i v_k = \partial_i v_k + \Gamma^l_{ik} v_l$, where v_i is a vector and Γ^l_{ik} are the Christoffel symbols. The Γ^l_{ik} symbols contain only derivatives from the components of the metric tensor $g_{\mu\nu}$ and we immediately conclude that there could be no δ_{ik} local term in the correlator of T_{0i} components. Thus, we see that introducing first gravitational background does allow to fix the subtraction term in a static correlator in the flat space.

Currents which we are considering now are somewhat similar to the standard superfluid current [41]. But there are important differences as well. In particular, the value of the superfluid current is not fixed in equilibrium while the current associated with the ChME has a unique value. Now, the statement is [17–19] that all the non-dissipative pieces in $\langle j_\mu \rangle$, $\langle T_{\mu\nu} \rangle$ can be conveniently determined by embedding the system into static electromagnetic plus gravitational background. We will outline briefly the proof following [17].

Static gravitational background in all the generality can be parameterized as follows:

$$ds^2 = -e^{2\sigma(x)} \left(dt + a_i(x) dx^i \right)^2 + g_{ij}(x) dx^i dx^j, \tag{11.33}$$

where x_i are spatial coordinates ($i = 1, 2, 3$ for definiteness), ∂_t is the Killing vector on this manifold, gravitational potentials σ, a_i, g_{ij} are smooth functions of the coordinates x_i. One assumes also presence of a static U(1) gauge field A,

$$A = A_0 dx^0 + A_i dx^i. \tag{11.34}$$

The A_0 component is related to the chemical potential, see (11.29).

Consider first the zeroth order in gradients. Then it is quite obvious that in equilibrium

$$u^\mu_{(0)} = e^{-\sigma(x)}(1, 0, 0, 0), \qquad T_{(0)}(x) = e^{-\sigma(x)} T_0, \qquad \mu_{(0)}(x) = e^{-\sigma(x)} A_0, \tag{11.35}$$

where $\sigma(x)$ enters the metric (11.33) and subscript (0) refers to the zeroth order in expansion in derivatives. Indeed, expressions for $T_{(0)}(x)$, $\mu_{(0)}(x)$ can be obtained directly from their invariant definition (11.29) while $u^\mu_{(0)}(x)$ is fixed by the normalization condition. Thus, the function W to this lowest order reduces to

$$W_{(0)} = \int \sqrt{g_3} \frac{e^\sigma}{T_0} P\left(T_0 e^{-\sigma}, A_0 e^{-\sigma}\right), \tag{11.36}$$

where $P(T, \mu)$ is pressure as function of temperature and chemical potential in flat space.

In higher orders in derivatives one expands u^μ, T, μ further:

$$u^\mu = u^\mu_{(0)} + u^\mu_{(1)} + u^\mu_{(2)} + \cdots, \tag{11.37}$$

$$T = T_{(0)} + T_{(1)} + T_{(2)} + \cdots, \tag{11.38}$$

$$\mu = \mu_{(0)} + \mu_{(1)} + \mu_{(2)} + \cdots, \tag{11.39}$$

where $u^\mu_{(n)}$, $T_{(n)}$, $\mu_{(n)}$ are expressions of n-th order in derivatives acting on the background fields σ, A_0, A_i, g_{ij}. It is important that both $T_{(n)}$ and $\mu_{(n)}$ are constructed on the same set of gauge- and diffeomorphic-invariant scalars. Let us denote the number of such scalars as $s_{(n)}$. As for the four velocity u^μ normalized to unit, δu^0 can be expressed in terms of δu^i and is not independent. Variations of the vector $u^i_{(n)}$ are expanded in the set of independent invariant vector combinations. The total number of such combinations is denoted as $v_{(n)}$.

Consider now a general decomposition of the energy-momentum tensor and current j_μ:

$$T^{\mu\nu} = E u^\mu u^\nu + P \Delta^{\mu\nu} + q^\mu u^\nu + q^\nu u^\mu + \tilde{\tau}^{\mu\nu}, \tag{11.40}$$

$$j^\mu = N u^\mu + v^\mu \tag{11.41}$$

where $q^\mu u_\mu = \tilde{\tau}^{\mu\nu} u_\mu = \Delta^{\mu\nu} u_\mu = 0$, $g_{\mu\nu} \tilde{\tau}^{\mu\nu} = 0$. Similar to the observations above, the scalars E, P, N can be written as expansions in independent gauge- and diffeomorphic-invariant scalars, vector q^μ is expanded in independent SO(3) vector structures while expansion of $\tau^{\mu\nu}$ would require introduction of a set of independent SO(3) tensor structures, with trace zero. The total number of the tensor structures is denoted by $t_{(n)}$. Note that within the standard relativistic hydrodynamics the energy-momentum tensor is defined somewhat different, in the so called Landau gauge, see (11.5). Namely, there is no vector q^μ and ϵ, P, n are those functions of temperature T and of chemical potential as determined by flat-space equilibrium thermodynamics. As a result, it is only $\tau_{\mu\nu}$ and v^μ which are to be expanded in gauge and diffeomorphic invariant structures introduced above. Therefore, the hydrodynamic tensors (11.5) are expanded in $s_{(n)} + v_{(n)} + t_{(n)}$ invariant structures.

The central point of the procedure invented in [17–19] is to equate the standard hydrodynamic tensors (11.5) to the expressions obtained by differentiating the functional W, see (11.28). This can be done only at the equilibrium point. In the equilibrium, not all the gauge- and diffeomorphic-invariant structures introduced above survive. The number of non-vanishing structures is called $s^e_{(n)}$, $v^e_{(n)}$, $t^e_{(n)}$. Moreover, these, surviving terms are non-dissipative since they exist in equilibrium. As for the dissipative terms, which correspond to the structures vanishing in equilibrium one gets no predictions or constraints concerning them.

By simple counting, the total number of the constraints is $3s^e_{(n)} + 2v^e_{(n)} + t^e_{(n)}$ and this number is exactly the same as needed to both find expansion corrections of the nth order to T, μ, u^i and determine the equilibrium stress tensor and current in the Landau gauge (11.5).[5] Indeed, the expansion of T, μ, u^i at the equilibrium point depends on $2s^e_{(n)} + v^e_{(n)}$ parameters while expansion of the energy-momentum tensor and of the current in the Landau gauge brings in $s^e_{(n)} + v^e_{(n)} + t^e_{(n)}$ terms. This completes the general proof that all the non-dissipative terms, like the chiral magnetic effect, are fixed uniquely. Actual details and explicit examples can be found in Refs. [17–19]. In particular, the results of Ref. [15] have been reproduced in this way.

The method of Refs. [17–19] outlined above allows to derive systematically chiral effects, like the ChME, for any number of conserved and anomalous currents and to any order in the derivative expansion. Remarkably, one avoids considering the entropy current s_μ altogether. The derivation makes it clear that one can fix only currents existing in the equilibrium. In other words, the currents are non-dissipative [40]. The issue will become central for us in Sect. 11.3.

11.2.3 Non-renormalization Theorems in Diagrammatic Approach

Very recently, there was a remarkable development [25, 42] in understanding the temperature-dependent chiral vortical effect, see the T^2 term in (11.9), (11.23). Namely, it was demonstrated that the bare-loop result (11.9) is modified in two-loop order in a well defined way.

As a preliminary remark, let us notice that the chiral vortical effect is determined [40] in terms of a static correlator, similar to the chiral magnetic effect, see (11.31). We recall the reader that the chiral vortical effect is defined in terms of the coefficient ξ_ω, see (11.16). In the non-relativistic limit we have the following piece in the axial-vector j^5_i:

$$\delta\left(j^5_i\right) = \xi_\omega \epsilon_{ijk} \partial_j v_k, \tag{11.42}$$

where v_k are components of the 3-velocity of an element of the liquid. Then for the coefficient ξ_ω one gets [40]:

$$\xi_\omega = \lim_{q_n \to 0} \sum_{ij} \epsilon_{ijn} \frac{i}{2q_n} \langle j^5_i, T_{0j} \rangle|_{\omega=0}, \tag{11.43}$$

(no summation over n). The argumentation is based on the well known analogy between the vector potential of a gauge field $\mathbf{A}$ and the metric-related vector $\mathbf{g}$, where $g_i \equiv g_{0i}$, or $ds^2 = dt^2 + 2g_i dt dx^i + dx^2_i$. In the rest-frame of the fluid but in the background of the gravitational potential $\mathbf{g}$ we have for the 4-velocity of the liquid $u_\mu = (-1, \mathbf{v}) = (-1, \mathbf{g})$. Therefore the "gravi-magnetic field" $\mathbf{B}_g \equiv \mathrm{curl}\,\mathbf{g}$ and we find, indeed, a complete analogy between the coefficient σ_B, see (11.31),

---

[5] Alternatively, this counting can be considered as a proof of the possibility to introduce the Landau gauge (11.5).

describing the chiral magnetic effect and the coefficient ξ_ω, see (11.42) and (11.43), describing the "chiral gravi-magnetic effect", or the chiral vortical effect as we call it.

Let us recall the reader that the non-renormalization theorem of Sect. 11.3.1 fixes the chiral vortical effect up to a temperature-dependent term, proportional to T^2 which can be evaluated at the vanishing chemical potential $\mu = 0$, see (11.23). We turn now to the problem of evaluating this missing term. In view of (11.43) we are interested then in the following term in the effective action:

$$S_{eff} = i\xi_\omega \int d^3x \epsilon_{ijk} A_i^5 \partial_j g_k \equiv iT^2 C_\omega \int d^4x a_i \partial_j b_k, \qquad (11.44)$$

where $a_i \equiv A_i^5$ is the gauge field coupled to j_i^5, $Tb_i \equiv g_{0i}$ is a component of the metric. We consider linearized gravity and the action is to be invariant under gauge and diffeomorphic transformations

$$a_i \rightarrow a_i + \partial_i \alpha, \qquad b_i \rightarrow T(\nabla_i \epsilon_0 + \nabla_0 \epsilon_i) \qquad (11.45)$$

where α is an arbitrary function, ϵ_μ is the diffeomorphism parameter, $x_\mu \rightarrow x_\mu + \epsilon_\mu$.

The Lagrangian density corresponding to the action (11.44) describes a 3d non-diagonal mass term mixing wave functions of the vectors a_i and b_i. The action (11.44) is gauge and diffeomorphic invariant, as it should be. However the mass term itself is not. This observation allows eventually to prove cancellation of a large class of radiative corrections [25]. The 3d nature of the action (11.44) is crucial for the proof. Note that although we started with a 4d gauge theory, reduction to a sum over a sequence of 3d theories is inherent to the problem since at finite temperature any 4d field theory reduces to a sum over Matsubara frequencies, with each frequency corresponding to a 3d theory.

The analysis of the radiative corrections to the 3d topological term (11.44) echoes the proof of non-renormalizability of a pure gauge-boson topological mass given about 30 years ago [27]. In that case the topological mass looks as [43, 44]:

$$S_{eff}^{gauge} = im_g \int d^3x \epsilon_{ijk} a_i \partial_j a_k. \qquad (11.46)$$

This topological mass arises on one-loop level in 3d gauge theories. The simplest Lagrangian of the matter field looks as

$$L_m = \bar{\psi}(D_\mu \gamma_\mu - m_0)\psi, \qquad (11.47)$$

where $D_\mu (\mu = 0, 1, 2)$ are covariant derivatives. The one-loop contribution is given by

$$(m_g)_{one\text{-}loop} = \frac{q^2}{4\pi} \frac{m_0}{|m_0|}, \qquad (11.48)$$

where q is the charge of the fermion. In higher orders of perturbation theory one-loop fermion graphs with a few photon exchanges arise. One can integrate first over the (massive) fermion and reduce any graph to a n-photon effective vertex. In the momentum space, it is denoted as $\Gamma^{(n)}_{\mu_1\dots\mu_n}(q_1,\dots,q_n)$ where q_i are photon momenta and the overall δ-function is factored out. The vertex is a function of $(n-1)$ independent momenta, $q_1,\dots,q_{n-1}$. Take now the limit of all the momenta $(q_1,\dots,q_{(n-1)})$ small. Then it is straightforward to prove that the effective vertex is to vanish in the limit of any independent momentum zero. In other words,

$$\Gamma^{(n)}(q_1,\dots,q_n) = O(q_1 \cdot q_2 \cdots q_{n-1}). \tag{11.49}$$

This relation is sufficient to prove that all the contributions to topological mass, beginning with two loops, vanish. Indeed, one can always choose the momenta corresponding to the external photon legs to be included into momenta $(q_1,\dots,q_{(n-1)})$ which are small. The one-loop graph is exceptional in this sense since there is only a single independent photon momentum and $\Gamma^{(1)}(q_1) = O(q_1)$.

Let us come back to discussion of the topological mass (11.44) relevant to the vortical effect. It is determined by the correlator (11.43). Let us split the momentum-density operator into the fermionic and gluonic parts:

$$T_{0j} = (T_{0j})_{fermionic} + (T_{0j})_{gluonic}. \tag{11.50}$$

As far as we keep only the fermionic part the main idea, that only 3d one-loop graphs can contribute to (11.44), remains the same. There is, however, an important change concerning infrared behavior of higher-loop graphs. Amusingly, the masslessness of the fermions in the original 4d field theory does not matter since in the 3d projection the fermions do have non-vanishing masses, $m_f^2 = 4\pi^2(n+1/2)^2$ where n, $(n = 0, 1, 2 \dots)$ enumerates Matsubara frequencies. The actual subtle point is that in non-Abelian 3d theory higher loops generically diverge badly in the infrared. According to Ref. [25] the infrared cut off is still provided by the non-perturbative gluon mass emerging [45] at finite temperatures. The effective gluon mass is of order $(m_g^2)_{eff} \sim g_{Y-M}^4(T)T^2$ and at external momenta much smaller than this effective mass one can expect that higher-loop contribution to the topological mass still vanishes. Finally, the evaluation of one-loop fermionic contributions to the topological mass (11.44) is more involved technically than in case of (11.46). Each 3d theory corresponding to a certain Matsubara frequency does contribute to (11.44). Indeed, according to (11.48) the one-loop contribution does not disappear with the growing fermion mass. As a result, one comes [25] to a divergent sum. The standard ζ-function regularization provides the final answer for the temperature-dependent vortical effect:

$$C_\omega^{ferm} T^2 = -T^2 \sum_{m=1}^{\infty} m \to \frac{1}{12} T^2, \tag{11.51}$$

where C_ω^{ferm} is the contribution of the fermionic loop to C_ω.

So far we discussed the fermionic part of T_{0j}, see (11.50), and argued that its contribution is exhausted by one-loop graphs. The argument does not apply, however,

311 Chiral Magnetic Effect in Hydrodynamic Approximation

to the gauge-field part, $(T_{0j})_{gluonic}$. It does generate a calculable two-loop radiative correction to the T^2 term in (11.9), see [25, 26]. We should not feel disappointed about this lack of complete cancellation of higher loops. Indeed, although the Adler-Bardeen theorem is commonly referred to as a proof of non-renormalization of the anomaly, it is known since long (see, e.g., [46]) that two-loop graphs corresponding to rescattering of gauge field do not vanish in fact. The proof of non-renormalizability of the ChME in Sect. 11.3.1 avoided this problem only because we treated the electromagnetic field as external (not dynamical). Note also that the T^2 term in the chiral vortical effect is present only in case of the singlet quark currents which is anomalous anyhow and the status of the corresponding ChME is not clear. Its evaluation, however, is an amusing demonstration that dissipation-free processes generically are not suppressed at all at high temperature.

11.2.4 Non-renormalization Theorems in Effective Field Theories

Here we develop another approach to the chiral effects based on an effective theory following mostly Ref. [30]. The basic idea is to treat chemical potentials as effective couplings. The basic rule can be memorized as an analogy between interaction with charge q and with chemical potential μ: $qA^\mu \to \mu u^\mu$, where A_μ is the vector-potential of an external field and u^μ is, as usual, the 4-velocity of an element of liquid, see (11.10).

Let us first substantiate the rule (11.10). Chemical potentials are introduced through the effective Hamiltonian:

$$\delta H = \mu Q + \mu_5 Q_5, \tag{11.52}$$

where $Q = \int d^3x \, \psi^\dagger \psi$ and $Q_5 = \int d^3x \, \psi^\dagger \gamma_5 \psi$. Note that this is indeed an effective, not fundamental interaction since any chemical potential is introduced thermodynamically, that is for a large number of particles. However, at least formally the Hamiltonian (11.52) looks exactly the same as the Hamiltonian for a fundamental (i.e., not effective) interaction of charges with the A_0 gauge field. Thus, we come to the analogy (11.10) in the particular case $u^0 = 1$, $u^i = 0$. This case was considered in many papers, see in particular [13, 14], and in the equilibrium the analogy between A_0 and μ is commonly used nowadays. We have already exploited this connection, see (11.29). In this sense the effective theory considered here can be viewed as a simplified version of a much more elaborated scheme outlined in Sect. 11.2.2.

Treating (11.52) as a small perturbation we can also fix the corresponding Lagrangian density:

$$S_{eff} = \int dx \left(i \bar\psi \gamma^\rho D_\rho \psi + \mu \bar\psi \gamma^0 \psi + \mu_5 \bar\psi \gamma^0 \gamma_5 \psi \right) + S_{int}, \tag{11.53}$$

where D_μ is the covariant derivative in external electromagnetic field and S_{int} is a fundamental interaction responsible for the formation of the liquid. For our discussion it is crucial that S_{int} does not induce any anomaly. We can drop then this interaction for our purposes.

So far we considered the whole of the liquid as being at rest and the chemical potential being constant through the whole of the volume. To study hydrodynamics one can use the standard trick of boosting the action into a local rest frame by utilizing the 4-vector u^μ:

$$S_{eff} = \int dx \left(i\bar{\psi}\gamma^\rho D_\rho \psi + \mu u_\mu \bar{\psi}\gamma^\mu \psi + \mu_5 u_\mu \bar{\psi}\gamma^\mu \gamma_5 \psi \right). \qquad (11.54)$$

The Lagrangian density (11.54) coincides in the rest frame with (11.53) but looks perfectly Lorentz invariant and can be used in covariant perturbative calculations. The boost velocity u^μ is treated then as a slowly varying external field, similar to A^μ.

The only difference between the effective action (11.54) and fundamental one is the presence of the terms proportional to the chemical potentials μ_5, μ. The fundamental theory is free from anomalies in the case considered now, $(E \cdot B) = 0$. Therefore the only possible source of anomalies in the effective theory are the triangle graphs with μ, μ_5 entering the vertices.

The presence of an anomaly in the effective theory can indeed be readily verified. For this purpose one can calculate triangle graphs or use Fujikawa-Vergeles path-integral considerations. According to the latter technique, anomalies emerge due to non-invariance of the path-integral measure under field transformations. Consider the following transformation

$$\psi \rightarrow e^{i\alpha\gamma_5 + i\beta}\psi. \qquad (11.55)$$

Then, by the standard technique one readily finds:[6]

$$\partial_\mu j_5^\mu = -\frac{1}{4\pi^2}\epsilon_{\mu\nu\alpha\beta}\left(\partial^\mu(A^\nu + \mu u^\nu)\partial^\alpha(A^\beta + \mu u^\beta) + \partial^\mu \mu_5 u^\nu \partial^\alpha \mu_5 u^\beta\right), \qquad (11.56)$$

$$\partial_\mu j^\mu = -\frac{1}{2\pi^2}\epsilon_{\mu\nu\alpha\beta}\partial^\mu(A^\nu + \mu u^\nu)\partial^\alpha \mu_5 u^\beta. \qquad (11.57)$$

Rewriting (11.56) and (11.57)

$$\partial_\mu\left(n_5 u^\mu + \frac{\mu^2 + \mu_5^2}{2\pi^2}\omega^\mu + \frac{\mu}{2\pi^2}B^\mu\right) = -\frac{1}{4\pi^2}\epsilon_{\mu\nu\alpha\beta}\partial^\mu A^\nu \partial^\alpha A^\beta,$$

$$\partial_\mu\left(n u^\mu + \frac{\mu\mu_5}{\pi^2}\omega^\mu + \frac{\mu_5}{2\pi^2}B^\mu\right) = 0 \qquad (11.58)$$

---

[6]We could have defined anomaly in such a way that it does not contribute to $\partial_\mu j^\mu$. However, in the presence of both chemical potentials μ and μ_5 there is no physical motivation for such a regularization.

allows for a straightforward comparison with results of the thermodynamic approach, see (11.23), (11.24). We find out that the effective theory does reproduce anomalous pieces of the transport coefficients obtained earlier in the leading order in the chemical potentials. As for the higher orders in the chemical potential there are apparent differences. Within the effective theory higher orders in the chemical potential belong to higher order perturbative terms. The triangle graphs which determine (11.58) have of course a very special status. First, they defy conservation of the currents and, second, they do not receive contributions due to the iteration of S_{int}, because of the Adler-Bardeen theorem.

As for higher in μ, μ_5 terms they are infrared sensitive and can be fixed only within a particular infrared-sensitive regularization scheme. Consider for example contribution of order μ^3 to the chiral vortical effect. It can be estimated as

$$\delta\xi \sim \frac{\mu^2}{2\pi^2} \frac{\mu}{\epsilon_{IR}}, \tag{11.59}$$

where ϵ_{IR} is an infrared cut off in the energy/momentum integration. Equation (11.59) can hardly be improved within the effective theory. On the other hand, the thermodynamic derivation of Sect. 11.2.3 does fix [15] terms of order μ^3 in the Landau gauge as:

$$\delta\xi = -\frac{\mu^2}{2\pi^2} \frac{2\mu n}{(\epsilon + p)}. \tag{11.60}$$

By comparing (11.59) and (11.60) we find

$$\epsilon_{IR} \sim (\epsilon + p)/n. \tag{11.61}$$

Note that the enthalpy $w = \epsilon + p$ is known to play the role of mass in relativistic hydrodynamics. Thus, the ratio $\mu \cdot n/(\epsilon + p)$ characterizes the contribution of the energy related to the chemical potential in units of the total energy, as far as μ is small. And it is quite natural to have an expansion in this parameter. The expansion coefficients within the effective theory are model dependent, however.

To summarize, the effective hydrodynamic theory defined through the substitution (11.10) allows for a straightforward and simple evaluation of the chiral effects in terms of anomalies of the effective theory. Terms of lowest order in expansion in the chemical potentials coincide with the results of other approaches. Higher orders, however, are infrared sensitive and model dependent within the effective theory. Within the thermodynamic approach of Sect. 11.2.1 these terms are apparently fixed by the procedure chosen to integrate the differential equations (11.21) beginning with small μ and keeping the pressure P constant. The use of the Landau frame is also needed.

Apart from the μ^2 term the vortical effect has a T^2 contribution, see Sect. 11.2.3. This term can be evaluated within the finite-temperature field theory at $\mu = 0$. For this reason the effective in chemical potentials theory introduced above does not help to evaluate the T^2 term.

11.2.5 Concluding Remarks

Powerful non-renormalization theorems have been proven in case of chiral magnetic and vortical effects in hydrodynamic approximation. Eventually, all the proofs go back to celebrated field theoretic non-renormalization theorems [12, 27]. To bridge them to hydrodynamics, it is crucial that chiral effects can be expressed in terms of certain spatial correlators at frequency $\omega = 0$. These static correlators are trivially continued to the Euclidean space. Which means, in turn, that we are dealing with thermodynamic observables. Continuation to the Euclidean space also allows to use the standard technique of Feynman graphs and utilize field-theoretic non-renormalization theorems. We also notice that all the theorems refer to topological terms in action. In other words, the action observes symmetries of the problem considered while the Lagrangian density does not. Chiral anomalies were put into such a context first in Ref. [47]. Moreover, it turns out that it is not crucial whether the corresponding one-loop graphs signify an anomaly or not. Probably, one and the same non-renormalization theorem can be proven either in terms of anomalies or non-anomalous graphs. The topological aspect, however, seems to be an indispensable ingredient to non-renormalizability.

11.3 Hydrodynamic Chiral Effects as Quantum Phenomena

11.3.1 Non-dissipative Currents

All the chiral effects which we are considering are non-dissipative. This was demonstrated in the thermodynamic language [17–20], see Sect. 11.2.2. Another line of reasoning is suggested in Ref. [33] and based on time-reversal invariance. Let us summarize the argumentation.[7] One compares, for example, the ordinary electric conductance, σ_E and "magnetic conductance" σ_B associated with the ChME:

$$\mathbf{j}_{el} = \sigma_E \mathbf{E}, \qquad \mathbf{j}_{el} = \sigma_B \mathbf{B}.$$

Moreover, $\sigma_E > 0$ since the work done by external electric field, $W = \mathbf{j}_{el} \cdot \mathbf{E} > 0$.

Under time reversal,

$$\mathbf{j}_{el}^T = -\mathbf{j}_{el}, \qquad \mathbf{E}^T = +\mathbf{E}, \qquad \mathbf{B}^T = -\mathbf{B}.$$

If we would try to obtain $\mathbf{j}_{el}^T = \sigma_E^T \mathbf{E}^T$ by time reversal of the relation $\mathbf{j}_{el} = \sigma_E \mathbf{E}$ then we would conclude that $\sigma_E^T = -\sigma_E$ which is in contradiction with the positivity of σ_E. There is no surprise of course in this failure. To the contrary, dissipation is indeed not a time-reversible process. (The time-reversal invariance is manifested instead in the Onsager relations which imply, in particular, $\sigma_E > 0$.) On the other

---

[7]The author is thankful to L. Stodolsky for a detailed discussion of the subject.

hand, the Hall relation $\mathbf{j}_{el} \sim \mathbf{E} \times \mathbf{B}$ is time-reversal invariant and this is possible only if there is no dissipation. Our chiral magnetic effect is of the same type as the Hall conductivity and, therefore, cannot be accompanied by dissipation because of the time-reversal invariance of theories considered. Note that there are no such symmetry-based arguments for, say, superfluid current. Therefore, the viscosity can be only dynamically suppressed and, indeed, according to the modern views [48] there is a universal lowest bound on the shear viscosity, $\eta/s \geq 1/4\pi$ where s is the entropy density. According to the logic outlined above no similar bound can exist for dissipation associated with the chiral magnetic effect since dissipation is forbidden by symmetry considerations. Most amusingly, no such bound can exist in case of superconductivity since the London equation $m_\gamma^2 \mathbf{A} = \mathbf{j}_{el}$ is invariant under time reversal and, therefore, dissipation is forbidden by symmetry considerations.

Furthermore, dissipation-free processes are usually quantum phenomena and in this section we discuss the ChME from this point of view. In fact, it is rather an open-end discussion since not much is known yet about the microscopic picture of the ChME in the hydrodynamic approximation.

There is no doubt that the ChME is rooted in the loop graphs of the underlying theory and represents a macroscopic manifestation of quantum phenomena. Calculation of a standard Feynman graph gets related to thermodynamics by a simple trick of identifying a constant piece in an external gauge field with the chemical potential, $A_0 \approx \mu$. Let us explain this point in more detail. The generating functional $W(sources(x))$, see (11.28), contains in fact a piece, $W_{anom}^{(1)}(sources(x))$ reproducing the chiral anomaly. Moreover, W_{anom} is uniquely fixed by the requirement that it does reproduce the anomaly within the formalism of Sect. 11.2.2. It turns out that W_{anom} in the static case considered can be written in a local form [17]. Explicitly to first order in derivatives:

$$W_{anom}^{(1)} = \frac{C}{2} \int d^3x \sqrt{g_3} \left(\frac{A_0}{3T_0} \epsilon^{ijk} A_i \partial_j A_k + \frac{A_0^2}{6T_0} \epsilon^{ijk} A_i \partial_j a_k \right), \qquad (11.62)$$

where the constant C determines the anomaly and further notations are specified in Sect. 11.2.2.

Let us emphasize that (11.62) is a pure field-theoretic input. However, once we identify A_0 with a macroscopic quantity, chemical potential, the action (11.62) determines macroscopic motions. Moreover, it is quite obvious that the two terms in the right-hand side of (11.62) do reproduce the chiral magnetic and chiral vortical effects, respectively. Comparison of (11.62) with (11.31) and (11.43) helps to check this.

So far, we discussed the exact chiral limit. If we would decide to estimate the effect of chiral symmetry violations, say, through finite fermionic masses, we have to address the field-theoretic calculations anew. In particular, the rate of production of massive fermions in parallel, constant electric and magnetic fields, E_z, B_z is given by the equation [49, 50]:

$$\frac{dN_5}{d^3x\,dt} = \frac{q^2 B_z E_z}{2\pi^2} \exp\left(-\frac{\pi m_f^2}{E_z} \right), \qquad (11.63)$$

which is replacing (11.26). We see that if we tend $E_z \to 0$ now we would not get any ChME, compare (11.27). Therefore, the formal static hydrodynamic limit is replaced for small fermionic masses by:

$$(\omega \equiv 0, q_i \to 0) \to (|q_i| \gg \omega, \omega \gg m_f), \tag{11.64}$$

and phenomenological implications of this modification are to be considered within a particular framework.

Quantum description of the chiral magnetic effect has another aspect which can be illustrated on the example of chiral fermions interacting with an external magnetic field [3–5, 9]. Namely, we should be able to obtain the same current (11.1) by evaluating the matrix element

$$j_\mu^{el} = q\langle \bar{\psi}(x)\gamma_\mu(x)\psi(x)\rangle, \tag{11.65}$$

where the averaging is over the thermodynamic ensemble. The simplest set up is non-interacting fermions and a non-vanishing chemical potential μ_5. In this case $\psi(x)$ represents solutions of the Dirac equation

$$(i\gamma^\mu D_\mu + \mu_5\gamma^0\gamma^5)\psi(x) = 0, \tag{11.66}$$

where $D_\mu = \partial_\mu - iA_\mu$ and the vector-potential A_μ corresponds to a constant magnetic field. The thermodynamic ensemble is described by the ideal gas of massless Fermi particles. Explicit calculation turns out to be feasible and the final result agrees with (11.1). We will give some details of the calculation in Sect. 11.3.4. Note that, at least from the technical point of view, this coincidence is not trivial at all. Indeed, (11.1) is based entirely on the evaluation of the anomalous triangle graph over perturbative vacuum state. The magnetic field is treated perturbatively. On the other hand, the matrix element (11.65) is saturated by the fermionic zero modes [9] which one finds explicitly, accounting for the magnetic field to all orders.

What is lacking in the case of the ChME in hydrodynamic approximation is a microscopic calculation similar to the direct evaluation of (11.65) for non-interacting particles just outlined. What makes such a calculation especially desirable is the observation that all the non-renormalization theorems reviewed in Sect. 11.2 do not indicate any crucial dependence on temperature. It seems highly non-trivial to keep up quantum coherence over an (infinitely) large length at finite temperature.

In the rest of this section we discuss microscopical picture for the chiral effects on the only example available so far [36], that is the case of superfluid.

11.3.2 Low-Dimensional Defects

A novel point brought by considering superfluid is the crucial role of low-dimensional defects, or singularities of hydrodynamic approximation. Such defects were considered in fact in many papers for various reasons, see in particular

[23, 34, 35, 51]. In case of rotating superfluid such defects have been known since long [41].

Consider first the chiral vortical effect. The crucial point is that the velocity field of the superfluid is known to be potential:

$$\mathbf{v}_s = \nabla\varphi, \tag{11.67}$$

and, naively, the vorticity vanishes identically since $\operatorname{curl} \mathbf{v}_s = 0$. If this were true, the chiral current (11.7) would disappear. But it is well known, of course, that the angular momentum is still transferred to the liquid through vortices [41]. The potential is singular on the linear defects, or vortices. The vortex is defined through circulation of velocity:

$$\oint \mathbf{v}_s d\mathbf{l} = 2\pi k, \tag{11.68}$$

where $d\mathbf{l}$ is an element of length and k is integer. The quantization condition (11.68) follows from the interpretation of φ as the phase of a wave function of identical particles. Equations (11.68), (11.67) imply that the velocity is singular. In three dimensions the singularity occupies a line, which can be called a defect of lower dimension. The energy of the vortex is logarithmically divergent, $E_{vortex} \sim l \ln(l/a)$ where l is the length of a (closed) vortex and a is of order distance between the constituents.

Note that the velocity v_s according to (11.68) falls off with increasing distance r to the singularity, $v_s = k/r$ while for a rotating solid body the velocity, to the contrary, grows with distance to the axis of rotation, $v = |\Omega|r$ where Ω is the angular velocity of rotation. Therefore, at first sight, distribution of velocities inside a rotating bucket with a superfluid is very different from ordinary liquid. This is not true, however, as far as the angular velocity is large enough. In this case, there are many defects and the distribution of velocities, once averaged over defects, is the same as for ordinary liquid. The proof [41] is based on the observation that thermodynamic equilibrium is determined by minimizing $E_{rotation} = E - \mathbf{M} \cdot \Omega$ where $\mathbf{M}$ is the angular momentum and velocity the averaged (over defects) is uniquely determined by this condition.

To summarize, vortices in superfluid represent an example of lower-dimensional defects. Although locally, or at small distances, defects look very different from the continuum picture, the thermodynamic results can be restored upon averaging over a large number of defects.

11.3.3 Relativistic Superfluidity

To consider vortex solutions in detail we need explicit examples of dynamical systems which exhibit relativistic superfluidity. The simplest and best understood example of this type seems to be the pionic medium at zero temperature and non-zero isospin chemical potential μ_I [52]. The system is described by the following chiral Lagrangian:

$$L = \frac{1}{4} f_\pi^2 \, \mathrm{Tr}\big[D^\mu U (D_\mu U)^\dagger\big], \qquad (11.69)$$

where U are 2×2 unitary matrices, functions of the pionic fields, see, e.g., [53]. Moreover, the chemical potential μ_I is switched on through the covariant derivatives, $D_0 U = \partial_0 U - \frac{\mu_I}{2}[\tau_3, U]$, $D_i U = \partial_i U$. To have the chiral current conserved we consider massless quarks. Then the chiral symmetry is spontaneously broken. In the common case of vanishing chemical potential the residual symmetry (realized linearly) is $SU(2)_{L+R}$. Furthermore, at non-zero μ_I this symmetry is broken to $U(1)_{L+R}$. The proof is straightforward. Namely, the potential energy corresponding to (11.69) equals to

$$V_{eff}(U) = \frac{f_\pi^2 \mu_I^2}{8} \, \mathrm{Tr}\big[\tau_3 U \tau_3 U^\dagger - 1\big], \qquad (11.70)$$

and the minima of that potential can be captured by substitution $U = \cos\alpha + i(\tau_1 \cos\phi + \tau_2 \sin\phi)\sin\alpha$:

$$V_{eff}(\alpha) = \frac{f_\pi^2 \mu_I^2}{4}(\cos 2\alpha - 1) \qquad (11.71)$$

and for the minimum one readily obtains $\cos\alpha = 0$. Then, depending on the sign of μ_I, squared mass of π^+ or π^- state becomes negative and the corresponding field is condensed. This means that the vacuum is described by $U = i(\tau_1 \cos\phi + \tau_2 \sin\phi)$ instead of the standard, i.e. $\mu = 0$ vacuum $U = I$. There emerges a new order parameter $\langle \overline{u}\gamma_5 d \rangle + h.c. = 2\langle \overline{\psi}\psi \rangle_{vac} \sin(\alpha) = 2\langle \overline{\psi}\psi \rangle_{vac}$. The system is thus a charged superfluid. It should be noted, that the degeneracy with respect to the angle ϕ above indicates that it can be identified as a 3d Goldstone field. In addition, there are two massive modes.

Because of the presence of a 3d Goldstone mode the superfluidity criterion (11.32) is satisfied:

$$\lim_{q \to 0} \int d^3 x \exp(i\mathbf{q} \cdot \mathbf{r})\langle T_{0i}(\mathbf{x}), T_{0k}(0)\rangle = \mu^2 \mathbf{q}_i \mathbf{q}_k / \mathbf{q}^2. \qquad (11.72)$$

Explicit evaluation of this correlator is based on the Josephson equation:

$$\partial_0 \phi = \mu \qquad (11.73)$$

which is satisfied now as an equation of motion following from (11.69). For $\mu = const$ (11.73) can be interpreted as condensation of $\partial_0\phi$, similar to the standard Higgs condensation but violating the Lorentz invariance [13, 14, 54].

As is argued, e.g., in [13, 14], $\partial_\mu\phi$ can be identified with non-normalized superfluid velocity. The vortex configuration is in principle determined by the value of the angular velocity of rotation of the liquid [16]. We address a generic case, when the quantum of circulation, n is rather high to consider defects thermodynamically but not high enough to ruin the superfluidity. As is known, an energetically preferable configuration is the uniform distribution of vortices with $n = 1$. Nearby any given vortex the Goldstone field is given by [54]:

$$\phi = \mu t + \varphi. \tag{11.74}$$

We will assume that vortices are well separated, $\delta x \gg a$, calculate the current for a single vortex and then sum it over all vortices, that is simply multiply by n. The effective Lagrangian for the interaction of fermions with the scalar field ϕ (we will limit ourselves to the case of a single fermion, and then sum up the result over colors and flavors) looks as

$$L = \overline{\psi} i (\partial_\mu + i \partial_\mu \phi) \gamma^\mu \psi. \tag{11.75}$$

Indeed because, of the Josephson equation (11.73) we reproduce the standard chemical potential term. Other components complete this term to a formally Lorentz invariant interaction, compare (11.10).

Using standard methods of evaluating the anomalous triangle diagrams, see, e.g., Sect. 11.2.4, one obtains for the axial current:

$$j_\mu^5 = \frac{1}{4\pi^2} \epsilon_{\mu\nu\alpha\beta} \partial^\nu \phi \partial^\alpha \partial^\beta \phi. \tag{11.76}$$

This current seems to vanish identically. However, for the vortex field, $\phi = \mu t + \varphi$, $[\partial_x, \partial_y]\phi = 2\pi \delta(x, y)$ and:

$$\left(j_3^5\right)_{vortex} = \frac{\mu}{2\pi} \delta(x, y). \tag{11.77}$$

The total current, or the sum over the vortices equals to

$$j_3^5 = \int d^2x j_3^5 = \frac{\mu}{2\pi} n. \tag{11.78}$$

It is worth noting that actually $n \sim \mu$ and the current (11.78) is quadratic in the chemical potential μ, as it should be.

So far we considered the chiral vortical effect. To evaluate the chiral magnetic effect, consider a charged superfluid and turn on a magnetic field. Then magnetic field would stream into tubes, or Abrikosov vortices. The vortex profile could be found by accounting for the finite photon mass. The chiral current can be obtained then by substituting the vortex configuration to the Dirac equation and solving it for the modes. The current is concentrated on the vortex center. In the hydrodynamic approximation we are considering the magnetic field is singular and the Dirac equation is poorly defined in this sense. However, using index theorems it is possible to evaluate number of zero modes, and the zero modes saturate the chiral current. We will give more details in the next subsection.

11.3.4 Zero Modes

We now proceed to microscopic picture based on the zero modes.[8] The Hamiltonian
has the form:

$$H = -i(\partial_i - i\partial_i\phi)\gamma^0\gamma^i + m\gamma^0 \tag{11.79}$$

and the Dirac equation decomposes as:

$$-H_R\psi_L = E\psi_L, \tag{11.80}$$

$$H_R\psi_R = E\psi_R, \tag{11.81}$$

where $H_R = (-i\partial_i + \partial_i\phi)\sigma_i$. Hence, any solution ψ_R of $H_R\psi_R = \epsilon\psi_R$ generates
both a solution with energy $E = \epsilon$, $\psi = \begin{pmatrix} 0 \\ \psi_R \end{pmatrix}$ and a solution with $E = -\epsilon$, $\psi = \begin{pmatrix} \psi_R \\ 0 \end{pmatrix}$.

Because of the invariance with the respect to translations in z direction, we decompose using the momentum eigenstates $-i\partial_3\psi_R = p_3\psi_R$. For each p_3,

$$H_R = p_3\sigma^3 + H_\perp, \tag{11.82}$$

$$H_\perp = (-i\partial_a - \partial_a\phi)\sigma^a, \quad a = 1, 2. \tag{11.83}$$

Notice that $\{\sigma^3, H_\perp\} = 0$. This means that if $|\lambda\rangle$ is an eigenstate of $H_\perp$ with non-
zero eigenvalue λ, $H_\perp|\lambda\rangle = \lambda|\lambda\rangle$, $\sigma_3|\lambda\rangle$ is an eigenstate with eigenvalue $-\lambda$. Also,
σ_3 maps zero eigenstates of $H_\perp$ to themselves, so that all the eigenstates of $H_\perp$ can
be classified with respect to σ_3.

We can now express eigenstates of H_R in terms of eigenstates of $H_\perp$. Since
$[H_R, H_\perp^2] = 0$, H_R will only mix states $|\lambda\rangle$, $|-\lambda\rangle$. For $\lambda > 0$, one can write,

$$\psi_R = c_1|\lambda\rangle + c_2\sigma_3|\lambda\rangle. \tag{11.84}$$

Solving (11.82) we find for the eigenvalues of energy:

$$\epsilon = \pm\sqrt{\lambda^2 + p_3^2}.$$

This means that every eigenstate of $H_\perp$ with eigenvalue $\lambda > 0$ produces two eigen-
states of H_R, while zero modes of $H_\perp$, $|\lambda = 0\rangle$ are eigenstates of H_R with eigenval-
ues:

$$\epsilon = \pm p_3, \tag{11.85}$$

where the sign corresponds to $\sigma^3|\lambda = 0\rangle = \pm|\lambda = 0\rangle$.

---

[8]This subsection is of rather technical nature and can be considered as an appendix. Moreover, the
presentation is close to that of Ref. [23], with a substitution $A_i \to \partial_i\phi$.

Therefore, the zero modes of $H_\perp$ are gapless modes of H, capable of travelling up or down the vortex, depending on the sign of σ_3 and chirality. These are precisely the carriers of the axial current along the vortex. Let $N_\pm+$ be the numbers of zero modes which are eigenstates of the matrix σ^3 with eigenvalues ± 1, respectively. Consider zero mode of $H_\perp$, $|\lambda\rangle = (u, v)$, where u and v are c-functions satisfying

$$\mathcal{D}v = 0, \qquad \mathcal{D}^\dagger u = 0. \tag{11.86}$$

Here

$$\mathcal{D} = -i\partial_1 - \partial_2 - (\partial_1\phi - i\partial_2\phi). \tag{11.87}$$

Define $N_+ = \dim(\ker(\mathcal{D}^\dagger))$, $N_- = \dim(\ker(\mathcal{D}))$, and

$$N = Index(H_\perp) = N_+ - N_- = \dim(\ker(\mathcal{D}^\dagger)) - \dim(\ker(\mathcal{D})). \tag{11.88}$$

Note that $H_\perp$ is an elliptic operator. Its index has been computed within various approaches in papers [55, 56]. In our case the index is given by

$$N = \frac{1}{2\pi} \int dx_i \partial_i\phi = n. \tag{11.89}$$

Moreover, for the most important case $n = 1$ the zero mode is easy to construct, see, e.g., Ref. [36]. The result (11.89) can be also obtained starting from the well known case of magnetic field parallel to z-axis and uniform in that direction. In the latter case the index is given by

$$N = \frac{e}{2\pi} \int dx_i A_i = \frac{e}{2\pi} \int d^2x B_z \tag{11.90}$$

and by substituting $eA_i \to \partial_i\phi$ we arrive at (11.89). Note, however, that in case of superfluid, which we discuss here, the index is an integer, whereas for non-superconducting case the flux is not quantized and the left-hand side of (11.90) is to be understood as the integer part of the right-hand side.

We now proceed to the computation of the axial current at a finite chemical potential μ. The axial current density in the third direction is given by:

$$j_5^3(x) = \overline{\psi}(x)\gamma^3\gamma^5\psi(x) = \psi_L^\dagger\sigma^3\psi_L(x) + \psi_R^\dagger\sigma^3\psi_R(x). \tag{11.91}$$

We are interested in the expectation value of the axial current along the vortex, $j_5^3 = \int d^2x \langle j_5^3(x)\rangle$. At finite chemical potential, we have:

$$\langle j_5^3(x)\rangle = \sum_E \theta(\mu - E)\psi_E^\dagger(x)\gamma^0\gamma^3\gamma^5\psi_E(x)$$

$$= \sum_\epsilon \left(\theta(\mu - \epsilon) + \theta(\mu + \epsilon)\right)\psi_{R\epsilon}^\dagger(x)\sigma^3\psi_{R\epsilon}(x). \tag{11.92}$$

Here, $\theta(\mu - E)$ is the Fermi-Dirac distribution (at zero temperature), ψ_E are eigenstates of H with eigenvalue E, $\psi_{R\epsilon}$ are eigenstates of H_R with eigenvalue ϵ. By substitution of the explicit form of $\psi_{R\epsilon}$ in terms of $H_\perp$ eigenstates, one obtains:

$$\langle j_5^3 \rangle = \frac{1}{L} \sum_{p_3} \sum_{\lambda \rangle 0} \sum_{s=\pm} \left(\theta(\mu - (\lambda^2 + p_3^2)^{\frac{1}{2}}) \right.$$

$$+ \theta(\mu + (\lambda^2 + p_3^2)^{\frac{1}{2}})) \langle \psi_R^s(\lambda, p_3) | \sigma^3 | \psi_R^s(\lambda, p_3) \rangle$$

$$+ \frac{1}{L} \sum_{p_3} \sum_{\lambda = 0} (\theta(\mu - p_3) + \theta(\mu + p_3)) \langle \lambda | \sigma^3 | \lambda \rangle. \qquad (11.93)$$

Here $\lambda > 0$ enumerate eigenstates of $H_\perp$, which generate eigenstates of H_R, $\psi_R^\pm(\lambda, p_3)$ with momentum p_3 and eigenvalue $\epsilon_\pm = \pm\sqrt{\lambda^2 + p_3^2}$, and $\lambda = 0$ label the zero modes of $H_\perp$. Moreover, the sum over all non-zero eigenstates vanishes, and only zero modes of $H_\perp$ generate $j_5^3 \neq 0$. For the zero modes, $\langle \lambda | \sigma^3 | \lambda \rangle = \pm 1$, and we obtain:

$$j_5^3 = (N_+ - N_-) \frac{1}{L} \sum_{p_3} (\theta(\mu - p_3) + \theta(\mu + p_3))$$

$$= n \int \frac{dp_3}{2\pi} (\theta(\mu - p_3) + \theta(\mu + p_3)) = \frac{\mu}{\pi} n. \qquad (11.94)$$

This result is similar to the macroscopic answer (11.7) for the vortical effect but there is inconsistency of a factor of two.

11.3.5 Concluding Remarks

Considering superfluid provides a unique possibility to develop a microscopic picture for chiral hydrodynamic effects. The calculations above demonstrate that the chiral currents are carried by quantum-mechanical zero modes and are indeed dissipation-free. This result is in agreement with the expectations.

However, this explicit example brings also new lessons. First of all, the prediction for the chiral vortical effect is changed by a factor of two and it is instructive to appreciate the reason for this change. Technically, the easiest way to trace the origin of this factor of two is to compare the calculation of the chiral vortical effect in this section with the calculation within effective theories, see Sect. 11.2.4. In the latter case, the chiral effects are described by anomalous triangle graphs, with vertices proportional to μu_μ or $q A_\mu$. In other words, the chemical potential μ plays the role similar to the electromagnetic coupling, or charge q while the field of fluid velocities, u_μ is similar to the electromagnetic field A_μ. The triangle graphs for the chiral magnetic and vortical effects looks very similar. The only difference is that vortical effect is quadratic in μ while the magnetic effect is linear both in q

and μ. Because of quantum mechanics, however, this difference results in a factor of two: in case of the vortical effect the corresponding graph has two identical vertices and this brings a factor of $1/2$, as usual. It is this factor which is absent from the calculation of the vortical effect in terms of defects. Indeed, there are two facets of the chemical potential. First, it plays the role of an effective coupling, as we have just explained. And, second, it limits the integration over the longitudinal momentum of zero modes, see Sect. 11.3.4. In the language of defects, these two roles are not interchangeable and the quantum-mechanical factor of $1/2$ of the effective theory is not reproduced.

Capitalizing on this technical explanation, we can say that in terms of defects we have a two-component picture. One component is superfluid with velocity field u_μ. The other component are zero modes responsible for the chiral magnetic and vortical effects. The zero modes are having speed of light and, therefore, are not equilibrated with the rest of the liquid. The two-component picture, however, does not necessarily differ in predictions from the one- component picture. Indeed, we did reproduce the standard answer in the case of chiral magnetic effect. Technically, the reason is the same as in Sect. 4.1 where we argued (following Ref. [41]) that vortices reproduce on average the velocity distribution of ordinary rotating liquid. Namely, linearity of the ChME effect in the chemical potential is the same crucial as linearity in the angular momentum of the energy E_{rot} in case of rotating liquid.

11.4 Conclusions

Theory of the chiral magnetic effect has been developing fast since the papers [3–5] put it into the actual context of the RHIC experiments. At the beginning, theory was focused on the mechanism of chirality production in heavy ion collisions. Already at this stage one has to turn to hydrodynamics describing the bulk of the RHIC data. Since the effective chiral chemical potential μ_5 vanishes on average and fluctuates from event to event in heavy ion collisions, it is mostly physics of fluctuations which—in the theoretical perspective—was studied at this stage.

Later, beginning with the paper [15] the interest shifted from phenomenology to more theoretical issues, such as unifying methods of field theory and thermodynamics to get exact results for chiral effects in hydrodynamic approximation. Very recently, we believe, the most exciting development is the emerging proof that chiral magnetic effect in hydrodynamics is a dissipation-free process. Moreover, examples known so far indicate that this ballistic-type of transport is provided by quantum-mechanical zero modes.

All these exciting results are valid, strictly speaking, in the exact chiral limit. This limitation might have rather severe phenomenological implications in case of realistic quantum chromodynamics. One could expect, therefore, in the near future a shift of interest to condensed-matter systems with fermionic excitations and linear dependence of the energy on the momentum. Of course, this classification of theoretical developments into various stages at the very best could be true only in its gross features. Nevertheless, it might be reasonable to present our conclusions within this, oversimplifying scheme.

Fluctuations of the Chiral Chemical Potential   Because of the space limitations we did not address the issue of fluctuations in the bulk of the review. If we start with $\langle \mu_5 \rangle = 0$ the current (11.1) vanishes. The chiral magnetic effect is still manifested through fluctuations. In particular, the spectrum of hydrodynamic excitations is sensitive to it. The so called chiral magnetic wave [57] corresponds to the following sequence of fluctuations:

$$(\delta n_Q) \to (\delta j^5) \to (\delta \mu_A) \to (\delta j^{el}) \to (\delta n_Q). \tag{11.95}$$

In more detail: first, a local fluctuation of electric charge density induces a fluctuation of axial current, see (11.8). Then the fluctuation of the axial current triggers a local fluctuation of the axial chemical potential. Finally and completing the cycle, the fluctuation of μ_5 results in a fluctuation of the electrical current, see (11.1). Thus, there should exist an excitation combining density waves of electric and chiral charges, the chiral magnetic wave.

Another result to be mentioned here is the observation [58] that there is a piece in the correlator of components of electric current which is uniquely determined in terms of the chiral anomaly squared:

$$F_{zz}(\omega) - F_{xx}(\omega) = \frac{(qB)^2}{4\pi^3} \frac{\omega}{e^{\beta\omega} - 1}, \tag{11.96}$$

where the z-axis points in the direction of an external magnetic field and $F_{ii}(\omega)$ is a correlator of the ith components of the electromagnetic current as a function of frequency ω, for a precise definition of the correlators see [58].

Non-renormalization Theorems and Non-dissipative Motions   The present review is focused on the non-renormalization theorems and non-dissipative, or quantum nature of the chiral effects, see Sects. 11.2 and 11.3, respectively. Non-renormalization theorems were proven within various approaches (thermodynamic, geometric, diagrammatic, effective field theories). There are no doubts that the non-renormalization theorems are valid within the approximations and assumptions made. The basic assumptions are exact chiral limit, hydrodynamic expansion in derivatives and equilibrium. The main message is that chiral currents are dissipation free and there is no suppression at high temperature.

The results reviewed in Sect. 11.2 imply that the dissipation-free motions considered are rather ballistic transports than superfluid-type phenomena (in agreement with the fact that the carriers of charges are fermions). Indeed, the entropy current associated with the chiral effects is not vanishing, unlike the superfluid case. According to (11.22), (11.24)

$$s_\mu \sim \frac{\mu}{T} \mu_5 B_\mu,$$

where s_μ is the entropy current in equilibrium associated with the magnetic field B_μ and μ, μ_5 are chemical potentials. The entropy current disappears for $\mu = 0$ but this is actually a kind of misleading. The point is that there is always a flow of degrees of freedom along the magnetic field and in the direction opposite to it. There is a cancellation, in terms of s_μ at $\mu = 0$ while $\mu \neq 0$ implies that the liquid is charged and there is a preferred direction of the flow of degrees of freedom. The electric current (11.1), on the other hand, counts the total number of degrees of freedom with charge $\pm q$.

Moreover, at least in case of noninteracting particles the carriers of the electromagnetic current (11.1) are identified as quantum-mechanical zero modes ψ_0 of Dirac particles in the external magnetic field B:

$$D_\mu \gamma^\mu \psi_0 = 0, \qquad N_0 \sim |q \cdot B \cdot \mu_5|,$$

where N_0 is the number of zero modes, see Sect. 11.3.4 or, e.g., [6] and references therein. Theoretically, no suppression of the current (11.1) is expected at non-zero temperatures. A straightforward conclusion would be that the number of zero modes does not go down with temperature. At first sight, it looks very unexpected that the quantum coherence could persist at high temperature. Let us mention, however, that something similar happens already at $T = 0$. Namely, measurements at small lattice spacings a (such that $a \cdot \Lambda_{QCD} \ll 1$) demonstrate that the number of zero modes survives wild quantum fluctuations which are of order $|A_\mu^{glue}| \sim 1/a$. However, the volume V_0 occupied by a zero mode goes to zero as a power of $(a \cdot \Lambda_{QCD})$, for a review see, e.g., [59, 60]. By analogy, one could expect for zero modes at high temperature

$$N_0(T) \approx const, \qquad V_0 \approx (\Lambda_{QCD}/T)^\gamma,$$

where the index $\gamma \sim (1\text{--}2)$, and zero modes at high temperature become lower-dimensional defects, somewhat similar to the vortices discussed in Sect. 11.3.3.

Technically, derivation of the dissipation-free hydrodynamics from the chiral symmetry of field theory is quite straightforward. Consider first the confining phase. In field theory, there is spontaneous symmetry breaking and light degrees of freedom are represented by massless Goldstone fields φ. This is the field theoretic input. The hydrodynamic output is superfluidity [52] associated with an extra thermodynamic potential (superfluid 3d velocity squared). The route from field theory to hydrodynamics is provided by replacing the ordinary time derivative with the covariant one, the chemical potential being a constant part of the gauge field A_0. As a result, a 4d massless Goldstone particle on the field-theoretic side becomes a 3d massless fields (plus the Josephson condition, $\partial_0 \phi = \mu$). Gradient of φ, $\nabla \varphi = \mathbf{v}_s$ is identified then with the (unrenormalized) superfluid velocity, or a new thermodynamic variable. Original fermionic degrees of freedom are counted as constituents of the normal component. Standard relativistic superfluidity is reproduced.

In this review, we are concerned with the case when there is no spontaneous breaking of the chiral symmetry. Still, there are massless particles, the chiral fermions themselves. Field theoretic input is then existence of polynomials in the effective action such that the action observes the symmetries while the density of the action does not respect the symmetries. The bridge to hydrodynamics is provided by the same inclusion of the chemical potential into the covariant derivative. As a result, there are new motions, or currents allowed in the equilibrium (similar to superfluidity). However, because we do not introduce new massless degrees of freedom, the dissipation-free motions are calculable now, and there are no new thermodynamic potentials. (For attempts to introduce chiral superfluidity in direct analogy with the ordinary superfluidity, i.e. through postulating a new light scalar see Refs. [61–64].)

In this sense, analogy between the ChME and other calculable macroscopic quantum effects, say, Casimir forces, (for review and references see, e.g., [65]) might be relevant. One considers first interaction of two fluctuating dipoles (atoms) with electric polarizabilities $\mathbf{p}_{1,2}(\omega) = \alpha_{1,2}(\omega)\mathbf{E}(\omega)$. At large distances retardation becomes important and the van der Waals potential is replaced by the Casimir-Polder static energy:

$$V_{CP} = -\frac{23}{4\pi}\frac{\alpha_1(0)\alpha_2(0)}{r^6}\frac{hc}{r}.$$

The Casimir-Polder potential refers to interaction of point-like sources due to two-photon exchange. Macroscopic interaction arises upon averaging over many point-like sources. Probably the best known example of this type is the force F acting on a unit area A of two conducting plates at distance a;

$$\frac{F}{A} = -\frac{\pi^2 hc}{240a^4}.$$

This is a macroscopic force of quantum origin, calculable from first principles. Similarly, the anomaly derived first for point-like particles becomes a macroscopic chiral magnetic effect upon averaging of a two-fermion exchange over many centers.

Towards Condensed-Matter Applications   A specific feature of the ChME is that even at equilibrium there is a non-vanishing current while Casimir forces are static. This difference is entirely due to the fact that chiral particles are massless and cannot be "stopped". Indeed the Casimir-Polder potential above is due to polarizabilities. As far as a constituent is massive we can have a static magnetic dipole $\mathbf{m} = \alpha_M \mathbf{H}$. However, for a massless, chiral particle there can be no static magnetic moment and, instead, we get a current, see (11.1).

Already from this simple reasoning we can conclude that transition to chiral effects is singular. If a small mass $m_f \neq 0$ is introduced the life-time T of the current (11.1) is finite and there is no motion in equilibrium, $t \to \infty$, see Sect. 11.3.1. Whether

$$T \sim \frac{1}{m_f} \quad \text{or} \quad T \sim m_f^2/\mu_5,$$

or else, remains, to our knowledge, an open question. The answer might depend on details of experiment.

It seems natural to assume that the proof of the non-dissipative nature of the chiral effects would be generalized to condensed matter systems, like, say, Weyl semimetals with chiral spectrum of excitations

$$\epsilon_f = v k_f,$$

where ϵ_f, k_f are energy and momentum of fermionic excitation. Indeed, many consequences from the chiral anomaly in relativistic field theory have close parallels in condensed-matter systems, see, in particular [66–69]. From the point of view of applications the condensed-matter systems seem of course more practical and one can start discussing principles of functioning of a new kind of devices [70].

Acknowledgements The author is grateful to T. Kalaydzhyan, D.E. Kharzeev, V.P Kirilin, A.V. Sadofyev, V.I. Shevchenko, L. Stodolsky and H. Verschelde for illuminating discussions. The work of the author was partially supported by grants PICS-12-02-91052, FEBR-11-02-01227-a and by the Federal Special-Purpose Program "Cadres" of the Russian Ministry of Science and Education.

References

1. E.V. Shuryak, What RHIC experiments and theory tell us about properties of quark-gluon plasma? Nucl. Phys. A **750**, 64–83 (2005). arXiv:hep-ph/0405066 [hep-ph]
2. D. Teaney, J. Laure, E.V. Shuryak, A hydrodynamic description of heavy ion collisions at the SPS and RHIC. arXiv:nucl-th/0110037 [nucl-th]
3. D.E. Kharzeev, Parity violation in hot QCD: why it can happen, and how to look for it. Phys. Lett. B **633**, 260–264 (2006). arXiv:0406125 [hep-ph]
4. D.E. Kharzeev, L.D. McLerran, H.J. Warringa, The effects of topological charge change in heavy ion collisions: 'Event by event P and CP violation'. Nucl. Phys. A **803**, 227–253 (2008). arXiv:0711.0950 [hep-ph]
5. K. Fukushima, D.E. Kharzeev, H.J. Warringa, The chiral magnetic effect. Phys. Rev. D **78**, 074033 (2008). arXiv:0808.3382 [hep-ph]
6. G. Basar, C.V. Dunne, The chiral magnetic effect and axial anomalies. arXiv:1207.4199 [hep-th]
7. D.E. Kharzeev, The chiral magnetohydrodynamics of QCD fluid at RHIC and LHC. J. Phys. G **38**, 124061 (2011). arXiv:1107.4004 [hep-ph]
8. A. Vilenkin, Equilibrium parity violating current in a magnetic field. Phys. Rev. D **22**, 3080 (1980)
9. H.B. Nielsen, M. Ninomiya, Adler-Bell-Jackiw anomaly and Weyl fermions in crystal. Phys. Lett. **130B**, 389 (1983)
10. S.L. Adler, Axial vector vertex in spinor electrodynamics. Phys. Rev. **177**, 2426–2438 (1969)
11. J.S. Bell, R. Jackiw, A PCAC puzzle: $\pi 0 \to \gamma\gamma$ in the sigma model. Nuovo Cimento A **60**, 47–61 (1969)
12. S.L. Adler, W.A. Bardeen, Absence of higher order corrections in the anomalous axial vector divergence equation. Phys. Rev. **182**, 1517 (1969)
13. D.T. Son, Hydrodynamics of relativistic systems with broken continuous symmetries. Int. J. Mod. Phys. A **16S1C**, 1284 (2001)
14. S. Dubovsky, L. Hui, A. Nicolis, D.T. Son, Effective field theory for hydrodynamics: thermodynamics, and the derivative expansion. Phys. Rev. D **85**, 085029 (2012). arXiv:1107.0731 [hep-th]

15. D.T. Son, P. Surowka, Hydrodynamics with triangle anomalies. Phys. Rev. Lett. **103**, 191601 (2009). arXiv:0906.5044 [hep-th]

16. L.D. Landau, E.M. Lifshitz, Course of Theoretical Physics, Fluid Mechanics, vol. 6, 2nd edn. ISBN 978-0-08-033933-7

17. N. Banerjee, J. Bhattacharya, S. Bhattacharyya, S. Jain, S. Minwalla, T. Sharma, Constraints on fluid dynamics from equilibrium partition functions. arXiv:1203.3544 [hep-th]

18. K. Jensen, M. Kaminski, P. Kovtun, R. Meyer, A. Ritz, A. Yarom, Towards hydrodynamics without an entropy current. Phys. Rev. Lett. **109**, 101601 (2012). arXiv:1203.3556 [hep-th]

19. K. Jensen, Triangle anomalies, thermodynamics, and hydrodynamics. arXiv:1203.3599 [hep-th]

20. A.Yu. Alekseev, V.V. Cheianov, J. Frohlich, Universality of transport properties in equilibrium, Goldstone theorem and chiral anomaly. Phys. Rev. Lett. **81**, 3503–3506 (1998). arXiv:9803346 [cond-mat]

21. J. Erdmenger, M. Haack, M. Kaminski, A. Yarom, Fluid dynamics of R-charged black holes. J. High Energy Phys. **0901**, 055 (2009). arXiv:0809.2488 [hep-th]

22. D.T. Son, A.R. Zhitnitsky, Quantum anomalies in dense matter. Phys. Rev. D **70**, 074018 (2004). arXiv:0405216 [hep-ph]

23. M.A. Metlitski, A.R. Zhitnitsky, Anomalous axion interactions and topological currents in dense matter. Phys. Rev. D **72**, 045011 (2005). arXiv:0505072 [hep-ph]

24. A. Vilenkin, Macroscopic parity violating effects: neutrino fluxes from rotating black holes and in rotating thermal radiation. Phys. Rev. D **20**, 1807 (1979)

25. S. Golkar, D.T. Son, (Non)-renormalization of the chiral vortical effect coefficient. arXiv:1207.5806 [hep-th]

26. D. Hou, H. Liu, H.-c. Ren, A possible higher order correction to the vortical conductivity in a gauge field plasma. Phys. Rev. D **86**, 121703(R) (2012). arXiv:1210.0969 [heh-th]

27. S.R. Coleman, B.R. Hill, No more corrections to the topological mass term in QED in three-dimensions. Phys. Lett. B **159**, 184 (1985)

28. M. Lublinsky, I. Zahed, Anomalous chiral superfluidity. Phys. Lett. B **684**, 119–122 (2010). arXiv:0910.1373 [hep-th]

29. M.I. Isachenkov, A.V. Sadofyev, The chiral magnetic effect in hydrodynamical approach. Phys. Lett. B **697**, 404–406 (2011). arXiv:1010.1550 [hep-th]

30. A.V. Sadofyev, V.I. Shevchenko, V.I. Zakharov, Notes on chiral hydrodynamics within effective theory approach. Phys. Rev. D **83**, 105025 (2011). arXiv:1012.1958 [hep-th]

31. M. Stone (ed.), *The Quantum Hall Effect* (World Scientific, Singapore, 1992)

32. T. Kimura, Hall and spin Hall viscosity ratio in topological insulators. arXiv:1004.2688 [cond-mat.mes-hall]

33. D.E. Kharzeev, H.-Y. Yee, Anomalies and time reversal invariance in relativistic hydrodynamics: the second order and higher dimensional formulations. Phys. Rev. D **84**, 045025 (2011). arXiv:1105.6360 [hep-th]

34. J. Goldstone, F. Wilczek, Fractional quantum numbers on solitons. Phys. Rev. Lett. **47**, 986–989 (1981)

35. C.G. Callan, J.A. Harvey, Anomalies and fermion zero modes on strings and domain walls. Nucl. Phys. B **250**, 427 (1985)

36. V.P. Kirilin, A.V. Sadofyev, V.I. Zakharov, Chiral vortical effect in superfluid. Phys. Rev. D **86**, 025021 (2012). arXiv:1203.6312 [hep-th]

37. K. Landsteiner, L. Melgar, Holographic flow of anomalous transport coefficients. arXiv:1206.4440 [hep-th]

38. Y. Neiman, Y. Oz, Relativistic Hydrodynamics with general anomalous charges. J. High Energy Phys. **1103**, 023 (2011). arXiv:1011.5107 [hep-th]

39. D.E. Kharzeev, H.J. Warringa, Chiral magnetic conductivity. Phys. Rev. D **80**, 034028 (2009). arXiv:0907.5007 [hep-ph]

40. K. Landsteiner, E. Megias, F. Pena-Benitez, Anomalies and transport coefficients: the chiral gravito-magnetic effect. arXiv:1110.3615 [hep-ph]

41. L.P. Pitaevskii, E.M. Lifshitz, *Statistical Physics*, part 2, vol. 9, 1st edn. (Butterworth-Heinemann, Oxford, 1980). ISBN 978-0-7506-2636-1
42. K. Jensen, R. Loganayagam, A. Yarom, Thermodynamics, gravitational anomalies and cones. arXiv:1207.5824 [hep-th]
43. R. Jackiw, S. Templeton, How super-renormalizable interactions cure their infrared divergences. Phys. Rev. D **23**, 2291–2304 (1981)
44. S. Deser, R. Jackiw, S. Templeton, Topologically massive gauge theories. Ann. Phys. **140**, 372–411 (1982)
45. D.J. Gross, R.D. Pisarski, L.G. Yaffe, QCD and instantons at finite temperature. Rev. Mod. Phys. **53**, 43 (1981)
46. A.A. Anselm, A.A. Johansen, Radiative corrections to the axial anomaly. JETP Lett. **49**, 214–218 (1989)
47. E. Witten, Global aspects of current algebra. Nucl. Phys. B **223**, 422–432 (1983)
48. P. Kovtun, D.T. Son, A.O. Starinets, Viscosity in strongly interacting quantum field theories from black hole physics. Phys. Rev. Lett. **94**, 111601 (2005). arXiv:0405231 [hep-th]
49. K. Fukushima, D.E. Kharzeev, H.J. Warringa, Real-time dynamics of the chiral magnetic effect. Phys. Rev. Lett. **104**, 212001 (2010). arXiv:1002.2495 [hep-ph]
50. H.J. Warringa, Dynamics of the chiral magnetic effect in a weak magnetic field. arXiv:1205.5679 [hep-th]
51. E. Witten, Cosmic superstrings. Phys. Lett. B **153**, 243 (1985)
52. D.T. Son, M.A. Stephanov, QCD at finite isospin density: from pion to quark–anti-quark condensation. Phys. At. Nucl. **64**, 834–842 (2001). arXiv:0011365 [hep-ph]
53. H. Leutwyler, On the foundations of chiral perturbation theory. Ann. Phys. **235**, 165–203 (1994)
54. A. Nicolis, Low-energy effective field theory for finite-temperature relativistic superfluids. arXiv:1108.2513 [hep-th]
55. Y. Aharonov, A. Casher, The ground state of a spin 1/2 charged particle in a two-dimensional magnetic field. Phys. Rev. A **19**, 2461–2462 (1979)
56. A.J. Niemi, G.W. Semenoff, Fermion number fractionization in quantum field theory. Phys. Rept. **135**, 99 (1986)
57. D.E. Kharzeev, H.-U. Yee, Chiral magnetic wave. Phys. Rev. D **83**, 085007 (2011). arXiv:1012.6026 [hep-th]
58. V. Shevchenko, Quantum measurements and chiral magnetic effect. arXiv:1208.0777 [hep-th]
59. F.V. Gubarev, S.M. Morozov, M.I. Polikarpov, V.I. Zakharov, Evidence for fine tuning of fermionic modes in lattice gluodynamics. JETP Lett. **82**, 343–349 (2005). arXiv:0505016 [hep-lat]
60. V.I. Zakharov, Dual string from lattice Yang-Mills theory. AIP Conf. Proc. **756**, 182–191 (2005). arXiv:0501011 [hep-ph]
61. V.P. Gusynin, V.A. Miransky, I.A. Shovkovy, Dimensional reduction and dynamical chiral symmetry breaking by a magnetic field in $(3+1)$-dimensions. Phys. Lett. B **349**, 477–483 (1995). arXiv:9412257 [hep-ph]
62. M.N. Chernodub, H. Verschelde, V.I. Zakharov, Two-component liquid model for the quark-gluon plasma. Theor. Math. Phys. **170**, 211–216 (2012). arXiv:1007.1879 [hep-ph]
63. H. Verschelde, V.I. Zakharov, Two-component quark-gluon plasma in stringy models. AIP Conf. Proc. **1343**, 137–139 (2011). arXiv:1012.4821 [hep-ph]
64. T. Kalaydzhyan, Chiral superfluidity of the quark-gluon plasma. arXiv:1208.0012 [hep-ph]
65. K.A. Milton, Van der Waals and Casimir-Polder forces. arXiv:1101.2238 [cond-mat.mes-hall]
66. D.T. Son, B.Z. Spivak, Chiral anomaly and classical negative magnetoresistance of Weyl metals. arXiv:1206.1627 [cond-mat.mes-hall]
67. I. Zahed, Anomalous chiral Fermi surface. Phys. Rev. Lett. **109**, 091603 (2012). arXiv:1204.1955 [hep-th]
68. D.T. Son, N. Yamamoto, Berry Curvature, triangle anomalies, and chiral magnetic effect in Fermi liquids. Phys. Rev. Lett. **109**, 181602 (2012). arXiv:1203.2697 [cond-mat.mes-hall]

69. E.V. Gorba, V.A. Miransky, I.A. Shovkovy, Surprises in relativistic matter in a magnetic field. Prog. Part. Nucl. Phys. **67**, 547–551 (2012). arXiv:1111.3401 [hep-ph]
70. D.E. Kharzeev, H.-U. Yee, Chiral electronics. arXiv:1207.0477 [cond-mat.mes-hall]

Chapter 12
Remarks on Decay of Defects with Internal Degrees of Freedom

A. Gorsky

12.1 Introduction

The chiral magnetic effect [1] seems to play an important role in the heave ions collision. The key question concerns the physical realization of the chiral chemical potential responsible for this phenomena. Usually it is attributed to the instanton effects however there are another sources for the effective chiral chemical potential In this paper we review three problems of such type: the decay of a metastable axion wall in an external magnetic field, the decay of walls in dense matter, and the decay of a nonabelian string with the excitation [2]. It is known that due to the one-loop Goldstone-Wilczek current [3] an axion wall in an external magnetic field develops a homogeneous density of the charged fermions [4–8]. The question we investigate here is whether the fermions stay localized on the remaining part of the domain wall during its decay or escape into the bulk? We find that in the leading approximation the fermions leave the domain wall and provide a very clear pattern of the chiral magnetic effect (CME) which currently attracts a considerable attention related to the observed charge separation effect at RHIC [1]. We shall thus argue that the decay of domain wall provides an effective chiral chemical potential responsible for the chiral magnetic effect.

The second problem is related to the decay of mesonic walls in QCD. The nontrivial density of the Skyrmions is generated in the external magnetic field on π^0 walls, and the field prevents the wall from nonperturbative decay above some B_{crit} [9]. There is a new point in the mesonic wall decay in dense matter. It is known [10] that in this case due to the anomalous term in the action the current is generated along the axion-like string at the boundary of the wall. In the decay problem the axion string is the boundary of the hole and therefore there is the circular boundary current during the decay. We analyze the impact of the created current on the magnetic field.

A. Gorsky (✉)
Institute of Theoretical and Experimental Physics, Moscow 117259, Russia
e-mail: gorsky@itep.ru

D. Kharzeev et al. (eds.), *Strongly Interacting Matter in Magnetic Fields*,
Lecture Notes in Physics 871, DOI 10.1007/978-3-642-37305-3_12,
© Springer-Verlag Berlin Heidelberg 2013

The third problem in our discussion concerns the decay process of a nonabelian string with excited internal CP_N degrees of freedom. The Lagrangian of the world-sheet theory supports a kink excitation [11–13] which can exist in isolation by itself in SUSY theories and as kink-antikink bound states in a generic non-SUSY case. We shall argue that in the nonabelian string decay the kink excitations on the string decrease the decay rate.

The paper is organized as follows. In Sect. 12.2 we consider the process of the axion domain wall decay. The decays of mesonic walls in QCD are considered in Sect. 12.3. In Sect. 12.4 we comment on the decay of nonabelian string with the emphasis on the role of kink excitations.

12.2 Decay of Axion-Like Domain Walls in $D = 3 + 1$ Theories

In this section we shall discuss the decay of the Abelian domain walls in the magnetic field. Let us discuss the decay of axion wall in the theory with the Lagrangian for the axion field $a(x)$ given by

$$L = f_a^{-2} \left[\frac{1}{2} (\partial a)^2 + m_a^2 \cos a(x) \right]. \tag{12.1}$$

The model also contains charged fermions interacting with the axion as

$$L_f = \bar{\psi} \left[i (\partial_v - i A_v) \gamma^v - m_f e^{i a(x) \gamma^5} \right] \psi \tag{12.2}$$

with A_v being the potential of the electromagnetic field. (The coupling e is absorbed into the normalization of A.) An integration over the fermionic field gives rise to the anomalous interaction between the axion and the electromagnetic field [3] described by the Lagrangian

$$L_{anom} = \frac{1}{16\pi^2} \varepsilon_{\mu\nu\lambda\sigma} A_\mu F_{\lambda\sigma} \partial_v a. \tag{12.3}$$

A derivation of this term through the analysis of the fermionic modes in the external field in the simplified model for the fermions with the Lagrangian

$$L = \bar{\psi} \left[i (\partial_v - i A_v) \gamma^v - \mu_1(z) - i \mu_2(z) \gamma^5 \right] \psi \tag{12.4}$$

can be found in [7]. The variation of μ_2 breaks the CP parity similarly to the axion field.

If the axion model admits a wall solution, the anomalous term yields the density of the electric charge at the domain wall [4]. The details of the domain wall solution are not important and the surface density of the induced charge in the background magnetic field is equal to

$$q = \frac{B \Delta a}{4\pi^2} \tag{12.5}$$

where B is the magnetic field perpendicular to the axion wall and Δa is the total variation of the axion field across the wall. Therefore a constant external magnetic field creates a homogeneous density of the induced electric charge on the wall. In the simplified model the distribution of the induced charge density in the domain wall background equals to

$$\rho(z) = \frac{B}{4\pi^2} \frac{d}{dz} \arctan \frac{\mu_2(z)}{\mu_1(z)}. \tag{12.6}$$

For any realistic magnetic field its total flux through an infinite wall is zero, so that the total charge of such wall is zero as well. This behavior however is a result of exact cancellation between areas with positive and negative surface charges.

An axion wall is metastable and decays through nucleation and subsequent expansion of a hole bounded by an axion string. When considering this process in a constant magnetic field, a natural questions arises concerning the fate of the induced electric charge during the decay as well as the back-reaction of the decay on the background magnetic field. In order to analyze this issue we consider different components of the GW current which are generated during the decay process. Let us assume for definiteness that the hole is created at the origin in the (x, y) plane.

During the decay the axion field can be approximated as

$$a(z, r, t) = f(z)\theta(r - r(t)) \tag{12.7}$$

with some profile function $f(z)$, and $r(t)$ is the time dependent radius of the expanding hole. Fermions are bounded by the domain wall and therefore there is no reason for them to stay at the same point when the hole is created. There are two logical possibilities: the fermions fly away from the domain wall plane or are captured by the axion string at the boundary of the hole. Using the GW expression we find that as the hole expands there arises a current perpendicular to the wall plane

$$J_z \propto (\partial_t a) B_z \propto f(z)\dot{r}\delta(r - r(t)) B_z \tag{12.8}$$

which is clearly localized near the boundary of the hole.

The current J_z is directed along the external magnetic field and provides the explicit example of the chiral magnetic effect resulting in the charge separation effect [1, 14] which was recently a subject of intensive theoretical and experimental studies. The domain wall decay process amounts to the effective time dependent chiral chemical potential

$$\mu_{5,eff} = \partial_t a.$$

The effective chiral chemical potential is localized at the boundary of the hole—the axion string. This is not a surprise since the axion string in magnetic field is chiral, i.e. there is an asymmetry between left and right modes of the fermions on the string.

We thus conclude that the fermions, initially localized at the wall, do fly away to the bulk in the decay process. One can perform a cross check of this conclusion by comparing the current 'to the bulk' J_z with the rate of the disappearance of the area of the wall that initially carried the fermions. Indeed, the rate at which the surface charge disappears through the growth of the hole is given by

$$dN = 2\pi q r \dot{r} \, dt \tag{12.9}$$

where q is the surface charge density on the wall, $r(t)$ radius of the hole and the speed $\dot{r}$ is fixed by the bounce solution. On the other hand if all fermions fly away we must have from the continuity equation

$$\frac{dN}{dt} = -\int d^3x \, \frac{dJ_z}{dz}. \tag{12.10}$$

Substituting the expression for the axion field corresponding to the undeformed $O(4)$ symmetric bounce solution into the anomalous current we obtain that the continuity equation is fulfilled. This implies that all fermions fly away from the wall plane during the decay process and there is no need for accumulating any fermions on the boundary of the hole.

It should be mentioned that in the considered case, where the area of the wall supporting the fermion charge changes and the fermions flow into the bulk, the charge conservation is implemented differently from the case of a fixed patch of the wall and varying magnetic field. In the latter situation the surface term of the GW current is not conserved by itself [15] and its divergence is localized at the string

$$\partial_\mu j_\mu \propto B\delta(r). \tag{12.11}$$

The apparent non-conservation of the current is compensated by the accumulation of the charge on the axionic string at the boundary of the patch.

The escape of the fermions to the bulk produces an effect on the probability of the wall decay in a magnetic field. Indeed, the tunneling process is described by a spherical Euclidean bounce configuration which is determined from the effective action

$$S_{eff} = 4\pi R^2 T_{string} - 4/3\pi R^3 T_{wall}, \tag{12.12}$$

where R is the radius of the bounce. At the extremum of this action the surface of the bounce is described by

$$R_{crit} = \frac{2T_{string}}{T_{wall}} \qquad x^2 + y^2 + t_E^2 = R_{crit}^2 \tag{12.13}$$

with t_E being the Euclidean time, so that the coordinate z transverse to the wall is not essential in the bounce solution.

The transverse direction however becomes of relevance when the magnetic field is switched on. In the leading approximation the effect of the magnetic field can be taken into account as follows. While the fermions are localized at the wall they are effectively massless and there is no energy associated with them. Once they escape in the bulk each fermion costs energy equal to its mass m_f. That is energetically the localization of the fermions on the wall suppresses the decay probability. From the $(2 + 1)$ dimensional viewpoint the decay proceeds with the energy loss since fermions escape from the wall. The fermion charge in a uniform magnetic field is proportional to the area of the wall (12.5). Therefore the effect an energy loss due to emission of fermions can be described by replacing the wall tension by an effective one

$$T_{wall} \to T_{wall,eff} = T_{wall} - \frac{B}{2\pi} m_f, \tag{12.14}$$

resulting in a suppression of the wall decay. Notably the decay is entirely suppressed, i.e. the wall is stabilized at the magnetic field exceeding the critical value

$$B_{crit} = 2\pi \frac{T_{wall}}{m_f}. \tag{12.15}$$

Let us remark that for the case of an induced wall decay at a non-vanishing energy the solution necessarily involves Minkowski part in the time evolution. An example of such two-step process involving a resonance behavior at particular values of the energy has been discussed in details in [16]. In the current problem this would happen if in a process, e.g. in a collision, an excited state of axionic string is produced, which then tunnels through the Euclidean region and eventually reaches the classical expansion regime.

12.3 Decays of Mesonic Walls

12.3.1 Decay of π^0 Domain Walls

In this section we consider the decay of walls in conventional QCD. One example of such object is provided by a wall built from π^0 mesons. A π^0 wall is not topological and can be 'unwound' inside the $SU(2)$ flavor group. Furthermore such walls are absolutely unstable in the absence of the magnetic field. However at $B > B_0 = 3m_\pi^2$ the wall becomes locally stable and at $B > B_1 = 16\pi f_\pi^2 m_\pi / m_N$ a patch of such wall carrying a baryon number becomes the lowest energy state with baryon number [9].

The tension of the domain wall calculated at the explicit solution [9] reads as

$$T_{pwall} = 8 f_\pi^2 m_\pi. \tag{12.16}$$

A magnetic field B applied perpendicularly to the wall generates a surface density of the baryon charge

$$q_B = \frac{B}{2\pi}, \tag{12.17}$$

which can be also viewed as a liquid of the Skyrmions on the surface [9].

The decay of the pionic wall implies a nontrivial baryonic current

$$J_\mu = \frac{1}{4\pi^2} \varepsilon_{\mu\nu\lambda\sigma} \partial_\nu \pi^0 F_{\lambda\sigma} \tag{12.18}$$

flowing into the bulk similar to the electric current in the axion example. While escaping from the wall the Skyrmions have mass m_N. Therefore the effective wall tension can be found as

$$T_{eff} = T_{pwall} - q_B m_N = 8 f_\pi^2 m_\pi - \frac{B}{2\pi} m_N. \tag{12.19}$$

One readily concludes from this expression that at $B > B_1$ the effective tension of the wall is negative, so that the decay is energetically impossible and the wall is absolutely stable. It can be noted that a somewhat similar behavior has been observed for the decay of electric strings in magnetic field [17].

12.3.2 Wall Decay in QCD at High Density

At high baryon density the ground state of QCD is in color-flavor locking (CFL) phase and the system develops color superconductivity (see [18] for a review). The theory at large baryon chemical potential μ is in the weak coupling regime and the dynamics of the low-energy degrees of freedom can be calculated perturbatively. In particular, the existence of a ϕ domain wall can be justified [19] from the effective Lagrangian for the Goldstone mode ϕ of $U(1)_A$ symmetry which is spontaneously broken by the condensate in the color-superconducting vacuum state.

The explicit form of the Lagrangian in two-flavor case reads as follows

$$L_{dense} = f^2 \left[(\partial_0 \phi)^2 - u^2 (\partial_i \phi)^2 \right] - a\mu^2 \Delta^2 \cos\phi \tag{12.20}$$

where a is dimensionless and vanishes in the limit $\mu \to \infty$, and u is the speed of sound: $u^2 = 1/3$. The parameters of the Lagrangian are

$$f^2 = \frac{\mu^2}{8\pi u^2}, \tag{12.21}$$

and Δ is the value of the gap. The tension of the wall can be derived immediately from the effective Lagrangian

$$T_{wall} = 8\sqrt{2a} u f \mu \Delta. \tag{12.22}$$

In the CFL phase the potential term in the Lagrangian of the lightest meson gets modified as

$$V_{CFL} = -\tilde{a} \left(\frac{m_s}{\mu} \right) \mu^2 \Delta^2 \cos\phi, \tag{12.23}$$

where m_s is the mass of the strange quark. Thus the tension of the wall in the CFL phase acquires additional m_s dependent factor.

The decay of the domain walls in the dense QCD matter has some peculiarities. In particular, one can notice that in the dense matter there is an anomalous

Chern-Simons term in the Lagrangian of the pseudoscalar meson proportional to the chemical potential μ [9, 10]:

$$\delta L = \frac{e}{24\pi^2} \mu \varepsilon_{0\nu\lambda\sigma} \partial_\nu \phi F_{\lambda\sigma}. \tag{12.24}$$

An immediate consequence of this term is that there is an electric current circulating along the axial string [10, 20, 21]

$$J = \frac{\mu e}{12\pi}. \tag{12.25}$$

It can be noted that the current does not depend on the value of external field. This current can be derived by the summation over the fermion modes [20, 21] similar to the calculation of the induced charge in the magnetic field [7].

The main difference from the wall decay discussed in the previous section is that in dense matter there necessarily is a current along the hole boundary identified with the axial ϕ string. The current plays a two-fold role. Firstly, its existence implies that during the decay process not all of the fermions populating the wall fly away from the plane. Some of them are captured by the axial string at the boundary instead. Secondly, the circular current induces magnetic field inside the hole and the direction of the field depends on the sign of the chemical potential. One thus concludes that in dense matter magnetic effects in the decay of the wall are necessarily essential since the field is generated by the induced current circulating along the hole boundary. The tunnelings starts at zero energy, so that there is no cusp at the bottom of the bounce configuration. However contrary to the axion and π^0-mesonic walls in the vacuum, the current along the boundary makes it impossible to describe the outflow of the fermion energy by an appropriately modified effective wall tension as in (12.19).

12.4 Nonabelian String Decay

The strings are quite common objects corresponding to effective solutions to the equations of motion in various models. The problem of their decays can be formulated, and a detailed analysis of the decay of an ANO string in the Abelian Higgs model can be found in [22]. As of yet only the decay of such ANO Abelian effective strings has been considered. However more general stringy solutions exist both in the SUSY and non-SUSY theories in the color-flavor locking phase. Their key feature is the existence of additional degrees of freedom due to the nontrivial embedding of the nonabelian string into the gauge group, which amounts to the orientational moduli providing CP_N degrees of freedom on the worldsheet [11–13]. Thus the problem that parallels the discussion in the previous section is that of the fate of the CP_N degrees of freedom living on the nonabelian string during the decay. The worldsheet theory is built from an N-component complex field n^i subject to the constraint

$$n_i^* n^i = 1. \tag{12.26}$$

The Lagrangian has the form

$$L = \frac{2}{g^2}\left[(\partial_\mu - iA_\mu)n_i^*(\partial_\mu + iA_\mu)n^i - \lambda(n_i^* n^i - 1)\right], \qquad (12.27)$$

where λ is the Lagrange multiplier enforcing the condition (12.26). At the quantum level this constraint is effectively eliminated and λ becomes dynamical. Moreover, A_μ is an auxiliary field which at the classical level enters the Lagrangian with no kinetic term. A kinetic term is generated, however, at the quantum level, so that the field A_μ becomes dynamical too.

One can also add to the Lagrangian a θ term of the form

$$L_\theta = \frac{\theta}{2\pi}\varepsilon_{\mu\nu}\partial^\mu A^\nu = \frac{\theta}{2\pi}\varepsilon_{\mu\nu}\partial^\mu\left(n_i^*\partial^\nu n^i\right). \qquad (12.28)$$

In non-SUSY $D = 4$ gauge theories there is only a single vacuum state, so that a single kink solution is impossible. However the spectrum involves a kink-antikink pair, which corresponds, from the four dimensional viewpoint, to a monopole-antimonopole pair localized at the nonabelian string [23]. In the non-SUSY case one can also introduce the θ-term in the bulk theory, which makes its way to the worldsheet theory of the nonabelian string as well [23]. In the worldsheet theory the θ term induces a constant Abelian electric field of the auxiliary gauge field $A(x)$ along string. In the string decay the electric field is completely screened in the emerging hole. One therefore concludes that the dyons have to be created at the ends of the string in this case.

Recently the interesting phenomena concerning the decay of the nonabelian string has been found [24]. It turns out that the CP_N model on the interval of length R undergoes the phase transition when R changes. It implies the peculiar features of the decay process. Indeed if we consider the bounce configuration for the decay process the contribution from the "hole" strongly depends on the radius of the hole which is defined by the parameters of the model. When the hole radius is small enough the CP_N model on the disc has no the mass gap while at large radius it is in the gapped phase. Hence one could expect the strong change of the decay probability near the critical value of the radius of the disc in the Euclidean space.

12.5 Summary

In this note we have reviewed decay of defects in external field. It was shown that in a magnetic field the axion domain wall evaporates all induced electric charge into the bulk. Such decay of the axion wall provides an explicit example of the chiral magnetic effect where the axion strings are responsible for the chiral chemical potential. The decay of the mesonic walls in magnetic field has some peculiarities. Namely, the decay probability of the pionic walls is suppressed by a magnetic field and above a critical value of the field the wall is non-perturbatively stable. In the CFL phase at high density the current along the boundary of the hole in the η-meson walls is generated decreasing the initial magnetic field.

Acknowledgements　　I thanks to M. Voloshin for the collaboration on these issues.

References

1. D.E. Kharzeev, L.D. McLerran, H.J. Warringa, The effects of topological charge change in heavy ion collisions: 'Event by event P and CP violation'. Nucl. Phys. A **803**, 227 (2008). arXiv:0711.0950 [hep-ph]
2. A. Gorsky, M.B. Voloshin, Phys. Rev. D **82**, 086008 (2010). arXiv:1006.5423 [hep-th]
3. J. Goldstone, F. Wilczek, Fractional quantum numbers on solitons. Phys. Rev. Lett. **47**, 986 (1981)
4. P. Sikivie, Experimental tests of the "invisible" axion. Phys. Rev. Lett. **51**, 1415 (1983). [Erratum. Phys. Rev. Lett. **52**, 695 (19840)]
5. I.I. Kogan, Kaluza-Klein and axion domain walls: induced charge and mass transmutation. Phys. Lett. B **299**, 16 (1993)
6. I.I. Kogan, Axions, monopoles and cosmic strings. arXiv:hep-ph/9305307
7. M.B. Voloshin, Once again on electromagnetic properties of a domain wall interacting with charged fermions. Phys. Rev. D **63**, 125012 (2001). arXiv:hep-ph/0102239
8. J.M. Izquierdo, P.K. Townsend, Axionic defect anomalies and their cancellation. Nucl. Phys. B **414**, 93 (1994). arXiv:hep-th/9307050
9. D.T. Son, M.A. Stephanov, Axial anomaly and magnetism of nuclear and quark matter. Phys. Rev. D **77**, 014021 (2008). arXiv:0710.1084 [hep-ph]
10. D.T. Son, A.R. Zhitnitsky, Quantum anomalies in dense matter. Phys. Rev. D **70**, 074018 (2004). arXiv:hep-ph/0405216
11. A. Hanany, D. Tong, Vortices, instantons and branes. J. High Energy Phys. **0307**, 037 (2003)
12. R. Auzzi, S. Bolognesi, J. Evslin, K. Konishi, A. Yung, Non-Abelian superconductors: vortices and confinement in N = 2 SQCD. Nucl. Phys. B **673**, 187 (2003)
13. M. Shifman, A. Yung, Non-Abelian string junctions as confined monopoles. Phys. Rev. D **70**, 045004 (2004). hep-th/0403149
14. A. Alekseev, V. Cheianov, J. Froehlich, cond-mat/9803346
15. C.G. Callan, J.A. Harvey, Anomalies and fermion zero modes on strings and domain walls. Nucl. Phys. B **250**, 427 (1985)
16. A.S. Gorsky, M.B. Voloshin, Nonperturbative production of multiboson states and quantum bubbles. Phys. Rev. D **48**, 3843 (1993). arXiv:hep-ph/9305219
17. M.N. Chernodub, Background magnetic field stabilizes QCD string against breaking. arXiv:1001.0570 [hep-ph]
18. M.G. Alford, A. Schmitt, K. Rajagopal, T. Schafer, Color superconductivity in dense quark matter. Rev. Mod. Phys. **80**, 1455 (2008). arXiv:0709.4635 [hep-ph]
19. D.T. Son, M.A. Stephanov, A.R. Zhitnitsky, Domain walls of high-density QCD. Phys. Rev. Lett. **86**, 3955 (2001). arXiv:hep-ph/0012041
20. M.A. Metlitski, Currents on superconducting strings at finite chemical potential and temperature. Phys. Lett. B **612**, 137 (2005). arXiv:hep-ph/0501144
21. M.A. Metlitski, A.R. Zhitnitsky, Anomalous axion interactions and topological currents in dense matter. Phys. Rev. D **72**, 045011 (2005). arXiv:hep-ph/0505072
22. M. Shifman, A. Yung, Metastable strings in Abelian Higgs models embedded in non-Abelian theories: calculating the decay rate. Phys. Rev. D **66**, 045012 (2002). arXiv:hep-th/0205025
23. A. Gorsky, M. Shifman, A. Yung, Non-Abelian Meissner effect in Yang-Mills theories at weak coupling. Phys. Rev. D **71**, 045010 (2005). arXiv:hep-th/0412082
24. A. Milekhin, Phys. Rev. D **86**, 105002 (2012). arXiv:1207.0417 [hep-th]

Chapter 13
A Chiral Magnetic Effect from AdS/CFT with Flavor

Carlos Hoyos, Tatsuma Nishioka, and Andy O'Bannon

13.1 Introduction

Consider a $(3 + 1)$-dimensional system of free, massless Dirac fermions ψ. The Lagrangian of such a system has two $U(1)$ symmetries, the vector one $U(1)_V$, with conserved current $\bar{\psi}\gamma^\mu\psi$, and the axial one $U(1)_A$, with conserved current $\bar{\psi}\gamma^\mu\gamma^5\psi$. $U(1)_A$ is anomalous, and can be explicitly broken by a nonzero Dirac mass. If we introduce an axial chemical potential μ_5 then we expect an imbalance in the number of left- and right-handed fermions. If we further introduce an external $U(1)_V$ magnetic field B then, assuming the fermions have positive charge, we expect their spins to align with B, and since they are massless their momenta will also align or anti-align depending on their chirality. Given the imbalance in chirality, we expect a net $U(1)_V$ current parallel to B. This is the simplest example of the chiral magnetic effect (CME) [1–3].

For free fermions, the axial anomaly determines the size of the chiral magnetic current as follows. An axial chemical potential is equivalent to a background $U(1)_A$ gauge field with constant time component, $A_t^5 = \mu_5$, or to a time-dependent phase $\psi \to e^{i\gamma^5\mu_5 t}\psi$. Via the axial anomaly such a phase shift can be traded for a θ-term of the form $a(t, \mathbf{x})F \wedge F$ with $a(t, \mathbf{x}) = \mu_5 t$ and F the $U(1)_V$ field strength.

C. Hoyos (✉)
Raymond and Beverly Sackler Faculty of Exact Sciences, School of Physics and Astronomy, Tel-Aviv University, Ramat-Aviv 69978, Israel
e-mail: choyos@post.tau.ac.il

T. Nishioka
Department of Physics, Princeton University, Princeton, NJ 08544, USA
e-mail: nishioka@princeton.edu

A. O'Bannon
Department of Applied Mathematics and Theoretical Physics, University of Cambridge, Cambridge CB3 0WA, UK
e-mail: A.OBannon@damtp.cam.ac.uk

D. Kharzeev et al. (eds.), *Strongly Interacting Matter in Magnetic Fields*,
Lecture Notes in Physics 871, DOI 10.1007/978-3-642-37305-3_13,
© Springer-Verlag Berlin Heidelberg 2013

The spacetime-dependent source $a(t, \mathbf{x})$ can be regarded as a background, non-dynamical axion field. Writing $F = dA$ and integrating by parts, we obtain an interaction of the form $da \wedge A \wedge F$. Varying the action with respect to A we find the chiral magnetic current, which is parallel to B and has magnitude

$$J = \frac{\mu_5}{2\pi^2} B. \tag{13.1}$$

Notice that the CME occurs only when parity P and charge conjugation times parity CP symmetries are broken. The quantity $\sigma \equiv J/B$ is called the chiral magnetic conductivity [4].

When the system has nontrivial time evolution J becomes a function of time, or in Fourier space a function of frequency, $J(\omega)$. For massless fermions, the $U(1)_A$ Ward identity fixes part of $J(\omega)$, namely the part linear in B in the DC limit $\omega \to 0$, to be the anomaly-determined value in (13.1): $\lim_{B,\omega \to 0} \frac{\partial}{\partial B} J(\omega) = \frac{\mu_5}{2\pi^2}$ [5]. This statement remains true for massless fermions even in the presence of interactions or of temperature, so long as the Ward identity is satisfied. Of course in $J(\omega)$, the finite-ω dependence and any other dependence on B [6] will be sensitive to the nature of the interactions and to the temperature. Crucially, a finite fermion mass violates the Ward identity, in which case the $\omega \to 0$, linear-in-B contribution to $J(\omega)$ is no longer "protected", i.e. can deviate from the value in (13.1).

A CME may occur in heavy-ion collisions such as those produced at the Relativistic Heavy-Ion Collider (RHIC) and the Large Hadron Collider (LHC) [2, 7, 8]. The dominant interaction in the early stages of collisions is the strong nuclear force, as described by Quantum Chromodynamics (QCD). Indeed, the quark-gluon plasma (QGP) created at RHIC appears to involve strongly-interacting degrees of freedom far from equilibrium: the plasma appears to thermalize quickly and have a short mean free path (both signs of strong interactions [9, 10]) but also expands and cools rapidly until hadronization occurs.

Two conditions must hold to produce a CME in a heavy-ion collision. First, the QCD vacuum apparently preserves P and CP, so some event-by-event violation of these is required. In a medium such as the QGP, one possible mechanism for such violations are fluctuations of the topological charge density [2]. Second, the collision must be non-central, i.e. the nuclei must not perfectly overlap upon impact. In that case, the net charge combined with the net angular momentum can produce large magnetic fields, although these may die quickly as the QGP expands [2].

Assuming that P and CP are broken and a magnetic field is present, we know of two mechanisms to produce a CME in finite-temperature QCD. The first occurs for sufficiently large temperatures, where the QCD plasma is deconfined and chiral symmetry is restored. In that case we may invoke a naïve picture of quarks as freely propagating fermions in a magnetic field, and apply the arguments above.

A second, more subtle, mechanism, discussed for example in Ref. [11], may occur at lower temperatures, when the QCD plasma is in a confined state with chiral symmetry broken. Here we expect a gas of hadrons rather than a QGP. The key observation is that an external electromagnetic field can convert a neutral pseudo-scalar meson, such as the π^0, η, or η', into a neutral vector meson, such as the ρ. More precisely, any effective action describing QCD and electromagnetic interactions will include for example a vertex of the form $B\pi^0\rho$. The vector meson so produced will be polarized in the direction of the magnetic field, and via interactions with charged mesons can induce a current parallel to B, thus producing a CME even in a confined phase. The same process may also occur in the late stages of QGP evolution, during hadronization when metastable domains with spontaneous breaking of P and CP could be formed [12].

To our knowledge, analysis of RHIC and LHC data appears to favor the presence of a CME in the QGP, although a better understanding of systematic errors and backgrounds is still needed before a firm conclusion can be made [13–17]. The strong interactions and far-from-equilibrium evolution of the QGP in a heavy-ion collision make a clean theoretical prediction for σ very difficult. Lattice simulations suggest that the CME occurs in thermal equilibrium [18–22]. Lattice simulations cannot yet reliably determine the time evolution of σ, or equivalently $\sigma(\omega)$, which is crucial for estimating the size of the chiral magnetic current in a heavy-ion collision.

An alternative approach to the CME is the anti-de Sitter/Conformal Field Theory correspondence (AdS/CFT) [23–25], or more generally gauge-gravity duality. Gauge-gravity duality equates a strongly-coupled non-Abelian gauge theory with a weakly-coupled theory of gravity on some background spacetime, such that the field theory lives on the boundary of the spacetime, hence the duality is holographic. In particular, a black hole spacetime is dual to a thermal equilibrium state in which the center symmetry is spontaneously broken, such as the high-temperature, deconfined phase of a confining theory, where the temperature of the field theory coincides with the Hawking temperature of the black hole [26].

Gauge-gravity duality has been most successful at describing out-of-equilibrium physics, especially near-equilibrium physics, *i.e.* hydrodynamics. Most importantly, *all* gauge theories with a gravity dual (in states with $SO(2)$ rotational symmetry [27]) have the same, very small, ratio of shear viscosity η to entropy density s, namely $\eta/s = 1/4\pi$ [28], which is surprisingly close to the value estimated for the QGP at RHIC [29, 30]. We take such universality, and indeed the universality of hydrodynamics in general, as encouragement to study the CME in many holographic systems, following Refs. [31–39], including systems without confinement or chiral symmetry breaking in vacuum.

A conserved $U(1)$ current in the field theory is dual to a $U(1)$ gauge field in the bulk and, roughly speaking, an anomaly for the current is dual to a $(4 + 1)$-dimensional Chern-Simons term for the bulk gauge field. The latter is thus typically a key ingredient in holographic descriptions of the CME [31–39]. More generally, holographic models dual to fluids with anomalous currents have been constructed for example in Refs. [37, 39–42]. These holographic studies are complementary to field theory studies of the effects of triangle anomalies on hydrodynamics [43–48], which themselves have been applied to study the CME in heavy-ion collisions [49–51].

One holographic model of QCD, the Sakai-Sugimoto model [52, 53] includes a bulk Chern-Simons term, although some confusion has arisen as to whether the

CME occurs in this model at all. The problem in this model is that the vector current is anomalous under $U(1)_V \times U(1)_A$ transformations and therefore is not conserved in the presence of arbitrary external sources. Modifying the vector current such that it is conserved, which in the gravity dual requires adding certain boundary counterterms, causes the chiral magnetic current J to vanish [31]. To our knowledge, no consensus has emerged on whether a CME occurs in the Sakai-Sugimoto model.[1]

The authors of Refs. [35, 36] argued that for any bulk theory with a Chern-Simons term both a conserved vector current and nonzero J are possible, but at a price: the bulk gravity solution becomes non-regular. More precisely, in Euclidean signature the bulk gauge field will not vanish at the horizon and hence will not be a regular one-form [55]. In Lorentzian signature the gauge field solution will be regular only on the future horizon. One conclusion is that no reliable holographic description of the CME in thermal equilibrium (regular on the past and future horizons) exists for a conserved vector current. Effectively, a source for the gauge field must be introduced at the black hole horizon, the meaning of which is unclear from a field theory point of view. We were thus motivated to study other models where such issues could be avoided or at least clarified.

We consider $\mathcal{N} = 4$ supersymmetric $SU(N_c)$ Yang-Mills (SYM) theory, in the 't Hooft large-N_c limit and with large 't Hooft coupling, coupled to a number $N_f \ll N_c$ of $\mathcal{N} = 2$ supersymmetric hypermultiplets in the fundamental representation of the gauge group, *i.e.* flavor fields. We introduce a complex mass $m = |m|e^{i\phi}$ for the flavor fields into the superpotential with a time-dependent phase $\phi = \omega t$, following Refs. [56–60]. For the fermions that effectively introduces an axial chemical potential $\mu_5 = \frac{1}{2}\omega$. The theory also has a $U(1)_V$ symmetry that we will call baryon number. We introduce a baryon number magnetic field B and compute (holographically) the resulting chiral magnetic current.

$\mathcal{N} = 4$ SYM at large N_c and large 't Hooft coupling is holographically dual to type IIB supergravity on $AdS_5 \times S^5$ [23]. The $N_f \ll N_c$ hypermultiplets are dual to a number $N_f \ll N_c$ of probe D7-branes extended along $AdS_5 \times S^3$ [61]. The phase of the mass corresponds to the position of the D7-branes in one of the transverse directions on the S^5, hence ω corresponds to the angular frequency of the D7-branes in that direction, and the axial charge density corresponds to the angular momentum of the D7-branes. The axial anomaly is realized holographically via the Wess-Zumino (WZ) coupling of D7-branes to the background Ramond-Ramond (RR) flux on the S^5. The $U(1)_V$ current is dual to the $U(1)$ gauge field on the worldvolume of the D7-branes, which thus encodes both the $U(1)_V$ magnetic field and the chiral magnetic current.

---

[1]An alternative way to fix the normalization of the currents in the Sakai-Sugimoto model is to demand that the bulk action be invariant under gauge transformations that are non-vanishing at spatial infinity (in field theory directions), which leads to different bulk counterterms [54] and produces a non-vanishing chiral magnetic current. For the sake of argument, here we are taking the phenomenological point of view that a $U(1)_A$ current in the presence of a $U(1)_V$ chemical potential should coincide with the weak-coupling result, which occurs with the bulk counterterms of Ref. [31] but not those of Ref. [54].

In short, our system is a D7-brane in $AdS_5 \times S^5$, rotating with angular frequency ω on the S^5 and with a worldvolume magnetic field B. At finite temperature we replace AdS_5 with AdS-Schwarzschild. When $|m| = 0$ the value of the chiral magnetic current agrees with the result from the calculation using the anomaly, (13.1). When $|m|$ is nonzero we find that a chiral magnetic current appears only when a certain $U(1)_V$-neutral pseudo-scalar operator has a nonzero expectation value, signaling the breaking of C times time reversal, CT. We interpret this as a neutral pseudo-scalar condensate being converted into a vector condensate by the magnetic field, in a manner somewhat similar to the CME in the low-temperature phase of QCD [11]. For nonzero $|m|$ the value of the chiral magnetic current is less than that in (13.1), and indeed both the current and the expectation value of the pseudo-scalar drop to zero for sufficiently large $|m|$ or B.

Although we were motivated to find a model describing a CME in equilibrium with regular bulk solutions, in states with a nonzero $|m|$ and a CME we can demonstrate that our system is out of equilibrium in two ways. First, we simply observe that the scalars in the $\mathcal{N} = 2$ hypermultiplet have the same mass, with the same phase, as the fermions, so when $|m|$ is nonzero the Lagrangian has explicit time dependence and hence the system cannot be in equilibrium. The explicit time dependence disappears in the limit $|m| \to 0$. Second, we observe that in our system the axial symmetry is part of the R-symmetry under which the adjoint fields of $\mathcal{N} = 4$ SYM are also charged, hence axial charge in the flavor sector can "leak" into the adjoint sector, also taking energy with it. We compute, both in the field theory and from holography, the rate at which that occurs, with perfect agreement. We find that the rate is nonzero only when the pseudo-scalar has nonzero expectation value. Our solutions are stationary because we inject an equal amount of charge from an external source coupled to the flavor fields. The corresponding supergravity statement is that we pump angular momentum into the D7-branes which then flows into a bulk horizon.

This paper is organized as follows. In Sect. 13.2 we describe the main characteristics of the field theory with a flavor mass that has a time-dependent phase. In Sect. 13.3 we describe the gravity dual and perform the holographic computation of the chiral magnetic current. In Sect. 13.4 we compute the loss rates of axial charge and of energy for our system. In Sect. 13.5 we summarize our results and discuss open questions for future research.

13.2 The Theory in Question

We study $\mathcal{N} = 4$ SYM theory in the 't Hooft limit of $N_c \to \infty$ with Yang-Mills coupling squared $g_{YM}^2 \to 0$ keeping the 't Hooft coupling $\lambda \equiv g_{YM}^2 N_c$ fixed, followed by the limit $\lambda \gg 1$. The theory has an $SO(6)_R$ R-symmetry. The field content of $\mathcal{N} = 4$ SYM theory is the gauge field, four Weyl fermions, and three complex scalars. The former are in the **4** representation and the latter in the **6** representation of

$SO(6)_R \simeq SU(4)_R$. We will also consider an $\mathcal{N} = 4$ SYM plasma with equilibrium temperature T.

We next introduce a number N_f of $\mathcal{N} = 2$ supersymmetric hypermultiplets in the fundamental representation of $SU(N_c)$, which in analogy with QCD we call flavor fields. In $\mathcal{N} = 1$ notation the field content of the hypermultiplet is two chiral superfields of opposite chirality, Q and $\tilde{Q}$ in the N_c and $\overline{N_c}$ representations, respectively. Each chiral superfield consists of a complex scalar and a Weyl fermion. We denote the scalars, the squarks, as q and $\tilde{q}$ and combine the Weyl fermions into a Dirac fermion ψ. The flavor fields' couplings break the $SO(6)_R$ symmetry down to $SO(4) \times U(1)_R$, of which an $SU(2)_R \times U(1)_R$ subgroup is the $\mathcal{N} = 2$ R-symmetry. The $U(1)_R$ does not affect the squarks but acts as an axial symmetry for the quarks. Given that our theory has only this Abelian chiral symmetry, we will use "axial symmetry" and "chiral symmetry" interchangeably. As in QCD, the axial $U(1)_R$ symmetry is anomalous. The flavor fields also have a $U(1)_V$ symmetry that simply rotates Q and $\tilde{Q}^\dagger$ by the same phase.

We will work in the probe limit, which consists of keeping N_f fixed when we take the 't Hooft $N_c \to \infty$ limit, and then working to leading order in the small parameter N_f/N_c. Physically that corresponds to neglecting quantum effects due to the flavor fields, such as the running of the coupling. For instance, when the 't Hooft coupling is small the probe limit consists of discarding diagrams with (s)quark loops.

In the probe limit some part of the $U(1)_R$ anomaly survives, as we now explain. Three types of triangle diagram contribute to the anomaly, each with a $U(1)_R$ current at one vertex and two other currents at the other vertices. For example one diagram has the $U(1)_R$ current and two gauge currents. We will denote that as the $U(1)_R \times SU(N_c) \times SU(N_c) \equiv U(1)_R SU(N_c)^2$ anomaly, with similar notation for other anomalies. Both adjoint and flavor fields will appear in the loop, hence that diagram will have an order N_c^2 contribution and an order $N_f N_c$ contribution. The next diagram is the $U(1)_R^3$ anomaly, which will similarly receive order N_c^2 and $N_f N_c$ contributions. The third diagram is the $U(1)_R U(1)_V^2$ anomaly. Only flavor fields carry the $U(1)_V$ charge, hence that diagram will be order $N_f N_c$, with no N_c^2 contribution. In the probe limit we neglect the order $N_f N_c$ contribution to the first two diagrams, since that is sub-leading. For the third diagram, however, the order $N_f N_c$ term is leading, hence we retain it.[2] That anomaly will give rise to the CME in our system.

$\mathcal{N} = 2$ supersymmetry allows for a constant, complex mass $m = |m|e^{i\phi}$ for the flavor fields. A nonzero $|m|$ explicitly breaks $U(1)_R$. We will introduce a mass with a time dependent phase: $|m|e^{i\phi} = |m|e^{i\omega t}$. Let us recall not only how $\phi = \omega t$ is equivalent to an axial chemical potential for the quarks, but also how $\phi = \omega t$ introduces explicit time dependence in the potential terms for q and $\tilde{q}$. Of the adjoint

---

[2]On the supergravity side the first two anomalies would appear in the type IIB supergravity sector, while the third will be associated with a WZ term on a probe D7-brane.

scalars, only one is charged under $U(1)_R$. We denote this scalar as Φ. The flavor couplings in the $\mathcal{N} = 1$ superpotential W are then

$$W \supset \tilde{Q} \Phi Q + |m| e^{i\phi} \tilde{Q} Q. \tag{13.2}$$

Integrating over superspace, we find the (normal) potential, from which we extract terms involving the squarks and terms involving the quarks. The terms involving q are [59]

$$V_q = q^\dagger |\Phi|^2 q - |m| e^{i\phi} q^\dagger \Phi^\dagger q - |m| e^{-i\phi} q^\dagger \Phi q + |m|^2 q^\dagger q. \tag{13.3}$$

The potential includes identical terms for $\tilde{q}$. The quark contribution is simply[3]

$$V_\psi = |m| \overline{\psi} e^{i\phi \gamma^5} \psi. \tag{13.4}$$

If we now perform a chiral rotation

$$\psi \to e^{-i\gamma^5 \phi/2} \psi, \tag{13.5}$$

then the derivative in ψ's kinetic term will act on ϕ, producing a new term that we may include in the potential,

$$V_\psi = |m| \overline{\psi} \psi - \frac{\partial_\mu \phi}{2} \overline{\psi} \gamma^\mu \gamma^5 \psi. \tag{13.6}$$

If we introduce $\phi = \omega t$ then clearly ω is equivalent to twice the axial chemical potential,

$$\omega = 2\mu_5. \tag{13.7}$$

Crucially, when $|m|$ is nonzero the squark terms of the form in (13.3) explicitly depend on t, so the Hamiltonian depends explicitly on time, energy is not conserved, and the system cannot be in equilibrium.

For the CME in low-temperature QCD the central players are the light pseudo-scalar and vector fields, the π^0, η, η' and the ρ, respectively. Excitations of these fields, the mesons, produce poles in the corresponding retarded two-point functions. Our theory has operators analogous to these which will play a role in our realization of the CME, so let us describe them in detail.

Our theory has gauge-invariant (s)quark bilinears, *i.e.* gauge-invariant operators built from two fields in the N_c and $\overline{N_c}$ representations. The two-point functions of these operators exhibit poles which we will call mesons in analogy with QCD. Unlike mesons in QCD, these modes are not associated with chiral symmetry breaking or confinement, rather they are deeply bound states with masses on the order of $|m|/\sqrt{\lambda}$ [62]. When $|m|$ is nonzero these are the lightest flavor degrees of freedom in our system.

---

[3] Our γ^5 is Hermitian and squares to the identity.

To determine the operators relevant for the CME, we treat $|m|$ and ϕ as external sources. We denote the associated operators as $\mathcal{O}_m$ and $\mathcal{O}_\phi$, respectively. For instance, varying minus the action with respect to $|m|$, we find a dimension three operator

$$\mathcal{O}_m = -\frac{\delta S}{\delta |m|}$$

$$= \overline{\psi} e^{i\phi\gamma^5}\psi - e^{i\phi}q^\dagger\Phi^\dagger q - e^{-i\phi}q^\dagger\Phi q - e^{i\phi}\tilde{q}^\dagger\Phi^\dagger\tilde{q}$$

$$- e^{-i\phi}\tilde{q}^\dagger\Phi\tilde{q} + 2|m|(q^\dagger q + \tilde{q}^\dagger\tilde{q}). \tag{13.8}$$

When ϕ is constant, $\mathcal{O}_m$ is just the $\mathcal{N}=2$ supersymmetric completion of the standard quark mass operator. Notice that $\mathcal{O}_m$ is charged under $U(1)_R$, and hence may serve as an order parameter for chiral symmetry breaking when $|m|=0$. Notice also that if $\phi = \omega t$ then $\mathcal{O}_m$ depends explicitly on time. Varying minus the action with respect to ϕ we find a dimension four operator,

$$\mathcal{O}_\phi = -\frac{\delta S}{\delta \phi}$$

$$= |m| i \overline{\psi} e^{i\phi\gamma^5}\gamma^5\psi + |m| q^\dagger i\left(e^{-i\phi}\Phi - e^{i\phi}\Phi^\dagger\right)q$$

$$+ |m|\tilde{q}^\dagger i\left(e^{-i\phi}\Phi - e^{i\phi}\Phi^\dagger\right)\tilde{q}. \tag{13.9}$$

Notice that $\mathcal{O}_\phi \propto |m|$, and again if $\phi = \omega t$ then $\mathcal{O}_\phi$ depends explicitly on time.

The $U(1)_V$ baryon number and $U(1)_R$ currents will also be involved in the CME. We denote the conserved $U(1)_V$ current as J^μ,

$$J^\mu = \overline{\psi}\gamma^\mu\psi - i\left(q^\dagger D^\mu q - (D^\mu q)^\dagger q\right) - i\left(\tilde{q}(D^\mu\tilde{q})^\dagger - (D^\mu\tilde{q})\tilde{q}^\dagger\right). \tag{13.10}$$

The contribution to the R-current from flavor fields is the same as half the axial current $J_R^\mu = \frac{1}{2}\overline{\psi}\gamma^\mu\gamma^5\psi$.[4] As mentioned in the introduction, adjoint fields also contribute to the R-current, hence the axial current will not be conserved even in the absence of anomalies. We discuss the non-conservation of quark axial charge in detail in Sect. 13.4.

Since discrete spacetime symmetries play a central role in the CME, we will also present the transformation properties of various operators under C, P and T, when $\phi = \omega t$:

	C	P	T
V_q	$\omega \to -\omega$	even	$\omega \to -\omega$
V_ψ	even	$\omega \to -\omega$	even
$i\overline{\psi}e^{i\omega t\gamma^5}\gamma^5\psi$	even	$\omega \to -\omega + \text{odd}$	odd
$q^\dagger i(e^{i\omega t}\Phi^\dagger - e^{-i\omega t}\Phi)q$	$\omega \to -\omega + \text{odd}$	even	$\omega \to -\omega$

---

[4] We are identifying R-charge transformations with shifts $\phi \to \phi + \delta\phi$, which for the quarks imply the $U(1)_A$ transformation $\psi \to e^{i\delta\phi\gamma^5/2}\psi$. With this convention the R-charge of the quarks is $1/2$.

where $\omega \rightarrow -\omega$ means that a sign flip of ω is the only change, and $\omega \rightarrow -\omega +$ odd means a sign flip of ω plus an overall sign flip are the only changes. The potential is not invariant under CPT, which is compatible with the breaking of Lorentz symmetry by the explicit time dependence. The only discrete spacetime symmetry under which the potential is invariant is CT. Notice that $\mathcal{O}_\phi$ is CT odd, so an expectation value $\langle \mathcal{O}_\phi \rangle$ may serve as an order parameter for spontaneous CT breaking.

Finally, let us explain the analogy between the CME in our system and in QCD. Our system has no dynamical electromagnetic $U(1)$, so to obtain a CME we will introduce a non-dynamical external $U(1)_V$ magnetic field $F_{xy} = B$. The phase ϕ will play the role of a non-dynamical external axion field $a(t, \mathbf{x})$. The operator $\mathcal{O}_\phi$ will play the role of a light, neutral pseudo-scalar, such as the π^0. At zero temperature and finite mass, we can then think of an expectation value $\langle \mathcal{O}_\phi \rangle$ as a condensate of pseudo-scalar mesons. The $U(1)_V$ current J^μ will play the role of a vector meson field, like the ρ, so we can think of the chiral magnetic current $\langle J^z \rangle$ as a condensate of vector mesons. Our holographic calculations will show that $\langle J^z \rangle$ is nonzero only when $\langle \mathcal{O}_\phi \rangle$ is nonzero, except in the chirally symmetric case $|m| = 0$. Our interpretation is that the mechanism for the CME in our system is similar to that of low-temperature QCD: the magnetic field converts pseudo-scalar mesons into vector mesons polarized in the direction of B. Notice that away from the chiral limit the CME occurs in our system only when CT is spontaneously broken, in contrast to the free fermion case which required P and CP breaking.

13.3   Chiral Magnetic Effect from Spinning Probe Branes

We begin in type IIB string theory with a supersymmetric intersection of N_c D3-branes and N_f D7-branes:

	x_0	x_1	x_2	x_3	x_4	x_5	x_6	x_7	x_8	x_9
D3	×	×	×	×						
D7	×	×	×	×	×	×	×	×		

$$(13.11)$$

Open strings with both ends on the D3-branes give rise at low energies to $\mathcal{N} = 4$ $SU(N_c)$ SYM theory, while open strings with one end on the D3-branes and one on the D7-branes give rise to $\mathcal{N} = 2$ hypermultiplets in the fundamental representation. The $SO(6)$ isometry in the directions $(x_4, \ldots, x_9)$ corresponds to the $SO(6)_R$ symmetry of $\mathcal{N} = 4$ SYM theory. Clearly the D7-branes break that to $SO(4) \times U(1)_R$, corresponding to rotations in $(x_4, \ldots, x_7)$ and (x_8, x_9) respectively. If we separate the D3- and D7-branes in the overall transverse directions, x_8 and x_9, then the 3–7 and 7–3 strings acquire a finite length, giving the hypermultiplets a mass. The complex mass $|m|e^{i\phi}$ thus corresponds simply to the relative positions of the D3- and D7-branes in that plane, with $|m|$ the separation distance and ϕ the angle in the plane. The breaking of $U(1)_R$ by a nonzero $|m|$ appears simply as the breaking of rotational symmetry in the (x_8, x_9)-plane. A time-dependent phase $\phi = \omega t$ corresponds to D7-branes spinning in the (x_8, x_9)-plane.

We take the usual limits for the D3-branes, $N_c \to \infty$ with $g_s N_c$ fixed, followed by taking $g_s N_c \gg 1$, where g_s is the string coupling and α' is the string length squared. We thus obtain type IIB supergravity in the near-horizon geometry of the D3-branes, $AdS_5 \times S^5$ where each factor has radius of curvature $L^4/\alpha'^2 = 4\pi g_s N_c \gg 1$. The solution includes N_c units of RR five-form flux on the S^5. AdS/CFT equates this theory with the low-energy theory on the D3-branes, $\mathcal{N} = 4$ SYM theory, with Yang-Mills coupling $g_{YM}^2 = 4\pi g_s$ and 't Hooft coupling $\lambda = g_{YM}^2 N_c$, so the theory is in the 't Hooft large-N_c limit with $\lambda \gg 1$.

We will use an $AdS_5 \times S^5$ metric of the form

$$ds^2 = -|g_{tt}| \, dt^2 + g_{xx} \, d\mathbf{x}^2 + g_{rr} \, dr^2 + g_{ss} \, ds_{S^3}^2 + g_{RR} \, dR^2 + g_{\phi\phi} \, d\phi^2 \quad (13.12)$$

$$= \frac{\rho^2}{L^2}\left(-dt^2 + d\mathbf{x}^2\right) + \frac{L^2}{\rho^2}\left(dr^2 + r^2 \, ds_{S^3}^2 + dR^2 + R^2 \, d\phi^2\right), \quad (13.13)$$

where ρ is the AdS_5 radial coordinate with the boundary at $\rho \to \infty$. The field theory lives in Minkowski space with coordinates $(t, \mathbf{x})$. We have split the six directions transverse to the D3-branes into $\mathbb{R}^4 \times \mathbb{R}^2$ where the latter $\mathbb{R}^2$ represents the (x_8, x_9)-plane. We have written the metric of both the $\mathbb{R}^4$ and the $\mathbb{R}^2$ in spherical coordinates. The former has radial coordinate r with $ds_{S^3}^2$ the metric of a unit S^3, while the latter has radial coordinate R and circle coordinate ϕ. Notice that $\rho^2 = r^2 + R^2$. The self-dual RR five-form can be derived from a four-form potential

$$C_4 = g_{xx}^2 \text{vol}_{\mathbb{R}^{3,1}} - g_{ss}^2 \, d\phi \wedge \text{vol}_{S^3}. \quad (13.14)$$

Starting now we will use units in which $L \equiv 1$. We can convert between string theory and supergravity quantities using $\alpha'^{-2} = \lambda$.

The $\mathcal{N} = 4$ SYM theory at finite temperature T is dual to supergravity in an AdS-Schwarzschild spacetime. In that case only $|g_{tt}|$ and g_{xx} change, becoming

$$|g_{tt}| = \rho^2 \frac{\gamma^2}{2} \frac{f^2(\rho)}{H(\rho)}, \qquad g_{xx} = \rho^2 \frac{\gamma^2}{2} H(\rho), \quad (13.15)$$

with

$$f(\rho) = 1 - \frac{1}{\rho^4}, \qquad H(\rho) = 1 + \frac{1}{\rho^4}. \quad (13.16)$$

In these coordinates the horizon is always at $\rho = 1$, but the Hawking temperature, which we identify with the $\mathcal{N} = 4$ SYM temperature, is $T = \gamma/\pi$, which we can vary by changing the parameter γ. We recover the $T = 0$ limit by first rescaling $\rho \to \sqrt{2}\rho/\gamma$, and the same for r and R, and then taking $\gamma \to 0$.

If we keep N_f finite as $N_c \to \infty$ we may treat the D7-branes as probes. The action describing the D7-brane's dynamics, S_{D7}, consists of two types of terms, a Dirac-Born-Infeld (DBI) term and Wess-Zumino (WZ) terms. We will consider only the $U(1)$ worldvolume theory of coincident D7-branes, so we will need only the Abelian D7-brane action,

$$S_{D7} = S_{DBI} + S_{WZ}, \quad (13.17)$$

$$S_{DBI} = -N_f T_{D7} \int d^8\xi \sqrt{-\det\left(g_{ab}^{D7} + (2\pi\alpha')\tilde{F}_{ab}\right)}, \qquad (13.18)$$

$$S_{WZ} = +\frac{1}{2} N_f T_{D7} (2\pi\alpha')^2 \int P[C_4] \wedge \tilde{F} \wedge \tilde{F}, \qquad (13.19)$$

where $T_{D7} = \frac{g_s^{-1}\alpha'^{-4}}{(2\pi)^7}$ is the D7-brane tension, ξ^a are the worldvolume coordinates, g_{ab}^{D7} is the induced metric on the brane, $\tilde{F}_{ab}$ is the $U(1)$ worldvolume field strength, and $P[C_4]$ is the pullback of the RR four-form to the D7-branes.

Let us introduce some convenient notation. First, we will absorb a factor of $(2\pi\alpha')$ into the field strength $(2\pi\alpha')\tilde{F}_{ab} \equiv F_{ab}$. Our D7-branes will be extended along $AdS_5 \times S^3$ inside $AdS_5 \times S^5$, that is, along the Minkowski coordinates $(t, \mathbf{x})$, the radial direction r, and the $S^3 \subset S^5$. In what follows we consider solutions for which the D7-brane Lagrangian will depend only on r, so we may trivially perform the integrations over the Minkowski and S^3 directions, producing factors of their respective volumes, $V_{\mathbb{R}^{3,1}}$ and $2\pi^2$. We will absorb the factor of the infinite volume of Minkowski space into the action, $S_{D7}/V_{\mathbb{R}^{3,1}} \to S_{D7}$. From now on we will refer to this rescaled action as the D7-brane action. We will absorb the factor of the S^3 volume into an overall factor

$$\mathcal{N} \equiv N_f T_{D7} 2\pi^2 = \frac{\lambda N_f N_c}{(2\pi)^4}, \qquad (13.20)$$

where in the second equality we converted to field theory quantities.

We now need an appropriate ansatz for the worldvolume fields to describe a CME in the field theory. The two scalars on the D7-brane worldvolume are R and ϕ. The former is dual to the operator $\mathcal{O}_m$ while the latter is dual to $\mathcal{O}_\phi$. More specifically, the asymptotic values of R and ϕ will be (proportional to) the modulus and phase of the complex mass, as is obvious from the initial D3/D7 intersection. We thus introduce $R(r)$ and $\phi(t, r) = \omega t + \varphi(r)$, which produces a time-dependent phase for the mass and allows for nonzero $|m|$, $\langle \mathcal{O}_m \rangle$ and $\langle \mathcal{O}_\phi \rangle$.

We can motivate the r-dependence in $\phi(t, r)$ from previous experience with probe branes in holographic spacetimes, following Ref. [63]. Suppose we introduce only $\phi(t) = \omega t$ and then perform a T-duality in the ϕ direction.[5] The background solution changes, and the D7-brane becomes a D8-brane extended in ϕ with worldvolume gauge field, $A_\phi(t) \propto \omega t$. The angular frequency ω becomes a constant electric field on the D8-brane, $F_{t\phi} \propto \omega$. A probe brane in a gravitational potential well, such as AdS, and with a constant worldvolume electric field will generally have a tachyonic instability: the gravitational potential reduces the effective tension of open strings, so at some point the constant electric field can rip strings apart. This tachyonic instability causes the Lorentzian action to become imaginary. Turning that around, an imaginary action signals a tachyon, since whenever a Lorentzian action

---

[5]Of course, T-duality is not a well-defined operation for an angular direction that can shrink to zero size, but here we are simply illustrating the similarities between spinning branes and branes with a worldvolume electric field.

S becomes a complex number, the weight factor e^{iS} in a path integral will have either an exponentially growing or exponentially decaying mode. The cure for the D8-brane's instability is to introduce r-dependence [63], *i.e.* $A_\phi(t,r) \propto \omega t + \varphi(r)$, producing a new constant of integration, associated with $\varphi(r)$, that we can adjust to maintain reality of the action and hence avoid the instability. T-duality back to a D7-brane produces the $\phi(t,r)$ above, although now the physical interpretation of the instability is very different. Now the instability occurs because in a gravitational potential well the local speed of light decreases, while the probe brane rotates at a constant angular frequency, so at some point the probe brane may have linear veloc-ity faster than the local speed of light. The D7-brane cures the problem by "twisting" in ϕ as a function of r [56–60].

For the CME we need a $U(1)_V$ magnetic field and we expect a current $\langle J^z \rangle$. The $U(1)_V$ current J^μ is dual to the $U(1)$ gauge field on the D7-branes, hence we include in our ansatz $F_{xy} = B$ and $A_z(r)$. In total, then, our ansatz includes[6] $R(r)$, $\phi(t,r) = \omega t + \varphi(r)$, $F_{xy} = B$ and $A_z(r)$.

We can argue that the probe D7-brane action must depend only on derivatives of ϕ and A_z as follows. The background solution has an isometry in ϕ: the met-ric and four-form in (13.12) and (13.14) are invariant under constant shifts of ϕ. Recalling that the scalars on the worldvolume of a D-brane are Goldstone bosons associated with the breaking of translation invariance in the transverse directions, and that Goldstone bosons can have only derivative interactions, we can conclude that the action S_{D7} will involve only derivatives of ϕ. For A_z we argue simply that the action depends only on the field strength and not A_z itself. Inserting our ansatz into the action, we find

$$S_{DBI} = -\mathcal{N} \int dr\, g_{SS}^{3/2} g_{xx}^{3/2} \sqrt{1 + \frac{B^2}{g_{xx}^2}}$$

$$\times \sqrt{\left(|g_{tt}| - g_{\pi\pi}\dot{\pi}^2\right)\left(g_{rr} + g_{RR}R'^2 + g^{xx}A_z'^2\right) + |g_{tt}|g_{\pi\pi}\pi'^2},$$

$$S_{WZ} = -\mathcal{N} B\omega \int dr\, g_{SS}^2 A_z'. \tag{13.21}$$

where primes denote $\frac{\partial}{\partial r}$ and dots denote $\frac{\partial}{\partial t}$. As advertised, the action depends only on derivatives of ϕ and A_z and hence produces two "constants of motion", $\frac{\delta S_{D7}}{\delta \phi'}$ and $\frac{\delta S_{D7}}{\delta A_z'}$. We may thus solve for ϕ' and A_z' and obtain an action for $R(r)$ only. We may do so in several ways. One is to solve for ϕ' and A_z', derive $R(r)$'s equation of motion from S_{D7}, and then plug the solutions for ϕ' and A_z' into that. Alternatively, we can plug the solutions directly into S_{D7}, perform a Legendre transform with respect to both ϕ' and A_z', and then derive $R(r)$'s equation of motion. We may also proceed in stages, for example by solving for and Legendre-transforming with respect to one only and then repeating the process for the second. The simplest approach turns out to be solving for ϕ' first and then for A_z'.

---

[6]We work in $A_r = 0$ gauge. Recall also that our B includes a factor of $(2\pi\alpha') \propto \lambda^{-1/2}$.

The equation of motion for ϕ' is

$$\frac{\delta S_{D7}}{\delta \phi'}$$

$$= -\mathcal{N} g_{SS}^{3/2} g_{xx}^{3/2} \sqrt{1 + \frac{B^2}{g_{xx}^2}}$$

$$\times \frac{|g_{tt}| g_{\pi\pi} \phi'}{\sqrt{(|g_{tt}| - g_{\pi\pi} \dot{\pi}^2)(g_{rr} + g_{RR} R'^2 + g^{xx} A_z'^2) + |g_{tt}| g_{\pi\pi} \pi'^2}}$$

$$\equiv \alpha, \tag{13.22}$$

where α is the first constant of motion. Solving for π' we get

$$\phi'^2 = \frac{\alpha^2}{|g_{tt}| g_{\pi\pi}} \frac{(|g_{tt}| - g_{\pi\pi} \dot{\pi}^2)(g_{rr} + g_{RR} R'^2 + g^{xx} A_z'^2)}{\mathcal{N}^2 g_{xx}^3 g_{SS}^3 |g_{tt}| g_{\pi\pi} (1 + \frac{B^2}{g_{xx}^2}) - \alpha^2}. \tag{13.23}$$

Next we Legendre transform with respect to ϕ',

$$\hat{S}_{D7} = \hat{S}_{DBI} + \hat{S}_{WZ} = S_{D7} - \int dr \, \phi' \frac{\delta S_{D7}}{\delta \phi'}. \tag{13.24}$$

Notice that S_{WZ} does not participate here, $\hat{S}_{WZ} = S_{WZ}$, so we focus on S_{DBI},

$$\hat{S}_{DBI} = S_{DBI} - \int dr \, \phi' \frac{\delta S_{DBI}}{\delta \phi'},$$

$$= -\mathcal{N} \int dr \, g_{SS}^{3/2} g_{xx}^{3/2} \sqrt{|g_{tt}| - g_{\pi\pi} \dot{\pi}^2} \sqrt{g_{rr} + g_{RR} R'^2 + g^{xx} A_z'^2}$$

$$\times \sqrt{1 + \frac{B^2}{g_{xx}^2} - \frac{\alpha^2 / \mathcal{N}^2}{|g_{tt}| g_{\pi\pi} g_{xx}^3 g_{SS}^3}}. \tag{13.25}$$

The equation of motion for A_z' is then

$$\frac{\delta \hat{S}_{D7}}{\delta A_z'} = \frac{\delta \hat{S}_{DBI}}{\delta A_z'} + \frac{\delta \hat{S}_{WZ}}{\delta A_z'} \equiv \beta, \tag{13.26}$$

where β is the second constant of motion. The two terms in β are

$$\frac{\delta \hat{S}_{DBI}}{\delta A_z'} = -\mathcal{N} g_{SS}^{3/2} g_{xx}^{3/2} \sqrt{|g_{tt}| - g_{\pi\pi} \dot{\pi}^2} \frac{g^{xx} A_z'}{\sqrt{g_{rr} + g_{RR} R'^2 + g^{xx} A_z'^2}}$$

$$\times \sqrt{1 + \frac{B^2}{g_{xx}^2} - \frac{\alpha^2 / \mathcal{N}^2}{|g_{tt}| g_{\pi\pi} g_{xx}^3 g_{SS}^3}}, \tag{13.27}$$

$$\frac{\delta \hat{S}_{WZ}}{\delta A'_z} = -\mathcal{N} B \omega g^2_{SS}.$$ (13.28)

We now solve for A'_z,

$$A'_z = \frac{(\beta + \mathcal{N} B \omega g^2_{SS}) g_{xx} \sqrt{g_{rr} + g_{RR} R'^2}}{\sqrt{\mathcal{N}^2 g^3_{xx} g^3_{SS}(|g_{tt}| - g_{\pi\pi} \dot{\pi}^2)(1 + \frac{B^2}{g^2_{xx}} - \frac{\alpha^2/\mathcal{N}^2}{|g_{tt}|g_{\pi\pi}g^3_{xx}g^3_{SS}}) - g_{xx}(\beta + \mathcal{N} B \omega g^2_{SS})^2}}.$$ (13.29)

Finally, we Legendre transform with respect to A'_z,

$$\hat{S}_{D7} = \hat{S}_{D7} - \int dr A'_z \frac{\delta \hat{S}_{D7}}{\delta A'_z} = -\mathcal{N} \int dr \sqrt{g_{rr} + g_{RR} R'^2}$$

$$\times \sqrt{g^3_{xx} g^3_{SS}(|g_{tt}| - g_{\pi\pi}\dot{\pi}^2)\left(1 + \frac{B^2}{g^2_{xx}} - \frac{\alpha^2/\mathcal{N}^2}{|g_{tt}|g_{\pi\pi}g^3_{xx}g^3_{SS}}\right) - g_{xx}\left(\frac{\beta}{\mathcal{N}} + B\omega g^2_{SS}\right)^2}.$$ (13.30)

We can derive $R(r)$'s equation of motion from this final form of the action, although is it cumbersome and unilluminating, so we will not present it.

We can now explain how to extract field theory information from bulk solutions. The fields have the following near-boundary asymptotic expansions:

$$R(r) = c_0 + \frac{c_2}{r^2} + \frac{1}{2}c_0\omega^2 \frac{\log r}{r^2} + O\left(\frac{\log r}{r^4}\right),$$ (13.31)

$$\phi(t,r) = \omega t + \frac{\alpha}{2\mathcal{N}c_0^3}\left(\frac{c_0}{r^2} - \frac{c_2 + \frac{1}{8}c_0\omega^2}{r^4} - \frac{1}{2}c_0\omega^2 \frac{\log r}{r^4}\right) + O\left(\frac{\log r}{r^6}\right),$$ (13.32)

$$A_z(r) = c_z + \frac{1}{2}\frac{\frac{\beta}{\mathcal{N}} + B\omega}{r^2} - \frac{1}{2}\frac{c_0^2 B\omega}{r^4} + O\left(\frac{1}{r^6}\right),$$ (13.33)

where c_0, c_2, and c_z are constants.

In each case the leading term acts as a source for the dual operator. c_0 is the asymptotic separation between the original D3-branes and the D7-branes, so the magnitude of the mass is c_0 times the string tension, $|m| = \frac{c_0}{2\pi\alpha'}$. The leading term in $\phi(t,r)$, in our case ωt, is the phase of the mass. c_z is a source for J^z, equivalent to the A_z component of an external gauge field. In our case we may safely set $c_z = 0$.

The coefficients of the sub-leading terms determine the expectation values of the dual operators. The exact relations follow from the holographic dictionary, which equates the on-shell bulk action with minus the generating functional of the field theory. Both the on-shell bulk action and the field theory generating functional are divergent, and require renormalization. Using standard techniques of holographic renormalization [64, 65], we find for $\langle \mathcal{O}_m \rangle$

$$\langle \mathcal{O}_m \rangle = \left(2\pi\alpha'\right)\mathcal{N}\left(-2c_2 - \frac{1}{2}\omega^2 c_0 - \frac{1}{2}\omega^2 c_0 \ln c_0^2\right). \tag{13.34}$$

Factors of the AdS radius, which we have set to one, make the argument of the logarithm dimensionless.[7] For the other operators we find

$$\langle \mathcal{O}_\phi \rangle = \alpha, \qquad \langle J^z \rangle = -\left(2\pi\alpha'\right)\beta. \tag{13.35}$$

As shown in the last section, when $|m|$ is nonzero the operators $\mathcal{O}_m$ and $\mathcal{O}_\phi$ depend explicitly on time. In our solutions c_0, c_2, and α will be time-independent constants, however, so the expressions above are only consistent for nonzero $|m|$ if the state in which we evaluate $\langle \mathcal{O}_m \rangle$ and $\langle \mathcal{O}_\phi \rangle$ has time dependence that cancels the time-dependence of the operators. Our configurations correspond then to a steady state.

Notice that $\langle \mathcal{O}_m \rangle$, $\langle \mathcal{O}_\phi \rangle$, and $\langle J^z \rangle$ are all of different orders in the large-N_c and large-λ counting. $\langle \mathcal{O}_m \rangle$ is of order $(2\pi\alpha')\mathcal{N} \propto \sqrt{\lambda}N_f N_c$ times factors of order one in the large-N_c and large-λ counting, such as c_0 and c_2. In contrast, $\langle \mathcal{O}_\phi \rangle = \alpha$, where from (13.22) we see that α is of order $\mathcal{N} \propto \lambda N_f N_c$ times factors of order one, so $\langle \mathcal{O}_\phi \rangle$ is bigger than $\langle \mathcal{O}_m \rangle$ by a factor of $\sqrt{\lambda}$. Recall, however, that $\mathcal{O}_\phi$ is $|m|$ times a dimension three operator. Using $|m| = \frac{c_0}{2\pi\alpha'} \propto \sqrt{\lambda}c_0$, we see that if $\langle \mathcal{O}_\phi \rangle$ scales as $\lambda N_f N_c$ then the expectation value of the dimension three operator must scale as $\langle \mathcal{O}_\phi \rangle / |m| \propto \sqrt{\lambda}N_f N_c$. The expectation values of the two dimension three operators, $\langle \mathcal{O}_m \rangle$ and $\langle \mathcal{O}_\phi \rangle / |m|$, thus have the same scaling. On the other hand we have $\langle J^z \rangle \propto \alpha'\beta$. From (13.27) we see that in terms of large-N_c and large-λ counting $\beta \propto \mathcal{N}A'_z$. Recall that we have absorbed a factor of $(2\pi\alpha')$ into A'_z. If we extract that factor, then we find $\beta \propto (2\pi\alpha')\mathcal{N} \propto \sqrt{\lambda}N_f N_c$ and hence $\langle J^z \rangle = -(2\pi\alpha')\beta \propto N_f N_c$. We thus find that $\langle J^z \rangle$'s normalization is independent of the coupling λ. That is not surprising. $\langle J^z \rangle$ is our chiral magnetic current, whose normalization is fixed by the $U(1)_A U(1)_V^2$ anomaly, and so is determined by the $U(1)_A$ and $U(1)_V$ charges, not the 't Hooft coupling.

The massless limit $|m| \to 0$, or equivalently $c_0 \to 0$, is subtle. For one thing, in that limit the phase of the mass becomes ill-defined. Moreover, in that limit we see from $\phi(t,r)$'s asymptotic expansion in (13.32) that α must also vanish, since otherwise the coefficients of the r-dependent terms in $\phi(t,r)$'s expansion would diverge. The vanishing of α in that limit makes sense, since $\alpha = \langle \mathcal{O}_\phi \rangle$ and we know from Sect. 13.2 that $\langle \mathcal{O}_\phi \rangle \propto |m|$. As mentioned in Sect. 13.2, in our system $\langle \mathcal{O}_\phi \rangle$ is an order parameter for spontaneous CT breaking, so the vanishing of $\langle \mathcal{O}_\phi \rangle$ as $|m| \to 0$ suggests that CT will always be restored in that limit. We must be cautious,

---

[7]At first glance, the $\langle \mathcal{O}_m \rangle$ in (13.34) appears to diverge in the flavor decoupling limit $c_0 \propto |m| \to \infty$, which is counter-intuitive. Both the analytic argument in the appendix of Ref. [57] and numerical calculations confirm that in fact $\langle \mathcal{O}_m \rangle \to 0$ in that limit.

however, since the expectation value of the dimension three operator $\langle \mathscr{O}_\phi \rangle / |m|$ need not vanish as $|m| \to 0$, so CT could still be broken. We will argue in Sect. 13.3.2 that it actually is restored in the states we consider.

Lastly, the R-charge density appears in the bulk as the angular momentum of the D7-brane. In what follows we will not compute the R-charge density or $\langle \mathscr{O}_m \rangle$, rather our focus will be on $\langle \mathscr{O}_\phi \rangle$ and $\langle J^z \rangle$. We will only care whether the R-charge density is zero or not. To determine that we only need to know whether the embedding $R(r)$ is nonzero or not: if $R(r) = 0$ then the D7-brane has no angular momentum, while if $R(r)$ is nonzero then the D7-brane has angular momentum. That is clear from the original D3/D7 construction, since if $R(r) = 0$ then the D7-brane is not extended at all in the (x_8, x_9)-plane and so cannot have angular momentum.

13.3.1 Solutions at Zero Temperature

Our goal now is to solve $R(r)$'s equation of motion, derived from (13.30), numerically in the pure AdS background dual to the zero-temperature vacuum of $\mathscr{N} = 4$ SYM.

Let us first quickly review what happens when B and ω are zero. Here we have no time-dependent phase for the flavor mass, and we expect no current, so we also set ϕ and A_z to zero. The induced metric on the D7-brane is then[8]

$$ds_{D7}^2 = \rho^2 \left(-dt^2 + dx^2 \right) + \frac{1}{\rho^2} \left(dr^2 \left(1 + R'^2 \right) + r^2 \, ds_{S^3}^2 \right), \tag{13.36}$$

where $\rho^2 = r^2 + R^2$, and the action becomes

$$S_{D7} = -\mathscr{N} \int dr \, r^3 \sqrt{1 + R'^2}. \tag{13.37}$$

Inside the square root factor appearing in the action is a sum of squares, hence the action will be extremized only when $R' = 0$, or in other words when the solution is constant $R = c_0$. These solutions describe flavor fields with an $\mathscr{N} = 2$ supersymmetry-preserving constant mass. $\mathscr{N} = 2$ supersymmetry demands that $\langle \mathscr{O}_m \rangle = 0$, which is indeed the case for these solutions, which have $c_2 = 0$ and hence via (13.34) $\langle \mathscr{O}_m \rangle = 0$.

The D7-brane is always extended along r from the boundary $r = \infty$ to $r = 0$, however for these constant solutions the D7-brane does not fill all of AdS_5. At the boundary the D7-brane wraps the maximum-volume equatorial S^3 inside the S^5, but as it extends into AdS_5, to smaller r, the S^3 shrinks and eventually collapses to zero size at the "North pole" of the S^5, which occurs when $r = 0$. Recalling that the radial coordinate of AdS_5 is not r but $\rho = \sqrt{r^2 + R^2}$, we see that at $r = 0$ the

---

[8] Starting now we suppress the r dependence in $R(r)$ for notational clarity, $R(r) \to R$, unless stated otherwise.

D7-brane has only reached $\rho = c_0$: from the perspective of an observer in AdS_5 the D7-brane simply ends at that point. The trivial solution $R = 0$ describes massless flavors. In that case the D7-brane fills all of AdS_5.

Notice that $R = c_0$ is a smooth solution because $R'(0) = 0$, that is, the slope of R is zero when the S^3 collapses at $r = 0$. If that does not occur then we see from (13.36) that the D7-brane will have a conical singularity at $r = 0$. The regularity condition $R'(0) = 0$ remains true when B and ω are nonzero. In what follows we will find solutions for which $R'(0)$ is nonzero and hence the D7-brane develops a conical singularity at $r = 0$.

Now let us introduce a nonzero B, with ϕ and A_z still zero [66–69]. Roughly speaking a nonzero B "pushes" the D7-brane toward the boundary. More precisely, suppose we fix $c_0 \propto |m| = 0$. Here an infinite number of solutions appear, of which only one is the trivial solution $R(r) = 0$. The key question then is which solution has the smallest on-shell action and hence is physically preferred? A numerical analysis reveals that the trivial solution is *not* the preferred one: that honor is reserved for a nontrivial $R(r)$ [66–69]. In fact, as we increase B the position where the D7-brane ends, $\rho = R(0)$, increases. Physically, the nonzero B causes the D7-brane to "bend", and increasing B pushes the endpoint of the D7-brane closer to the boundary. Notice what that means in the field theory: the preferred solution, being nontrivial, necessarily has a nonzero c_2, which from (13.34) indicates a nonzero $\langle \mathcal{O}_m \rangle$, hence chiral symmetry is spontaneously broken. The general lesson is that in our system a nonzero B promotes D7-brane bending, or in field theory language chiral symmetry breaking. The same remains true at nonzero c_0, although in that case c_0 explicitly breaks chiral symmetry.

Now let us return to nonzero B, ϕ, and A_z, and follow the arguments of Refs. [56–60]. The induced D7-brane metric is now

$$ds_{D7}^2 = g_{tt}^{D7} dt^2 + 2g_{tr}^{D7} dt\, dr + g_{rr}^{D7} dr^2 + \rho^2 d\mathbf{x}^2 + \frac{r^2}{\rho^2} ds_{S^3}^2, \qquad (13.38)$$

$$g_{tt}^{D7} = \rho^2 \left(-1 + \frac{\omega^2 R^2}{(r^2 + R^2)^2} \right), \qquad g_{tr}^{D7} = \frac{R^2 \omega \phi'}{r^2 + R^2},$$

$$g_{rr}^{D7} = \frac{1}{\rho^2} \left(1 + R'^2 + R^2 \pi'^2 \right), \qquad\qquad (13.39)$$

while the Legendre-transformed action in (13.30) is

$$\hat{S}_{D7} = -\mathcal{N} \int dr\, r^2 \sqrt{1 + R^2}$$

$$\times \sqrt{\left(1 - \frac{\omega^2 R^2}{(r^2 + R^2)^2} \right)\left(1 + \frac{B^2}{(r^2 + R^2)^2} - \frac{\alpha^2/\mathcal{N}^2}{R^2 r^6} \right) - \frac{1}{r^6}\left(\frac{\beta}{\mathcal{N}} + \frac{B\omega r^4}{(r^2 + R^2)^2} \right)^2 }.$$

$$\qquad (13.40)$$

The square root in the second line is of the form[9] $\sqrt{a(r)b(r) - c(r)^2}$ where

$$a(r) = 1 - \frac{\omega^2 R^2}{(r^2 + R^2)^2}, \qquad b(r) = 1 + \frac{B^2}{(r^2 + R^2)^2} - \frac{\alpha^2/\mathcal{N}^2}{R^2 r^6}, \qquad (13.41a)$$

$$c(r) = \frac{1}{r^3}\left(\frac{\beta}{\mathcal{N}} + \frac{B\omega r^4}{(r^2 + R^2)^2}\right). \qquad (13.41b)$$

Notice that $a(r)$ may change sign between $r \to \infty$ and $r \to 0$, but does not necessarily. More specifically, $a(r)$ is always positive at $r \to \infty$ and may become negative as $r \to 0$, depending on the behavior of $R(r)$. For the moment let us suppose that $a(r)$ does change sign. We will denote the value of r where $a(r)$ vanishes as r_*. Upon taking $a(r_*) = 0$ and doing some algebra we find the equation for a semicircle,

$$\left(R(r_*) - \frac{\omega}{2}\right)^2 + r_*^2 = \frac{\omega^2}{4}, \qquad (13.42)$$

where the radius[10] is $\omega/2$ and the center is at $(r_*, R(r_*)) = (0, \omega/2)$.

In fact, this semicircle is a horizon on the worldvolume of the D7-brane. If we change coordinates

$$d\hat{t} = dt + \frac{g_{tr}^{D7}}{g_{tt}^{D7}}\, dr, \qquad (13.43)$$

then the induced metric becomes

$$ds_{D7}^2 = \hat{g}_{\hat{t}\hat{t}}^{D7}\, d\hat{t}^2 + \hat{g}_{rr}^{D7}\, dr^2 + g_{xx}\, d\mathbf{x}^2 + g_{SS}\, ds_{S^3}^2, \qquad (13.44)$$

with

$$\hat{g}_{\hat{t}\hat{t}}^{D7} = g_{tt}^{D7}, \qquad \hat{g}_{rr}^{D7} = g_{rr}^{D7} - \frac{(g_{tr}^{D7})^2}{g_{tt}^{D7}}. \qquad (13.45)$$

We then have $\hat{g}_{\hat{t}\hat{t}}^{D7} = -\rho^2 a(r)$ and hence $a(r_*) = 0$ implies $g_{\hat{t}\hat{t}}^{D7}(r_*) = 0$. To understand the appearance of this horizon on the D7-brane, consider a light ray moving in the ϕ direction, at fixed values of all other coordinates. The line element for a light ray is null, hence $g_{tt}\, dt^2 + g_{\phi\phi}\, d\phi^2 = 0$, which gives us the local speed of light in the ϕ direction

$$\frac{d\phi}{dt} = \sqrt{\frac{|g_{tt}|}{g_{\pi\pi}}} = \frac{r^2 + R^2}{R}. \qquad (13.46)$$

---

[9]Notice that the Legendre-transformed action here has the same generic form as the Legendre-transformed D7-brane action with worldvolume electric and magnetic fields used in Ref. [70] for a holographic calculation of a Hall conductivity associated with the $U(1)_V$ symmetry. The similarity is not surprising, given the similarity between rotation and worldvolume electric fields due to T-duality, as explained above. Many of our arguments below are similar to those made in Ref. [70].

[10]Recall that we are using units in which the AdS_5 radius is $L \equiv 1$.

Clearly when ω is large enough to make $a(r) < 0$ the D7-brane is moving faster than the local speed of light at that value of r, and a worldvolume horizon appears. Formally we can associate a temperature with the worldvolume horizon, indicating that in the field theory the flavor sector has a finite temperature while the adjoint sector has $T = 0$, which is a clear signal that our system is not in equilibrium.

If $a(r)$ changes sign but $b(r)$ does not, then $a(r)b(r) - c(r)^2 < 0$ for some $r < r_*$ and because of the square root $\hat{S}_{D7}$ becomes imaginary, signaling a tachyonic instability as explained above. To avoid the instability we demand that $b(r_*) = 0$ also. Furthermore, as $a(r)$ and $b(r)$ approach zero, $c(r)$ must approach zero more quickly, otherwise we again encounter an instability. We thus also impose $c(r_*) = 0$.

The condition $a(r_*) = 0$ fixes the worldvolume horizon while the conditions $b(r_*) = 0$ and $c(r_*) = 0$ fix the two integration constants α and β, or equivalently via (13.35) $\langle \mathcal{O}_\phi \rangle$ and $\langle J^z \rangle$. In other words, for given values of $|m|$, ω, and B, regularity of the bulk solution determines unique values of the one-point functions $\langle \mathcal{O}_m \rangle$, $\langle \mathcal{O}_\phi \rangle$, and $\langle J^z \rangle$, as is standard in AdS/CFT. Explicitly, we find

$$\alpha = -\mathcal{N} R(r_*) r_*^3 \sqrt{1 + \frac{B^2}{\omega^2 R^2(r_*)}}, \qquad \beta = -\mathcal{N} \frac{B\omega r_*^4}{(R^2(r_*) + r_*^2)^2}. \qquad (13.47)$$

Using $a(r_*) = 0$ we can also express α and β in terms of $R(r_*)$ alone,

$$\alpha = -\mathcal{N} R(r_*)^{3/2} |R(r_*) - \omega|^{3/2} \sqrt{R(r_*)^2 + \frac{B^2}{\omega^2}}, \qquad \beta = -\mathcal{N} \frac{B}{\omega} (R(r_*) - \omega)^2. \qquad (13.48)$$

The D7-brane does not always develop a worldvolume horizon. What happens when it does not? In that case $a(r) > 0$ for all values of r. If α is nonzero then $b(r)$ will change sign, rendering the action imaginary, so we demand $\alpha = 0$. In addition if β is nonzero then again the action becomes imaginary because $a(r)b(r)$ goes as $1/r^4$ as $r \to 0$ while $c(r)^2$ goes like β^2/r^6, so clearly $\sqrt{a(r)b(r) - c(r)^2}$ will become imaginary at sufficiently small r. We thus also demand $\beta = 0$ for these solutions.

We thus have two classes of D7-brane embeddings, those with a worldvolume horizon and those without. The former have nonzero α and β, or in the field theory nonzero $\langle \mathcal{O}_\phi \rangle$ and $\langle J^z \rangle$, hence these solutions describe a CME with CT spontaneously broken. The latter have $\alpha = 0$ and $\beta = 0$ and hence no CME.

An exceptional case is the trivial solution $R(r) = 0$ and $\phi(t, r) = \omega t$, corresponding to a chirally-symmetric state with $|m| = 0$ and $\langle \mathcal{O}_m \rangle = 0$. That solution has no worldvolume horizon, yet from (13.48) we see that although α vanishes, β must be nonzero to maintain reality of the action. Recalling that $B = (2\pi\alpha')\tilde{B}$, where $\tilde{B}$ is the value of the magnetic field in the field theory (see above (13.20)), we find for the trivial solution

$$\beta = -(2\pi\alpha') \mathcal{N} \tilde{B} \omega. \qquad (13.49)$$

Using (13.7), (13.20), and (13.35) to translate to field theory quantities, we find

$$\langle J^z \rangle = \frac{N_f N_c}{2\pi^2} \mu_5 \tilde{B}, \tag{13.50}$$

which is the value fixed by the anomaly, as expected in the chirally-symmetric case. We hasten to add three things. First, we will see that (13.50) is unchanged at finite temperature, where the trivial solution remains a valid solution. Second, the trivial D7-brane has no angular momentum, so the corresponding field theory state has zero axial charge density, despite having nonzero μ_5 with zero mass gap, $|m| = 0$. Third, because the anomaly's contribution to $\langle J^z \rangle$ does not contain much dynamical information, will isolate the more interesting dynamical contributions by writing the $\langle J^z \rangle$ in (13.35) as

$$\langle J^z \rangle = -(2\pi\alpha')\beta = \frac{N_f N_c}{(2\pi)^2} \tilde{B}\omega \left(\frac{-\beta}{(2\pi\alpha')\mathcal{N} \tilde{B}\omega} \right). \tag{13.51}$$

The factor in parentheses in the final equality contains the non-trivial dynamical information. From (13.47) we see that $\beta \propto (2\pi\alpha')\mathcal{N} \tilde{B}\omega$, so the factor in parentheses also does not depend explicitly on the magnetic field, although it will depend implicitly through the embedding. Notice that the current is always proportional to B, as we expect for the CME.

To produce numerical solutions for the two classes of D7-brane embeddings, we must specify boundary conditions. The equation of motion for $R(r)$ is a second-order non-linear ordinary differential equation, for which we need two boundary conditions on $R(r)$. Solutions without worldvolume horizons are the simplest to produce. For these we set $\alpha = 0$ and $\beta = 0$, choose a value of $R(0)$ greater than $\omega/2$ to avoid a worldvolume horizon, and then impose $R'(0) = 0$ to guarantee regularity. Solutions with worldvolume horizons are trickier to obtain,[11] since the equation of motion itself depends on the values of α and β, or equivalently on r_* and $R(r_*)$, so we must choose these before we can solve the equation of motion. For these solutions we first choose a point on the semicircle in (13.42), which fixes the values of α and β via (13.48). We then obtain a condition on the first derivative at that point, $R'(r_*)$, from the equation of motion itself. We omit the explicit form, which is unilluminating. With these boundary conditions we can solve the equation of motion both inside the worldvolume horizon and outside. Notice that in these cases the value of $R'(0)$, which determines whether the D7-brane has a conical singularity at $r = 0$, is an *output* of the calculation. Figure 13.1 shows numerical solutions for $R(r)$ for various values of $|m|$, ω, and B.

Our first observation is that for all solutions with a worldvolume horizon $R'(0)$ is nonzero, so in our system at zero temperature all non-trivial solutions describing

---

[11]Solutions with nonzero B and worldvolume horizons but with $A_z(r) = 0$ were obtained in Ref. [57]. These solutions are in fact unphysical, since an ansatz with $A_z(r) = 0$ is inconsistent: in Ref. [57] the WZ term in (13.21) was omitted, but the presence of that term necessitates the introduction of $A_z(r)$. All other solutions in Ref. [57] besides these are consistent.

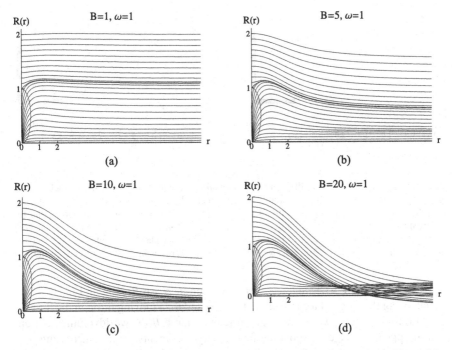

Fig. 13.1 Numerical D7-brane embeddings $R(r)$ for $T = 0$ and $\omega = 1$ for various values of B, in units of the AdS_5 radius. The *red semi-circle* denotes the worldvolume horizon of (13.42). (a) $B = 1$. (b) $B = 5$. (c) $B = 10$. (d) $B = 20$. The asymptotic value of $R(r)$ as $r \to \infty$ (*the far right in each plot*) is the coefficient c_0 in (13.31), which is proportional to the flavor mass $|m|$. The different classes of solutions, and their behavior as functions of B and $|m|$, are discussed in the accompanying text (Color figure online)

the CME have a conical singularity. In fact, for these solutions the on-shell action exhibits a divergence at $r = 0$, taking us outside of both the probe and supergravity limits, so strictly speaking we should not trust these solutions. Nevertheless, in Sect. 13.4 we will argue that the singularities are physically sensible, being intimately related with the time rates of change of axial charge and of energy in the field theory.

Our second observation is that all solutions with a worldvolume horizon have nonzero $|m|$. That makes sense, since these solutions have nonzero α, and hence must have nonzero $|m|$, as described above. Only solutions with $\alpha = 0$ can describe $|m| = 0$. That class of solutions includes the trivial one $R(r) = 0$ as well as nontrivial solutions without worldvolume horizons.

From Fig. 13.1 we can deduce the general behavior of solutions as we increase $|m|$ or B, as follows. Suppose we fix ω and B, *i.e.* we choose one of Figs. 13.1(a) through (d), and then begin with some $c_0 \propto |m|$ that is nonzero but much smaller than ω or $\sqrt{B}$, such that the solution is very close to the trivial $R(r) = 0$ solution. As we increase $|m|$, clearly the endpoint $R(0)$ also increases. For sufficiently large $|m|$ the worldvolume horizon will disappear, at which point α and β vanish. Alter-

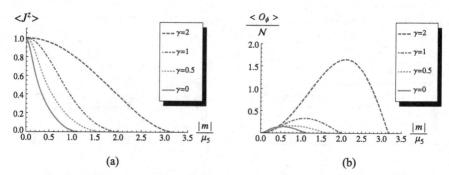

Fig. 13.2 (a) The value of $\langle J^z \rangle$, divided by the anomaly-determined value in (13.50), as a function of the flavor mass divided by the axial chemical potential, $|m|/\mu_5$. Here we set the magnetic field $B = 1$ and $\omega = 2\mu_5 = 1$ (in units of the AdS_5 radius). The different *curves* correspond to different temperatures $T = \gamma/\pi$, with *green solid, yellow dotted, red dot-dashed,* and *blue dashed* corresponding to $\gamma = 0, 0.5, 1, 2$, respectively. (b) The pseudo-scalar condensate $\langle \mathscr{O}_\phi \rangle / \mathscr{N}$ versus $|m|/\mu_5$, with $B = 1$ and $\omega = 2\mu_5 = 1$, for the same temperatures as in (a) (Color figure online)

natively, suppose we fix ω and $c_0 \propto |m|$, and then increase B. Now in Fig. 13.1 we are choosing the value of a curve at the far right and then moving through the figures from (a) to (d). Again we see that for sufficiently large B the worldvolume horizon will disappear.[12] The corresponding field theory statements are that increasing $|m|$ or B eventually restores CT and extinguishes the CME, since eventually $\langle \mathscr{O}_\phi \rangle = 0$ and $\langle J^z \rangle = 0$. The general lesson is that chiral symmetry breaking, whether explicit via $|m|$ or spontaneous via B, acts against the CME in our system.

Our main result in this section is Fig. 13.2. The green solid curve in Fig. 13.2(a) shows the exact behavior of $\langle J^z \rangle$, normalized to the value in (13.50), as we increase $|m|/\mu_5$, and the green curve in Fig. 13.2(b) shows the same for $\langle \mathscr{O}_\phi \rangle / \mathscr{N}$. At $|m| = 0$, $\langle J^z \rangle$ takes the value determined by the anomaly, while $\langle \mathscr{O}_\phi \rangle = 0$. Increasing $|m|/\mu_5$, $\langle J^z \rangle$ decreases monotonically and eventually reaches zero, while $\langle \mathscr{O}_\phi \rangle$ increases, reaches a maximum, and then drops to zero. We omit the curves for $\langle J^z \rangle$ and $\langle \mathscr{O}_\phi \rangle$ versus $\sqrt{B}/\mu_5$, which are qualitatively similar to those in Fig. 13.2.

13.3.2 Solutions at Finite Temperature

We now want to solve $R(r)$'s equation of motion numerically in the AdS-Schwarzschild background, corresponding to an $\mathscr{N} = 4$ SYM plasma at temperature T.

---

[12]From Fig. 13.1(d) we also see that for some D7-branes without a worldvolume horizon, $R(r)$ passes through zero for sufficiently large B. Such behavior has been observed many times for D7-branes with worldvolume magnetic field: see for example Refs. [66, 69]. As argued in Ref. [69], these solutions have a sensible interpretation in the field theory as a renormalization group flow, although in equilibrium they are not always the lowest-energy solutions. These D7-branes do not describe a CME and so are of less interest to us than D7-branes with worldvolume horizons.

Before doing so, let us briefly review what occurs when $\omega = 0$ and $B = 0$, *i.e.* when the worldvolume gauge field and ϕ vanish, summarizing Refs. [71–76]. The main difference from the zero-temperature, pure AdS_5 case is the presence of the AdS-Schwarzschild horizon, which divides D7-brane solutions into two categories. The first are similar to those in pure AdS_5, namely D7-branes for which the S^3 shrinks and eventually collapses to zero size at some value of ρ outside the AdS-Schwarzschild horizon. The second category consists of D7-branes for which the S^3 shrinks but does not reach zero size by the time the D7-brane intersects the AdS-Schwarzschild horizon. In the current context, solutions in the first category are called "Minkowski" embeddings while solutions in the second category are called "black hole" embeddings. In Euclidean signature, with compact time direction of period $1/T$, the time circle collapses to zero size at the horizon. The two types of D7-brane solution thus have distinct topology: Minkowski embeddings have a collapsing three-cycle, the S^3, while black hole embeddings have a collapsing one-cycle, the time circle. For Minkowski embeddings the condition to avoid a conical singularity when the S^3 collapses is $R'(0) = 0$, while the condition for black hole embeddings to avoid a singularity is that in the (r, R) plane the D7-brane must be perpendicular to the AdS-Schwarzschild horizon.

When ω and B both vanish, the theory has one physically meaningful dimensionless parameter, $T/|m|$. Suppose we fix $T/|m|$ such that we have a Minkowski embedding and then increase $T/|m|$, say by holding $|m|$ fixed but increasing T. In the bulk the AdS-Schwarzschild horizon will grow and move toward the boundary, eventually encountering the D7-brane. The D7-brane solution then becomes a black hole embedding. Such a process involves a change in topology, so we have reason to expect that in general any observable associated with the flavor fields in the field theory will exhibit discontinuous behavior. Indeed, the bulk transition from Minkowski to black hole embedding appears in the field theory as a first-order phase transition [71–76].

Perhaps the most dramatic change in that transition occurs in the spectrum of D7-brane excitations, dual to the spectrum of mesons. For a Minkowski embedding the fluctuations of worldvolume fields are normal modes, *i.e.* standing waves trapped between the AdS_5 boundary and the endpoint of the D7-brane. That translates into a field theory meson spectrum that is gapped and discrete [62]. For a black hole embedding the worldvolume fluctuations are quasi-normal modes, that is, the eigenfrequencies acquire an imaginary part. Physically, these fluctuations can leak energy into the AdS-Schwarzschild horizon and hence are damped. In the field theory the meson spectrum is gapless and continuous. The transition between the two is thus a kind of "meson melting" transition [77].

Introducing nonzero B, still keeping $\omega = 0$, qualitatively has the same effect as in the pure AdS_5 case: increasing B pushes the D7-brane toward the boundary. If we start with a black hole embedding, for example, and keep $T/|m|$ fixed while increasing $B/|m|^2$, eventually a transition occurs to a Minkowski embedding [66–69].

Now consider nonzero $|m|$, T, B, and ω. The induced D7-brane metric is then

$$ds_{D7}^2 = g_{tt}^{D7} \, dt^2 + 2g_{tr}^{D7} \, dt \, dr + g_{rr}^{D7} \, dr^2 + \rho^2 \frac{\gamma^2}{2} H(\rho) \, d\mathbf{x}^2 + \frac{r^2}{\rho^2} \, ds_{S^3}^2, \quad (13.52)$$

where g_{tr}^{D7} and g_{rr}^{D7} are the same as in (13.39) but

$$g_{tt}^{D7} = \rho^2 \left(-\frac{\gamma^2}{2} \frac{f(\rho)^2}{H(\rho)} + \frac{\omega^2 R^2}{(r^2 + R^2)^2} \right), \qquad f(\rho) = 1 - \frac{1}{\rho^4}, \qquad H(\rho) = 1 + \frac{1}{\rho^4},$$
(13.53)

and we recall that $\rho^2 = r^2 + R^2$, the AdS-Schwarzschild horizon is at $\rho = 1$, and the temperature is $T = \gamma/\pi$ in our conventions. The location of the D7-brane's worldvolume horizon r_* is now given by

$$\frac{\gamma^2}{2} \frac{f(\rho_*)^2}{H(\rho_*)} - \frac{\omega^2 R(r_*)^2}{(r_*^2 + R(r_*)^2)^2} = 0,$$
(13.54)

or equivalently

$$f(\rho_*)^2 = \frac{2H(\rho_*)}{\gamma^2} \frac{\omega^2 R(r_*)^2}{(r_*^2 + R(r_*)^2)^2}.$$
(13.55)

Clearly the D7-brane worldvolume horizon is always outside of the AdS-Schwarzschild horizon, since $f(\rho_*) > f(\rho = 1) = 0$. If $\omega/\gamma \propto \mu_5/T \to 0$ then $\rho_* \to 1$ and the two horizons coincide. At fixed mass, we may think of the $\mu_5/T \to 0$ limit either as small μ_5 at fixed T or large T at fixed μ_5.

The physical arguments of the last subsection for the reality of the action are unchanged. Applying those arguments to fix α and β we find

$$\alpha = -\mathcal{N} \frac{\gamma^4}{4} R(r_*) r_*^3 f(\rho_*) H(\rho_*) \sqrt{1 + \frac{4B^2}{\gamma^4 (R(r_*)^2 + r_*^2)^2 H(\rho_*)^2}},$$
$$\beta = -\mathcal{N} B \omega \frac{r_*^4}{(R(r_*)^2 + r_*^2)^2}.$$
(13.56)

In AdS-Schwarzschild our D7-brane solutions fall into three categories. The first two are the straightforward generalizations of the categories of the last subsection: Minkowski embeddings without worldvolume horizons and Minkowski embeddings with worldvolume horizons. The new category consists of black hole embeddings, which necessarily have worldvolume horizons, as explained above. As in the last subsection, solutions without a worldvolume horizon describe field theory states with no CME and no spontaneous breaking of CT, while solutions with a worldvolume horizon, whether Minkowski or black hole, describe field theory states with a CME and spontaneous breaking of CT.

The trivial solution $R(r) = 0$ falls into the third category of embeddings, since it necessarily intersects the AdS-Schwarzschild horizon. From (13.56) we see that for the trivial solution $\alpha = 0$ and β takes the value in (13.49), so $\langle J^z \rangle$ again takes the value in (13.50), hence we see that the value of $\langle J^z \rangle$ in the chirally-symmetric case, being fixed by the anomaly, is independent of temperature.

For Minkowski embeddings the procedure to generate numerical solutions is the same as in the last subsection. In particular, for solutions with a worldvolume horizon we first choose a point on the worldvolume horizon and then use the equation of

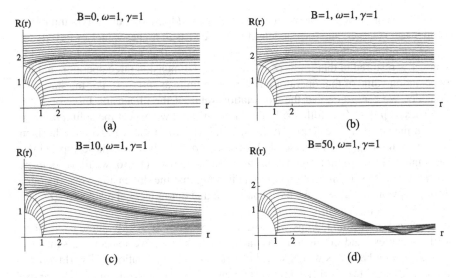

Fig. 13.3 Numerical D7-brane embeddings $R(r)$ for $\omega = 1$ and $\gamma = 1$, corresponding to a temperature $T = \gamma/\pi = 1/\pi$, for various values of B, in units of the AdS_5 radius. The *red quarter-circle* represents the AdS-Schwarzschild horizon while the other *red curve* (*the near-quarter-circle*) denotes the worldvolume horizon of (13.54). (**a**) $B = 0$. (**b**) $B = 1$. (**c**) $B = 10$. (**d**) $B = 50$. The different classes of solutions, and their behavior as functions of B, $|m|$, and T, are discussed in the accompanying text (Color figure online)

motion to determine the first derivative. We use the latter procedure for black hole embeddings too, since these necessarily have a worldvolume horizon. Notice that when we impose boundary conditions at the worldvolume horizon, the behavior of $R(r)$ and its derivative at $r = 0$ or at the AdS-Schwarzschild horizon, which determines whether the solution has a conical singularity, is an output of the calculation.

Figure 13.3 shows numerical solutions for $R(r)$ for various values of $|m|$, ω, B, and T. The results for Minkowski embeddings are similar to those of the last subsection. In particular, Minkowski embeddings with a worldvolume horizon have a nonzero $R'(0)$ and hence a conical singularity. The black hole embeddings, however, do not have such a conical singularity: as Fig. 13.3 suggests, and numerical analysis confirms, in the (r, R) plane depicted the D7-brane "hits" the black hole horizon perpendicularly.

Many of the conclusions from our $T = 0$ analysis remain valid at finite temperature. All solutions with nonzero α have nonzero $c_0 \propto |m|$. The worldvolume horizon eventually disappears as we increase $|m|$ or B: chiral symmetry breaking works against the CME in our system.

Figure 13.2(a) shows the chiral magnetic current $\langle J^z \rangle$, normalized to the anomaly-determined value in (13.50), versus $|m|/\mu_5$ for $B = 1$ and several values of T. At higher T the chiral magnetic current can persist to higher values of $|m|/\mu_5$ before dropping to zero. Figure 13.2(b) shows $\langle \mathcal{O}_\phi \rangle / \mathcal{N}$ versus $|m|/\mu_5$ for $B = 1$ and the same values of T as in Fig. 13.2(a). The qualitative behavior of the pseudoscalar condensate is similar to the $T = 0$ case, increasing, reaching a maximum, and then dropping to zero as we increase $|m|/\mu_5$. At higher T, the maximum of the condensate is larger, and the condensate also persists to higher values of $|m|/\mu_5$.

As mentioned above, seeing $\langle \mathscr{O}_\phi \rangle \to 0$ as $|m| \to 0$ is not enough to conclude that CT is restored in that limit. $\mathscr{O}_\phi$ is $|m|$ times a dimension three operator, and the expectation value of that operator could remain finite as $|m| \to 0$. For the states we consider, we can argue that the expectation value of the dimension three operator vanishes as $|m| \to 0$ as follows. In the limit $|m| \to 0$ we expect the solution to approach the constant one $R(r) \approx c_0 \propto |m|$, and we expect the worldvolume horizon to approach the AdS-Schwarzschild horizon, so $\rho_* \approx 1$. Inserting these approximations into (13.54) we find $f(\rho_*) \simeq |m|\omega/\gamma$, and then from (13.56) we find $\alpha \propto |m|^2$, so $\langle \mathscr{O}_\phi \rangle$ vanishes as $|m|^2$ as $|m| \to 0$, indicating that the dimension-three operator $\langle \mathscr{O}_\phi \rangle/|m|$ vanishes as $|m|$. We have confirmed that our numerical results for $\langle \mathscr{O}_\phi \rangle$ in Fig. 13.2(b) behave as $|m|^2$ as $|m| \to 0$.

For black hole embeddings we expect the spectrum of worldvolume excitations will be gapless and continuous, as in the $\omega = 0$ case. We expect the same for Minkowski embeddings with a worldvolume horizon: fluctuations of worldvolume fields will "see" the worldvolume horizon as a genuine horizon, and hence we expect them to behave in a fashion similar to those of black hole embeddings.[13] More specifically, since we can associate a temperature with the worldvolume horizon, we expect the solutions for linearized fluctuations to translate into field theory two-point functions with a form characteristic of thermal diffusion at that temperature [59]. Moreover, in the bulk the worldvolume and AdS-Schwarzschild horizons generally will not coincide, so the worldvolume temperature will generally be different from the $\mathscr{N} = 4$ SYM plasma temperature. As mentioned in Sect. 13.3.1, this is a clear signal that the system is not in equilibrium.

13.4 Loss Rates of Axial Charge and of Energy

As mentioned above, when $|m|$ is finite our system in not in equilibrium: axial charge can leak into the adjoint sector, taking energy with it. In this section we will compute the loss rate of axial charge in our system, as well as the loss rate of the energy density, or more precisely the expectation value of the "tt" component of the stress-energy tensor (density), $\langle T_{tt} \rangle$. We compute these loss rates both in field theory and from holography, following Refs. [59, 80], with perfect agreement. We will also explain how the states we study remain stationary despite these loss rates, and how the nonzero loss rates might be related to the singularities in some of the D7-brane solutions of Sects. 13.3.1 and 13.3.2.

In our system, axial current conservation can be violated in three ways. The first way is explicitly via a nonzero $|m|$. The second way is due to anomalies. The third way is due to the fact that the axial symmetry is part of the R-symmetry, so that an

---

[13]A gapless, continuous spectrum appears in the presence of a worldvolume electric field, which is closely related to rotation via T-duality as we argued above [78, 79].

axial charge density introduced in the flavor sector alone can leak into the adjoint sector. From the field theory point of view, axial charge can be lost to the adjoint sector in two ways, depending on whether mesons are melted or not. The equivalent bulk statement is that D7-brane angular momentum can be lost in two ways, depending on whether the D7-brane has a worldvolume horizon or not.

If mesons are not melted, then the loss of R-charge occurs when mesons radiate R-charged glueballs. In the 't Hooft limit the interactions of color singlet mesons and glueballs are suppressed, so we expect the loss of R-charge to the adjoint sector not to be apparent in that limit. The dual statement is that a D7-brane with angular momentum but no worldvolume horizon can only lose angular momentum via radiation of closed strings, but the relevant interaction is proportional to the string coupling $g_s \sim 1/N_c$ and so is suppressed in the classical supergravity limit, where $g_s \to 0$.

If mesons are melted, then the spectrum in the flavor sector is no longer just delta-functions representing color-singlet mesons, but a continuum of modes whose interactions with the adjoint fields experience no large-N_c suppression. The dual statement is that a D7-brane with angular momentum and a worldvolume horizon (whether a Minkowski or black hole embedding) can transfer angular momentum across the horizon even in the supergravity limit.

We will now confirm explicitly the above expectations by computing the time rate of change of axial/R-charge. We begin with the field theory calculations. For the rate of change of R-charge we just need to compute a Ward identity. Since an R-symmetry transformation is equivalent to a shift in the source, *i.e.* a shift in the phase of the flavor mass $\phi \to \phi + \delta\phi$, we have, using the definition of $\mathcal{O}_\phi$ in terms of a variation of minus the field theory action with respect to the phase of the flavor mass,[14]

$$\partial_\mu \langle J_R^\mu \rangle = \langle \mathcal{O}_\phi \rangle. \tag{13.57}$$

In our case nothing is changing in space, so we obtain for the time rate of change of the R-charge density $\partial_t \langle J_R^t \rangle = \langle \mathcal{O}_\phi \rangle$. Recall that $\mathcal{O}_\phi \propto |m|$, so a necessary condition for the rate of change to be nonzero is for $|m|$ to be nonzero. In that case the potential and therefore the Hamiltonian are time-dependent, so the energy density will also not be conserved. Recalling the definitions of the potential terms V_q and V_ψ from Sect. 13.2, the change in energy density is

$$\partial_t \langle T_{tt} \rangle = \partial_t \langle V_q + V_\psi \rangle = \langle \mathcal{O}_\phi \rangle \partial_t \phi, \tag{13.58}$$

where the second equality follows from the chain rule. In our case $\partial_t \phi = \omega$, and we just saw that $\langle \mathcal{O}_\phi \rangle = \partial_t \langle J_R^t \rangle$, so for our system

$$\partial_t \langle T_{tt} \rangle = \omega \partial_t \langle J_R^t \rangle. \tag{13.59}$$

---

[14]The variation of the path integral is $\int d^4x \langle -\frac{1}{2}\partial_\mu \delta\phi \overline{\psi}\gamma^\mu \gamma^5 \psi - \mathcal{O}_\phi \delta\phi \rangle$. Integrating by parts and demanding that the variation vanish for any $\delta\phi$, we find the relation in (13.57).

The two rates of change are directly proportional. As expected, when R-charge leaks into the adjoint sector, it takes energy with it.

We now turn to the bulk calculation of the same rates of change. The canonical momentum associated with the worldvolume scalar ϕ, which we denote π_ϕ^M, is[15]

$$\pi_\phi^M \equiv \frac{\delta S_{D7}}{\delta(\partial_M \phi)}, \qquad \partial_M \pi_\phi^M = 0. \qquad (13.60)$$

The second equation is just the ϕ equation of motion. The probe flavor contribution to the expectation value of the axial or R-symmetry current in the field theory directions (*i.e.* $M = \mu$), J_R^μ, is the integral of π_ϕ^μ over the worldvolume of the brane.[16]

$$\langle J_R^\mu \rangle = \int_{r_H}^\infty dr\, \pi_\phi^\mu. \qquad (13.61)$$

For concreteness we have written the lower endpoint of the r integration as r_H, as appropriate for a black hole embedding. For a Minkowski embedding the lower endpoint is $r = 0$. The R-charge density $\langle J_R^t \rangle$ is given by the angular momentum of the D7-brane, $\langle J_R^t \rangle = \int_{r_H}^\infty dr\, \pi_\phi^t$. Taking ∂_t of the charge density and using the ϕ equation of motion in (13.60), we find

$$\partial_t \langle J_R^t \rangle = \int_{r_H}^\infty dr\, \partial_t \pi_\phi^t = -\int_{r_H}^\infty dr\, \partial_r \pi_\phi^r = -\pi_\phi^r \big|_{r_H}^\infty, \qquad (13.62)$$

where in the second equality we assumed homogeneity, so the derivative of π_ϕ^M in any field theory spatial direction vanishes. Recalling from (13.22) that $\pi_\phi^r = \alpha$, which in our system is independent of r, the R-charge density appears to be constant in time, $\partial_t \langle J_R^t \rangle = 0$. That is indeed true since the states we study are stationary. The reason why is nontrivial, however: the two terms in the final equality above cancel one another. The contribution from the lower endpoint represents the angular momentum that the D7-brane is losing, or equivalently the R-charge that the flavors are dissipating into the adjoint sector, while the contribution from the upper endpoint represents angular momentum that we are pumping into the system by hand via a boundary condition on the D7-brane, since we are forcing the D7-brane to rotate at the boundary, or equivalently in the field theory R-charge that we are pumping into the system from an external source. Implicitly in our solutions we choose the latter precisely to cancel the former. The upshot is that the loss rate is the contribution from the lower endpoint,

$$\partial_t \langle J_R^t \rangle \big|_{\text{loss}} = \alpha = \langle \mathcal{O}_\phi \rangle, \qquad (13.63)$$

---

[15]Uppercase Latin letters $M, N, \ldots$ denote $(4+1)$-dimensional bulk coordinates, including the holographic radial direction, while lowercase Greek letters $\mu, \nu, \ldots$ denote $(3+1)$-dimensional boundary field theory directions.

[16]Here we are ignoring any potential divergences that may appear at the $r \to \infty$ endpoint of the integral, which can be cancelled with counterterms that do not affect our main results.

in perfect agreement with the field theory Ward identity. Only D7-branes with a worldvolume horizon have nonzero α, hence in the field theory only states with melted mesons have a nonzero rate of change for R-charge (in the 't Hooft limit), in conformity with our field theory intuition.

To compute the rate of change of the energy density, we need to compute the stress-energy tensor of the D7-brane. As explained in Ref. [80], we can do that in two equivalent ways. The first way is directly, by variation of the D7-brane action with respect to the background metric. The second way is via a Noether procedure. Although we have used both methods, we will only present the latter, which is more efficient. Defining a Lagrangian via $S_{D7} = \int dr \mathscr{L}$, the stress-energy tensor *density* of the D7-brane, in the AdS_5 directions, is

$$\Theta_N^M = \mathscr{L}\delta_N^M + 2F_{LN}\frac{\delta\mathscr{L}}{\delta F_{ML}} - \partial_N\theta\frac{\delta\mathscr{L}}{\delta\partial_M\theta} - \partial_N\pi\frac{\delta\mathscr{L}}{\delta\partial_M\pi}. \tag{13.64}$$

The D7-brane stress-energy tensor is then the integral of Θ_N^M over r. For values of M and N in field theory directions, we may equate the D7-brane stress-energy tensor's components with the flavor fields' contribution to the expectation value of the field theory stress-energy tensor [80],

$$\langle T_\nu^\mu \rangle = \int dr \, \Theta_\nu^\mu. \tag{13.65}$$

To compute the rate of change of energy density we will only need one component of the D7-brane stress-energy tensor density, $\Theta_t^r = -\omega\frac{\partial\mathscr{L}}{\partial\phi'} = -\omega\alpha$. We proceed in a similar manner to the calculation of the R-charge rate of change. Using conservation of the D7-brane stress-energy tensor density, $\partial_M\Theta_N^M = 0$, we find

$$\langle \partial_t T_{tt} \rangle = -\langle \partial_t T_t^t \rangle = -\int_{r_H}^\infty dr \, \partial_t\Theta_t^t = \int_{r_H}^\infty dr \, \partial_r\Theta_t^r = \Theta_t^r\Big|_{r_H}^\infty = -\omega\alpha\Big|_{r_H}^\infty. \tag{13.66}$$

The total energy is conserved, but the flux of energy at the boundary and the horizon is nonzero. The flux at the horizon corresponds to the rate of energy dissipation,

$$\partial_t \langle T_{tt} \rangle\big|_{\text{loss}} = \omega\alpha = \omega\langle \mathscr{O}_\phi \rangle, \tag{13.67}$$

where α is given by (13.56) and is negative. The holographic calculation reproduces the relation between energy and charge loss rates,[17]

$$\partial_t \langle T_{tt} \rangle\big|_{\text{loss}} = \omega\partial_t\langle J_R^t \rangle\big|_{\text{loss}}. \tag{13.68}$$

The dissipation rate is only of order $\alpha \sim \lambda N_f N_c$. That means that the flavor sector will only transfer an order N_c^2 amount of charge into the adjoint sector over a

---

[17]For black hole embeddings the loss rates can also be extracted from (suitably regulated) divergences in the angular momentum $\frac{\delta S_{D7}}{\delta\omega}$ and in the D7-brane stress-energy tensor at the AdS-Schwarzschild horizon, as explained in Ref. [80].

time of the order $N_c/\lambda \propto 1/g_{YM}^2 \gg 1$. For times parametrically shorter than N_c/λ, we can ignore the dissipation rate and treat the background as a reservoir, in which case the stationary solution in the probe limit is a reliable approximation to the actual solution. This is similar to what occurs with constant electric fields on the D7-brane, where both energy and momentum (but not angular momentum) are dissipated in the bulk [80].

Recall that the above analysis is valid for Minkowski embeddings with worldvolume horizon (or for black hole embeddings in the zero-temperature limit) simply by taking $r_H \to 0$. In Sect. 13.3 we saw that such D7-branes have a conical singularity at $r = 0$. We can understand this singularity as a consequence of the angular momentum and energy flux along the brane. When the angular momentum and energy flowing along the brane reach the "bottom" at $r = 0$, they must be dumped into some source, or really a sink. For black hole embeddings that source is hidden behind the AdS-Schwarzschild horizon, and the part of the D7-brane outside of that horizon is non-singular. In the absence of the AdS-Schwarzschild horizon, and neglecting the backreaction of the D7-brane, the source is manifested as the "naked" conical singularity of the embedding. Something very similar occurs for the holographic dual of $\mathcal{N} = 4$ SYM formulated on a spatial S^3 with R-charge chemical potentials. Bulk solutions, known as "superstars" (dual to zero-temperature BPS states), exhibit naked singularities that have a sensible physical interpretation in terms of charged sources, namely giant gravitons [81].

13.5 Summary and Discussion

We used AdS/CFT to study the CME in large-N_c, strongly-coupled $\mathcal{N} = 4$ SYM theory coupled to a number $N_f \ll N_c$ of $\mathcal{N} = 2$ supersymmetric flavor hypermultiplets. We introduced a time-dependent phase for the hypermultiplet mass, which for the hypermultiplet fermions is equivalent to an axial chemical potential, and we introduced an external, non-dynamical $U(1)_V$ magnetic field. When the magnitude of the hypermultiplet mass $|m|$ was zero, we found at both zero and finite temperature that the chiral magnetic current $\langle J^z \rangle$ coincided with the weak-coupling result in the chirally-symmetric limit, (13.50). When $|m|$ was nonzero we found that $\langle J^z \rangle$ had a smaller value than (13.50), and also that the $U(1)_V$-invariant and CT-odd pseudo-scalar operator $\mathcal{O}_\phi$ acquired a nonzero expectation value. Indeed, our main result was Fig. 13.2, which shows that for sufficiently large $|m|$ or B, compared to the axial chemical potential or the temperature, both $\langle J^z \rangle$ and $\langle \mathcal{O}_\phi \rangle$ drop to zero. In these cases we interpret the appearance of a chiral magnetic current as the conversion of the pseudo-scalar condensate to a vector condensate via the magnetic field.

Our holographic system describes the CME only in non-equilibrium states. Whenever $|m|$ is nonzero, the scalars in the hypermultiplet have masses with time-dependent phases, hence the Hamiltonian has explicit time dependence and energy is not conserved. Moreover, in our system the axial symmetry is part of the R-symmetry, so axial charge in the flavor sector can leak into the adjoint sector, also taking energy with it. We computed the associated loss rates, which we found to be proportional to $\langle \mathscr{O}_\phi \rangle$. When the CME occurs in our system at nonzero $|m|$, the flavor fields are losing axial charge and energy to the adjoint sector. In the probe limit these loss rates are negligible, however, and our states describing a CME were in fact stationary.[18]

The supergravity description of the above was N_f probe D7-branes extended along $AdS_5 \times S^3$ inside $AdS_5 \times S^5$, rotating on the S^5 and with worldvolume gauge fields that encode the magnetic field and chiral magnetic current. AdS space is effectively a gravitational potential well in which the local speed of light decreases as we move away from the boundary. A D7-brane rotating sufficiently quickly may at some point be rotating faster than the local speed of light and hence may develop a worldvolume horizon. We saw that indeed D7-brane solutions thus split into two categories, those that rotate quickly enough to develop a worldvolume horizon and those that don't. For the former the D7-brane becomes imaginary, signaling the presence of a tachyon, unless we introduce certain worldvolume fields and adjust their integration constants to maintain reality of the action. Via the holographic dictionary these integration constants were precisely the values of $\langle J^z \rangle$ and $\langle \mathscr{O}_\phi \rangle$. In the bulk the loss of axial charge and energy appear as the flow of angular momentum and energy across the worldvolume horizon. We found numerically that the angular momentum and energy flux produces a conical singularity in Minkowski embeddings with worldvolume horizon at the point where the S^3 collapses to zero size.

Although we focused on D7-branes, the CME can be realized in many similar flavor brane systems. The basic ingredients are a holographic spacetime with probe flavor D-branes satisfying two conditions: they describe $(3 + 1)$-dimensional flavor fields, and they have at least two transverse directions in which to rotate. An axial chemical potential, implemented as a time-dependent fermion mass, will be realized via rotation in a transverse plane, and the axial anomaly will be realized via a WZ coupling to RR flux in the internal space. A model relevant for applications to QCD is that of Ref. [83], with flavor D6-branes in the near-horizon geometry of D4-branes.

From a phenomenological point of view, models of the D3/D7 or D4/D6 type have advantages and disadvantages when compared to the Sakai-Sugimoto or other "AdS/QCD" models. Consider first the disadvantages. D3/D7-type models are typically less similar to large-N_c QCD, for example, non-Abelian chiral symmetries will generically be explicitly broken by (super)potential terms. Given that the general objective of holography is to uncover universal physics, this disadvantage may not be fatal. Beyond that we suspect that the problems with D7-branes will be generic: $U(1)_A$ charge may leak into the adjoint sector and Minkowski embeddings

---

[18] The loss rates are of order $\langle \mathscr{O}_\phi \rangle \propto \lambda N_c$, and so can be neglected for times shorter than N_c/λ. Taking into account the change in angular momentum and energy, *i.e.* computing the back-reaction of the D7-branes, would probably lead to an expanding horizon [59]. Gravity solutions exhibiting precisely that behavior have been constructed for external electric fields in Ref. [82].

describing a CME will likely be singular. On the other hand, the fact that Sakai-Sugimoto-like models require a source at the horizon to describe the CME suggests that the same (or similar) problems may appear in those models as well, once the meaning and effects of the source are clarified. Perhaps the principal advantage of D3/D7-type models is that the $U(1)_V$ current is conserved and gauge invariant under $U(1)_V$ transformations by construction, so among other things comparison with weak coupling calculations is straightforward. Another advantage is that quark masses are easy to introduce and the effects of chiral symmetry breaking are easy to study.

Although our focus was on the CME, our spinning D7-branes without world-volume horizon may have useful applications as well. When the solution describes massless flavors, the field theory is in a state with a finite charge density of fermions with $U(1)_A$ spontaneously broken, i.e. superfluid states [57]. These solutions have no loss of axial charge or energy and no obvious instabilities, at least to leading order in the $1/N_c$ expansion, and so deserve further study as models for strongly-coupled, many-body fermion physics.[19]

An important task for the future is a complete analysis of linearized fluctuations of worldvolume fields, to determine whether our solutions are stable. Although we avoided an obvious instability by demanding reality of the D7-brane action, more subtle instabilities may exist in the spectrum of worldvolume fluctuations, and indeed certain instabilities have been found in very similar systems [60]. Moreover, in our analysis we assumed homogeneity of the ground state, but in QCD with $U(1)_A$ or $U(1)_V$ chemical potentials and a strong magnetic field, in a state with chiral symmetry broken, the ground state may be the inhomogeneous "chiral magnetic spiral" [84]. Such a phase was indeed detected in the Sakai-Sugimoto model via analysis of linearized fluctuations [85], and a similar analysis should be done for the D3/D7 model.

Acknowledgements We thank O. Bergman, J. Charbonneau, S. Das, M. Kaminski, A. Karch, K.-Y. Kim, P. Kumar, K. Landsteiner, A. Rebhan, S. Ryu, D.T. Son, S. Sugimoto, Y. Tachikawa, T. Takayanagi, L. Yaffe, N. Yamamoto, A. Yarom, A. Zayakin, and A. Zhitnitsky for useful conversations. We also thank the Galileo Galilei Institute for Theoretical Physics for hospitality and the INFN for partial support during the completion of this work. C.H. also wants to thank the Erwin Schrödinger Institute in Vienna and T.N. would like to thank the IPMU in Tokyo for hospitality during the completion of this work. The work of T.N. was supported in part by the US NSF under Grants No. PHY-0844827 and PHY-0756966. The work of A.O'B. was supported by the European Research Council grant "Properties and Applications of the Gauge/Gravity Correspondence". This work was supported in part by DOE grant DE-FG02-96ER40956.

---

[19]These solutions may in fact also describe a CME, beyond leading order in the large-N_c approximation. In the presence of a black hole, Hawking radiation can induce thermal excitations on the probe brane which are dual to a gas of mesons in the field theory. The magnetic field can polarize pseudo-scalar mesons, producing an excess of vector mesons through the anomaly, hence a CME can occur. From the perspective of the gravitational theory, this is a quantum process, and as such is not captured by our classical calculation. From the perspective of the field theory, the meson gas produces only a $O(1/N_c^2)$ contribution to the total free energy of the system.

References

1. A.Y. Alekseev, V.V. Cheianov, J. Frohlich, Universality of transport properties in equilibrium, Goldstone theorem and chiral anomaly. cond-mat/9803346
2. D.E. Kharzeev, L.D. McLerran, H.J. Warringa, The effects of topological charge change in heavy ion collisions: 'Event by event P and CP violation'. Nucl. Phys. A **803**, 227–253 (2008). arXiv:0711.0950
3. K. Fukushima, D.E. Kharzeev, H.J. Warringa, The chiral magnetic effect. Phys. Rev. D **78**, 074033 (2008). arXiv:0808.3382
4. D.E. Kharzeev, H.J. Warringa, Chiral magnetic conductivity. Phys. Rev. D **80**, 034028 (2009). arXiv:0907.5007
5. K. Jensen, Triangle anomalies, thermodynamics, and hydrodynamics. arXiv:1203.3599
6. K. Fukushima, M. Ruggieri, Dielectric correction to the chiral magnetic effect. Phys. Rev. D **82**, 054001 (2010). arXiv:1004.2769
7. D. Kharzeev, Parity violation in hot QCD: Why it can happen, and how to look for it. Phys. Lett. B **633**, 260–264 (2006). hep-ph/0406125
8. D. Kharzeev, A. Zhitnitsky, Charge separation induced by P-odd bubbles in QCD matter. Nucl. Phys. A **797**, 67–79 (2007). arXiv:0706.1026
9. E. Shuryak, Why does the quark gluon plasma at rhic behave as a nearly ideal fluid? Prog. Part. Nucl. Phys. **53**, 273–303 (2004). hep-ph/0312227
10. E.V. Shuryak, What rhic experiments and theory tell us about properties of quark-gluon plasma? Nucl. Phys. A **750**, 64–83 (2005). hep-ph/0405066
11. M. Asakawa, A. Majumder, B. Muller, Electric charge separation in strong transient magnetic fields. Phys. Rev. C **81**, 064912 (2010). arXiv:1003.2436
12. D. Kharzeev, R.D. Pisarski, M.H.G. Tytgat, Possibility of spontaneous parity violation in hot QCD. Phys. Rev. Lett. **81**, 512–515 (1998). hep-ph/9804221
13. I.V. Selyuzhenkov (STAR Collaboration), Global polarization and parity violation study in Au + Au collisions. Rom. Rep. Phys. **58**, 049–054 (2006). nucl-ex/0510069
14. S.A. Voloshin (STAR Collaboration), Probe for the strong parity violation effects at rhic with three particle correlations. Indian J. Phys. **85**, 1103–1107 (2011). arXiv:0806.0029
15. B.I. Abelev et al. (STAR Collaboration), Observation of charge-dependent azimuthal correlations and possible local strong parity violation in heavy ion collisions. Phys. Rev. C **81**, 054908 (2010). arXiv:0909.1717
16. F. Wang, Effects of cluster particle correlations on local parity violation observables. Phys. Rev. C **81**, 064902 (2010). arXiv:0911.1482
17. P. Christakoglou, Charge dependent azimuthal correlations in Pb–Pb collisions at $\sqrt{s_{NN}} = 2.76$ TeV. J. Phys. G **38**, 124165 (2011). arXiv:1106.2826
18. P.V. Buividovich, E.V. Luschevskaya, M.I. Polikarpov, M.N. Chernodub, Chiral magnetic effect in $SU(2)$ lattice gluodynamics at zero temperature. JETP Lett. **90**, 412–416 (2009)
19. P.V. Buividovich, M.N. Chernodub, E.V. Luschevskaya, M.I. Polikarpov, Numerical study of chiral magnetic effect in quenched $SU(2)$ lattice gauge theory. PoS **LAT2009**, 080 (2009). arXiv:0910.4682
20. M. Abramczyk, T. Blum, G. Petropoulos, R. Zhou, Chiral magnetic effect in 2 + 1 flavor QCD + Qed. PoS **LAT2009**, 181 (2009). arXiv:0911.1348
21. A. Yamamoto, Chiral magnetic effect in lattice QCD with chiral chemical potential. Phys. Rev. Lett. **107**, 031601 (2011). arXiv:1105.0385
22. A. Yamamoto, Lattice study of the chiral magnetic effect in a chirally imbalanced matter. Phys. Rev. D **84**, 114504 (2011). arXiv:1111.4681
23. J.M. Maldacena, The large N limit of superconformal field theories and supergravity. Adv. Theor. Math. Phys. **2**, 231–252 (1998). hep-th/9711200
24. S.S. Gubser, I.R. Klebanov, A.M. Polyakov, Gauge theory correlators from non-critical string theory. Phys. Lett. B **428**, 105–114 (1998). hep-th/9802109
25. E. Witten, Anti-de Sitter space and holography. Adv. Theor. Math. Phys. **2**, 253–291 (1998). hep-th/9802150

26. E. Witten, Anti-de Sitter space, thermal phase transition, and confinement in gauge theories. Adv. Theor. Math. Phys. **2**, 505–532 (1998). hep-th/9803131

27. J. Erdmenger, P. Kerner, H. Zeller, Non-universal shear viscosity from Einstein gravity. Phys. Lett. B **699**, 301–304 (2011). arXiv:1011.5912

28. P. Kovtun, D.T. Son, A.O. Starinets, Viscosity in strongly interacting quantum field theories from black hole physics. Phys. Rev. Lett. **94**, 111601 (2005). hep-th/0405231

29. P. Romatschke, U. Romatschke, Viscosity information from relativistic nuclear collisions: How perfect is the fluid observed at rhic? Phys. Rev. Lett. **99**, 172301 (2007). arXiv:0706.1522

30. M. Luzum, P. Romatschke, Conformal relativistic viscous hydrodynamics: Applications to rhic results at $\sqrt{S_{NN}} = 200$ GeV. Phys. Rev. C **78**, 034915 (2008). arXiv:0804.4015

31. A. Rebhan, A. Schmitt, S.A. Stricker, Anomalies and the chiral magnetic effect in the Sakai-Sugimoto model. J. High Energy Phys. **1001**, 026 (2010). arXiv:0909.4782

32. H.-U. Yee, Holographic chiral magnetic conductivity. J. High Energy Phys. **0911**, 085 (2009). arXiv:0908.4189

33. A. Gorsky, P.N. Kopnin, A.V. Zayakin, On the chiral magnetic effect in soft-wall AdS/QCD. Phys. Rev. D **83**, 014023 (2011). arXiv:1003.2293

34. V.A. Rubakov, On chiral magnetic effect and holography. arXiv:1005.1888

35. A. Gynther, K. Landsteiner, F. Pena-Benitez, A. Rebhan, Holographic anomalous conductivities and the chiral magnetic effect. J. High Energy Phys. **1102**, 110 (2011). arXiv:1005.2587

36. L. Brits, J. Charbonneau, A constraint-based approach to the chiral magnetic effect. Phys. Rev. D **83**, 126013 (2011). arXiv:1009.4230

37. T. Kalaydzhyan, I. Kirsch, Fluid/gravity model for the chiral magnetic effect. Phys. Rev. Lett. **106**, 211601 (2011). arXiv:1102.4334

38. K. Landsteiner, E. Megias, L. Melgar, F. Pena-Benitez, Holographic gravitational anomaly and chiral vortical effect. J. High Energy Phys. **1109**, 121 (2011). arXiv:1107.0368

39. I. Gahramanov, T. Kalaydzhyan, I. Kirsch, Anisotropic hydrodynamics, holography and the chiral magnetic effect. Phys. Rev. D **85**, 126013 (2012). arXiv:1203.4259

40. J. Erdmenger, M. Haack, M. Kaminski, A. Yarom, Fluid dynamics of R-charged black holes. J. High Energy Phys. **0901**, 055 (2009). arXiv:0809.2488

41. N. Banerjee et al., Hydrodynamics from charged black branes. J. High Energy Phys. **1101**, 094 (2011). arXiv:0809.2596

42. C. Eling, Y. Neiman, Y. Oz, Holographic Non-Abelian charged hydrodynamics from the dynamics of null horizons. J. High Energy Phys. **1012**, 086 (2010). arXiv:1010.1290

43. D.T. Son, P. Surowka, Hydrodynamics with triangle anomalies. Phys. Rev. Lett. **103**, 191601 (2009). arXiv:0906.5044

44. A.V. Sadofyev, M.V. Isachenkov, The chiral magnetic effect in hydrodynamical approaxch. Phys. Lett. B **697**, 404–406 (2011). arXiv:1010.1550

45. Y. Neiman, Y. Oz, Relativistic hydrodynamics with general anomalous charges. J. High Energy Phys. **1103**, 023 (2011). arXiv:1011.5107

46. D.E. Kharzeev, H.-U. Yee, Chiral magnetic wave. Phys. Rev. D **83**, 085007 (2011). arXiv:1012.6026

47. S. Dubovsky, L. Hui, A. Nicolis, Effective field theory for hydrodynamics: Wess-Zumino term and anomalies in two spacetime dimensions. arXiv:1107.0732

48. S. Lin, An anomalous hydrodynamics for chiral superfluid. Phys. Rev. D **85**, 045015 (2012). arXiv:1112.3215

49. B. Keren-Zur, Y. Oz, Hydrodynamics and the detection of the QCD axial anomaly in heavy ion collisions. J. High Energy Phys. **1006**, 006 (2010). arXiv:1002.0804

50. D.E. Kharzeev, D.T. Son, Testing the chiral magnetic and chiral vortical effects in heavy ion collisions. Phys. Rev. Lett. **106**, 062301 (2011). arXiv:1010.0038

51. Y. Burnier, D.E. Kharzeev, J. Liao, H.-U. Yee, Chiral magnetic wave at finite baryon density and the electric quadrupole moment of quark-gluon plasma in heavy ion collisions. Phys. Rev. Lett. **107**, 052303 (2011). arXiv:1103.1307

52. T. Sakai, S. Sugimoto, Low energy hadron physics in holographic QCD. Prog. Theor. Phys. **113**, 843–882 (2005). hep-th/0412141
53. T. Sakai, S. Sugimoto, More on a holographic dual of QCD. Prog. Theor. Phys. **114**, 1083–1118 (2005). hep-th/0507073
54. O. Bergman, G. Lifschytz, M. Lippert, Magnetic properties of dense holographic QCD. Phys. Rev. D **79**, 105024 (2009). arXiv:0806.0366
55. S. Kobayashi, D. Mateos, S. Matsuura, R.C. Myers, R.M. Thomson, Holographic phase transitions at finite baryon density. J. High Energy Phys. **0702**, 016 (2007). hep-th/0611099
56. N. Evans, E. Threlfall, R-charge chemical potential in the M2–M5 system. arXiv:0807.3679
57. A. O'Bannon, Toward a holographic model of superconducting fermions. J. High Energy Phys. **0901**, 074 (2009). arXiv:0811.0198
58. N. Evans, E. Threlfall, Chemical potential in the gravity dual of a $2 + 1$ dimensional system. Phys. Rev. D **79**, 066008 (2009). arXiv:0812.3273
59. S.R. Das, T. Nishioka, T. Takayanagi, Probe branes, time-dependent couplings and thermalization in AdS/CFT. J. High Energy Phys. **1007**, 071 (2010). arXiv:1005.3348
60. S.P. Kumar, Spinning flavour branes and fermion pairing instabilities. Phys. Rev. D **84**, 026003 (2011). arXiv:1104.1405
61. A. Karch, E. Katz, Adding flavor to AdS/CFT. J. High Energy Phys. **0206**, 043 (2002). hep-th/0205236
62. M. Kruczenski, D. Mateos, R.C. Myers, D.J. Winters, Meson spectroscopy in AdS/CFT with flavour. J. High Energy Phys. **0307**, 049 (2003). hep-th/0304032
63. A. Karch, A. O'Bannon, Metallic AdS/CFT. J. High Energy Phys. **0709**, 024 (2007). arXiv:0705.3870
64. S. de Haro, S.N. Solodukhin, K. Skenderis, Holographic reconstruction of space-time and renormalization in the AdS/CFT correspondence. Commun. Math. Phys. **217**, 595–622 (2001). hep-th/0002230
65. K. Skenderis, Lecture notes on holographic renormalization. Class. Quantum Gravity **19**, 5849–5876 (2002). hep-th/0209067
66. V.G. Filev, C.V. Johnson, R.C. Rashkov, K.S. Viswanathan, Flavoured large N gauge theory in an external magnetic field. J. High Energy Phys. **0710**, 019 (2007). hep-th/0701001
67. V.G. Filev, Criticality, scaling and chiral symmetry breaking in external magnetic field. J. High Energy Phys. **0804**, 088 (2008). arXiv:0706.3811
68. T. Albash, V.G. Filev, C.V. Johnson, A. Kundu, Finite temperature large N gauge theory with quarks in an external magnetic field. J. High Energy Phys. **0807**, 080 (2008). arXiv:0709.1547
69. J. Erdmenger, R. Meyer, J.P. Shock, AdS/CFT with flavour in electric and magnetic Kalb-Ramond fields. J. High Energy Phys. **0712**, 091 (2007). arXiv:0709.1551
70. A. O'Bannon, Hall conductivity of flavor fields from AdS/CFT. Phys. Rev. D **76**, 086007 (2007). arXiv:0708.1994
71. J. Babington, J. Erdmenger, N.J. Evans, Z. Guralnik, I. Kirsch, Chiral symmetry breaking and pions in non-supersymmetric gauge/gravity duals. Phys. Rev. D **69**, 066007 (2004). hep-th/0306018
72. I. Kirsch, Generalizations of the AdS/CFT correspondence. Fortschr. Phys. **52**, 727–826 (2004). hep-th/0406274
73. K. Ghoroku, T. Sakaguchi, N. Uekusa, M. Yahiro, Flavor quark at high temperature from a holographic model. Phys. Rev. D **71**, 106002 (2005). hep-th/0502088
74. D. Mateos, R.C. Myers, R.M. Thomson, Holographic phase transitions with fundamental matter. Phys. Rev. Lett. **97**, 091601 (2006). hep-th/0605046
75. T. Albash, V.G. Filev, C.V. Johnson, A. Kundu, A topology-changing phase transition and the dynamics of flavour. Phys. Rev. D **77**, 066004 (2008). hep-th/0605088
76. D. Mateos, R.C. Myers, R.M. Thomson, Thermodynamics of the brane. J. High Energy Phys. **0705**, 067 (2007). hep-th/0701132
77. C. Hoyos-Badajoz, K. Landsteiner, S. Montero, Holographic meson melting. J. High Energy Phys. **0704**, 031 (2007). hep-th/0612169

78. T. Albash, V.G. Filev, C.V. Johnson, A. Kundu, Quarks in an external electric field in finite temperature large N gauge theory. J. High Energy Phys. **0708**, 092 (2008). arXiv:0709.1554

79. J. Mas, J.P. Shock, J. Tarrio, Holographic spectral functions in metallic AdS/CFT. J. High Energy Phys. **0909**, 032 (2009). arXiv:0904.3905

80. A. Karch, A. O'Bannon, E. Thompson, The stress-energy tensor of flavor fields from AdS/CFT. J. High Energy Phys. **0904**, 021 (2009). arXiv:0812.3629

81. R.C. Myers, O. Tafjord, Superstars and giant gravitons. J. High Energy Phys. **0111**, 009 (2001). hep-th/0109127

82. B. Sahoo, H.-U. Yee, Electrified plasma in AdS/CFT correspondence. J. High Energy Phys. **1011**, 095 (2010). arXiv:1004.3541

83. M. Kruczenski, D. Mateos, R.C. Myers, D.J. Winters, Towards a holographic dual of large-N_c QCD. J. High Energy Phys. **0405**, 041 (2004). hep-th/0311270

84. G. Basar, G.V. Dunne, D.E. Kharzeev, Chiral magnetic spiral. Phys. Rev. Lett. **104**, 232301 (2010). arXiv:1003.3464

85. K.-Y. Kim, B. Sahoo, H.-U. Yee, Holographic chiral magnetic spiral. J. High Energy Phys. **1010**, 005 (2010). arXiv:1007.1985

Chapter 14
Lattice Studies of Magnetic Phenomena in Heavy-Ion Collisions

P.V. Buividovich, M.I. Polikarpov, and O.V. Teryaev

14.1 Introduction

It has been realized recently that in heavy-ion collision experiments hadronic matter is affected not only by extremely high temperatures and densities, but also by superstrong magnetic fields with field strength being comparable to hadron masses squared. Such superstrong fields are created due to the relative motion of heavy ions themselves, since they carry large charge $Z \sim 100$ [1].

Obviously, the magnetic field is perpendicular to the collision plane, which can be reconstructed in experiment from the angular distribution of produced hadrons [2]. There is no direct experimental way to measure the absolute value of the field strength, but it can be estimated in some microscopic transport model, such as the Ultrarelativistic Quantum Molecular Dynamics model (UrQMD) [3].

Probably the most notable effect which arises due to magnetic fields in heavy-ion collisions is the so-called Chiral Magnetic Effect. The essence of the effect is the generation of electric current along the direction of the external magnetic field in the background of topologically nontrivial gauge field configurations [1, 4]. Such generation is not prohibited by $\mathscr{P}$-invariance, since topological charge density is a pseudoscalar field and thus nonzero topological charge explicitly breaks parity. However, since QCD is parity-invariant, the net current or the net electric charge should vanish when averaged over multiple collision events. Nevertheless, the non-

P.V. Buividovich (✉) · M.I. Polikarpov
ITEP, B. Cheremushkinskaya str. 25, 117218 Moscow, Russia
e-mail: buividovich@itep.ru

M.I. Polikarpov
e-mail: polykarp@itep.ru

P.V. Buividovich · O.V. Teryaev
JINR, Joliot-Curie str. 6, 141980 Dubna, Russia

O.V. Teryaev
e-mail: teryaev@theor.jinr.ru

D. Kharzeev et al. (eds.), *Strongly Interacting Matter in Magnetic Fields*,
Lecture Notes in Physics 871, DOI 10.1007/978-3-642-37305-3_14,
© Springer-Verlag Berlin Heidelberg 2013

trivial effect can still be detected if one considers dispersions of electric current or electric charge [5, 6]. Experimentally, the Chiral Magnetic Effect manifests itself as the dynamical enhancement of fluctuations of the numbers of charged hadrons emitted above and below the reaction plane [2, 7–9].

Another effect, which is closely related to the CME, but has different experimental signatures, is the anisotropic electric conductivity of hadronic matter in the strong magnetic field, which was discovered in lattice simulations [10]. Since the conductivity of the hadronic matter is directly related to the lepton emission rate [11, 12], such anisotropic conductivity should result in specific anisotropy of the dilepton emission rate w.r.t. the reaction plane. This anisotropy should grow with the centrality of the collision and with the charges of the colliding ions. This effect might also contribute to the observed abnormal dilepton yield in heavy-ion collisions [13]. Some theoretical considerations [14] as well as preliminary lattice data [15] suggest that at very strong magnetic fields with strength $eB > m_\rho^2$, where m_ρ is the mass of the charged ρ-meson, the anisotropic conductivity might even turn into the anisotropic superconductivity in the direction of the field. Unfortunately, such extremely large field strengths are hardly reachable with present-day heavy-ion colliders.

In this paper we give some estimates of the expected experimental signatures of superstrong magnetic fields, basing on the lattice data. In Sect. 14.2 we consider the Chiral Magnetic Effect and argue that its strength should decrease with increasing quark mass, which can be used to discriminate between the CME and other possible effects which might result in preferential emission of charged hadrons in the direction perpendicular to the reaction plane. In Sect. 14.3 we consider the dilepton emission rate, and estimate the contribution of the induced conductivity to the total dilepton yield and dilepton angular distribution in heavy-ion collisions.

14.2 Chiral Magnetic Effect

Chiral Magnetic Effect is usually characterized by the following experimental observables, suggested first in Ref. [8]:

$$a_{ab} = \frac{1}{N_e} \sum_{e=1}^{N_e} \frac{1}{N_a N_b} \sum_{i=1}^{N_a} \sum_{j=1}^{N_b} \cos(\phi_{ia} + \phi_{jb}), \qquad (14.1)$$

where $a, b = \pm$ denotes hadrons with positive or negative charges, respectively, N_e is the number of events used for data analysis, N_a and N_b are the total multiplicities of positively/negatively charged particles produced in each event, and ϕ_{ia}, ϕ_{jb} are the angles w.r.t. the reaction plane at which the hadrons with labels i and j are emitted. Summation in (14.1) goes over all produced hadrons. In practice, only sufficiently energetic particles are considered.

The main signatures of the CME is the growth of a_{ab} with impact parameter, as well as the negativity of a_{++} and a_{--} and the positivity of a_{+-} [2, 7–9]. However, it has been pointed out recently that these results can also be explained by other effects, such as the influence of nuclear medium on jet formation [16] and other in-medium effects [17]. It is therefore important to think about more refined experimental tests of the CME. Our main message in this paper is that the dependence of the charge fluctuations on quark mass can be used to discriminate between the CME and other possible phenomena which contribute to the observed asymmetry of charge fluctuations. Indeed, the CME emerges due to fluctuations of quark chirality [1], which are suppressed when the quark mass is increased.

The dependence of the observables (14.1) on the quark mass can be studied if one sums separately over charged mesons with different quark content, e.g. $u\bar{d}$ and $d\bar{u}$ (charged pions), $u\bar{s}$ and $\bar{u}s$ (charged kaons) or $\bar{u}c$, $\bar{c}u$, $\bar{d}c$, $d\bar{c}$ (D-mesons). The observables a_{ab} should then decrease with the meson mass. The dependence of a_{ab} on the centrality of the collision should also become weaker. Here we give a rough estimate of this effect basing on the results of lattice simulations.

Let us first note that the observables $a_{a,b}$ can be expressed in terms of the differences of multiplicities of charged hadrons emitted above and below the reaction plane [1]:

$$a_{ab} = \frac{c \langle \Delta_a \Delta_b \rangle}{\langle N_a \rangle \langle N_b \rangle}, \qquad (14.2)$$

where $a, b = \pm$, $\Delta_\pm$ are the differences of the multiplicities of hadrons with positive or negative charges above or below the reaction plane, respectively. The factor c depends on the hydrodynamical evolution of hadronic matter, and is usually close to unity. For multiplicities $\langle N_a \rangle \sim 1000$ one can also neglect the initial charge of heavy ions $Z \sim 100$ with a good precision, and assume that $\langle N_a \rangle = \langle N_b \rangle = N_q$, where N_q is the mean multiplicity of the same-charge hadrons per event.

Lattice results can be compared to experimental data by considering the quantity $a_{++} + a_{--} - 2a_{+-}$, which can be expressed solely in terms of the difference $\Delta Q = \Delta_+ - \Delta_-$ of net charges of hadrons emitted above and below the reaction plane:

$$a_{++} + a_{--} - 2a_{+-} = \frac{\langle (\Delta Q)^2 \rangle}{N_q^2} = \frac{\langle (\Delta_+ - \Delta_-)^2 \rangle}{N_q^2}. \qquad (14.3)$$

In turn, the dispersion of the charge difference $\langle (\Delta Q)^2 \rangle$ can be related to the vacuum expectation values of the squared current densities $\langle j_\mu^2(x) \rangle$ [5]. The contribution of each quark flavor $f = u, d, s, c$ to the total electromagnetic current is $j_\mu^f(x) = \bar{q}^f \gamma_\mu q^f$. We do not consider here the third-generation quarks, which are extremely rarely produced in heavy-ion collisions.

The simplest model which allows to express $\langle (\Delta Q)^2 \rangle$ in terms of $\langle j_\mu^{f\,2}(x) \rangle$ is the model of spherical fireball, which emits positively and negatively charged hadrons from its surface with intensity proportional to $\langle j_\mu^{f\,2}(x) \rangle$. This leads to the following relation [5]:

$$a_{++} + a_{--} - 2a_{+-} = \frac{4\pi \tau^2 \rho^2 r^2}{3N_q^2} \left(\langle (j_\parallel^f)^2 \rangle + 2\langle (j_\perp^f)^2 \rangle \right), \qquad (14.4)$$

Fig. 14.1 Schematic view of
the collision geometry. The
fireball is the hatched region
of volume
$V \sim \frac{4\pi}{3} (R - b/2)^3$ within
the intersection of two heavy
ions of radius R each

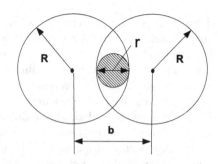

where $j_\parallel^f (x)$ and $j_\perp^f (x)$ are, respectively, the currents along the magnetic field and
perpendicular to it, τ is some characteristic collision time, r is the fireball radius
and ρ is some typical correlation length for electric charge density in the fireball. In
our estimates, we take $\tau \sim 0.3$ fm (this is a typical decay time for the magnetic field
in heavy-ion collisions [3]), $\rho \sim 0.2$ fm, which are reasonable parameters for, say,
gold-gold collisions at 60 GeV/nucleon. We also assume that the fireball is a sphere
with radius $r = R - b/2$ within the overlapping region between the two heavy ions
of radius R which collide at impact parameter b (see Fig. 14.1). The net multiplicity N_q and the impact parameter b as the functions of collision centrality can be
found in Table 1 in [1]. For simplicity, we assume that $\langle ((j_\mu)^f{}^2 \rangle$ are approximately
constant on the surface of the fireball. Note also, that in order to exclude the effects related to the dependence of multiplicities of strange and charmed mesons on
the collision centrality, which might be different from that of light mesons, we normalize the charge of emitted hadrons by the square of the total multiplicity of all
hadrons.

Several technical remarks are in order. First, we assume that the matter within the
fireball is in the state of thermal equilibrium, and thus the expectation values $\langle j_\mu^{f\,2} \rangle$
can be calculated from gauge theory in Euclidean space. We also assume that the
magnetic field is uniform and nearly time-independent. Of course, these are rather
rough approximations, but we are aiming here at qualitative rather than quantitative
estimates. We have calculated the currents in $SU(2)$ lattice gauge theory with background magnetic field both in the confinement phase, neglecting the contribution
from the virtual quark loops (quenched approximation). A comparison with $SU(3)$
gauge theory suggests that this is a reasonable approximation [18]. A more detailed
study has shown also that in the deconfinement phase the dispersions of local current densities are practically independent of the magnetic field [5], thus we do not
consider here this case. The expectation values $\langle j_\mu^{f\,2} \rangle$ contain also the ultraviolet divergent part, which we have removed by subtracting the corresponding expectation
values at zero temperature and zero magnetic field.

The masses of the valence quarks took the values $m_q = 50$ MeV (for u and d
quarks), $m_q = 110$ MeV (for s-quark) and $m_q = 1$ GeV (for c-quark). The lowest
value of the quark mass is dictated by the numerical stability of our algorithm. Thus
our calculations of $\langle j_\mu^{f\,2} \rangle$ at $m_q = 50$ MeV should be considered as only the lower
bound for $\langle j_\mu^{f\,2} \rangle$ at realistic masses of u and d quarks.

Fig. 14.2 Comparison of the quantity $a_{++} + a_{--} - 2a_{+-}$ for the experimental data by the STAR Collaboration [2] with the estimates (14.4) based on the results of lattice simulations at different quark masses

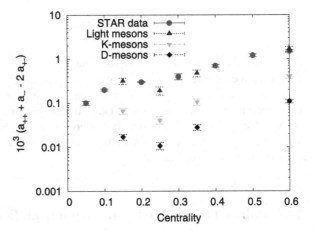

On Fig. 14.2 we compare the quantity $a_{++} + a_{--} - 2a_{+-}$ calculated for the experimental data by the STAR collaboration [9] with the estimate (14.4) based on the results of lattice simulations at different quark masses and different temperatures. In order to match the value of the magnetic field strength in experiment and in simulations, we take the rough estimate from Eq. (A.12) in [1]:

$$eB \sim (0.1b/R) \text{ GeV}^2, \qquad (14.5)$$

where b is the impact parameter. This rough fit also agrees by the order of magnitude with the results of more sophisticated calculations of the magnetic field within the UrQMD model [3].

One can see that the best agreement with the STAR data is obtained at the smallest quark mass. The combination $a_{++} + a_{--} - 2a_{+-}$ quickly decreases as the quark mass is increased—approximately by a factor of 5 as the quark mass changes from 50 MeV to 110 MeV, and by a factor of almost 20 as the quark mass further increases to 1 GeV. Since all observables a_{ab} are typically of the same order, one can expect that each such observable will also decrease with the quark mass. This dependence of asymmetry of angular distributions of mesons of different flavors on their mass can be used to discriminate between the CME and other phenomena which might cause such asymmetry.

The result may be compared with the perturbative analog of CME [19] resulting from straightforward generalization of Heisenberg-Euler Lagrangian depending on quark mass as m_q^{-4}

$$j_\mu^q = \frac{7\alpha\alpha_s}{45m_q^4} \tilde{F}_{\mu\nu}\partial^\nu(G\tilde{G}). \qquad (14.6)$$

The correspondence of this perturbative and Abelian effect to CME is manifested by the substitution

$$\frac{1}{m_q^4} \partial^\nu (G\tilde{G}) \rightarrow \partial^\nu \int d^4 z (G\tilde{G}) \rightarrow \partial^\nu \theta. \qquad (14.7)$$

As a result, the perturbative mechanism may become essential when quark mass is exceeding the inverse correlation length of topological charge density. One may expect that the transition point from non-perturbative to perturbative mass dependence is not too far from the strange quark mass, like it happen for vacuum quark condensates and strangeness polarization in nucleons (see [19] and references therein) so that the non perturbative lattice results are applicable for experimentally important case of strangeness separation.

14.3 Induced Conductivity and Abnormal Dilepton Yield

Another phenomenon which might be caused by superstrong magnetic fields acting on the hadronic matter is the induced anisotropic conductivity along the magnetic field [10]. While the Chiral Magnetic Effect is related to the local fluctuations of current density, the induced conductivity reflects the fact that these fluctuations also have long-range correlation in time. Indeed, by virtue of the Green-Kubo relations the conductivity is related to the zero-frequency limit of the spectral function $\rho_{\mu\nu}(w)$ which corresponds to the correlator $\langle \mathcal{T} j_\mu(x) j_\nu(y) \rangle$ in Minkowski space [20]:

$$\int d^3 \mathbf{x} \langle \mathcal{T} j_\mu(0,0) j_\nu(\mathbf{x},\tau) \rangle = \int\limits_0^{+\infty} \frac{dw}{2\pi} \frac{\cosh w \left(\tau - \frac{1}{2T}\right)}{\sinh(\frac{w}{2T})} \rho_{\mu\nu}(w). \qquad (14.8)$$

This spectral function can also be extracted from the results of lattice simulations using the so-called Maximal Entropy Method [21, 22].

The spectral function $\rho_{\mu\nu}(w)$ determines also the dilepton emission rate from either cold or hot hadronic matter [11, 23]:

$$\frac{R}{V} = -4e^4 \int \frac{d^3 p_1}{(2\pi)^3 2E_1} \frac{d^3 p_2}{(2\pi)^3 2E_2} L^{\mu\nu}(p_1, p_2) \frac{\rho_{\mu\nu}(q)}{q^4}, \qquad (14.9)$$

where p_1 and p_2 are the momenta of the leptons, $q = p_1 + p_2$,

$$L^{\mu\nu} = \left((p_1 \cdot p_2 + m^2)\eta^{\mu\nu} - p_1^\mu p_2^\nu - p_2^\mu p_1^\nu\right)$$

is the dilepton tensor ($\eta^{\mu\nu}$ is the Minkowski metric), m is the lepton mass. Thus the low-momentum limit of $\rho_{\mu\nu}(w)$ is related, on the one hand, to the emission rate of soft dileptons, and, on the other hand, to the conductivity of hadronic matter.

The enhancement of the conductivity due to the magnetic field should thus lead to the enhancement of the dilepton emission rate. This might provide a viable explanation of the abnormal soft dilepton yield observed in heavy-ion collisions [13]. Moreover, the anisotropy of the conductivity should lead to specific correlations between the momenta of the dileptons and the direction of the magnetic field (in other words, with the orientation of the reaction plane).

Lattice simulations show that the electric conductivity is nonzero only in the direction of the magnetic field and depends linearly on qB [10, 24]. For sufficiently small momenta p_1 and p_2 (and thus for small q) by virtue of the Green-Kubo theorem one has $\rho_{ij}(q) \approx \sim \sigma_{ij} q/T \sim B_i B_j/(|B|T)$ [20, 21]. Let us also neglect the lepton masses and go to the rest frame of the dilepton pair, where $\mathbf{p_1} = -\mathbf{p_2} \equiv p\mathbf{n}$, $q = (2p, \mathbf{0})$ and the spatial components of the dilepton tensor are $L^{ij} = p^2 (\delta_{ij} - n_i n_j)$. The dilepton emission rate is therefore proportional to

$$\frac{R}{V} \sim \int \frac{d^3 p}{(2\pi)^3 32 E B p^2} \left(\mathbf{B}^2 - (\mathbf{B} \cdot \mathbf{n})^2\right) \sim |B| \sin^2(\theta), \qquad (14.10)$$

where θ is the angle between the spatial momentum of the outgoing leptons and the magnetic field. Therefore, there should be more soft dileptons emitted perpendicular to the magnetic field than parallel to it. As a result, they are to large extend hidden inside the hadrons in the scattering plane which should lead to the difficulty in their experimental observation.

In order to estimate the effect of the magnetic field on the total dilepton yield in heavy-ion collisions, we normalize the conductivity induced by the magnetic field to the conductivity at zero magnetic field and at the temperature close to the deconfinement phase transition, $T = 1.12 T_c$. By virtue of (14.9), the ratio of these conductivities should be equal to the ratio of dilepton emission rates in the low-momentum region. The relevant lattice data is summarized in [10, 24]. In these works it was found that in the deconfinement phase the conductivity is practically independent of the magnetic field. Therefore here we will try to estimate possible contribution of the magnetic field to the dilepton emission rate from hadronic matter in the confinement phase. From the data presented in [10, 24], we estimate the ratio of the induced conductivity $\sigma(B, T < T_c)$ to the conductivity of quark-gluon plasma $\sigma(B = 0, T = 1.12T_c)$ as:

$$\sigma(B, T < T_c)/\sigma(B = 0, T = 1.12T_c) \approx \frac{eB}{(0.5 \text{ GeV})^2}.$$

We now use the estimate (14.5) for the magnetic field strength and take into account that the ratio of conductivities should be equal to the ratio of dilepton emission rates. We thus obtain for the contribution of the magnetic field to the dilepton emission rate:

$$R(B, T = 0)/R(B = 0, T = 1.12T_c) \approx 1/3\,0.4\,b/R, \qquad (14.11)$$

where the factor $1/3$ appears after averaging the expression (14.10) over the angle θ.

We conclude that at large impact parameters ($b \sim 2R$) the dilepton yield can increase by up to 20–30 % due to the influence of the magnetic field. This factor should be essentially reduced because the magnetic-induced dileptons are hidden inside the scattering plane. The observed abnormal dilepton yield is, however, maximal for central collisions (where it reaches several hundred percent as compared to the hadron resonance model) and decreases as the impact parameter grows [13]. Such behavior might be caused by several factors, such as the change of the temperature within the fireball or the change of the fireball volume with impact parameter. A proper investigation of such factors is out of the scope of this paper, and cannot be undertaken using the methods of lattice gauge theory.

14.4 Conclusions

In this paper we have summarized the main experimental signatures of the effects caused by superstrong magnetic fields in heavy ion collisions, namely, the Chiral Magnetic Effect and the abnormal dilepton yield.

The Chiral Magnetic Effect [1] results in preferential emission of charged hadrons in the direction perpendicular to the reaction plane. The origin of the Chiral Magnetic Effect is the fluctuations of chirality, which are suppressed as the quark mass grows. Thus this asymmetry in angular distributions of charged hadrons, characterized by the coefficients a_{ab} (14.1), should be strongly suppressed for strange or charmed hadrons.

The abnormal dilepton yield with specific angular dependence (14.10) is the consequence of electric conductivity of the hadronic matter induced by the magnetic field. In this case more dileptons are emitted in the direction perpendicular to the magnetic field. Let us also note that since according to our lattice data the magnetic field influences the conductivity only in the confinement phase [10], the significant change of dilepton yield in noncentral heavy-ion collisions might be a signature of the confinement-deconfinement phase transition.

More generally, lattice data suggests that the influence of the magnetic field on the properties of hadronic matter is stronger in the confinement phase. Thus heavy-ion collision experiments on colliders with lower beam energy but with larger luminosity (such as FAIR in Darmstadt, Germany or NICA in Dubna, Russia) might be more advantageous for studying magnetic phenomena.

Acknowledgements The authors are grateful to M.N. Chernodub, A.S. Gorsky, V.I. Shevchenko, M. Stephanov and A.V. Zayakin for interesting and useful discussions. We are also deeply indebted to D.E. Kharzeev for very valuable and enlightening remarks on the present work. The work was supported by the Russian Ministry of Science and Education under contract No. 07.514.12.4028. Numerical calculations were performed at the ITEP computer systems "Graphyn" and "Stakan" (authors are much obliged to A.V. Barylov, A.A. Golubev, V.A. Kolosov, I.E. Korolko, M.M. Sokolov for the help), the MVS 100K at Moscow Joint Supercomputer Center and at Supercomputing Center of the Moscow State University.

References

1. D.E. Kharzeev, L.D. McLerran, H.J. Warringa, Nucl. Phys. A **803**, 227 (2008). arXiv:0711.0950
2. B. Abelev et al. (STAR Collaboration), Phys. Rev. Lett. **103**, 251601 (2009). arXiv:0909.1739

3. V. Skokov, A. Illarionov, V. Toneev, Int. J. Mod. Phys. A **24**, 5925 (2009). arXiv:0907.1396
4. K. Fukushima, D.E. Kharzeev, H.J. Warringa, Phys. Rev. D **78**, 074033 (2008). arXiv:0808.3382
5. P.V. Buividovich, M.N. Chernodub, E.V. Luschevskaya, M.I. Polikarpov, Phys. Rev. D **80**, 054503 (2009). arXiv:0907.0494
6. V. Orlovsky, V. Shevchenko, Phys. Rev. D **82**, 094032 (2010). arXiv:1008.4977
7. I.V. Selyuzhenkov (STAR Collaboration), Rom. Rep. Phys. **58**, 049 (2006). arXiv:nucl-ex/0510069
8. S.A. Voloshin, Phys. Rev. C **70**, 057901 (2004). arXiv:hep-ph/0406311
9. S.A. Voloshin (STAR Collaboration), Probe for the strong parity violation effects at RHIC with three-particle correlations, in *Proceedings of Quark Matter, 2008* (2008). arXiv:0806.0029
10. P.V. Buividovich, M.N. Chernodub, D.E. Kharzeev, T. Kalaydzhyan, E.V. Luschevskaya, M.I. Polikarpov, Phys. Rev. Lett. **105**, 132001 (2010). arXiv:1003.2180
11. E.L. Bratkovskaya, O.V. Teryaev, V.D. Toneev, Phys. Lett. B **348**, 283 (1995)
12. S. Gupta, Phys. Lett. B **597**, 57 (2004). arXiv:hep-lat/0301006
13. A. Adare et al. (PHENIX Collaboration), Phys. Rev. C **81**, 034911 (2010). arXiv:0912.0244
14. M.N. Chernodub, Phys. Rev. D **82**, 085011 (2010). arXiv:1008.1055
15. V.V. Braguta, P.V. Buividovich, M.N. Chernodub, M.I. Polikarpov, Phys. Lett. B **718**, 667 (2012). arXiv:1104.3767
16. H. Petersen, T. Renk, S.A. Bass, Phys. Rev. C **83**, 014916 (2011). arXiv:1008.3846
17. G. Ma, B. Zhang, Phys. Lett. B **700**, 39 (2011). arXiv:1101.1701
18. V.V. Braguta, P.V. Buividovich, T.K. Kalaydzhyan, S.M. Kuznetsov, M.I. Polikarpov, PoS **LAT2010**, 190 (2010). arXiv:1011.3795
19. O.V. Teryaev, Nucl. Phys. B, Proc. Suppl. **86**, 219 (2011)
20. L.P. Kadanoff, P.C. Martin, Ann. Phys. **24**, 419 (1963)
21. G. Aarts, C. Allton, J. Foley, S. Hands, S. Kim, Phys. Rev. Lett. **99**, 022002 (2007). arXiv:hep-lat/0703008
22. M. Asakawa, T. Hatsuda, Y. Nakahara, Prog. Part. Nucl. Phys. **46**, 459 (2001). arXiv:hep-lat/0011040
23. L.D. McLerran, T. Toimela, Phys. Rev. D **31**, 545 (1985)
24. P.V. Buividovich, M.I. Polikarpov, Phys. Rev. D **83**, 094508 (2011). arXiv:1011.3001

Chapter 15
Chiral Magnetic Effect on the Lattice

Arata Yamamoto

15.1 Introduction

In the strong interaction, the gauge field forms nontrivial topology. The existence of the topology has been theoretically established, while its observation is difficult in experiments. *The chiral magnetic effect* is a possible candidate to detect the topological structure in heavy-ion collisions [1]. The chiral magnetic effect is the generation of an electric current in a strong magnetic field.

The essence of the chiral magnetic effect is the imbalance of the chirality, i.e., the number difference between the right-handed and left-handed quarks. The magnetic field induces the electric currents of the right-handed and left-handed quarks in opposite directions. If the chirality is imbalanced, a nonzero net electric current is induced. In a local domain of the QCD vacuum, the chiral imbalance is generated by the topological fluctuation and the axial anomaly. In the global QCD vacuum, the chirality is balanced as a whole. The strong theta parameter is experimentally zero, $\theta = 0$, although its reason is unknown. This is *the strong CP problem*. The chiral magnetic effect is regarded as the local violation of the P and CP symmetries.

Experimental facilities tried to measure the chiral magnetic effect through charged-particle correlations [2, 3]. However, the interpretation of the experimental data is not yet conclusive. On the theoretical side, the chiral magnetic effect has been studied in various frameworks, e.g., phenomenological models, the gauge-gravity duality, etc. The chiral magnetic effect has been also studied in the lattice simulations. The lattice simulation is a powerful framework to solve QCD nonperturbatively on computers. By means of the lattice simulation, we can study the chiral magnetic effect from first principles in QCD.

A. Yamamoto (✉)
Quantum Hadron Physics Laboratory, Theoretical Research Division, RIKEN Nishina Center, 2-1 Hirosawa, Wako, Saitama 351-0198, Japan
e-mail: arayamamoto@riken.jp

D. Kharzeev et al. (eds.), *Strongly Interacting Matter in Magnetic Fields*,
Lecture Notes in Physics 871, DOI 10.1007/978-3-642-37305-3_15,
© Springer-Verlag Berlin Heidelberg 2013

Fig. 15.1 A cartoon of how
to observe the chiral magnetic
effect on the lattice.
Left: A topological charge Q
of the gauge field induces a
nonuniform current density
distribution. *Right*: A chiral
chemical potential μ_5 induces
a uniform electric current

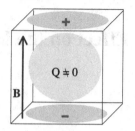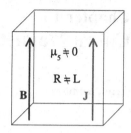

There are two approaches to analyze the chiral magnetic effect in the lattice simulation. In other words, there are two different ways to generate the chiral imbalance:

1. topological charge [4–9]
2. chiral chemical potential [10–12]

These concepts are schematically depicted in Fig. 15.1. In the first case, a topological charge of the background gauge field generates the chiral imbalance, which is spatially nonuniform. When an external magnetic field is applied, a current density distribution appears around the topological object. In the second case, a chiral chemical potential generates the chirally imbalanced matter, which is spatially uniform. A uniform electric current is induced by the external magnetic field.

In this chapter, we overview the theoretical background and the current status of the lattice studies of the chiral magnetic effect. Here we focus only on the lattice aspect of the chiral magnetic effect. For the theoretical and phenomenological aspects, see the corresponding chapters. We use the Euclidean metric and the lattice unit in the following sections.

15.2 Basics of the Lattice Simulation

The basic formalism of the lattice simulation has been well established. For the details, see the textbooks [13–16]. The formalism is based on the Euclidean QCD partition function

$$Z = \int DU\, D\bar{\psi}\, D\psi\, e^{-S_G[U] - S_F[\bar{\psi},\psi,U]}$$

$$= \int DU \det D[U] e^{-S_G[U]}. \tag{15.1}$$

The space-time is discretized as a hypercubic lattice. The gluon field is written as the SU(3) link variable

$$U_\mu(x) = \exp\big(igt^a A_\mu^a(x)\big). \tag{15.2}$$

The functional integral is numerically evaluated by the Monte Carlo simulation. We generate gauge configurations, which are sets of the link variable, and then calculate the expectation value of an operator as

$$\langle O[U] \rangle = \frac{1}{N_{\text{conf}}} \sum_{\{U\}} O[U]. \tag{15.3}$$

The gauge configurations are generated to satisfy the probability weight $P = \det D[U] e^{-S_G[U]}$. The simulation including the fermion determinant is called *the dynamical QCD simulation* or *the full QCD simulation*. *The quenched approximation* is often used to reduce the computational cost. In the quenched approximation, the fermion determinant is ignored and the probability weight is $P = e^{-S_G[U]}$. The quenched gauge configurations are independent of the fermion action.

The probability weight must be positive real, otherwise it cannot be interpreted as the probability weight. In QCD, the fermion action becomes complex at a finite quark chemical potential. The definition of the probability weight must be modified, e.g., by the reweighting method [17]. Even after the modification, the Monte Carlo simulation severely suffers from strong sign fluctuation. This is known as *the sign problem*. The sign problem at the quark chemical potential is an important unsolved problem in the lattice simulation [18]. As shown later, a chiral chemical potential does not cause the sign problem. This is similar to two-color QCD [19–21] and an isospin chemical potential [21–25].

The basic observable of the chiral magnetic effect is the local vector current density

$$j_\mu(x) = \bar{\psi}(x) \gamma_\mu \psi(x). \tag{15.4}$$

The fourth (zeroth) component corresponds to the local charge density. For calculating the local vector current density, we consider the Dirac eigenvalue problem

$$D[U]\phi_k(x) = (i\lambda_k + m)\phi_k(x), \tag{15.5}$$

and use the identity

$$\langle \bar{\psi}(x)\gamma_\mu \psi(x) \rangle = \langle \operatorname{tr} \gamma_\mu D[U]^{-1} \rangle = \left\langle \sum_k \frac{\bar{\phi}_k(x)\gamma_\mu \phi_k(x)}{i\lambda_k + m} \right\rangle. \tag{15.6}$$

Thus, its expectation value is obtained by inverting or diagonalizing the Dirac operator $D[U]$. In the case of diagonalizing, we can calculate the local vector current density of each Dirac eigenmode.

Chiral symmetry is a nontrivial problem on the lattice due to the Nielsen-Ninomiya no-go theorem [26, 27]. Most lattice fermions more or less break chiral symmetry. The lattice fermion with exact chiral symmetry has been known, although its computational cost is rather large in the dynamical simulation. We should select an appropriate lattice fermion, corresponding to the purpose of the simulation. For the details of the lattice fermions and chiral symmetry, see the reviews [28–30].

For a magnetic field, the QED gauge field is also introduced. When the magnetic field is external, i.e., not dynamical, the QED field strength term does not exist in the action. To couple the fermions to the magnetic field, the Dirac operator is replaced as

$$D[U] \rightarrow D[uU] \tag{15.7}$$

with the U(1) link variable

$$u_\mu(x) = \exp(iqA_\mu(x)). \tag{15.8}$$

We can apply a homogeneous magnetic field in a finite-volume box with periodic boundary conditions. For the homogeneous magnetic field in the z-direction, the U(1) link variables are set as

$$u_1(x) = \exp(-iqBN_s y) \quad \text{at } x = N_s, \tag{15.9}$$

$$u_2(x) = \exp(iqBx), \tag{15.10}$$

$$u_\mu(x) = 1 \quad \text{for other components} \tag{15.11}$$

in the lattice volume $N_s^3 \times N_t$ [31]. In this setup, the magnetic field is quantized as

$$qB = \frac{2\pi}{N_s^2} \times (\text{integer}). \tag{15.12}$$

This integer is the input parameter which controls the strength of the magnetic field in the simulation.

15.3 Lattice Simulation with a Topological Background

The gauge configuration possesses a topological charge. The topological charge of the gauge configuration is given as

$$Q = \frac{g^2}{64\pi^2} \int d^4x \, \varepsilon_{\mu\nu\lambda\rho} F_{\mu\nu}^a(x) F_{\lambda\rho}^a(x). \tag{15.13}$$

Euclidean topological objects, such as the instanton, can be reproduced on the lattice when the gauge configuration is smooth enough [32].

The fermion feels the background topology of the gauge configuration through *the zero mode*. The zero mode is defined as the eigenmode ϕ_k which has the zero eigenvalue $i\lambda_k = 0$ in (15.5). The topological charge and the zero mode are related through the Atiyah-Singer index theorem

$$N_R - N_L = N_f Q, \tag{15.14}$$

where N_R and N_L are the numbers of the right-handed zero modes and the left-handed zero modes, respectively [33]. The fermion zero mode is essential to generate the chiral imbalance in the topological background. We must use the lattice

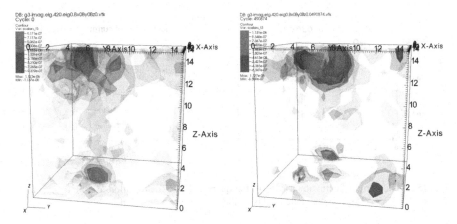

Fig. 15.2 The charge density distribution $\rho(x)$ in the $(2+1)$-flavor dynamical QCD + QED simulation at $a^2 q B = 0$ (*left*) and 0.0490874 (*right*) [8]. The magnetic field is applied along the z-axis. The temperature is above the critical temperature

fermion which is sensitive to the zero mode and satisfies the index theorem, e.g., the overlap fermion.

Naively, it is impossible to measure the local vector current density (15.4) in the topological background. The reason is as follows. In the QCD vacuum, the positive and negative topological charges appear with the same probability. In the simulation, the numbers of the gauge configurations with the positive and negative topological charges are the same, namely

$$\langle Q \rangle = 0. \tag{15.15}$$

The positive and negative topological charges induce the vector current in opposite directions. When all the Monte Carlo samples are averaged in all the topological sectors, the net vector current is zero. To measure the vector current, we must fix the topological sector by the lattice action which suppresses topology changing transitions [34]. Although the fixed-topology simulation cannot reproduce the $\theta = 0$ vacuum, we can obtain a finite expectation value of the vector current.

The fixed-topology analysis has been done in the $(2+1)$-flavor dynamical QCD simulation with the domain-wall fermion [8, 9]. This simulation includes not only the external magnetic field but also the dynamical QED effect. The domain-wall fermion does not have the exact zero mode due to small explicit chiral symmetry breaking, but has the "near" zero mode which becomes the exact zero mode in an ideal limit. In Fig. 15.2, we show the charge density distribution of one near zero mode in one typical gauge configuration [8]. The charge density of the kth eigenmode is defined as

$$\rho_k(x) = \frac{\phi_k^\dagger(x)\phi_k(x)}{i\lambda_k + m}. \tag{15.16}$$

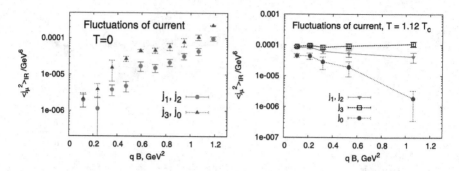

Fig. 15.3 The current fluctuation $\langle j_\mu^2 \rangle_{\text{IR}}$ in the quenched SU(2) simulation at $T = 0$ (*left*) and $T = 1.12 T_c$ (*right*) [5]. The magnetic field is applied in the $\mu = 3$ direction

The simulation was performed above the critical temperature. The charge density distribution at $B \neq 0$ differs from that at $B = 0$. This result suggests that some relation exists between the charge density and the magnetic field. However, the exact relation is not clear in this simulation. We need further investigation for evidence of the chiral magnetic effect.

Except for the fixed-topology simulation, the vector current itself is zero because of the parity oddness. In this case, a numerical observable is the parity-even quantity which reflects the topological fluctuation

$$\frac{\langle Q^2 \rangle}{V} \simeq (200 \,\text{MeV})^4. \tag{15.17}$$

For instance, the fluctuation of the vector current is parity-even. This situation is similar to the experimental observation. An experimental observable must be parity-even, although the chiral magnetic effect is a parity-odd process. We have to extract the parity-odd information from the parity-even particle correlation. This kind of analysis is not easy because the fluctuation can be easily induced by other irrelevant effects. The irrelevant contributions must be subtracted correctly.

The fluctuation $\langle j_\mu^2 \rangle$ of the vector current was calculated in the quenched SU(2) simulation at zero temperature [4], in the quenched SU(2) simulation at finite temperature [5, 6], and in the quenched SU(3) simulation [7]. The overlap Dirac operator was adopted in these simulations, although the zero modes were ignored. The vector currents are zero in all the directions because the topological sector is not fixed, but the current fluctuation is nonzero. In Fig. 15.3, we show the current fluctuation in the quenched SU(2) simulation below and above the critical temperature [5]. The ultraviolet part of the fluctuation is subtracted to obtain a clear signal as

$$\langle j_\mu^2 \rangle_{\text{IR}} = \frac{1}{V} \sum_{\text{site}} \langle j_\mu^2(x) \rangle_{B,T} - \frac{1}{V} \sum_{\text{site}} \langle j_\mu^2(x) \rangle_{B=0,T=0}, \tag{15.18}$$

where the index μ is not summed over. At zero temperature $T = 0$, all the fluctuations grow at stronger magnetic field. In particular, the longitudinal fluctuation $\langle j_3^2 \rangle$

grows faster than transverse fluctuations $\langle j_1^2 \rangle = \langle j_2^2 \rangle$. Above the critical temperature $T > T_c$, the longitudinal fluctuation is insensitive and the transverse fluctuations decrease at stronger magnetic field. As a consequence, the ratio of the longitudinal fluctuation to the transverse fluctuation is enhanced by the magnetic field in both cases.

As shown above, the magnetic field affects the charge density distribution and the current fluctuation. Note however that we must carefully check whether its origin is actually the chiral magnetic effect. In general, a strong magnetic field can induce a strong current fluctuation in the longitudinal direction, even if there is no topological object. This complication is the same as that in experiments. For identifying the chiral magnetic effect, we must distinguish a small topological contribution from other large contaminations in a high-precision simulation.

15.4 Lattice Simulation with a Chiral Chemical Potential

Another possible source of the chiral imbalance is a chiral chemical potential. The chiral chemical potential μ_5 is defined as

$$D(\mu_5) = \gamma_\mu \big(\partial_\mu + i g t^a A_\mu^a(x)\big) + m + \mu_5 \gamma_4 \gamma_5 \tag{15.19}$$

in the continuum space [35]. The chiral chemical potential directly couples to the chiral charge

$$N_5 \equiv N_R - N_L = -\int d^3x \, \langle \bar{\psi}(x)\gamma_4\gamma_5\psi(x)\rangle. \tag{15.20}$$

By using the chiral chemical potential, we can generate a chirally imbalanced QCD matter in equilibrium. The chiral chemical potential is the external parameter which tunes the chiral charge instead of the topological charge. The chiral chemical potential does not exist in the original QCD action because the chiral charge is not a conserved quantity. It is not a "chemical potential" in the exact sense.

Because the topological charge is not necessary in this approach, the sensitivity to the zero mode is not important for the choice of the fermion action. For example, the lattice Dirac operator of the Wilson fermion is

$$\frac{1}{m} D_W(\mu_5) = 1 - \kappa \sum_i \big[(1 - \gamma_i)T_{i+} + (1 + \gamma_i)T_{i-}\big]$$

$$- \kappa \big[\big(1 - \gamma_4 e^{\mu_5\gamma_5}\big)T_{4+} + \big(1 + \gamma_4 e^{-\mu_5\gamma_5}\big)T_{4-}\big], \tag{15.21}$$

with

$$\kappa \equiv \frac{1}{2m + 8}, \tag{15.22}$$

$$[T_{\mu+}]_{x,y} \equiv U_\mu(x)\delta_{x+\hat{\mu},y}, \tag{15.23}$$

$$[T_{\mu-}]_{x,y} \equiv U_\mu^\dagger(y)\delta_{x-\hat{\mu},y}. \tag{15.24}$$

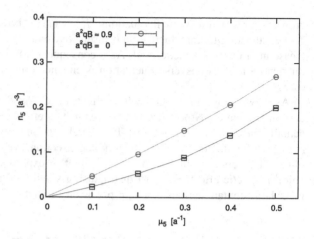

Fig. 15.4 The chiral charge density n_5 in the two-flavor dynamical QCD simulation [12]. The lattice spacing is $a \simeq 0.13$ fm and the temperature is $T \simeq 400$ MeV

The chiral chemical potential is introduced as the exponential matrix factor

$$e^{\pm \mu_5 \gamma_5} = \cosh \mu_5 \pm \gamma_5 \sinh \mu_5, \qquad (15.25)$$

which is the straightforward analogy to a quark chemical potential [36]. The Wilson-Dirac operator (15.21) reproduces the continuum form (15.19) in the continuum limit.

A notable feature of the chiral chemical potential is that it does not cause the sign problem unlike the quark chemical potential. The Wilson-Dirac operator (15.21) is "γ_5-Hermitian",

$$\gamma_5 D(\mu_5) = [\gamma_5 D(\mu_5)]^\dagger \quad \text{or} \quad \gamma_5 D(\mu_5)\gamma_5 = D^\dagger(\mu_5). \qquad (15.26)$$

In the two-flavor case, the fermion determinant is positive real,

$$\det \begin{pmatrix} D(\mu_5) & 0 \\ 0 & D(\mu_5) \end{pmatrix} = \det D(\mu_5) \det \gamma_5 D(\mu_5)\gamma_5 = |\det D(\mu_5)|^2 \geq 0. \quad (15.27)$$

Therefore there is no sign problem. We can exactly simulate a kind of finite density QCD matter by the chiral chemical potential.

In Fig. 15.4, we show the chiral charge density

$$n_5 = \frac{N_5}{V} = -\frac{1}{V} \sum_{\text{site}} \langle \bar{\psi}(x) \gamma_4 \gamma_5 \psi(x) \rangle \qquad (15.28)$$

of the Wilson fermion in the two-flavor dynamical QCD simulation [12]. The lattice spacing is $a \simeq 0.13$ fm. The physical temperature is $T \simeq 400$ MeV, which is above the critical temperature. The chiral charge density is finite at a finite chiral chemical potential. This means that the uniform chirally imbalanced matter is realized on the lattice. The total chiral charge in this lattice volume is $N_5 = n_5 V \simeq O(10^3)$. This number is much larger than a typical number of the topological charge. The

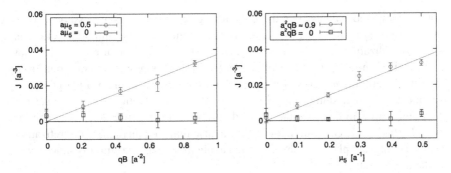

Fig. 15.5 The vector current density J in the two-flavor dynamical QCD simulation [12]. The data are plotted as a function of the magnetic field B (*left*) and of the chiral chemical potential μ_5 (*right*). The lattice spacing is $a \simeq 0.13$ fm and the temperature is $T \simeq 400$ MeV

typical number of the topological charge is $O(10)$ at most in the conventional lattice simulation. Owing to the large chiral imbalance, the analysis of the chiral magnetic effect becomes easy.

For the analysis of the chiral magnetic effect, the local vector current density (15.4) was measured. The vector current is induced only in the longitudinal direction of the magnetic field. The transverse components are exactly zero, $\langle j_1 \rangle = \langle j_2 \rangle = 0$. In Fig. 15.5, the induced current

$$J = \frac{1}{V} \sum_{\text{site}} \langle j_3(x) \rangle \tag{15.29}$$

is plotted as a function of the magnetic field and of the chiral chemical potential. This induced current is direct evidence of the chiral magnetic effect. The induced current is a linearly increasing function in both cases. Therefore, the functional form is

$$J = N_{\text{dof}} C \mu_5 q B. \tag{15.30}$$

Because all the fermions have the same charge in this simulation, the prefactor is $N_{\text{dof}} = 3(\text{color}) \times 2(\text{flavor}) = 6$. The overall coefficient C characterizes the strength of the induced current. This functional form is consistent with the analytical formula,

$$J = N_{\text{dof}} \frac{1}{2\pi^2} \mu_5 q B, \tag{15.31}$$

which was derived from the Dirac equation coupled with the background magnetic field [35]. Note that (15.31) is different from Ref. [35] by q due to the definition of the electric current, i.e., $J_{\text{EM}} = q J$. If there are several fermions with different charges, the total electric current is $J_{\text{EM}} = \sum_i q_i J_i = \sum_i q_i^2 C \mu_5 B$.

The overall coefficient is $C = 0.013 \pm 0.001$ in this lattice simulation and $C = 1/(2\pi^2) \simeq 0.05$ in the analytical formula. The induced current seems somehow smaller than the analytical formula. However, these overall coefficients should not be compared naively. For the quantitative argument, it is necessary to estimate several systematic effects in the lattice simulation. One important effect is the renormalization of the local vector current. The local vector current (15.4) is not renormalization-group invariant on the lattice [37]. This property is different from that in the continuum theory. The local vector current is renormalization-group invariant in the continuum theory because of the Ward identity. We must take the continuum limit to compare the induced currents on the lattice and in the continuum. By taking the continuum limit, we can also remove other lattice discretization artifacts. For example, the Wilson fermion explicitly breaks the chiral symmetry due to the discretization artifact.

The systematic errors were partly estimated in Ref. [12]. By varying simulation parameters, the dependences of the overall coefficient were examined in the quenched simulation. Although the dynamical QCD simulation is necessary for the quantitative argument, the quenched simulation is useful to understand which systematic effect is important. Actually, the quenched results were qualitatively similar to the dynamical QCD results. It turned out that the induced current is insensitive to the temperature, the quark mass, and the spatial volume. However, the overall coefficient strongly depends on the lattice spacing. The overall coefficient increases near the continuum limit. This systematic analysis indicates that the continuum extrapolation is necessary for the quantitative argument.

Another important effect is chiral symmetry. It is difficult to discuss chiral symmetry using the naive Wilson fermion. The Wilson fermion explicitly breaks chiral symmetry at a finite lattice spacing, while the explicit breaking vanishes in the continuum limit. One possible origin of the strong lattice spacing dependence might be this artificial chiral symmetry breaking. We should investigate the role of chiral symmetry in the chiral magnetic effect by performing the same analysis with a chiral lattice fermion, such as the domain-wall fermion or the overlap fermion.

15.5 Conclusion

In this review, we have overviewed the lattice studies of the chiral magnetic effect. The vector current and its fluctuation were measured in the chiral imbalance, which is generated by the topological charge or the chiral chemical potential. We should develop these pioneering works in future. In the future works, it is important to respect the essential pieces of the chiral magnetic effect, in particular, the fermion zero mode and chiral symmetry.

We see that the chiral magnetic effect is an observable phenomenon on the lattice. The lattice simulation is a hopeful approach to study the chiral magnetic effect in "numerical" experiments.

Acknowledgements   The author is supported by the Special Postdoctoral Research Program of RIKEN.

References

1. D.E. Kharzeev, L.D. McLerran, H.J. Warringa, Nucl. Phys. A **803**, 227 (2008). arXiv:0711. 0950 [hep-ph]
2. B.I. Abelev et al. (STAR Collaboration), Phys. Rev. Lett. **103**, 251601 (2009). arXiv:0909. 1739 [nucl-ex]
3. B.I. Abelev et al. (STAR Collaboration), Phys. Rev. C **81**, 054908 (2010). arXiv:0909.1717 [nucl-ex]
4. P.V. Buividovich, E.V. Luschevskaya, M.I. Polikarpov, M.N. Chernodub, JETP Lett. **90**, 412 (2009). [Pis'ma Zh. Eksp. Teor. Fiz. **90**, 456 (2009)]
5. P.V. Buividovich, M.N. Chernodub, E.V. Luschevskaya, M.I. Polikarpov, Phys. Rev. D **80**, 054503 (2009). arXiv:0907.0494 [hep-lat]
6. P.V. Buividovich, M.N. Chernodub, E.V. Luschevskaya, M.I. Polikarpov, PoS **LAT2009**, 080 (2009). arXiv:0910.4682 [hep-lat]
7. V.V. Braguta, P.V. Buividovich, T. Kalaydzhyan, S.V. Kuznetsov, M.I. Polikarpov, PoS **LAT2010**, 190 (2010). arXiv:1011.3795 [hep-lat]
8. M. Abramczyk, T. Blum, G. Petropoulos, R. Zhou, PoS **LAT2009**, 181 (2009). arXiv:0911. 1348 [hep-lat]
9. T. Ishikawa, T. Blum, PoS **LAT2011**, 196 (2011)
10. A. Yamamoto, Phys. Rev. Lett. **107**, 031601 (2011). arXiv:1105.0385 [hep-lat]
11. A. Yamamoto, PoS **LAT2011**, 220 (2011). arXiv:1108.0937 [hep-lat]
12. A. Yamamoto, Phys. Rev. D **84**, 114504 (2011). arXiv:1111.4681 [hep-lat]
13. M. Creutz, *Quarks, Gluons and Lattices*. Cambridge Monographs on Mathematical Physics (Cambridge University Press, Cambridge, 1983)
14. I. Montvay, G. Munster, *Quantum Fields on a Lattice*. Cambridge Monographs on Mathematical Physics (Cambridge University Press, Cambridge, 1994)
15. J. Smit, Introduction to quantum fields on a lattice: a robust mate. Camb. Lect. Notes Phys. **15**, 1 (2002)
16. H.J. Rothe, Lattice gauge theories: an introduction. World Sci. Lect. Notes Phys. **74**, 1 (2005)
17. A.M. Ferrenberg, R.H. Swendsen, Phys. Rev. Lett. **61**, 2635 (1988)
18. P. de Forcrand, PoS **LAT2009**, 010 (2009). arXiv:1005.0539 [hep-lat] (for a review)
19. S. Hands, J.B. Kogut, M.-P. Lombardo, S.E. Morrison, Nucl. Phys. B **558**, 327 (1999). hep-lat/ 9902034
20. J.B. Kogut, D.K. Sinclair, S.J. Hands, S.E. Morrison, Phys. Rev. D **64**, 094505 (2001). hep-lat/ 0105026
21. J.B. Kogut, D.K. Sinclair, Phys. Rev. D **66**, 014508 (2002). hep-lat/0201017
22. J.B. Kogut, D.K. Sinclair, Phys. Rev. D **66**, 034505 (2002). hep-lat/0202028
23. J.B. Kogut, D.K. Sinclair, Phys. Rev. D **70**, 094501 (2004). hep-lat/0407027
24. A. Nakamura, T. Takaishi, Nucl. Phys. Proc. Suppl. **129**, 629 (2004). hep-lat/0310052
25. P. de Forcrand, M.A. Stephanov, U. Wenger, PoS **LAT2007**, 237 (2007). arXiv:0711.0023 [hep-lat]
26. H.B. Nielsen, M. Ninomiya, Nucl. Phys. B **185**, 20 (1981). Erratum. Nucl. Phys. B **195**, 541 (1982)
27. H.B. Nielsen, M. Ninomiya, Nucl. Phys. B **193**, 173 (1981)
28. M. Creutz, Rev. Mod. Phys. **73**, 119 (2001). hep-lat/0007032
29. H. Neuberger, Annu. Rev. Nucl. Part. Sci. **51**, 23 (2001). hep-lat/0101006
30. S. Chandrasekharan, U.J. Wiese, Prog. Part. Nucl. Phys. **53**, 373 (2004). hep-lat/0405024
31. M.H. Al-Hashimi, U.-J. Wiese, Ann. Phys. **324**, 343 (2009). arXiv:0807.0630 [quant-ph]
32. M. Teper, Nucl. Phys. Proc. Suppl. **83**, 146 (2000). hep-lat/9909124 (for a review)
33. M.F. Atiyah, I.M. Singer, Ann. Math. **87**, 484 (1968)
34. H. Fukaya et al. (JLQCD Collaboration), Phys. Rev. D **74**, 094505 (2006). hep-lat/0607020
35. K. Fukushima, D.E. Kharzeev, H.J. Warringa, Phys. Rev. D **78**, 074033 (2008). arXiv:0808. 3382 [hep-ph]
36. P. Hasenfratz, F. Karsch, Phys. Lett. B **125**, 308 (1983)
37. L.H. Karsten, J. Smit, Nucl. Phys. B **183**, 103 (1981)

Chapter 16
Magnetism in Dense Quark Matter

Efrain J. Ferrer and Vivian de la Incera

16.1 Introduction

In this paper we review the status of the current knowledge about the magnetic field effects on color superconductivity (CS) at asymptotically high densities and discuss possible consequences of these effects for the physics of compact stars.

Contrary to what our naïve intuition might indicate, a magnetic field does not need to be of the order of the baryon chemical potential to produce a noticeable effect in a color superconductor. As discussed in [45, 54], a color superconductor can be characterized by various scales and different physics can occur at field strengths comparable to each of them. Specifically, for the so-called Color-Flavor-Locked (CFL) phase, the superconducting gap, the Meissner mass of the charged gluons, and the baryon chemical potential define three scales that determine the values of the magnetic field needed to produce different effects. Thus, the presence of sufficiently strong fields can modify the properties of the dense-matter phase which in turn might lead to observable signatures.

16.2 Magnetic Fields in Compact Stars

The density of matter in the core of compact stars is expected to exceed that of nuclear matter $\rho_{nuc} = 2.8 \times 10^{14}$ g/cm^3 [127]. At such densities, individual nucleons overlap substantially. Under such conditions matter might consist of weakly interacting quarks rather than of hadrons [36]. Due to the asymptotic freedom mechanism [35, 76, 118] one might think that at high baryon density QCD is amenable

E.J. Ferrer (✉) · V. de la Incera
Department of Physics, University of Texas at El Paso, El Paso, TX 79968, USA
e-mail: ejferrer@utep.edu

V. de la Incera
e-mail: vincera@utep.edu

D. Kharzeev et al. (eds.), *Strongly Interacting Matter in Magnetic Fields,*
Lecture Notes in Physics 871, DOI 10.1007/978-3-642-37305-3_16,
© Springer-Verlag Berlin Heidelberg 2013

to perturbative techniques [19, 22, 69, 70, 85, 87, 89, 106]. However, the ground state of the superdense quark system, a Fermi liquid of weakly interacting quarks, is unstable with respect to the formation of diquark condensates [17, 21, 68], a non-perturbative phenomenon essentially equivalent to the Cooper instability of BCS superconductivity. Given that in QCD one gluon exchange between two quarks is attractive in the color-antitriplet channel, at sufficiently high density and sufficiently small temperature quarks should condense into Cooper pairs, which are color an-titriplets. These color condensates break the $SU(3)$ color gauge symmetry of the ground state producing a color superconductor.

In the late 90's the interest in CS was regained after the finding, based on dif-ferent effective theories for low energy QCD [7, 121], that a color-breaking diquark condensate of much larger magnitude than originally thought may exist already at relatively moderate densities (of the order of a few times the nuclear matter density). At densities much higher than the masses of the u, d, and s quarks, one can assume the three quarks as massless. In this asymptotic region the most favored state is the CFL phase [7], characterized by a spin-zero diquark condensate antisymmetric in both color and flavor. Since the combination of high densities and relatively low temperatures can naturally exist in the dense cores of compact stars, it is expected that CS could be realized in that astrophysical environment.

Compact stars, on the other hand, are strongly magnetized objects. From the measured periods and spin down of soft-gamma repeaters (SGR) and anomalous X-ray pulsars (AXP), as well as the observed X-ray luminosities of AXP, it has been determined that certain class of neutron stars known as magnetars can have surface magnetic fields as large as 10^{14}–10^{16} G [83, 92, 109, 136, 137]. In addition, since the stellar medium has a very high electric conductivity, the magnetic flux should be conserved. Hence, it is natural to expect an increase of the magnetic field strength with increasing matter density, and consequently a much stronger magnetic field in the stars' core. Nevertheless, the interior magnetic fields of neutron stars are not directly accessible to observation, so one can only estimate their values with heuris-tic methods. Estimates based on macroscopic and microscopic analysis, for nuclear [24, 38], and quark matter considering both gravitationally bound and self-bound stars [65], have led to maximum fields within the range 10^{18}–10^{20} G, depending whether the inner medium is formed by neutrons [24, 38], or quarks [65].

For instance, from energy-conservation arguments we can estimate the maximum field strength for a quark star. One should expect that the magnetic energy density does not exceed the energy density of the self-bound quark matter, which is given as the energy density at zero pressure that can have a maximum value equal to that of the iron nucleus (roughly 939 MeV). Based on this reasoning, the maximum field allowed can be estimated as

$$B_{max} \simeq \frac{\varepsilon_{bind}^2}{e\hbar c} \leq \frac{(939 \text{ MeV})^2}{e\hbar c} \sim 1.5 \times 10^{20} \text{ G}. \qquad (16.1)$$

From this result, one notice that the inner field can reach values two orders of mag-nitude larger than the estimates done for gravitationally bound stars with nuclear matter [24, 38].

As we will see in Sect. 16.8, the magnetic field decreases the inner pressure along the field direction of the magnetized system. If the magnetic field that makes such a pressure component equal to zero is taken as the maximum value of the inner field allowable for a stable gravitational bound star, then a star with a quark matter core can have a maximum field $\sim 10^{19}$–10^{20} G [114], while one with nuclear matter can only have fields $\sim 10^{18}$ G [24, 38].

Therefore, the investigation of the properties of very dense matter in the presence of strong magnetic fields is of interest not just from a fundamental point of view, but it could be also closely connected to the physics of strongly magnetized neutron stars.

16.3 Magnetism in Spin-Zero Color Superconductivity

An important point to keep in mind in our analysis of the field effects in CS is that in spin-zero color superconductivity the electromagnetism is not the conventional one. In the color superconducting medium the conventional electromagnetic field is not an actual eigenfield, since it is mixed with one of the gluon fields, much like the mixing occurring between the hyper-field and the W-boson in the electroweak model in the presence of the Higgs condensate. Thus, even though the original electromagnetic $U(1)_{em}$ symmetry is broken by the formation of the charged quark Cooper pairs in the CFL phase [8], a residual $\tilde{U}(1)$ gauge symmetry still remains. The massless gauge field associated with this symmetry is given by the linear combination of the conventional photon field and the 8th gluon field [8, 9, 75],

$$\tilde{A}_\mu = \cos\theta A_\mu - \sin\theta G^8_\mu. \tag{16.2}$$

The corresponding orthogonal linear combination

$$\tilde{G}^8_\mu = \sin\theta A_\mu + \cos\theta G^8_\mu, \tag{16.3}$$

is massive. The field $\tilde{A}_\mu$ plays the role of an in-medium or rotated electromagnetic field. A magnetic field associated with $\tilde{A}_\mu$ can penetrate the CS without being subject to the Meissner effect, since the color condensate is neutral with respect to the rotated charge. However, the rotated electromagnetic field in the CFL superconductor is mostly formed by the original photon with only a small admixture of the 8th gluon since the mixing angle, $\cos\theta = g/\sqrt{e^2/3 + g^2}$, is sufficiently small.

The generator of the unbroken $\tilde{U}(1)$ symmetry, which corresponds to the long-range rotated photon in the CFL phase, is a matrix in $flavor_{(3\times3)} \otimes color_{(3\times3)}$ space given by $\tilde{Q}_{CFL} = Q \otimes I + I \otimes T_8/\sqrt{3}$, where Q is the conventional electromagnetic charge operator of quarks and T_8 is the 8th Gell-Mann matrix. Using the matrix representations, $Q = \text{diag}(-1/3, -1/3, 2/3)$ for (s, d, u) flavors, and $T_8 = \text{diag}(-1/\sqrt{3}, -1/\sqrt{3}, 2/\sqrt{3})$ for (b, g, r) colors, the $\tilde{Q}$ charges (in units of $\tilde{e} = e\cos\theta$) of different quarks are

$$
\begin{array}{ccccccccc}
s_b & s_g & s_r & d_b & d_g & d_r & u_b & u_g & u_r \\
\hline
0 & 0 & -1 & 0 & 0 & -1 & +1 & +1 & 0
\end{array}
\tag{16.4}
$$

For the $2SC$ color superconductor the ground state is formed by spin-zero diquarks, which are also neutral with respect to the rotated electromagnetic field associated with the remnant $\widetilde{U}(1)$ symmetry. In this system (see Refs. [7, 17, 30, 121]), the generator of the remnant $\widetilde{U}(1)$ symmetry is a matrix in $flavor_{(2\times2)} \otimes color_{(3\times3)}$ space given by $\widetilde{Q}_{2SC} = Q \otimes I - I \otimes T_8/\sqrt{3}$, with the usual matrix of electromagnetic charges of quarks in flavor space $Q = \mathrm{diag}(2/3, -1/3)$, and T_8 is the eighth generator of the $SU(3)_c$ gauge group in the adjoint representation. The rotated charges of the quarks in units of $\tilde{e} = e\cos\theta$, are given in this phase by

$$
\begin{array}{cccccc}
d_b & d_g & d_r & u_b & u_g & u_r \\
\hline
0 & -\frac{1}{2} & -\frac{1}{2} & 1 & \frac{1}{2} & \frac{1}{2}
\end{array}
\tag{16.5}
$$

and the massless rotated electromagnetic field and orthogonal massive field are defined in the same way as in the CFL case.

From now on, we will use "magnetic field" in short, when we refer to the "rotated magnetic field", since inside the superconductor only the rotated magnetic field is the physical long range field.

16.4 The Magnetic CFL Phase

The fact that the rotated magnetic field can penetrate the spin-zero color superconductor brings the possibility to look for possible field-interaction effects on the CS phase. An important consequence of this interaction was first studied in [61]. It is based in the following observation. Although the Cooper pairs have zero rotated charge, they can be formed either by neutral quarks or by quarks of opposite charges. If the magnetic field is strong enough so that the magnetic length $l_0 = 1/\sqrt{2eB}$ becomes smaller than the pairs' coherence length, then the magnetic field can interact with the pair constituents and significantly modify the pair structure of the condensate. As shown in [61–64], the presence of a magnetic field changes the CFL phase characterized by one single gap, producing a splitting of the CFL gap into a gap that gets contributions from both pairs of oppositely charged, as well as neutral, quarks, denoted by Δ_B, and one that only gets contributions from pairs of neutral quarks, denoted by Δ. The new phase that forms in the presence of the magnetic field also has color-flavor-locking, but with a smaller symmetry group $SU(2)_{C+L+R}$, a change that is reflected in the splitting of the Δ and Δ_B gaps. The less symmetric realization of the CFL pairing that occurs in the presence of a magnetic field, is known as the magnetic-CFL (MCFL) phase [61–64]. The MCFL phase has similarities, but also important differences with the CFL phase [54, 61–64, 71, 111].

In the strong-field limit, the gap formed by pairs of neutral and charged quarks satisfies the gap equation [61–64]

$$1 \approx \frac{g^2}{3\Lambda^2} \int_\Lambda \frac{d^3 q}{(2\pi)^3} \frac{1}{\sqrt{(q-\mu)^2 + 2(\Delta_B)^2}} + \frac{g^2 \tilde{e} \tilde{B}}{6\Lambda^2} \int_{-\Lambda}^\Lambda \frac{dq}{(2\pi)^2} \frac{1}{\sqrt{(q-\mu)^2 + (\Delta_B)^2}}, \tag{16.6}$$

The solution of (16.6) is given by

$$\Delta_B \sim 2\sqrt{\delta\mu} \exp\left(-\frac{3\Lambda^2 \pi^2}{g^2 (\mu^2 + \frac{\tilde{e}\tilde{B}}{2})}\right), \tag{16.7}$$

which can be compared with the CFL gap

$$\Delta_{CFL} \sim 2\sqrt{\delta\mu} \exp\left(-\frac{3\Lambda^2 \pi^2}{2g^2 \mu^2}\right). \tag{16.8}$$

Here we used $\delta \equiv \Lambda - \mu$, with Λ the ultraviolet cutoff of the NJL model that should be much larger than any of the typical energy scales of the system, and μ the baryon chemical potential.

The gap Δ, formed only by pairs of neutral quarks, should be found as the solution of the gap equation

$$1 \approx \frac{g^2}{4\Lambda^2} \int_\Lambda \frac{d^3 q}{(2\pi)^3} \left(\frac{17}{9} \frac{1}{\sqrt{(q-\mu)^2 - \Delta^2}} + \frac{7}{9} \frac{1}{\sqrt{(q-\mu)^2 + 2(\Delta_B)^2}}\right), \tag{16.9}$$

where it is apparent the interconnection with the gap Δ_B. This is how through Δ_B the magnetic field can affect Δ although it is formed only by neutral quarks as we already pointed out.

The solution of (16.9) is given by

$$\Delta \sim \frac{1}{2^{(7/34)}} \exp\left(-\frac{36}{17x} + \frac{21}{17} \frac{1}{x(1+y)} + \frac{3}{2x}\right) \Delta_{CFL}, \tag{16.10}$$

where $x \equiv g^2 \mu^2 / \Lambda^2 \pi^2$, and $y \equiv \tilde{e}\tilde{B}/\mu^2$

The exponent in (16.7) has the typical BCS form, but with different density of states for neutral and charged quarks, i.e. $\exp[1/(N_\mu + N_{\tilde{B}})\tilde{G}]$, where $N_\mu = \mu^2/\pi^2$ is the density of states at the Fermi surface of the neutral quarks with single chirality, $N_{\tilde{B}} = \tilde{e}\tilde{B}/2\pi^2$ is the density of states of the charged quarks lying at the zero Landau level at the Fermi surface, and $\tilde{G} = -g^2/3\Lambda^2$ is the characteristic effective coupling constant of the $\bar{3}$ channel [133]. The effect of the strong magnetic field $\tilde{e}\tilde{B}/2 \geq \mu^2$ is to increase the total density of states, thus producing a gap enhancement.

Although the situation here shares some similarities with the magnetic catalysis of chiral symmetry breaking [50, 51, 56, 57, 77, 78, 90, 91, 96, 97]; the way the field influences the pairing mechanism in the two cases is quite different. The particles participating in the chiral condensate are near the surface of the Dirac sea. The effect of a magnetic field there is to effectively reduce the space dimension where the particles are embedded at the lowest Landau level (LLL), which as a consequence strengthens their effective coupling, and so catalyzing the chiral condensate. Color superconductivity, on the other hand, involves quarks near the Fermi surface, with a pairing dynamics that is already $(1+1)$-dimensional. Therefore, the $\widetilde{B}$ field does not yield further dimensional reduction of the pairing dynamics near the Fermi surface and hence the LLL does not have a special significance here. Nevertheless, the field increases the density of states of the $\widetilde{Q}$-charged quarks, and it is through this effect, as shown in (16.7), that the pairing of the charged particles is reinforced by the penetrating magnetic field.

Note that our analytic solutions are only valid at strong magnetic fields. For fields of this order and larger, the Δ_B gap is larger than Δ_{CFL} at the same density values. How fast or slow the gaps do it depends very much on the values of the NJL couplings. For example, for $x \sim 0.3$, one finds $\Delta \sim 0.2\Delta_B$ for $y = 3/2$, while for $x \sim 1$ then $\Delta \sim 0.5\Delta_B$.

In a recent study [45], it was discovered that the MCFL phase actually contains one more condensate, which we will call Δ_M. This new condensate is associated with the magnetic moment of the Cooper pairs. Physically this is easy to understand. The presence of a uniform magnetic field explicitly breaks the spatial rotational symmetry $O(3)$ to the subgroup $O(2)$ of rotations about the axis along the field. As shown in [45], this symmetry reduction has non-trivial consequences for the ground state structure of the MCFL superconductor. When one performs the Fierz transformations in the quark system with both Lorentz and rotational $O(3)$ symmetries explicitly broken, various new pairing channels appear allowing in principle the formation of new condensates. Of particular interest is an attractive channel that leads to a spin-one condensate of Dirac structure $\Delta_M \sim C\gamma_5\gamma^1\gamma^2$. Such a gap does not break any symmetry that has not already been broken by the other condensates of the MCFL ground state, so it in principle is not forbidden. The new condensate corresponds to the zero spin projection of the average magnetic moment of the Cooper pairs in the medium.

From a physical point of view, it is natural to expect the formation of this extra condensate in the magnetized system because the diquarks formed by oppositely charged quarks with opposite spins will have a net magnetic moment that may point parallel or antiparallel to the magnetic field. Diquarks formed by quarks lying on any non-zero Landau level can have magnetic moments pointing in both directions, because each quark in the pair may have both spins. Hence the contribution of these diquarks to the net magnetic moment should tend to cancel out. On the other hand, diquarks from quarks in the LLL can only have one orientation of their magnetic moment with respect to the field, because the quarks in the LLL have only one possible spin projection. This implies that the main contribution to the new condensate should come from the quarks at the LLL, an expectation that is consistent with the numerical results found in [45], where the new gap was obtained to be negligibly small at weak magnetic fields, where the zero Landau level occupation is not significant. On the other hand, at strong magnetic fields, the condensate became comparable in magnitude to the original condensates, Δ and Δ_B, of the MCFL ground state [61], because the majority of the quarks occupy the LLL in that case.

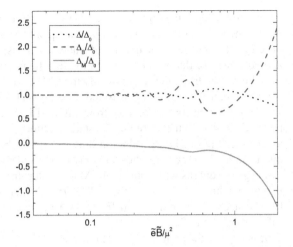

Fig. 16.1 The three gaps of the MCFL phase as a function of $\tilde{e}\tilde{B}/\mu^2$ for $\mu = 500$ MeV. They are scaled with respect to the CFL gap $\Delta_{CFL} = 25$ MeV

Although this new condensate is zero at zero magnetic field, we cannot ignore it even at very small magnetic fields because a self-consistent solution of the gap equations with $\Delta \neq 0$, and $\Delta_B \neq 0$, but $\Delta_M = 0$ is not possible. This is easy to understand since, as long as the magnetic field is not zero, there is always some occupation of the LLL. Thus, once a magnetic field is present, Δ_M has to be considered simultaneously with the spin-zero MCFL gaps.

The Δ_M condensate of the MCFL scenario described above shares a few similarities with the dynamical generation of an anomalous magnetic moment recently found in massless QED [56, 57]. Akin to the Cooper pairs of oppositely charged quarks in the MCFL phase, the fermion and antifermion that pair in massless QED also have opposite charges and spins and hence carries a net magnetic moment. A dynamical magnetic moment term in the QED Lagrangian does not break any symmetry that has not already been broken by the chiral condensate. Therefore, once the chiral condensate is formed due to the magnetic catalysis of chiral symmetry breaking [50, 51, 77, 78, 90, 91, 96, 97], the simultaneous formation of a dynamical mass and a dynamical magnetic moment is unavoidable [56, 57]. The realization of the anomalous magnetic moment condensate in magnetized massless QED produces a non-perturbative Zeeman effect [56, 57].

At moderate magnetic fields the energy gaps Δ and Δ_B exhibit oscillations when $\tilde{e}\tilde{B}/\mu^2$ is varied [71, 111], owed to the de Haas-van Alphen effect [80, 81] typical of charged fermion systems under magnetic fields (see for instance [40, 41, 71, 111]), while for Δ_M the oscillations are almost absent [45]. These features indicate, as already pointed out, that the main contribution to Δ_M should come from pairs whose charged quarks are at the LLL.

The previous discussion can be visualized in the plot of the gaps as functions of a dimensionless parameter $\tilde{e}\tilde{B}/\mu^2$ given in Fig. 16.1. Note that for small magnetic field, Δ and Δ_B are close to each other and approach the CFL gap $\Delta_{CFL} = 25$ MeV. As the magnetic field increases, Δ and Δ_B display oscillatory behaviors with respect to $\tilde{e}\tilde{B}/\mu^2$ as long as $\tilde{e}\tilde{B} < \mu^2$. As originally explained by Landau [93], these oscillations reveal the quantum nature of the interaction of the charged particles with the magnetic field (the well-known Landau quantization phenomenon), and are produced by the change in the density of states when passing from one Landau level to another. The oscillations cease when the first Landau level exceeds the Fermi surface. For ultra-strong fields, when only the LLL contributes to the gap equation, Δ_B is much larger than Δ, as it was found by analytical calculation in [61].

The only contribution to Δ_M from higher LLs can come when the number of particles is odd, so there are energy states occupied by a single particle, but that is a very small part. The cancellation does not occur, however, between the pairs of quarks in the LLL because they can only be formed by positive quarks with spin up and negative quarks with spin down. At low fields, the number of quarks in the LLL is scarce, while for fields of order $\tilde{e}\tilde{B} \geq \mu^2$, all the particles are constrained to the LLL, hence the variation of Δ_M from lower values at weak field, to higher values at sufficiently strong fields.

It is apparent from the graphical representation of Δ_M in Fig. 16.1, that its value remains relatively small up to magnetic-field values of the order of μ^2. In the field region between 10^{18}–10^{19} G, the magnitude of Δ_M grows from a few tenths of MeV to tens of MeV. It becomes comparable to the MCFL gap Δ_B when the field is strong enough to put all the quarks in the LLL, shown in the final segment of the plots in the figure.

Another important consequence of the gap Δ_M is the increment in the magnitude of Δ_B for any given value of the magnetic field in the strong field region, as compared to its own value found at the same field but ignoring the existence of Δ_M [71, 111]. This effect, combined with the increase of Δ_M at strong fields, will make the MCFL phase more stable than the regular CFL, a fact that could favor the realization of an MCFL core in magnetars.

Let us now discuss in more detail the difference between the CFL and MCFL phases from the point of view of symmetry. In the absence of a magnetic field, three-flavor massless quark matter at high baryonic density is in the energetically favored CFL phase. There, the diquark condensates lock the color and flavor transformations, breaking both symmetries. Thus, the symmetry breaking pattern in the CFL phase is

$$SU(3)_C \times SU(3)_L \times SU(3)_R \times U(1)_B \to SU(3)_{C+L+R}. \qquad (16.11)$$

In this case, there are only nine Goldstone bosons that survive to the Anderson-Higgs mechanism. One is a singlet, scalar mode, associated to the breaking of the baryonic symmetry, and the remaining octet is associated to the axial $SU(3)_A$ group, just like the octet of mesons in vacuum. At sufficiently high density, the anomaly is suppressed, and then one can as well consider the spontaneous breaking of an approximated $U(1)_A$ symmetry, and the additional pseudo Goldstone boson. We will ignore this effect, though.

Once electromagnetic effects are considered, the flavor symmetries of QCD are reduced, as only the d and s quarks have equal electromagnetic charges, $q = -e/3$, while the u quark has electromagnetic charge, $q = 2e/3$. However, because the electromagnetic structure constant $\alpha_{e.m.}$ is relatively small, this effect is considered to be really tiny, a small perturbation, and one can consider good approximated flavor symmetries. Nevertheless, in the presence of a strong magnetic field one cannot consider the effects of electromagnetism as a small perturbation. Flavor symmetries are explicitly reduced from $SU(3)_{L,R}$ to $SU(2)_{L,R}$. For sufficiently strong magnetic fields, in the MCFL phase the symmetry breaking pattern is then

$$SU(3)_C \times SU(2)_L \times SU(2)_R \times U(1)_A^{(1)} \times U(1)_B \times U(1)_{\text{e.m.}} \to SU(2)_{C+L+R}.$$
(16.12)

Here the symmetry group $U(1)_A^{(1)}$ is related to a current, which is an anomaly free linear combination of u, d and s axial currents, and such that $U(1)_A^{(1)} \subset SU(3)_A$. The locked $SU(2)$ group corresponds to the maximal unbroken symmetry, such that it maximizes the condensation energy. The counting of broken generators, after taking into account the Anderson-Higgs mechanism, leads to only five Goldstone bosons. As in the CFL case, one is associated with the breaking of the baryon symmetry; three Goldstone bosons are associated with the breaking of $SU(2)_A$, and another one with the breaking of $U(1)_A^{(1)}$. As before, if the effects of the anomaly could be neglected, there would be another pseudo Goldstone boson associated with the $U(1)_A$ symmetry. Thus, apart from modifying the value of the gaps, an applied strong magnetic field also affects the number of Goldstone bosons, reducing them from nine (neutral and charged) to five (neutral).

Once a magnetic field is present, the original symmetry group is reduced, and the low energy theory correspond to the breaking pattern (16.12), hence be described by five Goldstone bosons. In practice however, at weak magnetic fields, it is reasonable to treat the symmetry of the CFL phase as a good approximated symmetry, which means that at weak fields the low-energy excitations are essentially governed by nine approximately massless scalars (those of the breaking pattern (16.11)) instead of five.

A question of order here is: what do we exactly understand as a weak magnetic field? In other words, what is the threshold-field strength that effectively separates the CFL low energy behavior from the MCFL one? A fundamental clue in this direction was found in [54] by determining the term in the low-energy CFL Lagrangian that generates a field-induced mass for the charged Goldstone fields, so disconnecting them from the low-energy dynamics at some field strength and thereby effectively reducing the number of Goldstone bosons from the nine of the CFL phase, to the five neutral ones of the MCFL.

The threshold field $\tilde{B}_{MCFL}$ for the effective $CFL \to MCFL$ symmetry crossover was found to be [54]

$$\tilde{e}\tilde{B}_{MCFL} = \frac{4}{v_\perp^2} \Delta_{CFL}^2 \simeq 12\Delta_{CFL}^2,$$
(16.13)

where the weak-field approximation $v_\perp \simeq 1/\sqrt{3}$ was considered [134]. The threshold field does not depend on the decay constant f_π, therefore it depends on μ only through Δ_{CFL}. For $\Delta_{CFL} \sim 15$ MeV one gets $\tilde{e}\tilde{B}_{MCFL} \sim 10^{16}$ G. At these field strengths, the charged mesons decouple from the low-energy theory. When this decoupling occurs, the five neutral Goldstone bosons (including the one associated to the baryon symmetry breaking) that characterize the MCFL phase will drive the low-energy physics of the system. Therefore, coming from low to higher fields, the first magnetic phase that will effectively show up in the magnetized system for $\tilde{e}\tilde{B} \sim \Delta^2_{CFL}$ will be the MCFL [61–64].

Summarizing, in a color superconductor with three-flavor quarks at very high densities an increasing magnetic field produces a phase crossover from CFL to MCFL. During this phase transmutation no symmetry breaking occurs, since in principle once a magnetic field is present the symmetry is strictly speaking that of the MCFL, as discussed above. However, in practice for $\tilde{B} \sim \tilde{B}_{MCFL} \sim \Delta^2_{CFL}$ the main features of MCFL emerge through the low-energy behavior of the system [54]. At the threshold field $\tilde{B}_{MCFL}$, only five neutral Goldstone bosons remain out of the original nine characterizing the low-energy behavior of the CFL phase, because the charged Goldstone bosons acquire field dependent masses and can decay in lighter modes. For a meson to be stable in this system, its mass should be less than twice the gap, otherwise it will decay into a particle-antiparticle pair. That means that, as proved in Ref. [54], once the applied field produces a mass for the charged Goldstones of the order of the CFL gap it is reached the threshold field for the effective $CFL \rightarrow MCFL$ symmetry transmutation.

The existence of this phase transmutation is on the other hand manifested in the behavior of the gaps versus the magnetic field. At field strength smaller than the threshold field we find that $\Delta \approx \Delta_B \approx \Delta_{CFL}$, while for fields closer to $\tilde{B}_{MCFL}$ the gaps exhibit oscillations with respect to $\tilde{e}\tilde{B}/\mu^2$ [45, 71, 111], owed to the de Haas-van Alphen effect [80, 81].

It is worth to call attention to the analogy between the CFL-MCFL crossover and what could be called a "field-induced" Mott transition. Mott transitions were originally considered in condensed matter in the context of metal-insulator transitions in strongly-correlated systems [107, 108]. Later on, Mott transitions have been also discussed in QCD to describe delocalization of bound states into their constituents at a temperature defined as the Mott temperature [82]. By definition, the Mott temperature T_M is the temperature at which the mass of the bound state equals the mass of its constituents, so the bound state becomes a resonance at $T > T_M$. In the present work, the role of the Mott temperature is played by the threshold field $\tilde{B}_{MCFL}$. Mott transitions typically lead to the appearance of singularities at $T = T_M$ in a number of physically relevant observables. It is an open question, worth to be investigated whether similar singularities are or not present in the CFL-MCFL crossover at $\tilde{B}_{MCFL}$.

16.5 Magnetoelectric Effect in Cold-Dense Matter

It is well known that the phenomenon of CS shares many characteristics of condensed matter systems [13]. In this section, we discuss a new feature of CS that has its counterpart in magnetically ordered materials and has been known in the context of condensed matter for many years. It is the so called magnetoelectric (ME) effect, which establishes a relation between the electric and magnetic properties of certain materials. In general, it states that the electric polarization of such materials may depend on an applied magnetic field and/or that the magnetization may depend on an applied electric field. The first observations of magnetoelectricity took place when a moving dielectric was found to become polarized when placed in a magnetic field [124, 140]. In 1894, Pierre Curie [37] was the first in pointing out the possibility of an intrinsic ME effect for certain (non-moving) crystals on the basis of symmetry considerations. But it took many decades to be understood and proposed by Landau and Lifshitz [95] that the linear ME effect is only allowed in time-asymmetric systems. Recently the ME effect regained new interest in condensed matter thanks to new advancements in material science and with the development of the so-called multiferroic materials for which the ME effect is significant for practical applications [66].

As demonstrated in Refs. [46, 47] the ME effect also occurs in a highly magnetized CS medium like the MCFL phase. In particular, in [46, 47] it was shown how the electric susceptibility of this medium depends on an applied strong magnetic field.

Let us start by discussing the ME effect at weak fields. At weak fields this effect can be studied by taking into account the expansion of the system's free energy in powers of the electric $\widetilde{\mathbf{E}}$ and magnetic $\widetilde{\mathbf{B}}$ fields

$$F(\widetilde{\mathbf{E}}, \widetilde{\mathbf{B}}) = F_0 - \alpha_i \widetilde{E}_i - \beta_i \widetilde{B}_i - \gamma_{ij} \widetilde{E}_i \widetilde{B}_j - \eta_{ij} \widetilde{E}_i \widetilde{E}_j - \tau_{ij} \widetilde{B}_i \widetilde{B}_j$$
$$- \kappa_{ijk} \widetilde{E}_i \widetilde{E}_j \widetilde{B}_k - \lambda_{ijk} \widetilde{E}_i \widetilde{B}_j \widetilde{B}_k - \sigma_{ijkl} \widetilde{E}_i \widetilde{E}_j \widetilde{B}_k \widetilde{B}_l - \cdots. \quad (16.14)$$

In this weak-field expansion the coefficients α_i, γ_{ij} etc., which are the susceptibility tensors, can be found from the infinite set of one-loop polarization operator diagrams with external legs of the in-medium photon field $\widetilde{A}_\mu$ and internal lines of the full CFL quark propagator of the rotated charged quarks. Hence, these coefficients can only depend on the baryonic chemical potential, the temperature and the CFL gap. From (16.14), the electric polarization can be found as

$$P_i = -\frac{\partial F}{\partial \widetilde{E}_i} = \alpha_i + \gamma_{ij} \widetilde{B}_j + 2\eta_{ij} \widetilde{E}_j + 2\kappa_{ijk} \widetilde{E}_j \widetilde{B}_k + \lambda_{ijk} \widetilde{B}_j \widetilde{B}_k + 2\sigma_{ijkl} \widetilde{E}_j \widetilde{B}_k \widetilde{B}_l + \cdots.$$
$$(16.15)$$

If the tensor γ is different from zero the system exhibits the linear ME effect. From the free energy (16.14), we see that the linear ME effect can only exist if the time-reversal and parity symmetries are broken in the medium. In the CFL phase, the time-reversal symmetry is broken by the CFL gap [45], but parity is preserved. Thus, the linear ME effect cannot be present in this medium. The behavior under a time-reversal transformation underscores an important difference between the CFL color superconductivity and the conventional, electric superconductivity. While the CFL color superconductor is not invariant under time-reversal symmetry, the conventional superconductor is, since in the conventional superconductor the Cooper pairs are usually formed by time-reversed one-particle states [16]. In the conventional superconductor the violation of the T-invariance occurs only via some external perturbation which can lead in turn to pair breaking and to the so-called gapless superconductivity [2].

Higher-order ME terms are parameterized by the tensors κ, and λ. As it happens with γ, the coefficient $\lambda \neq 0$ is forbidden because it requires parity violation. On the other hand, although a $\kappa \neq 0$ term only requires time-reversal violation, to form a third-rank tensor independent of the momentum and parity invariant, the medium would need to have an extra spatial vector structure. However, the only tensor structures available to form such a third-rank tensor in the CFL phase are the metric tensor $g_{\mu\nu}$ and the medium fourth velocity u_μ, which in the rest frame is a temporal vector $u_\mu = (1, 0, 0, 0)$, so the coefficient κ should be zero too. Hence, we do not expect any ME effect associated with the lower terms in the weak-field expansion of the free energy (16.15).

At strong magnetic fields, the situation is quite different. In this case the expansion of the free energy can only be done in powers of a weak electric field, and the coefficients of each term can be found from the corresponding one-loop polarization operators, which now depend on the strong magnetic field in the MCFL phase. The free energy expansion in this case takes the form

$$F'(\widetilde{\mathbf{E}}, \widetilde{\mathbf{B}}) = F_0'(\widetilde{B}) - \alpha_i' \widetilde{E}_i - \eta_{ij}' \widetilde{E}_i \widetilde{E}_j - \cdots. \tag{16.16}$$

The tensors α' and η' can depend now on the baryonic chemical potential, temperature, magnetic field and gaps of the MCFL phase [45]. They can be found respectively by calculating the tadpole and the second rank polarization operator tensor of the MCFL phase in the strong field limit. An $\alpha' \neq 0$ would indicate that the MCFL medium behaves as a ferroelectric material [86, 99], but this is not the case because this phase is parity symmetric [45], hence $\alpha' = 0$. The tensor η', nevertheless, is not forbidden by any symmetry argument. If it is different from zero, η' would characterize the lowest order of the system dielectric response. More important, if η' results to be dependent on the magnetic field, this would imply that the electric polarization $P = \eta' E$ depends on the magnetic field through η', hence the MCFL phase would exhibit the ME effect.

To find the electric susceptibility η' in the strong-magnetic-field limit of the MCFL phase we start from [45]

$$F'(\widetilde{\mathbf{E}}, \widetilde{\mathbf{B}}) - F_0'(\widetilde{\mathbf{B}}) \sim \frac{1}{V} \int \widetilde{A}_0(x_3) \Pi_{00}(x_3 - x_3') \widetilde{A}_0(x_3') dx_3 dx_3' = -\eta' \widetilde{E}^2, \tag{16.17}$$

our task is then reduced to the calculation of the zero-zero component of the one-loop polarization operator at strong magnetic field in the infrared limit, $\Pi_{00}(p_0 = 0, p \to 0)$.

Now, the photon polarization operator should be gauge invariant. That is, in the strong-field approximation, it should satisfy the transversality condition in the reduced $(1+1)$-D space $(p_\mu^{\parallel} \Pi_{\mu\nu}^{\parallel}(p^{\parallel}) = 0)$. As known, the polarization operator tensor can be expanded in a superposition of independent transverse Lorentz tensors.

The number of these basic transverse tensors depends on the symmetries of the system under consideration. For example, in vacuum, where the only available tensorial structures are the four-momentum and the metric tensor, there is only one gauge invariant structure. When a medium is under consideration (i.e. at finite temperature or finite density), since the Lorentz symmetry is broken, there is an additional gauge invariant structure that can be formed by taking into account a new four-vector, the four-velocity of the medium center of mass, u_μ, [67]. When a magnetic field is applied on that medium, then the structure of the polarization operator is enriched by an additional tensor, $F_{\mu\nu}$. Then, at finite density and in the presence of a magnetic field, there are nine independent gauge-invariant tensorial structures [117]. At strong magnetic field, when the particles are confined to the LLL, due to the fact that the transverse momentum is zero, there is a dimensional reduction leaving only the tensors $g_{\mu\nu}^\parallel$, $p_\mu^\parallel$ and $u_\mu^\parallel = (1, 0)$ at our disposal. The original nine structures in [117] now reduce to only two

$$T_{\mu\nu}^{(1)} = \left(p^\parallel\right)^2 g_{\mu\nu}^\parallel - p_\mu^\parallel p_\nu^\parallel, \tag{16.18}$$

and

$$T_{\mu\nu}^{(2)} = \left[u_\mu^\parallel - \frac{p_\mu^\parallel (u^\parallel \cdot p^\parallel)}{(p^\parallel)^2}\right]\left[u_\nu^\parallel - \frac{p_\nu^\parallel (u^\parallel \cdot p^\parallel)}{p^2}\right]. \tag{16.19}$$

Moreover, one can readily check that the two tensors (16.18) and (16.19) are equivalent, which indicates that the rotated-photon polarization operator tensor, at strong magnetic field, only has one independent structure

$$\Pi_{\mu\nu}^\parallel(p^\parallel) = \Pi\left(p^\parallel, \mu, B\right)\left[(p^\parallel)^2 g_{\mu\nu}^\parallel - p_\mu^\parallel p_\nu^\parallel\right], \tag{16.20}$$

with $\Pi(p^\parallel, \mu, B)$ being a scalar coefficient depending on the photon longitudinal momentum, baryonic chemical potential and magnetic field.

At zero temperature, the regularized components of the polarization operator in powers of the photon momentum components p_0 and p_3, up to quadratic terms are given by

$$\Pi_{00R} = - \lim_{\Lambda \to \infty} \frac{\tilde{e}^2 |\tilde{e}\tilde{B}| p_3^2}{6\pi^2}\left(\frac{1}{\Delta_0^2} + \frac{1}{\Lambda^2}\right) = -\frac{\tilde{e}^2 |\tilde{e}\tilde{B}| p_3^2}{6\pi^2 \Delta_0^2}, \tag{16.21}$$

$$\Pi_{33R} = - \lim_{\Lambda \to \infty} \frac{\tilde{e}^2 |\tilde{e}\tilde{B}|}{6\pi^2}\left[\left(3 + \frac{p_0^2}{\Delta_0^2}\right) - \left(3 + \frac{p_0^2}{\Lambda^2}\right)\right] = -\frac{\tilde{e}^2 |\tilde{e}\tilde{B}| p_0^2}{6\pi^2 \Delta_0^2}, \tag{16.22}$$

and $\Pi_{30R} = \Pi_{03R} \simeq 0$. As should be expected, the regulator Λ introduced through the Pauli-Villars regularization scheme does not appear in the final results once we take $\Lambda \to \infty$.

Because Π_{00} has no constant contribution in the infrared limit $p_0 = 0$, $p_3 \to 0$, one immediately concludes that there is no Debye screening in the strong-field region, as it was the case at zero field in the CFL phase [100, 126]. This is simply because all quarks are bound within the rotated-charge neutral condensates. There is also no Meissner screening (i.e. Π_{33} is zero in the zero-momentum limit), as it should be expected from the remnant $\widetilde{U}(1)$ gauge symmetry. However, the condensates have electric dipole moments and could align themselves in an electric field. Hence, this should modify the dielectric constant of the medium. Since the quadratic term in the effective $\widetilde{U}(1)$ Lagrangian is given by $\tilde{A}_\mu(-p)[D_{\mu\nu}^{-1}(p) + \Pi_{\mu\nu}(p)]\tilde{A}_\nu(p)$, with D^{-1} being the bare rotated photon propagator, the effective action of the $\widetilde{U}(1)$ field in the strong-field region is given by

$$S_{eff} = \int d^4x \left[\frac{\varepsilon_\parallel}{2}\widetilde{\mathbf{E}}_\parallel \cdot \widetilde{\mathbf{E}}_\parallel + \frac{\varepsilon_\perp}{2}\widetilde{\mathbf{E}}_\perp \cdot \widetilde{\mathbf{E}}_\perp - \frac{1}{2\lambda_\parallel}\widetilde{\mathbf{H}}_\parallel \cdot \widetilde{\mathbf{H}}_\parallel - \frac{1}{2\lambda_\perp}\widetilde{\mathbf{H}}_\perp \cdot \widetilde{\mathbf{H}}_\perp \right], \quad (16.23)$$

where the separation between transverse and longitudinal parts is due to the $O(3) \rightarrow O(2)$ symmetry breaking produced by the strong magnetic field $\widetilde{B}$. In (16.23), $\widetilde{E}$, $\widetilde{H}$ are weak electric and magnetic field probes, respectively. In (16.23) the coefficients ε and λ denote the electric permittivity and magnetic permeability of the medium respectively.

From (16.21)–(16.22) it is straightforward that in the infrared limit the transverse and longitudinal components of the electric permittivity and magnetic permeability become

$$\lambda_\perp = \lambda_\parallel \simeq 1, \qquad \varepsilon_\perp = 1, \qquad \varepsilon_\parallel = 1 + \chi_{MCFL}^\parallel = 1 + \frac{\tilde{e}^2|\tilde{e}\widetilde{B}|}{6\pi^2\Delta_0^2}, \quad (16.24)$$

where $\chi_{MCFL}^\parallel$ is the longitudinal electric susceptibility. Notice that the longitudinal electric susceptibility is much larger than one because in the strong-magnetic-field limit $\tilde{e}\widetilde{B} \gg \Delta_0^2$ [45].

Although a static $\widetilde{U}(1)$ charge cannot be completely Debye screened by the $\widetilde{U}(1)$ neutral Cooper pairs, it can still be partially screened along the magnetic field direction because the medium is highly polarizable on that direction. This is due to the existence of Cooper pairs with opposite rotated charges $\widetilde{Q}$ that behave as electric dipoles with respect to the rotated electromagnetism of the MCFL phase. Moreover, the electric susceptibility depends on the magnetic field. When the magnetic field increases in the strong-field region, the susceptibility becomes smaller, because the coherence length $\xi \sim 1/\Delta_0$ decreases (i.e. Δ_0 increases) with the field at a quicker rate than $\sqrt{\tilde{e}\widetilde{B}}$ [45], and the pair's coherence length ξ plays the role of the dipole length. Hence, with increasing magnetic field the polarization effects weaken in the strong-field region. The tuning of the electric polarization by a magnetic field is what is called in condensed matter physics the magnetoelectric effect. From (16.24), we also see that at strong magnetic fields the medium turns out to be very anisotropic. The fact that the electric permittivity is only modified in the longitudinal direction is due to the confinement of the quarks to the LLL at high enough fields.

16.6 Paramagnetism in Color Superconductivity

Another nontrivial electromagnetic effect in cold-dense QCD is that an applied magnetic field can interact inside the color superconductor with the gluons, which as known, are neutral with respect to the conventional electromagnetism in vacuum.

Thus, we now analyze how the gluons are affected by an applied magnetic field in a CS state and how at sufficiently strong magnetic fields a new phase, that we call the Paramagnetic-CFL (PCFL) phase [52, 53], is created. In the color superconductor some of the gluons acquire rotated electric charges. In the CFL phase the $\tilde{Q}$-charge of the gluons in units of $\tilde{e}$ are

$$
\begin{array}{cccccccc}
G_\mu^1 & G_\mu^2 & G_\mu^3 & G_\mu^+ & G_\mu^- & I_\mu^+ & I_\mu^- & \tilde{G}_\mu^8 \\
0 & 0 & 0 & 1 & -1 & 1 & -1 & 0
\end{array}
\tag{16.25}
$$

The $\tilde{Q}$-charged fields in (16.25) correspond to the combinations $G_\mu^\pm \equiv \frac{1}{\sqrt{2}}[G_\mu^4 \mp iG_\mu^5]$ and $I_\mu^\pm \equiv \frac{1}{\sqrt{2}}[G_\mu^6 \mp iG_\mu^7]$.

To investigate the effect of the applied rotated magnetic field $\tilde{H}$ on the charged gluons, we should start from the effective action of the charged fields $G_\mu^\pm$ (the contribution of the other charges gluons $I_\mu^\pm$ is similar)

$$
\Gamma_{eff} = \int dx \left\{ -\frac{1}{4}(\tilde{f}_{\mu\nu})^2 + G_\mu^- \left[(\tilde{\Pi}_\mu \tilde{\Pi}_\mu)\delta_{\mu\nu} - 2i\tilde{e}\tilde{f}_{\mu\nu} \right.\right.
$$
$$
\left.\left. - (m_D^2 \delta_{\mu 0}\delta_{\nu 0} + m_M^2 \delta_{\mu i}\delta_{\nu i}) - \left(1 - \frac{1}{\varsigma}\tilde{\Pi}_\mu \tilde{\Pi}_\nu \right) \right] G_\nu^+ \right\}. \tag{16.26}
$$

Here, ς is the gauge fixing parameter, $\tilde{\Pi}_\mu = \partial_\mu - i\tilde{e}\tilde{A}_\mu$ is the covariant derivative in the presence of the external rotated field, m_D and m_M are the $G_\mu^\pm$-field Debye and Meissner masses respectively, and the field strength tensor for the rotated electromagnetic field if denoted by $\tilde{f}_{\mu\nu} = \partial_\mu \tilde{A}_\nu - \partial_\nu \tilde{A}_\mu$. The corresponding Debye and Meissner masses in (16.26) are given by [100, 126]

$$
m_D^2 = m_g^2 \frac{21 - 8\ln 2}{18}, \qquad m_M^2 = m_g^2 \frac{21 - 8\ln 2}{54}, \tag{16.27}
$$

with $m_g^2 = g^2(\mu^2/2\pi^2)$. We are neglecting the correction produced by the applied field to the gluon Meissner masses since it will be a second order effect. The effective action (16.26) is characteristic of a spin-1 charged field in a magnetic field (for details see for instance [42, 43]).

Assuming an applied magnetic field along the third spatial direction ($\tilde{f}_{12}^{ext} = \tilde{H}$), we find after diagonalizing the mass matrix of the field components (G_1^+, G_2^+) in (16.26)

$$
\begin{pmatrix} m_M^2 & i\tilde{e}\tilde{H} \\ -i\tilde{e}\tilde{H} & m_M^2 \end{pmatrix} \rightarrow \begin{pmatrix} m_M^2 + \tilde{e}\tilde{H} & 0 \\ 0 & m_M^2 - \tilde{e}\tilde{H} \end{pmatrix}, \tag{16.28}
$$

with corresponding eigenvectors $(G_1^+, G_2^+) \rightarrow (G, iG)$. We see that the lowest mass mode in (16.28) has a sort of "Higgs mass" above the critical field $\tilde{e}\tilde{H}_C = m_M^2$, indicating the setup of an instability for the G-field. This phenomenon is the well known "zero-mode problem" found in the presence of a magnetic field for Yang-Mills fields [110, 131], for the $W_\mu^\pm$ bosons in the electroweak theory [14, 15, 132], and even for higher-spin fields in the context of string theories [48, 49] and it is due to the presence of the gluon anomalous magnetic moment term $2i\tilde{e}\tilde{f}_{\mu\nu}G_\mu^- G_\nu^+$ in (16.26). Thus, to remove the instability it is needed the restructuring of the ground state through the condensate of the field bearing the tachyonic mode (i.e. the G-field).

It is worth to call attention that the gluon condensate under consideration is not the only charged spin-one condensate generated in a theory with a large fermion density. As known [58, 59, 98], a spin-one condensate of W-bosons can be originated at sufficiently high fermion density in the context of the electroweak theory at zero magnetic field. However, the physical implications of the gluon condensate induced by the magnetic field in the CS are fundamentally different from those associated to the homogeneous W-boson condensate of the dense electroweak theory [58, 59, 98]. One of the main physical differences is that the homogeneous W condensate, being electrically charged, so to compensate the excess of charge due to the finite density of electrons [58, 59, 98], breaks the electromagnetic $U(1)$ group producing a conventional superconducting state [60]; while the inhomogeneous gluon condensate in CS is formed with gluons of both charges, so keeping the condensate state neutral.

To find the G-field condensate and the induced magnetic field $\tilde{\mathbf{B}} = \nabla \times \tilde{\mathbf{A}}$, with $\tilde{\mathbf{A}}$ being the total rotated electromagnetic potential in the condensed phase in the presence of the external field $\tilde{H}$, we should start from the Gibbs free energy density $\mathscr{G} = \mathscr{F} - \tilde{H}\tilde{B}$, since it depends on both $\tilde{B}$ and $\tilde{H}$ ($\mathscr{F}$ is the system free energy density). Since specializing $\tilde{H}$ in the third direction the instability develops in the (x, y)-plane, we make the ansatz for the condensed field $\overline{G} = \overline{G}(x, y)$. Starting from (16.26) in the Feynman gauge $\varsigma = 1$, which in terms of the condensed field $\overline{G}$ implies $(\tilde{\Pi}_1 + i\tilde{\Pi}_2)\overline{G} = 0$, we have that the Gibbs free energy in the condensed phase is

$$\mathscr{G}_c = \mathscr{F}_{n0} + \tilde{\Pi}^2\overline{G}^2 - 2(\tilde{e}\tilde{B} - m_M^2)\overline{G}^2 + 2g^2\overline{G}^4 + \frac{1}{2}\tilde{B}^2 - \tilde{H}\tilde{B}, \qquad (16.29)$$

where $\mathscr{F}_{n0}$ is the system free energy in the normal phase ($\overline{G} = 0$) at zero magnetic field.

The minimum equations for the fields $\overline{G}$ and $\tilde{B}$ are respectively obtained from (16.29) as

$$\tilde{\Pi}^2\overline{G} + 2(m_M^2 - \tilde{e}\tilde{B})\overline{G} + 8g^2\overline{G}^2\overline{G} = 0, \qquad (16.30)$$

$$2\tilde{e}\overline{G}^2 - \tilde{B} + \tilde{H} = 0. \qquad (16.31)$$

Identifying $\overline{G}$ with the complex order parameter, (16.30)–(16.31) become analogous to the Ginzburg-Landau equations for a conventional superconductor except by the negative sign in front of the $\widetilde{B}$ field in (16.30) and the positive sign in the first term of the LHS of (16.31) [52]. The fact that those signs turn the opposite of those appearing in conventional superconductivity is due to the different nature of the condensates in both cases. While in conventional superconductivity the Cooper pair is a spin-zero condensate, here we have a condensate formed by spin-one charged particles interacting through their anomalous magnetic moment with the magnetic field (i.e. the term $2i\tilde{e}\tilde{f}_{\mu\nu}G_{\mu}^{-}G_{\nu}^{+}$ in (16.26)).

Notice that because of the different sign in the first term of (16.31), the resultant field $\widetilde{B}$ is stronger than the applied field $\widetilde{H}$, contrary to what occurs in conventional superconductivity. Thus, when a gluon condensate develops, the magnetic field will be antiscreened and the color superconductor will behave as a paramagnet. The antiscreening of a magnetic field has been also found in the context of the electroweak theory for magnetic fields $H \geq M_W^2/e \sim 10^{24}$ G [14]. Just as in the electroweak case, the antiscreening in the color superconductor is a direct consequence of the asymptotic freedom of the underlying theory [14].

Therefore, the magnetic field in the new phase is boosted to a higher value, which depends on the modulus of the $\overline{G}$-condensate. That is why the phase attained at $\widetilde{H} \geq \widetilde{H}_c$ is called paramagnetic CFL (PCFL) [52, 54]. It should be pointed out that at the scale of baryon densities typical of neutron-star cores ($\mu \simeq 400$ MeV, $g(\mu) \simeq 3$) the charged gluons magnetic mass in the CFL phase is $m_M^2 \simeq 16 \times 10^{-3}$ GeV^2. This implies a critical magnetic field of order $\widetilde{H}_c \simeq 0.7 \times 10^{17}$ G. Although it is a significant high value, it is in the expected range for the neutron star interiors with cold-dense quark matter [65, 114]. Let us underline that in our analysis we considered asymptotic densities where quark masses can be neglected. At lower densities where the Meissner masses of the charged gluons become smaller, the field values needed to develop the magnetic instability will be smaller.

To find the structure of the gluon condensate we should solve the non-linear differential equation (16.30). However, to get an analytic solution we can consider the approximation where $\widetilde{H} \approx \widetilde{H}_c = m_M^2$ and consequently $|\overline{G}| \approx 0$. In this approximation, (16.30) can be linearized as

$$\left[\partial_j^2 - \frac{4\pi i}{\widetilde{\Phi}_0}\widetilde{B}x\partial_y - 4\pi^2\frac{\widetilde{B}^2}{\widetilde{\Phi}_0^2}x^2 - \frac{1}{\xi^2}\right]\overline{G} = 0, \quad j = x, y \qquad (16.32)$$

where we fixed the gauge condition $\tilde{A}_2 = \widetilde{B}x_1$, and introduced the notations $\widetilde{\Phi}_0 = 2\pi/\tilde{e}$, and $\xi^2 = \frac{1}{2}(\tilde{e}\widetilde{B} - m_M^2)^{-1}$.

Equation (16.32) is formally similar to the Abrikosov's equation in type-II conventional superconductivity [2, 3, 73], with ξ playing the role of the coherence length and $\widetilde{\Phi}_0$ of the flux quantum per vortex cell. Then, following the Abrikosov's approach, a solution of (16.32) can be found as

$$\overline{G}(x, y) = \frac{1}{\sqrt{2\tilde{e}}\xi}e^{-\frac{x^2}{e\xi^2}}\vartheta_3(u/\tau), \qquad (16.33)$$

with $\vartheta_3(u/\tau)$ being the elliptic theta function with arguments

$$u = -i\pi b\left(\frac{x}{\xi^2} + \frac{y}{b^2}\right), \qquad \tau = -i\pi\frac{b^2}{\xi^2}. \qquad (16.34)$$

In (16.34) the parameter b is the periodic length in the y-direction ($b = \Delta y$). The double periodicity of the elliptic theta function also implies that there is a periodicity in the x-direction given by $\Delta x = \widetilde{\Phi}_0/b\widetilde{H}_c$. Therefore, the magnetic flux through each periodicity cell ($\Delta x \Delta y$) in the vortex lattice is quantized $\widetilde{H}_c \Delta x \Delta y = \widetilde{\Phi}_0$, with $\widetilde{\Phi}_0$ being the flux quantum per unit vortex cell. In this semi-qualitative analysis we considered the Abrikosov's ansatz of a rectangular lattice, but the lattice configuration should be carefully determined from a minimal energy analysis. For the rectangular lattice, we see that the area of the unit cell is $A = \Delta x \Delta y = \widetilde{\Phi}_0/\widetilde{H}_c$, so decreasing with $\widetilde{H}$.

In conclusion, to remove the instability created by an external uniform magnetic field in the z-direction, a periodic arrangement of vortices of charged gluon condensates is generated in the (x, y)-plane. The currents in the (x, y)-plane created by these vortices increase the magnitude of the net magnetic field in the direction of the original field, but since the magnitude of the resultant field varies in the (x, y)-plane, the vortex condensate leads to a net inhomogeneous magnetic field. Therefore, the presence of a supercritical magnetic field leads to the formation of a fluxoid along the z-direction and the appearance of a nontrivial topology on the perpendicular plane. From (16.31) we see that the resultant magnetic field can go from a minimum value $\widetilde{H}$ to a maximum at the core of the fluxoid that depends on the amplitude of the gluon condensate determined by the mismatch between the applied field and the gluon Meissner mass.

Summarizing, at low $\widetilde{H}$ field, the CFL phase behaves as an insulator, and the $\widetilde{H}$ field just penetrates through it without any change of strength. At sufficiently high field $\tilde{e}\widetilde{H} \sim m_M^2$, the condensation of $G^\pm$ is triggered inducing the formation of a lattice of magnetic flux tubes that breaks the translational and remaining rotational symmetries, creating the so-called paramagnetic phase. We stress that contrary to the situation in conventional type-II superconductivity, where the applied field only penetrates through the flux tubes and with a smaller strength, the vortex state in the color superconductor has the peculiarity that outside the flux tube the applied field $\widetilde{H}$ totally penetrates the sample, while inside the tubes the magnetic field becomes larger than $\widetilde{H}$ (this is the origin of the paramagnetic behavior of this CS phase). This effect provides an internal mechanism to increase the magnetic field of a compact star with a CS core.

16.7 Magnetic Phases in CFL Matter

From the discussions in the previous sections it is clear that in the three-flavor color superconductor at very high densities an increasing magnetic field produces a crossover from CFL to MCFL first, and then a phase transition from MCFL to PCFL.

During the crossover, no symmetry breaking occurs, since in principle once a magnetic field is present the symmetry is already that of the MCFL. At very weak magnetic fields, the color superconducting state is practically described by the CFL phase, because the charged mesons corresponding to the Goldstone modes, although massive, are so light that they cannot decay in pairs of quark-antiquark. When the field strength is of the order of the quarks' energy gap Δ_{CFL}, the charged mesons become heavy enough to decouple and the low-energy physics is indeed that of the MCFL phase, where five neutral massless mesons drive the low-energy behavior.

Going from MCFL to PCFL is, on the other hand, a real phase transition [52, 53], as the translational symmetry, as well as the remaining rotational symmetry in the plane perpendicular to the applied magnetic field are broken by the vortex state. This phase transition is driven by fields whose strengths are comparable to the magnetic masses m_M of the charged gluons, so creating a chromomagnetic instability that leads to the formation of a vortex state and the antiscreening of the magnetic field [52, 53].

This magnetic instability is characteristic of systems of charged bosons with higher spins ($s \geq 1$). Taking into account that at zero momentum the energy spectrum in a magnetic field H of a charged boson of spin s, charge e, gyromagnetic ratio g, and mass m is

$$E_n^2 = (2n + 1)eH - geH \cdot s + m^2, \tag{16.35}$$

it is evident that for spin-one particles, for which g = 2, the energy becomes imaginary, i.e. $E^2 < 0$, if the field satisfies $H > H_{cr} = m^2/e$), implying that when the field surpasses the critical value H_{cr}, one of the modes of the charged gauge field becomes tachyonic inducing the vortex formation.

Within a NJL model, for fields comparable to the baryon chemical potential, the ground state is that of the MCFL phase with sizable values of the three condensates Δ_B, Δ_M, and Δ. However, once the gluon effects are taken into account, the PCFL vortex state generated at lower fields is unavoidable and the picture becomes much more complicated due to the inhomogeneities of the gluon condensate and net magnetic field. From a physical point of view, it is natural to expect that in this situation the three fermion gaps will remain, because their physical origin, is still the same. That is, an inhomogeneous magnetic field will also distinguish between pairs of opposite charged quarks and pairs of neutral quarks, and those of opposite charged quarks will still have a magnetic moment contributing to the condensate Δ_M. However, all these condensates should become inhomogeneous in the (x, y)-plane.

16.8 Equation of State of the MCFL Phase

At present, some of the best-known characteristics of stellar objects are their masses and radii. The relation between the mass and the radius of a star is determined by the equation of state (EoS) of the inner phase of the matter in the star. If one can identify some features connecting the star's internal state (nuclear, strange, color superconducting, etc.) to its mass/radius relation, one would have an observational tool to discriminate among the actual realization of different star inner phases in nature. From previous theoretical studies [4, 6, 11, 12, 18, 23, 39, 101, 102, 112, 113, 125, 128] the mass-radius relationship predicted for neutron stars with different quark-matter phases (CS or unpaired) at the core are very similar to those having hadronic phases, at least for the observed mass/radius range. As a consequence, it is very difficult to find a clear observational signature that can distinguish among them. Nevertheless, an important ingredient was ignored in these studies: the magnetic field, which in some compact stars could reach very high values in the inner regions.

As pointed out in [65], a strong magnetic field can create a significant anisotropy in the longitudinal and transverse pressures. One would expect then, that the EoS, and consequently, the mass-radius ratio, become affected by sufficiently strong core fields. Given that we are beginning to obtain real observational constraints on the EoS of neutron stars [5], it is important to investigate the EoS in the presence of a magnetic field for different inner star phases to be able to discard those that do not agree with observations.

In order to understand the relevance of the magnetic field to tell apart neutron stars from stars with paired quark matter, it is convenient to recall that when the pressure exerted by the central matter density of neutron stars (which is about 200–600 MeV/fm^3) is contrasted with that exerted by an electromagnetic field, the field strength needed for these two contributions to be of comparable order results of order $\sim 10^{18}$ G [25]. It is worth to notice that even these very strong fields are not enough to produce quantum effects like the Landau quantization of the protons, because these effects only show up when the particles' cyclotron energy ehB/mc becomes comparable to its rest energy mc^2, which for protons means a field $\sim 10^{20}$ G.

However, for stars with paired quark matter, the situation is rather different. Naively, one might think that comparable matter and field pressures in this case would occur only at much larger fields, since the quark matter can only exist at even larger densities to ensure deconfinement. In reality, though, the situation is more subtle. As argued in [6], the leading term in the matter pressure coming from the contribution of the particles in the Fermi sea, $\sim \mu^4$, could be (almost) canceled out by the negative pressure of the bag constant and in such a case, the next-to-leading term would play a more relevant role than initially expected. Consequently, the magnetic pressure might only need to be of the order of that produced by the particles close to the Fermi surface, which becomes the next-to-leading contribution, $\sim \mu^2 \Delta^2$, with Δ the superconducting gap and μ the baryonic chemical potential. For typical values of these parameters in paired quark matter one obtains a field strength $\sim 10^{18}$ G. Moreover, the magnetic field can affect the pressure in a less obvious way too, since as shown in [62–64], it modifies the structure and magnitude of the superconductor's gap, an effect that, as found in [52, 53], starts to become relevant already at fields of order 10^{16} G and leads to de Haas van-Alphen oscillations of the gap magnitude [71, 111]. It is therefore quite plausible that the effects of moderately strong magnetic fields in the EoS of compact stars with color superconducting matter will be more noticeable than in stars made up only of nucleons, where quantum effects starts to be significant for field four orders of magnitude larger. This is why an evaluation of the EoS in magnetized quark phases is necessary and relevant.

In [114], a self-consistent analysis of the EoS of MCFL matter, was performed taking into consideration the solution of the gap equations and the anisotropy of the pressures in a magnetic field. In that study a uniform and constant magnetic field was assumed. The reliability of this assumption for neutron stars, where the magnetic field strength is expected to vary from the core to the surface in several orders, is based on the fact that the scale of the field variation in the stellar medium is much larger than the microscopic magnetic scale for both weak and strong magnetic fields [25]. Hence, when investigating the field effects in the EoS, it is consistent to take a magnetic field that is locally constant and uniform. This is the reason why such an approximation has been systematically used in all the previous works on magnetized nuclear [1, 20, 25, 27, 28, 33, 34, 72, 79, 88, 115, 119, 123, 135, 139, 142] and quark matter [31, 32, 74, 104, 105, 120].

16.8.1 Covariant Structure of the Energy-Momentum Tensor in a Magnetized System

In the reference frame comoving with the many-particle system, the system normal stresses (pressures) can be obtained from the diagonal spatial components of the average energy-momentum tensor $\langle \tau^{ii} \rangle$; the system energy, from its zeroth diagonal component $\langle \tau^{00} \rangle$; and the shear stresses (which are absent for the case of a uniform magnetic field) from the off-diagonal spatial components $\langle \tau^{ij} \rangle$ [94]. Then, to find the energy density and pressures of the dense magnetized system we need to calculate the quantum-statistical averages of the corresponding components of the energy-momentum tensor of the fermion system in the presence of a magnetic field.

These calculations were carried out long time ago in Ref. [26], using a QFT second-quantization approach. There, a quantum-mechanical average of the energy-momentum tensor in the eigen-states of the Dirac equation in the presence of the uniform magnetic field was first performed to get the corresponding quantum operator in the occupation-number space. The macroscopic stress-energy tensor was then found by averaging its quantum operator in the statistical ensemble using the many-particle density matrix. Similar calculations were performed in Ref. [65], but using a functional-method approach that makes it easier to recognize the thermodynamical quantities entering in the final results. An advantage of the procedure followed in [65], as compared with that of [26], is that it does not assume that the fermion fields entering in the definitions of the energy and pressures satisfy the classical equation of motions (i.e. the Dirac equations for ψ and $\overline{\psi}$), but the functional integrals integrate in all field configurations. Hence, the terms depending on the Lagrangian density $\mathscr{L}_\psi$ in $\tau_{\mu\nu}$ were kept, while in Ref. [26] the condition $\mathscr{L}_\psi = 0$ was considered as a constraint.

Let us then consider a hot and dense system of fermions in a constant and uniform magnetic field B. At this point it is convenient to introduce the covariant decomposition for the energy momentum tensor of the whole system containing the matter and field contributions. In order to accomplish this goal, we define the system thermodynamic potential as the sum of the matter, Ω_f, and field, $B^2/2$, contributions

$$\Omega = \Omega_f + \frac{B^2}{2}. \tag{16.36}$$

Taking into account the symmetries of the magnetized dense system, we can write the statistical average of the energy-momentum tensor as a combination of all the available independent structures

$$\frac{1}{\beta V}\langle\tilde{\tau}^{\mu\nu}\rangle = \Omega\eta^{\mu\nu} + (\mu N + TS)u^\mu u^\nu + BM\eta_\perp^{\mu\nu}, \tag{16.37}$$

where $N = -(\partial\Omega/\partial\mu)$ is the particle number density, $S = -(\partial\Omega/\partial T)$ is the system entropy, $M = -(\partial\Omega/\partial B)$ is the system magnetization and $\eta_\perp^{\mu\nu} = \widehat{F}^{\mu\rho}\widehat{F}_\rho^\nu$ (where $\widehat{F}^{\mu\rho} = F^{\mu\rho}/B$ denotes the normalized electromagnetic strength tensor).

To understand the origin of the covariant decomposition (16.37), notice that as a consequence of the breaking of the rotational symmetry $O(3)$ produced by the external magnetic field, the Minkowskian metric splits in transverse $\eta_\perp^{\mu\nu}$ and longitudinal $\eta_\parallel^{\mu\nu} = \eta^{\mu\nu} - \widehat{F}^{\mu\rho}\widehat{F}_\rho^\nu$ structures. Considering the quantum field limit with no magnetic field, i.e. when $T = \mu = B = 0$, the only term different from zero is the first one in the RHS of (16.37). In that case the system has Lorentz symmetry and the energy density, ε, and pressure, p, are given by $\varepsilon = -p = \Omega_f$. If temperature and/or density are switched on, then the Lorentz symmetry is broken specializing a particular reference frame comoving with the medium center of mass and having four velocity $u_\mu = (1, \vec{0})$. This is reflected in the second term of the RHS of (16.37). In this case, at $T = 0$ for instance, $\varepsilon = \Omega_f + \mu N$ and $p = -\Omega_f$. Finally, when there is an external uniform magnetic field acting on the system, the additional symmetry breaking $O(3) \rightarrow O(2)$ takes place, and $\langle\tilde{\tau}^{\mu\nu}\rangle$ get an anisotropy reflected in the appearance of the transverse metric structure $\eta_\perp^{\mu\nu}$ in (16.37). At $T = 0$ we then have

$$\varepsilon = \Omega_f - \mu\frac{\partial\Omega_f}{\partial\mu} + \frac{B^2}{2}, \tag{16.38}$$

$$p^\parallel = -\Omega_f - \frac{B^2}{2}, \qquad p^\perp = -\Omega_f + H\frac{\partial\Omega_f}{\partial B} + \frac{B^2}{2}. \tag{16.39}$$

See Ref. [65] for detailed derivations of the formulas for the pressures and energy density in a magnetic field.

16.8.2 MCFL Thermodynamic Potential

Let us turn our attention now to densities large enough for the fermion system to be in the MCFL phase. Our ultimate goal is to find the EoS of this superconducting phase. To find the density and pressure of this phase, we first need, as seen from (16.38)–(16.39), to obtain the contribution of the quarks to the thermodynamic potential. We can express the MCFL thermodynamic potential as the sum of the contributions coming from charged (Ω_C) and neutral (Ω_N) quarks [114]

$$\Omega_{MCFL} = \Omega_C + \Omega_N \tag{16.40}$$

with

$$\Omega_C = -\frac{\tilde{e}\tilde{B}}{4\pi^2} \sum_{n=0}^{\infty} \left(1 - \frac{\delta_{n0}}{2}\right) \int_0^{\infty} dp_3 e^{-(p_3^2 + 2\tilde{e}\tilde{B}n)/\Lambda^2} \left[8|\varepsilon^{(c)}| + 8|\bar{\varepsilon}^{(c)}|\right], \tag{16.41}$$

$$\Omega_N = -\frac{1}{4\pi^2} \int_0^{\infty} dp\, p^2 e^{-p^2/\Lambda^2} \left[6|\varepsilon^{(0)}| + 6|\bar{\varepsilon}^{(0)}|\right]$$

$$-\frac{1}{4\pi^2} \int_0^{\infty} dp\, p^2 e^{-p^2/\Lambda^2} \sum_{j=1}^{2} \left[2|\varepsilon_j^{(0)}| + 2|\bar{\varepsilon}_j^{(0)}|\right] + \frac{\Delta^2}{G} + \frac{2\Delta_B^2}{G}, \tag{16.42}$$

and

$$\varepsilon^{(c)} = \pm\sqrt{\left(\sqrt{p_3^2 + 2\tilde{e}\tilde{B}n} - \mu\right)^2 + \Delta_B^2},$$

$$\bar{\varepsilon}^{(c)} = \pm\sqrt{\left(\sqrt{p_3^2 + 2\tilde{e}\tilde{B}n} + \mu\right)^2 + \Delta_B^2}, \tag{16.43}$$

$$\varepsilon^{(0)} = \pm\sqrt{(p-\mu)^2 + \Delta^2}, \qquad \bar{\varepsilon}^{(0)} = \pm\sqrt{(p+\mu)^2 + \Delta^2},$$

$$\varepsilon_1^{(0)} = \pm\sqrt{(p-\mu)^2 + \Delta_a^2}, \qquad \bar{\varepsilon}_1^{(0)} = \pm\sqrt{(p+\mu)^2 + \Delta_a^2}, \tag{16.44}$$

$$\varepsilon_2^{(0)} = \pm\sqrt{(p-\mu)^2 + \Delta_b^2}, \qquad \bar{\varepsilon}_2^{(0)} = \pm\sqrt{(p+\mu)^2 + \Delta_b^2},$$

being the dispersion relations of the charged (c) and neutral (0) quarks. In the above we used the notation

$$\Delta_{a/b}^2 = \frac{1}{4}\left(\Delta \pm \sqrt{\Delta^2 + 8\Delta_B^2}\right)^2. \tag{16.45}$$

The MCFL gaps Δ and Δ_B were introduced in Sect. 16.4. In the integrals (16.41) and (16.42), Λ-dependent smooth cutoffs for the NJL model are used.

The effects of confinement can be incorporated by adding a bag constant $\mathscr{B}$ to Ω_{MCFL}. Besides the bag constant and the quark contributions, the thermodynamic

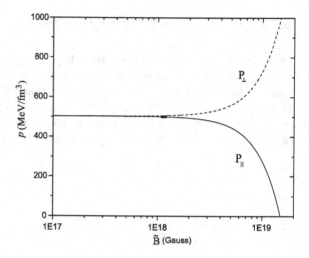

Fig. 16.2 Parallel and perpendicular pressures as a function of the magnetic field intensity for $\mu = 500$ MeV and bag constant $\mathscr{B} = 58$ MeV/fm^3

potential of the system also includes the pure Maxwell contribution, $\widetilde{B}^2/2$ [65]. Hence, the thermodynamic potential of the MCFL phase is given by

$$\Omega_B = \Omega_{MCFL} + \mathscr{B} + \frac{\widetilde{B}^2}{2}. \tag{16.46}$$

The gaps Δ, and Δ_B have to be found from their respective gap equations

$$\frac{\partial \Omega_{MCFL}}{\partial \Delta} = 0, \qquad \frac{\partial \Omega_{MCFL}}{\partial \Delta_B} = 0. \tag{16.47}$$

16.8.3 EoS in a Magnetic Field

The pressure and energy density of the MCFL phase are given by

$$\varepsilon_{MCFL} = \Omega_B - \mu \frac{\partial \Omega_B}{\partial \mu}, \tag{16.48}$$

$$p^{\parallel}_{MCFL} = -\Omega_B, \qquad p^{\perp}_{MCFL} = -\Omega_B + \widetilde{B} \frac{\partial \Omega_B}{\partial \widetilde{B}}. \tag{16.49}$$

Note the splitting between parallel $p^{\parallel}_{MCFL}$ (i.e. along the field) and transverse $p^{\perp}_{MCFL}$ (i.e. perpendicular to the field) pressures due to the magnetic field.

The magnetic field dependencies of the parallel and transverse pressures in (16.49) were studied in Ref. [114], and are plotted in Fig. 16.2. Similarly to what occurs in the case of a magnetized uncoupled fermion system at finite density [65], the transverse pressure in the MCFL phase increases with the field, while the parallel pressure decreases and reaches a zero value at field strength of order $\geq 10^{19}$ G

Fig. 16.3 Splitting of the
parallel and perpendicular
pressures, normalized to the
zero value pressure
$(p(\widetilde{B} = 0))$, as a function of
the magnetic field intensity
for $\mu = 500$ MeV and
$\mathscr{B} = 58$ MeV/fm³

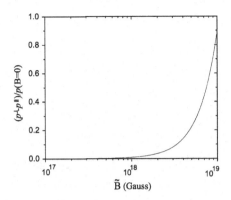

for the density under consideration ($\mu = 500$ MeV). We see from Fig. 16.2 that Ω_H and $\partial \Omega_B / \partial \widetilde{B}$ do not exhibit the Hass-van Alphen oscillations as happens with other physical quantities in the presence of a magnetic field [40, 41, 45, 71, 111]. This is due to the high contribution of the pure Maxwell term in Ω_B and $\partial \Omega_B / \partial \widetilde{B}$, which makes the oscillations of the matter part negligible in comparison.

The splitting between parallel and perpendicular pressures, shown in the vertical axis of Fig. 16.3, grows with the magnetic field strength. Comparing the found splitting with the pressure of the (isotropic) CFL phase, we can address how important this effect is for the EoS. Notice that for 3×10^{18} G the pressures splitting is ~10 % of their isotropic value at zero field (i.e. the one corresponding to the CFL phase).

In the graphical representation of the EoS in Fig. 16.4 the highly anisotropic behavior of the magnetized medium is explicitly shown. While the magnetic-field effect is significant for the $\varepsilon-p^{\parallel}$ relationship at $\widetilde{B} \sim 10^{18}$ G, with a shift in the energy density with respect to the zero-field value of ~200 MeV/fm³ for the same pressure, the field effect in the $\varepsilon-p^{\perp}$ relationship is smaller for the same range of field values.

The most important application of the EoS is to construct stellar models for compact stars composed of quark matter. This goal can be archived by using the relativistic equations of stellar structure, that is, the well known Tolman-Oppenheimer-Volkoff (TOV) and mass continuity equations.

$$\frac{dm}{dr} = 4\pi r^2 \varepsilon, \tag{16.50}$$

$$\frac{dP}{dr} = -\frac{\varepsilon m}{r^2}\left(1 + \frac{P}{\varepsilon}\right)\left(1 + \frac{4\pi r^3 P}{m}\right)\left(1 - \frac{2m}{r}\right)^{-1} \tag{16.51}$$

written in natural units, $c = G = 1$. However, it is clear that this set of differential equations apply only to isotropic EoS for systems with spherical symmetry.

If the magnetic field in the MCFL phase is high enough for the anisotropy in the pressure to be significant (i.e., expressed in terms of the pressure splitting to be $(\Delta p / p_{CFL}) \sim (\widetilde{B}^2 / \mu^2 \Delta^2) \sim \mathscr{O}(1)$) the spherically symmetric TOV equations become inappropriate, because the deviations lead to significant differences with respect of realistic axi-symmetric models, yet to be constructed [114].

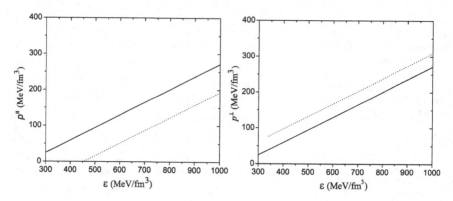

Fig. 16.4 Equation of state for MCFL matter considering parallel (*right panel*) and perpendicular (*left panel*) pressures for different values of $\widetilde{B}$: zero field (*solid line*), 10^{17} G (*dashed line*) and 5×10^{18} G (*dotted line*). Note that the low value of $\widetilde{B} = 10^{17}$ G is not distinguishable in the plots, being merged with the zero-field curve. The value of the bag constant was fixed to $\mathscr{B} = 58$ MeV/fm^3

16.9 Astrophysical Implications

As we have stressed, an important characteristic of neutron stars is that they typically possess very strong magnetic fields. Unveiling the interconnection between the star's magnetic field and the dense phases is important to understand the interplay between QCD and neutron star phenomenology. As discussed above, in recent years much interesting work has been done on the properties of the different nuclear phases that can be reached in dense astrophysical objects in the presence of strong magnetic fields. An important new step in this context would be to consider the consequences for the star's phenomenology of the possible new phases. Although at this point we do not know yet the quantitative details of the potential consequences, no doubt exploring them will shed new light on the important question of how we can infer the presence of a color-superconducting core from astronomical observations, and whether such observations can distinguish among different color superconducting phases. In what follow we discuss some of these related tasks.

16.9.1 Low-Energy Physics

The main challenge in determining the phases of matter inside a neutron star is to provide observables signatures of the presence of those phases. Currently there are many proposals to connect observations to the inner phases of the star. We want to discuss in general terms here, those connected to transport properties as conductivities, viscosities, etc. As known, these transport properties are determined by the low-energy spectrum of the phase, that is, by the lowest-energy modes as Goldstone bosons and gapless quark excitations, which as already shown, can be affected by the presence of a sufficiently high magnetic field. Let's briefly mention some examples of transport properties and how they could affect the star's observables.

Viscosity The viscosity of the interior of a star can be probed by observing how a rapidly spinning neutron star slows down. If the star slows down very quickly this indicates that it is unstable with respect to bulk flows (r-modes) that transfer the star angular momentum into gravitational radiation. But this can only occur if damping is sufficiently small. Based upon these arguments the possibility of pure CFL-quark-matter pulsars has been ruled out [59] since in the CFL phase viscous damping is negligible [103]. If the intermediate-density CS phase happens to have a large viscosity, it will not be restricted by r-modes arguments tough.

As we showed in [61], in a three-flavor theory the spectrum of the NG bosons of the CS is affected by the restructuring of the gap produced by the magnetic field. As a consequence, instead of the 9 Goldstone bosons that exist in the CFL phase, in the MCFL only 5 remain. In contrast to the CFL case where several Goldstone bosons are Q-charged, in the MCFL all are Q-neutral. Therefore, the scattering rate of the low-energy bosons should be different in the magnetic background, and this will be reflected in turn in the transport properties of the star. By investigating transport properties as thermal conductivity and viscosity in the MCFL phase (or in the extension of the MCFL phase when one takes into account the strange quark mass and neutrality effects) one could look for new observational effects that will allow us to distinguish between nuclear-core stars and quark-core stars.

Thermal Conductivity Given that neutrino emission rates and heat capacity generally rise with density, neutron star cooling is likely preferentially sensitive to the properties of matter in the core. Investigations [129, 130] on the impact of the thermal conductivity of dense quark matter on the star cooling process indicates that any CFL quark matter within the star will cool by conduction, not by neutrino emission. In this direction, to investigate how the magnetic field can affect the medium thermal conductivity is of interest.

Neutrino Emission and Detection Neutrino emission is the dominant heat loss mechanism of the stars in their first million years. In [84, 122] the neutrino emission from Nambu-Goldstone modes of the CFL phase has been investigated. These studies showed that the scattering of massless Goldstone modes, associated with the breaking of the baryon $U(1)_B$ symmetry, is not exponentially suppressed, and so, these modes dominate neutrino emission at late times. On the other hand, the time-of-arrival distribution of supernova neutrinos could be connected to possible phase transitions to and in quark matter [10, 29], but a detailed analysis of this suggestion requires a better understanding of both supernova itself and of the properties of quark matter at MeV temperatures.

If the CS phase at intermediate densities results to be a variety of gapless phase, the gapless modes could play a significant role in the transport properties of the star. A recent study [10] has shown that even a relatively small region of gCFL matter in a star would dominate the heat capacity and the heat loss by neutrino emission. However, we already know that the gCFL is not stable, so it is unlikely that this phase will occur within the star.

However, none of these studies have taken into account the presence of the in-medium magnetic field that penetrates the star's superconducting core. Nevertheless, a total understanding of the transport mechanism of a compact star with a quark core will not be complete without considering the modification of the color-superconducting gap by the strong in-medium magnetic field, as well as by the modification of the remaining Goldstone modes.

This effect can be relevant for the low energy physics of a color superconducting star's core and hence for its transport properties. In particular, the cooling of a compact star is determined by the particles with the lowest energy; so a star with a core of quark matter and sufficiently large magnetic field can have a distinctive cooling process. This study is a pending task that is worth to be undertaken.

16.9.2 Boosting Stellar Magnetic Fields via an Internal Mechanism

The standard model [136, 137] to explain the origin of the strong magnetic fields observed in the surface of magnetars is based on a magnetohydrodynamic dynamo mechanism that amplifies a seed magnetic field due to a rapidly rotating protoneutron star. This model requires a spin period <3 ms. Nevertheless, this mechanism cannot explain all the observed features of the supernova remnants surrounding these objects [138, 141].

As has been found recently, in color superconductors magnetic fields can be reinforced [52] and even generated [55]. It is natural to expect that if a color superconducting state exists in the core of neutron stars, it may have implications for the magnetic properties of such compact objects. At the moderately high densities that exist in the cores of neutron stars the most probable color superconducting state is not the CFL, but either an inhomogeneous phase or perhaps a strongly coupled 2SC phase. In the 2SC phase, the Meissner mass of the charged gluons decreases with decreasing density to values which are close to zero. For such small charged gluon masses, a magnetic field does not need to be too large to induce a vortex state. Fields with values $\check{H} > m_M^2$ will trigger the spontaneous generation of vortices of charged gluons, which in turn will enhance the existing magnetic field. Hence, CS could contribute to boosting the large magnetic fields observed in some stellar objects as magnetars, without having to rely only on the quick spinning assumed in the standard model of magnetars that, as known, are associated with some of the conflict of this model with the observations. These induced gluon vortices could produce a magnetic field of the order of the Meissner mass scale, which implies a magnitude in the range $\sim 10^{16}$–10^{17} G. Hence, the possibility of generating a magnetic field of such a large magnitude in the core of a compact star without relying on a magnetohydrodynamics effect, can be an interesting alternative to address the main criticism [138, 141] to the observational conundrum of the standard magnetar's paradigm [83, 92, 109, 136, 137]. On the other hand, to have a mechanism that associates the existence of high magnetic fields with CS at moderate densities can serve to single out the magnetars as the most probable astronomical objects for the realization of this high-dense state of matter.

16.9.3 Stability of Magnetized Quark Stars

It is now our goal to analyze the conditions for MCFL matter to become absolutely stable. This is done by comparing the energy density at zero pressure condition with that of the iron nucleus (~930 MeV). Depending on whether the energy density of the MCFL phase is higher or smaller than this value, the content of a magnetized strange quark could be or not made of MCFL matter.

The stability criterion for MCFL matter can be derived in a simple way. Following Farhi and Jaffe's [44] approach, we can determine the maximum value of the bag constant that satisfies the stability condition at zero pressure for each magnetic field value. We call reader's attention that in all these derivations we work within a self-consistent approach, in which the solutions of the gap equations are substituted in the pressures and energies of each phase before imposing the conditions of equilibrium and stability.

After imposing the zero pressure condition in the EoS for the MCFL phase, both the parallel and perpendicular pressures in the MCFL EoS need to vanish simultaneously. Therefore, the two equilibrium conditions become

$$p^{\parallel}_{MCFL} = -\Omega_{MCFL} - \mathscr{B} - \frac{\tilde{B}^2}{2} = 0, \tag{16.52}$$

$$p^{\perp}_{MCFL} = \tilde{B}\frac{\partial \Omega_{MCFL}}{\partial \tilde{B}} + \tilde{B}\frac{\partial \mathscr{B}}{\partial \tilde{B}} + \tilde{B}^2 = 0. \tag{16.53}$$

Where we are assuming that the bag constant depends on the magnetic field. It is not unnatural to expect that the applied magnetic field could modify the QCD vacuum, hence producing a field-dependent bag constant. One can readily verify that (16.52)–(16.53) are equivalent to require $p^{\parallel}_{MCFL} = 0$ and $\partial p^{\parallel}_{MCFL}/\partial \tilde{B} = 0$ at the equilibrium point.

Equation (16.53) can be rewritten as

$$\tilde{H} = M - \frac{\partial \mathscr{B}}{\partial \tilde{B}} \tag{16.54}$$

where $M = -\partial \Omega_{MCFL}/\partial \tilde{B}$ is the system magnetization. If we were to consider that the vacuum energy $\mathscr{B}$ does not depend on the magnetic field, we would need

$$M = \tilde{B}, \tag{16.55}$$

to ensure the equilibrium of the self-bound matter, a condition difficult to satisfy since it would imply that the medium response to the applied magnetic field (i.e. the medium magnetization M) is of the order of the applied field that produces it. Only if the MCFL matter were a ferromagnet this would be viable, but as known, the MCFL matter is on the contrary, an insulator. The other possibility for the equilibrium conditions (16.52) and (16.53) to hold simultaneously is to have a field-dependent bag constant capable to yield nonzero vacuum magnetization $M_0 = -\frac{\partial \mathscr{B}}{\partial \tilde{B}} \simeq \tilde{B}$.

Fig. 16.5 Stability window
for MCFL matter in the plane
$\tilde{B}$ vs. $\mathscr{B}$. The *curve* shown
corresponds to the borderline
value $\varepsilon/A = 930$ MeV

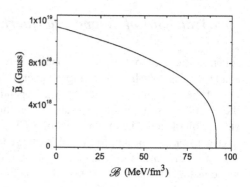

The following comment is in order. The fact that the bag constant needs to be field-dependent for self-bound stars in a strong magnetic field is a direct consequence of the lack of a compensating effect for the internal pressure produced by the magnetic field other than that applied by the vacuum (an exception could be of course if the paired quark matter would exhibit ferromagnetism). For gravitationally bound stars, on the other hand, the situation is different, since the own gravitational field can supply the pressure to compensate the one due to the field. For such systems, keeping $\mathscr{B}$ constant in the EoS is in principle possible. Under this assumption we considered a fixed $\mathscr{B}$-value in Fig. 16.4.

Taking into account that the matter energy density ε'_{MCFL} (i.e. the energy density that does not include the pure Maxwell contribution) divided by the baryon number is given by

$$\frac{\varepsilon'_{MCFL}}{n_A} = \frac{\Omega_{MCFL} + \mathscr{B}}{n_A} - \frac{\mu}{n_A}\frac{\partial \Omega_{MCFL}}{\partial \mu}. \tag{16.56}$$

We can write it under the zero parallel pressure condition ($\Omega_{MCFL} + \mathscr{B} = -\tilde{B}/2$) as

$$\left.\frac{\varepsilon'_{MCFL}}{n_A}\right|_{\mu_B} = 2\frac{\tilde{B}^2}{2n_A} + \frac{\mu_B}{n_A}N, \tag{16.57}$$

and the absolute stability condition becomes

$$\left.\frac{\varepsilon'_{MCFL}}{n_A}\right|_{\mu_B} = 2\frac{\tilde{B}^2}{2n_A} + \frac{\mu_B}{n_A}N \leq \varepsilon_0\left(Fe^{56}\right). \tag{16.58}$$

Then, finding μ_B as a function of $\tilde{B}$ from (16.58), and substituting it back in (16.52), we can numerically solve

$$\mathscr{B}(\tilde{B}) = -\Omega_{MCFL}(\mu_B, \tilde{B}) - \tilde{B}^2/2, \tag{16.59}$$

to determine the stability window in the plane $\tilde{B}$ versus $\mathscr{B}$ for the MCFL matter to be absolute stable (Fig. 16.5). The inner region, which corresponds to smaller bag constants for each given $\tilde{B}$, is the absolutely stable region.

Note that, contrary to Farhi and Jaffe [44], we did not impose a *minimum* value for the bag constant because we have no clear indication from experiments of the possible behavior of this parameter when a magnetic field is applied to a system.

In summary, our results indicate that a condition for the MCFL matter to be absolutely stable is a field-dependent bag constant that can give rise to a large vacuum magnetization at moderately strong fields (see Fig. 16.5). Under these circumstances, increasing the magnetic field tends to destabilize the self-bound MCFL matter. This result differs from that found in [116] where it was used a CFL model at $eB \neq 0$ with only one gap that was fixed by hand.

Acknowledgements This work has been supported in part by DOE Nuclear Theory grant DE-SC0002179.

References

1. A.M. Abrahams, S.L. Shapiro, Astrophys. J. **374**, 652 (1991)
2. A.A. Abrikosov, Sov. Phys. JETP **5**, 1174 (1957)
3. A.A. Abrikosov, L.P. Gorkov, Zh. Eksp. Teor. Fiz. **39**, 1781 (1960). [Sov. Phys. JETP **12**, 1961 (1243)]
4. B.K. Agrawal, S.K. Dhiman, Phys. Rev. D **79**, 103006 (2009)
5. M. Alford, Physics **3**, 44 (2010)
6. M. Alford, S. Reddy, Phys. Rev. D **67**, 074024 (2003)
7. M. Alford, K. Rajagopal, F. Wilczek, Phys. Lett. B **422**, 247 (1998)
8. M. Alford, K. Rajagopal, F. Wilczek, Nucl. Phys. B **537**, 443 (1999)
9. M. Alford, J. Berges, K. Rajagopal, Nucl. Phys. B **571**, 269 (2000)
10. M. Alford, P. Jotwani, C. Kouvaris, J. Kundu, K. Rajagopla, Phys. Rev. D **71**, 114011 (2005)
11. M. Alford, M. Brady, M.W. Paris, S. Reddy, Astrophys. J. **629**, 969 (2005)
12. M. Alford et al., Nature (London) **445**, E7 (2007)
13. M. Alford, A. Schmitt, K. Rajagopal, T. Schäfer, Rev. Mod. Phys. **80**, 1455 (2008) (and references therein)
14. J. Ambjorn, P. Olesen, Nucl. Phys. B **315**, 606 (1989)
15. J. Ambjorn, P. Olesen, Phys. Lett. B **218**, 67 (1989)
16. P.W. Anderson, Phys. Rev. Lett. **3**, 325 (1959)
17. D. Bailin, A. Love, Phys. Rep. **107**, 325 (1984)
18. M. Baldo et al., Phys. Lett. B **562**, 153 (2003)
19. V. Baluni, Phys. Rev. D **17**, 2092 (1978)
20. D. Bandyopadhyay et al., Phys. Rev. D **58**, 121301 (1998)
21. B.C. Barrois, Nucl. Phys. B **129**, 390 (1977)
22. G. Baym, S. Chin, Phys. Lett. B **62**, 241 (1976)
23. D. Blaschke, S. Fredriksson, H. Grigorian, A.M. Oztas, Nucl. Phys. A **736**, 203 (2004)
24. M. Bocquet, S. Bonazzola, E. Gourgoulhon, J. Novak, Astron. Astrophys. **301**, 757 (1995)
25. A. Broderick, M. Prakash, J.M. Lattimer, Astrophys. J. **537**, 351 (2000)
26. V. Canuto, H.-Y. Chiu, Phys. Rev. **173**, 1210 (1968)
27. V. Canuto, J. Ventura, Fundam. Cosm. Phys. **2**, 203 (1977)
28. C.Y. Cardall, M. Prakash, J.M. Lattimer, Astrophys. J. **554**, 332 (2001)
29. G.W. Carter, S. Reddy, Phys. Rev. D **62**, 103002 (2000)
30. R. Cassalbuoni, Z. Duan, F. Sannino, Phys. Rev. D **63**, 114026 (2001)
31. M. Chaichian et al., Phys. Rev. Lett. **84**, 5261 (2000)
32. S. Chakrabarty, Phys. Rev. D **54**, 1306 (1996)
33. S. Chakrabarty, D. Bandyopadhyay, S. Pal, Phys. Rev. Lett. **78**, 2898 (1997)

34. W. Chen, P.-Q. Zhang, L.-G. Liu, Mod. Phys. Lett. A **22**, 623 (2007)
35. S. Coleman, D. Gross, Phys. Rev. Lett. **31**, 851 (1973)
36. J.C. Collins, M.J. Perry, Phys. Rev. Lett. **30**, 1353 (1975)
37. P. Curie, J. Phys. **3**, 393 (1894)
38. L. Dong, S.L. Shapiro, APJ **383**, 745 (1991)
39. A. Drago, A. Lavagno, G. Pagliara, Phys. Rev. D **69**, 057505 (2004)
40. D. Ebert, K.G. Klimenko, Nucl. Phys. A **728**, 203 (2003)
41. D. Ebert, K.G. Klimenko, M.A. Vdovichenko, A.S. Vshivtsev, Phys. Rev. D **61**, 025005 (2000)
42. E. Elizalde, E.J. Ferrer, V. de la Incera, Ann. Phys. **295**, 33 (2002)
43. E. Elizalde, E.J. Ferrer, V. de la Incera, Phys. Rev. D **70**, 043012 (2004)
44. E. Farhi, R.L. Jaffe, Phys. Rev. D **30**, 2379 (1984)
45. B. Feng, E.J. Ferrer, V. de la Incera, Nucl. Phys. B **853**, 213 (2011)
46. B. Feng, E.J. Ferrer, V. de la Incera, Phys. Lett. B **706**, 232 (2011)
47. B. Feng, E.J. Ferrer, V. de la Incera, Phys. Rev. D **85**, 103529 (2012)
48. S. Ferrara, M. Porrati, Mod. Phys. Lett. A **8**, 2497 (1993)
49. E.J. Ferrer, V. de la Incera, Int. J. Mod. Phys. A **11**, 3875 (1996)
50. E.J. Ferrer, V. de la Incera, Phys. Rev. D **58**, 065008 (1998)
51. E.J. Ferrer, V. de la Incera, Phys. Lett. B **481**, 287 (2000)
52. E.J. Ferrer, V. de la Incera, Phys. Rev. Lett. **97**, 122301 (2006)
53. E.J. Ferrer, V. de la Incera, J. Phys. A, Math. Theor. **40**, 6913 (2007)
54. E.J. Ferrer, V. de la Incera, Phys. Rev. D **76**, 045011 (2007)
55. E.J. Ferrer, V. de la Incera, Phys. Rev. D **76**, 114012 (2007)
56. E.J. Ferrer, V. de la Incera, Phys. Rev. Lett. **102**, 050402 (2009)
57. E.J. Ferrer, V. de la Incera, Nucl. Phys. B **824**, 217 (2010)
58. E.J. Ferrer, V. de la Incera, A.E. Shabad, Phys. Lett. B **185**, 407 (1987)
59. E.J. Ferrer, V. de la Incera, A.E. Shabad, Nucl. Phys. B **309**, 120 (1988)
60. E.J. Ferrer, V. de la Incera, A.E. Shabad, Ann. Phys. **201**, 51 (1990)
61. E.J. Ferrer, V. de la Incera, C. Manuel, Phys. Rev. Lett. **95**, 152002 (2005)
62. E.J. Ferrer, V. de la Incera, C. Manuel, Nucl. Phys. B **747**, 88 (2006)
63. E.J. Ferrer, V. de la Incera, C. Manuel, PoS **JHW2005**, 022 (2006)
64. E.J. Ferrer, V. de la Incera, C. Manuel, J. Phys. A **39**, 6349 (2006)
65. E.J. Ferrer, V. de la Incera, J.P. Keith, I. Portillo, P.L. Springsteen, Phys. Rev. C **82**, 065802 (2010)
66. M. Fiebig, J. Phys. D, Appl. Phys. **38**, R123 (2005)
67. E.S. Fradkin, Quantum field theory and hydrodynamics, in *Proceedings of P. N. Lebedev Physical Institute*, vol. 29 (Nauka, Moscow, 1965), p. 7. (Engl. transl., Consultant Bureau, New York, 1967)
68. S. Frautschi, in *Proceedings of the Workshop on Hadronic Matter at Extreme Energy Density*, ed. by N. Cabibbo, Erice, Italy (1978)
69. B. Freedman, L. McLerran, Phys. Rev. D **16**, 1166 (1977)
70. B. Freedman, L. McLerran, Phys. Rev. D **17**, 1109 (1978)
71. K. Fukushima, H.J. Warringa, Phys. Rev. Lett. **100**, 03200 (2008)
72. I. Fushiki, E.H. Gudmundsson, C.J. Pethick, Astrophys. J. **342**, 958 (1989)
73. P.G. de Gennes, *Superconductivity of Metals and Alloys* (Benjamin, New York, 1966)
74. R. Gonzalez-Felipe, A. Perez-Martinez, H. Perez-Rojas, M. Orsaria, Phys. Rev. C **77**, 015807 (2008)
75. E.V. Gorbar, Phys. Rev. D **62**, 014007 (2000)
76. D. Gross, F. Wilczek, Phys. Rev. Lett. **30**, 1343 (1973)
77. V.P. Gusynin, V.A. Miransky, I.A. Shovkovy, Phys. Rev. Lett. **73**, 3499 (1994)
78. V.P. Gusynin, V.A. Miransky, I.A. Shovkovy, Nucl. Phys. B **563**, 361 (1999)
79. A.K. Hardings, D. Lai, Rep. Prog. Phys. **69**, 2631 (2006)
80. W.J. de Hass, P.M. van Alphen, Leiden Commun. A **212** (1930)
81. W.J. de Hass, P.M. van Alphen, Proc. R. Acad. Sci. Amsterdam **33**, 1106 (1930)

82. J. Hüfner, S.P. Klevansky, P. Rehberg, Nucl. Phys. A **606**, 260 (1996)
83. A.I. Ibrahim et al., Astrophys. J. **609**, L21 (2004)
84. P. Jaikumar, M. Prakash, T. Schafer, Phys. Rev. D **66**, 063003 (2002)
85. O. Kalashnikov, V. Klimov, Phys. Lett. B **88**, 328 (1979)
86. W. Känzig, in *Ferroelectrics and Antiferroelectrics*, ed. by F. Seitz, T.P. Das, D. Turnbull, E.L. Hahn. Solid State Physics, vol. 4 (Academic Press, San Diego, 1957), p. 5
87. J. Kapusta, Nucl. Phys. B **148**, 461 (1979)
88. V.R. Khalilov, Phys. Rev. D **65**, 056001 (2002)
89. B. Keister, C. Kisslinger, Phys. Lett. D **64**, 117 (1976)
90. K.G. Klimenko, Z. Phys. C **54**, 323 (1992)
91. K.G. Klimenko, Teor. Mat. Fiz. **90**, 3 (1992)
92. S. Kulkarni, D. Frail, Nature **365**, 33 (1993)
93. L.D. Landau, Z. Phys. **64**, 629 (1930)
94. L.D. Landau, E.M. Lifshitz, *The Classical Theory of Fields*, 4th edn. (Elsevier Butterworth-Heinemann, Waltham, 1975), Ch. 11, pp. 290–293
95. L.D. Landau, E.M. Lifshitz, *Electrodynamics of Continuous Media* (Pergamon, Oxford, 1960)
96. D.S. Lee, C.N. Leung, Y.J. Ng, Phys. Rev. D **55**, 6504 (1997)
97. C.N. Leung, Y.J. Ng, A.W. Ackley, Phys. Rev. D **54**, 4181 (1996)
98. A.D. Linde, Phys. Lett. B **86**, 39 (1979)
99. M. Lines, A. Glass, *Principles and Applications of Ferroelectrics and Related Materials* (Clarendon Press, Oxford, 1979)
100. D.F. Litim, C. Manuel, Phys. Rev. D **64**, 094013 (2001)
101. G. Lugones, I. Bombaci, Phys. Rev. D **72**, 065021 (2005)
102. G. Lugones, J.E. Horvath, Phys. Rev. D **66**, 074017 (2002)
103. C. Manuel, A. Dobado, F.J. LLanes-Estrada, J. High Energy Phys. **0509**, 076 (2005)
104. D.P. Menezes et al., Phys. Rev. C **79**, 035807 (2009)
105. D.P. Menezes et al., Phys. Rev. C **80**, 065805 (2009)
106. P. Morley, M. Kisslinger, Phys. Rep. C **36**, 3 (1979)
107. N.F. Mott, Rev. Mod. Phys. **40**, 677 (1968)
108. N.F. Mott, *Metal-Insulator Transitions* (Taylor and Francis, London, 1974)
109. T. Murakami et al., Nature **368**, 127 (1994)
110. N.K. Nielsen, P. Olesen, Nucl. Phys. B **144**, 376 (1978)
111. J.L. Noronha, I.A. Shovkovy, Phys. Rev. D **76**, 105030 (2007)
112. D. Page, S. Reddy, Annu. Rev. Nucl. Part. Sci. **56**, 327 (2006)
113. G. Pagliara, J. Schaffner-Bielich, Phys. Rev. D **77**, 063004 (2008)
114. L. Paulucci, E.J. Ferrer, V. de la Incera, J.E. Horvath, Phys. Rev. D **83**, 043009 (2011)
115. A. Perez-Martinez, H. Perez-Rojas, H.J. Mosquera-Cuesta, Eur. Phys. J. C **29**, 111 (2003)
116. A. Perez Martinez, R. Gonzalez Felipe, D. Manreza Paret, Int. J. Mod. Phys. E **20**, 84 (2011)
117. H. Perez Rojas, A.E. Shabad, Ann. Phys. **121**, 432 (1979)
118. H. Politzer, Phys. Rev. Lett. **30**, 1346 (1973)
119. A. Rabhi, C. Providencia, J. Da Providencia, J. Phys. G **35**, 125201 (2008)
120. A. Rabhi et al., J. Phys. G **36**, 115204 (2009)
121. R. Rapp, T. Schafer, E.V. Shuryak, M. Velkovsky, Phys. Rev. Lett. **81**, 53 (1998)
122. S. Reddy, M. Sadzikowski, M. Tachibana, Nucl. Phys. A **714**, 337 (2003)
123. Ö.E. Rögnvaldsson et al., Astrophys. J. **416**, 276 (1993)
124. W.C. Rontgen, Ann. Phys. **35**, 264 (1888)
125. S.B. Ruester, D.H. Rischke, Phys. Rev. D **69**, 045011 (2004)
126. A. Schmitt, Q. Wang, D.H. Rischke, Phys. Rev. D **69**, 094017 (2004)
127. S.L. Shapiro, S.A. Teukolsky, *Black Holes, White Duarfs, and Neutron Stars* (Wiley, New York, 1983)
128. B.K. Sharma, P.K. Panda, S.K. Patra, Phys. Rev. C **75**, 035808 (2007)
129. I. Shovkovy, P.J. Ellis, Phys. Rev. C **66**, 015802 (2002)
130. I. Shovkovy, P.J. Ellis, Phys. Rev. C **67**, 048801 (2003)

131. V.V. Skalozub, Sov. J. Nucl. Phys. **23**, 113 (1978)
132. V.V. Skalozub, Sov. J. Nucl. Phys. **43**, 665 (1986)
133. D.T. Son, Phys. Rev. D **59**, 094019 (1999)
134. D.T. Son, M.A. Stephanov, Phys. Rev. D **61**, 074012 (2000)
135. I.-S. Suh, G.J. Mathews, Astrophys. J. **546**, 1126 (2001)
136. C. Thompson, R.C. Duncan, Astrophys. J. **392**, L9 (1992)
137. C. Thompson, R.C. Duncan, Astrophys. J. **473**, 322 (1996)
138. J. Vink, L. Kuiper, Mon. Not. R. Astron. Soc. Lett. **370**, L14 (2006)
139. F.X. Wei et al., J. Phys. G **32**, 47 (2006)
140. H.A. Wilson, Philos. Trans. R. Soc. A **204**, 129 (1905)
141. R.-X. Xu. Adv. Space Res. **40**, 1453 (2007). astro-ph/0611608
142. F. Yang, H. Shen, Phys. Rev. C **79**, 025803 (2009)

Chapter 17
Anomalous Transport from Kubo Formulae

Karl Landsteiner, Eugenio Megías, and Francisco Peña-Benitez

17.1 Introduction

Anomalies in relativistic field theories of chiral fermions belong to the most intriguing properties of quantum field theory. Comprehensive reviews on anomalies can be found in the textbooks [1–3].

Hydrodynamics is an ancient subject. Even in its relativistic form it appeared that everything relevant to its formulation could be found in [4]. Apart from stability issues that were addressed in the 1960s and 1970s [5–7] leading to a second order formalism there seemed little room for new discoveries. The last years witnessed however an unexpected and profound development of the formulation of relativistic hydrodynamics. The second order contributions have been put on a much more systematic basis applying effective field theory reasoning [8, 9]. The lessons learned from applying the AdS/CFT correspondence [10–12] to the plasma phase of strongly coupled non-abelian gauge theories [13–15] played a major role (see [16] for a recent review).

The presence of chiral anomalies in otherwise conserved currents has profound implications for the formulation of relativistic hydrodynamics. The transport processes related to anomalies have surfaced several times and independently [17–22].

K. Landsteiner (✉) · F. Peña-Benitez
Instituto de Física Teórica UAM-CISC, Univ. Autónoma de Madrid, C/ Nicolás Cabrera 13-15, 28049 Madrid, Spain
e-mail: karl.landsteiner@csic.es

E. Megías
Grup de Física Teòrica and IFAE, Departament de Física, Universitat Autònoma de Barcelona, Bellaterra E-08193, Barcelona, Spain
e-mail: emegias@ifae.es

F. Peña-Benitez
Departamento de Física Teórica, Univ. Autónoma de Madrid, 28049 Madrid, Spain
e-mail: fran.penna@uam.es

D. Kharzeev et al. (eds.), *Strongly Interacting Matter in Magnetic Fields*,
Lecture Notes in Physics 871, DOI 10.1007/978-3-642-37305-3_17,
© Springer-Verlag Berlin Heidelberg 2013

The axial current was the focus in [23] and the first application of the AdS/CFT correspondence to anomalous hydrodynamics can be found already in [24]. The full impact anomalies have on the formulation of relativistic hydrodynamics was however not fully appreciated until recently.

The renewed interest in the formulation of relativistic hydrodynamics has its origin mostly in the spectacular experimental evidence for collective flow phenomena taking place in the physics of heavy ion collisions at RHIC and LHC. These experiments indicate the creation of a deconfined quark gluon plasma in a strongly coupled regime. In the context of heavy ion collisions it was argued in [25, 26] that the excitation of topologically non-trivial gluon field configurations in the early non-equilibrium stages of a heavy ion collision might lead to an imbalance in the number of left- and right-handed quarks. This situation can be modeled by an axial chemical potential and it was shown that an external magnetic field leads to an electric current parallel to the magnetic field. This chiral magnetic effect leads then to a charge separation perpendicular to the reaction plane in heavy ion collisions. The introduction of an axial chemical potential also allows to define a chiral magnetic conductivity which is simply the factor of proportionality between the magnetic field and the induced electric current. This effect is a direct consequence of the axial anomaly.

The application of the fluid/gravity correspondence to theories including chiral anomalies lead to another surprise: it was found that not only a magnetic field induces a current but that also a vortex in the fluid leads to an induced current [27, 28]. This is the chiral vortical effect. Again it is a consequence of the presence of a chiral anomaly. It was later realized that the chiral magnetic and vortical conductivities are almost completely fixed in the hydrodynamic framework by demanding the existence of an entropy current with positive definite divergence [29]. That this criterion did not fix the anomalous transport coefficients completely was noted in [30] and various terms depending on the temperature instead of the chemical potentials were shown to be allowed as undetermined integration constants. See also [31, 32] for a recent discussion of these anomaly coefficients with applications to heavy ion physics.

In the meanwhile Kubo formulae for the chiral magnetic conductivity [33] and the chiral vortical conductivity [34] had been developed. Up to this point only pure gauge anomalies had been considered to be relevant since the mixed gauge-gravitational anomaly in four dimensions is of higher order in derivatives and was thought not to be able to contribute to hydrodynamics at first order in derivatives. Therefore it came as a surprise that in the application of the Kubo formula for the chiral vortical conductivity to a system of free chiral fermions a purely temperature dependent contribution was found. This contribution was consistent with some the earlier found integration constants and it was shown to arise if and only if the system of chiral fermions features a mixed gauge-gravitational anomaly [35]. In fact these contributions had been found already very early on in [17–20]. The connection to the presence of anomalies was however not made at that time. The gravitational anomaly contribution to the chiral vortical effect was also established in a strongly coupled AdS/CFT approach and precisely the same result as at weak coupling was found [36].

The argument based on a positive definite divergence of the entropy current allows to fix the contributions form pure gauge anomalies uniquely and provides therefore a non-renormalization theorem. No such result is known thus far for the contributions of the gauge-gravitational anomaly.[1]

A gas of weakly coupled Weyl fermions in arbitrary dimensions has been studied in [39] and confirmed that the anomalous conductivities can be obtained directly from the anomaly polynomial under substitution of the field strength with the chemical potential and the first Pontryagin density by the negative of the temperature squared [40]. Recently the anomalous conductivities have also been obtained in effective action approaches [41, 42]. The contribution of the mixed gauge-gravitational anomaly appear in all these approaches as undetermined integrations constants.

We will review here what can be learned from the calculation of the anomalous conductivities via Kubo formulae. The advantage of the usage of Kubo formulae is that they capture all contributions stemming either from pure gauge or from mixed gauge-gravitational anomalies. The disadvantage is that the calculations can be performed only within a particular model and only in the weak or in the gravity dual of the strong coupling regime. Along the way we will explain our point of view on some subtle issues concerning the definition of currents and of chemical potentials when anomalies are present. These subtleties lead indeed to some ambiguous results [43] and [44]. A first step to clarify these issues was done in [45] and a more general exposition of the relevant issues has appeared in [46].

The review is organized as follows. In Sect. 17.2 we will briefly summarize the relevant issues concerning anomalies. We recall how vector like symmetries can always be restored by adding suitable finite counterterms to the effective action [47]. A related but different issue is the fact that currents can be defined either as consistent or as covariant currents. The hydrodynamic constitutive relations depend on what definition of current is used. We discuss subtleties in the definition of the chemical potential in the presence of an anomaly and define our preferred scheme. We discuss the hydrodynamic constitutive relations and derive the Kubo formulae that allow the calculation of the anomalous transport coefficients from two point correlators of an underlying quantum field theory.

In Sect. 17.3 we apply the Kubo formulae to a theory of free Weyl fermions and show that two different contributions arise. They are clearly identifiable as being related to the presence of pure gauge and mixed gauge-gravitational anomalies.

In Sect. 17.4 we define a holographic model that implements the mixed gauge-gravitational anomaly via a mixed gauge-gravitational Chern-Simons term. We calculate the same Kubo formulae as at weak coupling, obtaining the same values for chiral axi-magnetic and chiral vortical conductivities as in the weak coupling model.

We conclude this review with some discussions and outlook to further developments.

---

[1] See however the very recent attempts to establish non-renormalization theorems in [37] and [38].

17.2 Anomalies and Hydrodynamics

In this section we will review briefly anomalies. We compare the consistent with the covariant form of the anomaly and we introduce the Bardeen counterterm that allows to restore current conservation for vector like symmetries. Then we turn to the question of what we mean when we talk about the chemical potential. Two ways of introducing chemical potential, either through a deformation of the Hamiltonian or by demanding twisted boundary conditions along the thermal circle are shown to be in-equivalent in presence of an anomaly. Equivalence can still be achieved by introduction of a spurious axion field. We explain the implications for holography. The constitutive relations for anomalous currents are introduced in Landau and Laboratory frame. We discuss how they differ if we use the consistent instead of the covariant currents and derive the Kubo formulae for the anomalous conductivities.

17.2.1 Anomalies

Anomalies arise by integrating over chiral fermions in the path integral. They signal a fundamental incompatibility between the symmetries present in the classical theory and the quantum theory.

Unless otherwise stated we will always think of the symmetries as global symmetries. But we still will introduce gauge fields. These gauge fields serve as classical sources coupled to the currents. As a side effect their presence promotes the global symmetry to a local gauge symmetry. It is still justified to think of it as a global symmetry as long as we do not introduce a kinetic Maxwell or Yang-Mills term in the action.

In a theory with chiral fermions we define an effective action depending on these gauge fields by the path integral

$$e^{iW_{eff}[A_\mu]/\hbar} := \int \mathscr{D}\Psi \, \mathscr{D}\bar{\Psi} \, e^{iS[\psi, A_\mu]/\hbar}. \qquad (17.1)$$

The vector field $A_\mu(x)$ couples to a classically conserved current $J^\mu = \bar{\Psi}\gamma^\mu Q\Psi$. The charge operator Q can be the generator of a Lie group combined with chiral projectors $\mathscr{P}_\pm = \frac{1}{2}(1 \pm \gamma_5)$. General combinations are allowed although in the following we will mostly concentrate on a simple chiral $U(1)$ symmetry for which we can take $Q = \mathscr{P}_+$. The fermions are minimally coupled to the gauge field and the classical action has an underlying gauge symmetry

$$\delta\Psi = -i\lambda(x)Q\Psi, \qquad \delta A_\mu(x) = D_\mu\lambda(x), \qquad (17.2)$$

with D_μ denoting the gauge covariant derivative. Assuming that the theory has a classical limit the effective action in terms of the gauge fields allows for an expansion in $\hbar$

$$W_{eff} = W_0 + \hbar W_1 + \hbar^2 W_2 + \cdots. \qquad (17.3)$$

We find it convenient to use the language of BRST symmetry by promoting the gauge parameter to a ghost field $c(x)$.[2] The BRST symmetry is generated by

$$s A_\mu = D_\mu c, \qquad sc = -ic^2. \tag{17.4}$$

It is nilpotent $s^2 = 0$. The statement that the theory has an anomaly can now be neatly formalized. Since on gauge fields the BRST symmetry acts just as the gauge symmetry, gauge invariance translates into BRST invariance. An anomaly is present if

$$s W_{eff} = \mathcal{A} \quad \text{and} \quad \mathcal{A} \neq s B. \tag{17.5}$$

Because of the nilpotency of the BRST operator the anomaly has to fulfill the Wess-Zumino consistency condition

$$s \mathcal{A} = 0. \tag{17.6}$$

As indicated in (17.5) this has a possible trivial solution if there exists a local functional $B[A_\mu]$ such that $sB = \mathcal{A}$. An anomaly is present if no such B exists. The anomaly is a quantum effect. If it is of order $\hbar^n$ and if a suitable local functional B exists we could simply redefine the effective action as $\widetilde{W}_{eff} = W_{eff} - B$ and the new effective action would be BRST and therefore gauge invariant. The form and even the necessity to introduce a functional B might depend on the particular regularization scheme chosen. As we will explain in the case of an axial and vector symmetry a suitable B can be found that always allows to restore the vectorlike symmetry, this is the so-called Bardeen counterterm [47]. The necessity to introduce the Bardeen counterterm relies however on the regularization scheme chosen. In schemes that automatically preserve vectorlike symmetries, such as dimensional regularization, the vector symmetries are automatically preserved and no counterterm has to be added. Furthermore the Adler-Bardeen theorem guarantees that chiral anomalies appear only at order $\hbar$. Their presence can therefore by detected in one loop diagrams such as the triangle diagram of three currents.

We have introduced the gauge fields as sources for the currents

$$\frac{\delta}{\delta A_\mu(x)} W_{eff}[A] = \langle J^\mu \rangle. \tag{17.7}$$

For chiral fermions transforming under a general Lie group generated by T^a the chiral anomaly takes the form [1]

$$s W_{eff}[A] = - \int d^4 x \, c^a \left(D_\mu J^\mu \right)^a$$

$$= -\frac{\eta}{24\pi^2} \int d^4 x \, c^a \varepsilon^{\mu\nu\rho\sigma} \, \text{tr} \left[T^a \partial_\mu \left(A_\nu \partial_\rho A_\sigma + \frac{1}{2} A_\nu A_\rho A_\sigma \right) \right]. \tag{17.8}$$

---

[2] A recent comprehensive review on BRST symmetry is [48].

Where $\eta = +1$ for left-handed fermions and $\eta = -1$ for right-handed fermions. Differentiating with respect to the ghost field (the gauge parameter) we can derive a local form. To simplify the formulas we specialize this to the case of a single chiral $U(1)$ symmetry taking $T^a = 1$

$$\partial_\mu J^\mu = \frac{\eta}{96\pi^2} \varepsilon^{\mu\nu\rho\sigma} F_{\mu\nu} F_{\rho\sigma}. \tag{17.9}$$

This is to be understood as an operator equation. Sandwiching it between the vacuum state $|0\rangle$ and further differentiating with respect to the gauge fields we can generate the famous triangle form of the anomaly

$$\langle \partial_\mu J^\mu(x) J^\sigma(y) J^\kappa(z) \rangle = \frac{1}{12\pi^2} \varepsilon^{\mu\sigma\rho\kappa} \partial_\mu^x \delta(x-y) \partial_\rho^x \delta(x-z). \tag{17.10}$$

The one point function of the divergence of the current is non-conserved only in the background of parallel electric and magnetic fields whereas the non-conservation of the current as an operator becomes apparent in the triangle diagram even in vacuum.

By construction this form of the anomaly fulfills the Wess-Zumino consistency condition and is therefore called the *consistent anomaly*. In analogy we call the current defined by (17.7) the *consistent current*.

For a $U(1)$ symmetry the functional differentiation with respect to the gauge field and the BRST operator s commute,

$$\left[s, \frac{\delta}{\delta A_\mu(x)} \right] = 0. \tag{17.11}$$

An immediate consequence is that the consistent current is not BRST invariant but rather obeys

$$s J^\mu = \frac{1}{24\pi^2} \varepsilon^{\mu\nu\rho\lambda} \partial_\nu c F_{\rho\lambda} = -\frac{1}{24\pi^2} s K^\mu, \tag{17.12}$$

where we introduced the Chern-Simons current $K^\mu = \varepsilon^{\mu\nu\rho\lambda} A_\nu F_{\rho\lambda}$ with $\partial_\mu K^\mu = \frac{1}{2} \varepsilon^{\mu\nu\rho\lambda} F_{\mu\nu} F_{\rho\lambda}$.

With the help of the Chern-Simons current it is possible to define the so-called *covariant current* (in the case of a $U(1)$ symmetry rather the *invariant current*)

$$\tilde{J}^\mu = J^\mu + \frac{1}{24\pi^2} K^\mu \tag{17.13}$$

fulfilling

$$s \tilde{J}^\mu = 0. \tag{17.14}$$

The divergence of the covariant current defines the *covariant anomaly*

$$\partial_\mu \tilde{J}^\mu = \frac{1}{32\pi^2} \varepsilon^{\mu\nu\rho\sigma} F_{\mu\nu} F_{\rho\sigma}. \tag{17.15}$$

Notice that the Chern-Simons current cannot be obtained as the variation with respect to the gauge field of any functional. It is therefore not possible to define an effective action whose derivation with respect to the gauge field gives the covariant current.

Let us suppose now that we have one left-handed and one right-handed fermion with the corresponding left- and right-handed anomalies. Instead of the left-right basis it is more convenient to introduce a vector-axial basis by defining the vectorlike current $J_V^\mu = J_L^\mu + J_R^\mu$ and the axial current $J_A^\mu = J_L^\mu - J_R^\mu$. Let $V_\mu(x)$ be the gauge field that couples to the vectorlike current and $A_\mu(x)$ be the gauge field coupling to the axial current. The (consistent) anomalies for the vector and axial current turn out to be

$$\partial_\mu J_V^\mu = \frac{1}{24\pi^2}\varepsilon^{\mu\nu\rho\lambda}F_{\mu\nu}^V F_{\rho\lambda}^A, \tag{17.16}$$

$$\partial_\mu J_A^\mu = \frac{1}{48\pi^2}\varepsilon^{\mu\nu\rho\lambda}\left(F_{\mu\nu}^V F_{\rho\lambda}^V + F_{\mu\nu}^A F_{\rho\lambda}^A\right). \tag{17.17}$$

As long as the vectorlike current corresponds to a global symmetry nothing has gone wrong so far. If we want to identify the vectorlike current with the electric-magnetic current in nature we need to couple it to a dynamical photon gauge field and now the non-conservation of the vector current is worrisome to say the least. The problem arises because implicitly we presumed a regularization scheme that treats left-handed and right-handed fermions on the same footing. As pointed out first by Bardeen this flaw can be repaired by adding a finite counterterm (of order $\hbar$) to the effective action. This is the so-called Bardeen counterterm and has the form

$$B_{ct} = -\frac{1}{12\pi^2}\int \mathrm{d}^4x\,\varepsilon^{\mu\nu\rho\lambda}V_\mu A_\nu F_{\rho\lambda}^V. \tag{17.18}$$

Adding this counterterm to the effective action gives additional contributions of Chern-Simons form to the consistent vector and axial currents. With the particular coefficient chosen it turns out that the anomaly in the vector current is canceled whereas the axial current picks up an additional contribution such that after adding the Bardeen counterterm the anomalies are

$$\partial_\mu J_V^\mu = 0, \tag{17.19}$$

$$\partial_\mu J_A^\mu = \frac{1}{48\pi^2}\varepsilon^{\mu\nu\rho\lambda}\left(3F_{\mu\nu}^V F_{\rho\lambda}^V + F_{\mu\nu}^A F_{\rho\lambda}^A\right). \tag{17.20}$$

This definition of currents is mandatory if we want to identify the vector current with the usual electromagnetic current in nature! It is furthermore worth to point out that both currents are now invariant under the vectorlike $U(1)$ symmetry. The currents are not invariant under axial transformation, but these are anomalous anyway.

Generalizations of the covariant anomaly and the Bardeen counterterm to the non-abelian case can be found e.g. in [1].

There is one more anomaly that will play a major role in this review, the mixed gauge-gravitational anomaly [49–51].[3] So far we have considered only spin one currents and have coupled them to gauge fields. Now we also want to introduce the energy-momentum tensor through its coupling to a fiducial background metric $g_{\mu\nu}$. Just as the gauge fields, the metric serves primarily as the source for insertions of the energy momentum tensor in correlation functions. Just as in the case of vector and axial currents, the mixed gauge-gravitational anomaly is the statement that it is impossible in the quantum theory to preserve at the same time the vanishing of the divergence of the energy-momentum tensor and of chiral (or axial) $U(1)$ currents. It is however possible to add Bardeen counterterms to shift the anomaly always in the sector of the spin one currents and preserve translational (or diffeomorphism) symmetry. If we have a set of left-handed and right-handed chiral fermions transforming under a Lie Group generated by $(T_a)_L$ and $(T_a)_R$ in the background of arbitrary gauge fields and metric, the anomaly is conveniently expressed through the non-conservation of the *covariant* current as

$$\left(D_\mu J^\mu\right)_a = \frac{d_{abc}}{32\pi^2}\varepsilon^{\mu\nu\rho\lambda}F^b_{\mu\nu}F^c_{\rho\lambda} + \frac{b_a}{768\pi^2}\varepsilon^{\mu\nu\rho\lambda}R^\alpha{}_{\beta\mu\nu}R^\beta{}_{\alpha\rho\lambda}. \quad (17.21)$$

The purely group theoretic factors are

$$d_{abc} = \frac{1}{2}\mathrm{tr}\left(T_a\{T_b, T_c\}\right)_L - \frac{1}{2}\mathrm{tr}\left(T_a\{T_b, T_c\}\right)_R, \quad (17.22)$$

$$b_a = \mathrm{tr}(T_a)_L - \mathrm{tr}(T_a)_R. \quad (17.23)$$

Chiral anomalies are completely absent if and only if $d_{abc} = 0$ and $b_a = 0$.

17.2.2 Chemical Potentials for Anomalous Symmetries

Thermodynamics of systems with conserved charges can be described in a grand canonical ensemble where a Lagrange multiplier μ ensures that the partition function fulfills

$$T\frac{\partial \log(Z)}{\partial \mu} = \langle Q\rangle. \quad (17.24)$$

The textbook approach is to consider a deformation of the Hamiltonian

$$H \to H - \mu Q, \quad (17.25)$$

where Q is the charge in question. We can think of this as arising from the coupling of the (fiducial) gauge field A_μ to the current J^μ and giving a vacuum expectation value to $A_0 = \mu$. Since the fiducial gauge field leads to local gauge invariance we can remove the μQ coupling in the Hamiltonian by the gauge transformation $A_0 \to A_0 + \partial_0\chi$ with $\chi = -\mu t$.

---

[3]In $D = 4k + 2$ dimensions also purely gravitational anomalies can appear [52].

Fig. 17.1 At finite temperature field theories are defined on the Keldysh-Schwinger contour in the complexified time plane. The initial state at t_i is specified through the boundary conditions on the fields. The endpoint of the contour is at $t_i - i\beta$ where $\beta = 1/T$

Table 17.1 Two formalisms for the chemical potential	Formalism	Hamiltonian	Boundary condition
	(A)	$H - \mu Q$	$\Psi(t_i - i\beta) = \pm\Psi(t_i)$
	(B)	H	$\Psi(t_i - i\beta) = \pm e^{q\beta\mu}\Psi(t_i)$

In the context of finite temperature field theory such a gauge transformation is however not really allowed. One needs to define the field theory on the Keldysh-Schwinger contour in the complexified time plane as shown in Fig. 17.1. Fields are taken to be periodic or anti-periodic along the imaginary time direction $t = -i\tau$ with period $\beta = 1/T$ where T is the temperature

$$\Psi(t_i - i\beta) = \pm\Psi(t_i), \tag{17.26}$$

with the plus sign for bosons and the minus sign for fermions. The gauge transformation that removes the constant zero component of the gauge field is not periodic along the contour and therefore changes the boundary conditions on the fields. After the gauge transformation with $\chi = -\mu t$ the fields obey the boundary conditions

$$\Psi(t_i - i\beta) = \pm e^{q\mu\beta}\Psi(t_i). \tag{17.27}$$

Demanding these "twisted" boundary conditions is of course completely equivalent to having $A_0 = \mu$. The gauge invariant statement is that a charged field parallel transported around the Keldysh-Schwinger contour picks up a factor of $\exp(q\mu\beta)$. As long as we have honest non-anomalous symmetries under consideration we have therefore two (gauge-)equivalent formalisms of how to introduce the chemical potential summarized in Table 17.1 [53].

One convenient point of view on formalism (B) is the following. In a real time Keldysh-Schwinger setup we demand some initial conditions at initial (real) time $t = t_i$. These initial conditions are given by the boundary conditions in (B). From then on we do the (real) time development with the microscopic Hamiltonian H. In principle there is no need for the Hamiltonian H to preserve the symmetry present at times $t < t_i$. This seems an especially suited approach to situations where the charge in question is not conserved by the real time dynamics. In the case of an anomalous symmetry we can start at $t = t_i$ with a state of certain charge. As long as we have only external gauge fields present the one-point function of the divergence of the current vanishes and the charge is conserved. This is not true on the full theory since even in vacuum the three-point correlators are sensitive to the anomaly. For the formulation of hydrodynamics in external fields the condition that the one-point functions of the currents are conserved as long as there are no parallel electric and magnetic external fields (or a metric that has non-vanishing Pontryagin density) is sufficient.[4]

Let us assume now that Q is an anomalous charge, i.e. its associated current suffers from chiral anomalies. We first consider formalism (B) and ask what happens if we do now the gauge transformation that would bring us to formalism (A). Since the symmetry is anomalous the action transforms as

$$S[A + \partial \chi] = S[A] + \int d^4x \, \chi \varepsilon^{\mu\nu\rho\lambda} \left(C_1 F_{\mu\nu} F_{\rho\lambda} + C_2 R^\alpha{}_{\beta\mu\nu} R^\beta{}_{\alpha\rho\lambda} \right), \quad (17.28)$$

with the anomaly coefficients C_1 and C_2 depending on the chiral fermion content. It follows that formalisms (A) and (B) are physically inequivalent now, because of the anomaly. However, we would like to still come as close as possible to the formalism of (A) but in a form that is physically equivalent to the formalism (B). To achieve this we proceed by introducing a non-dynamical axion field $\Theta(x)$ and the vertex

$$S_\Theta[A, \Theta] = \int d^4x \, \Theta \varepsilon^{\mu\nu\rho\lambda} \left(C_1 F_{\mu\nu} F_{\rho\lambda} + C_2 R^\alpha{}_{\beta\mu\nu} R^\beta{}_{\alpha\rho\lambda} \right). \quad (17.29)$$

If we demand now that the "axion" transforms as $\Theta \to \Theta - \chi$ under gauge transformations we see that the action

$$S_{tot}[A, \Theta] = S[A] + S_\Theta[A, \Theta] \quad (17.30)$$

is gauge invariant. Note that this does not mean that the theory is not anomalous now. We introduce it solely for the purpose to make clear how the action has to be modified such that two field configurations related by a gauge transformation are physically equivalent. It is better to consider Θ as *coupling* and not a field, i.e. we consider it a spurion field. The gauge field configuration that corresponds to formalism (B) is simply $A_0 = 0$. A gauge transformation with $\chi = \mu t$ on the gauge invariant action S_{tot} makes clear that a physically equivalent theory is obtained by choosing the field configuration $A_0 = \mu$ and the time dependent coupling $\Theta = -\mu t$. If we define the current through the variation of the action with respect to the gauge field we get an additional contribution from S_Θ,

$$J_\Theta^\mu = 4C_1 \varepsilon^{\mu\nu\rho\lambda} \partial_\nu \Theta F_{\rho\lambda}, \quad (17.31)$$

---

[4]If dynamical gauge fields are present, such as the gluon fields in QCD even the one point function of the charge does decay over (real) time due to non-perturbative processes (instantons) or at finite temperature due to thermal sphaleron processes [54]. Even in this case in the limit of large number of colors these processes are suppressed and can e.g. not be seen in holographic models in the supergravity approximation.

and evaluating this for $\Theta = -\mu t$ we get the spatial current

$$J_\Theta^m = 8C_1 \mu B^m. \tag{17.32}$$

We do not consider this to be the chiral magnetic effect! This is only the contribution to the current that comes from the new coupling that we are forced to introduce by going to formalism (A) from (B) in a (gauge-)equivalent way. As we will see in the following chapters the chiral magnetic and vortical effect are on the contrary non-trivial results of dynamical one-loop calculations.

What is the Hamiltonian now based on the modified formalism (A)? We have to take of course the new coupling generated by the non-zero Θ. The Hamiltonian now is therefore

$$H - \mu \left(Q + 4C_1 \int d^3x \, \varepsilon^{0ijk} A_i \partial_j A_k \right), \tag{17.33}$$

where for simplicity we have ignored the contributions from the metric terms.

For explicit computations in Sects. 17.3 and 17.4 we will introduce the chemical potential through the formalism (B) by demanding twisted boundary conditions. It seems the most natural choice since the dynamics is described by the microscopic Hamiltonian H. The modified (A) based on the Hamiltonian (17.33) is however not without merits. It is convenient in holography where it allows vanishing temporal gauge field on the black hole horizon and therefore a non-singular Euclidean black hole geometry.[5]

17.2.2.1 Hydrodynamics and Kubo Formulae

The modern understanding of hydrodynamics is as an effective field theory. The equations of motion are the (anomalous) conservation laws of the energy-momentum tensor and spin one currents. These are supplemented by expression for the energy-momentum tensor and the current which are organized in a derivative expansion, the so-called constitutive relations. Symmetries constrain the possible terms. In the presence of chiral anomalies the constitutive relations for the energy-momentum tensor and the currents in the Landau frame are

$$T^{\mu\nu} = \varepsilon u^\mu u^\nu + p P^{\mu\nu} - \eta P^{\mu\alpha} P^{\nu\beta} \sigma_{\alpha\beta} - \zeta P^{\mu\nu} \partial_\alpha u^\alpha, \tag{17.34}$$

$$\tilde{J}_a^\mu = \rho_a u^\mu + \Sigma_{ab} \left(E_b^\mu - T P^{\mu\alpha} D_\alpha \frac{\mu_a}{T} \right) + \xi_{ab}^B B_a^\mu + \xi_a^V \omega^\mu. \tag{17.35}$$

It is important to specify that these are the constitutive relations for the *covariant* currents! Here ε is the energy density, p the pressure density, u^μ the local fluid velocity. $P^{\mu\nu} = g^{\mu\nu} + u^\mu u^\nu$ is the transverse projector to the fluid velocity.

---

[5]It is possible to define a generalized formalism to make any choice for the gauge field $A_0 = \nu$, so that one recovers formalism (A) when $\nu = \mu$ and formalism (B) when $\nu = 0$ as particular cases (see [55] for details).

$\sigma_{\mu\nu}$ is the symmetric traceless shear tensor. The non-anomalous transport coefficients are the shear viscosity η, the bulk viscosity ζ and the electric conductivities Σ_{ab}. External electric and magnetic fields are covariantized via $E_a^\mu = F_a^{\mu\nu} u_\nu$ and $B_a^\mu = \frac{1}{2} \varepsilon^{\mu\nu\rho\lambda} u_\nu F_{a,\rho\lambda}$. The vorticity of the fluid is $\omega^\mu = \varepsilon^{\mu\nu\rho\lambda} u_\nu \partial_\rho u_\lambda$.

The anomalous transport coefficients are the chiral magnetic conductivities ξ_{ab}^B and the chiral vortical conductivities ξ_a^V. At first order in derivatives the notion of fluid velocity is ambiguous and needs to be fixed by prescribing a choice of frame. We remark that the constitutive relations (17.34) and (17.35) are valid in the Landau frame where $T^{\mu\nu} u_\nu = \varepsilon u^\mu$.

To compute the Kubo formulae for the anomalous transport coefficients it turns out that the Landau frame is not the most convenient one. It fixes the definition of the fluid velocity through energy transport. Transport phenomena related to the generation of an energy current are therefore not directly visible, rather they are absorbed in the definition of the fluid velocity. It is therefore more convenient to go to another frame in which we demand that the definition of the fluid velocity is not influenced when switching on an external magnetic field or having a vortex in the fluid. In such a frame the constitutive relations take the form

$$T^{\mu\nu} = (\varepsilon + p) u^\mu u^\nu + p g^{\mu\nu} - \eta P^{\mu\alpha} P^{\nu\beta} \sigma_{\alpha\beta} - \zeta P^{\mu\nu} \partial_\alpha u^\alpha$$
$$+ Q^\mu u^\nu + Q^\nu u^\mu, \tag{17.36}$$

$$Q^\mu = \sigma_B^\varepsilon B^\mu + \sigma_V^\varepsilon \omega^\mu, \tag{17.37}$$

$$\tilde{J}^\mu = \rho u^\mu + \Sigma \left(E^\mu - T P^{\mu\alpha} D_\alpha \left(\frac{\mu}{T} \right) \right) + \sigma_B B^\mu + \sigma_V \omega^\mu. \tag{17.38}$$

In order to avoid unnecessary clutter in the equations we have specialized now to a single $U(1)$ charge. Notice that now there is a sort of "heat" current present in the constitutive relation for the energy-momentum tensor.

The derivation of Kubo formulae is better based on the usage of the *consistent* currents. Since the covariant and consistent currents are related by adding suitable Chern-Simons currents, the constitutive relations for the consistent current receive additional contribution from the Chern-Simons current

$$J^\mu = \tilde{J}^\mu - \frac{1}{24\pi^2} K^\mu. \tag{17.39}$$

If we were to introduce the chemical potential according to formalism (A) via a background for the temporal gauge field we would get an additional contribution to the consistent current from the Chern-Simons current. In this case it is better to go to the modified formalism (A') that also introduces a spurious axion field and another contribution to the current J_Θ (17.32) has to be added

$$J^\mu = \tilde{J}^\mu - \frac{1}{24\pi^2} K^\mu + J_\Theta^\mu. \tag{17.40}$$

For the derivation of the Kubo formulae it is therefore more convenient to work with formalism (B) in which $A_0 = 0$ and the chemical potential is introduced via the boundary conditions (17.27). Otherwise there arise additional contributions to the two point functions. We will briefly discuss them in the next subsection.

From the microscopic view the constitutive relations should be interpreted as the one-point functions of the operators $T^{\mu\nu}$ and J^μ in a near equilibrium situation, i.e. gradients in the fluid velocity, the temperature or the chemical potentials are assumed to be small. From this point of view Kubo formulae can be derived. In the microscopic theory the one-point function of an operator near equilibrium is given by linear response theory whose basic ingredient are the retarded two-point functions. If we consider a situation with only an external electric field in z-direction and all other sources switched off, i.e. the fluid being at rest $u^\mu = (1,0,0,0)$ and no gradients in temperature or chemical potentials the constitutive relations are simplified to

$$J^z = \Sigma E^z. \tag{17.41}$$

The electric field is $E^z = i\omega A^z$ in terms of the vector potential and using linear response theory the induced current is given through the retarded two-point function by

$$J^z = \langle J^z J^z \rangle A^z. \tag{17.42}$$

Equating the two expressions for the current we find the Kubo formula for the electric conductivity

$$\Sigma = \lim_{\omega \to 0} \frac{-i}{\omega} \langle J^z J^z \rangle. \tag{17.43}$$

This has to be evaluated at zero momentum. The limit in the frequency follows because the constitutive relation are supposed to be valid only to lowest order in the derivative expansion, therefore one needs to isolate the first non-trivial term.

Now we want to find some simple special cases that allow the derivation of Kubo formulae for the anomalous conductivities. A very convenient choice is to go to the restframe $u^\mu = (1,0,0,0)$, switch on a vector potential in the y-direction that depends only on the z direction and at the same time a metric deformation $g_{\mu\nu} = \eta_{\mu\nu} + h_{\mu\nu}$ with the only non-vanishing component h_{0y} depending on z only. To linear order in the background fields the non-vanishing components of the energy-momentum tensor and the current are

$$T^{0x} = -\sigma_B^\varepsilon \partial_z A_y - \sigma_V^\varepsilon \partial_z h_{0y}, \tag{17.44}$$

$$J^x = -\sigma_B \partial_z A_y - \sigma_V \partial_z h_{0y}, \tag{17.45}$$

since in the formalism (B) neither the Chern-Simons term nor the Θ coupling contribute. Going to momentum space and differentiating with respect to the sources A_y and h_{0y} we find therefore the Kubo formulae [26, 34]

$$\boxed{\begin{array}{ll} \sigma_B = \lim_{k_z \to 0} \frac{i}{k_z} \langle J^x J^y \rangle & \sigma_V = \lim_{k_z \to 0} \frac{i}{k_z} \langle J^x T^{0y} \rangle \\[2mm] \sigma_B^\varepsilon = \lim_{k_z \to 0} \frac{i}{k_z} \langle T^{0x} J^y \rangle & \sigma_V^\varepsilon = \lim_{k_z \to 0} \frac{i}{k_z} \langle T^{0x} T^{0y} \rangle \end{array}} \tag{17.46}$$

All these correlators are to be taken at precisely zero frequency. As these formulas are based on linear response theory the correlators should be understood as retarded ones. They have to be evaluated however at zero frequency and therefore the order of the operators can be reversed. From this it follows that the chiral vortical conductivity coincides with the chiral magnetic conductivity for the energy flux $\sigma_V = \sigma_B^\varepsilon$.[6]

We also want to discuss how these transport coefficients are related to the ones in the more commonly used Landau frame. They are connected by a redefinition of the fluid velocity of the form

$$u^\mu \to u^\mu - \frac{1}{\varepsilon + p} Q^\mu, \tag{17.47}$$

to go from (17.36)–(17.38) to (17.34)–(17.35). The corresponding transport coefficients of the Landau frame are therefore

$$\xi_B = \lim_{k_n \to 0} \frac{-i}{2k_n} \sum_{k,l} \varepsilon_{nkl} \left(\langle J^k J^l \rangle - \frac{\rho}{\varepsilon + p} \langle T^{0k} J^l \rangle \right), \tag{17.48}$$

$$\xi_V = \lim_{k_n \to 0} \frac{-i}{2k_n} \sum_{k,l} \varepsilon_{nkl} \left(\langle J^k T^{0l} \rangle - \frac{\rho}{\varepsilon + p} \langle T^{0k} T^{0l} \rangle \right), \tag{17.49}$$

where we have employed a slightly more covariant notation. The generalization to the non-abelian case is straightforward.

It is also worth to compare to the Kubo formulae for the dissipative transport coefficients as the electric conductivity (17.43). In the dissipative cases one first goes to zero momentum and then takes the zero frequency limit. In the anomalous conductivities this is the other way around, one first goes to zero frequency and then takes the zero momentum limit. Another observation is that the dissipative transport coefficients sit in the anti-Hermitean part of the retarded correlators, i.e. the spectral function whereas the anomalous conductivities sit in the Hermitean part. The rate at which an external source f_I does work on a system is given in terms of the spectral function of the operator O^I coupling to that source as

$$\frac{dW}{dt} = \frac{1}{2} \omega f_I(-\omega) \rho^{IJ}(\omega) f_J(\omega). \tag{17.50}$$

The anomalous transport phenomena therefore do no work on the system, first they take place at zero frequency and second they are not contained in the spectral function $\rho = \frac{-i}{2}(G_r - G_r^\dagger)$.

---

[6]Notice that h_{0y} can also be understood as the so-called gravito-magnetic vector potential $\mathbf{A}_g$, which is related to the gravito-magnetic field by $\mathbf{B}_g = \nabla \times \mathbf{A}_g$. This allows to interpret σ_V not only as the generation of a current due to a vortex in the fluid, i.e. the chiral vortical effect, but also as a *chiral gravito-magnetic* conductivity giving rise to a *chiral gravito-magnetic effect*, see [56] for details.

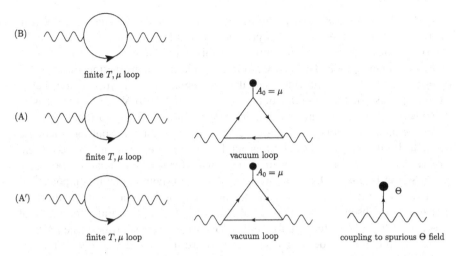

Fig. 17.2 Contributions to the Kubo formula for the chiral magnetic conductivity in the different formalisms for the chemical potential

17.2.3 Contributions to the Kubo Formulae

Now we want to give a detailed analysis of the different Feynman graphs that contribute to the Kubo formulae in the different formalisms for the chemical potentials. The simplest and most economic formalism is certainly the one labeled (B) in which we introduce the chemical potentials via twisted boundary conditions. The Hamiltonian is simply the microscopic Hamiltonian H. Relevant contributions arise only at first order in the momentum and at zero frequency and in this kinematic limit only the Kubo formulae for the chiral magnetic conductivity is affected. In Fig. 17.2 we summarize the different contributions to the Kubo formulae in the three ways to introduce the chemical potential.

The first of the Feynman graphs is the same in all formalisms. It is the genuine finite temperature and finite density one-loop contribution. This graph is finite because the Fermi-Dirac distributions cutoff the UV momentum modes in the loop. In the formalism (A) we need to take into account that there is also a contribution from the triangle graph with the fermions going around the loop in vacuum, i.e. without the Fermi-Dirac distributions in the loop integrals. For a non-anomalous symmetry this graph vanishes simply because on the upper vertex of the triangle sits a field configuration that is a pure gauge. If the symmetry under consideration is however anomalous the triangle diagram picks up just the anomaly. Even pure gauge field configurations become physically distinct from the vacuum and therefore this diagram gives a non-trivial contribution. On the level of the constitutive relations this contribution corresponds to the Chern-Simons current in (17.39). We consider this contribution to be unwanted. After all the anomaly would make even a constant value of the temporal gauge field A_0 observable in vacuum. An example is provided for a putative axial gauge field A_μ^5. If present the absolute value of its temporal component would be observable through the axial anomaly. We can be sure that in nature no such background field is present. The third line (A') introduces also the spurious axion field Θ the only purpose of this field is to cancel the contribution from the triangle graph. This cancellation takes place by construction since (A') is gauge equivalent to (B) in which only the first genuine finite T, μ part contributes. It corresponds to the contribution of the current J_{Θ}^{μ} in (17.40). We further emphasize that these considerations are based on the usage of the consistent currents.

In the interplay between axial and vector currents additional contributions arise from the Bardeen counterterm. It turns out that the triangle or Chern-Simons current contribution to the consistent vector current in the formalism (A) cancels precisely the first one [44, 45]. Our take on this is that a constant temporal component of the axial gauge field $A_0 = \mu_5$ would be observable in nature and can therefore be assumed to be absent. The correct way of evaluating the Kubo formulae for the chiral magnetic effect is therefore the formalism (B) or the gauge equivalent one (A').

At this point the reader might wonder why we introduced yet another formalism (A') which achieves apparently nothing but being equivalent to formalism (B). At least from the perspective of holography there is a good reason for doing so. In holography the strong coupling duals of gauge theories at finite temperature in the plasma phase are represented by five dimensional asymptotically Anti- de Sitter black holes. Finite charge density translates to charged black holes. These black holes have some non-trivial gauge flux along the holographic direction represented by a temporal gauge field configuration of the form $A_0(r)$ where r is the fifth, holographic dimension. It is often claimed that for consistency reasons the gauge field has to vanish on the horizon of the black hole and that its value on the boundary can be identified with the chemical potential

$$A_0(r_H) = 0 \quad \text{and} \quad A_0(r \to \infty) = \mu. \tag{17.51}$$

According to the usual holographic dictionary the gauge field values on the boundary correspond to the sources for currents. A non-vanishing value of the temporal component of the gauge field at the boundary is therefore dual to a coupling that modifies the Hamiltonian of the theory just as in (17.25). Thus with the boundary conditions (17.51) we have the holographic dual of the formalism (A). If anomalies are present they are represented in the holographic dual by five-dimensional Chern-Simons terms of the form $A \wedge F \wedge F$. The two point correlator of the (consistent) currents receives now contributions from the Chern-Simons term that is precisely of the form of the second graph in (A) in Fig. 17.2. As we have argued this is an a priori unwanted contribution. We can however cure that by introducing an additional term in the action of the form (17.29) living only on the boundary of the holographic space-time. In this way we can implement the formalism (A'), cancel the unwanted triangle contribution with the third graph in (A') in Fig. 17.2 and maintain $A_0(r_H) = 0$!

The claim that the temporal component of the gauge field has to vanish at the horizon is of course not unsubstantiated. The reasoning goes as follows. The Euclidean section of the black-hole space time has the topology of a disc in the r, τ

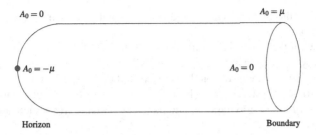

$A_0 = 0$

$A_0 = \mu$

$A_0 = -\mu$

$A_0 = 0$

Horizon

Boundary

Fig. 17.3 A sketch of the Euclidean black hole topology. A singularity at the horizon arises if we do not choose the temporal component of the gauge field to vanish there. On the other hand allowing the singularity to be present changes the topology to the one of a cylinder and this in turn allows twisted boundary conditions

directions, where τ is the Euclidean time (see Fig. 17.3). This is a periodic variable with period $\beta = 1/T$ where T is the (Hawking) temperature of the black hole and at the same time the temperature in the dual field theory. Using Stoke's law we have

$$\int_{\partial D} A_0 \, d\tau = \int_D F_{r0} \, dr \, d\tau, \tag{17.52}$$

where F_{r0} is the electric field strength in the holographic direction and D is a Disc with origin at $r = r_H$ reaching out to some finite value of r_f. If we shrink this disc to zero size, i.e. let $r_f \to r_H$ the r.h.s. of the last equation vanishes and so must the l.h.s. which approaches the value $\beta A_0(r_H)$. This implies that $A_0(r_H) = 0$. If on the other hand we assume that $A_0(r_H) \neq 0$ then the field strength must have a delta type singularity there in order to satisfy Stokes theorem. Strictly speaking the topology of the Euclidean section of the black hole is not anymore that of a disc since now there is a puncture at the horizon. It is therefore more appropriate to think of this as having the topology of a cylinder. Now if we want to implement the formalism (B) in holography we would find the boundary conditions

$$A_0(r_H) = \mu \quad \text{and} \quad A_0(r \to \infty) = 0, \tag{17.53}$$

and precisely such a singularity at the horizon would arise. In addition we would need to impose twisted boundary conditions around the Euclidean time τ for the fields just as in (17.27). Now the presence of the singularity seems to be a good thing: if the space time would still be smooth at the horizon it would be impossible to demand these twisted boundary conditions since the circle in τ shrinks to zero size there. If this is however a singular point of the geometry we can not really shrink the circle to zero size. The topology being rather a cylinder than a disc allows now for the presence of the twisted boundary conditions.

It is also important to note that in all formalisms the potential difference between the boundary and the horizon is given by μ. This has a very nice intuitive interpretation. If we bring a unit test charge from the boundary to the horizon we need the energy $\Delta E = \mu$. In the dual field theory this is just the energy cost of adding one unit of charge to the thermalized system and coincides with the elementary definition of the chemical potential.

From now on we will always only consider the genuine finite T, μ contribution that is the only one that arises in formalism (B).

The rest of this review is devoted to the explicit evaluation of these Kubo formulae in two different systems: free chiral fermions and a holographic model implementing the chiral and gravitational anomalies by suitable five dimensional Chern-Simons terms.

17.3 Weyl Fermions

We will now evaluate the Kubo formulae for the chiral magnetic, chiral vortical and energy flux conductivities (17.46) for a theory of N free chiral fermions Ψ^f transforming under a global symmetry group G generated by matrices $(T_a)^f{}_g$.

We denote the generators in the Cartan subalgebra by H_a. Chemical potentials μ_a can be switched on only in the Cartan subalgebra. Furthermore the presence of the chemical potentials breaks the group G to a subgroup $\widehat{G}$. Only the currents that lie in the unbroken subgroup are conserved (up to anomalies) and participate in the hydrodynamics. The chemical potential for the fermion Ψ^f is given by $\mu^f = \sum_a q_a^f \mu_a$, where we write the Cartan generator $H_a = q_a^f \delta^f{}_g$ in terms of its eigenvalues, the charges q_a^f. The unbroken symmetry group $\widehat{G}$ is generated by those matrices $T_a^f{}_g$ fulfilling

$$T_a^f{}_g \mu^g = \mu^f T_a^f{}_g. \tag{17.54}$$

There is no summation over indices in the last expression. From now on we will assume that all currents J_a lie in directions indicated in (17.54). We define the chemical potential through the boundary condition on the fermion fields around the thermal circle, i.e. we adopt the formalism (B) discussed in previous section,

$$\Psi^f(\tau - \beta) = -e^{\beta \mu^f} \Psi^f(\tau). \tag{17.55}$$

Therefore the eigenvalues of ∂_τ are $i\tilde{\omega}_n + \mu^f$ for the fermion spiecies f with $\tilde{\omega}_n = \pi T(2n+1)$ the fermionic Matsubara frequencies. A convenient way of expressing the current and the energy-momentum tensor is in terms of Dirac fermions and writing

$$J_a^i = \sum_{f,g=1}^{N} T_a^g{}_f \bar{\Psi}_g \gamma^i \mathscr{P}_+ \Psi^f, \qquad T^{0i} = \frac{i}{2} \sum_{f=1}^{N} \bar{\Psi}_f \left(\gamma^0 \partial^i + \gamma^i \partial^0\right) \mathscr{P}_+ \Psi^f,$$

$$\tag{17.56}$$

where we used the chiral projector $\mathscr{P}_\pm = \frac{1}{2}(1 \pm \gamma_5)$. The fermion propagator is

$$S(q)^f{}_g = \frac{\delta^f{}_g}{2} \sum_{t=\pm} \Delta_t\left(i\tilde{\omega}^f, \mathbf{q}\right) \mathscr{P}_+ \gamma_\mu \hat{q}_t^\mu, \qquad \Delta_t\left(i\tilde{\omega}^f, q\right) = \frac{1}{i\tilde{\omega}^f - t E_q},$$

$$\tag{17.57}$$

Fig. 17.4 1 loop diagram
contributing to the vortical
conductivity (17.58)

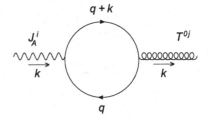

with $i\tilde{\omega}^f = i\tilde{\omega}_n + \mu^f$, $\hat{q}_t^\mu = (1, t\hat{q})$, $\hat{q} = \frac{\mathbf{q}}{E_q}$ and $E_q = |\mathbf{q}|$. For simplicity in the expressions we consider only left-handed fermions, but one can easily include right-handed fermions as well as they contribute in all our calculations in the same way as the left-handed ones up to a relative minus sign.

We will address in detail the computation of the vortical conductivities and sketch only the calculation of the magnetic conductivities since the latter one is a trivial extension of the calculation of the chiral magnetic conductivity in [33]. Then we show the results for the other conductivity coefficients.

17.3.1 Chiral Vortical Conductivity

The vortical conductivity is defined from the retarded correlation function of the current $J_a^i(x)$ and the energy momentum tensor or energy current $T^{0j}(x')$ (17.56), i.e.

$$G_a^V(x - x') = \frac{1}{2}\varepsilon_{ijn}\, i\theta(t - t')\langle[J_a^i(x), T^{0j}(x')]\rangle. \tag{17.58}$$

Going to Fourier space, one can evaluate this quantity as

$$G_a^V(k) = \frac{1}{4}\sum_{f=1}^{N} T_a^f \,_f \frac{1}{\beta}\sum_{\tilde{\omega}^f}\int \frac{d^3 q}{(2\pi)^3}\varepsilon_{ijn}\,\mathrm{tr}[S^f \,_f(q)\gamma^i S^f \,_f(q+k)(\gamma^0 q^j + \gamma^j i\tilde{\omega}^f)], \tag{17.59}$$

which corresponds to the one loop diagram of Fig. 17.4. The vertex of the two quarks with the graviton is $\sim \delta^f \,_g$, and therefore we find only contributions from the diagonal part of the group $\widehat{G}$. The metric we use through this section is the usual one in field theory computations, $g_{\mu\nu} = \mathrm{diag}(1, -1, -1, -1)$. We can split G_a^V into two contributions, i.e.

$$G_a^V(k) = G_{a,(0j)}^V(k) + G_{a,(j0)}^V(k), \tag{17.60}$$

which correspond to the terms $\gamma^0 q^j$ and $\gamma^j i\tilde{\omega}^f$ in (17.59) respectively. We will focus first on the computation of $G_{a,(0j)}^V$. After computation of the Dirac trace in (17.59), this term writes

$$G_{a,(0j)}^V(k) = \frac{1}{8} \sum_{f=1}^{N} T_d^f \; _f \frac{1}{\beta} \sum_{\tilde{\omega}^f} \int \frac{d^3q}{(2\pi)^3} q^j \sum_{t,u=\pm} \left[\varepsilon_{ijn}\left(t\frac{q^i}{E_q} + u\frac{k^i + q^i}{E_{q+k}} \right) \right.$$

$$\left. + i\frac{tu}{E_q E_{q+k}}(q_j k_n - q_n k_j) \right] \Delta_t \left(i\tilde{\omega}^f, \mathbf{k} \right) \Delta_u \left(i\tilde{\omega}^f + i\omega_n, \mathbf{q} + \mathbf{k} \right).$$

$$(17.61)$$

At this point one can make a few simplifications. Note that due to the antisymmetric tensor ε_{ijn}, the two terms proportional to q^i inside the bracket in (17.61) vanish. Regarding the term $\varepsilon_{ijn}q^j k^i$, it leads to a contribution $\sim \varepsilon_{ijn}k^j k^i$ after integration in d^3q, which is zero. Then the only term which remains is the one not involving ε_{ijn}. We can now perform the sum over fermionic Matsubara frequencies. One has

$$\frac{1}{\beta} \sum_{\tilde{\omega}^f} \Delta_t \left(i\tilde{\omega}^f, \mathbf{q} \right) \Delta_u \left(i\tilde{\omega}^f + i\omega_n, \mathbf{q} + \mathbf{p} \right)$$

$$= \frac{tn(E_q - t\mu^f) - un(E_{q+k} - u\mu^f) + \frac{1}{2}(u - t)}{i\omega_n + tE_q - uE_{q+k}}, \qquad (17.62)$$

where $n(x) = 1/(e^{\beta x} + 1)$ is the Fermi-Dirac distribution function. In (17.62) we have considered that $\omega_n = 2\pi T n$ is a bosonic Matsubara frequency. This result is also obtained in Ref. [33]. After doing the analytic continuation, which amounts to replacing $i\omega_n$ by $k_0 + i\varepsilon$ in (17.62), one gets

$$G_{a,(0j)}^V(k) = -\frac{i}{8} \sum_{f=1}^{N} T_d^f \; _f \int \frac{d^3q}{(2\pi)^3} \frac{\mathbf{q}^2 k_n - (\mathbf{q} \cdot \mathbf{k})q_n}{E_q E_{q+k}}$$

$$\times \sum_{t,u=\pm} \frac{un(E_q - t\mu^f) - tn(E_{q+k} - u\mu^f) + \frac{1}{2}(t - u)}{k_0 + i\varepsilon + tE_q - uE_{q+k}}. \qquad (17.63)$$

The term proportional to $\sim \frac{1}{2}(t - u)$ corresponds to the vacuum contribution, and it is ultraviolet divergent. By removing this term the finite temperature and chemical potential behavior is not affected, and the result becomes ultraviolet finite because the Fermi-Dirac distribution function exponentially suppresses high momenta. By making both the change of variable $\mathbf{q} \to -\mathbf{q} - \mathbf{k}$ and the interchange $u \to -t$ and $t \to -u$ in the part of the integrand involving the term $-tn(E_{q+k} - u\mu^f)$, one can express the vacuum substracted contribution of (17.63) as

$$\widehat{G}_{a,(0j)}^V(k) = \frac{i}{8} k_n \sum_{f=1}^{N} T_d^f \; _f \int \frac{d^3q}{(2\pi)^3} \frac{1}{E_q E_{q+k}} \left(\mathbf{q}^2 - \frac{(\mathbf{q} \cdot \mathbf{k})^2}{\mathbf{k}^2} \right)$$

$$\times \sum_{t,u=\pm} u \frac{n(E_q - \mu^f) + n(E_q + \mu^f)}{k_0 + i\varepsilon + tE_q + uE_{q+k}}, \qquad (17.64)$$

where we have used that $n(E_q - t\mu^f) + n(E_q + t\mu^f) = n(E_q - \mu^f) + n(E_q + \mu^f)$ since $t = \pm 1$. The result has to be proportional to k_n, so to reach this expression we have replaced q_n by $(\mathbf{q} \cdot \mathbf{k}) k_n / \mathbf{k}^2$ in (17.63). At this point one can perform the sum over u by using $\sum_{u=\pm} u / (a_1 + u a_2) = -2 a_2 / (a_1^2 - a_2^2)$, and the integration over angles by considering $\mathbf{q} \cdot \mathbf{k} = E_q E_k x$ and $E_{q+k}^2 = E_q^2 + E_k^2 + 2 E_q E_k x$, where $x := \cos(\theta)$ and θ is the angle between $\mathbf{q}$ and $\mathbf{k}$. Then one gets the final result

$$\widehat{G}_{a,(0j)}^V(k) = \frac{i}{16\pi^2} \frac{k_n}{\mathbf{k}^2} (k^2 - k_0^2) \sum_{f=1}^N T_a^f \,_f \int_0^\infty dq \, q f^V(q)$$

$$\times \left[1 + \frac{1}{8qk} \sum_{t=\pm} [k_0^2 - k^2 + 4q(q + tk_0)] \log\left(\frac{\Omega_t^2 - (q+k)^2}{\Omega_t^2 - (q-k)^2} \right) \right], \tag{17.65}$$

where $\Omega_t = k_0 + i\varepsilon + t E_q$, and

$$f^V(q) = n(E_q - \mu^f) + n(E_q + \mu^f). \tag{17.66}$$

The steps to compute $G_{a,(j0)}^V$ in (17.60) are similar. In this case the Dirac trace leads to a different tensor structure, in which the only contribution comes from the trace involving γ_5. The sum over fermionic Matsubara frequencies involves an extra $i\tilde{\omega}^f$, i.e.

$$\frac{1}{\beta} \sum_{\tilde{\omega}^f} i\tilde{\omega}^f \Delta_t(i\tilde{\omega}^f, \mathbf{q}) \Delta_u(i\tilde{\omega}^f + i\omega_n, \mathbf{q} + \mathbf{k})$$

$$= \frac{1}{i\omega_n + t E_q - u E_{q+k}} \left[E_q n(E_q - t\mu^f) - (E_{q+k} - u i\omega_n) n(E_{q+k} - u\mu^f) \right.$$

$$\left. - \frac{1}{2}(E_q - E_{q+k} + u i\omega_n) \right]. \tag{17.67}$$

The last term inside the bracket in the r.h.s. of (17.67) corresponds to the vacuum contribution which we choose to remove, as it leads to an ultraviolet divergent contribution after integration in d^3q. Making similar steps as for $\widehat{G}_{a,(0j)}^V$, i.e. performing the sum over u and integrating over angles, one gets the final result

$$\widehat{G}_{a,(j0)}^V(k) = -\frac{i}{32\pi^2} \frac{k_n}{k^3} \sum_{f=1}^N T_a^f \,_f \int_0^\infty dq \sum_{t=\pm} f_t^V(q, k_0)$$

$$\times \left[4tqkk_0 - (k^2 - k_0^2)(2q + tk_0) \log\left(\frac{\Omega_t^2 - (q+k)^2}{\Omega_t^2 - (q-k)^2} \right) \right], \tag{17.68}$$

where

$$f_t^V(q, k_0) = q f^V(q) + tk_0 n(E_q + t\mu^f). \tag{17.69}$$

The result for the vacuum substracted contribution of the retarded correlation function of the current and the energy momentum tensor, $\widehat{G}_a^V(k)$, writes as a sum of (17.65) and (17.68), according to (17.60). From these expressions one can compute the zero frequency, zero momentum, limit. Since

$$\lim_{k \to 0} \lim_{k_0 \to 0} \sum_{t=\pm} \log\left(\frac{\Omega_t^2 - (q+k)^2}{\Omega_t^2 - (q-k)^2}\right) = \frac{2k}{q}, \qquad (17.70)$$

the relevant integrals are

$$\int_0^\infty dq \, q f^V(q) = \int_0^\infty dq \, f_t^V(q, k_0 = 0) = \frac{(\mu^f)^2}{2} + \frac{\pi^2}{6} T^2. \qquad (17.71)$$

Finally it follows from (17.65) and (17.68) that the zero frequency, zero momentum, vortical conductivity writes

$$(\sigma_V)_a = \frac{1}{8\pi^2} \sum_{f=1}^N T_a^f \, _f \left[(\mu^f)^2 + \frac{\pi^2}{3} T^2\right]$$

$$= \frac{1}{16\pi^2} \left[\sum_{b,c} \mathrm{tr}(T_a\{H_b, H_c\})\mu_b\mu_c + \frac{2\pi^2}{3} T^2 \, \mathrm{tr}(T_a)\right]. \qquad (17.72)$$

Both $\widehat{G}_{a,(0j)}^V$ and $\widehat{G}_{a,(j0)}^V$ lead to the same contribution in $(\sigma_V)_a$. Equation (17.72) was first derived in [35], and it constitutes our main result in this section. The term involving the chemical potentials is induced by the chiral anomaly. More interesting is the term $\sim T^2$ which is proportional to the gravitational anomaly coefficient b_a [49–52]. This means that a non-zero value of this term has to be attributed to the presence of a gravitational anomaly. The Matsubara frequencies $\tilde{\omega}_n = \pi T(2n+1)$ generate a dependence on πT in the final result as compared to the chemical potentials, and then no factors of π show up for the term $\sim T^2$ in (17.72). Right-handed fermions contribute in the same way but with a relative minus sign. Therefore the $\sim T^2$ term appears only when the current in (17.58) has an axial component. The correlator with a vector current does not have this gravitational anomaly contribution.

17.3.2 Chiral Magnetic Conductivity

The chiral magnetic conductivity in the case of a vector and an axial $U(1)$ symmetry was computed at weak coupling in [33]. The corresponding Kubo formula involves the two point function of the current, see first expression in (17.46). Following the same method, we have computed it for the unbroken (non-abelian) symmetry group $\widehat{G}$. The relevant Green function is [35]

$$G^B_{ab}(k) = \frac{1}{2}\sum_{f,g} T^f_{a\ g}T^g_{b\ f}\frac{1}{\beta}\sum_{\tilde{\omega}^f}\int\frac{d^3q}{(2\pi)^3}\varepsilon_{ijn}\,\mathrm{tr}\big[S^f{}_f(q)\gamma^i S^f{}_f(q+k)\gamma^j\big].$$

(17.73)

The evaluation of this expression is exactly as in [33] so we skip the details. The zero frequency, zero momentum, limit of the magnetic conductivity is

$$(\sigma_B)_{ab} = \frac{1}{4\pi^2}\sum_{f,g=1}^{N} T^f_{a\ g}T^g_{b\ f}\,\mu^f = \frac{1}{8\pi^2}\sum_{c}\mathrm{tr}\big(T_a\{T_b, H_c\}\big)\mu_c.$$

(17.74)

In the second equality of (17.74) we have made use of (17.54). No contribution proportional to the gravitational anomaly coefficient is found in this case.

17.3.3 Conductivities for the Energy Flux

We will include for completeness the result of the chiral magnetic and vortical conductivities for the energy flux, corresponding to the last two expressions in (17.46).

The chiral magnetic conductivity for energy flux, σ^ε_B, follows from the correlation function of the energy momentum tensor and the current, and so it computes in the same way as the vortical conductivity in Sect. 17.3.1. From an evaluation of the corresponding Feynman diagram one finds that the result is the same as (17.59). Then one concludes that

$$\big(\sigma^\varepsilon_B\big)_a = (\sigma_V)_a,$$

(17.75)

where $(\sigma_V)_a$ is given by (17.72). Although these coefficients are equal, they describe different transport phenomena. Whereas $(\sigma^\varepsilon_B)_a$ describes the generation of an energy flux due to an external magnetic field $\mathbf{B}_a$, $(\sigma_V)_a$ describes the generation of the current $\mathbf{J}_a$ due to an external field that sources the energy-momentum tensor T^{0i}.

Finally the chiral vortical conductivity for the energy flux, σ^ε_V, follows from the correlation function of two energy momentum tensors. There are three contributions out of the four possible terms. One of these terms involves a sum over fermionic Matsubara frequencies of the form

$$\frac{1}{\beta}\sum_{\tilde{\omega}^f}(i\tilde{\omega}^f)^2\Delta_t(i\tilde{\omega}^f,\mathbf{q})\Delta_u(i\tilde{\omega}^f + i\omega_n, \mathbf{q}+\mathbf{k})$$

$$= \mathscr{F}(i\omega_n, E_q, E_{q+k}, t, u) + \frac{1}{i\omega_n + tE_q - uE_{q+k}}$$

$$\times\big[tE_q^2 n(E_q - t\mu^f) - u(E_{q+k} - ui\omega_n)^2 n(E_{q+k} - u\mu^f)\big],$$

(17.76)

where $\mathscr{F}$ corresponds to the ultraviolet divergent vacuum contribution which we choose to remove. The zero frequency, zero momentum, limit of the chiral vortical conductivity for the energy flux writes

$$\sigma_V^\varepsilon = \frac{1}{12\pi^2} \sum_{f=1}^N [(\mu^f)^3 + \pi^2 T^2 \mu^f]$$

$$= \frac{1}{24\pi^2} \left[\sum_{a,b,c} \mathrm{tr}(H_a\{H_b, H_c\})\mu_a\mu_b\mu_c + 2\pi^2 T^2 \sum_a \mathrm{tr}(H_a)\mu_a \right]. \quad (17.77)$$

This coefficient describes the generation of an energy flux due to a vortex (or a gravito-magnetic field). The correlators (17.75) and (17.77) enter the chiral magnetic and vortical conductivities in the Landau frame, respectively, as defined in [27–29], see (17.48)–(17.49). We have also checked that to lowest order in ω and k one has $\langle T^{0z} T^{0z} \rangle = p$, where p is the pressure of a free gas of massless fermions, and $\langle T^{0z} J^z \rangle = 0$ [34].

17.3.4 Summary and Specialization to the Group $U(1)_V \times U(1)_A$

The results for the different conductivities are neatly summarized as

$$(\dot{\sigma}_B)_{ab} = \frac{1}{4\pi^2} d_{abc} \mu^c, \quad (17.78)$$

$$(\sigma_V)_a = (\sigma_B^\varepsilon)_a = \frac{1}{8\pi^2} d_{abc} \mu^b \mu^c + \frac{T^2}{24} b_a, \quad (17.79)$$

$$\sigma_V^\varepsilon = \frac{1}{12\pi^2} d_{abc} \mu^a \mu^b \mu^c + \frac{T^2}{12} b_a \mu^a. \quad (17.80)$$

The axial and mixed gauge-gravitational anomaly coefficients are defined by

$$d_{abc} = \frac{1}{2} \left[\mathrm{tr}(T_a\{T_b, T_c\})_L - \mathrm{tr}(T_a\{T_b, T_c\})_R \right], \quad (17.81)$$

$$b_a = \mathrm{tr}(T_a)_L - \mathrm{tr}(T_a)_R, \quad (17.82)$$

where the subscripts L, R stand for the contributions of left-handed and right-handed fermions. The result shows that these conductivities are non-zero if and only if the theory features anomalies.

For phenomenological reasons it is interesting to specialize these results to the symmetry group $U(1)_V \times U(1)_A$, i.e. one vector and one axial current with chemical potentials $\mu_L = \mu + \mu_A$, $\mu_R = \mu - \mu_A$, charges $q_{V,A}^L = (1, 1)$ and $q_{V,A}^R = (1, -1)$ for one left-handed and one right-handed fermion. We find (for a vector magnetic field)

$$(\sigma_B)_{VV} = \frac{\mu_A}{2\pi^2}, \qquad (\sigma_B)_{AV} = \frac{\mu}{2\pi^2}, \quad (17.83)$$

$$(\sigma_V)_V = (\sigma_B^\varepsilon)_V = \frac{\mu \mu_A}{2\pi^2}, \qquad (\sigma_V)_A = (\sigma_B^\varepsilon)_A = \frac{\mu^2 + \mu_A^2}{4\pi^2} + \frac{T^2}{12}, \quad (17.84)$$

$$\sigma_V^\varepsilon = \frac{\mu_A}{6\pi^2}(3\mu^2 + \mu_A^2) + \frac{\mu_A}{6}T^2. \tag{17.85}$$

Here $(\sigma_B)_{VV}$ is the chiral magnetic conductivity [33], $(\sigma_B)_{AV}$ describes the generation of an axial current due to a vector magnetic field [57], $(\sigma_V)_V$ is the vector vortical conductivity, $(\sigma_V)_A$ is the axial vortical conductivity, and σ_V^ε is the vortical conductivity for the energy flux. The vector and axial magnetic conductivities for energy flux $(\sigma_B^\varepsilon)_V$ and $(\sigma_B^\varepsilon)_A$ coincide with the chiral vortical conductivities.

17.4 Holographic Model

In this section for simplicity we will consider a holographic system which realizes a single chiral $U(1)$ symmetry with a gauge and mixed gauge-gravitational anomaly [36]. As we saw in the previous section in a more realistic model $U(1)_V \times U(1)_A$ the transport coefficients receive contribution from the gravitational part only in the axial sector. For a study of such a system with a pure gauge anomaly using Kubo formulae, see [45].

17.4.1 Notation and Holographic Anomalies

Let us fix some conventions we will use in the Gravity Theory. We choose the five dimensional metric to be of signature $(-, +, +, +, +)$. Five dimensional indices are denoted with upper case Latin letters. The epsilon tensor has to be distinguished from the epsilon symbol by $\varepsilon_{ABCDE} = \sqrt{-g}\,\varepsilon(ABCDE)$. The symbol is defined by $\varepsilon(rtxyz) = +1$. We assume the metric can be decomposed in ADM like way and define an outward pointing normal vector to the holographic boundary of an asymptotically AdS space $n_A \propto g^{AB}\frac{\partial r}{\partial x^B}$ with unit norm $n_A n^A = 1$. So that the induced metric takes the form

$$h_{AB} = g_{AB} - n_A n_B. \tag{17.86}$$

In general a foliation of the space-time M with timelike surfaces defined through $r(x) = \text{const}$ can be written as

$$ds^2 = \left(N^2 + N_A N^A\right)dr^2 + 2N_A\,dx^A\,dr + h_{AB}\,dx^A\,dx^B. \tag{17.87}$$

The Christoffel symbols, Riemann tensor and extrinsic curvature are given by

$$\Gamma^M_{NP} = \frac{1}{2}g^{MK}(\partial_N g_{KP} + \partial_P g_{KM} - \partial_K g_{NP}), \tag{17.88}$$

$$R^M{}_{NPQ} = \partial_P \Gamma^M_{NQ} - \partial_Q \Gamma^M_{NP} + \Gamma^M_{PK}\Gamma^K_{NQ} - \Gamma^M_{QK}\Gamma^K_{NP}, \tag{17.89}$$

$$K_{AV} = h_A^C \nabla_C n_V = \frac{1}{2} \pounds_n h_{AB}, \tag{17.90}$$

where $\pounds_n$ denotes the Lie derivative in direction of n_A. Finally we can define our model. The action is given by

$$S = \frac{1}{16\pi G} \int_M d^5 x \sqrt{-g} \left[R + 2\Lambda - \frac{1}{4} F_{MN} F^{MN} + \varepsilon^{MNPQR} A_M \right.$$

$$\left. \times \left(\frac{\kappa}{3} F_{NP} F_{QR} + \lambda R^A{}_{BNP} R^B{}_{AQR} \right) \right] + S_{GH} + S_{CSK}, \tag{17.91}$$

$$S_{GH} = \frac{1}{8\pi G} \int_{\partial M} d^4 x \sqrt{-h} K, \tag{17.92}$$

$$S_{CSK} = -\frac{1}{2\pi G} \int_{\partial M} d^4 x \sqrt{-h} \, \lambda n_M \varepsilon^{MNPQR} A_N K_{PL} D_Q K_R^L, \tag{17.93}$$

where S_{GH} is the usual Gibbons-Hawking boundary term and D_A is the induced covariant derivative on the four dimensional surface. The second boundary term S_{CSK} is introduced to reproduce the gravitational anomaly at general hypersurface.

Let us study now the gauge symmetries of our model. We note that the action is diffeomorphism invariant, but they do depend explicitly on the gauge connection A_M. Under gauge transformations $\delta A_M = \nabla_M \xi$ they are therefore invariant only up to a boundary term. We have

$$\delta S = \frac{1}{16\pi G} \int_{\partial M} d^4 x \sqrt{-h} \, \xi \varepsilon^{MNPQR} \left(\frac{\kappa}{3} n_M F_{NP} F_{QR} + \lambda n_M R^A{}_{BNP} R^B{}_{AQR} \right)$$

$$- \frac{\lambda}{4\pi G} \int_{\partial M} d^4 x \sqrt{-h} \, n_M \varepsilon^{MNPQR} D_N \xi K_{PL} D_Q K_R^L. \tag{17.94}$$

Now without loss of generality we can choose the gauge $N = 1$ and $N_A = 0$ which defines the so called Gaussian normal coordinates, and the metric takes the form $ds^2 = dr^2 + \gamma_{ij} dx^i dx^j$. After doing the decomposition in terms of induced surface and orthogonal fields, all the terms depending on the extrinsic curvature cancel thanks to the contributions from S_{CSK}! The gauge variation of the action depends only on the intrinsic four dimensional curvature of the boundary and is given by

$$\delta S = \frac{1}{16\pi G} \int_{\partial M} d^4 x \sqrt{-h} \varepsilon^{mnkl} \left(\frac{\kappa}{3} \widehat{F}_{mn} \widehat{F}_{kl} + \lambda \widehat{R}^i{}_{jmn} \widehat{R}^j{}_{ikl} \right). \tag{17.95}$$

This has to be interpreted as the anomalous variation of the effective quantum action of the dual field theory. As consequence of the discussion in Sect. 17.2.1 we can recognize the form of the consistent anomaly and use (17.9) to fix κ for a single fermion transforming under a $U(1)_L$ symmetry. Similarly we can fix λ by matching to the gravitational anomaly of a single left-handed fermion (17.21) and find

$$-\frac{\kappa}{48\pi G} = \frac{1}{96\pi^2}, \qquad -\frac{\lambda}{16\pi G} = \frac{1}{768\pi^2}. \tag{17.96}$$

The bulk equations of motion are

$$G_{MN} - \Lambda g_{MN} = \frac{1}{2} F_{ML} F_N{}^L - \frac{1}{8} F^2 g_{MN} + 2\lambda \varepsilon_{LPQR(M} \nabla_B \left(F^{PL} R^B{}_{N)}{}^{QR} \right), \tag{17.97}$$

$$\nabla_N F^{NM} = -\varepsilon^{MNPQR} \left(\kappa F_{NP} F_{QR} + \lambda R^A{}_{BNP} R^B{}_{AQR} \right). \tag{17.98}$$

A remarkable fact is that the mixed Chern-Simons term does not introduce new singularities into the on-shell action for any asymptotically *AdS* solution, i.e. no new counterterm is needed to renormalize the theory. See [36] for a detailed discussion of the renormalization of the model and Appendix 1 to see the counterterms.

17.4.2 Applying Kubo Formulae and Linear Response

In order to compute the conductivities under study using the Kubo formulae (17.46), we will use tools of linear response theory. To do so we introduce metric and gauge fluctuations over a charged black hole background and use the AdS/CFT dictionary to compute the retarded propagators [58, 59]. Therefore we split the backgrounds and fluctuations as,

$$g_{MN} = g_{MN}^{(0)} + \varepsilon h_{MN}, \tag{17.99}$$

$$A_M = A_M^{(0)} + \varepsilon a_M. \tag{17.100}$$

After the insertion of these fluctuations and background fields in the action and expanding up to second order in ε we can read the on-shell boundary second order action which is needed to get the desired propagators [60],

$$\delta S_{ren}^{(2)} = \int \frac{d^d k}{(2\pi)^d} \left\{ \Phi_{-k}^I \mathscr{A}_{IJ} \Phi_k^{\prime J} + \Phi_{-k}^I \mathscr{B}_{IJ} \Phi_k^J \right\} \Big|_{r \to \infty}, \tag{17.101}$$

where prime means derivative with respect to the radial coordinate, Φ_k^I is a vector constructed with the Fourier transformed components of a_M and h_{MN},

$$\Phi^I \left(r, x^\mu \right) = \int \frac{d^d k}{(2\pi)^d} \Phi_k^I(r) e^{-i\omega t + i\mathbf{k}\mathbf{x}}, \tag{17.102}$$

and $\mathscr{A}$ and $\mathscr{B}$ are two matrices extracted from the boundary action and that we will show below.

For a coupled system the holographic computation of the correlators consists in finding a maximal set of linearly independent solutions that satisfy infalling boundary conditions on the horizon and that source a single operator at the AdS boundary [58–61]. To do so we can construct a matrix of solutions $F^I{}_J(k, r)$ such that each of its columns corresponds to one of the independent solutions and normalize it to the unit matrix at the boundary. Therefore, given a set of boundary values for the perturbations, φ_k^I, the bulk solutions are

$$\Phi_k^I(r) = F^I{}_J(k, r)\varphi_k^J. \qquad (17.103)$$

Finally using this decomposition we obtain the matrix of retarded Green's functions

$$G_{IJ}(k) = -2 \lim_{r \to \infty} \left(\mathscr{A}_{IM}\left(F^M{}_J(k, r)\right)' + \mathscr{B}_{IJ}\right). \qquad (17.104)$$

The system of equations (17.97)–(17.98) admit the following exact background AdS Reissner-Nordström black-brane solution

$$ds^2 = \frac{r^2}{L^2}\left(-f(r)\, dt^2 + d\mathbf{x}^2\right) + \frac{L^2}{r^2 f(r)}\, dr^2, \qquad (17.105)$$

$$A^{(0)} = \phi(r)\, dt = \left(v - \frac{\mu r_H^2}{r^2}\right) dt, \qquad (17.106)$$

where the horizon of the black hole is located at $r = r_H$ and the blackening factor of the metric is

$$f(r) = 1 - \frac{ML^2}{r^4} + \frac{Q^2 L^2}{r^6}. \qquad (17.107)$$

The parameters M and Q of the RN black hole are related to the chemical potential μ and the horizon r_H by[7]

$$M = \frac{r_H^4}{L^2} + \frac{Q^2}{r_H^2}, \qquad Q = \frac{\mu r_H^2}{\sqrt{3}}. \qquad (17.108)$$

The Hawking temperature is given in terms of these black hole parameters as

$$T = \frac{r_H^2}{4\pi L^2} f'(r_H) = \frac{(2r_H^2 M - 3Q^2)}{2\pi r_H^5}. \qquad (17.109)$$

The pressure of the gauge theory is $P = \frac{M}{16\pi G L^3}$ and its energy density is $\varepsilon = 3P$ due to the underlying conformal symmetry.

---

[7]The chemical potential is introduced as the energy needed to introduce an unit of charge from the boundary to behind the horizon $A(\infty) - A(r_H)$ which corresponds to the prescription (B) in Table 17.1. Observe that we have left the source value $A(\infty) = v$ as an arbitrary constant for reasons we will explain later.

To study the effect of anomalies we just turned on the shear sector (transverse momentum fluctuations) a_α and h_t^α and set without loss of generality the momentum k in the y-direction at zero frequency, so $\alpha = x, z$. Since we are interested in the hydrodynamical regime ($k, \omega \ll T$), it is just necessary to find solutions up to first order in momentum. So that we expand the fields in terms of the dimensionless momentum $p = k/4\pi T$ such as

$$h_t^\alpha(r) = h_t^{(0)\alpha}(r) + p h_t^{(1)\alpha}(r), \tag{17.110}$$

$$B_\alpha(r) = B_\alpha^{(0)}(r) + p B_\alpha^{(1)}(r), \tag{17.111}$$

with the gauge field redefined as $B_\alpha = a_\alpha/\mu$. For convenience we redefine new parameters and radial coordinate

$$\bar\lambda = \frac{4\mu\lambda L}{r_H^2}; \qquad \bar\kappa = \frac{4\mu\kappa L^3}{r_H^2}; \qquad a = \frac{\mu^2 L^2}{3r_H^2}; \qquad u = \frac{r_H^2}{r^2}. \tag{17.112}$$

In this new radial coordinate the horizon sits at $u = 1$ and the AdS boundary at $u = 0$. At zero frequency the system of differential equations consists on four second order equations.[8] The relevant physical boundary conditions on fields are: $h_t^\alpha(0) = \tilde{H}^\alpha$, $B_\alpha(0) = \tilde{B}_\alpha$; where the 'tilde' parameters are the sources of the boundary operators. The second condition compatible with the ingoing one at the horizon is regularity for the gauge field and vanishing for the metric fluctuation [34].

After solving the system perturbatively (see [36] for solutions), we can go back to the formula (17.104) and compute the corresponding holographic Green's functions. If we consider the vector of fields to be

$$\Phi_k^{\mathsf{T}}(u) = \left(B_x(u), h_t^x(u), B_z(u), h_t^z(u)\right), \tag{17.113}$$

the $\mathscr{A}$ and $\mathscr{B}$ matrices for that setup take the following form

$$\mathscr{A} = \frac{r_H^4}{16\pi G L^5} \, \mathrm{Diag}\left(-3af, \frac{1}{u}, -3af, \frac{1}{u}\right), \tag{17.114}$$

$$\mathscr{B}_{AdS+\partial} = \frac{r_H^4}{16\pi G L^5} \begin{pmatrix} 0 & -3a & \frac{4\kappa ik\mu^2\phi L^5}{3r_H^4} & 0 \\ 0 & -\frac{3}{u^2} & 0 & 0 \\ -\frac{4\kappa ik\mu^2\phi L^5}{3r_H^4} & 0 & 0 & -3a \\ 0 & 0 & 0 & -\frac{3}{u^2} \end{pmatrix}, \tag{17.115}$$

---

[8]The complete system of equations depending on frequency and momentum is showed in Appendix 2. The system consists of six dynamical equations and two constraints.

$$\mathscr{B}_{CT} = \frac{r_H^4}{16\pi G L^5} \begin{pmatrix} 0 & 0 & 0 & 0 \\ 0 & \frac{3}{u^2\sqrt{f}} & 0 & 0 \\ 0 & 0 & 0 & 0 \\ 0 & 0 & 0 & \frac{3}{u^2\sqrt{f}} \end{pmatrix}, \qquad (17.116)$$

where $\mathscr{B} = \mathscr{B}_{AdS+\partial} + \mathscr{B}_{CT}$.[9] Notice that there is no contribution to the matrices coming from the Chern-Simons gravity part, because the corresponding contributions vanish at the boundary. These matrices and the perturbative solutions are the ingredients to compute the matrix of propagators. Undoing the vector field redefinition introduced in (17.111) the non-vanishing retarded correlation functions at zero frequency are then

$$G_{x,tx} = G_{z,tz} = \frac{\sqrt{3}Q}{4\pi G L^3}, \qquad (17.117)$$

$$G_{x,z} = -G_{z,x} = \frac{i\sqrt{3}kQ\kappa}{2\pi G r_H^2} + \frac{ik\nu\kappa}{6\pi G}, \qquad (17.118)$$

$$G_{x,tz} = G_{tx,z} = -G_{z,tx} = -G_{tz,x} = \frac{3ikQ^2\kappa}{4\pi G r_H^4} + \frac{2ik\lambda\pi T^2}{G}, \qquad (17.119)$$

$$G_{tx,tx} = G_{tz,tz} = \frac{M}{16\pi G L^3}, \qquad (17.120)$$

$$G_{tx,tz} = -G_{tz,tx} = +\frac{i\sqrt{3}kQ^3\kappa}{2\pi G r_H^6} + \frac{4\pi i\sqrt{3}kQT^2\lambda}{G r_H^2}. \qquad (17.121)$$

We can do an important remark observing (17.118). Remember that we left the boundary value of the background gauge field (17.106) arbitrary as a constant ν. But as the $U(1)$ symmetry is anomalous in the Field Theory side, physical quantities have to be sensitive to the source A_0,[10] indeed as we can check they are. In particular if we choose the value $\nu = \mu$ which corresponds to formalism (A) in Table 17.1, we need to include the counterterm (17.29) in order to get the same propagator as at weak coupling. In fact in [44, 45] it has been shown that in the case of a propagator between two vector currents, choosing this specific value for ν the propagator would be zero, giving us in consequence a zero value for the chiral magnetic conductivity. Hence to be consistent with the scheme we are working at, let us just consider ν as a source in the field theory. Therefore the real propagator is the one with $\nu = 0$ because as is well known we have to set all sources to zero after taking the second

---

[9] $\mathscr{B}_{CT}$ is coming from the counterterms of the theory.

[10] In principle A_0 could be gauged away for the symmetric case and in consequence observables should not depend on its value. For example look at [45] to see how in presence of a $U(1)_V \times U(1)_A$ symmetry with only the $U(1)_V$ conserved, propagators do not depend on the specific value of the zero component of the vector gauge source V_0.

functional derivative of the effective action. Finally using the Kubo formulae (17.46) we recover the vortical and axial-magnetic conductivities

$$\sigma_B = -\frac{\sqrt{3}\,Q\kappa}{2\pi\,Gr_{\mathrm{H}}^2} = \frac{\mu}{4\pi^2}, \tag{17.122}$$

$$\sigma_V = \sigma_B^{\varepsilon} = -\frac{3Q^2\kappa}{4\pi\,G\bar{r}_{\mathrm{H}}^4} - \frac{2\lambda\pi\,T^2}{G} = \frac{\mu^2}{8\pi^2} + \frac{T^2}{24}, \tag{17.123}$$

$$\sigma_V^{\varepsilon} = -\frac{\sqrt{3}\,Q^3\kappa}{2\pi\,Gr_{\mathrm{H}}^6} - \frac{4\pi\sqrt{3}\,QT^2\lambda}{Gr_{\mathrm{H}}^2} = \frac{\mu^3}{12\pi^2} + \frac{\mu T^2}{12}. \tag{17.124}$$

All these expressions coincide with the results in Sect. 17.3, (17.78), (17.79) and (17.80) if we specialize to $d_{abc} = 1$ and $b_a = 1$. They are in perfect agreement with the literature [27–29, 34] except for the contribution coming from the gravitational anomaly which is manifest by the presence of the extra λT^2. All the numerical coefficients coincide precisely with the ones obtained at weak coupling; this we take as a strong hint that the anomalous conductivities are indeed completely determined by the anomalies and are not renormalized beyond one loop. Evidence for non-renormalization comes also from [62] where a holographic renormalization group running of the conductivities showed the same result at any value of the holographic cut-off. We also point out that the T^3 term that appears as undetermined integration constant in the hydrodynamic considerations in [63] should make its appearance in σ_V^{ε}. We do not find any such term which is consistent with the argument that this term is absent due to CPT invariance.

It is also interesting to write down the magnetic and vortical conductivities using (17.48) and (17.49) as they appear in the Landau frame to compare with the Son and Surowka form [29]

$$\xi_B = -\frac{\sqrt{3}\,Q(ML^2 + 3r_{\mathrm{H}}^4)\kappa}{8\pi\,GML^2r_{\mathrm{H}}^2} + \frac{\sqrt{3}\,Q\lambda\pi\,T^2}{GM} = \frac{1}{4\pi^2}\left(\mu - \frac{1}{2}\frac{n(\mu^2 + \frac{\pi^2T^2}{3})}{\varepsilon + P}\right), \tag{17.125}$$

$$\xi_V = -\frac{3Q^2\kappa}{4\pi\,GML^2} - \frac{2\pi\lambda T^2(r_{\mathrm{H}}^6 - 2L^2Q^2)}{GML^2r_{\mathrm{H}}^2}$$

$$= \frac{\mu^2}{8\pi^2}\left(1 - \frac{2}{3}\frac{n\mu}{\varepsilon + P}\right) + \frac{T^2}{24}\left(1 - \frac{2n\mu}{\varepsilon + P}\right). \tag{17.126}$$

These expressions agree with the literature except for the λT^2 term. A last comment can be done, the shear viscosity entropy ratio is not modified by the presence of the gravitational anomaly. We know that $\eta \propto \lim_{\omega \to 0} \frac{1}{\omega}\langle T^{xy}T^{xy}\rangle_{k=0}$, so we should solve the system at $k = 0$ for the fluctuations h_y^i, but the anomalous coefficients always appear with a momentum k (see Appendix 2), therefore if we switch off the momentum, the system looks precisely as the theory without anomalies. In [64] it has been shown that the black hole entropy doesn't depend on the extra Chern-Simons term.[11]

17.5 Conclusion and Outlook

In the presence of external sources for the energy momentum tensor and the currents, the anomaly is responsible for a non conservation of the latter. This is conveniently expressed through [52]

$$D_\mu J_a^\mu = \varepsilon^{\mu\nu\rho\lambda} \left(\frac{d_{abc}}{32\pi^2} F_{\mu\nu}^b F_{\rho\lambda}^c + \frac{b_a}{768\pi^2} R^\alpha_{\ \beta\mu\nu} R^\beta_{\ \alpha\rho\lambda} \right), \qquad (17.127)$$

where the axial and mixed gauge-gravitational anomaly coefficients, d_{abc} and b_a, are given by (17.22) and (17.23) respectively.

We have discussed in Sect. 17.2 the constitutive relations and derived the Kubo formulae that allow the calculation of transport coefficients at first order in the hydrodynamic expansion. We explained also subtleties in the definition of the chemical potential in the presence of anomalies. The explicit evaluation of these Kubo formulae in quantum field theory has been performed in Sect. 17.3 for the chiral magnetic, chiral vortical and energy flux conductivities of a relativistic fluid at weak coupling, and we found contributions proportional to the anomaly coefficients d_{abc} and b_a. Non-zero values of these coefficients are a necessary and sufficient condition for the presence of anomalies [52]. Therefore the non-vanishing values of the transport coefficients have to be attributed to the presence of chiral and gravitational anomalies.

In order to perform the analysis at strong coupling via AdS/CFT methods, we have defined in Sect. 17.4 a holographic model implementing both type of anomalies via gauge and mixed gauge-gravitational Chern-Simons terms. We have computed the anomalous magnetic and vortical conductivities from a charged black hole background and have found a non-vanishing vortical conductivity proportional to $\sim T^2$. These terms are characteristic for the contribution of the gravitational anomaly and they even appear in an uncharged fluid. The T^2 behavior had appeared already previously in neutrino physics [17–20]. In [30] similar terms in the vortical conductivities have been argued for, but just in terms of undetermined integration constants without any relation to the gravitational anomaly. Very recently a generalization of the results (17.122)–(17.124) to any even space-time dimension as a polynomial in μ and T [39] has been proposed. Finally, the consequences of this anomaly in hydrodynamics have been studied using a group theoretic approach, which seems to suggest that their effects could be present even at $T = 0$ [66]. The numerical values of the anomalous conductivities at strong coupling are in perfect agreement with weak coupling calculations, and this suggests the existence of a non-renormalization theorem including the contributions from the gravitational anomaly.

---

[11]For a four dimensional holographic model with gravitational Chern-Simons term and a scalar field this has also been shown in [65].

There are important phenomenological consequences of the present study to heavy ion physics. In [67] enhanced production of high spin hadrons (especially Ω^- baryons) perpendicular to the reaction plane in heavy ion collisions has been proposed as an observational signature for the chiral separation effect. Three sources of chiral separation have been identified: the anomaly in vacuum, the magnetic and the vortical conductivities of the axial current J_A^μ. Of these the contribution of the vortical effect was judged to be subleading by a relative factor of 10^{-4}. The T^2 term in (17.123) leads however to a significant enhancement. If we take μ to be the baryon chemical potential $\mu \approx 10$ MeV, neglect μ_A as in [67] and take a typical RHIC temperature of $T = 350$ MeV, we see that the temperature enhances the axial chiral vortical conductivity by a factor of the order of 10^4. We expect the enhancement at the LHC to be even higher due to the higher temperature.

In this review we have presented the computation of the transport coefficients, and in particular their gravitational anomaly contributions, via Kubo formulae. It would be interesting to calculate directly the constitutive relations of the hydrodynamics of anomalous currents via the fluid/gravity correspondence within the holographic model of Sect. 17.4, [27, 28, 68]. This approach will allow us to compute transport coefficients at higher orders [69, 70]. This study is currently in progress [71].

Acknowledgements This work has been supported by Plan Nacional de Altas Energías FPA2009-07908, FPA2008-01430 and FPA2011-25948, CPAN (CSD2007-00042), Comunidad de Madrid HEP-HACOS S2009/ESP-1473. E. Megías would like to thank the Institute for Nuclear Theory at the University of Washington, USA, and the Institut für Theoretische Physik at the Technische Universität Wien, Austria, for their hospitality and partial support during the completion of this work. The research of E.M. is supported by the Juan de la Cierva Program of the Spanish MICINN. F.P. has been supported by fellowship FPI Comunidad de Madrid.

Appendix 1: Boundary Counterterms

The result one gets for the counterterm coming from the regularization of the boundary action of the holographic model in Sect. 17.4 is

$$S_{ct} = -\frac{3}{8\pi G} \int_{\partial M} d^4 x \sqrt{-h} \left[1 + \frac{1}{2} P - \frac{1}{12} \left(P_j^i P_i^j - P^2 - \frac{1}{4} \widehat{F}_{ij} \widehat{F}^{ij} \right) \log e^{-2\rho} \right], \tag{17.128}$$

where hat on the fields means the induced field on the cut-off surface and

$$P = \frac{1}{6} \widehat{R}, \qquad P_j^i = \frac{1}{2} [\widehat{R}_j^i - P \delta_j^i]. \tag{17.129}$$

As a remarkable fact there is no contribution in the counterterm coming from the gauge-gravitational Chern-Simons term. This has also been derived in [72] in a similar model that does however not contain S_{CSK}.

Appendix 2: Equations of Motion for the Shear Sector

These are the complete linearized set of six dynamical equations of motion,

$$0 = B_\alpha''(u) + \frac{f'(u)}{f(u)} B_\alpha'(u) + \frac{b^2}{u f(u)^2}\left(\omega^2 - f(u)k^2\right) B_\alpha(u) - \frac{h_t^{\alpha'}(u)}{f(u)}$$

$$+ ik\varepsilon_{\alpha\beta}\left(\frac{3}{u f(u)}\bar{\lambda}\left(\frac{2}{3a}\left(f(u) - 1\right) + u^3\right)h_t^{\beta'}(u) + \bar{\kappa}\frac{B_\beta(u)}{f(u)}\right), \qquad (17.130)$$

$$0 = h_t^{\alpha''}(u) - \frac{h_t^{\alpha'}(u)}{u} - \frac{b^2}{u f(u)}\left(k^2 h_t^\alpha(u) + h_y^\alpha(u)\omega k\right) - 3au B_\alpha'(u)$$

$$\times i\bar{\lambda}k\varepsilon_{\alpha\beta}\left[\left(24au^3 - 6(1 - f(u))\right)\frac{B_\beta(u)}{u} + \left(9au^3 - 6(1 - f(u))\right)B_\beta'(u)\right.$$

$$\left. + 2u\left(uh_t^{\beta'}(u)\right)' - \frac{2ub^2}{f(u)}\left(h_y^\beta(u)\omega k + h_t^\beta(u)k^2\right)\right], \qquad (17.131)$$

$$0 = h_y^{\alpha''}(u) + \frac{(f/u)'}{f/u}h_y^{\alpha'}(u) + \frac{b^2}{u f(u)^2}\left(\omega^2 h_y^\alpha(u) + \omega k h_t^\alpha(u)\right) + 2uik\bar{\lambda}\varepsilon_{\alpha\beta}\left[uh_y^{\beta''}(u)\right.$$

$$\left. + \left(9f(u) - 6 + 3au^3\right)\frac{h_y^{\beta'}(u)}{f(u)} + \frac{b^2}{f(u)^2}\left(\omega k h_t^\beta(u) + w^2 h_y^\beta(u)\right)\right], \qquad (17.132)$$

and two constraints for the fluctuations at $\omega, k \neq 0$

$$0 = \omega\left(h_t^{\alpha'}(u) - 3au B_\alpha(u)\right) + f(u)kh_y^{\alpha'}(u) + ik\bar{\lambda}\varepsilon_{\alpha\beta}\left[2u^2\left(\omega h_t^{\beta'} + f(u)kh_y^{\beta'}(u)\right)\right.$$

$$\left. + \left(9au^3 - 6(1 - f(u))\right)B_\beta(u)\right]. \qquad (17.133)$$

References

1. R.A. Bertlmann, International Series of Monographs on Physics, vol. 91 (Clarendon, Oxford, 1996), p. 566
2. F. Bastianelli, P. van Nieuwenhuizen (Cambridge University Press, Cambridge, 2006), p. 379
3. K. Fujikawa, H. Suzuki (Clarendon, Oxford, 2004), p. 284
4. L.D. Landau, E.M. Lifshitz, 2nd edn. (Butterworth-Heinemann, Offord, January 15, 1987), p. 556
5. I. Müller, Z. Phys. **198**, 329 (1967)
6. W. Israel, Ann. Phys. **100**, 310 (1976)
7. W. Israel, J.M. Stewart, Ann. Phys. **118**, 341 (1979)
8. R. Baier, P. Romatschke, D.T. Son, A.O. Starinets, M.A. Stephanov, J. High Energy Phys. **0804**, 100 (2008). arXiv:0712.2451 [hep-th]
9. S. Bhattacharyya, V.E. Hubeny, S. Minwalla, M. Rangamani, J. High Energy Phys. **0802**, 045 (2008). arXiv:0712.2456 [hep-th]

10. J.M. Maldacena, Adv. Theor. Math. Phys. **2**, 231 (1998) [Int. J. Theor. Phys. **38**, 1113 (1999)]. hep-th/9711200
11. E. Witten, Adv. Theor. Math. Phys. **2**, 253 (1998). hep-th/9802150
12. S.S. Gubser, I.R. Klebanov, A.M. Polyakov, Phys. Lett. B **428**, 105 (1998). hep-th/9802109
13. S.S. Gubser, I.R. Klebanov, A.W. Peet, Phys. Rev. D **54**, 3915 (1996). hep-th/9602135
14. E. Witten, Adv. Theor. Math. Phys. **2**, 505 (1998). hep-th/9803131
15. P. Kovtun, D.T. Son, A.O. Starinets, Phys. Rev. Lett. **94**, 111601 (2005). hep-th/0405231
16. J. Casalderrey-Solana, H. Liu, D. Mateos, K. Rajagopal, U.A. Wiedemann, arXiv:1101.0618 [hep-th]
17. A. Vilenkin, Phys. Rev. D **20**, 1807 (1979)
18. A. Vilenkin, Phys. Rev. D **21**, 2260 (1980)
19. A. Vilenkin, Phys. Rev. D **22**, 3067 (1980)
20. A. Vilenkin, Phys. Rev. D **22**, 3080 (1980)
21. M. Giovannini, M.E. Shaposhnikov, Phys. Rev. D **57**, 2186 (1998). hep-ph/9710234
22. A.Y. Alekseev, V.V. Cheianov, J. Frohlich, Phys. Rev. Lett. **81**, 3503 (1998). cond-mat/9803346
23. G.M. Newman, D.T. Son, Phys. Rev. D **73**, 045006 (2006). hep-ph/0510049
24. G.M. Newman, J. High Energy Phys. **0601**, 158 (2006). hep-ph/0511236
25. D.E. Kharzeev, L.D. McLerran, H.J. Warringa, Nucl. Phys. A **803**, 227 (2008). arXiv:0711.0950 [hep-ph]
26. K. Fukushima, D.E. Kharzeev, H.J. Warringa, Phys. Rev. D **78**, 074033 (2008). arXiv:0808.3382 [hep-ph]
27. J. Erdmenger, M. Haack, M. Kaminski, A. Yarom, J. High Energy Phys. **0901**, 055 (2009). arXiv:0809.2488 [hep-th]
28. N. Banerjee, J. Bhattacharya, S. Bhattacharyya, S. Dutta, R. Loganayagam, P. Surowka, J. High Energy Phys. **1101**, 094 (2011). arXiv:0809.2596 [hep-th]
29. D.T. Son, P. Surowka, Phys. Rev. Lett. **103**, 191601 (2009). arXiv:0906.5044 [hep-th]
30. Y. Neiman, Y. Oz, J. High Energy Phys. **1103**, 023 (2011). arXiv:1011.5107 [hep-th]
31. T. Kalaydzhyan, I. Kirsch, Phys. Rev. Lett. **106**, 211601 (2011). arXiv:1102.4334 [hep-th]
32. I. Gahramanov, T. Kalaydzhyan, I. Kirsch, Phys. Rev. D **85**, 126013 (2012). arXiv:1203.4259 [hep-th]
33. D.E. Kharzeev, H.J. Warringa, Phys. Rev. D **80**, 034028 (2009). arXiv:0907.5007 [hep-ph]
34. I. Amado, K. Landsteiner, F. Pena-Benitez, J. High Energy Phys. **1105**, 081 (2011). arXiv:1102.4577 [hep-th]
35. K. Landsteiner, E. Megias, F. Pena-Benitez, Phys. Rev. Lett. **107**, 021601 (2011). arXiv:1103.5006 [hep-ph]
36. K. Landsteiner, E. Megias, L. Melgar, F. Pena-Benitez, J. High Energy Phys. **1109**, 121 (2011). arXiv:1107.0368 [hep-th]
37. S. Golkar, D.T. Son, arXiv:1207.5806 [hep-th]
38. K. Jensen, R. Loganayagam, A. Yarom, J. High Energy Phys. **1302**, 088 (2013). arXiv:1207.5824 [hep-th]
39. R. Loganayagam, P. Surowka, J. High Energy Phys. **1204**, 097 (2012). arXiv:1201.2812 [hep-th]
40. R. Loganayagam, arXiv:1106.0277 [hep-th]
41. K. Jensen, Phys. Rev. D **85**, 125017 (2012). arXiv:1203.3599 [hep-th]
42. N. Banerjee, J. Bhattacharya, S. Bhattacharyya, S. Jain, S. Minwalla, T. Sharma, J. High Energy Phys. **1209**, 046 (2012). arXiv:1203.3544 [hep-th]
43. H.-U. Yee, J. High Energy Phys. **0911**, 085 (2009). arXiv:0908.4189 [hep-th]
44. A. Rebhan, A. Schmitt, S.A. Stricker, J. High Energy Phys. **1001**, 026 (2010). arXiv:0909.4782 [hep-th]
45. A. Gynther, K. Landsteiner, F. Pena-Benitez, A. Rebhan, J. High Energy Phys. **1102**, 110 (2011). arXiv:1005.2587 [hep-th]
46. K. Landsteiner, E. Megias, L. Melgar, F. Pena-Benitez, J. Phys. Conf. Ser. **343**, 012073 (2012). arXiv:1111.2823 [hep-th]

47. W.A. Bardeen, Phys. Rev. **184**, 1848 (1969)
48. N. Dragon, F. Brandt, arXiv:1205.3293 [hep-th]
49. T. Kimura, Prog. Theor. Phys. **42**(5), 1191 (1969)
50. R. Delbourgo, A. Salam, Phys. Lett. B **40**, 381 (1972)
51. T. Eguchi, P.G.O. Freund, Phys. Rev. Lett. **37**, 1251 (1976)
52. L. Alvarez-Gaume, E. Witten, Nucl. Phys. B **234**, 269 (1984)
53. T.S. Evans, hep-ph/9510298
54. G.D. Moore, M. Tassler, J. High Energy Phys. **1102**, 105 (2011). arXiv:1011.1167 [hep-ph]
55. K. Landsteiner, E. Megias, L. Melgar, F. Pena-Benitez, Fortschr. Phys. **60**(9–10), 1064–1070 (2012). doi:10.1002/prop.201200021
56. K. Landsteiner, E. Megias, F. Pena-Benitez, arXiv:1110.3615 [hep-ph]
57. D.T. Son, A.R. Zhitnitsky, Phys. Rev. D **70**, 074018 (2004). hep-ph/0405216
58. D.T. Son, A.O. Starinets, J. High Energy Phys. **0209**, 042 (2002). hep-th/0205051
59. C.P. Herzog, D.T. Son, J. High Energy Phys. **0303**, 046 (2003). hep-th/0212072
60. M. Kaminski, K. Landsteiner, J. Mas, J.P. Shock, J. Tarrio, J. High Energy Phys. **1002**, 021 (2010). arXiv:0911.3610 [hep-th]
61. I. Amado, M. Kaminski, K. Landsteiner, J. High Energy Phys. **0905**, 021 (2009). arXiv:0903.2209 [hep-th]
62. K. Landsteiner, L. Melgar, J. High Energy Phys. **1210**, 131 (2012). arXiv:1206.4440 [hep-th]
63. Y. Neiman, Y. Oz, J. High Energy Phys. **1109**, 011 (2011). arXiv:1106.3576 [hep-th]
64. L. Bonora, M. Cvitan, P.D. Prester, S. Pallua, I. Smolic, Class. Quant. Grav. **28**, 195009 (2011). arXiv:1105.4792 [hep-th]
65. T. Delsate, V. Cardoso, P. Pani, J. High Energy Phys. **1106**, 055 (2011). arXiv:1103.5756 [hep-th]
66. V.P. Nair, R. Ray, S. Roy, Phys. Rev. D **86**, 025012 (2012). arXiv:1112.4022 [hep-th]
67. B. Keren-Zur, Y. Oz, J. High Energy Phys. **1006**, 006 (2010). arXiv:1002.0804 [hep-ph]
68. S. Bhattacharyya, V.E. Hubeny, S. Minwalla, M. Rangamani, J. High Energy Phys. **0802**, 045 (2008). arXiv:0712.2456 [hep-th]
69. D.E. Kharzeev, H.-U. Yee, Phys. Rev. D **84**, 045025 (2011). arXiv:1105.6360 [hep-th]
70. S. Bhattacharyya, J. High Energy Phys. **1207**, 104 (2012). arXiv:1201.4654 [hep-th]
71. E. Megias, F. Pena-Benitez, Holographic gravitational anomaly in first and second order hydrodynamics, Preprint IFT-UAM/CSIC-13-039
72. T.E. Clark, S.T. Love, T. ter Veldhuis, Phys. Rev. D **82**, 106004 (2010). arXiv:1006.2400 [hep-th]

Chapter 18
Quantum Criticality via Magnetic Branes

Eric D'Hoker and Per Kraus

18.1 Introduction

A statistical mechanics system undergoes a quantum phase transition when its ground state suffers a macroscopic rearrangement as an external parameter is varied. While a quantum phase transition takes place strictly at zero temperature, its presence governs quantum critical behavior in a small region of low temperature surrounding the quantum critical point. The existence of such a quantum critical region is believed to influence physics also at intermediate temperatures, and to have relevance to the phase structure of high-T_c superconductors and strange metals [58].

Holography provides concrete tools for studying non-Abelian gauge dynamics in terms of classical solutions to Einstein's equations of gravity, provided the number of colors N and the 't Hooft coupling $\lambda = Ng_{YM}^2$ are both large [34, 50, 62]. The classic example of this gauge/gravity duality relates $\mathcal{N} = 4$ super Yang-Mills theory in $3 + 1$-dimensional Minkowski space-time to Type IIB supergravity on $AdS_5 \times S^5$. Global symmetries match under the duality: the space-time isometry group $SO(4, 2) \times SO(6)$ on the gravity side maps to the conformal group $SO(4, 2)$ and the R-symmetry group $SU(4) \sim SO(6)$ on the gauge theory side, while the number of supersymmetries is the maximal allowed 32 on both sides (for general reviews on the AdS/CFT correspondence, see for example [2, 15, 46, 55]).

The gauge/gravity duality continues to apply to systems with fewer or no supersymmetries, and with broken conformal and Poincaré invariance. For example, renormalization group flow away from a conformal invariant gauge theory is dual to a space-time which is *asymptotically* AdS_5 but deviates from AdS_5 away

E. D'Hoker (✉) · P. Kraus
Department of Physics and Astronomy, University of California, Los Angeles, CA 90095, USA
e-mail: dhoker@physics.ucla.edu

P. Kraus
e-mail: pkraus@physics.ucla.edu

D. Kharzeev et al. (eds.), *Strongly Interacting Matter in Magnetic Fields*,
Lecture Notes in Physics 871, DOI 10.1007/978-3-642-37305-3_18,
© Springer-Verlag Berlin Heidelberg 2013

from the asymptotic limit. The dual to a gauge theory at finite temperature T is a space-time containing a black hole or black brane whose Hawking temperature is T. A charge density or chemical potential, and a background magnetic field may all be incorporated in the gauge theory as well via precise dual gravity prescriptions, thereby setting the stage for applying holographic methods to a wide variety of interesting strongly coupled systems in statistical mechanics. Reviews on the applications of holographic methods to condensed matter problems may be found in [35, 40].

Whether holography will ever be able to model reliably the condensed state of a specific compound remains to be seen. What has become clear, however, is that gauge/gravity duality can provide quantitative information on universal behavior, such as critical phenomena, critical exponents, transport properties, and the like for certain classes of strongly coupled systems. One of the first examples derived in this spirit is the ratio of the shear viscosity to the entropy density [56]; a more recent one gives a bound on the specific heat exponent [61].

In the present paper, we shall review recent holographic investigations into the critical behavior of gauge theory at finite temperature T, with electric charge density ρ, and subjected to an external magnetic field B. Supersymmetry will play no significant role here, as the gauge theory in question may be supersymmetric or not. We shall be interested in the thermodynamic properties of this system especially at low temperatures, as well as in the behavior of correlators at long distances.

The holographic dual to this system, in the large N and large λ approximations, is provided by a theory of gravity in $4 + 1$ space-time dimensions, plus an Abelian gauge field. Anomalies in the gauge current of the $3 + 1$-dimensional field theory side force the presence of a Chern-Simons term in the $4 + 1$-dimensional theory. The CS coupling is the only dimensionless free parameter in the theory, and the holographic dynamics will depend crucially on its value. All our results will be derived using this holographic model [16, 17].

Using holographic methods, a rich structure is found which exhibits quantum critical behavior [18, 19, 21], and the emergence of a $1 + 1$-dimensional CFT in IR correlators [20, 22]. Specifically, the system exhibits a quantum phase transition as the magnetic field B crosses a critical value B_c, the scale for which is set by the charge density ρ by $B_c \sim \rho^{2/3}$. Quantum critical behavior governs a region, depicted schematically in Fig. 18.1, where temperature is the largest scale. The mechanism underlying this transition on the gravity side is also illustrated in Fig. 18.1: it is driven by the expulsion of electric charge from within the horizon to the outside. More specifically, for vanishing magnetic field, the gravity solution is the AdS Reissner-Nordstrom black brane which has non-zero charge density and entropy density at $T = 0$. As B is increased electric charge gets expelled from within the black brane horizon to the outside, up till $B = B_c$ at which value the black brane carries no more charge or entropy density. This charge expulsion mechanism is realized in other holographic systems as well [36]. Other examples of magnetic field driven holographic phase transitions include [43, 47].

Fig. 18.1 The holographic picture dual to a meta-magnetic quantum phase transition is given by the gradual expulsion of electric charge from the region interior to the event horizon of a black brane to the region outside the horizon. The quantum critical point corresponds to the transition point at which all electric charge resides outside the horizon of the black brane

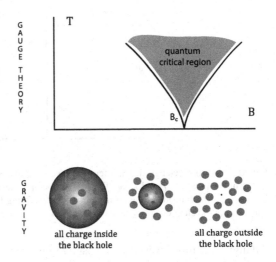

18.2 Basic Gauge Theory Dynamics

Before embarking on the study of strongly coupled gauge theory with the tools of holography, we shall summarize here some basic results on the dynamics of gauge theory in the presence of an external magnetic field.

18.2.1 Effective Low Energy Degrees of Freedom

In the presence of a constant magnetic field B, the energy levels of massless free bosons and fermions of electric charge q are given as follows,

$$\text{bosons} \quad E = \sqrt{p^2 + (2n+1)|qB|} \quad n = 0, 1, 2, \ldots$$

$$\text{fermions} \quad E = \sqrt{p^2 + 2n|qB|} \quad n = 0, 1, 2, \ldots$$

Here p is the momentum component parallel to the magnetic field B. The energy levels for bosons and fermions clearly do not match, so that supersymmetry is manifestly broken. Simple modifications in which supersymmetry is restored do exist, however, and were studied in [5].

For large B, only fermions in the lowest Landau level remain massless. More precisely, the fermions in the lowest Landau level are $1+1$-dimensional Weyl fermions, moving along the direction of the magnetic field, with momentum p, their chirality being correlated with their charge,

$$Bq > 0 \quad p > 0 \quad \text{field of right-movers } \psi_R$$

$$Bq < 0 \quad p < 0 \quad \text{field of left-movers } \psi_L$$

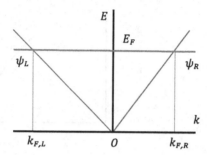

Fig. 18.2 At finite charge density, fermion levels fill up to a Fermi energy E_F, corresponding to Fermi momenta $k_{F,L}$ and $k_{F,R}$. The low temperature degrees of freedom are chiral fermions ψ_L and ψ_R which reside at the Fermi "surface", consisting here of only two points $k = k_{F,L}$ and $k = k_{F,R}$. In the charge symmetric case we have $k_{F,L} = -k_{F,R}$

All higher fermion levels and all boson levels acquire a large effective mass and will decouple from the spectrum.

In $\mathcal{N} = 4$ super-Yang-Mills theory, the operator of charges Q to which the magnetic field couples takes values in the R-symmetry algebra $SU(4)_R$. Although the gauge fields A_μ, gauginos λ_α, and scalars ϕ are now all strongly interacting, the above decoupling of charged fields will persist. Thus, the low energy effective degrees of freedom will be Weyl fermions ψ_L, ψ_R, whose coupling is induced by the remaining neutral gauge dynamics of the A_μ fields, and any remaining components of λ and ϕ which are neutral under Q. This system of Weyl fermions should underlie an interacting conformal field theory in $1 + 1$ dimensions. Any *asymmetry* in the spectrum of charges Q will give rise to the *chiral magnetic effect*, the observation of which is being considered in heavy ion collision experiments at RHIC [45].

At finite density and low or zero temperature, fermionic levels will fill up to a certain Fermi energy E_F. For the effective $1 + 1$-dimensional CFT discussed here, the Fermi surface consists of just two points corresponding to two values of the Fermi momentum. The left- and right-movers ψ_L and ψ_R live separately at $k = k_{F,L}$ and $k = k_{F,R}$ respectively. If the spectrum of charges is symmetric then $k_{F,L} = -k_{F,R}$, as is illustrated in Fig. 18.2. But in general, $k_{F,L} \neq -k_{F,R}$, and the ground state of the system will carry a nonzero total momentum. As the charge density ρ is being increased, E_F will increase. When E_F reaches the next Landau level, or the energies characteristic of the fully 3-dimensional excitations, a large number of degrees of freedom are being excited, and we may expect a quantum phase transition, at a critical charge density ρ_c, whose scale is set by the only possible dimensional scale in the problem, namely $\rho_c \sim B^{3/2}$.

18.2.2 Luttinger Liquids

The standard quantum field theory approach to systems of $1 + 1$-dimensional interacting chiral fermions is provided by the Luttinger approach to quantum liquids. One begins by identifying the excitations near the Fermi surface, in this case the Weyl fermions ψ_L, ψ_R introduced above. The Hamiltonian consists of bilinear terms which result from linearization around the Fermi surface, as well as all possible local four-Fermi interactions compatible with the symmetries of the system,

$$H_{\text{int}} = g_2\left(\psi_L^\dagger \psi_L\right)\left(\psi_R^\dagger \psi_R\right) + \frac{g_4}{2}\left(\psi_L^\dagger \psi_L\right)^2 + \frac{g_4}{2}\left(\psi_R^\dagger \psi_R\right)^2 \qquad (18.1)$$

Although the system was first solved in terms of fermionic fields by Dzyaloshinski and Larkin, modern methods based on bosonization provide a powerful reformulation in terms of two non-interacting boson fields [32]. A key thermodynamic relation for the entropy density s_{gauge} as a function of the temperature T is given by,

$$s_{\text{gauge}} = \frac{\pi}{3v} T \qquad v = v_F \sqrt{\left(1 + \frac{g_4}{2\pi v_F}\right)^2 - \left(\frac{g_2}{2\pi v_F}\right)^2} \qquad (18.2)$$

where v_F is the Fermi velocity, and v the actual velocity of the chiral excitations. Correlators may be obtained as well. For example, the two-point function of the charge density $\rho(x)$ takes the following form,

$$\langle \rho(x)\rho(0)\rangle = \frac{c_0}{x^2} + c_\Delta \frac{\cos(2k_F x)}{x^{2\Delta}} + \cdots \qquad (18.3)$$

where c_0 and c_Δ are constants, and Δ is the scaling dimension of the lowest dimensional operator which exchanges charge between ψ_L and ψ_R.

18.3 Holographic Dual Set-Up

In this section, we shall discuss the basic set up for the holographic dual in the supergravity approximation. Since we shall concentrate on thermodynamics, as well as on correlators of energy density, momentum density, and charge density, we may limit the quantum field operators to the stress tensor $\mathcal{T}^{\mu\nu}$ and the Maxwell current $\mathcal{J}^\mu$ of the 3 + 1-dimensional gauge theory. The holographic dual fields to these operators are respectively the metric g_{MN} and the Maxwell field A_M of the 4 + 1-dimensional Einstein-Maxwell-Chern-Simons theory with action,[1]

$$S = -\frac{1}{16\pi G_5} \int d^5x \sqrt{g}\left(R - \frac{12}{\ell^2} + F_{MN}F^{MN}\right) + \frac{k}{12\pi G_5} \int A \wedge F \wedge F \qquad (18.4)$$

Boundary as well as counterterm contributions to the action have not been exhibited here. Furthermore, G_5 is Newton's constant in 4 + 1 dimensions, $-12/\ell^2$ is the cosmological constant, and k is the dimensionless Chern-Simons coupling. The

---

[1]Einstein indices $\mu, \nu = 0, 1, 2, 3$ will be used in 3 + 1-dimensions, while Einstein indices $M, N = 0, 1, 2, 3, 4$ will be used in 4 + 1-dimensions. Our conventions are $g = -\det(g_{MN})$, as well as $R^L{}_{MNK} = \partial_K \Gamma^L_{MN} - \partial_N \Gamma^L_{MK} + \Gamma^P_{MN}\Gamma^L_{KP} - \Gamma^P_{MK}\Gamma^L_{NP}$ with $R_{MN} = R^L{}_{MLN}$ and $R = g^{MN} R_{MN}$.

anomaly of the chiral current $\mathscr{J}^\mu$ in the gauge theory is proportional to k, and we have $\partial_\mu \mathscr{J}^\mu \propto k\,\mathbf{E} \cdot \mathbf{B}$. The action is invariant under simultaneous reversal of the sign of A and k, allowing us to restrict attention to $k \geq 0$, without loss of generality.

For the special value $k = k_{\text{susy}} = 2/\sqrt{3}$, the action S coincides with the bosonic part of minimal supergravity in $4 + 1$ dimensions, and as such corresponds to a consistent truncation of all supersymmetric asymptotically AdS_5 compactifications of either Type IIB supergravity or M-theory [31]. Here, however, we shall leave k a free parameter, and investigate the phase diagram as a function of k.

18.3.1 Field Equations and Structure of the Solutions

The Bianchi identity is $dF = 0$, while the field equations are given as follows,

$$0 = d * F + kF \wedge F$$

$$R_{MN} = \frac{4}{\ell^2} g_{MN} + \frac{1}{3} g_{MN} F^{PQ} F_{PQ} - 2 F_{MP} F_N{}^P \qquad (18.5)$$

For vanishing Maxwell field $F = 0$, the field equations admit the AdS_5 solution of radius ℓ. Henceforth, we shall set $\ell = 1$. Denoting the coordinates of 4-dimensional space-time by $x^\mu = (t, x_1, x_2, x_3)$, and the holographic coordinate by r, the AdS_5 solution takes the form,

$$ds^2 = g_{MN} dx^M dx^N = \frac{dr^2}{4r^2} + 2r\left(-dt^2 + dx_1^2 + dx_2^2 + dx_3^2\right) \qquad (18.6)$$

Introducing a constant uniform magnetic field B along the direction x_3, and anticipating also the inclusion of finite temperature T, and constant uniform charge density ρ, we see that the symmetries to be imposed on the solutions should include,

1. Translation invariance in the coordinates t, x_1, x_2, x_3;
2. Rotation invariance in the x_1, x_2-plane.

The most general Ansatz consistent with these requirements is given by,

$$F = Bdx_1 \wedge dx_2 + Edr \wedge dt - Pdr \wedge dx_3 + \tilde{P}dt \wedge dx_3$$

$$ds^2 = f^{-1}dr^2 + Mdt^2 + 2Ldtdx_3 + Ndx_3^2 + K\left(dx_1^2 + dx_2^2\right) \qquad (18.7)$$

where all coefficient functions $B, E, P, \tilde{P}, f, K, L, M, N$ depend only on r. In view of the Bianchi identities, B and $\tilde{P}$ must be independent of r, and in view of the field equations, we have $\tilde{P} = 0$. This constant B is nothing but the constant magnetic background field. Finally, the residual reparametrization invariance in the variable r allows us to choose a coordinate r such that

$$f = L^2 - MN \qquad (18.8)$$

a choice which will prove convenient throughout.

18.3.2 Boundary Stress Tensor and Current

We will be considering asymptotically AdS_5 solutions, for which the metric and gauge field admit a Fefferman-Graham expansion [29]. Introducing a radial coordinate ρ, defined such that the AdS_5 boundary is located at $\rho = \infty$, the Fefferman-Graham gauge choice puts the fields in the following form,

$$A = A_\mu(\rho, x)dx^\mu$$

$$ds^2 = \frac{d\rho^2}{4\rho^2} + g_{\mu\nu}(\rho, x)dx^\mu dx^\nu \tag{18.9}$$

and their expansion in large ρ takes the following form,

$$A_\mu(\rho, x) = A_\mu^{(0)}(x) + \frac{1}{\rho}A_\mu^{(2)}(x) + \cdots$$

$$g_{\mu\nu}(\rho, x) = \rho g_{\mu\nu}^{(0)}(x) + g_{\mu\nu}^{(2)}(x) + \frac{1}{\rho}g_{\mu\nu}^{(4)}(x) + \frac{\ln\rho}{\rho}g_{\mu\nu}^{(\ln)}(x) + \cdots \tag{18.10}$$

The coefficients $g_{\mu\nu}^{(4)}$, $g_{\mu\nu}^{(\ln)}$, and the trace of $g_{\mu\nu}^{(4)}$ are fixed by the Einstein equations to be local functionals of the conformal boundary metric $g_{\mu\nu}^{(0)}$. The boundary stress tensor $T^{\mu\nu}$ and current J^μ of [8, 14] are defined in terms of the variation of the on-shell action with respect to $g_{\mu\nu}^{(0)}$ and $A_\mu^{(0)}$ respectively (see [22]). In terms of the Fefferman-Graham data the result is,

$$4\pi G_5 T_{\mu\nu}(x) = g_{\mu\nu}^{(4)}(x) + \text{local}$$

$$2\pi G_5 J_\mu(x) = A_\mu^{(2)}(x) + \text{local} \tag{18.11}$$

Indices are raised and lowered using the conformal boundary metric $g_{\mu\nu}^{(0)}$. The local terms denote tensors constructed locally from $g_{\mu\nu}^{(0)}$ and $A_\mu^{(0)}$, which may be dropped when computing correlators at non-coincident points.

18.4 The Purely Magnetic Brane: Zero Charge Density

The case of vanishing charge density, with zero or non-zero temperature, provides a physically interesting system, which lends itself to much simpler treatment than the charged case. For this reason, we shall investigate it first here, in its own right; see also [3–5].

18.4.1 The Purely Magnetic Brane at $T = 0$

We begin with the case of zero temperature. For vanishing charge density and temperature, Lorentz invariance in the t, x_3 directions is restored. To exhibit this symmetry, it will be convenient to re-interpret t, x_3 as light-cone coordinates by substituting $t \to x^+$ and $x_3 \to x^-$. In this space-time coordinate system, Lorentz transformations act by $x^\pm \to \lambda^{\pm 1} x^\pm$, and further restrict the Ansatz to $E = P = M = N = 0$. In terms of the remaining functions K, L, the Einstein equations reduce to,

$$0 = \left(L^2 K\right)'' - 24K$$

$$0 = K^2 L'' + K K' L' + 2K K'' L - L\left(K'\right)^2 \qquad (18.12)$$

$$-4B^2 = \left(K'\right)^2 L^2 + 4K K' L L' + K^2 \left(L'\right)^2 - 24K^2$$

Here, the prime stands for the derivative with respect to r. The last equation of (18.12) is a constraint, whose derivative with respect to r is linear in the first two equations, and vanishes, as soon as these equations are satisfied. The first equation gives L as a function of K by quadrature only. Substituting this form of L into the constraint then gives an equation for K which may be solved numerically.

The reduced field equations (18.12) admit an exact solution for which K is independent of r, which is given by,

$$K(r) = \frac{B}{\sqrt{3}} \qquad L(r) = 2\sqrt{3}\, r \qquad (18.13)$$

Substitution into the metric of (18.7) reveals that the corresponding space-time has the form $AdS_3 \times \mathbf{R}^2$, with an AdS_3 radius given by $\ell_3 = 1/\sqrt{3}$. The AdS_5 vacuum, with $K, L \sim r$ is not a solution to the reduced equations (18.12) when $B \neq 0$, but it does become an approximate solution in the limit of large r, namely when $B/r \to 0$. The $T = 0$ purely magnetic brane is the solution to (18.12) for which the functions K and L tend to the $AdS_3 \times \mathbf{R}^2$ solution of (18.13) when $r \to 0$, and is asymptotic to AdS_5 of (18.1) when $r \to \infty$,

$$K(r) \sim c_V r \qquad L(r) \sim 2r \qquad (18.14)$$

Numerical analysis confirms that such a purely magnetic brane solution exists and is regular for all $r > 0$, and gives the numerical value $c_V = 2.797$.

18.4.2 RG Flow and Thermodynamics

The holographic dual to the purely magnetic brane solution is a renormalization group flow from $3 + 1$-dimensional $\mathcal{N} = 4$ super-Yang-Mills theory in the UV (for large r) to a $1 + 1$-dim. CFT in the IR (for small r). This flow is schematically represented in Fig. 18.3. The holographic picture is consistent with the qualitative gauge dynamics behavior of strongly interacting Weyl fermions discussed in Sect. 18.2. The central charge c of this CFT may be derived using the Brown-Henneaux formula [12],

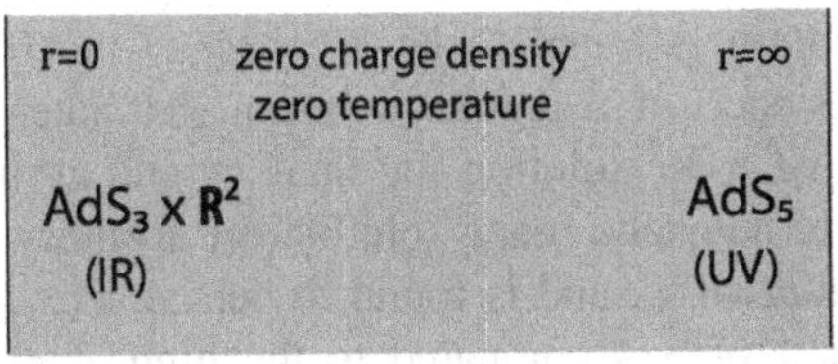

Fig. 18.3 The purely magnetic brane solution interpolates between an AdS_5 space-time with magnetic field for large r, and an $AdS_3 \times R^2$ space-time with magnetic field for small r. The holographic dual field theory has zero charge density and temperature

$$c = \frac{3\ell_3}{2G_3} \qquad \frac{1}{G_3} = \frac{B\,V_2}{G_5} \tag{18.15}$$

applied to an AdS_3 of radius $\ell_3 = 1/\sqrt{3}$, and where we have taken $x_{1,2}$ to be compactified on a $\mathbf{T}^2$ with area V_2.

The specific heat coefficient at low temperature may be expressed in terms of the entropy density s by s/T. In turn, the Cardy formula gives the entropy density s in terms of the central charge of a $1+1$-dimensional CFT. It may be used here to extract the holographic specific heat coefficient s_{grav}/T, and the entropy density s_{grav} in terms of the zero temperature purely magnetic brane, and we find,

$$\frac{s_{grav}}{T} = \frac{\pi}{3}c = \frac{\pi\,B\,V_2}{2\sqrt{3}G_5} = \sqrt{\frac{4}{3}}\frac{s_{gauge}}{T} \tag{18.16}$$

In the last equality, we have included the comparison with the entropy density s_{gauge} evaluated earlier for free fermions in the lowest Landau level. To exhibit this relation, we have used the AdS/CFT relation $G_5 = \pi/(2N^2)$. The fact that the gravity and gauge theory central charges do not agree can be understood as follows [5]. Comparing the central charges at small (but finite) and large values of the 't Hooft coupling should show agreement, because the central charge of a $D = 1+1$ CFT is unchanged under marginal deformation. However, the passage from zero to small 't Hooft coupling can be a discontinuous change if the CFT has a relevant operator that is either absent or present in the two cases. The $\sqrt{4/3}$ factor is presumably a result of the appearance of this relevant operator.

To derive the thermodynamics of the purely magnetic brane at all T, we replace the near-horizon $AdS_3 \times \mathbf{R}^2$ space by a BTZ $\times \mathbf{R}^2$ black brane. The latter is expected to solve the field equations (18.5) as well since BTZ may be obtained as a quotient of AdS_3 by a discrete group. Concretely, the absence of electric charge allow us to set $E = P = 0$ in (18.7), but the fields M, N need to be retained at finite temperature. The gauge field is still $F = Bdx^1 \wedge dx^2$, while the metric takes the following form,

$$K(r) = \frac{B}{\sqrt{3}} \qquad L(r) = 4\sqrt{3}(r - r_+)$$
$$N(r) = 1 \qquad M(r) = -12(r - r_+)(r_+ - r_-) \tag{18.17}$$

The Hawking temperature is found to be $T = 3(r_+ - r_-)/\pi$. Numerical analysis confirms the existence of a regular solution that interpolates between the above near-horizon BTZ $\times$ $\mathbf{R}^2$ solution for small r, and asymptotically AdS_5 for large r. Using this pure magnetic brane solution for arbitrary T, the entropy density may be calculated at all T, and is found to behave as $s_{\text{grav}} \sim T^3$ for high T, with the standard factor of $3/4$ compared to the high T gauge theory calculation.

18.4.3 Calculation of Current-Current Correlators at $T = 0$

The boundary current formalism discussed in Sect. 18.3.2 may be used to evaluate the various two-point functions of the gauge current $\mathscr{J}^\mu$ and the stress tensor $\mathscr{T}^{\mu\nu}$. In the absence of charge density for the purely magnetic brane, the cross correlators $\langle \mathscr{J}^\mu(x)\mathscr{T}^{\mu\nu}(y)\rangle$ will vanish identically. We begin by evaluating the current-current correlators, by combining linear response theory with the formulas of (18.11),

$$J^\mu(x) = i \int d^4y \sqrt{-\det\!\big(g^{(0)}_{\mu\nu}\big)} \langle \mathscr{J}^\mu(x) \mathscr{J}^\nu(y)\rangle \delta A^{(0)}_\nu(y) \qquad (18.18)$$

Here, the expectation value J^μ of the current $\mathscr{J}^\mu$ is sourced by a linear variation in the source $A^{(0)}_\nu$ of the gauge potential, using (18.11). As no variation of the metric is imposed, the current J^μ may be obtained by linearizing the Maxwell-Chern-Simons equations on the first line of (18.5) around the purely magnetic brane. We shall be interested in correlators in the $1+1$-dimensional effective low energy CFT only, and thus restrict to excitations carrying momentum along the magnetic field direction. The gauge potential for definite momentum is then given by,

$$A = A_B + \big(a_+(r, p)dx^+ + a_-(r, p)dx^-\big) e^{ipx} \qquad (18.19)$$

where $dA_B = Bdx_1 \wedge dx_2$, and we shall use the notations $px = p_+x^+ + p_-x^-$ and $p^2 = p_+p_-$ throughout. Denoting the metric fields of the purely magnetic brane solution at $T = 0$ by K and L, we find that the Maxwell-Chern-Simons equations may be decoupled in terms of the variables $\varepsilon_\pm = p_-a_+ \pm p_+a_-$ for which we obtain the following equations,

$$0 = KL\big(KL\varepsilon'_-\big)' - 4k^2B^2\varepsilon_- - 2K^2\frac{p^2}{L}\varepsilon_-$$
$$0 = KL\varepsilon'_+ - 2kB\varepsilon_- \qquad (18.20)$$

Given that the functions K and L are known only numerically, solving the above equations for general p^2 can only be achieved numerically. If we restrict attention to the regime of small p^2, however, then the linearized equations can be solved essentially analytically using the method of overlapping expansions.

18.4.4 Method of Overlapping Expansions

The characteristic scale of the purely magnetic brane solution, namely where the functional dependence of K and L transits from the behavior of (18.14) at large r to the behavior of (18.13) at small r, is set by $r \sim 1$.

In the *near-region*, defined by $r \ll 1$, the purely magnetic brane solution may be approximated by the behavior in (18.13), so that (18.20) becomes,

$$
0 = r^2 \varepsilon_-'' + r\varepsilon_-' - k^2 \varepsilon_- - \frac{p^2}{12\sqrt{3}r} \varepsilon_-
$$
$$
0 = r\varepsilon_+' - k\varepsilon_-
$$
(18.21)

The first equation is of the modified Bessel type and is solved by the Bessel functions $I_{2k}(p/\sqrt{r})$ and $K_{2k}(p/\sqrt{r})$. Only the solution $\varepsilon_-(r) \sim K_{2k}(p/\sqrt{r})$ is regular as $r \to 0$, which leads us to reject the solution I_{2k}.

In the *far-region*, defined by $p^2 \ll r$, we may neglect the last term of the first equation in (18.20), and solve the remaining equation in terms of the function,

$$
\psi(r) \equiv \int_\infty^r \frac{dr'}{K(r')L(r')}
$$
(18.22)

Note that $\psi(r)$ depends only on the data of the purely magnetic brane solution. Expressing the solution directly in terms of the original variables $a_\pm(r, p)$, we find,

$$
a_\pm = p_\pm \tilde{a}_0 + \left(a_\pm^{(0)} - p_\pm \tilde{a}_0\right)e^{\pm 2kB\psi(r)}
$$
(18.23)

where $a_\pm^{(0)}$ and $\tilde{a}_0$ are integration constants.

An *overlap-region*, in which the near-region and the far-region overlap in a finite interval, will exist provided $p^2 \ll 1$. Assuming that $p^2 \ll 1$, there will exist an overlap region in which we may match the $p^2/r \ll 1$ behavior of the Bessel function in the near-region solution,

$$
a_\pm = C \frac{(p^2/12)^{\mp k}}{\Gamma(1 \mp 2k)p_\mp} r^{\pm k} + p_\pm a_0
$$
(18.24)

with the $r \ll 1$ behavior of the far-region solution. The latter may be derived from the asymptotic behavior of the function $\psi(r)$, which is found to be for $r \ll 1$,

$$
\psi(r) \sim \frac{1}{2B} \ln r + \psi_0
$$
(18.25)

Numerical evaluation gives $\psi_0 \approx 0.2625$. Comparing the r-dependence in (18.24) and (18.23) using (18.25), we see that the near-region and far-region functional behaviors are indeed the same, and given by a constant term, as well as by $r^{\pm k}$ terms. Matching these functional dependences produces the full solution.

18.4.5 Current Two-Point Correlators

To complete the calculation of the current-current correlators in the long-distance approximation $p^2 \ll 1$ we use the overlapping expansion results obtained earlier. The large r approximation for the function $\psi(r)$ is obtained analytically from the asymptotic behavior of K and L in (18.14),

$$\psi(r) \sim -\frac{1}{2c_V r} \tag{18.26}$$

and used to derive the asymptotic behavior for the gauge potential for $r \to \infty$,

$$a_\pm(r, p) = a_\pm^{(0)} + \frac{1}{4r}a_\pm^{(2)} \qquad a_\pm^{(2)} = \mp\frac{4kB}{c_V}\left(a_\pm^{(0)} - p_\pm\tilde{a}_0\right) \tag{18.27}$$

To obtain $\tilde{a}_0$ and $a_\pm^{(2)}$ in terms of a_0 and $a_\pm^{(0)}$, we match the near-region solution of (18.24) with the far-region solution of (18.23). Including proper normalizations [22], we obtain the current-current correlators in the limit $p^2 \ll 1$,

$$\left\langle \mathscr{J}_+(p)\,\mathscr{J}_+(-p)\right\rangle = \frac{kc}{2\pi}\frac{p_+}{p_-}\frac{1}{1-\zeta p^{4k}}$$

$$\left\langle \mathscr{J}_-(p)\,\mathscr{J}_-(-p)\right\rangle = \frac{kc}{2\pi}\frac{p_-}{p_+}\frac{\zeta p^{4k}}{1-\zeta p^{4k}} \tag{18.28}$$

$$\left\langle \mathscr{J}_+(p)\,\mathscr{J}_-(-p)\right\rangle = -\frac{kc}{2\pi}\frac{\zeta p^{4k}}{1-\zeta p^{4k}}$$

where c is the Brown-Henneaux central charge derived in (18.15), and $\zeta = \zeta(k)$ is a k-dependent function whose precise form will not be needed here. We note that the above correlators saturate the chiral anomaly relation independently of ζ,

$$p_+\mathscr{J}_- + p_-\mathscr{J}_+ = \frac{kc}{\pi}\left(p_+a_-^{(0)} - p_-a_+^{(0)}\right) \tag{18.29}$$

To leading order in small p^2 the correlators involving $\mathscr{J}_-$ both vanish, while the correlator involving only $\mathscr{J}_+$ takes the following form in position space,

$$\left\langle \mathscr{J}_+(x)\,\mathscr{J}_+(0)\right\rangle = -\frac{kc}{2\pi^2}\frac{1}{(x^+)^2} \tag{18.30}$$

With our conventions, the above sign in the central term corresponds to a unitary Abelian Kac-Moody algebra for $k > 0$ and $c > 0$, as is the case here.

18.4.6 Maxwell-Chern-Simons Holography in AdS$_3$

Attempts to formulate Maxwell-Chern-Simons (MCS) holography directly in AdS_3 space-time are fraught with subtleties [1]. This circumstance may be investigated directly with the help of the near-region solutions derived in Sect. 18.4.4, and in particular the large r asymptotics of this solution given in (18.24),

$$a_\pm(r) = \alpha_\pm + \beta_\pm r^{\pm k} \tag{18.31}$$

From the point of view directly of AdS_3, it is unclear which coefficients should be used as sources, and which ones correspond to expectation values. That is, it is not clear which boundary conditions lead to a consistent theory, and indeed most boundary conditions lead to problems with instabilities and/or ghosts [1]. Symptoms of this may be detected in the current-current correlators derived in (18.29), by taking the limit $p^2 \gg 1$. In this limit, the correlators involving the component $\mathscr{J}_+$ vanish, while the correlator of $\mathscr{J}_-$ becomes in position space,

$$\langle \mathscr{J}_-(x)\,\mathscr{J}_-(0)\rangle = +\frac{kc}{2\pi^2}\frac{1}{(x^-)^2} \tag{18.32}$$

Although this correlator by itself saturates the chiral anomaly, its sign corresponds to that of a *non-unitary* Abelian Kac-Moody algebra. This violation of unitarity is a symptom of the disease which besets certain choices of boundary conditions for Maxwell-Chern-Simons directly in AdS_3.

However, by the same token we see that when the MCS theory is obtained as the IR limit of the holographic RG flow provided by the purely magnetic brane solution from an asymptotic AdS_5 completion, then the MCS theory makes perfect sense. The key point is that the IR theory comes with a built in UV cutoff, given by the scale at which the AdS_3 factor goes over to AdS_5. All the would-be inconsistencies are removed by the presence of the UV cutoff.

18.4.7 Effective Conformal Field Theory and Double-Trace Operators

The leading IR contribution to the two-point function of $\mathscr{J}_+$ in (18.28) may be parametrized by an effective free scalar field ϕ with canonical Lagrangian $\mathscr{L}_\phi \sim \partial_+\phi\partial_-\phi$. The subdominant p^{4k} terms in the correlators of (18.28) may be understood in terms of contributions to the Lagrangian from double-trace operators $\mathscr{L}_\mathcal{O} \sim \partial_+\mathcal{O}\partial_-\mathcal{O}$, where $\mathcal{O}$ is a conformal primary field of dimension (k, k). The expressions for the currents and two-point functions (in momentum space) of these operators are given as follows,

$$\begin{aligned}
\mathscr{J}_+ &= \partial_+\phi + \partial_+\mathcal{O} & \langle\phi(p)\phi(-p)\rangle &= p^{-2}\\
\mathscr{J}_- &= \partial_-\mathcal{O} & \langle\mathcal{O}(p)\mathcal{O}(-p)\rangle &= \zeta p^{4k-2}
\end{aligned} \tag{18.33}$$

These two fields summarize the entire IR behavior of the correlators, within our approximations. The importance of double trace operators in the holographic renormalization group [7, 10] has been stressed recently in [28, 38].

We note that the structure of the $\mathscr{J}_+$ correlators is quite reminiscent of the structure of the charge density two-point function in the Luttinger liquid model. Clearly there is a leading inverse square contribution both in (18.30) and in (18.3), while the p^{4k} higher order corrections in (18.28) are analogous to the $x^{-2\Delta}$ corrections of (18.3).

18.4.8 Stress Tensor Correlators and Emergent Virasoro Symmetry

Since the purely magnetic brane produces a flow from AdS_5 towards a space-time containing an AdS_3 factor, its holographic dual is expected to be a full-fledged CFT in the IR limit, endowed with left- and right-moving Virasoro algebras. This structure, and the value of the associated Brown-Henneaux central charge c of (18.15) dictate the structure of the two-point function of two stress tensor components. All correlators involving the component $\mathscr{T}_{+-}$ vanish at non-coincident points, as does the mixed correlator $\langle \mathscr{T}_{++}(x) \mathscr{T}_{--}(y) \rangle$. The remaining correlators are given by,

$$\langle \mathscr{T}_{\pm\pm}(x) \mathscr{T}_{\pm\pm}(0) \rangle = \frac{c}{8\pi^2 (x^\pm)^4} \tag{18.34}$$

These two-point correlators may be checked by explicit calculation using the method of overlapping expansions along the same lines as for the current correlators, and agree.

The existence of two Virasoro symmetry algebras in the IR brings to light the holographic realization of the *emergence of symmetries*. Indeed, the Brown-Henneaux coordinate transformations on the near-horizon AdS_3, which produce these Virasoro asymptotic symmetry algebras, correspond to pure gauge transformations. This is as expected, since gravity in three space-time dimensions is (locally) trivial. But these coordinate transformations on AdS_3 extend to perturbative deformations of the pure magnetic brane solution and interpolate to the AdS_5 boundary where they correspond to physical deformations which are not merely gauge transformations. Indeed, the asymptotic symmetry algebra at the asymptotically AdS_5 boundary of the purely magnetic brane is $SO(4, 2)$, a finite-dimensional Lie algebra of which the infinite-dimensional Virasoros are certainly not subalgebras. Therefore, we conclude that the Virasoro symmetries present in the IR are *emergent symmetries*, not present in the UV theory.

18.5 Holographic Dual Solutions for Non-zero Charge Density

A non-zero charge density gives rise to a wealth of interesting physics. As discussed in the introduction, the physical location of the charge, namely either inside or outside the event horizon and a mixture thereof, will to a large extent govern the phase diagram of the dual field theory. Remarkably, it will be possible to understand most of the low temperature dynamics, for large enough magnetic field, using the analytical methods of overlapping expansions, supplemented by a few numerical constants determined from the purely magnetic brane solution. In this section, we shall proceed analytically, and fill in the regions of the phase diagram not accessible through analytical results with the help of numerical results.

18.5.1 Reduced Field Equations

Investigating thermodynamics in the presence of charge density and a magnetic field in the x^3 direction will involve gravitational solutions which are invariant under translations in x^μ, and rotations in the x_1, x_2 plane. Thus, we need the full Ansatz of (18.7). The corresponding reduced field equations are as follows,

M1 $\left((NE + LP)e^{2V}\right)' + 2kbP = 0$

M2 $\left((LE + MP)e^{2V}\right)' - 2kbE = 0$

E1 $L'' + 2V'L' + 4\left(V'' + V'^2\right)L - 4PE = 0$

E2 $M'' + 2V'M' + 4\left(V'' + V'^2\right)M + 4E^2 = 0$ (18.35)

E3 $N'' + 2V'N' + 4\left(V'' + V'^2\right)N + 4P^2 = 0$

E4 $f\left(V'\right)^2 + f'V' + \dfrac{1}{4}\left(L'\right)^2 - \dfrac{1}{4}M'N' + b^2e^{-4V} + MP^2 + 2LEP + NE^2 = 6$

fV $\left(fe^{2V}\right)'' = 24\,e^{2V}$

We have used the notation $f = L^2 - MN$ of (18.8), and changed variables to $K = e^{2V}$. Also, we now denote the magnetic field as b, reserving the use of B for the value of the magnetic field in a canonical coordinate system.

The reduced field equations admit a number of first integrals. Using the potentials A and C for $E = A'$ and $P = -C'$, equations M1 and M2 admit obvious first integrals,

$$\left(NA' - LC'\right)e^{2V} + 2kbC = 0$$
$$\left(LA' - MC'\right)e^{2V} - 2kbA = 0$$ (18.36)

The integration constants to A and C that arise here have been absorbed into the definition of these functions. Forming combinations of equations E1, E2, and E3, and using (18.36), we find the following further first integrals,

$$\lambda e^{2V} - 4kbAC = \lambda_0 \qquad 2\lambda = NM' - MN'$$
$$\mu e^{2V} + 4kbA^2 = \mu_0 \qquad \mu = LM' - ML'$$ (18.37)
$$\nu e^{2V} + 4kbC^2 = \nu_0 \qquad \nu = NL' - LN'$$

where λ_0, μ_0, and ν_0 are the constant values of the corresponding first integrals. Equation fV is linear in f and may be solved for as a function of V. Finally, λ, μ,

ν satisfy a purely kinematic relation,

$$(f')^2 = 4(\lambda^2 - \mu\nu) + 4f(L')^2 - 4fM'N' \tag{18.38}$$

Under constant $\Lambda \in SL(2, \mathbf{R})$ transformations of the coordinates $x^\pm$,

$$\begin{pmatrix} \tilde{x}^+ \\ \tilde{x}^- \end{pmatrix} = \Lambda^{-1} \begin{pmatrix} x^+ \\ x^- \end{pmatrix} \tag{18.39}$$

the Ansatz (18.7), the reduced field equations (18.35), and the first integrals (18.36) and (18.37) are invariant provided the fields transform as,

$$\begin{pmatrix} \tilde{A} \\ -\tilde{C} \end{pmatrix} = \Lambda^t \begin{pmatrix} A \\ -C \end{pmatrix} \qquad \begin{pmatrix} \tilde{M} & \tilde{L} \\ \tilde{L} & \tilde{N} \end{pmatrix} = \Lambda^t \begin{pmatrix} M & L \\ L & N \end{pmatrix} \Lambda \tag{18.40}$$

while the field V and the combination f are invariant. The triplet (λ, μ, ν) is the $SL(2, \mathbf{R})$ analogue of angular momentum and transforms under the vector representation of $SL(2, \mathbf{R})$, just as their first integral values $(\lambda_0, \mu_0, \nu_0)$ do.

18.5.2 Near-Horizon Schrödinger Geometry

Introducing charge requires that $E \neq 0$ in the Ansatz of (18.7). Re-interpreting t as x^+ and x_3 as x^-, we see that turning on a charge corresponds to a deformation which is null in the $x^\pm$ coordinate system. This suggests the existence of a solution in which deformations in the x^- directions vanish. We begin by exhibiting an exact charged near-horizon solution in which the x_1, x_2-directions are frozen out by the presence of a magnetic field, so that $K = e^{2V}$ is constant. The gauge potential and electric field are found as follows,

$$A(r) = \frac{e_0}{k} r^k \qquad E(r) = e_0 r^{k-1} \tag{18.41}$$

The near-horizon metric takes the form,

$$ds^2 = \frac{dr^2}{12r^2} + 4\sqrt{3}rdtdx_3 - \left(\alpha_0 r + \frac{2e_0^2 r^{2k}}{k(2k-1)}\right)dt^2 + dx_1^2 + dx_2^2 \tag{18.42}$$

In these coordinates we have $b = \sqrt{3}$. This metric coincides with the Schrödinger space-time of [9, 60] and the null-warped solution of [6].

18.5.3 The Charged Magnetic Brane Solution

The near-horizon Schrödinger geometry at $r \to 0$ extends to a regular *charged magnetic brane solution* to the full reduced field equations (18.35) with asymptotic AdS_5

behavior. In this respect, the role played by the Schrödinger near-horizon geometry for the charged magnetic brane is parallel to the role played by the $AdS_3 \times \mathbf{R}^2$ near-horizon geometry of the purely magnetic brane. The full solution has vanishing deformations in the x^- direction, so that we can set $C = N = 0$, and the functions $K = e^{2V}$ and L remain those of the purely magnetic brane. The remaining fields A and M may be obtained by quadrature from $K = e^{2V}$ and L, and we find,

$$A(r) = A_\infty e^{2kb\psi(r)}$$

$$M(r) = L(r)\left(-\frac{\alpha_\infty}{2\sqrt{3}} - 4kb \int_\infty^r \frac{dr' \, A(r')^2}{K(r')L(r')^2}\right)$$

(18.43)

The function $\psi(r)$ was defined in (18.22). Using the $r \to 0$ asymptotics of $\psi(r)$ given in (18.25), we see that the gauge potential satisfies the standard regularity condition $A(0) = 0$ at the horizon. Using the $r \to \infty$ asymptotics of (18.26), we see that the integration constant $A_\infty = A(\infty)$ is the chemical potential. The integration constant α_∞ introduces a relative tilt between the light-cones in the UV and the IR. Solutions for different values of A_∞ and α_∞ are related to one another by $SL(2, \mathbf{R})$ transformations which preserve the restrictions $C = N = 0$, and we have,

$$\Lambda = \begin{pmatrix} \lambda_1 & 0 \\ \lambda_2 & \lambda_1^{-1} \end{pmatrix} \qquad \begin{aligned} \tilde{A}_\infty &= \lambda_1 A_\infty \\ \tilde{\alpha}_\infty &= \lambda_1^2 \alpha_\infty - 2\lambda_1\lambda_2 \end{aligned}$$

(18.44)

Therefore, all solutions with $A_\infty \neq 0$ are equivalent to one another under $SL(2, \mathbf{R})$. The asymptotic behavior near the boundary of AdS_5 is given as follows,

$$A(r) \sim A_\infty - \frac{c_E}{r} \qquad c_E = \frac{kb \, A_\infty}{c_V}$$

$$M(r) \sim -\frac{\alpha_\infty}{\sqrt{3}} r$$

(18.45)

The integration constants e_0 and α_0 of the near-horizon Schrödinger geometry may be related to the parameters α_∞ and A_∞ of the boundary, and we find,

$$A_\infty = e_0 \, e^{-2kb\psi_0}$$

$$\alpha_\infty = \alpha_0 + 16c_V^2 c_E^2 J(k) \qquad J(k) = \frac{1}{2k} \int_0^\infty dr \, \frac{e^{4kb\psi(r)}}{K(r)L(r)^2}$$

(18.46)

where the constant ψ_0 was defined in (18.25).

18.5.4 Regularity of the Solutions

In anticipation of extending the charged magnetic brane solution to finite temperature, we must require regularity of the solution as a black brane. Thus, the coefficient function of dt^2 in the metric must remain negative throughout the space-time region outside the (outer) horizon, which leads us to require,

$$M(r) \leq 0 \qquad (18.47)$$

with equality only at the horizon $r = 0$. At infinity, this imposes the condition $\alpha_\infty > 0$. The solution near the horizon of (18.42) imposes further conditions which depend on the value of k. In the parameter region $0 \leq k < 1/2$, the r^{2k} term dominates over the $\alpha_0 r$ term, and leads to $M(r) > 0$ as soon as $e_0 \neq 0$. Thus, the charged magnetic brane solution in the region $0 \leq k < 1/2$ is excluded, as it cannot arise as the zero temperature limit of a nonsingular finite temperature black brane.

In the parameter region $1/2 < k$, it is the $\alpha_0 r$ term that dominates, which requires $\alpha_0 \geq 0$. The value $\alpha_0 = 0$ is actually regular as well, since the r^{2k} term contributes negatively for $1/2 < k$. It is straightforward to see from (18.43) that these conditions are also sufficient to make the charged magnetic brane solution regular for all $0 < r < \infty$.

18.5.5 Existence of a Critical Magnetic Field

The regularity conditions derived in the preceding section on the parameters α_0 and α_∞ may be translated into conditions on physically observable parameters in the dual field theory. Since the boundary field theory is conformal invariant, only dimensionless combinations of data can enjoy physical meaning. The magnetic field B and the charge density ρ have non-trivial dimension, but the ratio defined by,

$$\hat{B} \equiv \frac{B}{\rho^{2/3}} \qquad (18.48)$$

is dimensionless, and physically observable. Expressions for B and ρ, which denote the values of the magnetic field and charge density in coordinates such that the AdS_5 metric takes a canonical for, may themselves may be read off from the boundary behavior of the solution, and are given by,

$$B = \frac{2b}{c_V} \qquad \rho = 4c_E \sqrt{\frac{2b}{\alpha_\infty}} \qquad (18.49)$$

where c_E was defined in (18.45). Thus, $\hat{B}^3$ may be cast in the following form,

$$\hat{B}^3 = \frac{3\alpha_\infty}{4c_V^3 c_E^2} \qquad \hat{B}_c^3 \equiv \frac{3(\alpha_\infty - \alpha_0)}{4c_V^3 c_E^2} = \frac{12J(k)}{c_V} \qquad (18.50)$$

Here, we have also defined the combination $\hat{B}_c$ in terms of which we obtain the following final expression for $\hat{B}$,

$$\frac{\hat{B}_c^3}{\hat{B}^3} = 1 - \frac{\alpha_0}{\alpha_\infty} \qquad (18.51)$$

Positivity of $J(k)$ for $k > 0$ implies $\alpha_0 < \alpha_\infty$, while regularity required $0 \leq \alpha_0$. Thus, we conclude that the charged magnetic brane solution obtained above is regular if and only if $\hat{B}_c \leq \hat{B}$. In this sense, $\hat{B}_c$ represents a critical magnetic field. Its value depends only on the CS coupling k and the data of the purely magnetic brane solution. Inspection of the behavior of L and ψ in the integral for $J(k)$ shows that $J(k)$, and hence $\hat{B}_c$ diverges as $k \to 1/2$, thus providing a natural physical end point for the validity of the charged magnetic brane solution.

18.5.6 Low T Thermodynamics for $\hat{B} > \hat{B}_c$

The low T behavior dual to the charged magnetic black brane solution must be investigated separately for magnetic fields $\hat{B} > \hat{B}_c$ and $\hat{B} \sim \hat{B}_c$. We begin here with the study of the former. The presence of a low non-zero temperature induces only small changes to the charged magnetic brane solution for large r, while substantially altering its near-horizon behavior. The corresponding leading T-dependent near-horizon behavior needs to be treated exactly to incorporate these effects.

Our starting point is the purely magnetic BTZ $\times \mathbf{R}^2$ solution already discussed in Sect. 18.4.2. Its metric is given by (18.17), but it will be convenient here to choose the outer horizon at $r = 0$, so that $r_+ = 0$, and to parametrize the solution as follows,

$$F = b dx^1 \wedge dx^2$$

$$ds^2 = \frac{dr^2}{12r^2 + mnr} - mr dt^2 + 4\sqrt{3} dt dx_3 + n dx_3^2 + dx_1^2 + dx_2^2 \tag{18.52}$$

For an asymptotically AdS_5 space-time given by,

$$ds^2 = \frac{dr^2}{4r^2} - \frac{\alpha_\infty}{\sqrt{3}} r dt^2 + 4r dt dx_3 + c_V \left(dx_1^2 + dx_2^2 \right) \tag{18.53}$$

the dimensionless form of the entropy density $\hat{s}$, and of the temperature $\hat{T}$ may be expressed as follows,

$$\hat{s} \equiv \frac{s}{B^{3/2}} = \frac{\sqrt{n c_V \alpha_\infty}}{24} \qquad \hat{T} \equiv \frac{T}{B^{1/2}} = \frac{m \sqrt{n c_V}}{4\pi \sqrt{\alpha_\infty}} \tag{18.54}$$

In their ratio all reference to n cancels out,

$$\frac{\hat{s}}{\hat{T}} = \frac{\pi}{6} \frac{\alpha_\infty}{m} \tag{18.55}$$

This ratio has a finite limit as $T \to 0$, and may be evaluated in terms of the data of the $T = 0$ charged magnetic solution, for which we have $m = \alpha_0$ and $n = 0$ by (18.42). Along with the result for α_0/α_∞ from (18.51), we find a remarkably simple formula,

$$\frac{\hat{s}}{\hat{T}} = \frac{\pi}{6} \frac{\hat{B}^3}{\hat{B}^3 - \hat{B}_c^3} \tag{18.56}$$

A number of remarks are in order.

1. In our system, the physical entropy density vanishes at zero temperature (in contrast with the non-vanishing entropy density used in [13, 27, 48] in $2+1$ dimensions.
2. The limit $\hat{B} \to \infty$ corresponds to vanishing charge density ρ at fixed B, and reproduces the zero charge density result of (18.16).
3. The dependence on $\hat{B}/\hat{B}_c$ manifested in (18.56) is reminiscent of the dependence on the excitation velocity and couplings in the Luttinger liquid theory in (18.3).
4. The divergence of $\hat{s}/\hat{T}$ at zero temperature as $\hat{B} \to \hat{B}_c$ signals the presence of a quantum critical point at $\hat{B}_c$. Therefore, in a small region around $T = 0$ and $\hat{B} = \hat{B}_c$ in the $\hat{T}, \hat{B}$ plane, we should expect to find quantum critical behavior, to be explored in the subsequent subsections.
5. The phase for $\hat{B} < \hat{B}_c$ has non-zero entropy density at $\hat{T} = 0$, and may be thought of as a deformation of the Reissner-Nordstrom solution for zero magnetic field.
6. Numerical solutions perfectly reproduce the above analytical approximations, as will be explained in Sect. 18.5.9.

18.5.7 Low T Thermodynamics for $\hat{B} = \hat{B}_c$

Precisely at the quantum critical point, we have $\hat{B} = \hat{B}_c$, or equivalently $\alpha_0 = 0$. The resulting near-horizon metric of (18.42) then simplifies slightly,

$$ds^2 = \frac{dr^2}{12r^2} + 4\sqrt{3}r dt dx_3 - \frac{2e_0^2 r^{2k}}{k(2k-1)} dt^2 + dx_1^2 + dx_2^2 \tag{18.57}$$

More importantly, however, the metric now is invariant under the following scaling transformations,

$$r \to \lambda r \qquad t \to \lambda^{-k} t \qquad x_3 \to \lambda^{k-1} x_3 \tag{18.58}$$

with x_1, x_2 unchanged. The associated dynamical scaling exponent is given by,

$$z = \frac{k}{1-k} \tag{18.59}$$

General arguments show that, for fixed $\hat{B} = \hat{B}_c$, the entropy density scales with temperature according to the relation $\hat{s} \sim \hat{T}^{d/z} = T^{1/z}$ given that the scaling theory here has space dimension $d = 1$.

Numerical analysis shows that the above prediction, based on the structure of the near-horizon metric and its scaling symmetry, is borne out in only a limited range for k. The actual behavior is as follows,

$$\hat{s} \sim \hat{T}^{(1-k)/k} \quad 1/2 < k < 3/4$$

$$\hat{s} \sim \hat{T}^{1/3} \quad 3/4 < k \tag{18.60}$$

The numerical accuracy for the exponents is better than 1 % for the points we have checked. Note that the exponent matches continuously across the value $k = 3/4$. Inspection of the numerically obtained metric functions shows that the near-horizon region for finite $\hat{T}$ holds an electrically charged black brane whose space-time metric differs from that of BTZ.

To understand why the arguments based on the scaling symmetry of the near-horizon metric fail for $k > 3/4$, we consider a scaling transformation which leaves the AdS_3 part of the metric invariant, but not necessarily the term in e_0^2. For example, applying the following scaling,

$$r \to \lambda r \qquad t \to t/\sqrt{\lambda} \qquad x_3 \to x_3/\sqrt{\lambda} \tag{18.61}$$

with x_1, x_2 unchanged, will scale the term in e_0^2 by a factor of λ^{2k-1}. Scaling towards the IR corresponds to $\lambda < 1$, and we see that the term in e_0^2 naively becomes irrelevant. Whether the term indeed is irrelevant becomes a dynamical question, which is not easy to settle. Detailed arguments were given in [19] that the separation point is indeed $k = 3/4$. The scaling exponent of $1/3$ may be reproduced for the range $k > 1$ with the help of the method of overlapping expansions. The calculations are technically involved, and will not be reproduced here.

In Fig. 18.4 we display numerical data illustrating the crossover from the behavior $\hat{s} \sim \hat{T}$ to the behavior $\hat{s} \sim \hat{T}^{1/3}$ for $k = k_{susy} = 2/\sqrt{3}$. Qualitatively, the cross-over behavior of Fig. 18.4 persists for all $k > 3/4$.

18.5.8 Scaling Function in the Quantum Critical Region

In a small region surrounding the quantum critical point $T = 0$ and $\hat{B} = \hat{B}_c$, a critical scaling regime sets in. For the range $k > 1$, we have been able to derive this scaling behavior with the help of the method of overlapping expansions, and we find,

$$\hat{s} = \hat{T}^{1/3} f\left(\frac{\hat{B} - \hat{B}_c}{\hat{T}^{2/3}}\right) \tag{18.62}$$

for a certain scaling function f (which is not to be confused with the metric function f introduced in (18.35)). At $\hat{B} = \hat{B}_c$, this formula reproduces the scaling behavior discussed in (18.60) of the previous section for $k > 1$. For $\hat{B} > \hat{B}_c$, and low temperature, namely $\hat{T}^{2/3} \ll (\hat{B} - \hat{B}_c)$, we should recover (18.56), so that we should have $f(x) \sim \pi \hat{B}_c/(18x)$ for large x. Actually, the method of overlapping expansions allows one to compute $f(x)$ for the range $k > 1$, and we shall quote here the result without reproducing the derivation given in [19],

$$f(x)\left(f(x)^2 + \frac{x}{32k\hat{B}_c^4}\right) = \frac{\pi}{576k\hat{B}_c^3} \tag{18.63}$$

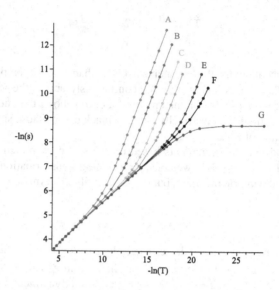

Fig. 18.4 The cross-over behavior of $\hat{s}$ versus $\hat{T}$ for $k = k_{susy} = 2/\sqrt{3}$ at values of $\hat{B} > \hat{B}_c$ corresponding to the curves A: $\hat{B}^3 = 0.125$, B: $\hat{B}^3 = 0.1247$, C: $\hat{B}^3 = 0.1246$, D: $\hat{B}^3 = 0.12458$, E: $\hat{B}^3 = 0.12457$, F: $\hat{B}^3 = 0.124569$, and G: $\hat{B}^3 = 0.124568$. To lighten notations, hats on the variables $\hat{T}$, $\hat{s}$ have not been exhibited in labeling the figure. At moderately low temperatures, $\hat{s}$ scales as $\hat{T}^{1/3}$ (*lower left corner*), while at ultra-low temperatures $\hat{s}$ scales as $\hat{T}$ for $\hat{B} > \hat{B}_c$ (*curves A, B, C, D, E, F*) and tends to a non-zero constant for $\hat{B} < \hat{B}_c$ (*curve G*). The *dots* represent numerical data points, while the *solid interpolating lines* are included to guide the eye

The scaling function f continues to apply for $\hat{B} < \hat{B}_c$, and we find the following behavior for the entropy density as a function of the magnetic field,

$$\hat{s} = \frac{\sqrt{\hat{B}_c - \hat{B}}}{4\sqrt{2k}\,\hat{B}_c^2} \tag{18.64}$$

The value of the exponent is reproduced to approximately 0.2 % accuracy by numerical simulations, and the prefactor to approximately 20 %.

18.5.9 Numerical Completion of the Holographic Phase Diagram

The full holographic phase diagram in the variables $\hat{B}$, $\hat{T}$ is presented in Fig. 18.5. Here, the various asymptotic behaviors are combined onto a single graph for the range $k > 3/4$. For the range $1/2 < k < 3/4$, the scaling exponent $1/3$ at $\hat{B} = \hat{B}_c$ must be replaced by $(1 - k)/k$, and the behavior of the entropy density for $\hat{B} < \hat{B}_c$ is altered though we have not systematically studied the corresponding modifications.

Fig. 18.5 The full
holographic phase diagram in
terms of the variables $\hat{B}$ and
$\hat{T}$ for $k > 3/4$. Hats on $\hat{B}$, $\hat{T}$,
$\hat{s}$, and $\hat{B}_c$ have not been
exhibited in the figure

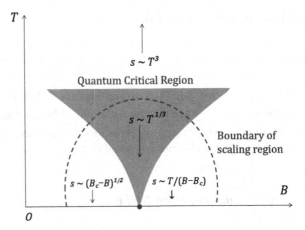

18.5.10 Correlators at Non-zero Charge Density

Correlators of the Maxwell current $\mathscr{J}^\mu$ and of the stress tensor $\mathscr{T}^{\mu\nu}$ may be evaluated, in the long distance approximation, in the presence of a magnetic field at zero temperature, but now with non-vanishing background electric charge density ρ, or equivalently, with chemical potential μ. As in the case with vanishing charge density studied in Sects. 18.4.3, 18.4.5, and 18.4.8, we use the method of overlapping expansions of 18.4.4 valid for $k > 1$. The calculations for the charged case proceed in analogy with the ones for the neutral case, but are now considerably more delicate and technically involved. We refer to the original paper [19] for their detailed derivation, and restrict here to quoting and explaining the results.

Correlators involving the operators with minus chirality are *unmodified* from the zero charge case. In particular, the two point function of $\mathscr{J}_-$ has no singularities, while the two point function of $\mathscr{T}_{--}$ continues to be given by (18.34). The correlators with plus chirality are found as follows,

$$\langle \mathscr{J}_+(x)\,\mathscr{J}_+(0)\rangle = -\frac{kc}{2\pi^2}\frac{1}{(x^+)^2}$$

$$\langle \mathscr{J}_+(x)\,\mathscr{T}_{++}(0)\rangle = +\frac{kc\mu}{2\pi^2}\frac{1}{(x^+)^2} \tag{18.65}$$

$$\langle \mathscr{T}_{++}(x)\,\mathscr{T}_{++}(0)\rangle = -\frac{kc\mu^2}{2\pi^2}\frac{1}{(x^+)^2} + \frac{c}{8\pi^2(x^+)^4}$$

where μ is the chemical potential, related to the charge density by,

$$\mu = A_\infty = \frac{\rho c_V \sqrt{\alpha_\infty}}{4kb\sqrt{2b}} \tag{18.66}$$

The system of correlators in the presence of charge may be related to the system of correlators of operators $\mathscr{J}_+^{(0)}$ and $\mathscr{T}_{++}^{(0)}$ at zero charge density by the following simple operator mixing relations,

$$\mathcal{J}_+ = \mathcal{J}_+^{(0)}$$
$$\mathcal{T}_{++} = \mathcal{T}_{++}^{(0)} - \mu_+ \mathcal{J}_+^{(0)} \tag{18.67}$$

Here, we have exhibited the natural Lorentz weight of the chemical potential by setting $\mu = \mu_+$. We see that the underlying Abelian Kac-Moody algebra for $\mathcal{J}_+^{(0)}$ and the underlying Virasoro algebras for $\mathcal{T}_{\pm\pm}^{(0)}$ are unmodified, with unchanged Kac-Moody level kc, and Virasoro central charge c.

18.5.11 Comments on Stability

The solutions studied here can, at least for special values of k, be uplifted to full solutions of higher dimensional supergravity and string theory, but nothing guarantees that they are stable solutions. There are two types of potential instabilities to be aware of: those coming from fields already included in our analysis, and those required by a consistent embedding into supergravity/string theory. Regarding the former, it has been observed in several contexts that the combination of electric charge and Chern-Simons terms can lead to instabilities towards spatially modulated phases [23, 53]. In some cases new solutions with reduced symmetry can be found [24, 25, 41, 42]. As for the latter, a supergravity/string theory embedding will typically bring along a variety of charged fields, and these may be unstable towards forming a condensate, as in holographic superconductors [37]. It is clearly an important challenge to determine when our solutions are unstable, and if not, to characterize the nature of the true ground state.

18.6 Quantum Criticality in 2 + 1 Dimensions

The critical theories studied so far originate from an underlying $3 + 1$-dimensional gauge theory in the UV which flows towards an effective $1 + 1$-dimensional strongly interacting CFT in the IR. Low temperature thermodynamics and long-distance correlators all signal massless propagation along the direction parallel to the magnetic field only. In the gravity dual, this IR behavior results from the existence of a near-horizon Schrödinger geometry of the form $WAdS_3 \times \mathbf{R}^2$, where $WAdS_3$ is a *null-warped* deformation of AdS_3 space-time. The physical mechanism driving the quantum critical transition on the gravity side is the gradual expulsion of electric charge from the inside of the black brane horizon to the outside of the horizon as the magnetic field $\hat{B} = B/\rho^{2/3}$ is being increased; see [36] for another example of this phenomenon.

While quantum criticality in $1 + 1$ dimensions is certainly of considerable physical interest, as was pointed out in the preceding section, it is probably even more urgent to extend the study to higher dimensions. Quantum criticality in $2 + 1$ dimensions is relevant to the physics of layered materials, such as cuprates, and graphene.

In the present section, we shall exhibit quantum critical behavior in $2+1$ dimensions systems in the presence of a magnetic field, and a non-vanishing electric charge density by holographic methods. Criticality here is driven by the same holographic mechanism that governed the $1 + 1$-dimensional case, namely charge expulsion from the black brane horizon. This time, however, the IR behavior in the gravity dual results from a flow from AdS_6 in the UV to a near-horizon Lifshitz geometry [44] in the IR which is a deformation of AdS_4. See [11, 33] for other examples of holographic RG flows involving Lifshitz spacetime. The AdS_6 geometry in the UV should be thought of as being dual to some $5 + 1$-dimensional CFT, examples of which do exist, and have been identified in [59].

18.6.1 Field Equations and Structure of the Solutions

The charge expulsion mechanism operating in the flow from AdS_5 to deformations of AdS_3 is made possible by the presence of the Chern-Simons interaction for the Maxwell field, and the existence of the transition crucially depends upon the strength of the associated Chern-Simons coupling k. This is because the Chern-Simons term provides the mechanism by which the bulk gauge field can carry its own charge.

The charge expulsion mechanism in higher dimensions that we shall focus on will also be made possible by the presence of Chern-Simons terms. Starting with AdS_6 in the UV does not support a Chern-Simons terms for the bulk gauge field all by itself. Thus, we are led to introducing further form fields. In the simplest extension, we add a single two-form potential C with field strength $G = dC$. The corresponding Einstein-Maxwell-Chern-Simons action then becomes,

$$
S = -\frac{1}{16\pi G_6} \int d^6x \sqrt{g} \left(R - \frac{20}{\ell^2} + F_{MN} F^{MN} + \frac{1}{3} G_{MNP} G^{MNP} \right) + S_{CS}
$$

$$
S_{CS} = \frac{k}{4\pi G_6} \int C \wedge F \wedge F
$$

(18.68)

where G_6 is the $5 + 1$-dimensional Newton constant, $F = dA$ is the Maxwell field strength, and $-20/\ell^2$ stands for the cosmological constant for an asymptotic AdS_6 vacuum solution of radius ℓ, which we shall set to 1. Boundary and counter term contributions to the action are not being exhibited here.

The Maxwell-Chern-Simons field equations are,

$$
d * F - 2k F \wedge G = 0
$$

$$
d * G + k F \wedge F = 0
$$

(18.69)

while the Einstein equations are,

$$
R_{MN} = -2 F_{MP} F_N{}^P - G_{MPQ} G_N{}^{PQ} + g_{MN} \left(5 + \frac{1}{4} F_{PQ} F^{PQ} + \frac{1}{6} G_{PQR} G^{PQR} \right)
$$

(18.70)

Clearly, the charge densities for both the F and G fields are proportional to the Chern-Simons coupling k.

As we focus here on thermodynamic questions, we shall be interested in solutions which are invariant under translations in $x^\mu = (t, x_1, x_2, x_3, x_4)$. The flow from AdS_6 in the UV to AdS_4 and its deformations in the IR will be generated by a constant magnetic field, which we shall choose in the direction $F_{34} = B$. It is natural to require rotation invariance in the x_3, x_4 plane, as well as in the remaining space directions x_1, x_2. A general Ansatz invariant under these symmetries was constructed in [21], and is given by,

$$F = Edr \wedge dt + \tilde{B}dx_1 \wedge dx_2 + Bdx_3 \wedge dx_4$$

$$G = (G_1 dr + G_2 dt) \wedge dx_1 \wedge dx_2 + (G_3 dr + G_4 dt) \wedge dx_3 \wedge dx_4 \quad (18.71)$$

$$ds^2 = \frac{dr^2}{U} - Udt^2 + e^{2V_1}\left(dx_1^2 + dx_2^2\right) + e^{2V_2}\left(dx_3^2 + dx_4^2\right)$$

By translation invariance in x^μ, the coefficients B, $\tilde{B}$, E, G_1, G_2, G_3, G_4, U, V_1, V_2 depend only on r. By the Bianchi identities for F and G, the quantities B, $\tilde{B}$, G_2, G_4 must actually be independent of r. The magnetic field $\tilde{B}$ plays the role of a magnetic field living in the IR $2 + 1$-dimensional field theory, and will be set to zero here for simplicity, $\tilde{B} = 0$. For $kB \neq 0$, the field equations then imply that $G_2 = 0$ and $G_3 G_4 = 0$. Solutions with either $G_3 \neq 0$ or $G_4 \neq 0$ do not have regular horizons, so we set also $G_3 = G_4 = 0$. The remaining reduced field equations were derived in [21], and will not be repeated here as they are reasonably involved.

18.6.2 Horizon and Asymptotic Data, Physical Quantities

We choose a coordinate r such that the horizon is at $r = 0$. We normalize the scales of the coordinates x^μ by setting,

$$U(0) = V_1(0) = V_2(0) = 0 \qquad U'(0) = 1 \quad (18.72)$$

The field equations relate the horizon values $G_1(0) = -2kbE(0)$. The asymptotic behavior for $r \to \infty$ may be parametrized analogously,

$$U(r) \sim r^2 \qquad e^{2V_1(r)} \sim v_1 r^2 \qquad e^{2V_2(r)} \sim v_2 r^2 \quad (18.73)$$

The asymptotics of the gauge field strength fixes the physical charge density ρ of the boundary theory by $r^4 E(r) \to \rho$. The dimensionless magnetic field $\hat{B} = B/\sqrt{\rho}$, temperature $\hat{T} = T/\sqrt{B}$, and entropy density $\hat{s} = s/B^2$ are then given by,

$$\hat{B} = \frac{b}{v_2\sqrt{\rho}} \qquad \hat{T} = \frac{\sqrt{v_2}}{4\pi\sqrt{b}} \qquad \hat{s} = \frac{v_2}{4v_1 b^2} \quad (18.74)$$

The equation of state corresponds to the relation $\hat{s} = \hat{s}(k, \hat{T}, \hat{B})$. We shall begin by discussing below analytical solutions available in various limits. Obtaining the function $\hat{s}$ throughout parameter space will, however, require numerical analysis.

18.6.3 Flows Towards the Electric IR Fixed Point

In the absence of a magnetic field, $B = 0$, the purely electric solution is given by the standard Reissner-Nordstrom form,

$$U = r^2 + \frac{q^2}{6r^6} - \frac{M}{r^3} \qquad V_1 = V_2 = \ln r \qquad E = \frac{\rho}{r^4} \qquad (18.75)$$

In the extremal limit, the location of the horizon r_+ is determined by $U(r_+) = U'(r_+) = 0$, and the entropy density $s \sim \rho \sim r_+^4$ does not vanish at $T = 0$. The near-horizon geometry of the purely electric solution is $AdS_2 \times \mathbf{R}^4$.

For $B \neq 0$, numerical analysis confirms the existence of a charged magnetic brane solution whose near-horizon behavior coincides with that of the purely electric solution, provided the Chern-Simons coupling k remains below a critical value k_c which will be determined shortly.

18.6.4 Flows Towards the Magnetic IR Fixed Point

The near-horizon behavior of the purely magnetic solution is given by $AdS_4 \times \mathbf{R}^2$ space-time at $T = 0$, or an AdS_4 Schwarzschild solution at $T \neq 0$. The two cases may be described together by,

$$U = \frac{20}{9}\left(r^2 - \frac{r_+^3}{r}\right) \qquad e^{2V_1} = \frac{20}{9}r^2 \qquad e^{2V_2} = \sqrt{\frac{3}{10}}B \qquad (18.76)$$

The temperature behaves as $T \sim r_+$, and the entropy density may be computed exactly in the low T approximation,

$$s = \frac{\pi^2}{5}\sqrt{\frac{3}{10}}BT^2 \qquad (18.77)$$

This T-dependence is precisely as expected of a $2 + 1$-dimensional CFT associated with the near-horizon AdS_4 space-time.

For $\rho \neq 0$, numerical analysis confirms the existence of a charged magnetic brane solution whose near-horizon behavior is that of the purely magnetic solution, provided k is larger than the critical value k_c already identified to end the purely electric flow. The T^2-dependence of the entropy density which is characteristic of $2 + 1$-dimensional CFT behavior, persists as long as $k > k_c$, but the coefficient is now found to be a non-trivial function,

$$s = A(k, \hat{B}) \, B \, T^2 \qquad (18.78)$$

We expect that the problem of calculating the function $A(k, \hat{B})$ may be amenable to analytical treatment, especially since numerical evaluation indicates the following behavior for intermediate and small $\hat{B}$,

$$A(k, \hat{B}) \sim c(k) \exp\!\left(d(k)\hat{B}^{-2}\right) \qquad (18.79)$$

with the characteristic behavior $d(k) \sim (k^2 - k_c^2)^{-1}$ obeyed to remarkable accuracy.

18.6.5 Flows Towards the Lifshitz IR Fixed Point

The near-horizon geometries, $AdS_2 \times \mathbf{R}^4$ for $k < k_c$, and $AdS_4 \times \mathbf{R}^2$ for $k > k_c$, are separated by a Lifshitz near-horizon geometry at the critical point $k = k_c$. Seeking near-horizon solutions which are invariant under space-time scalings, $r \to \lambda r$, $t \to t/\lambda$, $x_{1,2} \to \lambda^{-1/z}x_{1,2}$, and $x_{3,4} \to \lambda^{-\beta}x_{3,4}$, for real constants z and β, we find that the existence of such a solution requires $\beta = 0$, as well as

$$k = k_c = \frac{1}{\sqrt{3}} \qquad (18.80)$$

The dynamical scaling exponent z is constrained to the range $z > 1$, but otherwise arbitrary. From this scaling behavior in the near-horizon region, the scaling behavior of the low temperature behavior of the entropy density may be deduced, and we find,

$$\hat{s} \sim T^{\frac{2}{z}} \qquad (18.81)$$

in accord with the space-dimension of the IR theory being $d = 2$, and the parameter z standing for the dynamical critical exponent. As z runs through the range $1 < z < \infty$, the entropy density is being interpolated from its IR behavior for the purely electric fixed point at $z = \infty$ to its IR behavior for the purely magnetic fixed point at $z = 1$. This interpolating behavior is reflected in the functional dependence of z on the magnetic field $\hat{B}$, which interpolates between the following asymptotic behaviors,

$$\frac{1}{z} = \begin{cases} 0.105 \, \hat{B}^2 & \hat{B} \ll 1 \\ 1 - 0.894 \, \hat{B}^{-4} & \hat{B} \gg 1 \end{cases} \qquad (18.82)$$

18.6.6 The Full Phase Diagram

The full phase diagram may now be assembled from the behavior in the various regimes that we have examined in the preceding sections; see Fig. 18.6. For large T, the entropy density is dominated by temperature alone, and charge density ρ as well

Fig. 18.6 The full
holographic phase diagram in
terms of the variables k and
$\hat{T}$. To lighten notations, hats
on $\hat{T}$, $\hat{s}$ have not been
exhibited in the figure

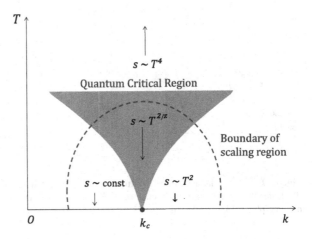

as magnetic field B have negligible effects. Thus, we have $s \sim T^4$, as expected form
the dual field theory by scaling.

For $k < k_c$, the flow from AdS_6 in the UV is towards the Reissner-Nordstrom
type near-horizon geometry $AdS_2 \times \mathbf{R}^4$ in the IR with $s \neq 0$ at $T = 0$. Without
doubt, this holographic solution will become unstable once charged scalar fields,
and/or space-dependent modulations are allowed. For $k > k_c$, the flow from AdS_6
in the UV is towards the $AdS_4 \times \mathbf{R}^2$ near-horizon geometry of the purely magnetic
brane, with its entropy density behaving as $s \sim BT^2$, characteristic of scaling in a
$2 + 1$-dimensional CFT. Finally, in the critical region, where $k \sim k_c$, the entropy
density at low $\hat{T}$ is governed by a scaling function,

$$\hat{s} = f(k - k_c, \hat{B}, \hat{T}) \tag{18.83}$$

In the absence to date of a (semi-)analytical solution connecting the near-horizon
Lifshitz geometry to the asymptotic AdS_6 region, the scaling function $f(k - k_c,$
$\hat{B}, \hat{T})$ is accessible only through numerical analysis.

18.7 Relation with Quantum Criticality in Condensed Matter

In this section, we shall point towards some exciting, though still speculative, applications of our holographic results to problems in condensed matter physics. One
application of the $1 + 1$-dimensional quantum criticality problem studied above is
to Strontium Ruthenates.

18.7.1 Meta-Magnetic Transitions in Strontium Ruthenates

The phase transitions exhibited by our holographic systems occur at a finite value
of the magnetic field and involve no change of symmetry. These are referred to as

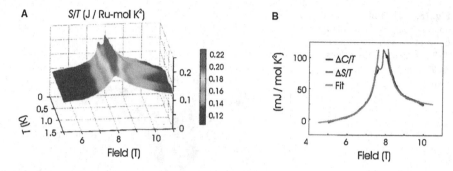

Fig. 18.7 Entropy landscape of $Sr_3Ru_2O_7$ near the meta-magnetic transition. The *right panel* illustrates the $1/(B - B_c)$ divergence of S/T, which is ultimately cut off by the appearance of a nematic phase. Figure taken from [57]

meta-magnetic phase transitions. A typical experimental situation is the following. A material has a line of first order phase transitions at finite temperature, reached by dialing the magnetic field, and with the line ending at a finite temperature critical point. By tuning some other control parameter, one can attempt to bring this critical point down to zero temperature, resulting in a quantum critical meta-magnetic transition [52], analogous to what we have found holographically. In particular, thermodynamic quantities such as the specific heat will diverge as the critical magnetic field is approached.

A version of this behavior, with some interesting twists, occurs in the Strontium Ruthenate compound $Sr_3Ru_2O_7$, and has been the subject of much experimental and theoretical interest in the past few years; e.g. [30, 57]. $Sr_3Ru_2O_7$ is a layered material, which, for a large magnetic field perpendicular to the 2-dimensional layers (around 8 T) exhibits meta-magnetic behavior with a "shrouded" quantum critical point. Notably, as the magnetic field reaches its critical value the entropy density behaves as $s/T \sim 1/(B - B_c)$, just as we found in our AdS_5 system. There appears to be no satisfactory theoretical understanding of this behavior. While this divergence in the entropy density appears to signal the onset of a quantum critical point, what actually seems to happen [30, 54] is that the system evolves into a nematic phase for $7.8T < B < 8.1T$. Spatial anisotropy in the nematic phase can be detected by applying a small in-plane component of magnetic field, which acts to align the domains, and then looking for anisotropic behavior of transport coefficients. The nematic phase seems to shroud the quantum critical point in a manner analogous to what occurs in high temperature superconductors. The fact that the would be divergence in s/T is cutoff by the appearance of a nematic phase has been described as nature's solution to the problem of avoiding a non-zero entropy density at zero temperature.

In [57], the complete *entropy landscape* of $Sr_3Ru_2O_7$ at finite temperature and magnetic field has been mapped out. The parallels with our system are clear, namely the $1/(B - B_c)$ divergence in the entropy density to temperature ratio; see Fig. 18.7. The most obvious difference is that our system is effectively $1 + 1$-dimensional at the critical point, while $Sr_3Ru_2O_7$ is strongly $2 + 1$-dimensional. But it is interesting to speculate whether a nematic phase will occur also in our holographic setup. In our ansatz we have assumed full translational and rotational invariance, but recent results indicate that there are frequently instabilities towards anisotropic phases [23–25, 41, 42, 53].

18.7.2 Relation to Hertz-Millis Theory

A standard approach to modeling magnetically tuned quantum phase transitions is based on the Hertz-Millis theory [39, 49, 51]. In $d + 1$ dimensions we consider the effective action

$$ S = \int d\omega d^d k \left(\frac{|\omega|}{|k|} + k^2 + (\hat{B} - \hat{B}_c) \right) |\phi(\omega, k)|^2 + \cdots \qquad (18.84) $$

The bosonic field ϕ represents the local magnetization, and one is supposed to think of this action as arising from integrating out gapless fermions at one-loop. There is no controlled approximation that justifies this approach, and indeed it is known to sometimes lead to predictions in conflict with experiment. Let us nonetheless make the following suggestive observations. We consider the action (18.84) with $d = 1$, and compare to our asymptotically AdS_5 critical theory. At $\hat{B} = \hat{B}_c$ the Hertz-Millis action is scale invariant, with k and ω assigned scaling dimensions 1 and 3 respectively. The dynamical critical exponent is therefore $z = 3$, which matches our AdS_5 result for $k > 3/4$, and will lead to the scaling law for the entropy, $s \sim T^{1/3}$, in one spatial dimension, as we found. Furthermore, $\hat{B} - \hat{B}_c$ plays the role of a relevant coupling of scaling dimension 2. This agrees with (18.62); to see this note that the argument of f in (18.62) has vanishing scale dimension, and $\hat{T}$ shares the same scaling dimension as ω, namely 3. Therefore, the scaling predictions of the Hertz-Millis theory can be identified in our holographic setup. Of course there are also differences; for instance, there is no analog of our finite ground state entropy density branch for $\hat{B} \le \hat{B}_c$.

Acknowledgements This work was supported in part by NSF grant PHY-07-57702.

During the course of this entire project, we have benefited from helpful conversations and correspondence with several colleagues, and we wish to thank here Vijay Balasubramanian, David Berenstein, Sudip Chakravarty, Geoffrey Compère, Jan de Boer, Frédéric Denef, Stéphane Detournay, Tom Faulkner, Jerome Gauntlett, Sean Hartnoll, Gary Horowitz, Finn Larsen, Alex Maloney, Eric Perlmutter, Joe Polchinski, Matt Roberts, Joan Simon, and especially Akhil Shah who collaborated on one of our papers. During parts of this work, we have enjoyed the hospitality of the KITP during the "Quantum Criticality and AdS/CFT Correspondence" program in 2009, and of the Aspen Center for Physics in 2011. One of us (E.D.) wishes to thank the Laboratoire de Physique Théorique de l'Ecole Normale Supérieure, and the Laboratoire de Physqiue Théorique et Hautes Energies, CNRS and Université Pierre et Marie Curie—Paris 6, and especially Constantin Bachas and Jean-Bernard Zuber for their warm hospitality while part of this work was being completed.

References

1. T. Andrade, J.I. Jottar, R.G. Leigh, Boundary conditions and unitarity: the Maxwell-Chern-Simons system in AdS_3/CFT_2. J. High Energy Phys. **1205**, 071 (2012). arXiv:1111.5054 [hep-th]
2. O. Aharony, S.S. Gubser, J.M. Maldacena, H. Ooguri, Y. Oz, Large N field theories, string theory and gravity. Phys. Rep. **323**, 183 (2000). arXiv:hep-th/9905111
3. A. Almheiri, Magnetic $AdS_2 \times R_2$ at weak and strong coupling. arXiv:1112.4820 [hep-th]
4. A. Almuhairi, AdS_3 and AdS_2 magnetic brane solutions. arXiv:1011.1266 [hep-th]
5. A. Almuhairi, J. Polchinski, Magnetic $AdS \times R^2$: supersymmetry and stability. arXiv:1108.1213 [hep-th]
6. D. Anninos, G. Compere, S. de Buyl, S. Detournay, M. Guica, The curious case of null warped space. J. High Energy Phys. **1011**, 119 (2010). arXiv:1005.4072 [hep-th]
7. V. Balasubramanian, P. Kraus, Space-time and the holographic renormalization group. Phys. Rev. Lett. **83**, 3605 (1999). hep-th/9903190
8. V. Balasubramanian, P. Kraus, A stress tensor for anti-de Sitter gravity. Commun. Math. Phys. **208**, 413 (1999). arXiv:hep-th/9902121
9. K. Balasubramanian, J. McGreevy, Gravity duals for non-relativistic CFTs. Phys. Rev. Lett. **101**, 061601 (2008). arXiv:0804.4053 [hep-th]
10. J. de Boer, E.P. Verlinde, H.L. Verlinde, On the holographic renormalization group. J. High Energy Phys. **0008**, 003 (2000). hep-th/9912012
11. H. Braviner, R. Gregory, S.F. Ross, Flows involving Lifshitz solutions. Class. Quant. Grav. **28**, 225028 (2011). arXiv:1108.3067 [hep-th]
12. J.D. Brown, M. Henneaux, Central charges in the canonical realization of asymptotic symmetries: an example from three-dimensional gravity. Commun. Math. Phys. **104**, 207 (1986)
13. M. Cubrovic, J. Zaanen, K. Schalm, Fermions and the AdS/CFT correspondence: quantum phase transitions and the emergent Fermi-liquid. arXiv:0904.1993 [hep-th]
14. S. de Haro, S.N. Solodukhin, K. Skenderis, Holographic reconstruction of space-time and renormalization in the AdS/CFT correspondence. Commun. Math. Phys. **217**, 595 (2001). arXiv:hep-th/0002230
15. E. D'Hoker, D.Z. Freedman, Supersymmetric gauge theories and the AdS/CFT correspondence. hep-th/0201253
16. E. D'Hoker, P. Kraus, Magnetic brane solutions in AdS. J. High Energy Phys. **0910**, 088 (2009). arXiv:0908.3875 [hep-th]
17. E. D'Hoker, P. Kraus, Charged magnetic brane solutions in AdS_5 and the fate of the third law of thermodynamics. arXiv:0911.4518 [hep-th]
18. E. D'Hoker, P. Kraus, Holographic metamagnetism, quantum criticality, and crossover behavior. J. High Energy Phys. **1005**, 083 (2010). arXiv:1003.1302 [hep-th]
19. E. D'Hoker, P. Kraus, Magnetic field induced quantum criticality via new asymptotically AdS_5 solutions. Class. Quant. Grav. **27**, 215022 (2010). arXiv:1006.2573 [hep-th]
20. E. D'Hoker, P. Kraus, Charged magnetic brane correlators and twisted Virasoro algebras. Phys. Rev. D **84**, 065010 (2011). arXiv:1105.3998 [hep-th]
21. E. D'Hoker, P. Kraus, Charge expulsion from black brane horizons, and holographic quantum criticality in the plane. arXiv:1202.2085 [hep-th]
22. E. D'Hoker, P. Kraus, A. Shah, RG flow of magnetic brane correlators. J. High Energy Phys. **1104**, 039 (2011). arXiv:1012.5072 [hep-th]
23. S.K. Domokos, J.A. Harvey, Baryon number-induced Chern-Simons couplings of vector and axial-vector mesons in holographic QCD. Phys. Rev. Lett. **99**, 141602 (2007). arXiv:0704.1604 [hep-ph]
24. A. Donos, J.P. Gauntlett, Helical superconducting black holes. Phys. Rev. Lett. **108**, 211601 (2012). arXiv:1203.0533 [hep-th]
25. A. Donos, J.P. Gauntlett, C. Pantelidou, Spatially modulated instabilities of magnetic black branes. J. High Energy Phys. **1201**, 061 (2012). arXiv:1109.0471 [hep-th]

26. A. Donos, J.P. Gauntlett, C. Pantelidou, Magnetic and electric AdS solutions in string- and M-theory. arXiv:1112.4195 [hep-th]

27. T. Faulkner, H. Liu, J. McGreevy, D. Vegh, Emergent quantum criticality, Fermi surfaces, and AdS$_2$. Phys. Rev. D **83**, 125002 (2011). arXiv:0907.2694 [hep-th]

28. T. Faulkner, H. Liu, M. Rangamani, Integrating out geometry: holographic Wilsonian RG and the membrane paradigm. J. High Energy Phys. **1108**, 051 (2011). arXiv:1010.4036

29. C. Fefferman, C.R. Graham, Conformal invariance, in *Elie Cartan et les Mathématiques d'aujourd'hui*, Astérisque (1985), p. 95

30. E. Fradkin, S.A. Kivelson, M.J. Lawler, J.P. Eisenstein, A.P. Mackenzie, Nematic Fermi fluids in condensed matter physics. Ann. Rev. Condens. Matter Phys. **1**, 153 (2010). arXiv:0910.4166

31. J.P. Gauntlett, O. Varela, Consistent Kaluza-Klein reductions for general supersymmetric AdS solutions. Phys. Rev. D **76**, 126007 (2007). arXiv:0707.2315 [hep-th]

32. T. Giamarchi, *Quantum Physics in One Dimension* (Oxford University Press, New York, 2004)

33. K. Goldstein, S. Kachru, S. Prakash, S.P. Trivedi, Holography of charged dilaton black holes. J. High Energy Phys. **1008** 078 (2010). arXiv:0911.3586 [hep-th]

34. S.S. Gubser, I.R. Klebanov, A.M. Polyakov, Gauge theory correlators from non-critical string theory. Phys. Lett. B **428**, 105 (1998). arXiv:hep-th/9802109

35. S.A. Hartnoll, Lectures on holographic methods for condensed matter physics. Class. Quant. Grav. **26**, 224002 (2009). arXiv:0903.3246 [hep-th]

36. S.A. Hartnoll, L. Huijse, Fractionalization of holographic Fermi surfaces. arXiv:1111.2606 [hep-th]

37. S.A. Hartnoll, C.P. Herzog, G.T. Horowitz, Building a holographic superconductor. Phys. Rev. Lett. **101**, 031601 (2008). arXiv:0803.3295 [hep-th]

38. I. Heemskerk, J. Polchinski, Holographic and Wilsonian renormalization groups. J. High Energy Phys. **1106**, 031 (2011). arXiv:1010.1264

39. J.A. Hertz, Quantum critical phenomena. Phys. Rev. B **14**, 1165 (1976)

40. C.P. Herzog, Lectures on holographic superfluidity and superconductivity. J. Phys. A **42**, 343001 (2009). arXiv:0904.1975 [hep-th]

41. N. Iizuka, K. Maeda, Study of anisotropic black branes in asymptotically anti-de Sitter. arXiv:1204.3008 [hep-th]

42. N. Iizuka, S. Kachru, N. Kundu, P. Narayan, N. Sircar, S.P. Trivedi, Bianchi attractors: a classification of extremal black brane geometries. arXiv:1201.4861 [hep-th]

43. K. Jensen, A. Karch, E.G. Thompson, A holographic quantum critical point at finite magnetic field and finite density. arXiv:1002.2447 [hep-th]

44. S. Kachru, X. Liu, M. Mulligan, Gravity duals of Lifshitz-like fixed points. Phys. Rev. D **78**, 106005 (2008). arXiv:0808.1725 [hep-th]

45. D.E. Kharzeev, L.D. McLerran, H.J. Warringa, The effects of topological charge change in heavy ion collisions: 'Event by event P and CP violation'. Nucl. Phys. A **803**, 227 (2008). arXiv:0711.0950 [hep-ph]

46. I.R. Klebanov, TASI lectures: introduction to the AdS/CFT correspondence. hep-th/0009139

47. G. Lifschytz, M. Lippert, Holographic magnetic phase transition. Phys. Rev. D **80**, 066007 (2009). arXiv:0906.3892 [hep-th]

48. H. Liu, J. McGreevy, D. Vegh, Non-Fermi liquids from holography. Phys. Rev. D **83**, 065029 (2011). arXiv:0903.2477 [hep-th]

49. H.v. Lohneysen, A. Rosch, M. Mojta, P. Wolfle, Fermi-liquid instabilities at magnetic quantum phase transitions. Rev. Mod. Phys. **79**, 1015–1075 (2007)

50. J.M. Maldacena, The large N limit of superconformal field theories and supergravity. Adv. Theor. Math. Phys. **2**, 231 (1998) [Int. J. Theor. Phys. 38, 1113 (1999)]. arXiv:hep-th/9711200

51. A.J. Millis, Effect of a nonzero temperature on quantum critical points in itinerant fermion systems. Phys. Rev. B **48**, 7183–7196 (1993)

52. A.J. Millis, A.J. Schofield, G.G. Lonzarich, S.A. Grigera, Metamagnetic quantum criticality in metals. Phys. Rev. Lett. **88**, 217204 (2002)
53. S. Nakamura, H. Ooguri, C.S. Park, Gravity dual of spatially modulated phase. Phys. Rev. D **81**, 044018 (2010). arXiv:0911.0679 [hep-th]
54. V. Oganesyan, S.A. Kivelson, E. Fradkin, Quantum theory of a nematic Fermi fluid. Phys. Rev. B **64**, 195109 (2001)
55. J. Polchinski, Introduction to gauge/gravity duality. arXiv:1010.6134 [hep-th]
56. G. Policastro, D.T. Son, A.O. Starinets, The shear viscosity of strongly coupled $N = 4$ supersymmetric Yang-Mills plasma. Phys. Rev. Lett. **87**, 081601 (2001). hep-th/0104066
57. A.W. Rost, R.S. Perry, J.-F. Mercure, A.P. Mackenzie, S.A. Grigera, Entropy landscape of phase formation associated with quantum criticality in $Sr_3Ru_2O_7$. Science **325**(5946), 1360–1363 (2009)
58. S. Sachdev, *Quantum Phase Transitions* (Cambridge University Press, New York, 2011)
59. N. Seiberg, Five-dimensional SUSY field theories, nontrivial fixed points and string dynamics. Phys. Lett. B **388**, 753 (1996). hep-th/9608111
60. D.T. Son, Toward an AdS/cold atoms correspondence: a geometric realization of the Schroedinger symmetry. Phys. Rev. D **78**, 046003 (2008). arXiv:0804.3972 [hep-th]
61. N. Ogawa, T. Takayanagi, T. Ugajin, Holographic Fermi surfaces and entanglement entropy. J. High Energy Phys. **1201**, 125 (2012). arXiv:1111.1023 [hep-th]
62. E. Witten, Anti-de Sitter space and holography. Adv. Theor. Math. Phys. **2**, 253 (1998). arXiv: hep-th/9802150

Chapter 19
Charge-Dependent Correlations in Relativistic Heavy Ion Collisions and the Chiral Magnetic Effect

Adam Bzdak, Volker Koch, and Jinfeng Liao

19.1 Introduction

The theoretical study of topological solitons in field theories has a long history. Quite generically, these objects arise as solutions to the classical equations of motion for field theories due to the nonlinearity of the equations as well as due to specific boundary conditions. They are found in field theories of various dimensions (2D kinks, 3D monopoles, 4D instantons), and are known to be particularly important in the non-perturbative domain where the theories are strongly coupled. For a recent review, see e.g. [1].

Topological objects in Quantum Chromodynamics (QCD) are known to play important roles in many fundamental aspects of QCD [1]. For example, instantons are responsible for various properties of the QCD vacuum, such as spontaneous breaking of chiral symmetry and the $U_A(1)$ anomaly (see e.g. [2, 3]). Magnetic monopoles, on the other hand, are speculated to be present in the QCD vacuum in a Bose-condensed form which then enforce the color confinement, known as the dual superconductor model for QCD confinement, which is strongly supported by lattice QCD calculations (see e.g. [4, 5]). Alternatively vortices are believed to describe the chromo-electric flux configuration (i.e. flux tube) between a quark-anti-quark

A. Bzdak (✉) · J. Liao
RIKEN BNL Research Center, Brookhaven National Laboratory, Upton, NY 11973, USA
e-mail: abzdak@bnl.gov

V. Koch
Nuclear Science Division, Lawrence Berkeley National Laboratory, MS70R0319, 1 Cyclotron Road, Berkeley, CA 94720, USA
e-mail: vkoch@lbl.gov

J. Liao
Physics Department and Center for Exploration of Energy and Matter, Indiana University, 2401 N Milo B. Sampson Lane, Bloomington, IN 47408, USA
e-mail: liaoji@indiana.edu

D. Kharzeev et al. (eds.), *Strongly Interacting Matter in Magnetic Fields*,
Lecture Notes in Physics 871, DOI 10.1007/978-3-642-37305-3_19,
© Springer-Verlag Berlin Heidelberg 2013

pair in the QCD vacuum which in turn gives rise to the confining linear potential (see e.g. reviews in [5, 6]). Some of these objects, such as monopoles [7–12] and flux tubes [13–15], may also be important degrees of freedom in the hot and deconfined QCD matter close to the transition temperature T_c, and may be responsible for the observed properties of the so called strongly coupled quark-gluon plasma [16–22].

Since the existence of such topological objects is theoretically well motivated and their effects on the dynamics are deemed to be important, a direct experimental detection of such objects or at least of certain unique imprints by them, would be a highly desirable goal. This review will discuss recent efforts and progress toward that goal, specifically in the context of relativistic heavy ion collisions through the measurement and analysis of charge-dependent correlations.

19.1.1 The Chiral Magnetic Effect in Brief

An interesting suggestion by Kharzeev and collaborators [23–31] on the direct manifestation of effects from topological objects is the possible occurrence of $\mathcal{P}$- and $\mathcal{CP}$-odd (local) domains due to the so-called sphaleron or anti-sphaleron transitions in the hot dense QCD matter created in relativistic heavy ion collisions. Imagine that in a single event created in a heavy ion collision the gauge field configurations in the space-time zone of the created hot dense matter experience a single sphaleron transition. As a result this local zone acquires a non-zero topological charge which is parity-odd. This non-zero topological charge, when coupled with light quarks through the triangle anomaly, induces a non-zero chirality for the quarks. In other words it generates an imbalance between left- and right-handed quark numbers, or a non-zero axial charge density. To be precise, there is no violation of parity at the interaction level, but rather a local creation of matter with non-zero axial charge density, which is a $\mathcal{P}$- and $\mathcal{CP}$-odd quantity.

A concrete proposal for experimental detection is the so-called Chiral Magnetic Effect (CME) [25]. The effect itself states that in the presence of external electromagnetic (EM) magnetic field **B**, a nonzero axial charge density will lead to an EM electric current along the direction of the magnetic field **B**:

$$\mathbf{j}_V = \frac{N_c\, e}{2\pi^2} \mu_A \mathbf{B} \tag{19.1}$$

where μ_A is the axial chemical potential associated with the non-zero axial charge density present in the system, and N_c is the number of colors. This elegant relation is theoretically well established in both the weakly-coupled and the strongly-coupled regimes of the theory as will be discussed in several contributions to this volume.

At first sight, it might seem that the above relation is violating parity: under spatial rotation and inversion the EM electric current $\mathbf{j}_V$ transforms like a vector, while the magnetic field, **B**, transforms like an axial- or pseudo-vector. Therefore, the factor in (19.1) relating the two will have to be parity-odd. This is indeed the case, since μ_A that enters the above relation is a pseudo-scalar quantity which changes sign under parity transformation. Thus the CME relation, (19.1), is invariant under parity transformation. However, in a region with nonzero, either positive or negative, μ_A certain parity-odd observables, e.g. the pseudoscalar quantity $\langle \mathbf{j}_V \cdot \mathbf{B} \rangle$, may acquire nonzero expectation values. It is only in this sense that one may refer to it as "local parity violation".

In addition there is a complimentary relation, as one might have guessed from the "duality" by interchanging the roles of V (vector) and A (axial), that has been called the Chiral Separation Effect (CSE). The CSE refers to the separation of chiral (or axial) charge along the axis of the external EM magnetic field at finite density of the vector charge, for example at finite baryon number density [32, 33]. The resulting axial current is given by

$$\mathbf{j}_A = \frac{N_c\, e}{2\pi^2} \mu_V \mathbf{B} \qquad (19.2)$$

with the μ_V here being the baryon number chemical potential. Furthermore the combination of the two effects, CME and CSE, gives rise to an interesting propagating collective mode: the vector density induces an axial current which transports and creates a locally nonzero axial charge density, which in turn leads to a vector current that further transports and creates a locally nonzero vector density, and so on. This is called Chiral Magnetic Wave (CMW) [34], just like Maxwell's electromagnetic waves represent the coupled evolution of the electric and magnetic fields. The CMW is a general concept that includes both the CME and CSE effects. It is robust in the sense that it takes the form of a collective excitation like the sound wave without relying on a quasi-particle picture.

We end the general introduction with two comments: first, the CME in the language of CMW induces a charged dipole (of the vector density distribution) that results from an initial nonzero axial charge density; second it has been recently pointed [35–37] that an initial vector charge density via CMW will lead to a charged quadrupole distribution that may be observable in heavy ion collisions. For the rest of this contribution we will focus on the charged dipole signal for the CME phenomenon.

19.1.2 Hunting for the CME in Heavy Ion Collisions

Now we turn to two key questions: can the Chiral Magnetic Effect occur in heavy ion collisions, and if so, what observables serve as unambiguous signals for the CME?

The answer to the first question seems to be positive. Two elements are needed for the CME to occur: an external magnetic field and a locally nonzero axial charge density. The relativistically moving heavy ions, typically with large positive charges (e.g. $+79e$ for Au), carry strong magnetic (and electric) fields with them. In the short moments before/during/after the impact of two ions in non-central collisions, there is a very strong magnetic field in the reaction zone [25, 38]. In fact, such a magnetic field is estimated to be of the order of $m_\pi^2 \approx 10^{18}$ Gauss [39, 40] (see also [41]), probably the strongest, albeit transient, magnetic field in the present Universe. The other required element, a locally non-vanishing axial charge density, can also be created in the reaction zone during the collision process through sphaleron transitions (see e.g. [31] for discussions and references therein). As such, it appears at least during the very early stage of a heavy ion collision, there can be both strong magnetic field and nonzero axial charge density in the created hot matter. Therefore, the CME should take place, that is, an electric current will be generated either parallel or anti-parallel to the magnetic field **B** depending on the axial charge density is positive (due to sphaleron) or negative (due to anti-sphaleron). How large this current is, is of course another question, see e.g. [42, 43].

The answer to the second question is much more difficult. Extracting the effects of the CME, which most likely occur at the very early stage of the collision, from the final observed hadrons, involves many uncertainties. First, it is quite unclear how long the magnetic field could remain strong: while the peak value is large, it decays very rapidly with time (if the only source of such field is from the protons in the ions) [44]. Second, if the CME current is generated mostly at very early time, it is not clear to which extent this current could survive without significant modifications, since we know that the created quark-gluon plasma behaves like a strongly interacting fluid. Furthermore, even if this current survives, one has to find the right observable for its detection. At present, there is no satisfactory resolution on the first two issues. This will likely require comprehensive and quantitative model studies. In this review we will only focus on the third issue—the observables to be used for measuring the possible CME current and related "background" effects.

In a simplistic view, one may consider the ultimate manifestation of the CME as a separation of charged hadrons along the direction of the initial magnetic field: more positive hadrons moving in one direction while more negative hadrons in the opposite direction. As a result, the momentum distribution of the final hadrons will have a charged dipole moment. The direction of such a momentum space dipole is expected to be along the **B** field, parallel or anti-parallel, depending on the sign of the initial axial charge density in a given event. Since the initial axial charge may be positive and negative with equal probability, the event average of the momentum space dipole vanishes, $\langle \mathbf{j}_V \cdot \mathbf{B} \rangle = 0$. This reflects the fact that parity is not broken globally by the strong interaction, so that any pseudo-scalar quantity, such as $\langle \mathbf{j}_V \cdot \mathbf{B} \rangle$, will have to vanish. What one can hope for, however, is to measure the fluctuation or variance of this charge separation, i.e. $\langle (\mathbf{j}_V \cdot \mathbf{B})^2 \rangle$, which is a parity even quantity. As we will discuss later, the prize one has to pay is that other, conventional correlations, not related to the CME, may contribute to observables which are sensitive to the variance of charged dipole moment.

Recently the STAR Collaboration at the Brookhaven's Relativistic Heavy Ion Collider (RHIC) has reported [45, 46] first measurements of a charge dependent correlation function in heavy ion collisions, which may by sensitive to the Chiral Magnetic Effect. The essential idea of the measurement, proposed by Voloshin [47], is based on two important features: first, in non-central heavy ion collisions, the direction of initial strong magnetic field is strongly correlated with the so-called reaction plane, which is spanned by the impact parameter and the beam direction. The **B** field is pointing (mostly) along the normal of reaction plane, albeit with random up/down orientation; second, the CME-induced current, or the charged dipole in momentum space, implies particular charge-dependent correlation patterns. The same-sign charged hadrons will prefer moving together while the opposite-sign charged hadrons moving back-to-back along the **B** field direction, and thus perpendicular to the reaction plane, which is commonly referred to as the out-of-plane direction.[1] While these measurements and their implications will be discussed in detail in Sect. 19.3, let us briefly summarize the present status: the STAR (later PHENIX, and also ALICE) data show very interesting charge dependent azimuthal correlation patterns, and some features are in line with the CME predictions. Other aspects of the data, on the other hand, are very hard to understand within the framework of the CME. At present, therefore, the observation of the Chiral Magnetic Effect in heavy ion collisions, and the local parity violation in the aforementioned sense, has not been established experimentally, and additional measurements as well as further theoretical analysis are required before definitive conclusions can be drawn.

This review is organized as follows: in Sect. 19.2, we will present a general discussion on the charge-dependent correlation measurements in heavy ion collisions, with the emphasis on the CME related observables; in Sect. 19.3, the presently available data from heavy ion collisions at a variety of collision energies will be examined and their interpretations will be critically evaluated; in Sect. 19.4, various possible "background" effects and their manifestation in various observables will be quantitatively analyzed; finally in Sect. 19.5 we summarize and conclude.

19.2 The Charge-Dependent Correlation Measurements

In this section, we focus on various charge-dependent correlation measurements in heavy ion collisions and what can be learned from these observables. The emphasis will not be on the data themselves, which will be the subject of the next section. Instead we will set up the conceptual framework for studying the azimuthal correlations, discuss possible complications in the design of the observables, and examine the connection between physical effects and the measurements.

---

[1] As a note of caution, the strong correlation between the **B** field direction and the participant-plane are considerably modified when the strong fluctuations in the initial condition are properly taken into account. As a result the two are rather weakly correlated in very central and very peripheral collisions [40, 41, 48].

19.2.1 General Considerations Concerning Azimuthal Correlation Measurements

The basic experimental information about the (hadronic) final state of a heavy ion collision consists of the momenta and the identity—the electric charge, mass and possibly other quantum numbers—of all hadrons observed in the acceptance of a given experiment. Customarily, the three-momentum $\mathbf{p}$ is represented by the (longitudinal) rapidity, y, the transverse momentum p_t as well as the azimuthal angle ϕ. Events may further be grouped according to the charged particle multiplicity, which is a good measure of the centrality or impact parameter of a collision. From a given sample of events one can then extract the single particle distributions, $d^3N/dydp_t^2d\phi$ either for all charged hadrons or, more selectively, for identified pions, kaons, protons, etc. In order to study possible correlations one analyses two-particle, three-particle and multi-particle distributions of various kinds. Most of the discussion in this review will focus on the dependence of various measurements on the azimuthal angle. The rapidity y and the transverse momentum p_t will either be in specific bins or integrated over.

The analysis of azimuthal distributions has to deal with the fact that the azimuthal direction of each collision, characterized by either the direction of the angular momentum or the impact parameter, is randomly distributed in the laboratory frame. Therefore, a single particle azimuthal distribution, $dN/d\phi$ will always be uniform and, thus, rather meaningless. To learn something about azimuthal distributions, one either measures distributions of the difference of the azimuthal angles of two particles, $dN/d(\phi_1 - \phi_2)$, or one determines the azimuthal orientation of a given event and studies distributions with respect to this direction. Commonly the azimuthal direction of the so-called reaction plane is used to characterize the orientation of an event. As already discussed in the Introduction, the reaction plane is spanned by the beam direction and the impact parameter of the collision. Its orientation in the laboratory frame is given by the so-called reaction plane angle, Ψ_{RP}, which measures the direction of the impact parameter in the laboratory frame. Given the reaction plane angle, one then can study azimuthal angular distributions with respect to the reaction plane angle, $f(\phi - \Psi_{RP}) = dN/d(\phi - \Psi_{RP})$. Clearly the determination of the reaction plane requires the measurement of other particles in addition to that used for the angular distribution (for a comprehensive review, see [49]). Therefore, the extraction of azimuthal distributions will require the measurement of two-particle (for the angular difference distribution $dN/d(\phi_1 - \phi_2)$) or even higher particle distributions.

However, it is important to distinguish between the need to measure two- or many-particle distributions to study azimuthal distributions, and the presence of true dynamical two- or many- particle correlations. To make this distinction more transparent, it is useful to introduce an *intrinsic* frame or coordinate system where the x-direction is given by the direction of the impact parameter, which is typically referred to as the so-called "in-plane-direction", and the y direction is defined by the

Fig. 19.1 A schematic demonstration of the proposed simultaneous analysis of $\hat{Q}_1^c$ and $\hat{Q}_2$ vectors in the same event

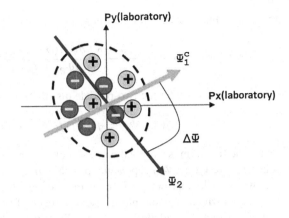

angular momentum, or the so-called "out-of plane direction". The relative angle of the x-axis of the intrinsic frame and that of the laboratory frame is then given by the reaction plane angle Ψ_{RP}, as illustrated in Fig. 19.1. In theoretical considerations and model calculations the orientation of the reaction plane is assumed to be known, or in other words, these calculations take place in the intrinsic frame. Finally, the azimuthal angle Φ in the intrinsic frame is related to the laboratory angle ϕ by

$$\Phi = \phi - \Psi_{RP}. \tag{19.3}$$

To continue, let us, as an example, consider a single particle distribution in the intrinsic frame

$$f_1(\Phi) = f_1(\phi - \Psi_{RP}) \propto 1 + 2v_2 \cos\left[2(\Phi)\right] = 1 + 2v_2 \cos\left[2(\phi - \Psi_{RP})\right] \tag{19.4}$$

which has an azimuthal asymmetry, characterized by the second Fourier component of strength v_2. This kind of distribution, which will be relevant for the subsequent discussion, is important in the context of the observed azimuthal asymmetries in heavy ion collisions, which are generally attributed to the hydrodynamics evolution of the system in non-central collisions. The parameter v_2 is commonly referred to as the elliptic flow coefficient. For a detailed discussion see [49]. The value for the elliptic flow parameter, v_2, may be obtained by measuring the second moment of the angular distribution, $\langle \cos 2(\phi - \Psi_{RP}) \rangle$. To this end we have to determine the reaction plane angle in each event, calculate the average moment in the intrinsic frame of each event and then average over events:

$$\langle \cos\left[2(\phi - \Psi_{RP})\right] \rangle$$

$$= \frac{1}{N_{events}} \sum_{event\ i=1}^{N_{events}} \left\{ \frac{1}{N(i)} \sum_{particle\ k=1}^{N(i)} \cos\left[2\left(\phi_k - \Psi_{RP}(i)\right)\right] \right\}. \tag{19.5}$$

In terms of the distribution function f_1 this can be expressed as[2]

$$\langle\cos[2(\phi - \Psi_{RP})]\rangle = \frac{\int d\Psi_{RP} \int d\phi f_1(\phi - \Psi_{RP})\cos[2(\phi - \Psi_{RP})]}{\int d\Psi_{RP} \int d\phi f_1(\phi - \Psi_{RP})}. \tag{19.6}$$

Let us next consider the two-particle distribution

$$f_2(\Phi_1, \Phi_2) = f_1(\Phi_1)f_1(\Phi_2) + C(\Phi_1, \Phi_2) \tag{19.7}$$

where the first term is simply the product of the single particle distributions, and the second term, $C(\Phi_1, \Phi_2)$ represents possible, *true*, two-particle correlations. Since the two-particle distribution depends on two angles, Φ_1 and Φ_2, in general it will have terms which depend only on the difference of the angle $\sim(\Phi_1 - \Phi_2) = (\phi_1 - \phi_2)$, and which are independent of the direction of the reaction plane. It will also have terms which depend on the sum of the angles, $\sim(\Phi_1 + \Phi_2) = (\phi_1 + \phi_2 - 2\Psi_{RP})$ which are dependent on the reaction plane direction. This may be illustrated by inserting into (19.7) the single particle distribution, (19.4), and neglecting the correlation term, i.e., setting $C(\Phi_1, \Phi_2) = 0$. In this case

$$\begin{aligned} f_2(\Phi_1, \Phi_2) &= f_1(\Phi_1)f_1(\Phi_2) \\ &\sim 2v_2^2\cos[2(\Phi_1 - \Phi_2)] + 2v_2^2\cos[2(\Phi_1 + \Phi_2)] \\ &= 2v_2^2\cos[2(\phi_1 - \phi_2)] + 2v_2^2\cos[2(\phi_1 + \phi_2 - 2\Psi_{RP})]. \end{aligned} \tag{19.8}$$

The term $\sim\cos[2(\phi_1 - \phi_2)]$ which depends on the difference of the angles can then be extracted by the measurement of the two-particle correlation

$$\langle\cos[2(\phi_1 - \phi_2)]\rangle \sim \int_{\Phi_1}\int_{\Phi_2} f_2(\Phi_1, \Phi_2)\cos[2(\phi_1 - \phi_2)]. \tag{19.9}$$

The measurement of the term $\sim\cos[2(\phi_1 + \phi_2 - 2\Psi_{RP})]$ requires the determination of the reaction plane, or at least a three-particle correlation measurement. For our example, (19.8), $\langle\cos[2(\phi_1 - \phi_2)]\rangle \sim v_2^2$, and in fact this is one of the frequently used (and the simplest) methods for measuring the elliptic flow. However, this method suffers from the so-called "non-flow" [49] contributions, which are due to the correlation term we have neglected in our example. Our simple example also demonstrates a very important fact: single particle distributions, such as f_1 do contribute to multi-particle azimuthal correlations. This will be essential for the subsequent discussion where one of the tasks will be to disentangle the effects from true correlations and contributions from the single particle distributions.

The above discussion can be easily extended to three- (and more) particle densities with the same basic conclusions:

---

[2]In reality the ability to express the actual measurement, as described in (19.5), in terms of an average of moments of the intrinsic distribution over the reaction plane angle requires a detailed analysis of all non-flow effects and flow fluctuations, as discussed in detail in Ref. [49].

- The n-particle density will have terms which do not depend on the reaction plane, and thus may be extracted by the measurement of appropriate n-particle correlations. It will also have reaction plane dependent terms, which require the measurement of at least $n + 1$ particle correlations or the determination of the reaction plane.
- Unless not very carefully designed, multi-particle correlations will contain contributions from the single particle distribution.

Finally, the measurement of angular correlations is of course not restricted to the second Fourier moment. Recently the harmonic moments, $\langle \cos[n(\phi_1 - \phi_2)] \rangle$, have been measured in order to study flow fluctuations [50–52]. These correlations may also be measured in a more selective way, such as correlations for particles with same or opposite electric charges (the charge-dependent correlations), correlations for particles with certain quantum numbers (e.g. baryon-strangeness [53]), or correlations for particles within or between certain kinematic regions (e.g. the soft-hard correlations [54, 55]), etc.

19.2.2 Measuring the Charge Separation Through Azimuthal Correlations

Let us turn to possible azimuthal correlation measurements as the signal for the Chiral Magnetic Effect. Specifically, as discussed in the Introduction, we have to find azimuthal correlations which are sensitive to a possible out-of-plane "charge separation".

We begin by defining what we mean by "charge separation effect". Consider the distribution of final state hadrons in the transverse momentum space as schematically shown in the Fig. 19.1. If the "center" of the positive charges happens to be different from that of the negative charges, then there is a separation between two types of charges which may be quantified by an "electric dipole moment" in the transverse momentum space. Such a separation may arise either simply from statistical fluctuations or may be due to specific dynamical effect such as the CME. We note that such a charge separation occurs already at the single-particle distribution level in the *intrinsic* frame. Let us, therefore, define a charge-dependent single-particle azimuthal distribution, which, besides a possible momentum-space electric dipole moment, also includes the presence of elliptic flow:

$$f_\chi(\phi, q) \propto 1 + 2v_2 \cos[2(\phi - \Psi_{RP})] + 2q\chi d_1 \cos(\phi - \Psi_{CS}). \quad (19.10)$$

Here q and ϕ represent the charge and the azimuthal angle of a particle, respectively. The parameters v_2 and d_1 quantify the elliptic flow and the charge separation effect, while Ψ_{CS} specifies the azimuthal orientation of the electric-dipole and Ψ_{RP} the direction of the reaction plane (see Fig. 19.1). It is important to notice that an additional random variable $\chi = \pm 1$ is introduced. This accounts for the fact that in a given event we may have sphaleron or anti-sphaleron transitions resulting in charge separation parallel or anti-parallel to the magnetic field. Consequently the sampling over all events with a given reaction plane angle, Ψ_{RP}, corresponds to averaging the intrinsic distribution f_χ over χ, namely $f_1 = \langle f_\chi \rangle_\chi \propto 1 + 2v_2 \cos(2\phi - 2\Psi_{RP})$. Physically speaking this means that the charge separation (or electric dipole, being $\mathcal{P}$-odd) flips sign randomly and averages to zero, thus causing the expectation value of any parity-odd operator to vanish. However, since $\langle \chi^2 \rangle = 1$ the presence of an event-by-event electric dipole may be observable through its variance.

For measurements related to heavy ion collisions one may reasonably assume particle charges to be $|q| = 1$ which is the case for almost all charged particles, e.g., charged pions and kaons, protons, etc. We note, that the above distribution does *not* contain a directed flow term $\sim \cos(\phi - \Psi_{RP})$ for either type of charges, which is reasonable if the distribution is measured in a symmetric rapidity bin.

Finally one may also consider a p_t-differential formulation of the charge separation effect or charge separation effects associated with higher harmonics in the azimuthal angle ϕ. We note that the charge separation term considered in (19.10) is actually the lowest harmonic in a more general charge-dependent Fourier series expansion in terms of the azimuthal angle. Various higher harmonics may be present due to e.g. the occurrence of multiple topological objects and their distributions over the entire transverse plane, the influence of transverse flow as well as the re-scattering of the CME current with medium. Here we concentrate the discussion on a possible measurement of the lowest harmonic that is most relevant to the CME current.

Let us next discuss how the above defined charge-dependent intrinsic single-particle distribution contributes to the charge-dependent azimuthal correlations recently measured by the STAR Collaboration in [45, 46]. Note that here we are only considering the contribution from the charge separation term in (19.10), while there are certainly additional contributions from two- and multi-particle correlations which we will discuss later in Sect. 19.4. Specifically the STAR Collaboration has measured the following two- and three-particle correlations [45, 46].

(i) The two-particle correlation $\langle \cos(\phi_i - \phi_j) \rangle$ for same-charge pairs $(++/--)$ and opposite-charge pairs $(+-)$. The contribution to this correlator due to the charge-dependent intrinsic single-particle distribution, (19.10) is:

$$\delta_{++/--} \equiv \langle \cos(\phi_i - \phi_j) \rangle_{++/--} = d_1^2, \tag{19.11}$$

$$\delta_{+-} \equiv \langle \cos(\phi_i - \phi_j) \rangle_{+-} = -d_1^2. \tag{19.12}$$

(ii) The three-particle correlation $\langle \cos(\phi_i + \phi_j - 2\phi_k) \rangle$ for same-charge pairs $(i, j = ++/--)$ and opposite-charge pairs $(i, j = +-)$ with the third particle, denoted by index k, having any charge. The contribution to these correlators due to the distribution, (19.10), turns out to be

$$\langle \cos(\phi_i + \phi_j - 2\phi_k) \rangle_{++/--,k\text{-any}} = v_2 d_1^2 \cos(2\Delta\Psi_{CS}), \tag{19.13}$$

$$\langle \cos(\phi_i + \phi_j - 2\phi_k) \rangle_{+-,k\text{-any}} = -v_2 d_1^2 \cos(2\Delta\Psi_{CS}) \tag{19.14}$$

where "k-any" indicates that the charge of the 3-rd particle may assume any value/sign. We have also defined the relative angle of the charged dipole with respect to the reaction plane, $\Delta\Psi_{CS} \equiv \Psi_{CS} - \Psi_{RP}$. The purpose of correlating the charged pair with the third particle is to address the reaction plane dependence of the pair-distribution, as discussed in the previous section, Sect. 19.2.1. Indeed, the STAR Collaboration has demonstrated [45, 46] that the above three particle correlator is dominated by the reaction plane dependent two-particle correlation function $\langle\cos(\phi_i + \phi_j - 2\Psi_{RP})\rangle$ and within errors they have found that

$$\langle\cos(\phi_i + \phi_j - 2\phi_k)\rangle = v_2\langle\cos(\phi_i + \phi_j - 2\Psi_{RP})\rangle. \tag{19.15}$$

Based on the distribution (19.10) we find the same relation between these correlation functions, since the reaction-plane dependent two-particle correlation is given by

$$\gamma_{++/--} \equiv \langle\cos(\phi_i + \phi_j - 2\Psi_{RP})\rangle_{++/--} = d_1^2 \cos(2\Delta\Psi_{CS}) \tag{19.16}$$

for same-charge pairs, and

$$\gamma_{+-} \equiv \langle\cos(\phi_i + \phi_j - 2\Psi_{RP})\rangle_{+-} = -d_1^2 \cos(2\Delta\Psi_{CS}) \tag{19.17}$$

for opposite-charge pairs.

To make contact with the predictions of the CME for the above correlation functions, let us assume for the moment that an accurate identification of the reaction plane could be achieved. In this case we may rotate all events such that $\Psi_{RP} = 0$. Furthermore the CME predicts $\Delta\Psi_{CS} = \pi/2$, and thus, for $\Psi_{RP} = 0$ the charge separation term will take the form of $\sim d\sin(\phi)$ [31, 45–47]. If the only contribution to the above correlations would be due to the CME, a very specific pattern arises:

$$\gamma_{++/--} = \langle\cos(\phi_i + \phi_j - 2\Psi_{RP})\rangle_{++/--} = -d_1^2 < 0, \tag{19.18}$$

$$\delta_{++/--} = \langle\cos(\phi_i - \phi_j)\rangle_{++/--} = +d_1^2 > 0, \tag{19.19}$$

while

$$\gamma_{+-} = \langle\cos(\phi_i + \phi_j - 2\Psi_{RP})\rangle_{+-} = +d_1^2 > 0, \tag{19.20}$$

$$\delta_{+-} = \langle\cos(\phi_i - \phi_j)\rangle_{+-} = -d_1^2 < 0. \tag{19.21}$$

This pattern for the correlations γ and δ, if seen in the data, would constitute a very strong evidence for occurrence of the CME in these collisions. However, as pointed out in [56], and as we shall discuss in more detail in Sect. 19.3 the STAR measurements do not show the above pattern. For example, while in the above analysis for the same-charge pairs the correlators γ and δ are expected to be equal in magnitude but *opposite* in sign, i.e., $\gamma_{++} = -\delta_{++}$ the STAR data finds them approximately equal in magnitude but with the *same* (negative) sign.

19.2.3 The $\hat{Q}_1^c$ Vector Analysis for Measuring the Charge Separation

When exploring an important phenomenon such a local parity violation, it is very useful to develop multiple observables which test its predictions, such as the Chiral Magnetic Effect. This is particularly the case in the present situation. The signals due to the CME are expected to be rather weak and the observables are not free from various backgrounds due to "conventional" physics, such as two-particle correlations. In addition, the interpretation of the STAR data is rather ambiguous. Therefore, it will be very helpful to have an alternative observable which is sensitive to a possible charge separation with specific azimuthal orientation. Currently there are a few proposals, for example the $\hat{Q}_1^c$ vector analysis [57], the charge multiplicity asymmetry correlations [58, 59], and the out-of-plane charge asymmetry distribution [60]. Here we focus on a detailed discussion of the $\hat{Q}_1^c$ vector analysis [57].

The $\hat{Q}_1^c$ vector analysis aims at a direct measurement of the intrinsic charge-dependent distribution in (19.10) by identifying the charged dipole moment vector $\hat{Q}_1^c$ of the final-state hadron distribution in the transverse momentum space. The magnitude Q_1^c and azimuthal angle Ψ_1^c of this vector can be determined in a given event by the following:

$$Q_1^c \cos \Psi_1^c \equiv \sum_i q_i \cos \phi_i$$
$$Q_1^c \sin \Psi_1^c \equiv \sum_i q_i \sin \phi_i$$

(19.22)

where the summation is over all charged particles in the event, with q_i the electric charge and ϕ_i the azimuthal angle of each particle. This method is in close analogy to the $\hat{Q}_1$ and $\hat{Q}_2$ vector analysis used for directed and elliptic flow (see e.g. [49]). In the $\hat{Q}_2$ analysis one evaluates the charge independent quadrupole moment Q_2 and its direction Ψ_2 in a similar fashion

$$Q_2 \cos 2\Psi_2 \equiv \sum_i \cos 2\phi_i$$
$$Q_2 \sin 2\Psi_2 \equiv \sum_i \sin 2\phi_i.$$

(19.23)

The angle Ψ_2 is a measure for the reaction plane angle, Ψ_{RP} such that for a system with infinite many particles $\Psi_2 \to \Psi_{RP}$.

Contrary to $\hat{Q}_2$, the charge dipole vector, $\hat{Q}_1^c$, incorporates the *electric charge* q_i of the particles. The mathematical details regarding the observable $\hat{Q}_1^c$ and its relation to multi-particle correlations can be found in [57].

In each event, both angles Ψ_1^c and Ψ_2 are determined from a finite number of final state hadrons (see Fig. 19.1). While these angles correspond to their idealized expectations Ψ_{CS} and Ψ_{RP} only in the limit of infinite multiplicity, their distribution

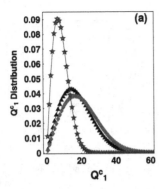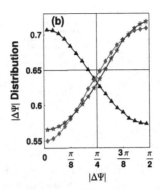

Fig. 19.2 (a) The Q_1^c and (b) $|\Delta\Psi|$ distributions for the four different scenarios described in the text (Color figure online)

and in particular the distribution of their difference, $\Delta\Psi = \Psi_1^c - \Psi_2$ should provide a good estimator for the magnitude of the charged dipole angle with respect to the reaction plane, $\Delta\Psi_{CS} = \Psi_{CS} - \Psi_{RP}$.

The combined Q_1^c- and Q_2-analysis will then provide distributions for the magnitude of the electric dipole, Q_1^c, and its relative angle with respect to Ψ_2, $\Delta\Psi$. This is demonstrated in Fig. 19.2 where we show the distributions for various scenarios calculated in a Monte Carlo simulation [57].

- The black triangles correspond to a "benchmark" scenario, where we have only elliptic flow but neither a charged dipole nor any true pair correlations. Therefore the resulting distributions for Q_1^c and $|\Delta\Psi|$ arise only from pure statistical fluctuations.
- The red diamonds have been obtained by adding a physical dipole along the out-of-plane direction with a magnitude of $d_1 = 0.025$, to the benchmark scenario.
- The green boxes are based on the case where back-to-back angular correlation for about 1 % of the same-charge pairs but no dipole have been added to the benchmark scenario.
- The blue stars result the case where same-side angular correlation for about 1 % of the opposite-charge pairs but no dipole have been added to the benchmark scenario.

As can be seen from the comparison in Fig. 19.2 and a more detailed discussion in [57], only the combined analysis of the distributions of angle and magnitude, is able to distinguish between scenarios based on conventional two-particle correlations and those involving a true charged momentum space dipole as predicted by the CME. As further discussed in [57] the final conclusion on the possible existence of an electric dipole will likely require a joint analysis of all three types of measurements, discussed in this section: the Q_1^c distribution, the $\Delta\Psi$ distribution, as well as the charge-dependent azimuthal correlations γ and δ.

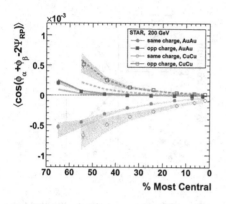
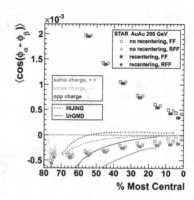

Fig. 19.3 The data from the STAR Collaboration for the reaction plane dependent correlation function $\langle\cos(\phi_\alpha + \phi_\beta - 2\Psi_{RP})\rangle$ (*left*) and the reaction-plane independent correlation function $\langle\cos(\phi_\alpha - \phi_\beta)\rangle$ (*right*) for like-sign and unlike-sign pairs. Also shown (*lines*) are results from various model calculations. The figures are from [45, 46]

19.3 Interpretation of the Available Data

After having discussed the general aspects of charge dependent correlation functions in Sect. 19.2 we will now turn our attention to the actual measurements of such correlation function. Following the proposal by Voloshin [47] the STAR Collaboration [45, 46] presented the first measurement of the reaction-plane dependent charged-pair correlation function

$$\gamma_{\alpha,\beta} = \langle\cos(\phi_\alpha + \phi_\beta - 2\Psi_{RP})\rangle \tag{19.24}$$

for pairs of particles with same, $(\alpha, \beta) = (+, +), (-, -)$, and opposite charge, $(\alpha, \beta) = (+, -)$. As already discussed in the previous section, in order to obtain the correlator $\gamma_{\alpha,\beta}$ STAR measured three-particle correlation functions, and demonstrated rather convincingly that, within errors, they are related to the reaction plane dependent two-particle charged-pair correlation function by

$$\langle\cos(\phi_\alpha + \phi_\beta - 2\phi_k)\rangle = v_2\langle\cos(\phi_\alpha + \phi_\beta - 2\Psi_{RP})\rangle = v_2\gamma_{\alpha,\beta} \tag{19.25}$$

where v_2 denotes the measured elliptic flow parameter characterizing the elliptic azimuthal asymmetry. The results of the STAR measurement for $\gamma_{\alpha,\beta}$ are shown in the left panel of Fig. 19.3.

Since the relation, (19.25), has been established experimentally, we will concentrate our discussion on the charge dependent pair correlation function, $\gamma_{\alpha,\beta}$, (19.24). Furthermore, we will choose a frame where the reaction plane angle is set to zero, $\Psi_{RP} = 0$, so that

$$\gamma_{\alpha,\beta} = \langle\cos(\phi_\alpha + \phi_\beta)\rangle. \tag{19.26}$$

In this frame the in-plane direction coincides with the x-axis and the out-of-plane direction points along the y-axis. Also the average direction of the magnetic field will be along the y-axis.

Before we examine the STAR data more carefully let us recall what the prediction for the charge separation due to the Chiral Magnetic Effects are. As discussed in the previous section, the electric momentum space dipole induced by the CME will point (in an ideal situation) either parallel or anti-parallel to the direction of the magnetic field, which in the frame where $\Psi_{RP} = 0$ points along the y-axis (neglecting fluctuations of the magnetic field [40]). Therefore, the charge separation due to the CME predicts to have pairs of same charge preferably moving together along the positive or negative y-direction. Pairs with opposite charge, on the other hand, are predicted to move away from each other along the y-axis. In terms of the azimuthal angles, ϕ_α, ϕ_β this means

$$(\phi_\alpha, \phi_\beta) = \left(\frac{\pi}{2}, \frac{\pi}{2}\right) \quad \text{or} \quad \left(\frac{3\pi}{2}, \frac{3\pi}{2}\right) \tag{19.27}$$

for same-charge pairs, and

$$(\phi_\alpha, \phi_\beta) = \left(\frac{\pi}{2}, \frac{3\pi}{2}\right) \quad \text{or} \quad \left(\frac{3\pi}{2}, \frac{\pi}{2}\right) \tag{19.28}$$

for opposite-charge pairs. Since

$$\cos\left(\frac{\pi}{2} + \frac{\pi}{2}\right) = \cos\left(\frac{3\pi}{2} + \frac{3\pi}{2}\right) = -1, \tag{19.29}$$

$$\cos\left(\frac{3\pi}{2} + \frac{\pi}{2}\right) = 1 \tag{19.30}$$

the correlation function $\gamma_{\alpha,\beta}$, (19.26), is expected to be negative for same-charge pairs and positive for opposite-charge pairs. While the STAR data, shown in Fig. 19.3, indeed show a negative value for same-charge pairs, the result for opposite-charge pairs is, at best, only mildly positive and, within errors, compatible with zero. Since opposite charged pairs are predicted to move away from each other, one may argue their (anti-)correlation should be weakened as these particles will have to traverse the entire fireball [25]. Therefore, at first sight, the STAR data may indeed show a first evidence for the charge separation pattern as predicted by the CME. However, the interpretation of the data is more difficult.

The complication arises from the fact that the correlation function $\gamma_{\alpha,\beta}$ does not unambiguously determine the angular correlation of the pair. To see this, consider a same-charge pair with angles $(\phi_\alpha, \phi_\beta) = (0, \pi)$. In this case the particles move away from each other in the in-plane direction. This is just the opposite of the correlation predicted by the CME, where the two particles are moving with each other in the out-of-plane direction. For both cases we get

$$\cos(\phi_\alpha, \phi_\beta) = \cos(0 + \pi) = \cos\left(\frac{\pi}{2} + \frac{\pi}{2}\right) = -1. \tag{19.31}$$

Thus, the correlation function $\gamma_{\alpha,\beta}$ is *not* able to distinguish between same-side out-of-plane correlations and back-to-back in-plane correlations. However, this ambiguity can easily be resolved by considering the reaction plane independent correlation function

$$\delta_{\alpha,\beta} = \langle \cos(\phi_\alpha - \phi_\beta) \rangle \tag{19.32}$$

which STAR has also measured, and we show their results in the right panel of Fig. 19.3. In the frame, where $\Psi_{RP} = 0$, the two correlation functions may be decomposed in the in-plane $\sim \langle \cos(\phi_\alpha) \cos(\phi_\beta) \rangle$ and out-of-plane $\sim \langle \sin(\phi_\alpha) \sin(\phi_\beta) \rangle$ components:

$$\begin{aligned}
\gamma_{\alpha,\beta} &= \langle \cos(\phi_\alpha + \phi_\beta) \rangle = \langle \cos(\phi_\alpha) \cos(\phi_\beta) \rangle - \langle \sin(\phi_\alpha) \sin(\phi_\beta) \rangle, \\
\delta_{\alpha,\beta} &= \langle \cos(\phi_\alpha - \phi_\beta) \rangle = \langle \cos(\phi_\alpha) \cos(\phi_\beta) \rangle + \langle \sin(\phi_\alpha) \sin(\phi_\beta) \rangle.
\end{aligned} \tag{19.33}$$

Qualitatively the STAR measurement in Au+Au collisions for both these correlation functions, $\gamma_{\alpha,\beta}$ and $\delta_{\alpha,\beta}$ for same-sign and opposite-sign pairs of charged particles, may be characterized as follows (see Fig. 19.3):

- For same-sign pairs:

$$\langle \cos(\phi_\alpha + \phi_\beta) \rangle_{\text{same}} \simeq \langle \cos(\phi_\alpha - \phi_\beta) \rangle_{\text{same}} < 0. \tag{19.34}$$

Using (19.34) this implies

$$\begin{aligned}
\langle \sin(\phi_\alpha) \sin(\phi_\beta) \rangle_{\text{same}} &\simeq 0, \\
\langle \cos(\phi_\alpha) \cos(\phi_\beta) \rangle_{\text{same}} &< 0.
\end{aligned} \tag{19.35}$$

- For opposite-sign pairs we find that

$$\begin{aligned}
\langle \cos(\phi_\alpha + \phi_\beta) \rangle_{\text{opposite}} &\simeq 0, \\
\langle \cos(\phi_\alpha - \phi_\beta) \rangle_{\text{opposite}} &> 0.
\end{aligned} \tag{19.36}$$

Again, using (19.34), this means

$$\langle \sin(\phi_\alpha) \sin(\phi_\beta) \rangle_{\text{opposite}} \simeq \langle \cos(\phi_\alpha) \cos(\phi_\beta) \rangle_{\text{opposite}} > 0. \tag{19.37}$$

The decomposition of the actual data into the in-plane and out-of-plane components is shown in Fig. 19.4. Obviously the correlations for same-charge pairs are predominantly in-plane and back-to-back. This is exactly the *opposite* of what has been predicted by the Chiral Magnetic Effect. This is illustrated in Fig. 19.5 were we have sketched the experimental situation for same-charge pairs based on the STAR data. For pairs with opposite charge, both in-plane and out-of-plane correlations have the same (positive) sign and magnitude. This implies that opposite-charged pairs move together equally likely in the in-plane and out-of-plane directions. This behavior can at least qualitatively be understood by resonance/cluster decays [58, 59] or local charge conservation [61].

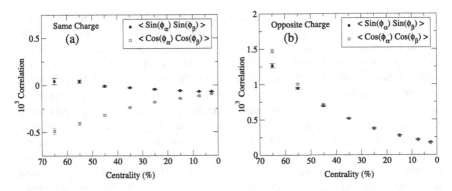

Fig. 19.4 Correlations in-plane $\langle\cos(\phi_\alpha)\cos(\phi_\beta)\rangle$ and out-of-plane $\langle\sin(\phi_\alpha)\sin(\phi_\beta)\rangle$ for same- and opposite-charge pairs in $Au + Au$ collisions as seen in the STAR data

Fig. 19.5 Schematic illustration of the actual STAR measurement (*red*) together with the predictions from the Chiral Magnetic Effect (*black*) for same-charge pairs (Color figure online)

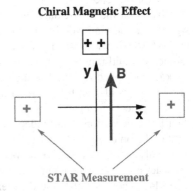

In addition to the data shown in Fig. 19.3, STAR has also analyzed the reaction plane dependent correlation function $\gamma_{\alpha,\beta}$ differentially as a function of the pair transverse momentum (sum and difference) and rapidity difference. Both these results are within qualitative expectations for a charge separation effect due to the CME [56]. Unfortunately, similar differential information is not available for the reaction plane independent correlation function, $\delta_{\alpha,\beta}$. Therefore a differential decomposition into in-plane and out-of-plane components, as we have done here, unfortunately is not possible at this time. Such information may help to further constrain possible background effects as well as predictions from the CME.

Recently, the ALICE Collaboration reported [62] the measurement of the same correlation functions for Pb + Pb collisions at a center of mass energy of $\sqrt{s} = 2.76\,\text{TeV}$, about ten times that of the STAR measurement. Just like STAR, ALICE determined the reaction plane dependent correlation function $\gamma_{\alpha,\beta}$ integrated over transverse momentum and rapidity as well as differentially. Within errors, the data for the integrated correlation function $\gamma_{\alpha,\beta}$ agree with those of the STAR measurement, and the differential measurement show the same qualitative features.

For the reaction plane independent correlation function, $\delta_{\alpha,\beta}$, on the other hand, the ALICE date differ from those by STAR. In particular, ALICE finds this correlation function to be positive for *both* opposite- and same-charge pairs. ALICE also provides the in-plane and out-of-plane pair correlations, $\langle\cos(\phi_\alpha)\cos(\phi_\beta)\rangle$ and $\langle\sin(\phi_\alpha)\sin(\phi_\beta)\rangle$, respectively. Similar to the STAR measurement ALICE finds that for opposite-charge pairs the in- and out-of-plane correlations are nearly identical and positive. For the same-charge pairs, however, ALICE finds both in- and out-of-plane projections to be positive, with the out-of-plane correlation slightly larger than the in-plane projection. This finding would be in qualitative agreement with the expectations from the CME. Amusingly, early predictions [63] for the collision energy dependence of the CME expected a smaller effect at the very high energies where ALICE has been measuring, largely due to the shorter duration of the magnetic field. Of course the complex dynamics of heavy ion collisions and the various background contributions turn quantitative predictions for these correlation functions into a very difficult task, and a final resolution will require a systematic analysis of all available data at various energies.

Given that the STAR data show an in-plane back-to-back correlation for same-sign pairs, one may wonder if there is still room for a charge separation effect due to the CME. This has been analyzed in [56] with the result that for the transverse momentum and rapidity integrated data, which the above analysis is based on, the backgrounds need to exactly cancel the CME induced charge separation. This may be just a coincidence, however. After all the data by ALICE show a different trend for the same-sign correlations. For this situation to be clarified further differential data for the reaction plane independent correlation function, $\delta_{\alpha,\beta}$ are required for both collision energies.

Finally, as part of the RHIC beam energy scan program, STAR has measured the reaction-plane dependent correlator $\gamma_{\alpha,\beta}$ for various collision energies [64], see also [44]. They find that the difference for the correlator between same-sign and opposite-sign pairs decreases with decreasing beam energy. Such a behavior is expected from the CME. However, as we will discuss in the next section, all background terms scale with the elliptic flow parameter, v_2, which is known to decrease with decreasing collision energy as well.

To conclude this section, presently available experimental results concerning the CME are inconclusive. While the integrated STAR data disfavor the presence of the CME, the ALICE data allow for more positive conclusions. Clearly, progress requires data at lower energies as well as, and most importantly, differential measurements of both the reaction plane dependent and reaction plane independent correlation functions. In addition, given the rather unsettled state of affairs, measurements of other observables, such as the one proposed in the previous section, would be very welcome.

19.4 Discussion of Various Background Contributions

As discussed in the previous sections the Chiral Magnetic Effect (CME)—if it exists—contributes to the reaction plane dependent two-particle correlator, first introduced in [47]. As in the previous section we denote the reaction plane dependent two-particle correlator by

$$\gamma \equiv \langle \cos(\phi_1 + \phi_2 - 2\Psi_{RP}) \rangle, \qquad (19.38)$$

where ϕ_1 and ϕ_2 are the azimuthal angles of two particles, and Ψ_{RP} is the reaction plane angle. In the following we will distinguish between $\gamma_{++/--}$, γ_{+-} and γ denoting respectively the correlator (19.38) for same-sign pairs, opposite-sign pairs, and the correlator without specifying the sign of measured particles.

As discussed in Sect. 19.3 a detailed measurement of γ was performed both at RHIC by the STAR Collaboration [45, 46] and at the LHC by the ALICE Collaboration [62]. However, as already discussed in the previous section, the interpretation of experimental data is not straightforward since various effects can contribute to γ.

The presence of elliptic flow allows for practically all "conventional" two-particle correlations to contribute to the reaction-plane dependent correlation function, γ. This can be easily seen from the decomposition of γ into in-plane and out-of-plane projections, (19.34)

$$\gamma = \langle \cos(\phi_1)\cos(\phi_2) \rangle - \langle \sin(\phi_1)\sin(\phi_2) \rangle. \qquad (19.39)$$

It is quite obvious that even if the underlying correlation mechanism does not depend on the reaction plane it will contribute to γ in the presence of the elliptic anisotropy v_2. This can be seen in an extreme, though unrealistic, situation where all particles are produced exactly in-plane. In this case $\langle \sin(\phi_1)\sin(\phi_2) \rangle = 0$ simply because there are no particles in the out-of-plane direction and $\gamma = \langle \cos(\phi_1)\cos(\phi_2) \rangle$. Obviously in this case, the presence of any two-particle angular correlation mechanism will result in a non-zero value of γ.

In this section we will focus exclusively on the contribution to γ driven by the non-vanishing elliptic anisotropy v_2. First we will derive the general expression which relates the elliptic anisotropy and the correlator γ in the presence of arbitrary two-particle correlations. Next we will discuss a few explicit mechanisms that need to be understood quantitatively, before any conclusions about the existence of the CME can be made. In particular we will address corrections due to transverse momentum conservation (TMC) [65–67] and the local charge conservation [61], both of which appear to contribute significantly to γ. In the last part of the paper we will discuss the possibility of removing, in the model independent way, the elliptic-flow-related background from γ.

19.4.1 General Relation

In this part we derive the general relation between the elliptic anisotropy v_2 and the two-particle correlator γ in the presence of an arbitrary reaction plane independent two-particle correlations.

By definition, the two-particle correlator γ is

$$\gamma = \frac{\int \rho_2(\phi_1, \phi_2, x_1, x_2, \Psi_{RP}) \cos(\phi_1 + \phi_2 - 2\Psi_{RP}) d\phi_1 d\phi_2 dx_1 dx_2}{\int \rho_2(\phi_1, \phi_2, x_1, x_2, \Psi_{RP}) d\phi_1 d\phi_2 dx_1 dx_2}, \quad (19.40)$$

where, to simplify our notation, we denote: $x = (p_t, \eta)$ and $dx = p_t dp_t d\eta$. Here p_t is the absolute value of transverse-momentum, while η is pseudorapidity (or rapidity). ρ_2 is the two-particle distribution in the intrinsic frame with the reaction plane angle Ψ_{RP}. It can be expressed in terms of the single-particle distributions, and the underlying correlation function C (see Sect. 19.2)

$$\rho_2(\phi_1, \phi_2, x_1, x_2, \Psi_{RP}) = \rho(\phi_1, x_1, \Psi_{RP}) \rho(\phi_2, x_2, \Psi_{RP}) \big[1 + C(\phi_1, \phi_2, x_1, x_2)\big]. \quad (19.41)$$

To simplify our calculation we assume the single-particle distribution to be

$$\rho(\phi, x, \Psi_{RP}) = \frac{\rho_0(x)}{2\pi} \big[1 + 2v_2(x) \cos(2\phi - 2\Psi_{RP})\big], \quad (19.42)$$

where $\rho_0(x)$ and $v_2(x)$ depend solely on $x = (p_t, \eta)$. We neglect higher moments v_n since their contribution to γ turns out to be proportional to $v_n v_m$ which is much smaller then the leading term $\sim v_2$, see Ref. [68].

If $v_2(x) \neq 0$, the single particle distributions depend on the reaction plane. Therefore, the part of the two-particle density (19.41) involving the two-particle correlation function C depends on the reaction plane even if C depends only on $\phi_1 - \phi_2$.

Here we want to concentrate on those correlations that depend only on $\Delta\phi = \phi_1 - \phi_2$, namely the underlying correlation mechanism is insensitive to the reaction plane orientation. The correlation function may be expanded in a Fourier series

$$C(\Delta\phi, x_1, x_2) = \sum_{n=0}^{\infty} a_n(x_1, x_2) \cos(n\Delta\phi), \quad (19.43)$$

where $a_n(x_1, x_2)$ does not depend on ϕ_1 and ϕ_2. Substituting (19.43) and (19.41) into (19.40), we obtain

$$\gamma = \frac{1}{2N^2} \int \rho_0(x_1) \rho_0(x_2) a_1(x_1, x_2) \big[v_2(x_1) + v_2(x_2)\big] dx_1 dx_2, \quad (19.44)$$

where $N = \int \rho_0(x) dx$, and we have assumed that $a_n \ll 1$.

Equation (19.44) explains why all correlation mechanisms with a non-zero $a_1(x_1, x_2)$ contribute to γ. For instance, it has been shown that transverse momentum conservation (TMC) leads to a correlation function which depends on $\cos(\Delta\phi)/N_{tot}$ [69], where N_{tot} is the total number of produced particles. In this case $a_1(x_1, x_2) \propto 1/N_{tot}$. Let us also emphasize that all correlations that depend on the momentum difference between particles $\Delta k = |\mathbf{k}_1 - \mathbf{k}_2|$ also contribute to γ. In this case:

$$C(\Delta k) = C\big(k_1^2 + k_2^2 - 2k_1 k_2 \cos(\Delta\phi)\big), \quad (19.45)$$

which naturally leads to a non-vanishing a_1 term.

To summarize, (19.44) explains why transverse-momentum conservation [65–67], local charge-conservation [61], resonance- (cluster-)decay [58, 59], and all other correlations with $\Delta\phi$ dependence contribute to γ.

19.4.2 Transverse Momentum Conservation

Very soon after publication of the experimental data by the STAR Collaboration it was realized that transverse momentum conservation convoluted with the non-zero elliptic anisotropy can lead to a substantial corrections for γ [65, 66, 70]. This can be easily seen for the simplified situation where all particles are measured (in the full phase-space), and where they all have exactly the same magnitude of transverse momentum $|\mathbf{p}_{i,t}| = |\mathbf{p}_t|$, $i = 1, \ldots, N_{\text{tot}}$.

In the frame where $\Psi_{RP} = 0$ the correlator γ may be written as

$$\gamma = \left\langle \frac{\sum_{i\neq j} \cos(\phi_i + \phi_j)}{\sum_{i\neq j} 1} \right\rangle, \qquad (19.46)$$

or, alternatively,

$$\gamma = \left\langle \frac{(\sum_i \cos(\phi_i))^2 - (\sum_i \sin(\phi_i))^2 - \sum_i \cos(2\phi_i)}{\sum_{i\neq j} 1} \right\rangle, \qquad (19.47)$$

where i and j are summed over all particles in the full phase-space. In the simplified scenario, where $|\mathbf{p}_{i,t}| = |\mathbf{p}_t|$, the conservation of transverse momentum implies

$$\sum_i \cos(\phi_i) = \sum_i \sin(\phi_i) = 0. \qquad (19.48)$$

Consequently we obtain

$$\gamma = -\left\langle \frac{\sum_i \cos(2\phi_i)}{N_{\text{tot}}(N_{\text{tot}} - 1)} \right\rangle \approx \frac{-v_2}{N_{\text{tot}}}, \qquad (19.49)$$

where N_{tot} is the total number of particles. Taking, for example, the centrality class 40–50 % we approximately have $v_2 \approx 0.1$ and $N_{\text{tot}} \approx 1500$ leading to $\gamma \approx -0.7 \cdot 10^{-4}$ from TMC. This is roughly a factor $3-4$ smaller than the experimental data for the same-charge pairs. This is only a simple estimation and a more realistic AMPT calculations [67] suggest that the TMC contribution is roughly factor 2 smaller than the STAR data.

In a similar way we obtain for the reaction plane independent correlation function

$$\delta = \langle \cos(\phi_1 - \phi_2) \rangle \approx -\frac{1}{N_{\text{tot}}}. \qquad (19.50)$$

In this case for $N_{\text{tot}} \approx 1500$ we obtain $\delta \approx -0.7 \cdot 10^{-3}$ which is comparable or slightly larger in magnitude than same-charge data by the STAR Collaboration.

Similar results hold also in a more realistic situation, where only a small fraction of all particles is measured, and the magnitudes of transverse momenta are distributed according to the thermal distribution. This has been discussed in detail in [66], and we will only show the most important results.

Using the central limit theorem and implying the global conservation of transverse momentum, the two-particle distribution function reads [66, 69, 71]

$$\rho_2(\mathbf{p_1}, \mathbf{p_2}) \simeq \rho(\mathbf{p_1})\rho(\mathbf{p_2})\left(1 + \frac{2}{N_{tot}} - \frac{(p_{1,x} + p_{2,x})^2}{2N_{tot}\langle p_x^2 \rangle_F} - \frac{(p_{1,y} + p_{2,y})^2}{2N_{tot}\langle p_y^2 \rangle_F}\right), \quad (19.51)$$

where x and y denote the two components of transverse momentum. F denotes that the appropriate average is calculated for all particles in the full phase-space. The single particle distribution, $\rho(\mathbf{p_1})$, is given by (19.42).

Before we continue let us clarify one subtle point. Equation (19.51) is derived assuming that we first sample particles with a given v_2 and next we conserve transverse momentum for all particles. In reality the opposite scenario should be considered. First we should sample partons/particles with a conserved transverse momentum, and after this the elliptic anisotropy v_2 should be generated according to some dynamical model. Of course the second approach is much more challenging and renders analytical calculations difficult. At the end of this section we will show that both procedures lead to comparable results for γ and δ and their transverse momentum distributions, however the rapidity distributions are quite different. We will come back to this point later.

Using (19.51) and (19.40) we can derive the following relations for γ and δ:

$$\gamma = -\frac{1}{N_{tot}}\frac{\langle p_t \rangle_\Omega^2}{\langle p_t^2 \rangle_F}\frac{2\bar{v}_{2,\Omega} - \bar{\bar{v}}_{2,F} - \bar{\bar{v}}_{2,F}(\bar{v}_{2,\Omega})^2}{1 - (\bar{v}_{2,F})^2}, \quad (19.52)$$

and

$$\delta = -\frac{1}{N_{tot}}\frac{\langle p_t \rangle_\Omega^2}{\langle p_t^2 \rangle_F}\frac{1 + (\bar{v}_{2,\Omega})^2 - 2\bar{\bar{v}}_{2,F}\bar{v}_{2,\Omega}}{1 - (\bar{v}_{2,F})^2}, \quad (19.53)$$

where we have introduced certain weighted moments of v_2:

$$\bar{v}_2 = \frac{\langle v_2(p_t, \eta)p_t \rangle}{\langle p_t \rangle}, \qquad \bar{\bar{v}}_2 = \frac{\langle v_2(p_t, \eta)p_t^2 \rangle}{\langle p_t^2 \rangle}. \quad (19.54)$$

In the above equations F and Ω denote averages that are calculated for all particles in the full phase-space, or for all actually measured particles in the restricted phase-space, respectively. Performing explicit calculations with reasonable assumptions about p_t and η dependence of the single-particle distribution $\rho(p_t, \eta)$ and elliptic flow $v_2(p_t, \eta)$, we have found that for mid-central and peripheral collisions [66]

$$\gamma \cdot N_{part} \approx -0.005, \qquad \delta \cdot N_{part} \approx -0.05, \quad (19.55)$$

where N_{part} is the number of participants, also referred to as wounded nucleons [72].

Fig. 19.6 The two-particle
azimuthal correlator
$\langle \cos(\phi_1 + \phi_2) \rangle$ vs
$p_+ = (p_{1,t} + p_{2,t})/2$ (*blue
line*) and $p_- = |p_{1,t} - p_{2,t}|$
(*red line*) for $N_{part} = 100$.
The results are in qualitative
and partly quantitative
agreement with the STAR
data for the same-charge
correlator. Figure from
Ref. [66] (Color figure
online)

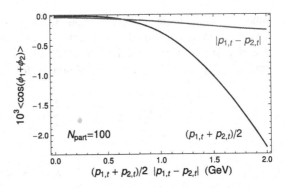

To summarize, transverse momentum conservation results in a negative contributions to both γ and δ, and they are of the same order of magnitude as the experimental measurement for like-sign pairs. More precisely they are a factor of 3–5 (very peripheral—mid-central) less in magnitude for γ, and a factor 1.5–4 (mid-central—very peripheral) larger for δ than the STAR data for the same-sign correlator. While it is rather difficult to understand the data with only transverse momentum conservation it is interesting to notice that the STAR experiment is sensitive enough to the effect of global transverse momentum conservation.

Using (19.40), (19.51) we can easily calculate the dependence of γ and δ on the sum $p_+ = (p_{1,t} + p_{2,t})/2$ and difference $p_- = |p_{1,t} - p_{2,t}|/2$ of the transverse momenta of the pair. For the STAR data [45, 46], γ is growing roughly linearly with p_+ and is approximately constant as a function of p_-. Interestingly a very similar behavior is obtained in the scenario with only global transverse momentum conservation. As seen in Fig. 19.6, the contribution of TMC to γ is consistent with the data for $p_+ > 1$ GeV and underestimates the data for $p_+ < 1$ GeV. As expected, for δ very similar dependence on p_+ and p_- is obtained but rescaled by a value of v_2.

Recently, very similar results were obtained in the AMPT model calculation [67], where the transverse momentum is conserved on the event-by-event basis, and we will come back to this point later.

19.4.2.1 Pseudorapidity Dependence

Since the contribution to γ due to transverse momentum conservation is proportional to v_2, one would naively expect that its (pseudo)rapidity dependence trace the rather mild (pseudo)rapidity dependence of v_2.

However, as shown in Ref. [65], under quite reasonable assumptions we can obtain very similar rapidity dependence as in the experimental data. In the STAR measurement [45, 46] the correlator γ is maximum for $|\eta_1 - \eta_2| = 0$ and is approximately linearly decreasing to values consistent with zero at $|\eta_1 - \eta_2| \approx 2$.

For simplicity of the argument let us assume that at the time $t = 0$ all produced partons/particles are distributed between two separate bins in rapidity with enforced global transverse momentum conservation. It means that if in the first bin the total transverse momentum equals $\mathbf{K}_{1,t}$, in the second bin $\mathbf{K}_{2,t} = -\mathbf{K}_{1,t}$ in each event. Assuming further that all particles have the same magnitude of transverse momentum $|\mathbf{p}_{i,t}| = |\mathbf{p}_t|$, $i = 1, \ldots, N_{\text{tot}}$, we obtain the following relation

$$\gamma \propto \left\langle \sum_{i \in 1} \cos(\phi_i) \sum_{k \in 2} \cos(\phi_k) - \sum_{i \in 1} \sin(\phi_i) \sum_{k \in 2} \sin(\phi_k) \right\rangle, \tag{19.56}$$

and

$$\gamma \propto \langle K_{x,1} K_{x,2} - K_{y,1} K_{y,2} \rangle = \langle K_{y,1}^2 - K_{x,1}^2 \rangle, \tag{19.57}$$

where x and y denote the components of transverse momentum. It should be noted that here we calculate the two-particle correlator where one particle is taken from a bin number 1, and the second particle from a bin number 2.

Now let us evaluate γ at time t=0. In this case it is reasonable to assume that there is no elliptic anisotropy v_2 and $\gamma = 0$ by definition. If we assume that bins are separated enough in rapidity so that there is no momentum exchange between two bins during the fireball evolution, i.e., the total transverse momentum $\mathbf{K}_{1,t}$ is constant, then $\gamma = 0$ also in the final state, even if subsequently a non-zero v_2 is generated. Of course this simple argument demonstrates only that having the global TMC we can still obtain a nontrivial dependence of γ as a function of $\eta_1 - \eta_2$. In Ref. [65] this problem was studied in detail in the cascade model, where it was shown that the satisfactory description of the data can be obtained.

We conclude that although the TMC probably cannot explain the data completely it gives rise to contributions which are of the same order of magnitude and which exhibit both the transverse momentum and rapidity dependence in qualitative and partly quantitative agreement with the STAR data.

19.4.3 AMPT Model

It is desirable to study the correlators γ and δ also in an advanced Monte Carlo model, which allows for the conservation of transverse momentum on the event-by-event basis, and which generates reasonable values for the elliptic anisotropy v_2. Recently such calculation was performed in the AMPT model and the results are presented in Ref. [67] with the conclusion that the signal coming form AMPT is dominated by TMC. Indeed, the obtained results, summarized in Figs. 19.7 and 19.8, are in very good agreement with the previous discussion. As seen in Fig. 19.7, the calculated correlator γ is in a good qualitative agreement with the data, however it underestimates the STAR data by a factor of $\sim$2. It was also shown that initializing the AMPT calculation with the charge dipole leads to a better description of the data.

In Fig. 19.8 the correlator γ is plotted as a function of $p_+ = (p_{1,t} + p_{2,t})/2$ and $\Delta\eta = \eta_1 - \eta_2$. Again, qualitatively the AMPT model reproduce the data but underestimate it by a factor of 2.

Fig. 19.7 The two-particle azimuthal correlator γ as a function of centrality in the AMPT model with different values of initial charge separation

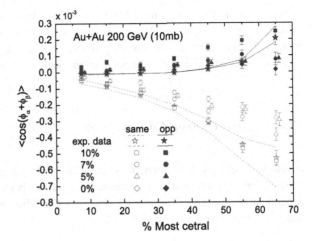

Fig. 19.8 The two-particle azimuthal correlator γ as a function of $p_+ = (p_{1,t} + p_{2,t})/2$ and $\Delta\eta = \eta_1 - \eta_2$ in the AMPT model with different initial values of charge separation

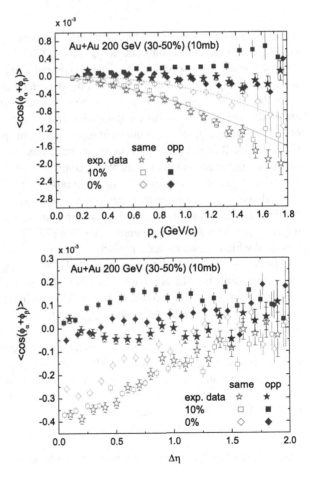

As seen from the figures the AMPT model calculation (without initial dipole) is consistent with the global TMC and allows to understand the behavior of the STAR data. It clearly demonstrates that all possible background effects must be studied very thoroughly before any conclusion about local parity violation can be reached.

19.4.4 Local Charge Conservation

Given the previous discussion, it is useful to construct a two-particle correlator which is insensitive to transverse momentum conservation and other charge independent correlations. The natural choice is the difference of opposite-sign and same-sign pair correlator (19.38), see [61]:

$$\gamma_P \equiv \frac{1}{2}(2\gamma_{+-} - \gamma_{++} - \gamma_{--}). \tag{19.58}$$

It is clear that only correlations that are charge sensitive will contribute to γ_P. While the CME, if present, will contribute to γ_P, global TMC, for example will not. Therefore the successful description of γ_P with conventional physics would constitute a serious challenge for the interpretation of the data in terms of the CME.

In Ref. [61] it was argued that γ_P can be fully understood assuming that charges are produced later in the collision (delayed hadronization). Indeed, in the calculation of Ref. [61] the charges are produced in pairs in the same point in space-time. Due to the collective flow the initial correlation in space-time is translated into correlations in momentum space, and consequently it contributes to γ_P. In this approach particles are emitted according to the blast-wave model with the additional requirement of local charge conservation at freeze-out. Local charge balance is enforced within the finite range in rapidity σ_η and the azimuthal angle σ_ϕ. By comparing the model with experimental data on the balance function [73], the values of σ_η and σ_ϕ can be extracted which allows to make prediction for γ_P.

The details of this calculation are presented in Ref. [61]. Here we summarize only the main results. It turns out that the model with delayed charge creation and local charge conservation can provide a successful description of the balance function both in the relative angle $\Delta\phi = \phi_1 - \phi_2$ and the relative pseudorapidity $\Delta\eta$, see Fig. 19.9 as an example.[3]

Using the best values of parameters σ_η and σ_ϕ the contribution of local charge conservation to γ_P can be calculated. The results are presented in Fig. 19.10.

As seen in Fig. 19.10 the agreement of the model with the STAR data is very good. It suggests that the two-particle charge sensitive correlations may be dominated by the local charge conservation.

---

[3]Recently a similar model was proposed to explain the fall-off of the same-side ridge in $\Delta\eta$ [74].

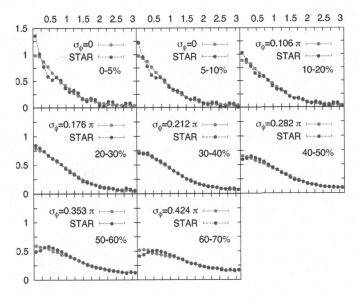

Fig. 19.9 Model with a delayed charge production and the local charge conservation vs the STAR data on the balance function in the relative azimuthal angle $\Delta\phi = \phi_1 - \phi_2$

19.4.5 Decomposition of Flow-Induced and Flow-Independent Contributions

From the analysis of the data in Sect. 19.3 as well as the discussion of various "background" effects, it appears rather plausible that the observed charge-dependent correlation patterns in γ and δ contain contributions from more than one source. In particular there are effects whose contributions to these correlations are flow-dependent, for example the transverse momentum conservation (TMC) or the local charge conservation (LCC). On the other hand the CME, if present, is flow independent. Let us, therefore, attempt a decomposition of flow-induced and flow-independent contributions.

We first consider correlation effects where the underlying correlation function C is independent of the reaction plane orientation:

$$C(\phi_1, \phi_2) \propto \rho(\phi_1, \Psi_{RP})\rho(\phi_2, \Psi_{RP})C(\phi_1 - \phi_2), \tag{19.59}$$

where ρ is a single particle distribution. Note that the above is true for both TMC and LCC effects. A correlation effect of this type will contribute to the measured correlators as follows:

$$\gamma_{\alpha,\beta} \sim v_2 F_{\alpha,\beta}, \qquad \delta_{\alpha,\beta} \sim F_{\alpha,\beta}, \tag{19.60}$$

with the factor $F_{\alpha,\beta}$ representing the strength of the effects, and (α, β) is either $++/--$ or $+-$. Both the TMC and the LCC follow this pattern albeit with opposite contributions. Thus F represents the total of all effects of this type.

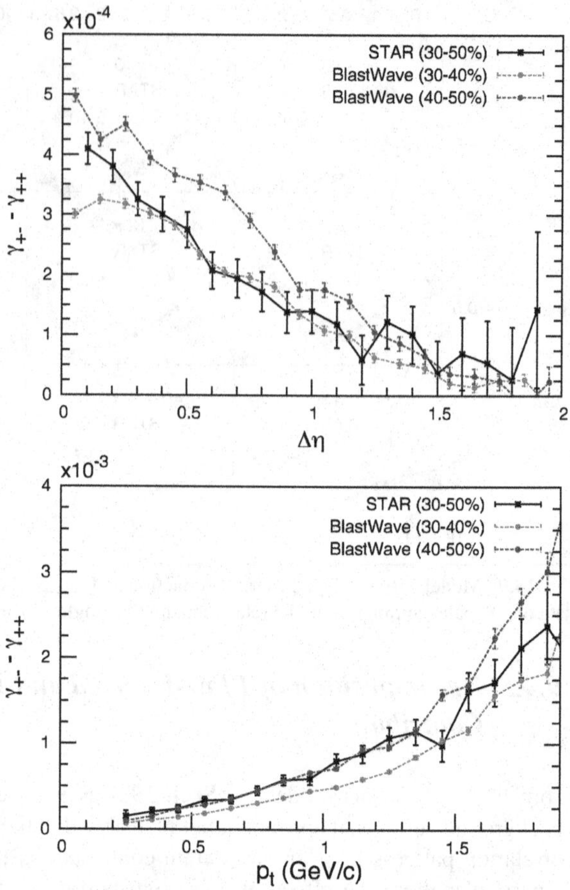

Fig. 19.10 Model with a delayed charge production and local charge conservation vs the STAR data on γ_P as a function of p_+ and $\Delta\eta$

We should note, however, that the above relation, (19.60) is a simplification of the exact relation between γ and $v_2(p_t, \eta)$, which is given in (19.44). Since the purpose of the present discussion is to gain some qualitative insight into the various contributions, we assume here that γ is approximately proportional to the integrated v_2.

Next we consider possible contributions of the CME type. They would appear in the two-particle density in the following form

$$\rho_2(\phi_1, \phi_2) \propto \sin(\phi_1 - \Psi_{RP}) \sin(\phi_2 - \Psi_{RP}), \tag{19.61}$$

which explicitly involves the reaction plane. This term will contribute to the measured correlators as follows:

$$\gamma_{\alpha,\beta} \sim -H_{\alpha,\beta}, \qquad \delta_{\alpha,\beta} \sim H_{\alpha,\beta}, \tag{19.62}$$

with the factor H representing the strength of the effects. It should be pointed out that besides the CME, there are possibly other effects that may also contribute to

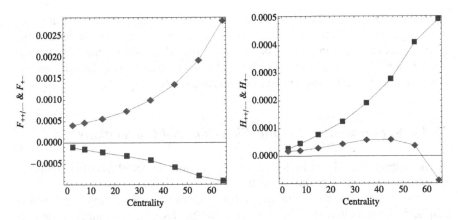

Fig. 19.11 The strength factors $F_{\alpha,\beta}$ (*left*) and $H_{\alpha,\beta}$ (*right*) extracted from the decomposition analysis (see text for details). The *blue boxes* and *red diamonds* are for $++/--$ and $+-$, respectively (Color figure online)

the correlators with the above pattern. One example is a possible dipole asymmetry from initial condition fluctuations that preferably aligns with the out-of-plane direction, see Ref. [75] for details. However, this effect will be charge independent, i.e., $H_{++/--} = H_{+-} > 0$, whereas the CME predicts a charge dependence, $H_{++/--} = -H_{+-} > 0$.

Combining the two types of contributions we arrive at the following decomposition for the reaction plane dependent and independent correlation functions,

$$\gamma_{\alpha,\beta} \sim v_2 F_{\alpha,\beta} - H_{\alpha,\beta}, \qquad \delta_{\alpha,\beta} \sim F_{\alpha,\beta} + H_{\alpha,\beta}. \qquad (19.63)$$

Given these relations, we may use the STAR data for $\gamma_{\alpha,\beta}$, $\delta_{\alpha,\beta}$, and v_2 to extract the strength factors $F_{\alpha,\beta}$ and $H_{\alpha,\beta}$ as a function of centrality. The result of such a decomposition is shown in Fig. 19.11.

Given this analysis we make the following observations: (a) Both components are charge dependent, i.e. there is significant difference between $++/--$ and $+-$; (b) In both cases, however, the same-charge and opposite-charge signals are not symmetric with respect to zero. This may indicate that in each category there are likely more than one source of correlations; (c) There is a strong residual centrality dependence for both types component, although the dependence on centrality from v_2 has been removed. This may indicate that the correlations depend also on the multiplicity, which changes from central to peripheral collisions.

Although, as already noted, the above analysis is qualitative, let us entertain a possible scenario, which would be consistent with the above observations: The flow-induced signals may have two sources, the TMC with $F^{TMC}_{++/--} = F^{TMC}_{+-} < 0$ and the LCC with $F^{LCC}_{++/--} = 0$ and $F^{LCC}_{+-} > |F^{TMC}_{+-}| > 0$. The flow-independent signals may be from two different sources, the CME with $H^{CME}_{++/--} > 0$ and $H^{CME}_{+-} < 0$ and the dipole asymmetry from fluctuations (DAF) with $H^{DAF}_{++/--} = H^{DAF}_{+-} > 0$. Such a combination would indeed lead to correlations with magnitude and sign in qualitative agreement with the data. However, a quantitative analysis would have to be based on the exact decomposition based on (19.44). Alternatively, one may attempt a separation of flow dependent and flow independent contributions in experiment. How this could be achieved will be discussed in the following.

19.4.6 Suppression of Elliptic-Flow-Induced Correlations

In this chapter we will discuss the possibility of removing the elliptic-flow-induced background from the experimental data.

As seen in (19.44) all contributions due to correlations are proportional to the elliptic flow parameter, v_2. Therefore, it would be desirable to control or remove this contribution by a suitable measurement. There are essentially two ways to go about this.

First, as proposed in Ref. [76], is to study collisions of deformed nuclei, such as $U + U$. By selecting very central, "face on face" collisions where the deformation of the nuclei is imprinted on the fireball, elliptic flow should be generated while at the same time the magnetic field will be very small. Should one observe correlations of the same magnitude and structure as ones already reported by STAR, this would identify their origin as being due to conventional two-particle correlations. The observation of considerably smaller correlations combined with a sizable v_2, on the other hand, would lend support for the existence of the CME. This approach, while challenging to analyze, is at present being attempted at RHIC, were first $U + U$ collisions were made available.

Alternatively, as proposed in Ref. [68] one may make use of the large event-by-event fluctuations of v_2. By selecting events with different v_2 in a given centrality class we can control this background. In principle the measurement can be extrapolated to $v_2 = 0$ (and consequently $v_2(p_t, \eta) = 0$) which will allow to extract correlations that only depend on the reaction plane orientation. Indeed, as presented in Fig. 19.12 even at $b = 10$ fm we expect a large fluctuation of initial eccentricity that will translate to large fluctuations of elliptic flow v_2.

Of course it is important to remove this background under the condition that the contribution from the Chiral Magnetic Effect is approximately unchanged. In Fig. 19.13 we demonstrate that indeed it is the case. Both the wounded nucleons and spectators' contribution to the magnetic field weekly depends on ϵ_2.

We believe this analysis should help to clarify the situation. Observation of non-zero γ_{++} or γ_{+-} at vanishing value of elliptic anisotropy v_2 will suggest the existence of the correlation mechanism that is sensitive to the average direction of the magnetic field—possibly the Chiral Magnetic Effect.

To summarize this section, clearly there are many contributions based on conventional physics which contribute to the azimuthal correlations analyzed by the various experiments. In addition to those discussed in more detail in this section other mechanisms, such as the decay of multi-particle clusters [58, 59], have been proposed. While it is more difficult to asses their contribution quantitatively, their

Fig. 19.12 The distribution of initial eccentricity ϵ_2 calculated in the Glauber Monte-Carlo at the impact parameter $b = 10$ fm

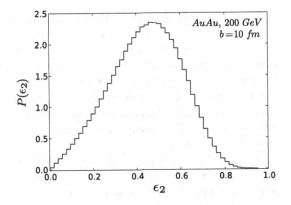

Fig. 19.13 The out-of-plane component of the magnetic field from wounded protons and spectator protons as a function of initial eccentricity ϵ_2 at a given impact parameter $b = 10$ fm

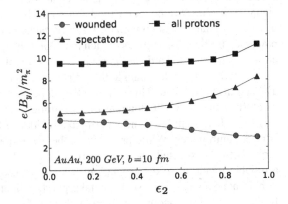

influence cannot a priory be ignored. Therefore, it seems the best way forward is to separate the influence of elliptic flow and magnetic field experimentally as discussed in the last part of this section.

19.5 Summary and Conclusions

In this review we have concentrated on the observational aspects of the search for phenomena related to local parity violation of the strong interaction. Specifically we have discussed various observables and their measurement for the charge separation, which is a predicted consequence of induced currents due to sphaleron and anti-sphaleron transitions in an external magnetic field. This phenomenon is often referred to as the Chiral Magnetic Effect (CME).

We have discussed various properties and aspects of azimuthal angle correlations, and we have emphasized that, due to the elliptic flow observed in heavy ion collisions, virtually any two-particle correlations contribute to the azimuthal correlations. We have further discussed an alternative observable, which in our view may be better suited in discriminating between the backgrounds and the CME.

We have examined the presently available data on reaction plane dependent and independent correlation functions of same- and opposite-charged pairs. Our phenomenological analysis of the data by the STAR Collaboration showed that the measured correlations of same-charge pairs are predominantly back-to-back, and in-plane. This is opposite to the predictions from the CME, where same-side out-of-plane correlations for pairs of the same charge are expected. The data by the ALICE Collaboration taken at about ten times the STAR collision energy, on the other hand, show a correlation, albeit small, which is qualitatively consistent with the CME expectations.

However, before any conclusion on the CME can be drawn, the contributions due to "conventional" correlations need to be accounted for. As we have discussed in some detail, both the conservation of transverse momentum as well as local charge conservation give rise to corrections which are of the same order as the experimental signal. These need to be understood and properly subtracted from the data in order to see if a signal consistent with the CME remains. Since there are conceivably many other two-particle correlations, which may enter due to the presence of elliptic flow, an important step towards answering the question about the existence of the CME is to experimentally disentangle the elliptic flow phenomenon from the creation of a strong magnetic field. This can be either done by colliding deformed nuclei or by carefully utilizing the fluctuations of the elliptic flow.

In conclusion, the present experimental evidence for the existence of the CME is rather ambiguous. While progress on the assessment of the various background terms is to be expected, the sheer variety of possible correlations will likely limit a reliable quantitative determination of all the backgrounds. Therefore, the most important next step is the experimental separation of elliptic flow and magnetic field. In this context it is encouraging to note, that the first $U + U$ collisions at RHIC have just been recorded.

Finally let us close with a note of caution. Besides the CME there are other phenomena related to local non-vanishing topological charge fluctuations, such as the Chiral Magnetic Wave. One of the predictions in this case is the difference of elliptic flow between positively and negatively charged pions for collisions at lower energies [35, 36]. However, again there may be other, more mundane effects which lead to similar phenomena, such as an increased stopping of baryon number and isospin at lower energies [77].

Acknowledgements A.B. was supported by Contract No. DE-AC02-98CH10886 with the US Department of Energy. V.K. was supported by the Office of Nuclear Physics in the US Department of Energy's Office of Science under Contract No. DE-AC02-05CH11231. J.L. is grateful to the RIKEN BNL Research Center for partial support. A.B. also acknowledges the grant No. N202 125437 of the Polish Ministry of Science and Higher Education (2009–2012).

References

1. G. 't Hooft, arXiv:hep-th/0010225
2. T. Schafer, E.V. Shuryak, Rev. Mod. Phys. **70**, 323 (1998)

3. E. Shuryak, *The QCD Vacuum, Hadrons and Superdense Matter*, 2nd edn. (World Scientific, Singapore, 2004)
4. G. Ripka, arXiv:hep-ph/0310102
5. G.S. Bali, arXiv:hep-ph/9809351
6. J. Greensite, Prog. Part. Nucl. Phys. **51**, 1 (2003)
7. J. Liao, E. Shuryak, Phys. Rev. C **75**, 054907 (2007)
8. J. Liao, E. Shuryak, Phys. Rev. Lett. **101**, 162302 (2008)
9. M.N. Chernodub, V.I. Zakharov, Phys. Rev. Lett. **98**, 082002 (2007)
10. A. D'Alessandro, M. D'Elia, Nucl. Phys. B **799**, 241 (2008)
11. M. Cristoforetti, E. Shuryak, Phys. Rev. D **80**, 054013 (2009)
12. A. D'Alessandro, M. D'Elia, E. Shuryak, Phys. Rev. D **81**, 094501 (2010)
13. J. Liao, E. Shuryak, Phys. Rev. C **77**, 064905 (2008)
14. J. Liao, E. Shuryak, Phys. Rev. D **73**, 014509 (2006)
15. J. Liao, E. Shuryak, Nucl. Phys. A **775**, 224 (2006)
16. E. Shuryak, Prog. Part. Nucl. Phys. **62**, 48 (2009)
17. D.E. Kharzeev, Nucl. Phys. A **827**, 118C (2009)
18. J. Liao, E. Shuryak, Phys. Rev. Lett. **102**, 202302 (2009)
19. E. Shuryak, Phys. Rev. C **80**, 054908 (2009) [Erratum. Phys. Rev. C **80**, 069902 (2009)]
20. C. Ratti, E. Shuryak, Phys. Rev. D **80**, 034004 (2009)
21. M. Lublinsky, C. Ratti, E. Shuryak, Phys. Rev. D **81**, 014008 (2010)
22. D.M. Ostrovsky, G.W. Carter, E.V. Shuryak, Phys. Rev. C **66**, 036004 (2002)
23. D. Kharzeev, Phys. Lett. B **633**, 260 (2006)
24. D. Kharzeev, A. Zhitnitsky, Nucl. Phys. A **797**, 67 (2007)
25. D.E. Kharzeev, L.D. McLerran, H.J. Warringa, Nucl. Phys. A **803**, 227 (2008)
26. K. Fukushima, D.E. Kharzeev, H.J. Warringa, Phys. Rev. D **78**, 074033 (2008)
27. K. Fukushima, D.E. Kharzeev, H.J. Warringa, Nucl. Phys. A **836**, 311 (2010)
28. D.E. Kharzeev, H.J. Warringa, Phys. Rev. D **80**, 034028 (2009)
29. P.V. Buividovich, M.N. Chernodub, E.V. Luschevskaya, M.I. Polikarpov, Phys. Rev. D **80**, 054503 (2009)
30. P.V. Buividovich, M.N. Chernodub, E.V. Luschevskaya, M.I. Polikarpov, Nucl. Phys. B **826**, 313 (2010)
31. D.E. Kharzeev, Ann. Phys. **325**, 205 (2010)
32. D.T. Son, A.R. Zhitnitsky, Phys. Rev. D **70**, 074018 (2004). arXiv:hep-ph/0405216
33. M.A. Metlitski, A.R. Zhitnitsky, Phys. Rev. D **72**, 045011 (2005). arXiv:hep-ph/0505072
34. D.E. Kharzeev, H.-U. Yee, Phys. Rev. D **83**, 085007 (2011). arXiv:1012.6026 [hep-th]
35. Y. Burnier, D.E. Kharzeev, J. Liao, H.-U. Yee, Phys. Rev. Lett. **107**, 052303 (2011). arXiv:1103.1307 [hep-ph]
36. Y. Burnier, D.E. Kharzeev, J. Liao, H.-U. Yee, arXiv:1208.2537 [hep-ph]
37. E.V. Gorbar, V.A. Miransky, I.A. Shovkovy, Phys. Rev. D **83**, 085003 (2011). arXiv:1101.4954 [hep-ph]
38. J. Rafelski, B. Muller, Phys. Rev. Lett. **36**, 517 (1976)
39. V. Skokov, A. Illarionov, V. Toneev, Int. J. Mod. Phys. A **24**, 5925 (2009)
40. A. Bzdak, V. Skokov, Phys. Lett. B **710**, 171 (2012). arXiv:1111.1949 [hep-ph]
41. W.-T. Deng, X.-G. Huang, Phys. Rev. C **85**, 044907 (2012). arXiv:1201.5108 [nucl-th]
42. B. Muller, A. Schafer, Phys. Rev. C **82**, 057902 (2010). arXiv:1009.1053 [hep-ph]
43. M. Asakawa, A. Majumder, B. Muller, Phys. Rev. C **81**, 064912 (2010). arXiv:1003.2436 [hep-ph]
44. V.D. Toneev, V. Voronyuk, E.L. Bratkovskaya, W. Cassing, V.P. Konchakovski, S.A. Voloshin, Phys. Rev. C **85**, 034910 (2012). arXiv:1112.2595 [hep-ph]
45. B.I. Abelev et al. (STAR Collaboration), Phys. Rev. Lett. **103**, 251601 (2009)
46. B.I. Abelev et al. (STAR Collaboration), Phys. Rev. C **81**, 054908 (2010)
47. S.A. Voloshin, Phys. Rev. C **70**, 057901 (2004)
48. J. Bloczynski, X.-G. Huang, X. Zhang, J. Liao, Phys. Lett. B **718**, 1529 (2013). arXiv:1209.6594 [nucl-th]

49. S.A. Voloshin, A.M. Poskanzer, R. Snellings, arXiv:0809.2949 [nucl-ex]
50. S. Chatrchyan et al. (CMS Collaboration), J. High Energy Phys. **1107**, 076 (2011). arXiv:1105.2438 [nucl-ex]
51. G. Aad et al. (ATLAS Collaboration), arXiv:1203.3087 [hep-ex]
52. B. Abelev et al. (ALICE Collaboration), arXiv:1205.5761 [nucl-ex]
53. V. Koch, A. Majumder, J. Randrup, Phys. Rev. Lett. **95**, 182301 (2005). nucl-th/0505052
54. X. Zhang, J. Liao, Phys. Lett. B **713**, 35 (2012). arXiv:1202.1047 [nucl-th]
55. J. Liao, AIP Conf. Proc. **1441**, 874 (2012). arXiv:1109.0271 [nucl-th]
56. A. Bzdak, V. Koch, J. Liao, Phys. Rev. C **81**, 031901 (2010)
57. J. Liao, V. Koch, A. Bzdak, Phys. Rev. C **82**, 054902 (2010). arXiv:1005.5380 [nucl-th]
58. F. Wang, Phys. Rev. C **81**, 064902 (2010). arXiv:0911.1482 [nucl-ex]
59. Q. Wang, arXiv:1205.4638 [nucl-ex]
60. N.N. Ajitanand, R.A. Lacey, A. Taranenko, J.M. Alexander, Phys. Rev. C **83**, 011901 (2011). arXiv:1009.5624 [nucl-ex]
61. S. Schlichting, S. Pratt, Phys. Rev. C **83**, 014913 (2011)
62. B. Abelev et al. (ALICE Collaboration), arXiv:1207.0900 [nucl-ex]
63. V.D. Toneev, V. Voronyuk, Phys. Atom. Nucl. **75**, 607 (2012). arXiv:1012.1508 [nucl-th]
64. B. Mohanty, [STAR Collaboration], J. Phys. G **38**, 124023 (2011). arXiv:1106.5902 [nucl-ex]
65. S. Pratt, S. Schlichting, S. Gavin, Phys. Rev. C **84**, 024909 (2011)
66. A. Bzdak, V. Koch, J. Liao, Phys. Rev. C **83**, 014905 (2011)
67. G.-L. Ma, B. Zhang, Phys. Lett. B **700**, 39 (2011)
68. A. Bzdak, Phys. Rev. C **85**, 044919 (2012)
69. N. Borghini, P.M. Dinh, J.-Y. Ollitrault, Phys. Rev. C **62**, 034902 (2000)
70. S. Pratt, arXiv:1002.1758 [nucl-th]
71. Z. Chajecki, M. Lisa, Phys. Rev. C **78**, 064903 (2008)
72. A. Bialas, M. Bleszynski, W. Czyz, Nucl. Phys. B **111**, 461 (1976)
73. M.M. Aggarwal et al. (STAR Collaboration), Phys. Rev. C **82**, 024905 (2010)
74. P. Bozek, W. Broniowski, arXiv:1204.3580 [nucl-th]
75. D. Teaney, L. Yan, Phys. Rev. C **83**, 064904 (2011)
76. S.A. Voloshin, Phys. Rev. Lett. **105**, 172301 (2010)
77. J. Steinheimer, V. Koch, M. Bleicher, arXiv:1207.2791 [nucl-th]

Chapter 20
Holography, Fractionalization and Magnetic Fields

Tameem Albash, Clifford V. Johnson, and Scott McDonald

Aspects of the low energy physics of matter charged under a global $U(1)$ at finite density can be studied at strong coupling using the AdS/CFT correspondence and deformations thereof [1–5] (for recent reviews, see Refs. [6–10]). While it remains unclear just how far-reaching the tools of holography will be in helping understand experimentally accessible physics of various condensed matter systems, it is already apparent that potentially powerful new ways of characterizing several classes of important behavior may be emerging from the lines of research underway. The language is that of a dual gravitational system, which has the utility that it is often very geometrical in character, while also being in terms of quantities that are gauge invariant. The dual effective field theories are usually formulated perturbatively as gauge theories (or generalizations thereof) for gauge group of rank N, where N is large. The natural gauge invariant variables to use at strong coupling are usually less easy to work with. It is in this strong coupling regime where the gravitational language is most effective.

At finite charge density, there is a non-zero gauge field A_t switched on in the gravitational ("bulk") background. (The gauged $U(1)$ there is the global $U(1)$ of the dual theory, according to the usual dictionary [1–3]). A most natural circumstance is to have an event horizon present, with a Reissner–Nordstrom black hole sourcing the electric flux, as first explored in this context in Refs. [11, 12]. Already there is interesting physics to be learned from such systems, giving insights into such phenomena as the holographic Hall and Nernst effects and other transport properties [13–15], holographic Fermi surfaces [16–19], and so forth (see for example the

T. Albash (✉) · C.V. Johnson · S. McDonald
Department of Physics and Astronomy, University of Southern California, Los Angeles,
CA 90089-0484, USA
e-mail: talbash@usc.edu

C.V. Johnson
e-mail: johnson1@usc.edu

S. McDonald
e-mail: smacdona@usc.edu

D. Kharzeev et al. (eds.), *Strongly Interacting Matter in Magnetic Fields*,
Lecture Notes in Physics 871, DOI 10.1007/978-3-642-37305-3_20,
© Springer-Verlag Berlin Heidelberg 2013

reviews in Refs. [6–8]), but more recent insights have shown a wider context into which such horizon-endowed charged spacetimes fit nicely: The electric flux can be sourced by electrically charged fermionic matter in the bulk of spacetime *it instead* of a horizon, giving a kind of charged "star", where the geometry in the infra-red (IR) ceases to be $AdS_2 \times \mathbb{R}^2$ (we stay with infinite volume henceforth in this discussion), and becomes of Lifshitz form [20]. The heuristic argument here is that [20] the matter in the bulk effectively screens the electric field in the IR, and the equations of motion yield the Lifshitz geometry, known to be appropriate when there is a massive gauge field [21]. One supplementary way to think about this solution is that it is a result of the black hole becoming unstable due to pair creation induced by the electric field. The charges separate in the field, one escaping to infinity while the other falls into the horizon (for suitable particle mass). The black hole loses charge and energy and ultimately a gas of charged matter, the star, is all that is left.

Naturally, the question arises as to what these two different situations (horizon source *vs.* matter source) represent for the dual low energy physics. Since at large N the entropy of a spacetime with horizon has an extra factor of some power of N (typically N^2 for ordinary gauge theory) compared to the entropy for spacetimes without, it is clear that the degrees of freedom in each case are organized very differently, as noted very early on in holographic studies in the context of thermally driven confinement–deconfinement phase transitions (modeled by the Hawking–Page transition [22]) [5].

The observation in the present context is that the difference here is between fully fractionalized and fully unfractionalized (or "mesonic") phases of the low energy finite density system.[1] This is of some considerable interest in condensed matter physics, since descriptions of charge and/or spin separation (where the charge or spin degrees of freedom of an electron may move separately) are relatively common for getting access to certain types of physics often associated with the low energy dynamics of lattice models. The idea (aspects of it are reviewed in Ref. [10]) is that the N^2 degrees of freedom accessible when the horizon is present are typical of the larger number of fractionalized variables available (the analogue of the electron being split into separate charge and spin degrees of freedom plus a Lagrange multiplier field that becomes a dynamical gauge field at low energy), while the fewer degrees of freedom of the non-horizon systems are more like the unfractionalized "electron" (a composite particle in this picture). This fractionalization leads to a violation of the Luttinger theorem, where the fractionalized variables carry the missing charge density [25–28].

This rather compelling picture is certainly worth exploring, since we have a fully non-perturbative tool for getting access to fractionalization. One exploration at zero

---

[1]The idea that these holographic systems may capture the dynamics of fractionalized phases seems to have first begun to emerge in the work of Ref. [23] in their semi-holographic approach to the low energy physics. There, they used the term "quasiunparticle" for the effective particles in the unfractionalized phase. In work dedicated to addressing the issue, Ref. [24] further elucidated the connection between holographic physics and fractionalized phases. See Refs. [9, 10] for a review of some of these ideas.

temperature (of interest to us in the current work, as we shall see) is Ref. [29], where it was shown how to tune an operator (that is relevant in the ultraviolet (UV)) in the system that allows for a mixture of fully fractionalized and fully unfraction-alized phases. This is dual to turning on a dilaton-like scalar field in the bulk theory. The dilaton scalar field couples to gravity, has a non-trivial potential, and due to a term of the form: $-\exp(\frac{2}{\sqrt{3}}\Phi)F^2$, (where F is the gauge field strength) it effec-tively spatially modulates the coupling strength of the bulk gauge field, allowing for different configurations of matter and gravity to minimize the action for given asymptotic fields. If the dilaton diverges positively in the IR, the effective Maxwell coupling vanishes in the IR, allowing for a geometry with a horizon again, while still having the matter back-react enough on the bulk geometry to support a star that carries some of the electric charge. This gives a mixture of fractionalized and par-tially fractionalized phases. If the dilaton diverges negatively in the IR, the effective Maxwell coupling diverges there and we have that the favored situation is the purely "mesonic" case where there is no horizon and only charged matter. The negatively diverging dilaton serves to enhance the Maxwell sector's tendency to destabilize a charged black hole due to pair production, forcing it to evaporate completely in favor of a charged star.

It is worth noting that the same way that the Einstein–Maxwell–AdS system in various dimensions may be consistently uplifted to 10 and 11 dimensional su-pergravity as combinations of spins in the compact directions (actually, using equal spins [11]), the systems with a dilaton sector can be uplifted, but now all the spins are equal except one [30]. This is worth noting that since this means that the Einstein–Maxwell-dilaton–AdS system's consistent uplift may allow the fate of the regions with diverging dilaton to be studied in M-theory.

In each case, the reduced gravity system is coupled to a gas of charged fermions in the bulk, represented by a fluid stress energy tensor with a given pressure and density, and an appropriate equation of state is input in order to characterize the system. This mirrors the standard construction of neutron stars using the Tolman–Oppenheimer–Volkoff method [31, 32] which was first generalized to asymptoti-cally AdS geometries in Ref. [33].

In this paper, we begin the exploration of the important situation of having a magnetic field present in the system. We learn some very interesting lessons from this study. One of them is that we need a dilaton present to achieve unfractionalized phases of the physics. Since we are including no magnetically charged matter in our system, and in the limit we take there are no current sources of magnetic field, for a smooth gravitational dual it is intuitively clear that there must be a horizon to source the magnetic field. One can imagine therefore that there might be configurations where such a magnetic source is surrounded by a gas of matter that forms a star that sources all the electric flux. This would be the fully unfractionalized magnetic case, with the purely dyonic Reissner–Nordstrom black hole being the fractionalized counterpart.

In fact, it turns out that the matter alone is unable to back-react enough on the magnetic geometry that we find to support a star. In other words, the magnetic field's presence stops the electric field from giving the local chemical potential (as seen by the charged bulk matter) the profile needed to support a star. Put differently, the magnetic field suppresses the pair-creation channel by which a black hole can leak its charge into a surrounding cloud of fermions to form a star. This is where the dilaton can come in. A negatively diverging dilaton in the IR will send the Maxwell coupling to infinity. This makes the pair-creation channel viable again, and allow for the system to seek stable solutions that have some of the electric flux sourced by fermions outside the horizon.

In summary we find that in the limit we are working, when there is a magnetic field in the system there is necessarily an event horizon, and it is essential to have a dilaton present to not have all the electric flux sourced behind the horizon (at least for the zero temperature case we study here). Introducing the dilaton in the theory allows us to construct a mesonic phase as well as a partially fractionalized phase (the fully fractionalized phase is simply the already known dilaton–dyon solution).

The work we report on in this paper is but a first step in characterizing these systems in the presence of magnetic field. Our goal here is to exhibit the solutions we find.[2] We will postpone the analysis of the fluctuations of the solutions to examine various transport properties of the system, and also leave for a later time the comparison of the actions of the various solutions for given magnetic field and chemical potential, which allow a full exploration of the phase diagram.

20.1 Charged Ideal Fluid with a Magnetic Field

20.1.1 Gravity Background

We consider a simple model of Einstein–Maxwell theory coupled to a dilaton Φ and an ideal fluid source (akin to that of Ref. [29]). The Einstein–Maxwell sector has action given by:

$$S_{EM} = \int d^4x \sqrt{-G}\left[\frac{1}{2\kappa^2}\left(R - 2\partial_\mu\Phi\partial^\mu\Phi - \frac{V(\Phi)}{L^2}\right) - \frac{Z(\Phi)}{4e^2}F_{\mu\nu}F^{\mu\nu}\right]. \quad (20.1)$$

The Einstein equations of motion are given by:

$$R_{\mu\nu} - \frac{1}{2}\left(R - 2\partial_\lambda\Phi\partial^\lambda\Phi - \frac{V(\Phi)}{L^2}\right)G_{\mu\nu} - 2\partial_\mu\Phi\partial_\nu\Phi$$

$$= \kappa^2\left(\frac{Z(\Phi)}{e^2}\left(F_{\mu\rho}F_\nu^\rho - \frac{1}{4}G_{\mu\nu}F_{\lambda\rho}F^{\lambda\rho}\right) + T_{\mu\nu}^{\text{fluid}}\right), \quad (20.2)$$

---

[2]We emphasize that our work is both qualitatively and quantitatively different from Ref. [34]. Their work is in five dimensions, for a spherical star, and their TOV treatment involves an electrically neutral star.

with the following stress-energy tensor and external current for a perfect (charged) fluid (the action that gives rise to this stress energy tensor is given in Sect. 20.1.3):

$$T^{\text{fluid}}_{\mu\nu} = \frac{1}{L^2\kappa^2}\left((\tilde{P}(r)+\tilde{\rho}(r))u_\mu u_\nu + \tilde{P}(r)G_{\mu\nu}\right), \qquad J_\mu = \frac{1}{eL^2\kappa}\tilde{\sigma}(r)u_\mu. \quad (20.3)$$

The Maxwell equations are given by:

$$\partial_\nu\left(\sqrt{-G}Z(\Phi)F^{\mu\nu}\right) = e^2\sqrt{-G}J^\mu. \quad (20.4)$$

We begin with the following metric ansatz:

$$ds^2 = L^2\left(-f(r)dt^2 + a(r)d\mathbf{x}^2 + g(r)dr^2\right), \quad (20.5)$$

and we take $u_t = -\sqrt{-G_{tt}}$. For the Maxwell field, we take as ansatz:

$$F_{rt} = \frac{eL}{\kappa}h'(r), \qquad F_{xy} = \frac{eL}{\kappa}\tilde{B}. \quad (20.6)$$

The resulting equations of motion can be reduced (after some work) to the following six equations:

$$\tilde{P}'(r) + \frac{f'(r)}{2f(r)}\left(\tilde{P}(r)+\tilde{\rho}(r)\right) - \tilde{\sigma}(r)\frac{h'(r)}{\sqrt{f(r)}} = 0, \quad (20.7)$$

$$\frac{a''(r)}{a(r)} - \frac{a'(r)}{a(r)}\left(\frac{g'(r)}{2g(r)} + \frac{f'(r)}{2f(r)} + \frac{a'(r)}{2a(r)}\right)$$
$$+ \left(g(r)(\tilde{P}(r)+\tilde{\rho}(r)) + 2\Phi'(r)^2\right) = 0, \quad (20.8)$$

$$\frac{f''(r)}{f(r)} - \frac{f'(r)}{f(r)}\left(\frac{g'(r)}{2g(r)} + \frac{f'(r)}{2f(r)} - \frac{2a'(r)}{a(r)}\right)$$
$$+ \left(\frac{a'(r)^2}{2a(r)^2} - 2\Phi'(r)^2 - g(r)\left(5\tilde{P}(r)+\tilde{\rho}(r)-2V(\Phi)\right)\right) = 0, \quad (20.9)$$

$$\Phi'(r)^2 + g(r)\left(-\frac{Z(\Phi)\tilde{B}^2}{2a(r)^2} + \left(\tilde{P}(r)-\frac{1}{2}V(\Phi)\right)\right)$$
$$- \frac{a'(r)^2}{4a(r)^2} - \frac{a'(r)f'(r)}{2a(r)f(r)} - \frac{Z(\Phi)h'(r)^2}{2f(r)} = 0, \quad (20.10)$$

$$h''(r) - h'(r)\left(\frac{g'(r)}{2g(r)} + \frac{f'(r)}{2f(r)} - \frac{a'(r)}{a(r)} - \frac{Z'(\Phi)\Phi'(r)}{Z(\Phi)}\right)$$
$$- \frac{\sqrt{f(r)}g(r)}{Z(\Phi)}\tilde{\sigma}(r) = 0. \quad (20.11)$$

$$\Phi''(r) + \Phi'(r)\left(\frac{f'(r)}{2f(r)} - \frac{g'(r)}{2g(r)} + \frac{a'(r)}{a(r)}\right)$$

$$-\frac{Z'(\Phi)}{4}\left(\frac{g(r)\tilde{B}^2}{a(r)^2} - \frac{h'(r)^2}{f(r)}\right) - \frac{g(r)V'(\Phi)}{4} = 0. \tag{20.12}$$

For the charge density $\tilde{\sigma}(r)$ and the energy density $\tilde{\rho}(r)$, we show in Sect. 20.1.2 that using the density of states of charged quanta in a magnetic field results in the free fermion gas result when using the appropriate electron star limits:

$$\tilde{\sigma}(r) = \frac{1}{3}\tilde{\beta}\left(\tilde{\mu}(r)^2 - \tilde{m}^2\right)^{3/2},$$

$$\tilde{\rho}(r) = \frac{\tilde{\beta}}{8}\left(\tilde{\mu}(r)\sqrt{\tilde{\mu}(r)^2 - \tilde{m}^2}\left(2\tilde{\mu}(r)^2 - \tilde{m}^2\right)\right. \tag{20.13}$$

$$\left. + \tilde{m}^4\ln\left(\frac{\tilde{m}}{\tilde{\mu}(r) + \sqrt{\tilde{\mu}(r)^2 - \tilde{m}^2}}\right)\right),$$

and (20.7) is satisfied with the following equation for the pressure:

$$\tilde{P}(r) = -\tilde{\rho}(r) + \tilde{\mu}(r)\tilde{\sigma}(r), \tag{20.14}$$

where we have defined $\tilde{\mu}(r) = h(r)/\sqrt{f(r)}$ as the local chemical potential. Here, $\tilde{m} = \kappa m/e$, where m is the fermion mass. Equations (20.8), (20.9), and (20.10) are derived from the Einstein equations of motion, and only two of them are dynamical equations; the remaining equation is a constraint. For concreteness, let us take from this point forward:

$$V(\Phi) = -6\cosh(2\Phi/\sqrt{3}), \qquad Z(\Phi) = e^{2\Phi/\sqrt{3}}. \tag{20.15}$$

This choice matches the choice in Ref. [35] that give rise to three-equal-charge dilatonic black holes in four dimensions. We want our solutions to asymptote to AdS$_4$ so we will require that the UV behavior of our fields is given by:

$$g(r) = \frac{1}{r^2}, \qquad f(r) = \frac{1}{r^2}, \qquad a(r) = \frac{1}{r^2}, \tag{20.16}$$

which in turn gives the following UV behavior for the remaining fields:

$$\Phi(r) = r\phi_1 + r^2\phi_2, \qquad h(r) = \mu - \rho r, \tag{20.17}$$

where ϕ_1 is proportional to the source of the operator dual to the dilaton, ϕ_2 to the vev of the same operator, μ the chemical potential in the dual field theory, and ρ the charge density.

20.1.2 *Density of States for* $(3 + 1)$*-Dimensional Fermions in a Magnetic Field*

The appropriate choice for the density of states should be that of the 3 (spatial) dimensional Landau levels. For a fermion of mass m with charge q in the nth Landau level with momentum k, the energy in magnetic field B is given by ($c = \hbar = 1$):

$$E_n(k) = \sqrt{2q B(n + 1) + k^2 + m^2}, \qquad (20.18)$$

where for simplicity, we are assuming that $q B > 0$. At level n and momentum k, the number of states is given by:

$$N_n(k) = g_s \left(\frac{k L_r}{2\pi} \right) \left(2q B \frac{L_x L_y}{4\pi} \right), \qquad (20.19)$$

where g_s is the spin degeneracy ($=2$ for spin $1/2$). In turn, at a given energy E, the number of states is given by:

$$
\begin{aligned}
N(E) &= g_s \sum_{n=0}^{\infty} \frac{L_x L_y L_r}{8\pi^2} 2q B \sqrt{E^2 - m^2 - 2q B(n + 1)}\, \Theta\left(E - \sqrt{m^2 + 2q B(n + 1)}\right) \\
&= V \frac{\beta}{2} 2q B \sum_n \sqrt{E^2 - m^2 - 2q B(n + 1)}\, \Theta\left(E - \sqrt{m^2 + 2q B(n + 1)}\right),
\end{aligned}
$$
$$(20.20)$$

where we have defined a dimensionless constant of proportionality which is $O(1)$. Therefore, we can write the density of states as:

$$
\begin{aligned}
g(E) = \frac{\beta}{2} 2q B \sum_{n=0}^{\infty} &\left[\frac{E}{\sqrt{E^2 - m^2 - 2q B(n + 1)}} \Theta\left(E - \sqrt{m^2 + 2q B(n + 1)}\right) \right. \\
&\left. + \sqrt{E^2 - m^2 - 2q B(n + 1)}\, \delta\left(E - \sqrt{m^2 + 2q B(n + 1)}\right) \right]. \quad (20.21)
\end{aligned}
$$

The second term will always give us zero contribution for the terms we are interested in, so we will drop it. The energy density ρ and charge density σ are given by:

$$\sigma = \int_0^\mu dE\, g(E), \qquad \rho = \int_0^\mu dE\, E g(E), \qquad (20.22)$$

where μ is the chemical potential. To proceed, we identify the local magnetic field in our gravity background with $q B$ and the local chemical potential with μ:

$$q B = \frac{F_{xy}}{L^2 A(r)} = \frac{e}{L\kappa} \frac{\tilde{B}}{A(r)}, \qquad \mu = \frac{A_t}{L\sqrt{f(r)}} = \frac{e}{\kappa} \frac{h(r)}{\sqrt{f(r)}} \equiv \frac{e}{k} \tilde{\mu}. \qquad (20.23)$$

For conciseness, let us define:

$$\tilde{q} \equiv \frac{\kappa}{Le^2} \frac{e}{A(r)}. \tag{20.24}$$

In terms of dimensionless variables $\tilde{\sigma}(r) \equiv eL^2\kappa\sigma(r)$ and $\tilde{\rho}(r) \equiv L^2\kappa^2\rho(r)$, we now have:

$$\tilde{\sigma}(r) = \frac{\tilde{\beta}}{2} 2\tilde{q}\tilde{B} \sum_{n=0}^{\infty} \Theta\left(\tilde{\mu}^2 - \tilde{m}^2 - 2\tilde{q}\tilde{B}(n+1)\right)$$

$$\times \int_{\sqrt{\tilde{m}^2+2\tilde{q}\tilde{B}(n+1)}}^{\tilde{\mu}} \frac{d\tilde{E}\,\tilde{E}}{\sqrt{\tilde{E}^2 - \tilde{m}^2 - 2\tilde{q}\tilde{B}(n+1)}},$$

$$\tilde{\rho}(r) = \frac{\tilde{\beta}}{2} 2\tilde{q}\tilde{B} \sum_{n=0}^{\infty} \Theta\left(\tilde{\mu}^2 - \tilde{m}^2 - 2\tilde{q}\tilde{B}(n+1)\right) \tag{20.25}$$

$$\times \int_{\sqrt{\tilde{m}^2+2\tilde{q}\tilde{B}(n+1)}}^{\tilde{\mu}} \frac{d\tilde{E}\,\tilde{E}^2}{\sqrt{\tilde{E}^2 - \tilde{m}^2 - 2\tilde{q}\tilde{B}(n+1)}}.$$

where $\tilde{\beta} = e^4L^2\beta/\kappa^2$ and $\tilde{m} = \kappa m/e$. To be in the classical gravity regime, we must have $\kappa/L \ll 1$. To be able to use the flat space physics, we must have a large density relative to the curvature scale of the geometry, we must have $\sigma L^3 \sim L\sigma(r)/e\kappa \gg 1$. To have $\tilde{\beta} \sim O(1)$, we must have that $e^2L/\kappa \sim O(1)$. Therefore by our previous requirements, we must have $e^2 \sim \kappa/L$. These are the requirements from Ref. [20]. With these requirements, we have the following result:

$$2\tilde{q}\tilde{B} \ll 1. \tag{20.26}$$

We can perform the integrals explicitly. We have that:

$$\int_{\sqrt{\tilde{m}^2+2\tilde{q}\tilde{B}(n+1)}}^{\tilde{\mu}} \frac{d\tilde{E}\,\tilde{E}}{\sqrt{\tilde{E}^2 - \tilde{m}^2 - 2\tilde{q}\tilde{B}(n+1)}} = \sqrt{\tilde{\mu}^2 - \tilde{m}^2 - 2\tilde{q}\tilde{B}(n+1)},$$

$$\int_{\sqrt{\tilde{m}^2+2\tilde{q}\tilde{B}(n+1)}}^{\tilde{\mu}} \frac{d\tilde{E}\,\tilde{E}^2}{\sqrt{\tilde{E}^2 - \tilde{m}^2 - 2\tilde{q}\tilde{B}(n+1)}} \tag{20.27}$$

$$= \frac{1}{2}\left[\tilde{\mu}\sqrt{\tilde{\mu}^2 - \tilde{m}^2 - 2\tilde{q}\tilde{B}(n+1)} \right.$$

$$\left. - (\tilde{m}^2 + 2\tilde{q}\tilde{B}(n+1))\ln\left(\frac{\sqrt{\tilde{m}^2 + 2\tilde{q}\tilde{B}(n+1)}}{\tilde{\mu} + \sqrt{\tilde{m}^2 + 2\tilde{q}\tilde{B}(n+1)}}\right) \right].$$

Furthermore, we see that the spacing between Landau levels is extremely small, and to zeroth order in $2\tilde{q}\tilde{B}$ we can approximate the sum by an integral (Euler–Maclaurin):

$$2\tilde{q}\tilde{B}\sum_{n=0}^{\infty}\Theta\left(\tilde{\mu}^2 - m^2 - 2\tilde{q}\tilde{B}(n+1)\right)F\left(2\tilde{q}\tilde{B}(n+1)\right) \approx \int_0^{\tilde{\mu}^2-\tilde{m}^2} F(x)dx + O(2\tilde{q}\tilde{B}).$$

(20.28)

We can perform this final integration to give us our final result for the charge density and energy density:

$$\tilde{\sigma} = \frac{1}{3}\tilde{\beta}\left(\tilde{\mu}^2 - \tilde{m}^2\right)^{3/2} + O(2\tilde{q}\tilde{B}),$$

$$\tilde{\rho} = \frac{1}{8}\tilde{\beta}\left(\tilde{\mu}\sqrt{\tilde{\mu}^2 - \tilde{m}^2}\left(2\tilde{\mu}^2 - \tilde{m}^2\right) + \tilde{m}^4\ln\left(\frac{\tilde{m}}{\tilde{\mu}+\sqrt{\tilde{\mu}^2 - \tilde{m}^2}}\right)\right) + O(2\tilde{q}\tilde{B}).$$

(20.29)

Interestingly enough, the zeroth order results are exactly the free fermion gas results in the absence of a magnetic field. Since the pressure equation (20.7) is the same as that in the free fermion in the absence of a magnetic field case, the ansatz:

$$\tilde{P} = -\tilde{\rho} + \tilde{\mu}\tilde{\sigma},$$

(20.30)

trivially satisfies that equation.

20.1.3 Action Calculation

We discuss in detail an action that recovers the equations of motion used for our background. Our calculation generalizes the calculation performed in Ref. [20] to the case with a magnetic field. We consider an action for the fluid along the lines of Refs. [36–38] given by:

$$\mathcal{L}_{\text{fluid}} = \sqrt{-G}\left(-\rho(\sigma) + \sigma u^\mu(\partial_\mu\phi + A_\mu + \alpha\partial_\mu\beta) + \lambda\left(u^\mu u_\mu + 1\right)\right), \quad (20.31)$$

where ϕ is a Clebsch potential variable, (α, β) are potential variables, and λ is a Lagrange multiplier. The equations of motion from this part of the action give:

$$\delta\sigma: \quad -\rho'(\sigma) + u^\mu(\partial_\mu\phi + A_\mu + \alpha\partial_\mu\beta) = 0, \quad (20.32)$$

$$\delta u_\mu: \quad G^{\mu\nu}\left(\sigma(\partial_\nu\phi + A_\nu + \alpha\partial_\nu\beta) + 2\lambda u_\nu\right) = 0, \quad (20.33)$$

$$\delta\lambda: \quad u_\mu u^\mu = -1, \quad (20.34)$$

$$\delta\phi: \quad \partial_\mu\left(\sqrt{-G}G^{\mu\nu}\sigma u_\nu\right) = 0, \quad (20.35)$$

$$\delta\alpha: \quad u^\mu\partial_\mu\beta = 0, \quad (20.36)$$

$$\delta\beta: \quad \partial_\mu\left(\sqrt{-G}\sigma u^\mu\alpha\right) = 0. \quad (20.37)$$

We define the local chemical potential $\mu \equiv \rho'(\sigma)$ such that:

$$\mu = u^{\mu}(\partial_{\mu}\phi + A_{\mu} + \alpha\partial_{\mu}\beta). \tag{20.38}$$

From the equation of the fluctuation of u_{μ}, we can multiply through by u_{μ} to fix the Lagrange multiplier:

$$\lambda = \frac{1}{2}\sigma\mu = \frac{1}{2}(P + \rho), \tag{20.39}$$

where we have used the thermodynamic relation $P = -\rho + \mu\sigma$. Now we recall that we wish to fix $u_t = -\sqrt{-G_{tt}}$, $u_x = u_y = 0$, $\mu = A_t/(-u_t)$ and choose a gauge where $A_x = 0$, $A_y = eL\tilde{B}x/\kappa$. Since our metric ansatz has $G^{xx} = G^{yy}$, we can satisfy the equation for the fluctuation of u_{μ} (and all the equations) with:

$$\phi = -\frac{eL}{\kappa}\tilde{B}xy, \qquad \alpha = \frac{eL}{\kappa}\tilde{B}y, \qquad \beta = x, \tag{20.40}$$

with these choices, we note that:

$$\frac{\delta\mathcal{L}_{\text{fluid}}}{\delta G^{\mu\nu}} = \sqrt{-G}\left(-\frac{1}{2}G_{\mu\nu}(-\rho + \mu\sigma) - \frac{1}{2}\delta_{\mu}^{t}\delta_{\nu}^{t}\mu\sigma u_t u_t\right), \tag{20.41}$$

which for us recovers the (on-shell) fluid energy–momentum tensor using in our equations of motion:

$$T_{\mu\nu}^{\text{fluid}} = \frac{-2}{\sqrt{-G}}\frac{\delta\mathcal{L}_{\text{fluid}}}{\delta G^{\mu\nu}} = (P + \rho)u_{\mu}u_{\nu} + PG_{\mu\nu} = \frac{1}{L^2\kappa^2}((\tilde{P} + \tilde{\rho})u_{\mu}u_{\nu} + \tilde{P}G_{\mu\nu}). \tag{20.42}$$

20.2 The Role of the Dilaton

We first remark on why Einstein–Maxwell theory alone (*i.e.* without a dilaton) is not sufficient for studying backgrounds with an electric star in the presence of a magnetic field. Let us consider the case where the dilaton is absent in the theory. We find the following series expansion in the IR ($r \to \infty$):

$$a(r) = \sum_{n=0}^{\infty}\frac{a_n}{r^n}, \quad f(r) = \frac{1}{r^2}\sum_{n=0}^{\infty}\frac{f_n}{r^n}, \quad g(r) = \frac{1}{r^2}\sum_{n=0}^{\infty}\frac{g_n}{r^n}, \quad h(r) = \frac{1}{r}\sum_{n=0}^{\infty}\frac{h_n}{r^n}, \tag{20.43}$$

where the leading coefficients fixed by the equations of motion are given by:

$$g_0 = \frac{1}{6}, \qquad \frac{h_0}{\sqrt{f_0}} = \sqrt{1 - \frac{\tilde{B}^2}{6a_0^2}}, \tag{20.44}$$

where we have picked the positively charged horizon (sign of h_0) without loss of generality. This is of course the well established near horizon geometry of the ex-

tremal dyonic black hole in AdS_4. We find that all the coefficients of $f(r)$ are undetermined by the equations of motion, as well as a_0 and a_1. The value of a_0 is fixed by
requiring that $r^2 a(r)$ be equal to one in the UV, so it is the ratio a_1/a_0 that is the free
parameter. We note that the near horizon behavior of the local chemical potential is:

$$\frac{h(r)}{\sqrt{f(r)}} = \frac{h_0}{\sqrt{f_0}}\left(1 - \frac{1}{r}\frac{a_1}{a_0}f_0 + O(r^{-2})\right). \tag{20.45}$$

This chemical potential *decreases* (this is because only for $a_1 > 0$ do we get an
asymptotically AdS_4 solution) from its IR value. This will be an important point for
what comes next.

Let us now consider the case where we turn on the source and look for a solution
where we have a star in the IR. We find that an identical expansion as (20.43) works,
but the leading coefficients are given by:

$$g_0 = \frac{1}{6}, \qquad \frac{h_0}{\sqrt{f_0}} = \tilde{m}, \qquad \tilde{B}^2 = 6a_0^2(1 - \tilde{m}^2). \tag{20.46}$$

However, the situation is more grave than this; the only solution allowed is:

$$\frac{h(r)}{\sqrt{f(r)}} = \tilde{m}, \qquad a(r) = a_0, \tag{20.47}$$

which is clearly not an asymptotically AdS_4 solution (one can attempt to generalize the IR expansion to having $f(r) \propto r^{-2z}$ and $g(r) \propto r^{-z}$ but the results remain
the same up to factors of z). For $f(r) = 1/r^2$, the solution is simply the near horizon geometry of a dyon where the electric charge has been set by $\tilde{m}$. Furthermore,
since the chemical potential is equal to $\tilde{m}$ everywhere, there is no star, so we cannot
construct a solution with a star in the IR.

We may consider building a solution where there is no star in the IR but instead
exists a finite distance away from the horizon. Therefore, we would have the same
IR expansion as (20.43), and hope that we are able to populate the star at some finite
distance $r < \infty$. This would require $\tilde{m} > h_0/\sqrt{f_0}$. but, as pointed out in (20.45), we
find that the chemical potential decreases as we move away from the horizon. In fact,
(as we know already from the exact solution) the chemical potential monotonically
decreases from the horizon to the UV, so a star can never form. Therefore, we find
that we cannot support a magnetic star in the simple Einstein–Maxwell setup.

One way forward is to introduce the dilaton in the theory, which changes the
effective coupling strength of the Maxwell field. As mentioned in the introduction,
a dilaton was first used in this context in Ref. [29]. Introducing the dilaton is dual
to using a relevant operator (in the UV) to induce the theory to flow to different
IR phases. Three different phases were identified: a "mesonic" phase, where all the
electric charge is *not* behind the horizon, a partially "fractionalized" phase where *a
fraction* of the charge is behind the horizon, and a fully fractionalized phase, where
all the charge is behind the horizon. We will use this nomenclature to label our
solutions.

The dilaton's behavior in the IR naturally provides us a classification scheme for
our solutions. It is useful to review this before introducing the star. The dilaton can

diverge positively or negatively in the IR, giving rise to a purely electric or purely magnetic horizon, or it can take a finite value, giving a dyon solution. Since we are only interested in geometries with a magnetic field, we will ignore the purely electric case in what follows. We briefly review the IR asymptotics of the two magnetic cases. The purely magnetic case has an IR ($r \to \infty$) expansion for the fields:

$$f(r) = \frac{1}{r^2}\left(\sum_{n=0}^{\infty} \frac{f_n}{r^{4n/3}}\right), \qquad g(r) = \frac{1}{r^{8/3}}\left(\sum_{n=0}^{\infty} \frac{g_n}{r^{4n/3}}\right), \qquad h(r) = 0,$$

$$\Phi(r) = -\frac{\sqrt{3}}{3}\ln r + \left(\sum_{n=0}^{\infty} \frac{\Phi_n}{r^{4n/3}}\right), \qquad a(r) = \frac{1}{r^{2/3}}\left(\sum_{n=0}^{\infty} \frac{a_n}{r^{4n/3}}\right), \qquad (20.48)$$

$$\tilde{B}^2 = \frac{3}{2}e^{-4\Phi_0/\sqrt{3}}a_0^2.$$

The coefficients f_n are completely undetermined by the equations of motion as well as Φ_0 and a_0 (note that a_0 is not really a free parameter since it determines what the coefficient for dx^2 is in the UV which needs to be 1). Therefore, we see that there are two free parameters in the IR, Φ_0 and the function $f(r)$ (subject to obeying the right asymptotics in the UV and IR). This translates to the freedom of choosing a particular magnetic field and source for the dilaton in the dual field theory. Finally, we note that the behavior of the dilaton in the IR means that the Maxwell coupling is diverging, which means that quantum loop effects would become important near the horizon.

The dyon solution has an IR expansion for the fields given by:

$$f(r) = \frac{1}{r^2}\sum_{n=0}^{\infty} \frac{f_n}{r^n}, \qquad g(r) = \frac{1}{r^2}\sum_{n=0}^{\infty} \frac{g_n}{r^n}, \qquad a(r) = \sum_{n=0}^{\infty} \frac{a_n}{r^n},$$

$$h(r) = \frac{1}{r}\sum_{n=0}^{\infty} \frac{h_n}{r^n}, \qquad \Phi(r) = \sum_{n=0}^{\infty} \frac{\Phi_n}{r^n}, \qquad g_0 = \frac{e^{2\Phi_0/\sqrt{3}}}{3(1+e^{4\Phi_0/\sqrt{3}})}, \qquad (20.49)$$

$$\frac{h_0}{\sqrt{f_0}} = \frac{1}{\sqrt{e^{2\Phi_0/\sqrt{3}} + e^{2\sqrt{3}\Phi_0}}}, \qquad \tilde{B}^2 = 3a_0^2.$$

The coefficients f_n are completely undetermined as well as Φ_0, a_0, and a_1. In the IR, there are three free parameters: a_1/a_0, Φ_0, and the function $f(r)$ (subject to obeying the right asymptotics). This translates to a choice of the magnetic field, the chemical potential, and the source for the dilaton in the dual field theory. Note that since we are studying the extremal case, the electric charge behind the horizon is fixed by the magnetic field:

$$\lim_{r \to \infty} Z(\Phi)(*F)_{xy} = e^{2\Phi_0/\sqrt{3}}\frac{a_0 h_0}{\sqrt{g_0 f_0}} = \tilde{B}. \qquad (20.50)$$

We now turn on the pressure, charge density, energy density due to the electron star and investigate the various gravitational solutions we can construct.

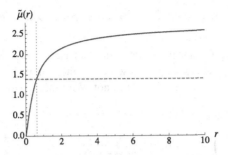

Fig. 20.1 Solution for the local chemical potential (*solid blue curve*) of a star in the IR using $f(r) = r^{-2}$ with $\tilde{\beta} = 1$, $\tilde{m} = 3 \cdot 3^{1/4}/(2(2\tilde{\beta})^{1/2})$, $\Phi_0 = -\sqrt{3}$. The *red dashed line* is the value of $\tilde{m}$ and the *black dotted line* is the position where the star ends. Corresponds to a solution with $\tilde{B} = 6.63$, $\mu = 3.08$, $\rho = 0.53$, $\phi_1 = -17.58$, $\phi_2 = 177.36$ (Color figure online)

20.3 Solutions with Stars

20.3.1 Mesonic Phase: Star in the Infra-Red

Consider an IR expansion for the fields:

$$f(r) = \frac{1}{r^2}\left(\sum_{n=0}^{\infty}\frac{f_n}{r^{2n/3}}\right), \qquad g(r) = \frac{1}{r^{8/3}}\left(\sum_{n=0}^{\infty}\frac{g_n}{r^{2n/3}}\right),$$

$$a(r) = \frac{1}{r^{2/3}}\left(\sum_{n=0}^{\infty}\frac{a_n}{r^{2n/3}}\right), \qquad \Phi(r) = -\frac{\sqrt{3}}{3}\ln r + \left(\sum_{n=0}^{\infty}\frac{\Phi_n}{r^{2n/3}}\right), \qquad (20.51)$$

$$h(r) = \frac{1}{r}\left(\sum_{n=0}^{\infty}\frac{h_n}{r^{2n/3}}\right).$$

This solution is a natural extension of the purely magnetic dilaton-black hole reviewed in the previous section, and the leading order behavior is not changed from the pure magnetic dilatonic black hole except that the introduction of the star now changes the power of the expansion as well as turns on the gauge field A_t:

$$\tilde{B}^2 = \frac{3}{2}e^{-4\Phi_0/\sqrt{3}}a_0^2, \qquad g_0 = \frac{16}{27}e^{2\Phi_0/\sqrt{3}}, \qquad \frac{h_0}{\sqrt{f_0}} = \frac{16}{81}\tilde{\beta}\left(\frac{h_0^2}{f_0} - \tilde{m}^2\right)^{3/2}.$$

$$(20.52)$$

There is no charge behind the horizon since $\lim_{r\to\infty} Z(\Phi)(*F)_{xy} \propto r^{-1/3}$. As in the sourceless case, the coefficients f_n are undetermined by the equations of motion. We can then solve for all remaining coefficients explicitly in terms of a_0, f_n, and Φ_0 order by order as an expansion in the IR. We show an example of a solution in Fig. 20.1.

20.3.2 Partially Fractionalized Phases: Star Outside Horizon

For the case of having no star in the IR with an electric charge behind the horizon, there are two cases to consider: the case where the star is a finite distance away from the horizon, and the case where it is not. We first consider the latter. The IR expansion is given by:

$$
f(r) = \frac{1}{r^2}\left(\sum_{n=0}^{\infty}\frac{f_n}{r^{n/2}}\right), \qquad g(r) = \frac{1}{r^2}\sum_{n=0}^{\infty}\frac{g_n}{r^{n/2}}, \qquad a(r) = \sum_{n=0}^{\infty}\frac{a_n}{r^{n/2}},
$$

$$
h(r) = \frac{1}{r}\sum_{n=0}^{\infty}\frac{h_n}{r^{n/2}}, \qquad \Phi(r) = \sum_{n=0}^{\infty}\frac{\Phi_n}{r^{n/2}}, \qquad B^2 = 3a_0^2, \qquad g_0 = \frac{e^{2\Phi_0/\sqrt{3}}}{3(1+e^{4\Phi_0/\sqrt{3}})},
$$

$$
\frac{h_0}{\sqrt{f_0}} = \pm\frac{1}{\sqrt{e^{2\Phi_0/\sqrt{3}}+e^{2\sqrt{3}\Phi_0}}}, \qquad \tilde{m} = \frac{1}{\sqrt{e^{2\Phi_0/\sqrt{3}}+e^{2\sqrt{3}\Phi_0}}},
$$

$$
f_1 = 0 = h_1 = 0 = g_1 = 0 = a_1 = 0. \tag{20.53}
$$

All higher $n \geq 2$ coefficients are non-zero with a_2/a_0 and $f_{n\geq2}$ as free parameters. The leading IR behavior is unchanged from the dyon, except that we find that the value of Φ_0 is fixed by the mass $\tilde{m}$, and the expansion is now in terms of fractional powers. An example is shown in Fig. 20.2(a).

The local chemical potential at the horizon is equal to $\tilde{m}$ meaning that at the horizon we have no star. However, we find that $\tilde{\mu}$ in the IR evolves as:

$$
\frac{h(r)}{\sqrt{f(r)}} = \tilde{m} + \frac{1}{r}\left(\frac{e^{-\Phi_0/\sqrt{3}}(-5 - 8e^{4\Phi_0/\sqrt{3}} + e^{8\Phi_0/\sqrt{3}})}{18(1+e^{4\Phi_0\sqrt{3}})^{5/2}}\frac{a_2}{a_0}\right) + O\left(r^{-2}\right). \tag{20.54}
$$

The second term in the expansion appears in the value of other coefficients to some fractional power, and therefore is required to be positive

$$
\left(-5 - 8e^{4\Phi_0/\sqrt{3}} + e^{8\Phi_0/\sqrt{3}}\right)\frac{a_2}{a_0} > 0. \tag{20.55}
$$

We are also required to take $a_2 > 0$ in order to get an asymptotically AdS solution (this observation is found numerically). Therefore, since for star formation we need the chemical potential to increase, we have a condition on Φ_0:

$$
\Phi_0 > \frac{\sqrt{3}}{4}\ln(4 + \sqrt{21}) \equiv \Phi_c. \tag{20.56}
$$

For $\Phi_0 < \Phi_c$, the chemical potential decreases monotonically from its value at the horizon and does not allow a star to form. For $\Phi_0 = \Phi_c$, the solution is AdS$_2 \times R^2$ everywhere (if we choose $f(r) = r^{-2}$). This solution therefore separates the partially fractionalized from the fully fractionalized solution for this class of star backgrounds.

Fig. 20.2 The local chemical potential (*solid blue curve*) for the partially fractionalized phases with (**a**) the star ending at the horizon and (**b**) the star a finite distance from the horizon. The *dashed red line* is the mass $\tilde{m}$ and the *dot dashed black line* is where the star starts/ends. For both $f(r) = r^{-2}$, $\tilde{\beta} = 10$, $\Phi_0 = \sqrt{3}$.
(**a**) $\tilde{m} = \frac{1}{\sqrt{e^{2\Phi_0/\sqrt{3}} + e^{2\sqrt{3}\Phi_0}}}$,
$\tilde{B} = 1.28$, $\mu = 0.3$, $\rho = 1.28$, $\phi_1 = 7.49$, $\phi_2 = -30.86$.
(**b**) $\tilde{m} = 0.0498$, $\tilde{B} = 1.28$, $\mu = 0.3$, $\rho = 1.28$, $\phi_1 = 7.47$, $\phi_2 = -30.82$. In these examples, the star has a negligible effect on the overall charge of the system (Color figure online)

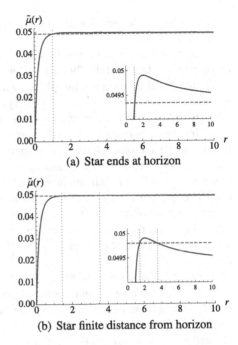

(a) Star ends at horizon

(b) Star finite distance from horizon

There is also the possibility of having a star that is a finite distance from the horizon. In this case, the IR expansion would be that of the dilaton–dyon in (20.49). It requires that

$$h_0 < \tilde{m} < \tilde{m}_{max}, \qquad \Phi_0 > \Phi_c, \tag{20.57}$$

where $\tilde{m}_{max}$ is the maximum mass after which the chemical potential never gets high enough to populate the star. An example is shown in Fig. 20.2(b).

20.4 Concluding Remarks

We have constructed three types of zero temperature electron star solutions with background magnetic field. Note that, including the solution without a star, our work does not address which of the four solution types is the thermodynamically preferred state for fixed chemical potential, dilaton source, and magnetic field. this is for future work. For the case of vanishing magnetic field, such a phase diagram at zero temperature was constructed [29], where it was observed that a Lifshitz fixed point separates the mesonic and partially fractionalized phases when the phase transition is continuous (the Lifshitz point is avoided in the case of the first order transition), and a third order transition separates the partially fractionalized phase from the fully fractionalized phase. Here, we have observed that an AdS$_2 \times \mathbb{R}^2$ geometry appears to separate the fractionalized phase from the fully fractionalized phase at finite magnetic field, and it would be interesting to study how this point affects the phase transition.

We note the absence of any Lifshitz solutions in our presentation of the various phases so far. In Ref. [29], the Lifshitz solution plays the important role of mediating between the mesonic phase (with no horizon) from the partially fractionalized phase (with an electrically charge horizon) at zero magnetic field. However, as indicated earlier, in the presence of a magnetic field, all the phases we study involve a horizon behind which the magnetic charge hides, and the transition from the mesonic phase to the fractionalized phase is a transition between an electrically neutral horizon and an electrically charged horizon. This transition may be softer, which might explain our inability to find a mediating solution. This picture is of course incomplete without an action calculation to study the exact nature of the transition between the two phases. (Note that since the dilaton diverges on the horizon, some care may be needed in the treatment of these solutions, perhaps by exploiting their M-theory uplift mentioned in the introduction.)

Another obvious generalization of our work would be to construct finite temperature solutions as was done for the electron star at zero magnetic field [39, 40]. Finally, it would be interesting to explore the transport properties of our systems, employing fluctuation analyses and computing the appropriate Green's functions. We will leave these matters for another day.

Acknowledgements   TA, CVJ, and SM would like to thank the US Department of Energy for support under grant DE-FG03-84ER-40168. TA is also supported by the USC Dornsife College of Letters, Arts and Sciences. We thank Kristan Jensen, Joe Polchinski, and Herman Verlinde for comments.

References

1. J.M. Maldacena, The large n limit of superconformal field theories and supergravity. Adv. Theor. Math. Phys. **2**, 231–252 (1998). hep-th/9711200
2. S.S. Gubser, I.R. Klebanov, A.M. Polyakov, Gauge theory correlators from non-critical string theory. Phys. Lett. B **428**, 105–114 (1998). hep-th/9802109
3. E. Witten, Anti-de Sitter space and holography. Adv. Theor. Math. Phys. **2**, 253–291 (1998). hep-th/9802150
4. O. Aharony, S.S. Gubser, J.M. Maldacena, H. Ooguri, Y. Oz, Large n field theories, string theory and gravity. Phys. Rept. **323**, 183–386 (2000). hep-th/9905111
5. E. Witten, Anti-de sitter space, thermal phase transition, and confinement in gauge theories. Adv. Theor. Math. Phys. **2**, 505–532 (1998). hep-th/9803131
6. J. McGreevy, Holographic duality with a view toward many-body physics. Adv. High Energy Phys. **2010**, 723105 (2010). arXiv:0909.0518 [hep-th]
7. S.A. Hartnoll, Lectures on holographic methods for condensed matter physics. Class. Quant. Grav. **26**, 224002 (2009). arXiv:0903.3246 [hep-th]
8. C.P. Herzog, Lectures on holographic superfluidity and superconductivity. J. Phys. A **42**, 343001 (2009). arXiv:0904.1975 [hep-th]
9. S. Sachdev, What can gauge-gravity duality teach us about condensed matter physics? Annu. Rev. Condens. Matter Phys. **3**(1), 9–33 (2012)

10. S.A. Hartnoll, Horizons, holography and condensed matter. arXiv:1106.4324 [hep-th]
11. A. Chamblin, R. Emparan, C.V. Johnson, R.C. Myers, Charged AdS black holes and catastrophic holography. Phys. Rev. D **60**, 064018 (1999). hep-th/9902170
12. A. Chamblin, R. Emparan, C.V. Johnson, R.C. Myers, Holography, thermodynamics and fluctuations of charged AdS black holes. Phys. Rev. D **60**, 104026 (1999). hep-th/9904197
13. S.A. Hartnoll, P. Kovtun, Hall conductivity from dyonic black holes. Phys. Rev. D **76**, 066001 (2007). arXiv:0704.1160 [hep-th]
14. S.A. Hartnoll, P.K. Kovtun, M. Muller, S. Sachdev, Theory of the Nernst effect near quantum phase transitions in condensed matter, and in dyonic black holes. Phys. Rev. B **76**, 144502 (2007). arXiv:0706.3215 [cond-mat.str-el]
15. S.A. Hartnoll, C.P. Herzog, Ohm's law at strong coupling: S duality and the cyclotron resonance. Phys. Rev. D **76**, 106012 (2007). arXiv:0706.3228 [hep-th]
16. S.-S. Lee, A Non-Fermi liquid from a charged black hole: a critical Fermi ball. Phys. Rev. D **79**, 086006 (2009). arXiv:0809.3402 [hep-th]
17. H. Liu, J. McGreevy, D. Vegh, Non-Fermi liquids from holography. arXiv:0903.2477 [hep-th]
18. M. Cubrovic, J. Zaanen, K. Schalm, String theory, quantum phase transitions, and the emergent Fermi liquid. Science **325**, 439–444 (2009). arXiv:0904.1993 [hep-th]
19. T. Faulkner, H. Liu, J. McGreevy, D. Vegh, Emergent quantum criticality, Fermi surfaces, and AdS2. Phys. Rev. D **83**, 125002 (2011). arXiv:0907.2694 [hep-th]
20. S.A. Hartnoll, A. Tavanfar, Electron stars for holographic metallic criticality. Phys. Rev. D **83**, 046003 (2011). arXiv:1008.2828 [hep-th]
21. S. Kachru, X. Liu, M. Mulligan, Gravity duals of Lifshitz-like fixed points. Phys. Rev. D **78**, 106005 (2008). arXiv:0808.1725 [hep-th]
22. S.W. Hawking, D.N. Page, Thermodynamics of black holes in anti-de Sitter space. Commun. Math. Phys. **87**, 577 (1983)
23. T. Faulkner, J. Polchinski, Semi-holographic Fermi liquids. arXiv:1001.5049 [hep-th]
24. S. Sachdev, Holographic metals and the fractionalized Fermi liquid. Phys. Rev. Lett. **105**, 151602 (2010). arXiv:1006.3794 [hep-th]
25. S.A. Hartnoll, D.M. Hofman, A. Tavanfar, Holographically smeared Fermi surface: quantum oscillations and Luttinger count in electron stars. Europhys. Lett. **95**, 31002 (2011). arXiv: 1011.2502 [hep-th]
26. N. Iqbal, H. Liu, M. Mezei, Semi-local quantum liquids. J. High Energy Phys. **1204**, 086 (2012). arXiv:1105.4621 [hep-th]
27. S. Sachdev, A model of a Fermi liquid using gauge-gravity duality. Phys. Rev. D **84**, 066009 (2011). arXiv:1107.5321 [hep-th]
28. L. Huijse, S. Sachdev, Fermi surfaces and gauge-gravity duality. Phys. Rev. D **84**, 026001 (2011). arXiv:1104.5022 [hep-th]
29. S.A. Hartnoll, L. Huijse, Fractionalization of holographic Fermi surfaces. arXiv:1111.2606 [hep-th]
30. M. Cvetic, S.S. Gubser, Thermodynamic stability and phases of general spinning branes. J. High Energy Phys. **07**, 010 (1999). hep-th/9903132
31. J.R. Oppenheimer, G.M. Volkoff, On massive neutron cores. Phys. Rev. **55**, 374–381 (1939). http://link.aps.org/doi/10.1103/PhysRev.55.374
32. R.C. Tolman, Static solutions of Einstein's field equations for spheres of fluid. Phys. Rev. **55**, 364–373 (1939). http://link.aps.org/doi/10.1103/PhysRev.55.364
33. J. de Boer, K. Papadodimas, E. Verlinde, Holographic neutron stars. J. High Energy Phys. **1010**, 020 (2010). arXiv:0907.2695 [hep-th]
34. P. Burikham, T. Chullaphan, Holographic magnetic star. J. High Energy Phys. **1206**, 021 (2012). arXiv:1203.0883 [hep-th]
35. S.S. Gubser, F.D. Rocha, Peculiar properties of a charged dilatonic black hole in AdS5. Phys. Rev. D **81**, 046001 (2010). arXiv:0911.2898 [hep-th]
36. J.D. Brown, Action functionals for relativistic perfect fluids. Class. Quant. Grav. **10**, 1579–1606 (1993). arXiv:gr-qc/9304026 [gr-qc]

37. L. Bombelli, R.J. Torrence, Perfect fluids and Ashtekar variables, with applications to Kantowski-Sachs models. Class. Quant. Grav. **7**(10), 1747 (1990). http://stacks.iop.org/0264-9381/7/i=10/a=008
38. R. de Ritis, M. Lavorgna, G. Platania, C. Stornaiolo, Charged spin fluid in the Einstein-Cartan theory. Phys. Rev. D **31**, 1854–1859 (1985). http://link.aps.org/doi/10.1103/PhysRevD.31.1854
39. V.G.M. Puletti, S. Nowling, L. Thorlacius, T. Zingg, Holographic metals at finite temperature. J. High Energy Phys. **1101**, 117 (2011). arXiv:1011.6261 [hep-th]
40. S.A. Hartnoll, P. Petrov, Electron star birth: a continuous phase transition at nonzero density. Phys. Rev. Lett. **106**, 121601 (2011). arXiv:1011.6469 [hep-th]

Chapter 21
Holographic Description of Strongly Correlated Electrons in External Magnetic Fields

E. Gubankova, J. Brill, M. Čubrović, K. Schalm, P. Schijven, and J. Zaanen

21.1 Introduction

The study of strongly interacting fermionic systems at finite density and temperature is a challenging task in condensed matter and high energy physics. Analytical methods are limited or not available for strongly coupled systems, and numerical simulation of fermions at finite density breaks down because of the sign problem [1, 2]. There has been an increased activity in describing finite density fermionic matter by a gravity dual using the holographic AdS/CFT correspondence [3]. The gravitational solution dual to the finite chemical potential system is the electrically charged AdS-Reissner-Nordström (RN) black hole, which provides a background where only the metric and Maxwell fields are nontrivial and all matter fields vanish.

E. Gubankova (✉)
ITP, J. W. Goethe-University, D-60438 Frankfurt am Main, Germany
e-mail: gubankova@th.physik.uni-frankfurt.de

E. Gubankova
ITEP, Moscow, Russia

J. Brill · M. Čubrović · K. Schalm · P. Schijven · J. Zaanen
Instituut Lorentz, Leiden University, Niels Bohrweg 2, 2300 RA Leiden, The Netherlands

J. Brill
e-mail: jellebrill@gmail.com

M. Čubrović
e-mail: cubrovic@lorentz.leidenuniv.nl

K. Schalm
e-mail: kschalm@lorentz.leidenuniv.nl

P. Schijven
e-mail: aphexedpiet@gmail.com

J. Zaanen
e-mail: jan@lorentz.leidenuniv.nl

D. Kharzeev et al. (eds.), *Strongly Interacting Matter in Magnetic Fields*,
Lecture Notes in Physics 871, DOI 10.1007/978-3-642-37305-3_21,
© Springer-Verlag Berlin Heidelberg 2013

In the classical gravity limit, the decoupling of the Einstein-Maxwell sector holds and leads to universal results, which is an appealing feature of applied holography. Indeed, the celebrated result for the ratio of the shear viscosity over the entropy density [4] is identical for many strongly interacting theories and has been considered a robust prediction of the AdS/CFT correspondence.

However, an extremal black hole alone is not enough to describe finite density systems as it does not source the matter fields. In holography, at leading order, the Fermi surfaces are not evident in the gravitational geometry, but can only be detected by external probes; either probe D-branes [3] or probe bulk fermions [5–8]. Here we shall consider the latter option, where the free Dirac field in the bulk carries a finite charge density [9]. We ignore electromagnetic and gravitational backreaction of the charged fermions on the bulk spacetime geometry (probe approximation). At large temperatures, $T \gg \mu$, this approach provides a reliable hydrodynamic description of transport at a quantum criticality (in the vicinity of superfluid-insulator transition) [10]. At small temperatures, $T \ll \mu$, in some cases sharp Fermi surfaces emerge with either conventional Fermi-liquid scaling [6] or of a non-Fermi liquid type [7] with scaling properties that differ significantly from those predicted by the Landau Fermi liquid theory. The non-trivial scaling behavior of these non-Fermi liquids has been studied semi-analytically in [8] and is of great interest as high-T_c superconductors and metals near the critical point are believed to represent non-Fermi liquids.

What we shall study is the effects of magnetic field on the holographic fermions. A magnetic field is a probe of finite density matter at low temperatures, where the Landau level physics reveals the Fermi level structure. The gravity dual system is described by a AdS dyonic black hole with electric and magnetic charges Q and H, respectively, corresponding to a $2 + 1$-dimensional field theory at finite chemical potential in an external magnetic field [11]. Probe fermions in the background of the dyonic black hole have been considered in [12–14]; and probe bosons in the same background have been studied in [15]. Quantum magnetism is considered in [16].

The Landau quantization of momenta due to the magnetic field found there, shows again that the AdS/CFT correspondence has a powerful capacity to unveil that certain quantum properties known from quantum gases have a much more ubiquitous status than could be anticipated theoretically. A first highlight is the demonstration [17] that the Fermi surface of the Fermi gas extends way beyond the realms of its perturbative extension in the form of the Fermi-liquid. In AdS/CFT it appears to be gravitationally encoded in the matching along the scaling direction between the 'bare' Dirac waves falling in from the 'UV' boundary, and the true IR excitations living near the black hole horizon. This IR physics can insist on the disappearance of the quasiparticle but, if so, this 'critical Fermi-liquid' is still organized 'around' a Fermi surface. The Landau quantization, the organization of quantum gaseous matter in quantized energy bands (Landau levels) in a system of two space dimensions pierced by a magnetic field oriented in the orthogonal spatial direction, is a second such quantum gas property. We shall describe here following [12], that despite the strong interactions in the system, the holographic computation reveals the same strict Landau-level quantization. Arguably, it is the mean-field nature imposed by large N limit inherent in AdS/CFT that explains this. The system is effectively non-interacting to first order in $1/N$. The Landau quantization is not manifest from the geometry, but as we show this statement is straightforwardly encoded in the symmetry correspondences associated with the conformal compactification of AdS on its flat boundary (i.e., in the UV CFT).

An interesting novel feature in strongly coupled systems arises from the fact that the background geometry is only sensitive to the total energy density $Q^2 + H^2$ contained in the electric and magnetic fields sourced by the dyonic black hole. Dialing up the magnetic field is effectively similar to a process where the dyonic black hole loses its electric charge. At the same time, the fermionic probe with charge q is essentially only sensitive to the Coulomb interaction gqQ. As shown in [12], one can therefore map a magnetic to a non-magnetic system with rescaled parameters (chemical potential, fermion charge) and same symmetries and equations of motion, as long as the Reissner-Nordström geometry is kept.

Translated to more experiment-compatible language, the above magnetic-electric mapping means that the spectral functions at nonzero magnetic field h are identical to the spectral function at $h = 0$ for a reduced value of the coupling constant (fermion charge) q, provided the probe fermion is in a Landau level eigenstate. A striking consequence is that the spectrum shows conformal invariance for arbitrarily high magnetic fields, as long as the system is at negligible to zero density. Specifically, a detailed analysis of the fermion spectral functions reveals that at strong magnetic fields the Fermi level structure changes qualitatively. There exists a critical magnetic field at which the Fermi velocity vanishes. Ignoring the Landau level quantization, we show that this corresponds to an effective tuning of the system from a regular Fermi liquid phase with linear dispersion and stable quasiparticles to a non-Fermi liquid with fractional power law dispersion and unstable excitations. This phenomenon can be interpreted as a transition from metallic phase to a "strange metal" at the critical magnetic field and corresponds to the change of the infrared conformal dimension from $\nu > 1/2$ to $\nu < 1/2$ while the Fermi momentum stays nonzero and the Fermi surface survives. Increasing the magnetic field further, this transition is followed by a "strange-metal"-conformal crossover and eventually, for very strong fields, the system always has near-conformal behavior where $k_F = 0$ and the Fermi surface disappears.

For some Fermi surfaces, this surprising metal-"strange metal" transition is not physically relevant as the system prefers to directly enter the conformal phase. Whether a fine tuned system exists that does show a quantum critical phase transition from a FL to a non-FL is determined by a Diophantine equation for the Landau quantized Fermi momentum as a function of the magnetic field. Perhaps these are connected to the magnetically driven phase transition found in AdS_5/CFT_4 [18]. We leave this subject for further work.

Overall, the findings of Landau quantization and "discharge" of the Fermi surface are in line with the expectations: both phenomena have been found in a vast array of systems [19] and are almost tautologically tied to the notion of a Fermi surface in a magnetic field. Thus we regard them also as a sanity check of the whole bottom-up approach of fermionic AdS/CFT [5–7, 17], giving further credit to the holographic Fermi surfaces as having to do with the real world.

Next we use the information of magnetic effects the Fermi surfaces extracted from holography to calculate the quantum Hall and longitudinal conductivities. Generally speaking, it is difficult to calculate conductivity holographically beyond the Einstein-Maxwell sector, and extract the contribution of holographic fermions. In the semiclassical approximation, one-loop corrections in the bulk setup involving charged fermions have been calculated [17]. In another approach, the backreaction of charged fermions on the gravity-Maxwell sector has been taken into account and incorporated in calculations of the electric conductivity [9]. We calculate the one-loop contribution on the CFT side, which is equivalent to the holographic one-loop calculations as long as vertex corrections do not modify physical dependencies of interest [17, 20]. As we dial the magnetic field, the Hall plateau transition happens when the Fermi surface moves through a Landau level. One can think of a difference between the Fermi energy and the energy of the Landau level as a gap, which vanishes at the transition point and the $2 + 1$-dimensional theory becomes scale invariant. In the holographic D3–D7 brane model of the quantum Hall effect, plateau transition occurs as D-branes move through one another [21, 22]. In the same model, a dissipation process has been observed as D-branes fall through the horizon of the black hole geometry, that is associated with the quantum Hall insulator transition. In the holographic fermion liquid setting, dissipation is present through interaction of fermions with the horizon of the black hole. We have also used the analysis of the conductivities to learn more about the metal-strange metal phase transition as well as the crossover back to the conformal regime at high magnetic fields.

We conclude with the remark that the findings summarized above are in fact somewhat puzzling when contrasted to the conventional picture of quantum Hall physics. It is usually stated that the quantum Hall effect requires three key ingredients: Landau quantization, quenched disorder[1] and (spatial) boundaries, i.e., a finite-sized sample [23]. The first brings about the quantization of conductivity, the second prevents the states from spilling between the Landau levels ensuring the existence of a gap and the last one in fact allows the charge transport to happen (as it is the boundary states that actually conduct). In our model, only the first condition is satisfied. The second is put by hand by assuming that the gap is automatically preserved, i.e. that there is no mixing between the Landau levels. There is, however, no physical explanation as to how the boundary states are implicitly taken into account by AdS/CFT.

We outline the holographic setting of the dyonic black hole geometry and bulk fermions in Sect. 21.2. In Sect. 21.3 we prove the conservation of conformal symmetry in the presence of the magnetic fields. Section 21.4 is devoted to the holographic fermion liquid, where we obtain the Landau level quantization, followed by a detailed study of the Fermi surface properties at zero temperature in Sect. 21.5. We calculate the DC conductivities in Sect. 21.6, and compare the results with available data in graphene.

---

[1]Quenched disorder means that the dynamics of the impurities is "frozen", i.e. they can be regarded as having infinite mass. When coupled to the Fermi liquid, they ensure that below some scale the system behaves as if consisting of non-interacting quasiparticles only.

21.2 Holographic Fermions in a Dyonic Black Hole

We first describe the holographic setup with the dyonic black hole, and the dynamics of Dirac fermions in this background. In this paper, we exclusively work in the probe limit, i.e., in the limit of large fermion charge q.

21.2.1 Dyonic Black Hole

We consider the gravity dual of 3-dimensional conformal field theory (CFT) with global $U(1)$ symmetry. At finite charge density and in the presence of magnetic field, the system can be described by a dyonic black hole in 4-dimensional anti-de Sitter space-time, AdS_4, with the current J_μ in the CFT mapped to a $U(1)$ gauge field A_M in AdS. We use $\mu, \nu, \rho, \ldots$ for the spacetime indices in the CFT and $M, N, \ldots$ for the global spacetime indices in AdS.

The action for a vector field A_M coupled to AdS_4 gravity can be written as

$$S_g = \frac{1}{2\kappa^2} \int d^4x \sqrt{-g} \left(\mathcal{R} + \frac{6}{R^2} - \frac{R^2}{g_F^2} F_{MN} F^{MN} \right), \qquad (21.1)$$

where g_F^2 is an effective dimensionless gauge coupling and R is the curvature radius of AdS_4. The equations of motion following from (21.1) are solved by the geometry corresponding to a dyonic black hole, having both electric and magnetic charge:

$$ds^2 = g_{MN} dx^M dx^N = \frac{r^2}{R^2} \left(-f dt^2 + dx^2 + dy^2 \right) + \frac{R^2}{r^2} \frac{dr^2}{f}. \qquad (21.2)$$

The redshift factor f and the vector field A_M reflect the fact that the system is at a finite charge density and in an external magnetic field:

$$f = 1 + \frac{Q^2 + H^2}{r^4} - \frac{M}{r^3},$$

$$A_t = \mu \left(1 - \frac{r_0}{r} \right), \qquad A_y = hx, \qquad A_x = A_r = 0, \qquad (21.3)$$

where Q and H are the electric and magnetic charge of the black hole, respectively. Here we chose the Landau gauge; the black hole chemical potential μ and the magnetic field h are given by

$$\mu = \frac{g_F Q}{R^2 r_0}, \qquad h = \frac{g_F H}{R^4}, \qquad (21.4)$$

with r_0 is the horizon radius determined by the largest positive root of the redshift factor $f(r_0) = 0$:

$$M = r_0^3 + \frac{Q^2 + H^2}{r_0}. \qquad (21.5)$$

The boundary of the *AdS* is reached for $r \to \infty$. The geometry described by (21.2)–(21.3) describes the boundary theory at finite density, i.e., a system in a charged medium at the chemical potential $\mu = \mu_{bh}$ and in transverse magnetic field $h = h_{bh}$, with charge, energy, and entropy densities given, respectively, by

$$\rho = 2\frac{Q}{\kappa^2 R^2 g_F}, \qquad \varepsilon = \frac{M}{\kappa^2 R^4}, \qquad s = \frac{2\pi}{\kappa^2}\frac{r_0^2}{R^2}. \tag{21.6}$$

The temperature of the system is identified with the Hawking temperature of the black hole, $T_H \sim |f'(r_0)|/4\pi$,

$$T = \frac{3r_0}{4\pi R^2}\left(1 - \frac{Q^2 + H^2}{3r_0^4}\right). \tag{21.7}$$

Since Q and H have dimensions of $[L]^2$, it is convenient to parametrize them as

$$Q^2 = 3r_*^4, \qquad Q^2 + H^2 = 3r_{**}^4. \tag{21.8}$$

In terms of r_0, r_* and r_{**} the above expressions become

$$f = 1 + \frac{3r_{**}^4}{r^4} - \frac{r_0^3 + 3r_{**}^4/r_0}{r^3}, \tag{21.9}$$

with

$$\mu = \sqrt{3}g_F\frac{r_*^2}{R^2 r_0}, \qquad h = \sqrt{3}g_F\frac{\sqrt{r_{**}^4 - r_*^4}}{R^4}. \tag{21.10}$$

The expressions for the charge, energy and entropy densities, as well as for the temperature are simplified as

$$\rho = \frac{2\sqrt{3}}{\kappa^2 g_F}\frac{r_*^2}{R^2}, \qquad \varepsilon = \frac{1}{\kappa^2}\frac{r_0^3 + 3r_{**}^4/r_0}{R^4}, \qquad s = \frac{2\pi}{\kappa^2}\frac{r_0^2}{R^2},$$
$$T = \frac{3}{4\pi}\frac{r_0}{R^2}\left(1 - \frac{r_{**}^4}{r_0^4}\right). \tag{21.11}$$

In the zero temperature limit, i.e., for an extremal black hole, we have

$$T = 0 \quad \to \quad r_0 = r_{**}, \tag{21.12}$$

which in the original variables reads $Q^2 + H^2 = 3r_0^4$. In the zero temperature limit (21.12), the redshift factor f as given by (21.9) develops a double zero at the horizon:

$$f = 6\frac{(r - r_{**})^2}{r_{**}^2} + \mathcal{O}\big((r - r_{**})^3\big). \tag{21.13}$$

As a result, near the horizon the AdS_4 metric reduces to $AdS_2 \times R^2$ with the curvature radius of AdS_2 given by

$$R_2 = \frac{1}{\sqrt{6}} R. \qquad (21.14)$$

This is a very important property of the metric, which considerably simplifies the calculations, in particular in the magnetic field.

In order to scale away the AdS_4 radius R and the horizon radius r_0, we introduce dimensionless variables

$$r \to r_0 r, \qquad r_* \to r_0 r_*, \qquad r_{**} \to r_0 r_{**},$$
$$M \to r_0^3 M, \qquad Q \to r_0^2 Q, \qquad H \to r_0^2 H, \qquad (21.15)$$

and

$$(t, \mathbf{x}) \to \frac{R^2}{r_0}(t, \mathbf{x}), \qquad A_M \to \frac{r_0}{R^2} A_M, \qquad \omega \to \frac{r_0}{R^2}\omega,$$

$$\mu \to \frac{r_0}{R^2}\mu, \qquad h \to \frac{r_0^2}{R^4}h, \qquad T \to \frac{r_0}{R^2}T, \qquad (21.16)$$

$$ds^2 \to R^2 ds^2.$$

Note that the scaling factors in the above equation that describes the quantities of the boundary field theory involve the curvature radius of AdS_4, not AdS_2.

In the new variables we have

$$T = \frac{3}{4\pi}(1 - r_{**}^4) = \frac{3}{4\pi}\left(1 - \frac{Q^2 + H^2}{3}\right), \qquad f = 1 + \frac{3r_{**}^4}{r^4} - \frac{1 + 3r_{**}^4}{r^3},$$
$$(21.17)$$

$$A_t = \mu\left(1 - \frac{1}{r}\right), \qquad \mu = \sqrt{3}g_F r_*^2 = g_F Q, \qquad h = g_F H,$$

and the metric is given by

$$ds^2 = r^2\left(-f dt^2 + dx^2 + dy^2\right) + \frac{1}{r^2}\frac{dr^2}{f}, \qquad (21.18)$$

with the horizon at $r = 1$, and the conformal boundary at $r \to \infty$.

At $T = 0$, r_{**} becomes unity, and the redshift factor develops the double zero near the horizon,

$$f = \frac{(r - 1)^2(r^2 + 2r + 3)}{r^4}. \qquad (21.19)$$

As mentioned before, due to this fact the metric near the horizon reduces to $AdS_2 \times R^2$ where the analytical calculations are possible for small frequencies [8]. However, in the chiral limit $m = 0$, analytical calculations are also possible in the bulk AdS_4 [24], which we utilize in this paper.

21.2.2 Holographic Fermions

To include the bulk fermions, we consider a spinor field ψ in the AdS_4 of charge q and mass m, which is dual to an operator $\mathcal{O}$ in the boundary CFT_3 of charge q and dimension

$$\Delta = \frac{3}{2} + mR, \tag{21.20}$$

with $mR \geq -\frac{1}{2}$ and in dimensionless units corresponds to $\Delta = \frac{3}{2} + m$. In the black hole geometry, (21.2), the quadratic action for ψ reads as

$$S_\psi = i \int d^4 x \sqrt{-g} (\bar{\psi} \Gamma^M \mathcal{D}_M \psi - m \bar{\psi} \psi), \tag{21.21}$$

where $\bar{\psi} = \psi^\dagger \Gamma^{\underline{t}}$, and

$$\mathcal{D}_M = \partial_M + \frac{1}{4} \omega_{abM} \Gamma^{ab} - iq A_M, \tag{21.22}$$

where ω_{abM} is the spin connection, and $\Gamma^{ab} = \frac{1}{2}[\Gamma^a, \Gamma^b]$. Here, M and a, b denote the bulk space-time and tangent space indices respectively, while μ, ν are indices along the boundary directions, i.e. $M = (r, \mu)$. Gamma matrix basis (Minkowski signature) is given in [8].

We will be interested in spectra and response functions of the boundary fermions in the presence of magnetic field. This requires solving the Dirac equation in the bulk [6, 7]:

$$(\Gamma^M \mathcal{D}_M - m) \psi = 0. \tag{21.23}$$

From the solution of the Dirac equation at small ω, an analytic expression for the retarded fermion Green's function of the boundary CFT at zero magnetic field has been obtained in [8]. Near the Fermi surface it reads as [8]:

$$G_R(\Omega, k) = \frac{(-h_1 v_F)}{\omega - v_F k_\perp - \Sigma(\omega, T)}, \tag{21.24}$$

where $k_\perp = k - k_F$ is the perpendicular distance from the Fermi surface in momentum space, h_1 and v_F are real constants calculated below, and the self-energy $\Sigma = \Sigma_1 + i \Sigma_2$ is given by [8]

$$\Sigma(\omega, T)/v_F = T^{2\nu} g\left(\frac{\omega}{T}\right) = (2\pi T)^{2\nu} h_2 e^{i\theta - i\pi\nu} \frac{\Gamma(\frac{1}{2} + \nu - \frac{i\omega}{2\pi T} + \frac{i\mu_q}{6})}{\Gamma(\frac{1}{2} - \nu - \frac{i\omega}{2\pi T} + \frac{i\mu_q}{6})}, \tag{21.25}$$

where ν is the zero temperature conformal dimension at the Fermi momentum, $\nu \equiv \nu_{k_F}$, given by (21.58), $\mu_q \equiv \mu q$, h_2 is a positive constant and the phase θ is such that the poles of the Green's function are located in the lower half of the complex frequency plane. These poles correspond to quasinormal modes of the Dirac equation (21.23) and they can be found numerically solving $F(\omega_*) = 0$ [25, 26], with

$$F(\omega) = \frac{k_\perp}{\Gamma(\frac{1}{2} + \nu - \frac{i\omega}{2\pi T} + \frac{i\mu_q}{6})} - \frac{h_2 e^{i\theta - i\pi\nu}(2\pi T)^{2\nu}}{\Gamma(\frac{1}{2} - \nu - \frac{i\omega}{2\pi T} + \frac{i\mu_q}{6})}, \qquad (21.26)$$

The solution gives the full motion of the quasinormal poles $\omega_*^{(n)}(k_\perp)$ in the complex ω plane as a function of $k_\perp$. It has been found in [8, 25, 26], that, if the charge of the fermion is large enough compared to its mass, the pole closest to the real ω axis bounces off the axis at $k_\perp = 0$ (and $\omega = 0$). Such behavior is identified with the existence of the Fermi momentum k_F indicative of an underlying strongly coupled Fermi surface.

At $T = 0$, the self-energy becomes $T^{2\nu} g(\omega/T) \to c_k \omega^{2\nu}$, and the Green's function obtained from the solution to the Dirac equation reads [8]

$$G_R(\Omega, k) = \frac{(-h_1 v_F)}{\omega - v_F k_\perp - h_2 v_F e^{i\theta - i\pi\nu} \omega^{2\nu}}, \qquad (21.27)$$

where $k_\perp = \sqrt{k^2 - k_F}$. The last term is determined by the IR AdS_2 physics near the horizon. Other terms are determined by the UV physics of the AdS_4 bulk.

The solutions to (21.23) have been studied in detail in [6–8]. Here we simply summarize the novel aspects due to the background magnetic field [27]

- The background magnetic field h introduces a discretization of the momentum:

$$k \to k_{\text{eff}} = \sqrt{2|qh|l}, \quad \text{with } l \in N, \qquad (21.28)$$

with Landau level index l [13, 14, 25, 26]. These discrete values of k are the analogue of the well-known Landau levels that occur in magnetic systems.
- There exists a (non-invertible) mapping on the level of Green's functions, from the magnetic system to the non-magnetic one by sending

$$(H, Q, q) \mapsto \left(0, \sqrt{Q^2 + H^2}, q\sqrt{1 - \frac{H^2}{Q^2 + H^2}}\right). \qquad (21.29)$$

The Green's functions in a magnetic system are thus equivalent to those in the absence of magnetic fields. To better appreciate that, we reformulate (21.29) in terms of the boundary quantities:

$$(h, \mu_q, T) \mapsto \left(0, \mu_q, T\left(1 - \frac{h^2}{12\mu^2}\right)\right), \qquad (21.30)$$

where we used dimensionless variables defined in (21.15), (21.17). The magnetic field thus effectively decreases the coupling constant q and increases the chemical potential $\mu = g_F Q$ such that the combination $\mu_q \equiv \mu q$ is preserved [12]. This is an important point as the equations of motion actually only depend on this combination and not on μ and q separately [12]. In other words, (21.30) implies that the additional scale brought about by the magnetic field can be understood as changing μ and T independently in the effective non-magnetic system instead of only tuning the ratio μ/T. This point is important when considering the thermo-dynamics.

- The discrete momentum $k_{\text{eff}} = \sqrt{2|qh|l}$ must be held fixed in the transformation (21.29). The bulk-boundary relation is particularly simple in this case, as the Landau levels can readily be seen in the bulk solution, only to remain identical in the boundary theory.

- Similar to the non-magnetic system [12], the IR physics is controlled by the near horizon $AdS_2 \times R^2$ geometry, which indicates the existence of an IR CFT, characterized by operators $\mathcal{O}_l, l \in N$ with operator dimensions $\delta = 1/2 + \nu_l$:

$$\nu_l = \frac{1}{6}\sqrt{6\left(m^2 + \frac{2|qh|l}{r_{**}^2}\right) - \frac{\mu_q^2}{r_{**}^4}}, \qquad (21.31)$$

in dimensionless notation, and $\mu_q \equiv \mu q$. At $T = 0$, when $r_{**} = 1$, it becomes

$$\nu_l = \frac{1}{6}\sqrt{6(m^2 + 2|qh|l) - \mu_q^2}. \qquad (21.32)$$

The Green's function for these operators $\mathcal{O}_l$ is found to be $\mathcal{G}_l^R(\omega) \sim \omega^{2\nu_l}$ and the exponents ν_l determines the dispersion properties of the quasiparticle excitations. For $\nu > 1/2$ the system has a stable quasiparticle and a linear dispersion, whereas for $\nu \le 1/2$ one has a non-Fermi liquid with power-law dispersion and an unstable quasiparticle.

21.3 Magnetic Fields and Conformal Invariance

Despite the fact that a magnetic field introduces a scale, in the absence of a chemical potential, all spectral functions are essentially still determined by conformal symmetry. To show this, we need to establish certain properties of the near-horizon geometry of a Reissner-Nordström black hole. This leads to the AdS_2 perspective that was developed in [8]. The result relies on the conformal algebra and its relation to the magnetic group, from the viewpoint of the infrared CFT that was studied in [8]. Later on we will see that the insensitivity to the magnetic field also carries over to AdS_4 and the UV CFT in some respects. To simplify the derivations, we consider the case $T = 0$.

21.3.1 The Near-Horizon Limit and Dirac Equation in AdS₂

It was established in [8] that an electrically charged extremal AdS-Reissner-Nordström black hole has an AdS_2 throat in the inner bulk region. This conclusion carries over to the magnetic case with some minor differences. We will now give a quick derivation of the AdS_2 formalism for a dyonic black hole, referring the reader to [8] for more details (that remain largely unchanged in the magnetic field).

Near the horizon $r = r_{**}$ of the black hole described by the metric (21.2), the redshift factor $f(r)$ develops a double zero:

$$f(r) = 6\frac{(r - r_{**})^2}{r_{**}^2} + \mathcal{O}\big((r - r_{**})^3\big). \tag{21.33}$$

Now consider the scaling limit

$$r - r_{**} = \lambda \frac{R_2^2}{\zeta}, \qquad t = \lambda^{-1}\tau, \qquad \lambda \to 0 \text{ with } \tau, \zeta \text{ finite}. \tag{21.34}$$

In this limit, the metric (21.2) and the gauge field reduce to

$$ds^2 = \frac{R_2^2}{\zeta^2}(-d\tau^2 + d\zeta^2) + \frac{r_{**}^2}{R^2}(dx^2 + dy^2),$$

$$A_\tau = \frac{\mu R_2^2 r_0}{r_{**}^2}\frac{1}{\zeta}, \qquad A_x = Hx \tag{21.35}$$

where $R_2 = \frac{R}{\sqrt{6}}$. The geometry described by this metric is indeed $AdS_2 \times R^2$. Physically, the scaling limit given in (21.34) with finite τ corresponds to the long time limit of the original time coordinate t, which translates to the low frequency limit of the boundary theory:

$$\frac{\omega}{\mu} \to 0, \tag{21.36}$$

where ω is the frequency conjugate to t. (One can think of λ as being the frequency ω.) Near the AdS_4 horizon, we expect the AdS_2 region of an extremal dyonic black hole to have a CFT_1 dual. We refer to [8] for an account of this AdS_2/CFT_1 duality. The horizon of AdS_2 region is at $\zeta \to \infty$ (coefficient in front of $d\tau$ vanishes at the horizon in (21.35)) and the infrared CFT (IR CFT) lives at the AdS_2 boundary at $\zeta = 0$. The scaling picture given by (21.34)–(21.35) suggests that in the low frequency limit, the 2-dimensional boundary theory is described by this IR CFT (which is a CFT_1). The Green's function for the operator $\mathcal{O}$ in the boundary theory is obtained through a small frequency expansion and a matching procedure between the two different regions (inner and outer) along the radial direction, and can be expressed through the Green's function of the IR CFT [8].

The explicit form for the Dirac equation in the magnetic field is of little interest for the analytical results that follow. It can be found in [27]. Of primary interest is its limit in the IR region with metric given by (21.35):

$$\left(-\frac{1}{\sqrt{g_{\zeta\zeta}}}\sigma^3\partial_\zeta - m + \frac{1}{\sqrt{-g_{\tau\tau}}}\sigma^1\left(\omega + \frac{\mu_q R_2^2 r_0}{r_{**}^2\zeta}\right) - \frac{1}{\sqrt{g_{ii}}}i\sigma^2\lambda_l\right)F^{(l)} = 0, \tag{21.37}$$

where the effective momentum of the lth Landau level is $\lambda_l = \sqrt{2|qh|l}$, $\mu_q \equiv \mu q$ and we omit the index of the spinor field. To obtain (21.37), it is convenient to pick the gamma matrix basis as $\Gamma^{\hat{\zeta}} = -\sigma_3$, $\Gamma^{\hat{t}} = i\sigma_1$ and $\Gamma^{\hat{i}} = -\sigma_2$. We can write explicitly:

$$\begin{pmatrix} \frac{\zeta}{R_2}\partial_\zeta + m & -\frac{\zeta}{R_2}\left(\omega + \frac{\mu_q R_2^2 r_0}{r_{**}^2 \zeta}\right) + \frac{R}{r_{**}}\lambda_l \\ \frac{\zeta}{R_2}\left(\omega + \frac{\mu_q R_2^2 r_0}{r_{**}^2 \zeta}\right) + \frac{R}{r_{**}}\lambda_l & \frac{\zeta}{R_2}\partial_\zeta - m \end{pmatrix}\begin{pmatrix} y \\ z \end{pmatrix} = 0. \quad (21.38)$$

Note that the AdS_2 radius R_2 enters for the (τ, ζ) directions. At the AdS_2 boundary, $\zeta \to 0$, the Dirac equation to the leading order is given by

$$\zeta\partial_\zeta F^{(l)} = -U F^{(l)}, \quad U = R_2\begin{pmatrix} m & -\frac{\mu_q R_2 r_0}{r_{**}^2} + \frac{R}{r_{**}}\lambda_l \\ \frac{\mu_q R_2 r_0}{r_{**}^2} + \frac{R}{r_{**}}\lambda_l & -m \end{pmatrix}. \quad (21.39)$$

The solution to this equation is given by the scaling function $F^{(l)} = Ae_+\zeta^{-\nu_l} + Be_-\zeta^{\nu_l}$ where $e_\pm$ are the real eigenvectors of U and the exponent is

$$\nu_l = \frac{1}{6}\sqrt{6\left(m^2 + \frac{R^2}{r_{**}^2}2|qh|l\right)R^2 - \frac{\mu_q^2 R^4 r_0^2}{r_{**}^4}}. \quad (21.40)$$

The conformal dimension of the operator $\mathcal{O}$ in the IR CFT is $\delta_l = \frac{1}{2} + \nu_l$. Comparing (21.40) to the expression for the scaling exponent in [8], we conclude that the scaling properties and the AdS_2 construction are unmodified by the magnetic field, except that the scaling exponents are now fixed by the Landau quantization. This "quantization rule" was already exploited in [25, 26] to study de Haas-van Alphen oscillations.

21.4 Spectral Functions

In this section we will explore some of the properties of the spectral function, in both plane wave and Landau level basis. We first consider some characteristic cases in the plane wave basis and make connection with the ARPES measurements.

21.4.1 Relating to the ARPES Measurements

In reality, ARPES measurements cannot be performed in magnetic fields so the holographic approach, allowing a direct insight into the propagator structure and the spectral function, is especially helpful. This follows from the observation that the spectral functions as measured in ARPES are always expressed in the plane wave basis of the photon, thus in a magnetic field, when the momentum is not a good quantum number anymore, it becomes impossible to perform the photoemission spectroscopy.

In order to compute the spectral function, we have to choose a particular fermionic plane wave as a probe. Since the separation of variables is valid throughout the bulk, the basis transformation can be performed at every constant r-slice. This means that only the x and y coordinates have to be taken into account (the plane wave probe lives only at the CFT side of the duality). We take a plane wave propagating in the $+x$ direction with spin up along the r-axis. In its rest frame such a particle can be described by

$$\Psi_{\text{probe}} = e^{i\omega t - i p_x x} \begin{pmatrix} \xi \\ \xi \end{pmatrix}, \quad \xi = \begin{pmatrix} 1 \\ 0 \end{pmatrix}. \tag{21.41}$$

Near the boundary (at $r_b \to \infty$) we can rescale our solutions of the Dirac equation, details can be found in [27]:

$$F_l = \begin{pmatrix} \zeta_l^{(1)}(\tilde{x}) \\ \xi_+^{(l)}(r_b)\zeta_l^{(1)}(\tilde{x}) \\ \zeta_l^{(2)}(\tilde{x}) \\ -\xi_+^{(l)}(r_b)\zeta_l^{(2)}(\tilde{x}) \end{pmatrix}, \quad \tilde{F}_l = \begin{pmatrix} \zeta_l^{(1)}(\tilde{x}) \\ \xi_-^{(l)}(r_b)\zeta_l^{(1)}(\tilde{x}) \\ -\zeta_l^{(2)}(\tilde{x}) \\ \xi_-^{(l)}(r_b)\zeta_l^{(2)}(\tilde{x}) \end{pmatrix}, \tag{21.42}$$

with rescaled $\tilde{x}$ defined in [27]. This representation is useful since we calculate the components $\xi_\pm(r_b)$ related to the retarded Green's function in our numerics (we keep the notation of [8]).

Let $\mathcal{O}_l$ and $\tilde{\mathcal{O}}_l$ be the CFT operators dual to respectively F_l and $\tilde{F}_l$, and $c_k^\dagger$, c_k be the creation and annihilation operators for the plane wave state Ψ_{probe}. Since the states F and $\tilde{F}$ form a complete set in the bulk, we can write

$$c_p^\dagger(\omega) = \sum_l (U_l^*, \tilde{U}_l^*) \begin{pmatrix} \mathcal{O}_l^\dagger(\omega) \\ \tilde{\mathcal{O}}_l^\dagger(\omega) \end{pmatrix} = \sum_l (U_l^* \mathcal{O}_l^\dagger(\omega) + \tilde{U}_l^* \tilde{\mathcal{O}}_l^\dagger(\omega)) \tag{21.43}$$

where the overlap coefficients $U_l(\omega)$ are given by the inner product between Ψ_{probe} and F:

$$U_l(p_x) = \int dx \bar{F}_l^\dagger i \Gamma^0 \Psi_{\text{probe}} = -\int dx e^{-i p_x x} \xi_+(r_b) \left(\zeta_l^{(1)\dagger}(\tilde{x}) - \zeta_l^{(2)\dagger}(\tilde{x}) \right), \tag{21.44}$$

with $\bar{F} = F^\dagger i \Gamma^0$, and similar expression for $\tilde{U}_l$ involving $\xi_-(r_b)$. The constants U_l can be calculated analytically using the numerical value of $\xi_\pm(r_b)$, and by noting that the Hermite functions are eigenfunctions of the Fourier transform. We are interested in the retarded Green's function, defined as

$$G_{\mathcal{O}_l}^R(\omega, p) = -i \int d^x dt e^{i\omega t - i p \cdot x} \theta(t) G_{\mathcal{O}_l}^R(t, x)$$

$$G_{\mathcal{O}_l}^R(t, x) = \langle 0 | [\mathcal{O}_l(t, x), \bar{\mathcal{O}}_l(0, 0)] | 0 \rangle \tag{21.45}$$

$$G^R = \begin{pmatrix} G_\mathcal{O} & 0 \\ 0 & \tilde{G}_\mathcal{O} \end{pmatrix},$$

where $\tilde{G}_\mathcal{O}$ is the retarded Green's function for the operator $\tilde{\mathcal{O}}$.

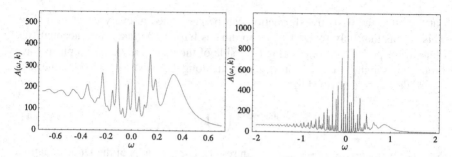

Fig. 21.1 Two examples of spectral functions in the plane wave basis for $\mu/T = 50$ and $h/T = 1$. The conformal dimension is $\Delta = 5/4$ (*left*) and $\Delta = 3/2$ (*right*). Frequency is in the units of effective temperature T_{eff}. The plane wave momentum is chosen to be $k = 1$. Despite the convolution of many Landau levels, the presence of the discrete levels is obvious

Exploiting the orthogonality of the spinors created by $\mathcal{O}$ and $\mathcal{O}^\dagger$ and using (21.43), the Green's function in the plane wave basis can be written as

$$G^R_{c_p}(\omega, p_x) = \sum_l \text{tr}\begin{pmatrix} U \\ \tilde{U} \end{pmatrix}(U^*, \tilde{U}^*)G^R$$

$$= \left(|U_l(p_x)|^2 G^R_{\mathcal{O}_l}(\omega, l) + |\tilde{U}_l(p_x)|^2 \tilde{G}^R_{\mathcal{O}_l}(\omega, l)\right). \qquad (21.46)$$

In practice, we cannot perform the sum in (21.46) all the way to infinity, so we have to introduce a cutoff Landau level l_{cut}. In most cases we are able to make l_{cut} large enough that the behavior of the spectral function is clear.

Using the above formalism, we have produced spectral functions for two different conformal dimensions and fixed chemical potential and magnetic field (Fig. 21.1). Using the plane wave basis allows us to directly detect the Landau levels. The unit used for plotting the spectra (here and later on in the paper) is the effective temperature T_{eff} [6]:

$$T_{\text{eff}} = \frac{T}{2}\left(1 + \sqrt{1 + \frac{3\mu^2}{(4\pi T)^2}}\right). \qquad (21.47)$$

This unit interpolates between μ at $T/\mu = 0$ and T and is of or $T/\mu \to \infty$, and is convenient for the reason that the relevant quantities (e.g., Fermi momentum) are of order unity for any value of μ and h.

21.4.2 Magnetic Crossover and Disappearance of the Quasiparticles

Theoretically, it is more convenient to consider the spectral functions in the Landau level basis. For definiteness let us pick a fixed conformal dimension $\Delta = \frac{5}{4}$ which corresponds to $m = -\frac{1}{4}$. In the limit of weak magnetic fields, $h/T \to 0$, we should reproduce the results that were found in [6].

In Fig. 21.2(A) we indeed see that the spectral function, corresponding to a low value of μ/T, behaves as expected for a nearly conformal system. The spectral function is approximately symmetric about $\omega = 0$, it vanishes for $|\omega| < k$, up to a small residual tail due to finite temperature, and for $|\omega| \gg k$ it scales as ω^{2m}.

In Fig. 21.2(B), which corresponds to a high value of μ/T, we see the emergence of a sharp quasiparticle peak. This peak becomes the sharpest when the Landau level l corresponding to an effective momentum $k_{\text{eff}} = \sqrt{2|qh|l}$ coincides with the Fermi momentum k_F. The peaks also broaden out when k_{eff} moves away from k_F. A more complete view of the Landau quantization in the quasiparticle regime is given in Fig. 21.3, where we plot the dispersion relation (ω–k map). Both the sharp peaks and the Landau levels can be visually identified.

Collectively, the spectra in Fig. 21.2 show that conformality is only broken by the chemical potential μ and not by the magnetic field. Naively, the magnetic field introduces a new scale in the system. However, this scale is absent from the spectral functions, visually validating the discussion in the previous section that the scale h can be removed by a rescaling of the temperature and chemical potential.

One thus concludes that there is some value h_c' of the magnetic field, depending on μ/T, such that the spectral function loses its quasiparticle peaks and displays near-conformal behavior for $h > h_c'$. The nature of the transition and the underlying mechanism depends on the parameters (μ_q, T, Δ). One mechanism, obvious from the rescaling in (21.29), is the reduction of the effective coupling q as h increases. This will make the influence of the scalar potential A_0 negligible and push the system back toward conformality. Generically, the spectral function shows no sharp change but is more indicative of a crossover.

A more interesting phenomenon is the disappearance of coherent quasiparticles at high effective chemical potentials. For the special case $m = 0$, we can go beyond numerics and study this transition analytically, combining the exact $T = 0$ solution found in [24] and the mapping (21.30). In the next section, we will show that the transition is controlled by the change in the dispersion of the quasiparticle and corresponds to a sharp phase transition. Increasing the magnetic field leads to a decrease in phenomenological control parameter ν_{k_F}. This can give rise to a transition to a non-Fermi liquid when $\nu_{k_F} \leq 1/2$, and finally to the conformal regime at $h = h_c'$ when $\nu_{k_F} = 0$ and the Fermi surface vanishes.

21.4.3 Density of States

As argued at the beginning of this section, the spectral function can look quite different depending on the particular basis chosen. Though the spectral function is an attractive quantity to consider due to connection with ARPES experiments, we will also direct our attention to basis-independent and manifestly gauge invariant quantities. One of them is the density of states (DOS), defined by

$$D(\omega) = \sum_l A(\omega, l), \tag{21.48}$$

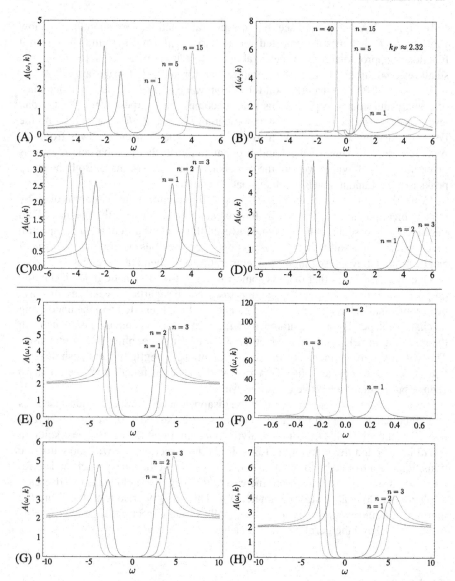

Fig. 21.2 Some typical examples of spectral functions $A(\omega, k_{\text{eff}})$ vs. ω in the Landau basis, $k_{\text{eff}} = \sqrt{2|qh|n}$. The *top four* correspond to a conformal dimension $\Delta = \frac{5}{4}$ $m = -\frac{1}{4}$ and the *bottom four* to $\Delta = \frac{3}{2}$ ($m = 0$). In each plot we show different Landau levels, labelled by index n, as a function of μ/T and h/T. The ratios take values $(\mu/T, h/T) = (1, 1), (50, 1), (1, 50), (50, 50)$ from left to right. Conformal case can be identified when μ/T is small regardless of h/T (plots in the *left panel*). Nearly conformal behavior is seen when both μ/T and h/T are large. This confirms our analytic result that the behavior of the system is primarily governed by μ. Departure from the conformality and sharp quasiparticle peaks are seen when μ/T is large and h/T is small in 21.2(B) and 21.2(F). Multiple quasiparticle peaks arise whenever $k_{\text{eff}} = k_F$. This suggests the existence of a critical magnetic field, beyond which the quasiparticle description becomes invalid and the system exhibits a conformal-like behavior. As before, the frequency ω is in units of T_{eff}

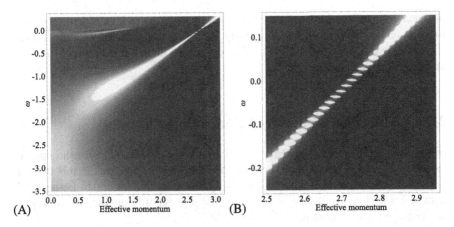

Fig. 21.3 Dispersion relation ω vs. k_{eff} for $\mu/T = 50$, $h/T = 1$ and $\Delta = \frac{5}{4}$ ($m = -\frac{1}{4}$). The spectral function $A(\omega, k_{\text{eff}})$ is displayed as a density plot. (**A**) On a large energy and momentum scale, we clearly sees that the peaks disperse almost linearly ($\omega \approx v_F k$), indicating that we are in the stable quasiparticle regime. (**B**) A zoom-in near the location of the Fermi surface shows clear Landau quantization

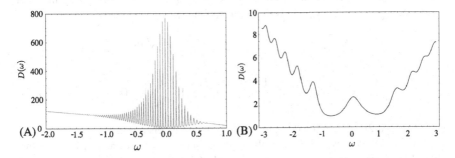

Fig. 21.4 Density of states $D(\omega)$ for $m = -\frac{1}{4}$ and (**A**) $\mu/T = 50$, $h/T = 1$, and (**B**) $\mu/T = 1$, $h/T = 1$. Sharp quasiparticle peaks from the splitting of the Fermi surface are clearly visible in (**A**). The case (**B**) shows square-root level spacing characteristic of a (nearly) Lorentz invariant spectrum such as that of graphene

where the usual integral over the momentum is replaced by a sum since only discrete values of the momentum are allowed.

In Fig. 21.4, we plot the density of states for two systems. We clearly see the Landau splitting of the Fermi surface. A peculiar feature of these plots is that the DOS seems to grow for negative values of ω. This, however, is an artefact of our calculation. Each individual spectrum in the sum (21.48) has a finite tail that scales as ω^{2m} for large ω, so each term has a finite contribution for large values of ω. When the full sum is performed, this fact implies that $\lim_{\omega \to \infty} D(\omega) \to \infty$. The relevant information on the density of states can be obtained by regularizing the sum, which in practice is done by summing over a finite number of terms only, and then considering the peaks that lie on top of the resulting finite-sized envelope. The physical point in Fig. 21.4(A) is the linear spacing of Landau levels, corresponding to a non-relativistic system at finite density. This is to be contrasted with Fig. 21.4B where the level spacing behaves as $\propto \sqrt{h}$, appropriate for a Lorentz invariant system and realized in graphene [28].

21.5 Fermi Level Structure at Zero Temperature

In this section, we solve the Dirac equation in the magnetic field for the special case $m = 0$ ($\Delta = \frac{3}{2}$). Although there are no additional symmetries in this case, it is possible to get an analytic solution. Using this solution, we obtain Fermi level parameters such as k_F and v_F and consider the process of filling the Landau levels as the magnetic field is varied.

21.5.1 Dirac Equation with $m = 0$

In the case $m = 0$, it is convenient to solve the Dirac equation including the spin connection (see details in [27]) rather than scaling it out:

$$\left(-\frac{\sqrt{g_{ii}}}{\sqrt{g_{rr}}} \sigma^1 \partial_r - \frac{\sqrt{g_{ii}}}{\sqrt{-g_{tt}}} \sigma^3 (\omega + q A_t) + \frac{\sqrt{g_{ii}}}{\sqrt{-g_{tt}}} \sigma^1 \frac{1}{2} \omega_{\hat{t}\hat{r}t} \right.$$
$$\left. - \sigma^1 \frac{1}{2} \omega_{\hat{x}\hat{r}x} - \sigma^1 \frac{1}{2} \omega_{\hat{y}\hat{r}y} - \lambda_l \right) \otimes 1 \begin{pmatrix} \psi_1 \\ \psi_2 \end{pmatrix} = 0, \qquad (21.49)$$

where $\lambda_l = \sqrt{2|qh|l}$ are the energies of the Landau levels $l = 0, 1, \ldots$, $g_{ii} \equiv g_{xx} = g_{yy}$, $A_t(r)$ is given by (21.3), and the gamma matrices are defined in [27]. In this basis the two components ψ_1 and ψ_2 decouple. Therefore, in what follows we solve for the first component only (we omit index 1). Substituting the spin connection, we have [20]:

$$\left(-\frac{r^2 \sqrt{f}}{R^2} \sigma^1 \partial_r - \frac{1}{\sqrt{f}} \sigma^3 (\omega + q A_t) - \sigma^1 \frac{r\sqrt{f}}{2R^2} \left(3 + \frac{rf'}{2f} \right) - \lambda_l \right) \psi = 0, \quad (21.50)$$

with $\psi = (y_1, y_2)$. It is convenient to change to the basis

$$\begin{pmatrix} \tilde{y}_1 \\ \tilde{y}_2 \end{pmatrix} = \begin{pmatrix} 1 & -i \\ -i & 1 \end{pmatrix} \begin{pmatrix} y_1 \\ y_2 \end{pmatrix}, \qquad (21.51)$$

which diagonalizes the system into a second order differential equation for each component. We introduce the dimensionless variables as in (21.15)–(21.17), and make a change of the dimensionless radial variable:

$$r = \frac{1}{1-z}, \qquad (21.52)$$

with the horizon now being at $z = 0$, and the conformal boundary at $z = 1$. Performing these transformations in (21.50), the second order differential equations for $\tilde{y}_1$ reads

$$\left(f \partial_z^2 + \left(\frac{3f}{1-z} + f' \right) \partial_z + \frac{15f}{4(1-z)^2} + \frac{3f'}{2(1-z)} + \frac{f''}{4} \right.$$

$$\left. + \frac{1}{f} \left((\omega + q\mu z) \pm \frac{if'}{4} \right)^2 - iq\mu - \lambda_i^2 \right) \tilde{y}_1 = 0. \qquad (21.53)$$

The second component $\tilde{y}_2$ obeys the same equation with $\mu \mapsto -\mu$.

At $T = 0$,

$$f = 3z^2(z - z_0)(z - \bar{z}_0), \quad z_0 = \frac{1}{3}(4 + i\sqrt{2}). \qquad (21.54)$$

The solution of this fermion system at zero magnetic field and zero temperature $T = 0$ has been found in [24]. To solve (21.53), we use the mapping to a zero magnetic field system (21.29). The combination $\mu_q \equiv \mu q$ at non-zero h maps to $\mu_{q,\mathrm{eff}} \equiv \mu_{\mathrm{eff}} q_{\mathrm{eff}}$ at zero h as follows:

$$\mu_q \mapsto q \sqrt{1 - \frac{H^2}{Q^2 + H^2}} \cdot g_F \sqrt{Q^2 + H^2} = \sqrt{3} q g_F \sqrt{1 - \frac{H^2}{3}} = \mu_{q,\mathrm{eff}} \qquad (21.55)$$

where at $T = 0$ we used $Q^2 + H^2 = 3$. We solve (21.53) for zero modes, i.e. $\omega = 0$, and at the Fermi surface $\lambda = k$, and implement (21.55).

Near the horizon ($z = 0$, $f = 6z^2$), we have

$$6z^2 \tilde{y}_{1;2}'' + 12z \tilde{y}_{1;2}' + \left(\frac{3}{2} + \frac{(\mu_{q,\mathrm{eff}})^2}{6} - k_F^2 \right) \tilde{y}_{1;2} = 0, \qquad (21.56)$$

which gives the following behavior:

$$\tilde{y}_{1;2} \sim z^{-\frac{1}{2} \pm \nu_k}, \qquad (21.57)$$

with the scaling exponent ν following from (21.32):

$$\nu = \frac{1}{6} \sqrt{6k^2 - (\mu_{q,\mathrm{eff}})^2}, \qquad (21.58)$$

at the momentum k. Using Maple, we find the zero mode solution of (21.53) with a regular behavior $z^{-\frac{1}{2} + \nu}$ at the horizon [20, 24]:

$$\tilde{y}_1^{(0)} = N_1 (z - 1)^{\frac{3}{2}} z^{-\frac{1}{2} + \nu} (z - \bar{z}_0)^{-\frac{1}{2} - \nu} \left(\frac{z - z_0}{z - \bar{z}_0} \right)^{\frac{1}{4}(-1 - \sqrt{2} \mu_{q,\mathrm{eff}}/z_0)}$$

$$\times {}_2F_1 \left(\frac{1}{2} + \nu - \frac{\sqrt{2}}{3} \mu_{q,\mathrm{eff}}, \nu + i \frac{\mu_{q,\mathrm{eff}}}{6}, 1 + 2\nu, \frac{2i\sqrt{2}z}{3z_0(z - \bar{z}_0)} \right), \qquad (21.59)$$

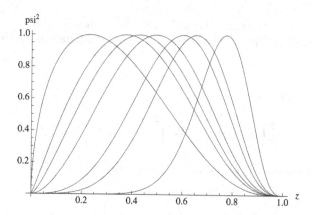

Fig. 21.5 Density of the zero mode $\psi^{0\dagger}\psi^0$ vs. the radial coordinate z (the horizon is at $z = 0$ and the boundary is at $z = 1$) for different values of the magnetic field h for the first (with the largest root for k_F) Fermi surface. We set $g_F = 1$ ($h \to H$) and $q = \frac{15}{\sqrt{3}}$ ($\mu_{q,\text{eff}} \to 15\sqrt{1 - \frac{H^2}{3}}$). From right to left the values of the magnetic field are $H = \{0, 1.40, 1.50, 1.60, 1.63, 1.65, 1.68\}$. The amplitudes of the *curves* are normalized to unity. At weak magnetic fields, the wave function is supported away from the horizon while at strong fields it is supported near the horizon

and

$$\tilde{y}_2^{(0)} = N_2(z - 1)^{\frac{3}{2}} z^{-\frac{1}{2}+\nu}(z - \bar{z}_0)^{-\frac{1}{2}-\nu}\left(\frac{z - z_0}{z - \bar{z}_0}\right)^{\frac{1}{4}(-1+\sqrt{2}\mu_{q,\text{eff}}/z_0)}$$

$$\times {}_2F_1\left(\frac{1}{2} + \nu + \frac{\sqrt{2}}{3}\mu_{q,\text{eff}}, \nu - i\frac{\mu_{q,\text{eff}}}{6}, 1 + 2\nu, \frac{2i\sqrt{2}z}{3z_0(z - \bar{z}_0)}\right), \quad (21.60)$$

where ${}_2F_1$ is the hypergeometric function and N_1, N_2 are normalization factors. Since normalization factors are constants, we find their relative weight by substituting solutions given in (21.59) back into the first order differential equations at $z \sim 0$,

$$\frac{N_1}{N_2} = -\frac{6i\nu + \mu_{q,\text{eff}}}{\sqrt{6}k}\left(\frac{z_0}{\bar{z}_0}\right)^{\mu_{q,\text{eff}}/\sqrt{2}z_0}. \quad (21.61)$$

The same relations are obtained when calculations are done for any z. The second solution $\tilde{\eta}_{1;2}^{(0)}$, with behavior $z^{-\frac{1}{2}-\nu}$ at the horizon, is obtained by replacing $\nu \to -\nu$ in (21.59).

To get insight into the zero-mode solution (21.59), we plot the radial profile for the density function $\psi^{(0)\dagger}\psi^{(0)}$ for different magnetic fields in Fig. 21.5. The momentum chosen is the Fermi momentum of the first Fermi surface (see the next section). The curves are normalized to have the same maxima. Magnetic field is increased from right to left. At small magnetic field, the zero modes are supported away from the horizon, while at large magnetic field, the zero modes are supported near the horizon. This means that at large magnetic field the influence of the black hole to the Fermi level structure becomes more important.

21.5.2 Magnetic Effects on the Fermi Momentum and Fermi Velocity

In the presence of a magnetic field there is only a true pole in the Green's function whenever the Landau level crosses the Fermi energy [25, 26]

$$2l|qh| = k_F^2. \tag{21.62}$$

As shown in Fig. 21.2, whenever the equation (21.62) is satisfied the spectral function $A(\omega)$ has a (sharp) peak. This is not surprising since quasiparticles can be easily excited from the Fermi surface. From (21.62), the spectral function $A(\omega)$ and the density of states on the Fermi surface $D(\omega)$ are periodic in $\frac{1}{h}$ with the period

$$\Delta\left(\frac{1}{h}\right) = \frac{2\pi q}{A_F}, \tag{21.63}$$

where $A_F = \pi k_F^2$ is the area of the Fermi surface [25, 26]. This is a manifestation of the de Haas-van Alphen quantum oscillations. At $T = 0$, the electronic properties of metals depend on the density of states on the Fermi surface. Therefore, an oscillatory behavior as a function of magnetic field should appear in any quantity that depends on the density of states on the Fermi energy. Magnetic susceptibility [25, 26] and magnetization together with the superconducting gap [29] have been shown to exhibit quantum oscillations. Every Landau level contributes an oscillating term and the period of the lth level oscillation is determined by the value of the magnetic field h that satisfies (21.62) for the given value of k_F. Quantum oscillations (and the quantum Hall effect which we consider later in the paper) are examples of phenomena in which Landau level physics reveals the presence of the Fermi surface. The superconducting gap found in the quark matter in magnetic fields [29] is another evidence for the existence of the (highly degenerate) Fermi surface and the corresponding Fermi momentum.

Generally, a Fermi surface controls the occupation of energy levels in the system: the energy levels below the Fermi surface are filled and those above are empty (or non-existent). Here, however, the association to the Fermi momentum can be obscured by the fact that the fermions form highly degenerate Landau levels. Thus, in two dimensions, in the presence of the magnetic field the corresponding effective Fermi surface is given by a single point in the phase space, that is determined by n_F, the Landau index of the highest occupied level, i.e., the highest Landau level below the chemical potential.[2] Increasing the magnetic field, Landau levels 'move up' in the phase space leaving only the lower levels occupied, so that the effective Fermi momentum scales roughly (excluding interactions) as a square root of the magnetic field, $k_F \sim \sqrt{n_F} \sim k_F^{\max}\sqrt{1 - h/h_{\max}}$. High magnetic fields drive the effective density of the charge carriers down, approaching the limit when the Fermi momentum coincides with the lowest Landau level.

---

[2]We would like to thank Igor Shovkovy for clarifying the issue with the Fermi momentum in the presence of the magnetic field.

Many phenomena observed in the paper can thus be qualitatively explained by Landau quantization. As discussed before, the notion of the Fermi momentum is lost at very high magnetic fields. In what follows, the quantitative Fermi level structure at zero temperature, described by k_F and v_F values, is obtained as a function of the magnetic field using the solution of the Dirac equation given by (21.59), (21.60). As in [12], we neglect first the discrete nature of the Fermi momentum and velocity in order to obtain general understanding. Upon taking the quantization into account, the smooth curves become combinations of step functions following the same trend as the smooth curves (without quantization). While usually the grand canonical ensemble is used, where the fixed chemical potential controls the occupation of the Landau levels [30], in our setup, the Fermi momentum is allowed to change as the magnetic field is varied, while we keep track of the IR conformal dimension v.

The Fermi momentum is defined by the matching between IR and UV physics [8], therefore it is enough to know the solution at $\omega = 0$, where the matching is performed. To obtain the Fermi momentum, we require that the zero mode solution is regular at the horizon ($\psi^{(0)} \sim z^{-\frac{1}{2}+v}$) and normalizable at the boundary. At the boundary $z \sim 1$, the wave function behaves as

$$a(1-z)^{\frac{3}{2}-m} \begin{pmatrix} 1 \\ 0 \end{pmatrix} + b(1-z)^{\frac{3}{2}+m} \begin{pmatrix} 0 \\ 1 \end{pmatrix}. \tag{21.64}$$

To require it to be normalizable is to set the first term $a = 0$; the wave function at $z \sim 1$ is then

$$\psi^{(0)} \sim (1-z)^{\frac{3}{2}+m} \begin{pmatrix} 0 \\ 1 \end{pmatrix}. \tag{21.65}$$

Equation (21.65) leads to the condition $\lim_{z \to 1}(z-1)^{-3/2}(\tilde{y}_2^{(0)} + i\tilde{y}_1^{(0)}) = 0$, which, together with (21.59), gives the following equation for the Fermi momentum as function of the magnetic field [20, 24]

$$\frac{{}_2F_1(1+v+\frac{i\mu_{q,\text{eff}}}{6}, \frac{1}{2}+v-\frac{\sqrt{2}\mu_{q,\text{eff}}}{3}, 1+2v, \frac{2}{3}(1-i\sqrt{2}))}{{}_2F_1(v+\frac{i\mu_{q,\text{eff}}}{6}, \frac{1}{2}+v-\frac{\sqrt{2}\mu_{q,\text{eff}}}{3}, 1+2v, \frac{2}{3}(1-i\sqrt{2}))} = \frac{6v - i\mu_{q,\text{eff}}}{k_F(-2i+\sqrt{2})}, \tag{21.66}$$

with $v \equiv v_{k_F}$ given by (21.58). Using Mathematica to evaluate the hypergeometric functions, we numerically solve the equation for the Fermi surface, which gives effective momentum as if it were continuous, i.e. when quantization is neglected. The solutions of (21.66) are given in Fig. 21.6. There are multiple Fermi surfaces for a given magnetic field h. Here and in all other plots we choose $g_F = 1$, therefore $h \to H$, and $q = \frac{15}{\sqrt{3}}$. In Fig. 21.6, positive and negative k_F correspond to the Fermi surfaces in the Green's functions G_1 and G_2. The relation between two components is $G_2(\omega, k) = G_1(\omega, -k)$ [7], therefore Fig. 21.6 is not symmetric with respect to the x-axis. Effective momenta terminate at the dashed line $v_{k_F} = 0$. Taking into account Landau quantization of $k_F \to \sqrt{2|qh|l}$ with $l = 1, 2\ldots$, the plot consists of stepwise functions tracing the existing curves (we depict only positive k_F). Indeed

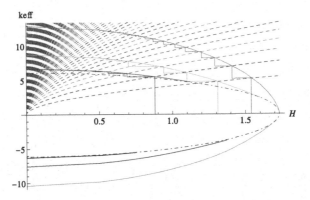

Fig. 21.6 Effective momentum k_{eff} vs. the magnetic field $h \rightarrow H$ (we set $g_F = 1$, $q = \frac{15}{\sqrt{3}}$). As we increase magnetic field the Fermi surface shrinks. *Smooth solid curves* represent situation as if momentum is a continuous parameter (for convenience), stepwise solid functions are the real Fermi momenta which are discretized due to the Landau level quantization: $k_F \rightarrow \sqrt{2|qh|l}$ with $l = 1, 2, \ldots$ where $\sqrt{2|qh|l}$ are Landau levels given by *dotted lines* (only positive discrete k_F are shown). At a given h there are multiple Fermi surfaces. From right to left are the first, second etc. Fermi surfaces. The *dashed-dotted line* is $v_{k_F} = 0$ where k_F is terminated. Positive and negative k_{eff} correspond to Fermi surfaces in two components of the Green's function

Fig. 21.7 Landau level numbers n corresponding to the quantized Fermi momenta vs. the magnetic field $h \rightarrow H$ for the three Fermi surfaces with positive k_F. We set $g_F = 1$, $q = \frac{15}{\sqrt{3}}$. From right to left are the first, second and third Fermi surfaces

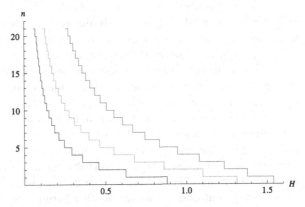

Landau quantization can be also seen from the dispersion relation at Fig. 21.3, where only discrete values of effective momentum are allowed and the Fermi surface has been chopped up as a result of it Fig. 21.3(B).

Our findings agree with the results for the (largest) Fermi momentum in a three-dimensional magnetic system considered in [31], compare the stepwise dependence $k_F(h)$ with Fig. 21.5 in [31].

In Fig. 21.7, the Landau level index l is obtained from $k_F(h) = \sqrt{2|qh|l}$ where $k_F(h)$ is a numerical solution of (21.66). Only those Landau levels which are below the Fermi surface are filled. In Fig. 21.6, as we decrease magnetic field first nothing happens until the next Landau level crosses the Fermi surface which corresponds to a jump up to the next step. Therefore, at strong magnetic fields, fewer states contribute

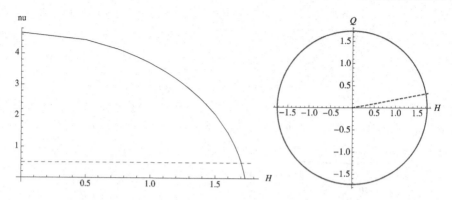

Fig. 21.8 *Left panel.* The IR conformal dimension $\nu \equiv \nu_{k_F}$ calculated at the Fermi momentum vs. the magnetic field $h \to H$ (we set $g_F = 1$, $q = \frac{15}{\sqrt{3}}$). Calculations are done for the first Fermi surface. *Dashed line* is for $\nu = \frac{1}{2}$ (at $H_c = 1.70$), which is the border between the Fermi liquids $\nu > \frac{1}{2}$ and non-Fermi liquids $\nu < \frac{1}{2}$. *Right panel.* Phase diagram in terms of the chemical potential and the magnetic field $\mu^2 + h^2 = 3$ (in dimensionless variables $h = g_F H$, $\mu = g_F Q$; we set $g_F = 1$). Fermi liquids are above the *dashed line* ($H < H_c$) and non-Fermi liquids are below the *dashed line* ($H > H_c$)

to transport properties and the lowest Landau level becomes more important (see the next section). At weak magnetic fields, the sum over many Landau levels has to be taken, ending with the continuous limit as $h \to 0$, when quantization can be ignored.

In Fig. 21.8, we show the IR conformal dimension as a function of the magnetic field. We have used the numerical solution for k_F. Fermi liquid regime takes place at magnetic fields $h < h_c$, while non-Fermi liquids exist in a narrow band at $h_c < h < h_c'$, and at h_c' the system becomes near-conformal.

In this figure we observe the pathway of the possible phase transition exhibited by the Fermi surface (ignoring Landau quantization): it can vanish at the line $\nu_{k_F} = 0$, undergoing a crossover to the conformal regime, or cross the line $\nu_{k_F} = 1/2$ and go through a non-Fermi liquid regime, and subsequently cross to the conformal phase. Note that the primary Fermi surface with the highest k_F and ν_{k_F} seems to directly cross over to conformality, while the other Fermi surfaces first exhibit a "strange metal" phase transition. Therefore, all the Fermi momenta with $\nu_{k_F} > 0$ contribute to the transport coefficients of the theory. In particular, at high magnetic fields when for the first (largest) Fermi surface $k_F^{(1)}$ is nonzero but small, the lowest Landau level $n = 0$ becomes increasingly important contributing to the transport with half degeneracy factor as compared to the higher Landau levels.

In Fig. 21.9, we plot the Fermi momentum k_F as a function of the magnetic field for the first Fermi surface (the largest root of (21.66)). Quantization is neglected here. At the left panel, the relatively small region between the dashed lines corresponds to non-Fermi liquids $0 < \nu < \frac{1}{2}$. At large magnetic field, the physics of the Fermi surface is captured by the near horizon region (see also Fig. 21.5) which is $AdS_2 \times R^2$. At the maximum magnetic field, $H_{max} = \sqrt{3} \approx 1.73$, when the black hole becomes pure magnetically charged, the Fermi momentum vanishes when it

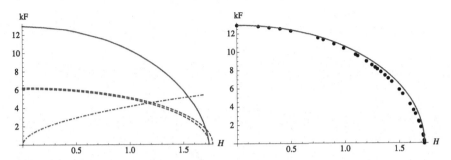

Fig. 21.9 Fermi momentum k_F vs. the magnetic field $h \to H$ (we set $g_F = 1, q = \frac{15}{\sqrt{3}}$) for the first Fermi surface. *Left panel.* The *inner* (closer to x-axis) *dashed line* is $v_{k_F} = 0$ and the *outer dashed line* is $v_{k_F} = \frac{1}{2}$, the region between these lines corresponds to non-Fermi liquids $0 < v_{k_F} < \frac{1}{2}$. The *dashed-dotted line* is for the first Landau level $k_1 = \sqrt{2qH}$. The first Fermi surface hits the *border-line* between a Fermi and non-Fermi liquids $v = \frac{1}{2}$ at $H_c \approx 1.70$, and it vanishes at $H_{\max} = \sqrt{3} = 1.73$. *Right panel.* Circles are the data points for the Fermi momentum calculated analytically, *solid line* is a fit function $k_F^{\max} \sqrt{1 - \frac{H^2}{3}}$ with $k_F^{\max} = 12.96$

crosses the line $v_{k_F} = 0$. This only happens for the first Fermi surface. For the higher Fermi surfaces the Fermi momenta terminate at the line $v_{k_F} = 0$, Fig. 21.6. Note the Fermi momentum for the first Fermi surface can be almost fully described by a function $k_F = k_F^{\max} \sqrt{1 - \frac{H^2}{3}}$. It is tempting to view the behavior $k_F \sim \sqrt{H_{\max} - H}$ as a phase transition in the system although it strictly follows from the linear scaling for $H = 0$ by using the mapping (21.29). (Note that also $\mu = g_F Q = g_F \sqrt{3 - H^2}$.) Taking into account the discretization of k_F, the plot will consist of an array of step functions tracing the existing curve. Our findings agree with the results for the Fermi momentum in a three dimensional magnetic system considered in [31], compare with Fig. 21.5 there.

The Fermi velocity given in (21.27) is defined by the UV physics; therefore solutions at non-zero ω are required. The Fermi velocity is extracted from matching two solutions in the inner and outer regions at the horizon. The Fermi velocity as function of the magnetic field for $v > \frac{1}{2}$ is [20, 24]

$$v_F = \frac{1}{h_1} \left(\int_0^1 dz \sqrt{g/g_{tt}} \, \psi^{(0)\dagger} \psi^{(0)} \right)^{-1} \lim_{z \to 1} \frac{|\tilde{y}_1^{(0)} + i \tilde{y}_2^{(0)}|^2}{(1-z)^3},$$

$$h_1 = \lim_{z \to 1} \frac{\tilde{y}_1^{(0)} + i \tilde{y}_2^{(0)}}{\partial_k (y_2^{(0)} + i \tilde{y}_1^{(0)})},$$

(21.67)

where the zero mode wavefunction is taken at k_F (21.59).

We plot the Fermi velocity for several Fermi surfaces in Fig. 21.10. Quantization is neglected here. The Fermi velocity is shown for $v > \frac{1}{2}$. It is interesting that the Fermi velocity vanishes when the IR conformal dimension is $v_{k_F} = \frac{1}{2}$. Formally, it follows from the fact that $v_F \sim (2v - 1)$ [8]. The first Fermi surface is at the

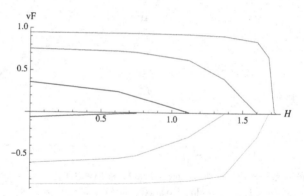

Fig. 21.10 Fermi velocity v_F vs. the magnetic field $h \to H$ (we set $g_F = 1$, $q = \frac{15}{\sqrt{3}}$) for the regime of Fermi liquids $v \geq \frac{1}{2}$. Fermi velocity vanishes at $v_{k_F} = \frac{1}{2}$ (x-axis). For the first Fermi surface, the *top curve*, Fermi velocity vanishes at $H_c \approx 1.70$. The region $H < H_c$ corresponds to the Fermi liquids and quasiparticle description. The *multiple lines* are for various Fermi surfaces in ascending order, with the first Fermi surface on the right. The Fermi velocity v_F has the same sign as the Fermi momentum k_F. As above, positive and negative v_F correspond to Fermi surfaces in the two components of the Green's function

far right. Positive and negative v_F correspond to the Fermi surfaces in the Green's functions G_1 and G_2, respectively. The Fermi velocity v_F has the same sign as the Fermi momentum k_F. At small magnetic field values, the Fermi velocity is very weakly dependent on H and it is close to the speed of light; at large magnetic field values, the Fermi velocity rapidly decreases and vanishes (at $H_c = 1.70$ for the first Fermi surface). Geometrically, this means that with increasing magnetic field the zero mode wavefunction is supported near the black hole horizon Fig. 21.5, where the gravitational redshift reduces the local speed of light as compared to the boundary value. It was also observed in [8, 24] at small fermion charge values.

21.6 Hall and Longitudinal Conductivities

In this section, we calculate the contributions to Hall σ_{xy} and the longitudinal σ_{xx} conductivities directly in the boundary theory. This should be contrasted with the standard holographic approach, where calculations are performed in the (bulk) gravity theory and then translated to the boundary field theory using the AdS/CFT dictionary. Specifically, the conductivity tensor has been obtained in [11] by calculating the on-shell renormalized action for the gauge field on the gravity side and using the gauge/gravity duality $A_M \to j_\mu$ to extract the R charge current-current correlator at the boundary. Here, the Kubo formula involving the current-current correlator is used directly by utilizing the fermion Green's functions extracted from holography in [8]. Therefore, the conductivity is obtained for the charge carriers described by the fermionic operators of the boundary field theory.

The use of the conventional Kubo formula to extract the contribution to the transport due to fermions is validated in that it also follows from a direct AdS/CFT computation of the one-loop correction to the on-shell renormalized AdS action [17]. We study in particular stable quasiparticles with $\nu > \frac{1}{2}$ and at zero temperature. This regime effectively reduces to the clean limit where the imaginary part of the self-energy vanishes Im $\Sigma \to 0$. We use the gravity-"dressed" fermion propagator from (21.27) and to make the calculations complete, the "dressed" vertex is necessary, to satisfy the Ward identities. As was argued in [17], the boundary vertex which is obtained from the bulk calculations can be approximated by a constant in the low temperature limit. Also, according to [32, 33], the vertex only contains singularities of the product of the Green's functions. Therefore, dressing the vertex will not change the dependence of the DC conductivity on the magnetic field [32, 33]. In addition, the zero magnetic field limit of the formulae for conductivity obtained from holography [17] and from direct boundary calculations [20] are identical.

21.6.1 Integer Quantum Hall Effect

Let us start from the "dressed" retarded and advanced fermion propagators [8]: G_R is given by (21.27) and $G_A = G_R^*$. To perform the Matsubara summation we use the spectral representation

$$
G(i\omega_n, \mathbf{k}) = \int \frac{d\omega}{2\pi} \frac{A(\omega, \mathbf{k})}{\omega - i\omega_n}, \tag{21.68}
$$

with the spectral function defined as $A(\omega, \mathbf{k}) = -\frac{1}{\pi} \operatorname{Im} G_R(\omega, \mathbf{k}) = \frac{1}{2\pi i}(G_R(\omega, \mathbf{k}) - G_A(\omega, \mathbf{k}))$. Generalizing to a non-zero magnetic field and spinor case [30], the spectral function [34] is

$$
A(\omega, \mathbf{k}) = \frac{1}{\pi} e^{-\frac{k^2}{|qh|}} \sum_{l=0}^{\infty} (-1)^l (-h_1 v_F)
$$
$$
\times \left(\frac{\Sigma_2(\omega, k_F) f(\mathbf{k}) \gamma^0}{(\omega + \varepsilon_F + \Sigma_1(\omega, k_F) - E_l)^2 + \Sigma_2(\omega, k_F)^2} + (E_l \to -E_l) \right), \tag{21.69}
$$

where $\varepsilon_F = v_F k_F$ is the Fermi energy, $E_l = v_F \sqrt{2|qh|l}$ is the energy of the Landau level, $f(\mathbf{k}) = P_- L_l(\frac{2k^2}{|qh|}) - P_+ L_{l-1}(\frac{2k^2}{|qh|})$ with spin projection operators $P_\pm = (1 \pm i\gamma^1 \gamma^2)/2$, we take $c = 1$, the generalized Laguerre polynomials are $L_n^\alpha(z)$ and by definition $L_n(z) = L_n^0(z)$, (we omit the vector part $\mathbf{k}\gamma$, it does not contribute to the DC conductivity), all γ's are the standard Dirac matrices, h_1, v_F and k_F are real constants (we keep the same notations for the constants as in [8]). The self-energy $\Sigma \sim \omega^{2\nu k_F}$ contains the real and imaginary parts, $\Sigma = \Sigma_1 + i\Sigma_2$. The imaginary part comes from scattering processes of a fermion in the bulk, e.g. from pair creation, and from the scattering into the black hole. It is exactly due to in-elastic/dissipative processes that we are able to obtain finite values for the transport coefficients, otherwise they are formally infinite.

Using the Kubo formula, the DC electrical conductivity tensor is

$$\sigma_{ij}(\Omega) = \lim_{\Omega \to 0} \frac{\operatorname{Im} \Pi_{ij}^R}{\Omega + i0^+}, \tag{21.70}$$

where $\Pi_{ij}(i\Omega_m \to \Omega + i0^+)$ is the retarded current-current correlation function; schematically the current density operator is $j^i(\tau, \mathbf{x}) = q v_F \sum_\sigma \bar{\psi}_\sigma(\tau, \mathbf{x}) \gamma^i \psi_\sigma(\tau, \mathbf{x})$. Neglecting the vertex correction, it is given by

$$\Pi_{ij}(i\Omega_m) = q^2 v_F^2 T \sum_{n=-\infty}^{\infty} \int \frac{d^2 k}{(2\pi)^2} \operatorname{tr}\big(\gamma^i G(i\omega_n, \mathbf{k}) \gamma^j G(i\omega_n + i\Omega_m, \mathbf{k})\big). \tag{21.71}$$

The sum over the Matsubara frequency is

$$T \sum_n \frac{1}{i\omega_n - \omega_1} \frac{1}{i\omega_n + i\Omega_m - \omega_2} = \frac{n(\omega_1) - n(\omega_2)}{i\Omega_m + \omega_1 - \omega_2}. \tag{21.72}$$

Taking $i\Omega_m \to \Omega + i0^+$, the polarization operator is now

$$\Pi_{ij}(\Omega) = \frac{d\omega_1}{2\pi} \frac{d\omega_2}{2\pi} \frac{n_{FD}(\omega_1) - n_{FD}(\omega_2)}{\Omega + \omega_1 - \omega_2} \int \frac{d^2 k}{(2\pi)^2} \operatorname{tr}\big(\gamma^i A(\omega_1, \mathbf{k}) \gamma^j A(\omega_2, \mathbf{k})\big), \tag{21.73}$$

where the spectral function $A(\omega, \mathbf{k})$ is given by (21.69) and $n_{FD}(\omega)$ is the Fermi-Dirac distribution function. Evaluating the traces, we have

$$\sigma_{ij} = -\frac{4q^2 v_F^2 (h_1 v_F)^2 |qh|}{\pi \Omega}$$

$$\times \operatorname{Re} \sum_{l,k=0}^{\infty} (-1)^{l+k+1} \big\{ \delta_{ij} (\delta_{l,k-1} + \delta_{l-1,k}) + i\varepsilon_{ij} \operatorname{sgn}(qh)(\delta_{l,k-1} - \delta_{l-1,k}) \big\}$$

$$\times \int \frac{d\omega_1}{2\pi} \left(\tanh \frac{\omega_1}{2T} - \tanh \frac{\omega_2}{2T} \right) \left(\frac{\Sigma_2(\omega_1)}{(\tilde{\omega}_1 - E_l)^2 + \Sigma_2^2(\omega_1)} + (E_l \to -E_l) \right)$$

$$\times \left(\frac{\Sigma_2(\omega_2)}{(\tilde{\omega}_2 - E_k)^2 + \Sigma_2^2(\omega_2)} + (E_k \to -E_k) \right), \tag{21.74}$$

with $\omega_2 = \omega_1 + \Omega$. We have also introduced $\tilde{\omega}_{1;2} \equiv \omega_{1;2} + \varepsilon_F + \Sigma_1(\omega_{1;2})$ with ε_{ij} being the antisymmetric tensor ($\varepsilon_{12} = 1$), and $\Sigma_{1;2}(\omega) \equiv \Sigma_{1;2}(\omega, k_F)$. In the momentum integral, we use the orthogonality condition for the Laguerre polynomials $\int_0^\infty dx e^x L_l(x) L_k(x) = \delta_{lk}$.

From (21.74), the term symmetric/antisymmetric with respect to exchange $\omega_1 \leftrightarrow \omega_2$ contributes to the diagonal/off-diagonal component of the conductivity (note the antisymmetric term $n_{FD}(\omega_1) - n_{FD}(\omega_2)$). The longitudinal and Hall DC conductivities ($\Omega \to 0$) are thus

$$
\sigma_{xx} = -\frac{2q^2(h_1v_F)^2|qh|}{\pi T} \int_{-\infty}^{\infty} \frac{d\omega}{2\pi} \frac{\Sigma_2^2(\omega)}{\cosh^2 \frac{\omega}{2T}}
$$

$$
\times \sum_{l=0}^{\infty} \left(\frac{1}{(\tilde{\omega} - E_l)^2 + \Sigma_2^2(\omega)} + (E_l \to -E_l) \right)
$$

$$
\times \left(\frac{1}{(\tilde{\omega} - E_{l+1})^2 + \Sigma_2^2(\omega)} + (E_{l+1} \to -E_{l+1}) \right), \tag{21.75}
$$

$$
\sigma_{xy} = -\frac{q^2(h_1v_F)^2 \, \mathrm{sgn}(qh)}{\pi} \nu_h,
$$

$$
\tag{21.76}
$$

$$
\nu_h = 2 \int_{-\infty}^{\infty} \frac{d\omega}{2\pi} \tanh \frac{\omega}{2T} \Sigma_2(\omega) \sum_{l=0}^{\infty} \alpha_l \left(\frac{1}{(\tilde{\omega} - E_l)^2 + \Sigma_2^2(\omega)} + (E_l \to -E_l) \right),
$$

where $\tilde{\omega} = \omega + \varepsilon_F + \Sigma_1(\omega)$. The filling factor ν_h is proportional to the density of carriers: $|\nu_h| = \frac{\pi}{|qh|h_1v_F}n$ (see derivation in [27]). The degeneracy factor of the Landau levels is α_l: $\alpha_0 = 1$ for the lowest Landau level and $\alpha_l = 2$ for $l = 1, 2 \ldots$. Substituting the filling factor ν_h back to (21.76), the Hall conductivity can be written as

$$
\sigma_{xy} = \frac{\rho}{h}, \tag{21.77}
$$

where ρ is the charge density in the boundary theory, and both the charge q and the magnetic field h carry a sign (the prefactor $(-h_1v_F)$ comes from the normalization choice in the fermion propagator (21.27), (21.69) as given in [8], which can be regarded as a factor contributing to the effective charge and is not important for further considerations). The Hall conductivity (21.77) has been obtained using the AdS/CFT duality for the Lorentz invariant $2+1$-dimensional boundary field theories in [11]. We recover this formula because in our case the translational invariance is maintained in the x and y directions of the boundary theory.

Low frequencies give the main contribution in the integrand of (21.76). Since the self-energy satisfies $\Sigma_1(\omega) \sim \Sigma_2(\omega) \sim \omega^{2\nu}$ and we consider the regime $\nu > \frac{1}{2}$, we have $\Sigma_1 \sim \Sigma_2 \to 0$ at $\omega \sim 0$ (self-energy goes to zero faster than the ω term). Therefore, only the simple poles in the upper half-plane $\omega_0 = -\varepsilon_F \pm E_l + \Sigma_1 + i\Sigma_2$ contribute to the conductivity where $\Sigma_1 \sim \Sigma_2 \sim (-\varepsilon_F \pm E_l)^{2\nu}$ are small. The same logic of calculation has been used in [30]. We obtain for the longitudinal and Hall conductivities

$$
\sigma_{xx} = \frac{2q^2(h_1v_F)^2 \Sigma_2}{\pi T} \times \left(\frac{1}{1 + \cosh \frac{\varepsilon_F}{T}} + \sum_{l=1}^{\infty} 4l \frac{1 + \cosh \frac{\varepsilon_F}{T} \cosh \frac{E_l}{T}}{(\cosh \frac{\varepsilon_F}{T} + \cosh \frac{E_l}{T})^2} \right), \tag{21.78}
$$

$$\sigma_{xy} = \frac{q^2(h_1 v_F)^2 \mathrm{sgn}(qh)}{\pi} \times 2\left(\tanh\frac{\varepsilon_F}{2T} + \sum_{l=1}^{\infty}\left(\tanh\frac{\varepsilon_F + E_l}{2T} + \tanh\frac{\varepsilon_F - E_l}{2T}\right)\right),$$

$$(21.79)$$

where the Fermi energy is $\varepsilon_F = v_F k_F$ and the energy of the Landau level is $E_l = v_F\sqrt{2|qh|l}$. Similar expressions were obtained in [30]. However, in our case the filling of the Landau levels is controlled by the magnetic field h through the field-dependent Fermi energy $v_F(h)k_F(h)$ instead of the chemical potential μ.

At $T = 0$, $\cosh\frac{\omega}{T} \to \frac{1}{2}e^{\frac{\omega}{T}}$ and $\tanh\frac{\omega}{2T} = 1 - 2n_{\mathrm{FD}}(\omega) \to \mathrm{sgn}\omega$. Therefore the longitudinal and Hall conductivities are

$$\sigma_{xx} = \frac{2q^2(h_1 v_F)^2 \Sigma_2}{\pi T}\sum_{l=1}^{\infty} l\delta_{\varepsilon_F, E_l} = \frac{2q^2(h_1 v_F)^2 \Sigma_2}{\pi T} \times n\delta_{\varepsilon_F, E_n}, \qquad (21.80)$$

$$\sigma_{xy} = \frac{q^2(h_1 v_F)^2 \mathrm{sgn}(qh)}{\pi} 2\left(1 + 2\sum_{l=1}^{\infty}\theta(\varepsilon_F - E_l)\right)$$

$$= \frac{q^2(h_1 v_F)^2 \mathrm{sgn}(qh)}{\pi} \times 2(1 + 2n)\theta(\varepsilon_F - E_n)\theta(E_{n+1} - \varepsilon_F), \quad (21.81)$$

where the Landau level index runs $n = 0, 1, \ldots$. It can be estimated as $n = [\frac{k_F^2}{2|qh|}]$ when $v_F \neq 0$ ([ ] denotes the integer part), with the average spacing between the Landau levels given by the Landau energy $v_F\sqrt{2|qh|}$. Note that $\varepsilon_F \equiv \varepsilon_F(h)$. We can see that (21.81) expresses the integer quantum Hall effect (IQHE). At zero temperature, as we dial the magnetic field, the Hall conductivity jumps from one quantized level to another, forming plateaus given by the filling factor

$$v_h = \pm 2(1 + 2n) = \pm 4\left(n + \frac{1}{2}\right), \qquad (21.82)$$

with $n = 0, 1, \ldots$. (Compare to the conventional Hall quantization $v_h = \pm 4n$, that appears in thick graphene.) Plateaus of the Hall conductivity at $T = 0$ follow from the stepwise behavior of the charge density ρ in (21.77):

$$\rho \sim 4\left(n + \frac{1}{2}\right)\theta(\varepsilon_F - E_n)\theta(E_{n+1} - \varepsilon_F), \qquad (21.83)$$

where n Landau levels are filled and contribute to ρ. The longitudinal conductivity vanishes except precisely at the transition point between the plateaus. In Fig. 21.11, we plot the longitudinal and Hall conductivities at $T = 0$, using only the terms after $\times$ sign in (21.79). In the Hall conductivity, plateau transition occurs when the Fermi level (in Fig. 21.11) of the first Fermi surface $\varepsilon_F = v_F(h)k_F(h)$ (Fig. 21.9) crosses the Landau level energy as we vary the magnetic field. By decreasing the magnetic field, the plateaus become shorter and increasingly more Landau levels contribute to the Hall conductivity. This happens because of two factors: the Fermi level moves

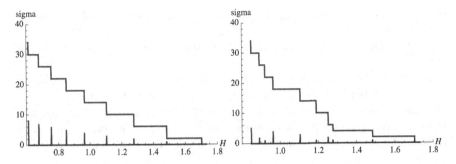

Fig. 21.11 Hall conductivity σ_{xy} and longitudinal conductivity σ_{xx} vs. the magnetic field $h \to H$ at $T = 0$ (we set $g_F = 1$, $q = \frac{15}{\sqrt{3}}$). *Left panel* is for IQHE. *Right panel* is for FQHE. At strong magnetic fields, the Hall conductivity plateau $\nu_h = 4$ appears together with plateaus $\nu_h = 2$ and $\nu_h = 6$ in FQHE (details are in [27]). Irregular pattern in the length of the plateaus for FQHE is observed in experiments on thin films of graphite at strong magnetic fields [28]

up and the spacing between the Landau levels becomes smaller. This picture does not depend on the Fermi velocity as long as it is nonzero.

21.6.2 Fractional Quantum Hall Effect

In [27], using the holographic description of fermions, we obtained the filling factor at strong magnetic fields

$$\nu_h = \pm 2j, \tag{21.84}$$

where j is the effective Landau level index. Equation (21.84) expresses the fractional quantum Hall effect (FQHE). In the quasiparticle picture, the effective index is integer $j = 0, 1, 2, \ldots$, but generally it may be fractional. In particular, the filling factors $\nu = 2/m$ where $m = 1, 2, 3, \ldots$ have been proposed by Halperin [35] for the case of bound electron pairs, i.e. $2e$-charge bosons. Indeed, QED becomes effectively confining in ultraquantum limit at strong magnetic field, and the electron pairing is driven by the Landau level quantization and gives rise to $2e$ bosons. In our holographic description, quasiparticles are valid degrees of freedom only for $\nu > 1/2$, i.e. for weak magnetic field. At strong magnetic field, poles of the fermion propagator should be taken into account in calculation of conductivity. This will probably result in a fractional filling factor. Our pattern for FQHE Fig. 21.11 resembles the one obtained by Kopelevich in Fig. 3 [36] which has been explained using the fractional filling factor of Halperin [35].

The somewhat regular pattern behind the irregular behavior can be understood as a consequence of the appearance of a new energy scale: the average distance between the Fermi levels. For the case of Fig. 21.11, we estimate it to be $\langle \varepsilon_F^{(m)} - \varepsilon_F^{(m+1)} \rangle = 4.9$ with $m = 1, 2$. The authors of [30] explain the FQHE through the opening of a gap in the quasiparticle spectrum, which acts as an order parameter related to the particle-hole pairing and is enhanced by the magnetic field (magnetic catalysis). Here, the energy gap arises due to the participation of multiple Fermi surfaces.

A pattern for the Hall conductivity that is strikingly similar to Fig. 21.11 arises in the AA and AB-stacked bilayer graphene, which has different transport properties from the monolayer graphene [37], compare with Figs. 2, 5 there. It is remarkable that the bilayer graphene also exhibits the insulating behavior in a certain parameter regime. This agrees with our findings of metal-insulating transition in our system.

21.7 Conclusions

We have studied strongly coupled electron systems in the magnetic field focussing on the Fermi level structure, using the AdS/CFT correspondence. These systems are dual to Dirac fermions placed in the background of the electrically and magnetically charged AdS-Reissner-Nordström black hole. At strong magnetic fields the dual system "lives" near the black hole horizon, which substantially modifies the Fermi level structure. As we dial the magnetic field higher, the system exhibits the non-Fermi liquid behavior and then crosses back to the conformal regime. In our analysis we have concentrated on the Fermi liquid regime and obtained the dependence of the Fermi momentum k_F and Fermi velocity v_F on the magnetic field. Remarkably, k_F exhibits the square root behavior, with v_F staying close to the speed of light in a wide range of magnetic fields, while it rapidly vanishes at a critical magnetic field which is relatively high. Such behavior indicates that the system may have a phase transition.

The magnetic system can be rescaled to a zero-field configuration which is thermodynamically equivalent to the original one. This simple result can actually be seen already at the level of field theory: the additional scale brought about by the magnetic field does not show up in thermodynamic quantities meaning, in particular, that the behavior in the vicinity of quantum critical points is expected to remain largely uninfluenced by the magnetic field, retaining its conformal invariance. In the light of current condensed matter knowledge, this is surprising and might in fact be a good opportunity to test the applicability of the probe limit in the real world: if this behavior is not seen, this suggests that one has to include the backreaction to metric to arrive at a realistic description.

In the field theory frame, we have calculated the DC conductivity using k_F and v_F values extracted from holography. The holographic calculation of conductivity that takes into account the fermions corresponds to the corrections of subleading order in $1/N$ in the field theory and is very involved [17]. As we are not interested in the vertex renormalization due to gravity (it does not change the magnetic field dependence of the conductivity), we have performed our calculations directly in the field theory with AdS gravity-dressed fermion propagators. Instead of controlling the occupancy of the Landau levels by changing the chemical potential (as is usual in non-holographic setups), we have controlled the filling of the Landau levels by varying the Fermi energy level through the magnetic field. At zero temperature, we have reproduced the integer QHE of the Hall conductivity, which is observed in graphene at moderate magnetic fields. While the findings on equilibrium physics (Landau quantization, magnetic phase transitions and crossovers) are within expectations and indeed corroborate the meaningfulness of the AdS/CFT approach as compared to the well-known facts, the detection of the QHE is somewhat surprising as the spatial boundary effects are ignored in our setup. We plan to address this question in further work.

Interestingly, at large magnetic fields we obtain the correct formula for the filling factor characteristic for FQHE. Moreover our pattern for FQHE resembles the one obtained in [36] which has been explained using the fractional filling factor of Halperin [35]. In the quasiparticle picture, which we have used to calculate Hall conductivity, the filling factor is integer. In our holographic description, quasiparticles are valid degrees of freedom only at weak magnetic field. At strong magnetic field, the system exhibits non-Fermi liquid behavior. In this case, the poles of the fermion propagator should be taken into account to calculate the Hall conductivity. This can probably result in a fractional filling factor. We leave it for future work.

Notably, the AdS-Reissner-Nordström black hole background gives a vanishing Fermi velocity at high magnetic fields. It happens at the point when the IR conformal dimension of the corresponding field theory is $\nu = \frac{1}{2}$, which is the borderline between the Fermi and non-Fermi liquids. Vanishing Fermi velocity was also observed at high enough fermion charge [24]. As in [24], it is explained by the red shift on the gravity side, because at strong magnetic fields the fermion wavefunction is supported near the black hole horizon modifying substantially the Fermi velocity. In our model, vanishing Fermi velocity leads to zero occupancy of the Landau levels by stable quasiparticles that results in vanishing regular Fermi liquid contribution to the Hall conductivity and the longitudinal conductivity. The dominant contribution to both now comes from the non-Fermi liquid and conformal contributions. We associate such change in the behavior of conductivities with a metal-"strange metal" phase transition. Experiments on highly oriented pyrolitic graphite support the existence of a finite "offset" magnetic field h_c at $T = 0$ where the resistivity qualitatively changes its behavior [38–41]. At $T \neq 0$, it has been associated with the metal-semiconducting phase transition [38–41]. It is worthwhile to study the temperature dependence of the conductivity in order to understand this phase transition better.

Acknowledgements The work was supported in part by the Alliance program of the Helmholtz Association, contract HA216/EMMI "Extremes of Density and Temperature: Cosmic Matter in the Laboratory" and by ITP of Goethe University, Frankfurt (E. Gubankova), by a VIDI Innovative Research Incentive Grant (K. Schalm) from the Netherlands Organization for Scientific Research (NWO), by a Spinoza Award (J. Zaanen) from the Netherlands Organization for Scientific Research (NWO) and the Dutch Foundation for Fundamental Research of Matter (FOM). K. Schalm thanks the Galileo Galilei Institute for Theoretical Physics for the hospitality and the INFN for partial support during the completion of this work.

References

1. J. Zaanen, Quantum critical electron systems: the uncharted sign worlds. Science **319**, 1205 (2008)
2. P. de Forcrand, Simulating QCD at finite density. PoS **LAT2009**, 010 (2009). arXiv:1005.0539 [hep-lat]
3. S.A. Hartnoll, J. Polchinski, E. Silverstein, D. Tong, Towards strange metallic holography. J. High Energy Phys. **1004**, 120 (2010). arXiv:0912.1061 [hep-th]
4. P. Kovtun, D.T. Son, A.O. Starinets, Viscosity in strongly interacting quantum field theories from black hole physics. Phys. Rev. Lett. **94**, 111601 (2005). arXiv:hep-th/0405231
5. S.-S. Lee, A non-Fermi liquid from a charged black hole: a critical Fermi ball. Phys. Rev. D **79**, 086006 (2009). arXiv:0809.3402 [hep-th]
6. M. Čubrović, J. Zaanen, K. Schalm, String theory, quantum phase transitions and the emergent Fermi-liquid. Science **325**, 439 (2009). arXiv:0904.1993 [hep-th]
7. H. Liu, J. McGreevy, D. Vegh, Non-Fermi liquids from holography. Phys. Rev. D **83**, 065029 (2011). arXiv:0903.2477 [hep-th]
8. T. Faulkner, H. Liu, J. McGreevy, D. Vegh, Emergent quantum criticality, Fermi surfaces, and AdS2. Phys. Rev. D **83**, 125002 (2011). arXiv:0907.2694 [hep-th]
9. S.A. Hartnoll, A. Tavanfar, Electron stars for holographic metallic criticality. Phys. Rev. D **83**, 046003 (2011). arXiv:1008.2828 [hep-th]
10. S.A. Hartnoll, P.K. Kovtun, M. Mueller, S. Sachdev, Theory of the Nernst effect near quantum phase transitions in condensed matter, and in dyonic black holes. Phys. Rev. B **76**, 144502 (2007). arXiv:0706.3215 [hep-th]
11. S.A. Hartnoll, P. Kovtun, Hall conductivity from dyonic black holes. Phys. Rev. D **76**, 066001 (2007). arXiv:0704.1160 [hep-th]
12. P. Basu, J.Y. He, A. Mukherjee, H.-H. Shieh, Holographic non-Fermi liquid in a background magnetic field. arxiv:0908.1436 [hep-th]
13. T. Albash, C.V. Johnson, Landau levels, magnetic fields and holographic Fermi liquids. J. Phys. A, Math. Theor. **43**, 345404 (2010). arXiv:1001.3700 [hep-th]
14. T. Albash, C.V. Johnson, Holographic aspects of Fermi liquids in a background magnetic field. J. Phys. A, Math. Theor. **43**, 345405 (2010). arXiv:0907.5406 [hep-th]
15. T. Albash, C.V. Johnson, A holographic superconductor in an external magnetic field. J. High Energy Phys. **0809**, 121 (2008). arXiv:0804.3466 [hep-th]
16. N. Iqbal, H. Liu, M. Mezei, Q. Si, Quantum phase transitions in holographic models of magnetism and superconductors. arXiv:1003.0010 [hep-th]
17. T. Faulkner, N. Iqbal, H. Liu, J. McGreevy, D. Vegh, From black holes to strange metals. arXiv:1003.1728 [hep-th]
18. E. D'Hoker, P. Kraus, J. High Energy Phys. **1005**, 083 (2010). arXiv:1003.1302 [hep-th]
19. A. Auerbach, Quantum magnetism approaches to strongly correlated electrons. arxiv: cond-mat/9801294
20. E. Gubankova, Particle-hole instability in the AdS_4 holography. arXiv:1006.4789 [hep-th]
21. J.L. Davis, P. Kraus, A. Shah, Gravity dual of a quantum hall plateau transition. J. High Energy Phys. **0811**, 020 (2008). arXiv:0809.1876 [hep-th]
22. E. Keski-Vakkuri, P. Kraus, Quantum Hall effect in AdS/CFT. J. High Energy Phys. **0809**, 130 (2008). arXiv:0805.4643 [hep-th]
23. A.H. MacDonald, Introduction to the physics of the quantum Hall regime. arXiv:cond-mat/9410047
24. T. Hartman, S.A. Hartnoll, Cooper pairing near charged black holes. arXiv:1003.1918 [hep-th]
25. F. Denef, S.A. Hartnoll, S. Sachdev, Quantum oscillations and black hole ringing. Phys. Rev. D **80**, 126016 (2009). arXiv:0908.1788 [hep-th]
26. F. Denef, S.A. Hartnoll, S. Sachdev, Black hole determinants and quasinormal modes. Class. Quant. Grav. **27**, 125001 (2010). arXiv:0908.2657 [hep-th]

27. E. Gubankova, J. Brill, M. Cubrovic, K. Schalm, P. Schijven, J. Zaanen, Holographic fermions in external magnetic fields. Phys. Rev. D **84**, 106003 (2011). arXiv:1011.4051 [hep-th]

28. Y. Zhang, Z. Jiang, J.P. Small, M.S. Purewal, Y.-W. Tan, M. Fazlollahi, J.D. Chudow, J.A. Jaszczak, H.L. Stormer, P. Kim, Landau level splitting in graphene in high magnetic fields. Phys. Rev. Lett. **96**, 136806 (2006). arXiv:cond-mat/0602649

29. J.L. Noronha, I.A. Shovkovy, Color-flavor locked superconductor in a magnetic field. Phys. Rev. D **76**, 105030 (2007). arXiv:0708.0307 [hep-ph]

30. V.P. Gusynin, S.G. Sharapov, Transport of Dirac quasiparticles in graphene: Hall and optical conductivities. Phys. Rev. B **73**, 245411 (2006). arXiv:cond-mat/0512157

31. E.V. Gorbar, V.A. Miransky, I.A. Shovkovy, Dynamics in the normal ground state of dense relativistic matter in a magnetic field. Phys. Rev. D **83**, 085003 (2011). arXiv:1101.4954 [hep-th]

32. M.A.V. Basagoiti, Transport coefficients and ladder summation in hot gauge theories. Phys. Rev. D **66**, 045005 (2002). arXiv:hep-ph/0204334

33. J.M.M. Resco, M.A.V. Basagoiti, Color conductivity and ladder summation in hot QCD. Phys. Rev. D **63**, 056008 (2001). arXiv:hep-ph/0009331

34. N. Iqbal, H. Liu, Real-time response in AdS/CFT with application to spinors. Fortschr. Phys. **57**, 367 (2009). arXiv:0903.2596 [hep-th]

35. B.I. Halperin, Helv. Phys. Acta **56**, 75 (1983)

36. Y. Kopelevich, B. Raquet, M. Goiran, W. Escoffier, R.R. da Silva, J.C. Medina Pantoja, I.A. Luk'yanchuk, A. Sinchenko, P. Monceau, Searching for the fractional quantum Hall effect in graphite. Phys. Rev. Lett. **103**, 116802 (2009)

37. Y.-F. Hsu, G.-Y. Guo, Anomalous integer quantum Hall effect in AA-stacked bilayer graphene. Phys. Rev. B **82**, 165404 (2010). arXiv:1008.0748 [cond-mat]

38. Y. Kopelevich, V.V. Lemanov, S. Moehlecke, J.H.S. Torrez, Landau level quantization and possible superconducting instabilities in highly oriented pyrolitic graphite. Fiz. Tverd. Tela **41**, 2135 (1999) [Phys. Solid State **41**, 1959 (1999)]

39. H. Kempa, Y. Kopelevich, F. Mrowka, A. Setzer, J.H.S. Torrez, R. Hoehne, P. Esquinazi, Solid State Commun. **115**, 539 (2000)

40. M.S. Sercheli, Y. Kopelevich, R.R. da Silva, J.H.S. Torrez, C. Rettori, Solid State Commun. **121**, 579 (2002)

41. Y. Kopelevich, P. Esquinazi, J.H.S. Torres, R.R. da Silva, H. Kempa, F. Mrowka, R. Ocana, Metal-insulator-metal transitions, superconductivity and magnetism in graphite. Stud. H-Temp. Supercond. **45**, 59 (2003). arXiv:cond-mat/0209442

Chapter 22
A Review of Magnetic Phenomena in Probe-Brane Holographic Matter

Oren Bergman, Johanna Erdmenger, and Gilad Lifschytz

22.1 Introduction

The behavior of strongly interacting matter subject to background magnetic fields is an interesting and physically relevant problem in many different scenarios, ranging from the effective 2d electron gas in graphene, to magnetars, which are neutron stars with a strong magnetic field. Magnetic fields give rise to a rich array of phenomena. Some examples in QCD are the magnetic catalysis of chiral symmetry breaking [1–3], anomaly-driven phases of baryonic matter [4], and the chiral magnetic effect [5]. It has also been suggested that magnetic fields induce ρ-meson condensation and superconductivity in the QCD vacuum [6, 7]. There are also many interesting examples in condensed matter physics, most notably the fractional quantum Hall effect [8, 9].

Gauge/gravity duality, also known as holographic duality, has emerged in recent years as a particularly useful approach to strong-coupling dynamics. Although it does not seem to be directly applicable to physical systems, this approach can be used to study theoretical systems that exhibit the same type of phenomena, and that capture some of the relevant physics. The techniques of holographic duality are especially efficient in addressing questions associated with finite temperature and density, background fields and transport properties, that are difficult to study

O. Bergman (✉)
Technion, Israel Institute of Technology, Haifa 32000, Israel
e-mail: bergman@physics.technion.ac.il

J. Erdmenger
Max Planck Institute for Physics, 80805 Munich, Germany
e-mail: jke@mppmu.mpg.de

G. Lifschytz
Department of Mathematics and Physics, University of Haifa at Oranim, Tivon 36006, Israel
e-mail: giladl@research.haifa.ac.il

D. Kharzeev et al. (eds.), *Strongly Interacting Matter in Magnetic Fields*,
Lecture Notes in Physics 871, DOI 10.1007/978-3-642-37305-3_22,
© Springer-Verlag Berlin Heidelberg 2013

using other non-perturbative methods. This approach can also lead to new ideas for constructing effective theories of the physical phenomena one is interested in.

Holographic models are divided into two main types, commonly referred to as top-down models and bottom-up models. In top-down models the bulk gravitational description of the system corresponds to a consistent solution of a well-defined quantum gravity theory, either in the context of the full string theory or in terms of the low-energy effective supergravity theory. This then defines some particular strong-coupling boundary dynamics. In bottom-up models, on the other hand, one builds into the description what one needs in order to produce the desired boundary dynamics. Each approach has advantages and disadvantages. Top-down models are more firmly grounded than bottom-up models, however they are more restrictive in terms of the variety and scope of phenomena they can exhibit.

Probe-brane models are a class of top-down holographic models, in which matter fields transforming in the fundamental representation of a gauge group are incorporated by embedding "flavor" D-branes in the gravitational background dual to the gauge theory [10]. These branes are treated as probes, in the sense that we neglect their backreaction on the background. (This corresponds to the 'quenched' approximation in the dual gauge theory, where matter loops are neglected in computing gluon amplitudes.) The matter fields are manifest in this construction: they correspond to the open strings between the flavor branes and the "color" branes that make up the background. In particular, one can easily design probe brane models in which the light matter degrees of freedom are purely fermionic, which is obviously a desirable feature for many physical systems, including QCD and condensed matter electron systems.

The matter dynamics is determined by the properties of the probe brane embedding. In particular, the fluctuations of the probe brane worldvolume fields correspond to gauge-invariant composite operators that describe the mesonic states of the matter system. There are generically two types of embeddings at finite temperature: "BH embeddings", in which the brane extends to the horizon of the background, and "MN embeddings", in which the brane terminates outside the horizon. The two embedding types describe different phases of the matter in the dual gauge theory. For example, in the MN phase the mesons are stable since they are associated with real eigenfrequencies of the probe brane fluctuations. In the BH phase, on the other hand, some of the energy of the fluctuations is dissipated into the black hole, leading to complex eigenfrequencies and damping. In this case the mesons have a finite lifetime. MN embeddings are favored at low temperature, and as the temperature is increased one generically observes a first order phase transition to a BH embedding.

The most extensively studied probe-brane models are the D3–D7 model [10], in which D7-branes are added to the D3-brane background, and the D4–D8 (or Sakai-Sugimoto) model [11], in which D8-branes and anti-D8-branes are added to the background of D4-branes compactified on a circle. Both models describe a strongly-coupled gauge theory in four dimensions with fundamental matter degrees of freedom, and both exhibit a number of phenomena similar to QCD. More recently, a different D3–D7 system, more closely related to the D4–D8 system, has been used as a model of strongly-interacting fermionic matter in three spacetime dimensions [12–15]. We will refer to this as the D3–D7' model. This model exhibits several interesting phenomena that are familiar in planar condensed matter systems.

Probe-brane models are especially well-designed to study the properties of the dual matter systems at non-zero density and in background electromagnetic fields. Both are implemented by turning on specific components of the probe-brane world-volume gauge field, and solving the resulting coupled differential equations for the embedding and the gauge fields. Here one observes another basic difference between the two types of embeddings in terms of their response to a background electric field. MN embeddings correspond to electrical insulators with a mass-gap to charged excitations, and BH embeddings describe gapless conductors.

Probe-brane models also exhibit a number of interesting phenomena in a background magnetic field, which are qualitatively similar to the phenomena listed in the beginning. In this paper we will review how each of the three models mentioned above respond at non-zero density to a background magnetic field in various situations. In particular, we will encounter the magnetic catalysis effect in both the D3–D7 and D4–D8 models. In the D3–D7 model we will also demonstrate the formation of a superfluid state. In the D4–D8 model we will describe anomaly-generated currents and baryonic states, as well as a metamagnetic-like transition. In the D3–D7' model we will see both quantum and anomalous Hall effects, as well as how the magnetic field influences the instability to the formation of stripes, and the zero-sound mode.

This paper is divided into three main sections, reviewing each of the probe-brane models in turn. For completeness let us mention that magnetic fields also play an important role in the related D3–D5 system, where the probe D5-brane corresponds to additional $(2 + 1)$-dimensional degrees of freedom in the dual gauge theory. In this model, the magnetic field leads to a phase transition of Berezinskii-Kosterlitz-Thouless (BKT) type [16, 17]. For brevity we do not discuss this model in this review.

22.2 The D3–D7 Model

22.2.1 Brane Construction

The starting point for this model is the usual configuration of the AdS/CFT correspondence [18] which involves a stack of N D3 branes. This has an open string interpretation in which the low-energy degrees of freedom are described by $U(N)$ $\mathcal{N} = 4$ super-Yang-Mills theory. On the other hand, in the closed string interpretation of N D3 branes, the low-energy *near-horizon* limit gives rise to the space $AdS_5 \times S^5$. Identifying the two pictures leads to the AdS/CFT correspondence.

Let us now add N_f probe D7-branes to this configuration, as first done in [10] and reviewed in detail in [19]. Within $(9 + 1)$-dimensional flat space, the D3-branes are extended along the 0123 directions, whereas the D7-branes are extended along the 01234567 directions. This configuration preserves $1/4$ of the total amount of supersymmetry in type IIB string theory (corresponding to 8 real supercharges) and has an $SO(4) \times SO(2)$ isometry in the directions transverse to the D3-branes. The $SO(4)$ rotates x^4, x^5, x^6, x^7, while the $SO(2)$ group acts on x^8, x^9. Separating the D3-branes from the D7-branes in the $(8, 9)$ directions by a distance l explicitly breaks the $SO(2)$ group. These geometrical symmetries are also present in the dual field theory: The dual field theory is an $\mathcal{N} = 2$ supersymmetric $(3 + 1)$-dimensional theory in which the degrees of freedom of $\mathcal{N} = 4$ super-Yang-Mills theory are coupled to N_f hypermultiplets of flavor fields with fermions and scalars (ψ_i, q^n), $i = 1, 2, n = 1, 2$, which transform in the fundamental representation of the gauge group. Separating the D7-branes from the D3-branes corresponds to giving a mass to the hypermultiplets.

For massless flavor fields, the Lagrangian is classically invariant under conformal transformations $SO(4, 2)$.[1] Moreover, the theory is invariant under the R-symmetries $SU(2)_R$ and $U(1)_R$ as well as under the global $SU(2)_\Phi$, which rotates the scalars in the adjoint hypermultiplet. Note that the mass term in the Lagrangian breaks the $U(1)_R$ symmetry explicitly. If all N_f flavor fields have the same mass m, the field theory is invariant under a global $U(N_f)$ flavor group. The baryonic $U(1)_B$ symmetry is a subgroup of the $U(N_f)$ flavor group. These symmetries of the field theory side may be identified with symmetries of the D3–D7 brane intersection and hence also with the dual gravity description.

For this field theory, gauge invariant composite operators may now be constructed which transform in suitable representations of the $SU(2) \times SU(2) \times U(1)$ symmetry group isomorphic to the geometrical $SO(4) \times SO(2)$. These operators are expected to be dual to the fluctuations of the D7-brane which transform in the same representation, as worked out in detail in [20]. An example of a meson operator is given by

$$\mathcal{M}^A = \bar{\psi}_i \sigma^A{}_{ij} \psi_j + \bar{q}^m X^A q^m, \quad (i, m = 1, 2), \tag{22.1}$$

with X^A the vector (X^8, X^9) of adjoint scalars associated with the $(8, 9)$ directions, and $\sigma^A \equiv (\sigma^1, \sigma^2)$ a doublet of Pauli matrices. Thus (22.1) has charge $+2$ under $U(1)_R$. It is a singlet under both $SU(2)_\Phi$ and $SU(2)_R$. The conformal dimension is $\Delta = 3$. This operator may be viewed as a supersymmetric generalization of a mesonic operator in QCD, with the index A labeling two scalar mesons.

The standard AdS/CFT duality relates the $\mathcal{N} = 4$ super-Yang-Mills degrees of freedom to supergravity on $AdS_5 \times S^5$. In addition, there are new degrees of freedom associated to the D7-brane worldvolume fields originating from the open strings on the D7-brane. The additional duality maps these to the mesonic operators in the field theory. This is an open-open string duality, as opposed to the standard AdS/CFT correspondence, which is an open-closed string duality. The dynamics of the D7-brane is described by the Dirac-Born-Infeld (DBI) action

---

[1] However note that the scale-invariance is broken at the quantum level since the beta function is proportional to N_f/N_c and therefore non-vanishing. In the limit $N_c \to \infty$ with N_f being fixed, the beta function is approximately zero, i.e. we may treat the theory as being scale invariant also at the quantum level.

$$S_{D7} = -\frac{\mu_7}{g_s} \int d^8\xi \sqrt{-\det\left(G_{ab} + B_{ab}^{(2)} + 2\pi\alpha' F_{ab}\right)}, \qquad (22.2)$$

where $\mu_7 = [(2\pi)^7 \alpha'^4]^{-1}$. G and $B^{(2)}$ are the induced metric and two-form field on the probe brane worldvolume, and F_{ab} is the worldvolume field strength. The D7-brane action also contains a fermionic term S_{D7}^f. In addition there may also be contributions of Wess-Zumino form. An example for this will be discussed below.

Let us write the $AdS_5 \times S^5$ metric in the form

$$ds^2 = \frac{r^2}{R^2} \eta_{ij}\, dx^i\, dx^j + \frac{R^2}{r^2}\left(d\rho^2 + \rho^2\, d\Omega_3^2 + dx_8^2 + dx_9^2\right), \qquad (22.3)$$

with $i, j = 0, 1, 2, 3$, $\rho^2 = x_4^2 + \cdots + x_7^2$, $r^2 = \rho^2 + x_8^2 + x_9^2$ and R the AdS radius. Since the D7-brane is transverse to x_8, x_9 in flat space, we see that it extends along AdS_5 and wraps an S^3 inside S^5 in the near-horizon background. The action for a static D7-brane embedding, for which F_{ab} may be consistently set to zero on its world-volume, is given from (22.2) up to angular factors by

$$S_{D7} = -\frac{\mu_7}{g_s} \int d^8\xi\, \rho^3 \sqrt{1 + \dot{x}_8^2 + \dot{x}_9^2}, \qquad (22.4)$$

where a dot indicates a ρ derivative (e.g. $\dot{x}_8 \equiv \partial_\rho x_8$). The ground state configuration of the D7-brane then corresponds to the solution of the equation of motion

$$\frac{d}{d\rho}\left[\frac{\rho^3}{\sqrt{1 + \dot{x}_8^2 + \dot{x}_9^2}}\frac{dx}{d\rho}\right] = 0, \qquad (22.5)$$

where x denotes either x_8 or x_9. Clearly the action is minimized by x_8, x_9 being any arbitrary constant. Therefore the embedded D7-brane is flat. According to string theory, the choice of the position in the x_8, x_9 plane corresponds to choosing the quark mass in the gauge theory action. The fact that x_8, x_9 are constant at all values of the radial coordinate ρ, which corresponds to the holographic renormalization scale, may be interpreted as non-renormalization of the mass in the dual field theory.

In general, the equations of motion have asymptotic ($\rho \to \infty$) solutions of the form

$$x = l + \frac{c}{\rho^2} + \cdots, \qquad (22.6)$$

where l is related to the quark mass m by

$$m = \frac{l}{2\pi\alpha'}. \qquad (22.7)$$

In agreement with the standard AdS/CFT result about the asymptotic behavior of supergravity fields near the boundary, the parameter c must correspond to the vev of an operator with the same symmetries as the mass and of dimension three, since ρ carries energy dimension. c is therefore a measure of the quark condensate $\bar{\psi}\psi$.

c is obtained from $\partial \mathcal{L} / \partial m$ which in addition to the fermion bilinear also includes scalar squark terms. We may consistently assume that the squarks have zero vev. Moreover, supersymmetry requires that a vev for c must be absent since c is an F-term of a chiral superfield: $\tilde{\psi} \psi$ is the F-term of $\tilde{Q} Q$. Supersymmetry is broken if $c = \langle \tilde{\psi} \psi \rangle \neq 0$. This is reflected also in the supergravity solution: The solutions to the supergravity equations of motion with c non-zero are not regular in AdS space and are therefore excluded.

We therefore consider the regular supersymmetric embeddings of the D7-brane for which the quark mass m may be non-zero, but the condensate c vanishes. For massive embeddings, the D7-brane is separated from the stack of D3-branes in either the x_8 or x_9 directions, where the indices refer to the coordinates given in (22.3). In this case the radius of the S^3 becomes a function of the radial coordinate r in AdS$_5$. At a radial distance from the deep interior of the AdS space given by the hypermultiplet mass, the radius of the S^3 shrinks to zero. From a five-dimensional AdS point of view, this gives a minimal value for the radial coordinate r beyond which the D7-brane cannot extend further. This is in agreement with the induced metric on the D7-brane world-volume, which is given by

$$ds^2 = \frac{\rho^2 + l^2}{R^2} \eta_{ij} \, dx^i \, dx^j + \frac{R^2}{\rho^2 + l^2} \, d\rho^2 + \frac{R^2 \rho^2}{\rho^2 + l^2} \, d\Omega_3{}^2, \tag{22.8}$$

$$d\Omega_3{}^2 = d\psi^2 + \cos^2 \psi \, d\beta^2 + \sin^2 \psi \, d\gamma^2, \tag{22.9}$$

where $\rho^2 = r^2 - l^2$ and Ω_3 are spherical coordinates in the 4567-space. For $\rho \to \infty$, this is the metric of $AdS_5 \times S^3$. When $\rho = 0$ (*i.e.* $r^2 = l^2$), the radius of the S^3 shrinks to zero.

22.2.2 Finite Temperature

The finite temperature system is realized holographically by placing the D7-brane in an AdS-Schwarzschild black hole background with metric given by

$$ds^2 = \frac{r^2}{2R^2} \left(\frac{dx^2(r_h{}^4 + r^4)}{r^4} - \frac{dt^2(r^4 - r_h{}^4)^2}{r^4(r_h{}^4 + r^4)} \right)$$

$$+ \frac{R^2}{r^2} (dL^2 + d\rho^2 + L^2 \, d\phi^2 + \rho^2 \, d\Omega_3^2), \tag{22.10}$$

where $r^2 = \rho^2 + L^2$. The temperature is given by $T = r_h / (\pi R^2)$.

In this metric we have introduced polar coordinates (L, ϕ) in the (x_8, x_9) plane and consider solutions for D7-brane embeddings with $L = L(\rho)$, $\phi = $ const. The asymptotic near-boundary behavior of these brane embeddings is given by

$$L(\rho) = l + \frac{c}{\rho^2} + \cdots, \tag{22.11}$$

where $m = 1/(2\pi\alpha')$ is the bare quark mass and c is proportional to the condensate $\langle\bar\psi\psi\rangle$. At finite temperature, supersymmetry is broken and brane embeddings with c non-zero are possible, in contrast to the supersymmetric case discussed above.

Depending on the quark mass, there are two different types of embeddings: Those which reach the black hole, and those which do not since the S^3 they wrap shrinks to zero outside the black hole horizon. The first type of branes is referred to as 'black hole' (BH) embeddings, while the second type is referred to as 'Minkowski' (MN) embeddings. In the BH case, fluctuations of the probe brane have complex eigenfrequencies or quasi-normal modes, which means that the mesons associated with these fluctuations decay. In the MN phase, the mesons are stable. The phase transition between the two types of embeddings is first order.

At finite temperature, the solution with $m = 0$ also has $c = 0$. However, in gravity backgrounds corresponding to confining field theories, brane embeddings with $m = 0$, $c \neq 0$ are possible. These realize spontaneous chiral symmetry breaking [21].

22.2.3 Magnetic Catalysis

Let us now consider, as was first done in [22], a magnetic field induced by a pure gauge B-field in the worldvolume direction of the D3-branes,

$$B^{(2)} = B\,dx^2 \wedge dx^3, \tag{22.12}$$

which satisfies $dB^{(2)} = 0$. This contributes to the DBI action (22.2). Since $B^{(2)}$ and $2\pi\alpha'F$ enter (22.2) in the same way, we may trade $B^{(2)}$ for a gauge field on the probe brane via $F = -B^{(2)}/2\pi\alpha'$, which justifies interpreting B of (22.12) as a magnetic field.

In addition, there is a non-trivial Wess-Zumino contribution to the action, which at first order in α' is of the form

$$S_{WZ} = 2\pi\alpha'\mu_7 \int F \wedge C^{(6)}. \tag{22.13}$$

In the presence of the B-field, this leads to an additional non-trivial contribution to the action, as explained in detail in [22]. This gives rise to a non-trivial $C^{(6)}$ which breaks supersymmetry on the worldvolume of the D7-brane.

Since supersymmetry is broken, the D7-brane now has a profile which depends on ρ as in (22.11), even at zero temperature. The Lagrangian corresponding to (22.2) takes the form

$$\mathcal{L} = -\frac{\mu_7}{g_s}\rho^3 \sin\psi \cos\psi \sqrt{1 + L'^2}\sqrt{1 + \frac{R^4 B^2}{(\rho^2 + L^2)^2}}. \tag{22.14}$$

For $m = 0$, the brane embedding solution obtained from this Lagrangian has non-zero $c \propto \langle\bar\psi\psi\rangle$ in (22.11). The magnetic field therefore induces spontaneous chiral

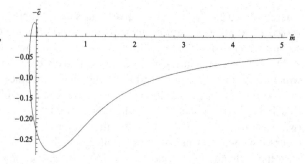

Fig. 22.1 $-\tilde{c}(\tilde{m})$ at vanishing temperature, with $-\tilde{c} = \langle \bar{\psi}\psi \rangle (2\pi\alpha')^3/(R^3 B^{3/2})$, $\tilde{m} = m(2\pi\alpha')/(R\sqrt{B})$. Reproduced from [23]

symmetry breaking, a phenomenon known as *magnetic catalysis* [1–3]. For large m, the condensate may be calculated analytically [22] and is found to be

$$\langle \bar{\psi}\psi \rangle \propto -c = -\frac{R^4}{4l}B^2. \tag{22.15}$$

For small m, c has to be evaluated numerically. The result is shown in Fig. 22.1. By evaluating the free energy, it has been shown that when there is more than one solution, the one with the larger condensate is preferred.

At small values of the magnetic field, it is possible to analytically evaluate the shift of the meson masses due to its presence. For fluctuations of the embedding scalar in particular, a Zeeman splitting is observed [22]. While in the absence of a magnetic field, in the supersymmetric case described here, the scalar meson mass obtained from the fluctuations is $M_0(n) = 2m/\sqrt{\lambda} \cdot \sqrt{(n+1)(n+2)}$ [20], for non-zero magnetic field there is a mass splitting

$$M_\pm = M_0 \pm \frac{1}{\sqrt{\lambda}}\frac{B}{m}. \tag{22.16}$$

A review of magnetic catalysis in probe D7-brane systems is given in [24]. Magnetic catalysis is also found in systems involving N_f D7 branes where the back-reaction of the metric on the background geometry is taken into account [25–27]. Out-of-equilibrium dynamics associated with the phase transition induced by magnetic catalysis has been investigated in [28].

In the finite temperature case, there is a competition between two mechanisms: The black hole attracts the D7-brane, while it is repelled at small radii by the magnetic field. This implies a phase transition between a phase where the D7-brane reaches the black hole and one where it does not. This is shown in Fig. 22.2 for different values of the magnetic field, where the dimensionless ratio B/T^2 is used. A detailed discussion of the normalization is found in [23].

There is a critical value for B/T^2 above which the probe brane is repelled from the black hole for all values of the bare quark mass: In this case, chiral symmetry is spontaneously broken and the mesons are stable. The phase diagram is shown in Fig. 22.3.

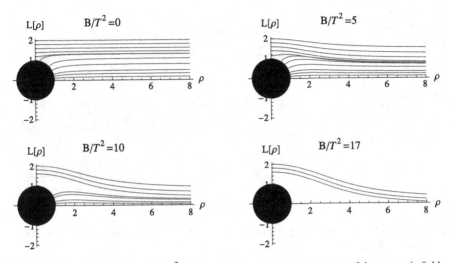

Fig. 22.2 Increasing values of B/T^2 for fixed T show the repulsive nature of the magnetic field. We see that for large enough B/T^2, the melted phase is never reached, and the chiral symmetry is spontaneously broken. Figure reproduced from [23]

Fig. 22.3 Phase diagram for the D3/D7 system in the $(B/T^2, T/m)$ plane. Figure from [23]

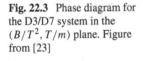

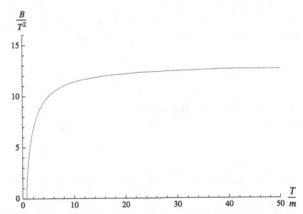

More involved phase diagrams are obtained if a $U(1)$ chemical potential and density are turned on in addition to the magnetic field by considering a non-trivial profile for the $U(1)$ gauge field on the D7-brane [29–31]. An example of a phase diagram is shown in Fig. 22.4.

22.2.4 Superfluid

At finite isospin density, the D3–D7 model realizes a holographic superfluid [32, 33]. Finite isospin density is obtained by considering two coincident D7-branes, and using an ansatz for solving the equations of motion, which involves a non-trivial

Fig. 22.4 Phase transitions at constant B field in the (T, μ) plane for the D3/D7 model. Figure reproduced from [31] by kind permission of the authors. c, d, J refer to the condensate, density and electric current, respectively

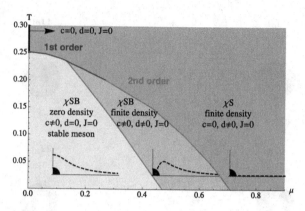

profile for the temporal component of the $SU(2)$ worldvolume gauge field, with asymptotic behavior

$$A^3{}_t(\rho) \propto \mu^3 + \frac{d^3}{\rho^2}. \tag{22.17}$$

μ^3 breaks the $SU(2)$ symmetry explicitly to a residual $U(1)_3$. In the presence of this background, the energetically favored solution also involves a non-trivial spatial component of the worldvolume gauge field,

$$A^1{}_x(\rho) \propto \frac{d^1{}_x}{\rho^2}. \tag{22.18}$$

Here the leading contribution is absent in the asymptotic behavior, so the $U(1)_3$ symmetry is spontaneously broken. $A^1{}_x$ is dual to a condensate of the form

$$d^1{}_x \propto \langle \bar{\psi} \sigma^1 \gamma_x \psi + \bar{\phi} \sigma^i \partial_x \phi \rangle, \tag{22.19}$$

which is the supersymmetric equivalent of the ρ meson. The calculation of the frequency-dependent conductivity $\sigma(\omega)$ for this solution shows that it describes a superfluid: $\sigma(\omega)$ displays a gap. For the Sakai-Sugimoto model discussed below, a similar condensation mechanism has been found in [34] and superfluidity has been discussed in [35].

As discussed in [36, 37], a similar condensation process also happens when the profile (22.17) for the *temporal* component of the $SU(2)$ gauge field is replaced by a non-trivial profile for a *spatial* component of the form

$$A^3{}_x = By, \tag{22.20}$$

which corresponds to a background magnetic field. In this case a similar condensation as above takes place. This has been demonstrated by analyzing the fluctuations about the magnetic field background [37]: The quasi-normal modes of particular fluctuations cross into the upper half of the complex frequency plane above a critical value of the magnetic field, indicating an instability. This is shown in Fig. 22.5.

Fig. 22.5 Quasi-normal modes cross into the upper half plane above a critical magnetic field, signaling an instability. Figure reproduced from [37]

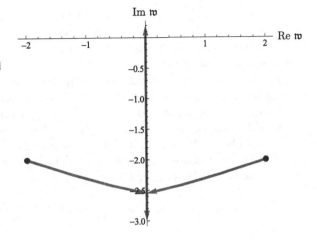

Finally let us note that the Hall conductivity has been calculated for the D3/D7 model in [38]. Unlike the isospin case, the ground state involving the ρ condensate is spatially modulated for the magnetic field background, leading to an Abrikosov lattice [39]. A similar ρ meson condensation mechanism in a background magnetic field has been found in the context of field theory in [6, 7, 40], based on similar earlier results in electroweak theory [41]. For the Sakai-Sugimoto model which we discuss below, a similar mechanism has been discussed in [42, 43].

22.3 The D4–D8 (Sakai-Sugimoto) Model

22.3.1 Basics

N_c D4-branes on $\mathbb{R}^{1,3} \times S^1$ with anti-periodic boundary conditions for fermions provide a holographic model for the low energy behavior of 4d $SU(N_c)$ Yang-Mills theory with $g_{YM}^2 = 4\pi g_s \sqrt{\alpha'}/R_4$ [44]. The near-horizon background at zero temperature is given by (we work with dimensionless coordinates rescaled by R)

$$ds_{con}^2 = u^{\frac{3}{2}}\left(-dx_0^2 + dx^2 + f(u)\,dx_4^2\right) + u^{-\frac{3}{2}}\left(\frac{du^2}{f(u)} + u^2\,d\Omega_4^2\right),$$

$$e^{\Phi} = g_s u^{3/4}, \qquad F_4 = 3\pi\left(\alpha'\right)^{3/2} N_c\,d\Omega_4, \tag{22.21}$$

where $f(u) = 1 - (u_{KK}^3/u^3)$, $u_{KK} = 4R^2/(9R_4^2)$ and $R = (\pi g_s N_c)^{1/3}\sqrt{\alpha'}$. The IR "wall" at $u = u_{KK}$ implies that the dual gauge theory is confining. At nonzero temperature there is another possible background with a metric

$$ds_{dec}^2 = u^{\frac{3}{2}}\left(-f(u)\,dx_0^2 + dx^2 + dx_4^2\right) + u^{-\frac{3}{2}}\left(\frac{du^2}{f(u)} + u^2\,d\Omega_4^2\right), \tag{22.22}$$

where $f(u) = 1 - (u_T^3/u^3)$ and $u_T = (4\pi/3)^2 R^2 T^2$. This background becomes the dominant one when $T > 1/(2\pi R_4)$. The presence of a horizon at $u = u_T$ in this background indicates that the gauge theory undergoes a (first order) deconfinement transition at this temperature.

Quarks are added to the model by including D8-branes and anti-D8-branes that are localized on the circle [11]. With N_f D8-branes at one point and N_f anti-D8-branes at another point, the model has N_f flavors of massless right-handed and left-handed fermions, and a $U(N_f)_R \times U(N_f)_L$ chiral symmetry. The 8-branes are treated as probes in the near horizon background of the D4-branes. The flavor dynamics is thus encoded in the 5d effective worldvolume theory of the D8-branes, which includes a DBI term and a CS term (in Lorentzian signature)[2]

$$S_{DBI} = -\mathcal{N} \int d^4x \, du \, u^{1/4} \sqrt{-\det(g_{MN} + f_{MN})}, \tag{22.23}$$

$$S_{CS} = -\frac{\mathcal{N}}{8} \int d^4x \, du \, \varepsilon^{MNPQR} a_M f_{NP} f_{QR}. \tag{22.24}$$

The dimensionless worldvolume gauge field a_M and field strength f_{MN} are defined as $a_M = (2\pi\alpha'/R)A_M$ and $f_{MN} = 2\pi\alpha' F_{MN}$, and the overall normalization is given by $\mathcal{N} = \mu_8 \Omega_4 R^9/g_s = (1/3)N_c R^6 (2\pi)^{-5}(\alpha')^{-3}$. The anti-D8-brane has a similar action in terms of its worldvolume gauge field $\bar{a}_M$. The DBI term is identical to that of the D8-brane, and the CS term has the opposite sign. We define the vector combination as $a_M^V = \frac{1}{2}(a_M + \bar{a}_M)$, and the axial combination as $a_M^A = \frac{1}{2}(\bar{a}_M - a_M)$.

In the low-temperature confining background (22.21) the D8-brane and anti-D8-brane connect at $u = u_0 \geq u_{KK}$ into a smooth U-shaped configuration (Fig. 22.6a), reflecting the spontaneous breaking of the $U(1)_R \times U(1)_L$ chiral symmetry to the diagonal $U(1)_V$. The embedding is determined by the DBI action (setting $f_{MN} = 0$)

$$S_{DBI}^{con} = -\mathcal{N} \int d^4x \, du \, u^4 \left[f(u)\big(x_4'(u)\big)^2 + \frac{1}{u^3 f(u)} \right]^{\frac{1}{2}}, \tag{22.25}$$

which implies an asymptotic behavior

$$x_4(u) \approx \frac{L}{2} - \frac{2}{9} \frac{u_0^4 \sqrt{f(u_0)}}{u^{9/2}}, \tag{22.26}$$

where L is the asymptotic brane-antibrane separation.

The normalizable fluctuations of the D8-brane worldvolume fields in this embedding correspond to the (low spin) mesons of the model. Their mass scale is set by u_0, which we can think of as the mass of a "constituent quark" described by an open string from u_0 to u_{KK}. There is one massless pseudoscalar field φ, precisely as one

---

[2] For simplicity, we will consider the single flavor case with one D8-brane and one anti-D8-brane. This does not affect any of the results qualitatively.

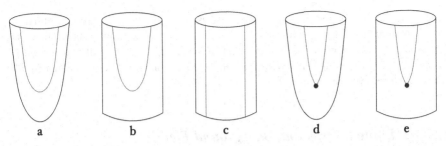

Fig. 22.6 D8-brane embeddings in the Sakai-Sugimoto model: (**a**) confined vacuum, (**b**) deconfined vacuum, (**c**) deconfined plasma, (**d**) confined nuclear matter, (**e**) deconfined nuclear matter

expects from the broken chiral symmetry. This is related to the η' meson in QCD. It appears (in a gauge with $a_u = 0$) as the zero mode a_μ^{A}:[3]

$$a_\mu^A\left(x^\mu, u\right) = -\partial_\mu \varphi\left(x^\mu\right)\psi_0(u) + \text{higher modes}, \tag{22.27}$$

where

$$\psi_0(u) = \frac{2}{\pi} \arctan \sqrt{\frac{u^3}{u_{KK}^3} - 1}. \tag{22.28}$$

Baryons are described by D4-branes wrapped on S^4 inside the D8-brane. Their charge comes from the N_c strings which must be attached to the wrapped D4-brane to cancel a tadpole due to the background RR field. These strings end on the D8-brane, giving N_c units of charge.[4]

In the high-temperature deconfining background (22.22) the D8-branes and anti-D8-branes can be either connected (Fig. 22.6b) or disconnected, with $x_4(u) = L/2$ (Fig. 22.6c), the latter corresponding to the restoration of the chiral symmetry [46]. The DBI action in the high-temperature deconfining background is very similar:

$$S_{DBI}^{dec} = -\mathcal{N} \int d^4x\, du\, u^4 \left[f(u)\left(x_4'(u)\right)^2 + \frac{1}{u^3} \right]^{\frac{1}{2}}. \tag{22.29}$$

Consequently the properties of the U embedding in this background are qualitatively similar to those of the embedding in the confining background, for example in terms of the spectrum of mesons. In the disconnected embedding there are no normaliz-

---

[3]Note that, although the boundary value is non-zero, this is a normalizable mode since the field strength is normalizable. Ordinarily, boundary values of bulk fields correspond to parameters in the boundary theory. But in this case there is a possible ambiguity, since the boundary value of a_μ^A can also describe a non-trivial gradient of the pseudoscalar field.

[4]For $N_f > 1$ the baryons correspond to instantons in the non-abelian D8-brane theory [11, 45]. This reproduces the known description of baryons as Skyrmions in the chiral Lagrangian. In this description the baryon charge comes from the CS term coupling the $U(1)_V$ field to the instanton density in the $SU(N_f)_V$ part.

able fluctuations corresponding to mesons, as one expects in a chiral-symmetric phase. Comparing the (Euclidean) actions of the two embeddings shows that the disconnected one becomes dominant when $T > 0.154/L$. In particular, for small L ($L < 0.97R_4$) the gauge theory has an intermediate phase of deconfinement with broken chiral symmetry.

22.3.2 Finite Density and Background Fields

The D8-brane worldvolume vector and axial gauge fields are dual to conserved vector and axial currents in the gauge theory, and therefore[5]

$$j_{V,A}^\mu = \frac{1}{\mathcal{N} V_4} \frac{\partial S_{D8}|_{on\text{-}shell}}{\partial a_\mu^{V,A}(u \to \infty)}. \qquad (22.30)$$

The chemical potentials are defined by[6]

$$\mu_V = a_0^V(u \to \infty) \quad \text{and} \quad \mu_A = a_0^A(u \to \infty). \qquad (22.31)$$

In our conventions quarks carry one unit of vector charge and baryons carry N_c units. Nevertheless we will refer to the vector current as the "baryon number current". We are also interested in studying the effects of background "electromagnetic" fields that couple to this current, which correspond to turning on spacetime dependent boundary values of the worldvolume gauge field, in particular

$$e_i = f_{0i}(u \to \infty), \qquad b_i = \varepsilon_{ijk} f_{jk}(u \to \infty). \qquad (22.32)$$

In some situations one may be required to add boundary terms to the action. These are especially relevant if there is a CS term in the bulk. In deriving the equations of motion from the variational principle one usually assumes that the surface terms vanish. However in some instances one has to be more careful. The surface terms (in the $a_u = 0$ gauge) are given in general by

$$\delta S|_{on\text{-}shell} = \int d^4x \frac{\partial \mathcal{L}}{\partial a_\nu'} \delta a_\nu \Big|_{u_{min}}^{\infty} + \int d^3x\, du\, \frac{\partial \mathcal{L}}{\partial(\partial_\mu a_\nu)} \delta a_\nu \Big|_{x_\mu \to -\infty}^{x_\mu \to \infty}. \qquad (22.33)$$

In holography the boundary values of the fields at $u \to \infty$ are fixed, so $\delta a_\mu(u \to \infty) = 0$. However $\delta a_\mu(u_{min})$ and $\delta a_\mu(x_\mu \to \pm\infty)$ need not vanish. Therefore a sur-

---

[5] The axial symmetry is broken by an anomaly. However this is a subleading effect at large N_c which we can neglect. In particular, we will assume that the one-flavor pseudoscalar η' is massless. For a discussion of the $U(1)_A$ anomaly and the η' mass in the context of the Sakai-Sugimoto model see [11, 47].

[6] We would like to stress that this is a gauge invariant definition. The standard boundary condition on the gauge field in AdS/CFT fixes the value of $a_M(u \to \infty)$. In this case only the transformations that vanish at $u \to \infty$ are gauged in the bulk. In particular, these transformations do not change the asymptotic value of a_0.

face term may be non-trivial if the fields extend to these boundaries. In order to have a well-defined variational principle we must therefore add boundary terms $S_\partial(u_{min}) + S_\partial(x_\mu \to \pm\infty)$, whose variation cancels the surface terms in the variation of the bulk action.

These boundary terms also allow one to derive an alternative and useful definition of the conserved currents. We do this by varying the *off-shell* action, now allowing $a_\mu(u \to \infty)$ to vary, and then going on-shell by applying the equations of motion. Due to the boundary terms, only the surface term at $u \to \infty$ remains. Thus

$$j^\mu = \frac{1}{\mathcal{N}} \frac{\partial \mathscr{L}}{\partial a'_\mu} (u \to \infty)\bigg|_{on\text{-}shell}. \tag{22.34}$$

In particular this relates the charge density in the boundary theory to the bulk radial electric field. The boundary term $S_\partial(u_{min})$ for a_0 should then be interpreted as a source term for this field. One could in principle also add boundary terms at $u \to \infty$. These have no effect on the derivation of the equations of motion, but may change the value of the on-shell action, and therefore may lead to additional contributions to (22.34).

A state with a non-zero baryon number density corresponds to an embedding with a radial electric field. In particular, for a U embedding (in both the confining and deconfining backgrounds) this requires the addition of a baryonic source at the tip, corresponding to a uniform spatial distribution of wrapped D4-branes (Figs. 22.6d, 22.6e) [48] (see also [49, 50]). We assume that the distribution is dilute enough so that we can ignore interactions between the D4-branes. We should therefore include the D4-brane action, which is given by (in the confining background)

$$S_{D4} = -\mathcal{N} V_4 n_{D4} N_c \left(\frac{1}{3} u_0 - a_0^V(u_0) \right), \tag{22.35}$$

where n_{D4} is the density of D4-branes. The first term is the D4-brane DBI action, and corresponds to the baryon mass, and the second term comes from the N_c strings that connect each D4-brane to the D8-brane. The second term is precisely the boundary term at $u_{min} = u_0$ that was discussed above.

The resulting asymptotic behavior of the gauge field is

$$a_0^V(u) \approx \mu_V - \frac{2}{3} \frac{d}{u^{3/2}}, \tag{22.36}$$

where $d = N_c n_{D4}$ is the baryon number density. On the other hand, extremizing the action with respect to n_{D4} fixes the value of the gauge field at the tip to $a_0^V(u_0) = u_0/3 = m_{baryon}/N_c$. This implies, as expected, that a non-zero density configuration exists only when the chemical potential is above the baryon mass. In fact the non-zero density state is always the dominant one. The transition to "nuclear matter" occurs at $\mu_V = m_{baryon}/N_c$. Near the critical point the density scales linearly with the chemical potential $d \sim \mu_V - m_{baryon}/N_c$. The D4-brane action also sources the embedding field $x_4(u)$, creating a cusp at $u = u_0$. This can be understood in terms of a force balance condition between the D4-branes pulling down and the D8-brane pulling up.

At high temperature the preferred embedding is parallel and there is a different finite density solution. In this case the gauge fields a_μ and $\bar{a}_\mu$ are independent, and the field theory has a conserved axial current as well as a baryon current. For a baryonic solution we take $a_0^V(\infty) = \mu_V$ and $a_0^A(\infty) = 0$. In addition, since the D8-brane and anti-D8-brane reach the horizon we must impose $a_0^V(u_T) = a_0^A(u_T) = 0$. Therefore the radial vector electric field, and thus the baryon number density, is non-zero when $\mu_V > 0$. In this phase $d \sim T^3 \mu_V$ for small μ_V.

22.3.3 Magnetic Catalysis of Chiral Symmetry Breaking

A strong magnetic field in QCD is believed to catalyze the spontaneous breaking of chiral symmetry [1–3]. The basic mechanism for this is that in a strong magnetic field all the quarks sit in the lowest Landau level, and the dynamics is effectively $1 + 1$-dimensional. This phenomenon has been exhibited in the Sakai-Sugimoto model in [51, 52].[7]

With a uniform background magnetic field b, the D8-brane action in the deconfining background becomes

$$S_{D8}^{dec} = -\mathcal{N} \int d^4x \, du \, u^4 \sqrt{\left(f(u)\left(x_4'(u)\right)^2 + \frac{1}{u^3} \right)\left(1 + \frac{b^2}{u^3} \right)}. \quad (22.37)$$

The U embedding has the same form as before (22.26), but now u_0 depends on the magnetic field, as shown in Fig. 22.7a. The mass scale associated with chiral symmetry breaking is seen to increase with the magnetic field. One therefore expects that chiral symmetry breaking becomes more favored as the magnetic field increases. This is indeed the case, as can be seen by comparing the Euclidean actions of the U and parallel embeddings as the temperature and magnetic field are varied. The resulting phase diagram is shown in Fig. 22.7b. We observe that in this model the critical temperature approaches a finite value at infinite magnetic field.

A qualitatively similar effect was observed in the D3–D7 model above. However in the D3–D7 model there is a critical value of B/T^2 above which the chiral symmetry is always broken, whereas in the D4–D8 there is a critical temperature above which the chiral symmetry is always broken.

It is also instructive to study the effect of the background magnetic field on the mesons. This was partly done in [55], in which the high spin mesons were studied. It was shown that the magnetic field enhances their stability by increasing their angular momentum, and thereby increasing the dissociation temperature at which they fall apart into their quark constituents. This is consistent with the above results.

---

[7]At non-zero baryon number density the magnetic field can actually induce an inverse magnetic catalysis in this model [53, 54].

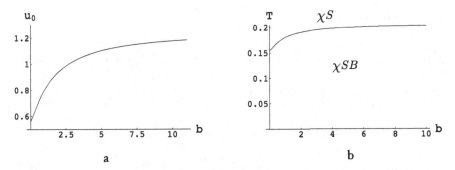

Fig. 22.7 Magnetic catalysis: (**a**) Chiral symmetry breaking mass scale. (**b**) Phase diagram

22.3.4 Anomalous Currents

The chiral anomaly leads to two interesting phenomena when both the magnetic field and chemical potential are non-zero. The first is the generation of anomalous currents in the chiral symmetric phase of QCD. In [56, 57] it was shown that the combination of a magnetic field and a non-zero baryon chemical potential generates an axial current

$$\mathbf{J}_A = \frac{e}{2\pi^2} \mu_B^{phys} \mathbf{B}. \tag{22.38}$$

Since the source for this current is the anomaly it is an exact result, and should be valid in particular at strong coupling. Similarly, an anomalous vector current is generated in a non-zero axial chemical potential:

$$\mathbf{J}_V = \frac{e}{2\pi^2} \mu_A^{phys} \mathbf{B}. \tag{22.39}$$

This is known as the "chiral magnetic effect", and may have some relevance to heavy ion physics at RHIC [5]. Within the D3/D7 model, this effect has been discussed in [58].

In the Sakai-Sugimoto model, the chiral-symmetric phase corresponds to the parallel D8–$\overline{\text{D8}}$ embedding in the deconfined background, and the chiral anomaly is encoded in the five-dimensional CS term (22.24). The background magnetic field and chemical potentials correspond to different components of the worldvolume gauge field. Through the five-dimensional CS term these source a third component, which corresponds to a current in the four-dimensional theory [59]. Let us review the calculation of the anomalous axial current in the Sakai-Sugimoto model. The calculation of the anomalous vector current is virtually identical.

To be specific, we will consider a background magnetic field in the x_1 direction by turning on a background gauge field $a_3^V = x_2 b$. A non-trivial boundary value of a_0^V will then source, via the CS term, a non-trivial a_1^A. In general a_0^V and a_1^A can depend on both u and x_2 in this case, although on-shell they will depend only on u. We will take $a_0^V(\infty) = \mu_V$ and $a_1^A(\infty) = 0$. The D8-brane DBI and CS actions in this case become

$$S_{DBI}^{dec} = -\mathscr{N} \int_{u_T}^{\infty} d^4x\, du\, u^{5/2} \sqrt{\left(1 - (a_0^{V\prime})^2 + f(u)(a_1^{A\prime})^2\right)\left(1 + \frac{b^2}{u^3}\right)}, \qquad (22.40)$$

$$S_{CS} = -\mathscr{N} \int d^4x\, du\, \left[b(a_0^V a_1^{A\prime} - a_0^{V\prime} a_1^A) + a_3^V (a_0^{V\prime} \partial_2 a_1^A - \partial_2 a_0^V a_1^{A\prime}) \right]. \qquad (22.41)$$

The necessary boundary terms are

$$S_\partial = -\frac{1}{2}\mathscr{N} \int d^3x\, du\, a_3^V \left(a_0^V a_1^{A\prime} - a_0^{V\prime} a_1^A \right) \Big|_{x_2 \to -\infty}^{x_2 \to \infty}. \qquad (22.42)$$

By integrating by parts the last two terms in the CS action one can show that up to a surface term at $u \to \infty$, the bulk CS and boundary actions combine into a bulk action

$$S_{CS} + S_\partial$$

$$= -\mathscr{N} \int d^4x\, du \left[\frac{3}{2} b(a_0^V a_1^{A\prime} - a_0^{V\prime} a_1^A) - \frac{1}{2} a_3^V (a_0^V \partial_2 a_1^A - \partial_2 a_0^V a_1^A) \right]. \qquad (22.43)$$

One can get rid of the remaining surface term by adding a boundary term at $u \to \infty$ [59], however this particular term does not contribute to the on-shell action, so we might as well ignore it.[8]

The equations of motion for $a_0^V(u)$ and $a_1^A(u)$ can be integrated once to yield

$$\frac{\sqrt{u^5 + b^2 u^2}\, a_0^{V\prime}(u)}{\sqrt{1 - (a_0^{V\prime}(u))^2 + f(u)(a_1^{A\prime}(u))^2}} = -3b a_1^A(u) + d, \qquad (22.44)$$

$$\frac{\sqrt{u^5 + b^2 u^2}\, f(u) a_1^{A\prime}(u)}{\sqrt{1 - (a_0^{V\prime}(u))^2 + f(u)(a_1^{A\prime}(u))^2}} = -3b a_0^V(u), \qquad (22.45)$$

where d is the baryon number charge density. The integration constant in the a_1^A equation vanishes since $a_0^V(u_T) = 0$ and $f(u_T) = 0$. Using (22.34) and (22.45) we can then evaluate the axial current:

$$j_A^1 = \frac{3}{2} b a_0^V(\infty) = \frac{3}{2} b \mu_V. \qquad (22.46)$$

The correctly normalized physical currents are given by $J = 2(2\pi\alpha'\mathscr{N}/R^5)j$, where the factor of 2 comes from adding the anti-D8-brane contribution, and the physical chemical potentials are $\mu^{phys} = (R/(2\pi\alpha'))\mu$. Thus in terms of the physical variables our result translates to

---

[8]Other boundary terms at $u \to \infty$ could affect the on-shell action, and therefore the currents. See for example [60, 61].

$$\mathbf{J}_A = \frac{N_c}{4\pi^2} \mu_V^{phys} \mathbf{B}. \tag{22.47}$$

Similarly, for an axial chemical potential we would get

$$\mathbf{J}_V = \frac{N_c}{4\pi^2} \mu_A^{phys} \mathbf{B}. \tag{22.48}$$

Interestingly, our results are half of the weak-coupling results (where $e = N_c$ in the holographic model). It has been argued that the discrepancy in the axial current is due to a different treatment of the triangle anomaly, consistent vs. covariant, and can be corrected by adding an appropriate Bardeen counterterm on the boundary [61]. However, the same counterterm leads to a vanishing vector current. This issue is still under investigation.

22.3.5 The Pion Gradient Phase

In the broken chiral symmetry phase the chiral anomaly leads to a novel finite density phase that dominates over nuclear matter at large magnetic fields [4]. In this phase the baryon charge is carried not by baryons but rather by a non-zero pion gradient background:

$$D = \frac{e}{4\pi^2 f_\pi} \mathbf{B} \cdot \nabla \pi^0, \tag{22.49}$$

where

$$\nabla \pi^0 = \frac{e}{4\pi^2 f_\pi} \mu_B^{phys} \mathbf{B}. \tag{22.50}$$

In the Sakai-Sugimoto model (in the $a_u = 0$ gauge) the pseudoscalar meson appears in the zero mode of a_μ^A (22.27), so

$$\partial_\mu \varphi(x^\mu) = -a_\mu^A(x^\mu, u \to \infty). \tag{22.51}$$

As in the chiral-symmetric phase, the presence of a vector chemical potential together with a background magnetic field sources a component of the axial gauge field, which in this case corresponds to a non-trivial gradient of the pseudoscalar field [59, 62]. Since there is only one flavor this field should really be thought as the η' meson. For simplicity, we will consider only the confined phase with the antipodal D8-brane embedding, namely $u_0 = u_{KK}$. (The results do not change qualitatively for more general U-shape embeddings, or for U-shape embeddings in the deconfined phase.)

Following [59], the D8-brane DBI action in this case is

$$S_{DBI}^{con} = -\mathcal{N} \int_{u_{KK}}^{\infty} d^4x \, du \, u^{5/2} \sqrt{\left(\frac{1}{f(u)} - (a_0^{V'})^2 + (a_1^{A'})^2\right)\left(1 + \frac{b^2}{u^3}\right)}, \tag{22.52}$$

and the CS plus boundary actions are the same as in the deconfined phase (22.43). The equations of motion integrate to

$$\frac{\sqrt{u^5 + b^2 u^2} a_0^{V\prime}(u)}{\sqrt{\frac{1}{f(u)} - (a_0^{V\prime}(u))^2 + (a_1^{A\prime}(u))^2}} = -3ba_1^A(u) + N_c n_{D4}, \qquad (22.53)$$

$$\frac{\sqrt{u^5 + b^2 u^2} a_1^{A\prime}(u)}{\sqrt{\frac{1}{f(u)} - (a_0^{V\prime}(u))^2 + (a_1^{A\prime}(u))^2}} = -3ba_0^V(u) + c, \qquad (22.54)$$

where we have explicitly included the baryon sources in the a_0^V equation. Our boundary conditions are now $a_0^V(\infty) = \mu_V$ and $a_1^A(\infty) = -\nabla\varphi(x^\mu)$. In particular $a_1^A(\infty)$ is a field rather than a parameter in the boundary theory, and we must minimize the action with respect to its value. This simply sets $j_A^1 = 0$ and therefore sets the integration constant in the a_1^A equation to $c = \frac{3}{2}b\mu_V$.[9]

We can now compute the total baryon number charge density d using the same procedure as in the previous section for the current. In the absence of sources $n_{D4} = 0$ and we find

$$d = -\frac{3}{2}ba_1^A(\infty) = \frac{3}{2}b\nabla\varphi. \qquad (22.55)$$

Let us express this in terms of the physical variables. First we must define a field with a canonically normalized kinetic term. Inserting (22.27) into the action (22.52) we find that the canonically normalized field is given by

$$\eta'(x^\mu) = \frac{R^2}{2\pi\alpha'} f_{\eta'}\varphi(x^\mu), \qquad f_{\eta'}^2 = \frac{N_c u_{KK}^{3/2}}{4\pi^4\alpha'}. \qquad (22.56)$$

Converting to physical variables we then find

$$D = \frac{N_c}{4\pi^2 f_{\eta'}} \mathbf{B} \cdot \nabla\eta', \qquad (22.57)$$

in agreement with (22.49). We would like to stress that this agreement did not depend on the specific value of $f_{\eta'}$ required for canonical normalization, since it cancels out when we express the result in terms of φ. The correct numerical factor of $1/(4\pi^2)$ is a direct consequence of including the proper boundary terms in the action, leading to the "3/2" in (22.55).

To find the value of the gradient $\nabla\varphi$ we need to solve (22.53) and (22.54). The result will not be as simple as (22.50). In particular it is not linear in the magnetic field, since we are using the full non-linear DBI action. It turns out that a closed form solution can be found in terms of a new variable

---

[9]This is also consistent with the fact that there are no quarks in this phase to carry such a current.

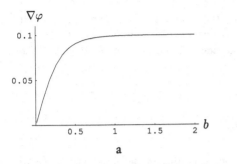

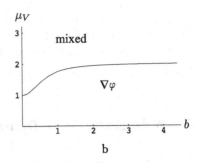

Fig. 22.8 (a) The pion gradient. (b) Phase diagram with magnetic field and baryon chemical potential

$$y = \int_{u_{KK}}^{u} \frac{3b\, d\tilde{u}}{\sqrt{f(\tilde{u})}\sqrt{\tilde{u}^5(1 + b^2\,\tilde{u}^{-3}) - (\frac{3}{2}b\mu_V)^2 + (3b\nabla\varphi)^2}}. \tag{22.58}$$

The solution is

$$a_0^V(y) = \frac{\mu_V}{2}\left(\frac{\cosh y}{\cosh y_\infty} + 1\right), \qquad a_1^A(y) = -\frac{\mu_V}{2}\frac{\sinh y}{\cosh y_\infty}, \tag{22.59}$$

where $y_\infty = y(u \to \infty)$. The pseudoscalar gradient is then given by

$$\nabla\varphi = -a_1^A(\infty) = \frac{\mu_V}{2}\tanh y_\infty. \tag{22.60}$$

The dependence on b is shown in Fig. 22.8a. For small b the behavior is linear in b:

$$\nabla\varphi \approx \frac{\pi}{2u_{KK}^{3/2}}\mu_V b, \tag{22.61}$$

and in terms of the physical quantities:

$$\nabla\eta' \approx \frac{N_c}{4\pi^2 f_{\eta'}}\mu_V^{phys}\mathbf{B}, \tag{22.62}$$

in agreement with the single flavor version of (22.50).

As in the case with no magnetic field, an embedding that includes sources is possible above a critical value of the chemical potential, which then becomes the dominant configuration. This describes a "mixed phase" that includes both "pion-gradient" matter and nuclear matter, with a total baryon number density

$$d = \frac{3}{2}b\nabla\varphi + N_c\, n_{D4}. \tag{22.63}$$

As before, one can find a closed form solution for a_0^V and a_1^A, and from it determine the values of $\nabla\varphi$ and n_{D4} in terms of the magnetic field b and baryon number chemical potential μ_V. The resulting phase diagram is shown in Fig. 22.8b. In particular, the critical value of μ_V is determined by setting $n_{D4} = 0$. The relative proportion of baryons in the mixed phase increases with μ_V and decreases with b.

22.3.6 Magnetic Phase Transition

The high-temperature chiral-symmetric phase of this model also exhibits an interesting magnetic phenomenon associated with the distribution of the baryonic charge [63]. In what follows we analyze the situation for the D8-brane. As in the broken chiral symmetry phase above, the distribution of baryonic charge along the radial direction u changes with b. We can therefore identify two types of baryonic charge, one originating from the horizon, and the other from outside the horizon. The latter d_*, corresponds to D4-branes that are radially smeared inside the D8-brane. This can be best seen from the longitudinal and transverse conductivities [64]

$$\sigma_L = \frac{\sqrt{u_T^8 + b^2 u_T^5 + u_T^3(d - d_*)^2}}{u_T^3 + b^2}, \tag{22.64}$$

$$\sigma_T = \frac{b(d - d_*)}{u_T^3 + b^2} + \frac{d_*}{b}, \tag{22.65}$$

where $d_* = -3ba_1(u_T)$. In particular only the horizon charge $d - d_*$ contributes to the longitudinal conductivity. In the transverse conductivity, the horizon charge contributes as an ordinary dissipative fluid, whereas the charge outside the horizon d_* behaves as a dissipation-free fluid. This is consistent with an interpretation of d_* as the charge filling the lowest Landau level. As the magnetic field increases more of the charge is "lifted" from the horizon, representing the transition to the lowest Landau level in the boundary theory.

In fact for a fixed density at low enough temperature this transition is a first order phase transition as a function of the magnetic field, in which the charges jump into the lowest Landau level. This is easiest to see in the zero temperature limit.[10] In this case one can solve the gauge field equations analytically in terms of a variable

$$z = \int_0^u \frac{3b\,d\tilde{u}}{\sqrt{\tilde{u}^5 + b^2\tilde{u}^2 + d^2\cosh^{-2}z_\infty}}, \tag{22.66}$$

where $z_\infty = z(u \to \infty)$. The solution is

---

[10]Strictly speaking, at zero temperature the theory is in the confining (and broken chiral symmetry) phase. We are considering the meta-stable state obtained by adiabatically reducing the temperature.

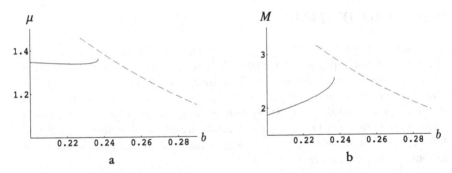

Fig. 22.9 (a) μ and (b) M as functions of b for $d = 1$ and $T = 0.09$, below the critical point. There are now two branches of stable solutions, and the phase transition between them occurs at $b = 0.235$

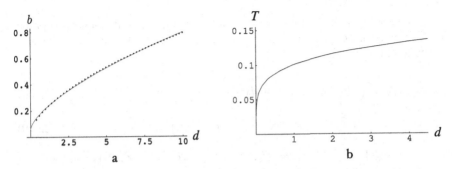

Fig. 22.10 (a) The phase diagram in the d–b plane at $T = 0.07$, and (b) the critical line in the T–d plane

$$a_0^V = \frac{d \sinh z}{3b \cosh z_\infty}, \qquad a_1^A = \frac{d \cosh z}{3b \cosh z_\infty} - \frac{d}{3b}. \qquad (22.67)$$

There are actually three solutions, representing two stable phases and an unstable phase. As the magnetic field b is increased, for a fixed total baryon number density d, one finds a first order phase transition between the two stable phases. Both the magnetization and the chemical potential are discontinuous in this transition (Fig. 22.9), which is reminiscent of a metamagnetic phase transition. The large magnetic field phase represents the situation where all the charge is in the lowest Landau level, with the chemical potential and free energy given by

$$\mu = \frac{d}{3b}, \qquad F = \frac{d^2}{6b}. \qquad (22.68)$$

The magnetic transition persists also at non-zero temperatures that are low relative to the density d. Too a very good approximation this happens when $b \sim d^{2/3}$ (Fig. 22.10a), which is the behavior expected for the lowest Landau level. At high temperature the transition disappears (Fig. 22.10b).

22.4 The D3–D7' Model

The study of magnetic properties of planar matter is a very active area of research in condensed matter physics. A simple holographic model for charged fermions in three dimensions can be obtained by T-dualizing the D4–D8 setup. This leads to a D3–D7 configuration with the two sets of branes intersecting on a plane [12, 14] (for a related model using a D2–D8 system see [65]). However, unlike in the D4–D8 configuration, here the branes have a mutually transverse coordinate. On the one hand this allows the fermions to be massive, but on the other hand it leads to an instability since the different branes repel.

22.4.1 Stable Embeddings

First we have to address the issue of stability. As before, we will employ the probe approximation and consider a single probe D7-brane. The background (at finite temperature) in this case is

$$L^{-2} ds_{10}^2 = r^2 \left(-h(r) dt^2 + dx^2 + dy^2 + dz^2\right) + r^{-2} \left(\frac{dr^2}{h(r)} + r^2 d\Omega_5^2\right), \quad (22.69)$$

$$F_5 = 4L^4 \left(r^3 dt \wedge dx \wedge dy \wedge dz \wedge dr + d\Omega_5\right), \quad (22.70)$$

where $h(r) = 1 - r_T^4/r^4$, $r_T = \pi L T$ and $L^2 = \sqrt{4\pi g_s N_c} \alpha'$. It is convenient to parameterize the five-sphere as an $S^2 \times S^2$ fibered over an interval:

$$d\Omega_5^2 = d\psi^2 + \cos^2 \psi \left(d\Omega_2^{(1)}\right)^2 + \sin^2 \psi \left(d\Omega_2^{(2)}\right)^2, \quad (22.71)$$

where $0 \leq \psi \leq \pi/2$. The first S^2 shrinks at the "south pole" $\psi = \pi/2$ and the second S^2 at the "north pole" $\psi = 0$. The D7-brane wraps the two S^2's and extends along (x, y), and has an embedding described by $z(r)$ and $\psi(r)$. In particular ψ is dual to the fermion bi-linear operator in the field theory corresponding to the fermion mass. The D7-brane DBI action in this background is given by

$$S_{DBI} = -4\mathcal{N} \int d^3x \, dr \, r^2 \cos^2 \psi \, \sin^2 \psi \sqrt{1 + r^4 h(r) z'^2 + r^2 h(r) \psi'^2}, \quad (22.72)$$

where $\mathcal{N} \equiv 4\pi^2 \mu_7 L^8/g_s$. A massless embedding would correspond to $\psi = \pi/4$. However the fluctuations contain a mode that violates the Breitenlohner-Freedman bound, and therefore the embedding is unstable. This can also be seen by trying a more general embedding with a large r behavior of

$$\psi(r) \sim \frac{\pi}{4} + cr^\Delta. \quad (22.73)$$

The equation of motion for ψ gives $\Delta(\Delta + 3) = -8$, which does not have a real solution.

Fortunately, the D7-brane can be stabilized by turning on some worldvolume flux [66], in this case on the two-spheres [15]:

$$2\pi\alpha' F = \frac{L^2}{2}\left(f_1 \, d\Omega_2^{(1)} + f_2 \, d\Omega_2^{(2)}\right), \quad f_i = \frac{2\pi\alpha'}{L^2}n_i \quad (n_i \in \mathbb{Z}). \quad (22.74)$$

This changes the DBI action,

$$S_{DBI}$$

$$= -\mathcal{N}\int d^3x \, dr \, r^2 \sqrt{\left(4\cos^4\psi + f_1^2\right)\left(4\sin^4\psi + f_2^2\right)\left(1 + r^4 h z'^2 + r^2 h \psi'^2\right)}, \quad (22.75)$$

and there is now also a CS term which gives,

$$S_{CS} = -\mathcal{N} f_1 f_2 \int d^3x \, dr \, r^4 z'(r). \quad (22.76)$$

The asymptotic behavior of $\psi(r)$ is now

$$\psi(r) \sim \psi_\infty + m r^{\Delta_+} - c_\psi r^{\Delta_-}, \quad (22.77)$$

where ψ_∞ is determined by the solution of

$$\left(f_1^2 + 4\cos^4\psi_\infty\right)\sin^2\psi_\infty = \left(f_2^2 + 4\sin^4\psi_\infty\right)\cos^2\psi_\infty, \quad (22.78)$$

and

$$\Delta_\pm = -\frac{3}{2} \pm \frac{1}{2}\sqrt{9 + 16\frac{f_1^2 + 16\cos^6\psi_\infty - 12\cos^4\psi_\infty}{f_1^2 + 4\cos^6\psi_\infty}}. \quad (22.79)$$

In particular, the embedding is stable for a large enough flux. The coefficient of the leading term is related to the fermion mass, and that of the subleading term corresponds to the bi-linear condensate. Note that the scaling dimension of the bi-linear operator is given by $-\Delta_-$. This represents a large anomalous dimension, which is not surprising given that the model is non-supersymmetric. We should require however that the operator be relevant, namely that $\Delta_- \geq -3$, and therefore that $\Delta_+ \leq 0$, in order to consider the leading term as a "mass deformation".

Generally there are two types of embeddings, that differ in their small r behavior: Minkowski-like (MN) embeddings, in which the D7-brane terminates smoothly outside the horizon (Fig. 22.11a), and black-hole (BH) embeddings, in which the D7-brane crosses the horizon (Fig. 22.11b). We refer the reader to [15] for the explicit embedding equations for $\psi(r)$ and $z(r)$, and for their numerical solutions.

In an MN embedding $\psi(r_0) = \pi/2$ or 0 for some $r_0 > r_T$, corresponding to one or the other S^2 shrinking. This indicates that the dual field theory has a mass-gap related to $r_0 - r_T$. An important condition for the existence of MN embeddings is the absence of sources for the worldvolume gauge field. Unlike in the model

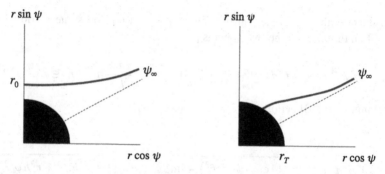

Fig. 22.11 (a) An MN embedding with $\psi(r_0) = \pi/2$. (b) A BH embedding

of the previous section, there are no localized sources in this model. All sources correspond to branes (or strings) connecting the D7-brane to the horizon. These inevitably pull the D7-brane down to the horizon, resulting in a BH embedding instead. This means that the flux on the S^2 that shrinks must vanish. For $\psi(r_0) = \pi/2$, which means that $f_1 = 0$, we find stable massive embeddings in the range

$$0.5235 \lesssim \psi_\infty \lesssim 0.6251. \tag{22.80}$$

(There are also the "mirror" embeddings with $\psi(r_0) = 0$ and $f_2 = 0$.) Note that the allowed values of ψ_∞ are quantized since the stabilizing flux is quantized. There are also two isolated MN embeddings with $\psi_\infty = 0$ and $\psi_\infty = \pi/2$.

BH embeddings describe gapless phases in the dual theory. These embeddings exist generically for any f_1, f_2 satisfying the stability condition.

22.4.2 Finite Density and Background Fields

For embeddings corresponding to finite density states in a background magnetic field we need to turn on the appropriate components of the worldvolume gauge field. As in the D4–D8 model we will work with the dimensionless field $a_\mu = (2\pi\alpha'/L)A_\mu$. There are additional terms in the DBI action,

$$S_{DBI} = -\mathcal{N} \int dr\, r^2 \sqrt{\left(4\cos^4\psi + f_1^2\right)\left(4\sin^4\psi + f_2^2\right)}$$

$$\times \sqrt{\left(1 + r^4 h(r)z'^2 + r^2 h(r)\psi'^2 - a_0'^2\right)\left(1 + \frac{b^2}{r^4}\right)}, \tag{22.81}$$

and also in the CS action,

$$S_{CS} = -\mathcal{N} f_1 f_2 \int dr\, r^4 z'(r) + 2\mathcal{N} \int dr\, c(r) b a_0'(r), \tag{22.82}$$

where

$$c(r) = \frac{1}{8\pi^2 L^4} \int_{S^2 \times S^2} C_4\big(\psi(r)\big) = \psi(r) - \frac{1}{4}\sin 4\psi(r) - \psi_\infty + \frac{1}{4}\sin 4\psi_\infty.$$

$$(22.83)$$

We have fixed a gauge for the RR field such that $c(\infty) = 0$. For MN embeddings we must also add a boundary term at $r = r_0$ (as explained in Sect. 22.3)[11]

$$S_\partial(r_0) = 2\mathcal{N} c(r_0) b a_0(r_0).$$

$$(22.84)$$

The quantity $c(r_0)$ has a nice physical interpretation: it is the total amount of 5-form flux captured by the D7-brane in the MN embedding. It is completely fixed by the asymptotic value of the embedding angle ψ_∞. For BH embeddings the boundary term vanishes since $a_0(r_T) = 0$.

The integrated equation of motion for $a_0(r)$ is given by

$$G(r) a_0'(r) = d - 2bc(r),$$

$$(22.85)$$

where

$$G(r) = r^2 \left(1 + \frac{b^2}{r^4}\right) \sqrt{\frac{(f_1^2 + 4\cos^4\psi)(f_2^2 + 4\sin^4\psi)}{Y(r)}},$$

$$(22.86)$$

$$Y(r) = \left(1 + \frac{b^2}{r^4}\right)\left(1 + hr^4 z'^2 + hr^2\psi'^2 - a_0'^2\right),$$

$$(22.87)$$

and d is the total charge density. As in Sect. 22.3, we are using (22.34) to define the conserved currents. The quantity on the RHS of (22.85), $\tilde{d}(r) \equiv d - 2bc(r)$, is the contribution to the charge density from radial positions below r.

We would also like to study the response of the system to a background electric field. To this end we should consider a more general ansatz for the gauge field with $a_x(t, r) = te + a_x(r)$, $a_y(x, r) = xb + a_y(r)$, in addition to $a_0(r)$. The current densities will be contained in the asymptotic behaviors of $a_x(r)$ and $a_y(r)$. The gauge field equations in this case become

$$G(r) a_0'(r) = \left[\tilde{d}(r)\left(1 - \frac{e^2}{r^4 h(r)}\right) + \tilde{j}_y(r)\frac{eb}{r^4 h(r)}\right]\frac{1 + \frac{b^2}{r^4}}{1 + \frac{b^2}{r^4} - \frac{e^2}{r^4 h(r)}},$$

$$(22.88)$$

$$G(r) a_y'(r) = \left[\tilde{d}(r)\frac{eb}{r^4 h(r)} - \frac{\tilde{j}_y(r)}{h(r)}\left(1 + \frac{b^2}{r^4}\right)\right]\frac{1 + \frac{b^2}{r^4}}{1 + \frac{b^2}{r^4} - \frac{e^2}{r^4 h(r)}},$$

$$(22.89)$$

---

[11] In [15] this term was derived by demanding invariance of the CS term $\int C_4 \wedge F \wedge F$ under gauge transformations of the RR field and then fixing $c(\infty) = 0$. However it can also be obtained by canceling the surface term in the variation of the CS term, when we present it as $\int F_5 \wedge A \wedge F$.

$$G(r)\,a_x'(r) = -\frac{j_x}{h(r)}\left(1 + \frac{b^2}{r^4}\right), \tag{22.90}$$

where j_x is the longitudinal current density and $\tilde{j}_y(r)$ is defined by analogy with $\tilde{d}(r)$ as $\tilde{j}_y(r) \equiv j_y - 2c(r)e$, where j_y is the transverse current density. The factor $G(r)$ is defined as before (22.86), now more generally with

$$Y(r) = \left(1 + \frac{b^2}{r^4} - \frac{e^2}{hr^4}\right)\left(1 + hr^4 z'^2 + hr^2 \psi'^2\right)$$

$$- \left(1 + \frac{b^2}{r^4}\right)a_0'^2 + ha_x'^2 + \left(1 - \frac{e^2}{hr^4}\right)ha_y'^2 - \frac{2eb}{r^4}a_0'a_y'. \tag{22.91}$$

22.4.3 Quantum Hall States

Let us consider first the response of the gapped MN embeddings. This is determined by the requirement that there are no sources, namely by regularity of the gauge field at $r = r_0$. This implies, in particular, that

$$\tilde{d}(r_0) = d - 2c(r_0)b = 0. \tag{22.92}$$

The entire charge in the MN embedding is thus due to the CS term and corresponds to a "fluid" of D5-branes inside the D7-brane. The charge density is proportional to the magnetic field, and the proportionality constant is fixed by the value of $c(r_0)$, and therefore of ψ_∞. This is the key property of a quantum Hall state, which is characterized by a specific quantized value of the Landau-level filling fraction $\nu \propto d/b$. In terms of the physical variables $D = (2\pi\alpha' \mathcal{N}/L^4)d$ and $B = b/(2\pi\alpha')$, the filling fraction is given by

$$\nu = \frac{2\pi D}{B} = \frac{2N_c}{\pi}c(r_0). \tag{22.93}$$

For the range of values of ψ_∞ needed for stability (22.80) we get

$$0.6972 \lesssim \frac{\nu}{N_c} \lesssim 0.8045. \tag{22.94}$$

Furthermore, the filling fractions are quantized according to the quantization of ψ_∞. The actual numbers can be obtained by solving (22.78), for a specific flux f_2 (with $f_1 = 0$), and plugging into (22.83) with $\psi(r_0) = \pi/2$, but they are not particularly illuminating (for example, they are not rational numbers). The isolated embeddings with $\psi_\infty = 0$ and $\psi_\infty = \pi/2$ correspond to $\nu/N_c = 1$ and 0, respectively.

The current densities can likewise be computed by requiring regularity of the spatial components of the gauge field. This condition implies that

$$j_x = 0 \quad \text{and} \quad \tilde{j}_y(r_0) = j_y - 2c(r_0)e = 0, \tag{22.95}$$

from which we can deduce the longitudinal and transverse conductivities:

$r_0 - r_T$

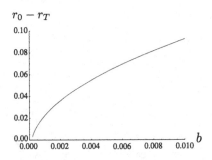

$\tilde{\omega}$

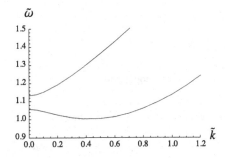

Fig. 22.12 (a) Mass-gap for charged states as a function of the magnetic field. (b) Dispersion relation of the two lowest neutral modes, showing the magneto-roton

$$\sigma_{xx} = 0, \qquad \sigma_{xy} = \frac{\nu}{2\pi}. \tag{22.96}$$

Thus the MN embeddings, when they exist, describe quantum Hall states with quantized transverse conductivities, and vanishing longitudinal conductivities. Furthermore, in the holographic description, the quantization is topological since it originates from the Dirac quantization of the magnetic fluxes on the S^2's. In particular, σ_{xy} in the MN embeddings is independent of the temperature.

Quantum Hall states are gapped to both charged and neutral excitations. In this model charged excitations are described by strings stretched from the D7-brane to the horizon, and therefore have a mass proportional to $r_0 - r_T$. This is seen to increase with the magnetic field, as shown in Fig. 22.12a. The neutral excitations correspond to fluctuations of the D7-brane worldvolume fields, and are also found to be massive [67] (see also [68]). The spectrum of neutral excitations includes a magneto-roton, which is a collective excitation whose dispersion relation has a minimum at non-zero momentum (Fig. 22.12b). A similar phenomenon is seen in real quantum Hall states [69].

22.4.4 Fermi-Like Liquid

BH embeddings describe gapless Fermi-like liquids. For a BH embedding, $\tilde{d}(r_T)$ corresponds to the horizon charge density carried by the "quarks", and it need not vanish. In particular, if we add sources to an MN embedding, thereby violating (22.92), it deforms continuously into a BH embedding with horizon charge.

To compute the electrical response in a BH embedding we have a couple of options. The standard approach is to extract the conductivities using linear response from the current-current correlators, computed holographically by studying fluctuations of the bulk gauge fields to quadratic order. The other option is to find a consistent solution in the presence of an external electric field [70]. The advantage of the second approach, when it is applicable, is that it gives the complete non-linear response. Using this method for the BH embeddings one finds

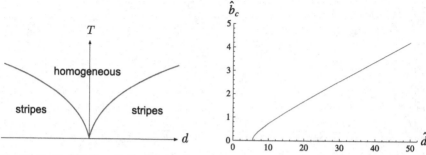

Fig. 22.13 (*Left*): Phase diagram in the T–d plane showing the quantum critical point. (*Right*): The phase diagram in the $\hat{b}$, $\hat{d}$ plane for $m = 0$. *Above the line* the system is in the homogeneous phase and *below the line* in the striped phase

$$\sigma_{xy} = \frac{N_c}{2\pi^2}\left(\frac{b}{b^2 + r_T^4}\tilde{d}(r_T) + 2c(r_T)\right),$$
(22.97)

$$\sigma_{xx} = \frac{N_c}{2\pi^2}\frac{r_T^2}{b^2 + r_T^4}\sqrt{\tilde{d}(r_T)^2 + \left(f_1^2 + 4\cos^4\psi(r_T)\right)\left(f_2^2 + 4\sin^4\psi(r_T)\right)\left(b^2 + r_T^4\right)}.$$
(22.98)

Note that the transverse conductivity has two components. The first involves the horizon charge, and resembles the contribution of an ordinary dissipative system of charges. The remaining charge, corresponding to the fluid of D5-branes inside the D7-brane, contributes like a dissipationless system. The longitudinal conductivity involves only the first component. This is basically the same separation that was seen in the D4–D8 model (see (22.64), (22.65)).

This state of holographic matter exhibits a variety of other interesting phenomena as a function of the charge density, temperature, background magnetic field, and mass.

Consider first the state at $T = 0$, $d = 0$, $b = 0$ and $m = 0$. In this case the D7-brane embedding is actually $AdS_4 \times S^2 \times S^2$, so this situation is described by a conformal field theory. Note that in this case $\sigma_{xx} \neq 0$. At non-zero density the system becomes unstable to the formation of stripes. The instability is signaled by the existence of a quasi-normal mode (in the transverse gauge field sector) with positive imaginary part in a finite range of momenta [71]. At a high enough temperature or high enough magnetic field this instability disappears. It is convenient to parametrize the situation with $\hat{b} = b/r_T^2$ and $\hat{d} = d/r_T^2$. Then the instability towards the striped phase at zero magnetic field and $m = 0$ happens for $\hat{d} > 5.5$. This is demonstrated in Fig. 22.13a. Above the quantum critical point ($T = m = d = 0$) there is a region which resembles a Fermi-like liquid, and on both sides there is a striped phase. For non-zero $\hat{b}$ and $m = 0$ the instability sets in at some other value of $\hat{d}$, as shown in Fig. 22.13b [72]. As m increases the instability sets in at a lower temperature.

The system also has a zero sound mode. At non-zero temperature the quasi-normal mode with the smallest imaginary part at low momentum is a purely dis-

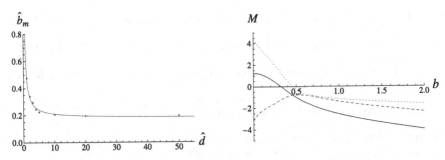

Fig. 22.14 (*Left*): The *line* in the $\hat{b}, \hat{d}$ plane for $m = 0$, separating the situation when the ze-ro-sound is gapped (*above the line*) and the situation when it is gapless (*below the line*). (*Right*): The magnetization as a function of magnetic field (*solid line*) and the individual contributions from the DBI (*dashed line*) and the CS term (*pointed line*) for $m = 1$ $d = 1$ and $r_T = 0.1$

sipative hydrodynamical mode ($\omega(k = 0) = 0$). At some non-zero momentum it meets another purely dissipative mode (ω purely imaginary) and crosses from a hy-drodynamical regime into a collisionless regime, where the resulting complex mode can be identified with the finite temperature zero-sound mode [71] (see also [73]). The zero sound mode becomes massive as the magnetic field crosses a critical value [72, 74] (Fig. 22.14a).

For $m \neq 0$ the system can have a non-zero transverse conductivity, even at zero magnetic field. This is due to having a non-trivial $c(r)$ for these embeddings. This is reminiscent of the anomalous Hall effect (AHE) that appears in ferromagnetic materials (for a review see [75]). Indeed for $m \neq 0$ the system is ferromagnetic (Fig. 22.14b) due to the second term in (22.82). Note that both the AHE and the ferromagnetic behavior, as well as the instability towards a striped phase, have a common origin in the Chern-Simon term $\int dr\, c(r) F \wedge F$ in the brane action.

Acknowledgements   O.B. and G.L. would like to thank Matt Lippert and Niko Jokela, who were an integral part in all our work on the models reviewed in Sects. 22.3 and 22.4. O.B. also thanks the Aspen Center for Physics, where this work was completed, for its hospitality. J.E. would like to thank her collaborators Martin Ammon, Matthias Kaminski, Patrick Kerner, René Meyer, Jonathan Shock and Migael Strydom, involved in the joint work presented in Sect. 22.2. This work was supported in part by the Israel Science Foundation under grant No. 392/09, and in part by the US-Israel Binational Science Foundation under grant No. 2008-072.

References

1. V.P. Gusynin, V.A. Miransky, I.A. Shovkovy, Catalysis of dynamical flavor symmetry break-ing by a magnetic field in $(2+1)$-dimensions. Phys. Rev. Lett. **73**, 3499–3502 (1994)
2. V.P. Gusynin, V.A. Miransky, I.A. Shovkovy, Dimensional reduction and dynamical chiral symmetry breaking by a magnetic field in $(3+1)$-dimensions. Phys. Lett. B **349**, 477–483 (1995)
3. V.A. Miransky, I.A. Shovkovy, Magnetic catalysis and anisotropic confinement in QCD. Phys. Rev. D **66**, 045006 (2002)

4. D.T. Son, M.A. Stephanov, Axial anomaly and magnetism of nuclear and quark matter. Phys. Rev. D **77**, 014021 (2008)
5. K. Fukushima, D.E. Kharzeev, H.J. Warringa, The chiral magnetic effect. Phys. Rev. D **78**, 074033 (2008)
6. M.N. Chernodub, Superconductivity of QCD vacuum in strong magnetic field. Phys. Rev. D **82**, 085011 (2010)
7. M.N. Chernodub, Spontaneous electromagnetic superconductivity of vacuum in strong magnetic field: evidence from the Nambu–Jona-Lasinio model. Phys. Rev. Lett. **106**, 142003 (2011)
8. D.C. Tsui, H.L. Stormer, A.C. Gossard, Two-dimensional magnetotransport in the extreme quantum limit. Phys. Rev. Lett. **48**, 1559–1562 (1982)
9. R.B. Laughlin, Anomalous quantum Hall effect—an incompressible quantum fluid with fractionally charged excitations. Phys. Rev. Lett. **50**, 1395–1398 (1983)
10. A. Karch, E. Katz, Adding flavor to AdS/CFT. J. High Energy Phys. **0206**, 043 (2002)
11. T. Sakai, S. Sugimoto, Low energy hadron physics in holographic QCD. Prog. Theor. Phys. **113**, 843–882 (2005)
12. S.-J. Rey, Talk given at Strings 2007 in Madrid, 2007
13. S.-J. Rey, String theory on thin semiconductors: holographic realization of Fermi points and surfaces. Prog. Theor. Phys. Suppl. **177**, 128–142 (2009)
14. J.L. Davis, P. Kraus, A. Shah, Gravity dual of a quantum Hall plateau transition. J. High Energy Phys. **0811**, 020 (2008)
15. O. Bergman, N. Jokela, G. Lifschytz, M. Lippert, Quantum Hall effect in a holographic model. J. High Energy Phys. **1010**, 063 (2010)
16. K. Jensen, A. Karch, D.T. Son, E.G. Thompson, Holographic Berezinskii-Kosterlitz-Thouless transitions. Phys. Rev. Lett. **105**, 041601 (2010)
17. N. Evans, A. Gebauer, K.-Y. Kim, M. Magou, Phase diagram of the D3/D5 system in a magnetic field and a BKT transition. Phys. Lett. B **698**, 91–95 (2011)
18. J.M. Maldacena, The large N limit of superconformal field theories and supergravity. Adv. Theor. Math. Phys. **2**, 231–252 (1998)
19. J. Erdmenger, N. Evans, I. Kirsch, E. Threlfall, Mesons in gauge/gravity duals—a review. Eur. Phys. J. A **35**, 81–133 (2008)
20. M. Kruczenski, D. Mateos, R.C. Myers, D.J. Winters, Meson spectroscopy in AdS/CFT with flavor. J. High Energy Phys. **0307**, 049 (2003)
21. J. Babington, J. Erdmenger, N.J. Evans, Z. Guralnik, I. Kirsch, Chiral symmetry breaking and pions in nonsupersymmetric gauge/gravity duals. Phys. Rev. D **69**, 066007 (2004)
22. V.G. Filev, C.V. Johnson, R.C. Rashkov, K.S. Viswanathan, Flavoured large N gauge theory in an external magnetic field. J. High Energy Phys. **0710**, 019 (2007)
23. J. Erdmenger, R. Meyer, J.P. Shock, AdS/CFT with flavour in electric and magnetic Kalb-Ramond fields. J. High Energy Phys. **0712**, 091 (2007)
24. V.G. Filev, R.C. Raskov, Magnetic catalysis of chiral symmetry breaking. A holographic prospective. Adv. High Energy Phys. **2010**, 473206 (2010)
25. V.G. Filev, D. Zoakos, Towards unquenched holographic magnetic catalysis. J. High Energy Phys. **1108**, 022 (2011)
26. J. Erdmenger, V.G. Filev, D. Zoakos, Magnetic catalysis with massive dynamical flavours (2011)
27. M. Ammon, V.G. Filev, J. Tarrio, D. Zoakos, D3/D7 quark-gluon plasma with magnetically induced anisotropy. J. High Energy Phys. **1209**, 039 (2012)
28. N. Evans, T. Kalaydzhyan, K.-y. Kim, I. Kirsch, Non-equilibrium physics at a holographic chiral phase transition. J. High Energy Phys. **1101**, 050 (2011)
29. N. Evans, A. Gebauer, K.-Y. Kim, M. Magou, Holographic description of the phase diagram of a chiral symmetry breaking gauge theory. J. High Energy Phys. **1003**, 132 (2010)
30. K. Jensen, A. Karch, E.G. Thompson, A holographic quantum critical point at finite magnetic field and finite density. J. High Energy Phys. **1005**, 015 (2010)

31. N. Evans, A. Gebauer, K.-Y. Kim, E, B, μ, T phase structure of the D3/D7 holographic dual. J. High Energy Phys. **1105**, 067 (2011)
32. M. Ammon, J. Erdmenger, M. Kaminski, P. Kerner, Superconductivity from gauge/gravity duality with flavor. Phys. Lett. B **680**, 516–520 (2009)
33. M. Ammon, J. Erdmenger, M. Kaminski, P. Kerner, Flavor superconductivity from gauge/gravity duality. J. High Energy Phys. **0910**, 067 (2009)
34. O. Aharony, K. Peeters, J. Sonnenschein, M. Zamaklar, Rho meson condensation at finite isospin chemical potential in a holographic model for QCD. J. High Energy Phys. **0802**, 071 (2008)
35. A. Rebhan, A. Schmitt, S.A. Stricker, Meson supercurrents and the Meissner effect in the Sakai-Sugimoto model. J. High Energy Phys. **0905**, 084 (2009)
36. J. Erdmenger, P. Kerner, M. Strydom, Holographic superconductors at finite isospin density or in an external magnetic field. PoS **FacesQCD**, 004 (2010)
37. M. Ammon, J. Erdmenger, P. Kerner, M. Strydom, Black hole instability induced by a magnetic field. Phys. Lett. B **706**, 94–99 (2011)
38. A. O'Bannon, Hall conductivity of flavor fields from AdS/CFT. Phys. Rev. D **76**, 086007 (2007)
39. Y.-Y. Bu, J. Erdmenger, J.P. Shock, M. Strydom, Magnetic field induced lattice ground states from holography. J. High Energy Phys. **1303**, 165 (2013)
40. V.V. Braguta, P.V. Buividovich, M.N. Chernodub, A.Yu. Kotov, M.I. Polikarpov, Electromagnetic superconductivity of vacuum induced by strong magnetic field: numerical evidence in lattice gauge theory. Phys. Lett. B **718**, 667–671 (2012)
41. J. Ambjorn, P. Olesen, Electroweak magnetism: theory and application. Int. J. Mod. Phys. A **5**, 4525–4558 (1990)
42. N. Callebaut, D. Dudal, H. Verschelde, Holographic study of rho meson mass in an external magnetic field: paving the road towards a magnetically induced superconducting QCD vacuum? PoS **FacesQCD**, 046 (2010)
43. N. Callebaut, D. Dudal, H. Verschelde, Holographic rho mesons in an external magnetic field. J. High Energy Phys. **1303**, 033 (2013)
44. E. Witten, Anti-de Sitter space, thermal phase transition, and confinement in gauge theories. Adv. Theor. Math. Phys. **2**, 505–532 (1998)
45. H. Hata, T. Sakai, S. Sugimoto, S. Yamato, Baryons from instantons in holographic QCD. Prog. Theor. Phys. **117**, 1157 (2007)
46. O. Aharony, J. Sonnenschein, S. Yankielowicz, A holographic model of deconfinement and chiral symmetry restoration. Ann. Phys. **322**, 1420–1443 (2007)
47. O. Bergman, G. Lifschytz, Holographic $U(1)_A$ and string creation. J. High Energy Phys. **0704**, 043 (2007)
48. O. Bergman, G. Lifschytz, M. Lippert, Holographic nuclear physics. J. High Energy Phys. **0711**, 056 (2007)
49. J.L. Davis, M. Gutperle, P. Kraus, I. Sachs, Stringy NJL and Gross-Neveu models at finite density and temperature. J. High Energy Phys. **0710**, 049 (2007)
50. M. Rozali, H.-H. Shieh, M. Van Raamsdonk, J. Wu, Cold nuclear matter in holographic QCD. J. High Energy Phys. **0801**, 053 (2008)
51. O. Bergman, G. Lifschytz, M. Lippert, Response of holographic QCD to electric and magnetic fields. J. High Energy Phys. **0805**, 007 (2008)
52. C.V. Johnson, A. Kundu, External fields and chiral symmetry breaking in the Sakai-Sugimoto model. J. High Energy Phys. **0812**, 053 (2008)
53. F. Preis, A. Rebhan, A. Schmitt, Inverse magnetic catalysis in dense holographic matter. J. High Energy Phys. **1103**, 033 (2011)
54. F. Preis, A. Rebhan, A. Schmitt, Holographic baryonic matter in a background magnetic field. J. Phys. G **39**, 054006 (2012)
55. C.V. Johnson, A. Kundu, Meson spectra and magnetic fields in the Sakai-Sugimoto model. J. High Energy Phys. **0907**, 103 (2009)

56. M.A. Metlitski, A.R. Zhitnitsky, Anomalous axion interactions and topological currents in dense matter. Phys. Rev. D **72**, 045011 (2005)
57. G.M. Newman, D.T. Son, Response of strongly-interacting matter to magnetic field: some exact results. Phys. Rev. D **73**, 045006 (2006)
58. C. Hoyos, T. Nishioka, A. O'Bannon, A chiral magnetic effect from AdS/CFT with flavor. J. High Energy Phys. **1110**, 084 (2011)
59. O. Bergman, G. Lifschytz, M. Lippert, Magnetic properties of dense holographic QCD. Phys. Rev. D **79**, 105024 (2009)
60. A. Rebhan, A. Schmitt, S.A. Stricker, Anomalies and the chiral magnetic effect in the Sakai-Sugimoto model. J. High Energy Phys. **1001**, 026 (2010)
61. A. Rebhan, A. Schmitt, S. Stricker, Holographic chiral currents in a magnetic field. Prog. Theor. Phys. Suppl. **186**, 463–470 (2010)
62. E.G. Thompson, D.T. Son, Magnetized baryonic matter in holographic QCD. Phys. Rev. D **78**, 066007 (2008)
63. G. Lifschytz, M. Lippert, Holographic magnetic phase transition. Phys. Rev. D **80**, 066007 (2009)
64. G. Lifschytz, M. Lippert, Anomalous conductivity in holographic QCD. Phys. Rev. D **80**, 066005 (2009)
65. N. Jokela, M. Jarvinen, M. Lippert, A holographic quantum Hall model at integer filling. J. High Energy Phys. **1105**, 101 (2011)
66. R.C. Myers, M.C. Wapler, Transport properties of holographic defects. J. High Energy Phys. **0812**, 115 (2008)
67. N. Jokela, G. Lifschytz, M. Lippert, Magneto-roton excitation in a holographic quantum Hall fluid. J. High Energy Phys. **1102**, 104 (2011)
68. N. Jokela, M. Jarvinen, M. Lippert, Fluctuations of a holographic quantum Hall fluid. J. High Energy Phys. **1201**, 072 (2012)
69. S.M. Girvin, A.H. MacDonald, P.M. Platzman, Magneto-roton theory of collective excitations in the fractional quantum Hall effect. Phys. Rev. B **33**, 2481–2494 (1986)
70. A. Karch, A. O'Bannon, Metallic AdS/CFT. J. High Energy Phys. **0709**, 024 (2007)
71. O. Bergman, N. Jokela, G. Lifschytz, M. Lippert, Striped instability of a holographic Fermi-like liquid. J. High Energy Phys. **1110**, 034 (2011)
72. N. Jokela, G. Lifschytz, M. Lippert, Magnetic effects in a holographic Fermi-like liquid. J. High Energy Phys. **1205**, 105 (2012)
73. R.A. Davison, A.O. Starinets, Holographic zero sound at finite temperature. Phys. Rev. D **85**, 026004 (2012)
74. M. Goykhman, A. Parnachev, J. Zaanen, Fluctuations in finite density holographic quantum liquids. J. High Energy Phys. **1210**, 045 (2012)
75. N. Nagaosa, J. Sinova, S. Onoda, A.H. MacDonald, N.P. Ong, Anomalous Hall effect. Rev. Mod. Phys. **82**, 1539–1592 (2010)